BUTTERWORTHS STUDENT

COMMERCIAL AND CONSUMER LAW

BUTTERWORTHS STUDENT STATUTES

Commercial and Consumer Law

Second edition

RICHARD HOOLEY

of the Middle Temple, Barrister-at-Law
Senior Tutor of Fitzwilliam College, Cambridge
University Lecturer in Law, University of Cambridge

Butterworths

London, Edinburgh, Dublin

1999

United Kingdom	Butterworths, a Division of Reed Elsevier (UK) Ltd, Halsbury House, 35 Chancery Lane, LONDON WC2A 1EL and 4 Hill Street, EDINBURGH EH2 3JZ
Australia	Butterworths, a Division of Reed International Books Australia Pty Ltd, CHATSWOOD, New South Wales
Canada	Butterworths Canada Ltd, MARKHAM, Ontario
Hong Kong	Butterworths Asia (Hong Kong), HONG KONG
India	Butterworths India, NEW DELHI
Ireland	Butterworth (Ireland) Ltd, DUBLIN
Malaysia	Malayan Law Journal Sdn Bhd, KUALA LUMPUR
New Zealand	Butterworths of New Zealand Ltd, WELLINGTON
Singapore	Butterworths Asia, SINGAPORE
South Africa	Butterworths Publishers (Pty) Ltd, DURBAN
USA	Lexis Law Publishing, CHARLOTTESVILLE, Virginia

© Reed Elsevier (UK) Ltd 1999

A CIP Catalogue record for this book is available from the British Library.

ISBN 0 406 98300 3

Printed and bound in Great Britain by Butler & Tanner Ltd, Frome and London

Visit us at our website: http://www.butterworths.co.uk

Preface

This book aims to provide students with the key legislative materials relevant to the core areas of commercial and consumer law. It also includes legislation from relevant areas of the general law of contract, tort and restitution. Since the first edition was published there have been a number of important legislative developments including the Sale of Goods (Amendment) Act 1995, the Arbitration Act 1996 and the Late Payment of Commercial Debts (Interest) Act 1998, as well as the Deregulation (Bills of Exchange) Order 1996, the Consumer Credit (Increase of Monetary Limits) (Amendment) Order 1998, the Consumer Credit (Further Increase of Monetary Amounts) Order 1998 and the Consumer Protection (Cancellation of Contracts Concluded away from Business Premises) (Amendment) Regulations 1998. All are incorporated in this new edition.

There is also a wider selection of international trade materials than in the previous edition. The inclusion of the ICC's *Uniform Customs and Practice for Documentary Credits* (UCP 500) should prove particularly useful for those students taking international trade modules. The coverage of consumer related material has also been strengthened to include the Property Misdescriptions Act 1991 and the Timeshare Act 1992, as well as important delegated legislation. The increased emphasis on consumer legislation is reflected in the new title for this second edition.

A feature of the first edition was that it took account of anticipated legislative developments. The second edition does the same by including not only the Contracts (Rights of Third Parties) Bill currently before Parliament, but also a number of proposed EC Directives dealing with such matters as credit transfers, distance contracts, settlement finality, consumer guarantees, late payment, electronic commerce and the distance marketing of consumer financial services. This is tomorrow's law, but we all need to be aware of it today. Proposed legislation is printed in its most up to date form at the time of going to press; however, students should note that changes may well be made to some of these proposals before they reach the statute book.

In order to keep the book to a manageable size and cost it has been necessary to be selective in choosing its contents. Not all material is produced in full but all material is produced in its most recent form, with supplementary notes providing useful details of its legislative history.

I am grateful to those who have suggested material for this new edition, and welcome further suggestions for the next one.

Richard Hooley
Cambridge, June 1999

Contents

Preface v

PART I
STATUTES

Statute of Frauds (1677), s 4 3
Life Assurance Act 1774, ss 1–4 3
Bills of Lading Act 1855, ss 1–3 5
Mercantile Law Amendment Act 1856, ss 5, 16 6
Bills of Sale Act 1878, ss 1, 3–16, 20–22, 24, Schs A, B 7
Bills of Sale Act (1878) Amendment Act 1882, ss 1, 3–18, Schedule 16
Bills of Exchange Act 1882, ss 1–81, 81A, 83–95, 97, 99, Sch 1 22
Factors Act 1889, ss 1–13, 16, 17 65
Bills of Sale Act 1890, ss 1–3 70
Partnership Act 1890, ss 5, 14, 50 71
Marine Insurance Act 1906, ss 1–91, 94, Sch 1 72
Marine Insurance (Gambling Policies) Act 1909, ss 1, 2 107
Law of Property Act 1925, ss 53, 136, 137, 209 109
Third Parties (Rights Against Insurers) Act 1930, ss 1–3, 5 113
Law Reform (Frustrated Contracts) Act 1943, ss 1–3 117
Cheques Act 1957, ss 1–8 120
Hire-Purchase Act 1964, ss 27–29, 37 123
Carriage of Goods by Road Act 1965, ss 1–8, 8A, 13, 14, Schedule 128
Misrepresentation Act 1967, ss 1–3, 5, 6 152
Trade Descriptions Act 1968, ss 1–10, 12–16, 18–39, 43 154
Carriage of Goods by Sea Act 1971, ss 1–6, Schedule, Arts I–IV,
    IVbis, V–X 177
Powers of Attorney Act 1971, ss 4, 5, 7, 11 189
Unsolicited Goods and Services Act 1971, ss 1–3, 3A, 4–7 192
Supply of Goods (Implied Terms) Act 1973, ss 8–11, 11A, 12, 14,
    15, 18 198
Fair Trading Act 1973, ss 1–3, 12–32, 34–42, 118–125, 129–132, 134,
    137, 138, 140, Schs 1, 4–6 206
Consumer Credit Act 1974, ss 1, 2, 4, 6–37, 39–41, 43–103, 105–121,
    123–142, 145–175, 177, 179–189, 192, 193, Schs 1, 2 255

Contents

Unsolicited Goods and Services (Amendment) Act 1975, ss 3, 4 395

Torts (Interference with Goods) Act 1977, ss 1–8, 10–14, 16, 17, Sch 1 396

Unfair Contract Terms Act 1977, ss 1–7, 9–14, 26, 27, 29, 31, 32, Schs 1, 2 409

Civil Liability (Contribution) Act 1978, ss 1–7, 10 422

Banking Act 1979, ss 47, 52 426

Sale of Goods Act 1979, ss 1–15, 15A, 16–20, 20A, 20B, 21–35, 35A, 36–39, 41–57, 59–62, 64 428

Supply of Goods and Services Act 1982, ss 1–5, 5A, 6–10, 10A, 11–16, 18–20 462

Companies Act 1985, ss 35, 35A, 35B, 36C, 196, 349, 395, 396, 747 476

Insolvency Act 1986, ss 8–11, 15, 17, 27, 29, 40, 43, 44, 86, 122, 123, 127–129, 175, 212–214, 238–241, 244, 245, 247, 249, 251, 344, 423–425, 435, 444 482

Consumer Protection Act 1987, ss 1–8, 10–16, 18–35, 37–47, 49, 50, Sch 2 517

Food Safety Act 1990, ss 2, 8, 14, 15, 60 575

Property Misdescriptions Act 1991, ss 1, 2, 7 579

Timeshare Act 1992, ss 1–9, 10A, 12, 13, Sch 1 582

Carriage of Goods by Sea Act 1992, ss 1–6 603

Arbitration Act 1996, ss 1–106, 108–110, Schs 1, 2 608

Late Payment of Commercial Debts (Interest) Act 1998, ss 1–17 679

Contracts (Rights of Third Parties) Bill, cll 1–8 691

PART II
STATUTORY INSTRUMENTS

Consumer Transactions (Restrictions on Statements) Order 1976, arts 1–5 701

Business Advertisements (Disclosure) Order 1977, arts 1, 2 704

Consumer Protection (Cancellation of Contracts Concluded away from Business Premises) Regulations 1987, regs 1–8, 10, 11, Schedule 705

Control of Misleading Advertisements Regulations 1988, regs 1–11 719

Consumer Protection (Code of Practice for Traders on Price Indications) Approval Order 1988, arts 1, 2, Schedule 727

Consumer Credit (Exempt Agreements) Order 1989, arts 1–6 744

Property Misdescriptions (Specified Matters) Order 1992, arts 1, 2, Schedule 750

Banking Coordination (Second Council Directive) Regulations 1992, regs 1, 2, 18, 19, 57–63, Schs 5, 10 753

Package Travel, Package Holidays and Package Tours Regulations 1992, regs 1–28, Schs 1–3 766

Contents

Commercial Agents (Council Directive) Regulations 1993, regs 1–23,
   Schedule 788
General Product Safety Regulations 1994, regs 1–5, 7–18 800
Unfair Terms in Consumer Contracts Regulations 1994, regs 1–8,
   Schs 1–3 810
Price Indications (Resale of Tickets) Regulations 1994, regs 1–8 817
Trading Schemes Regulations 1997, regs 1–11, Schs 1, 2.................... 821
Trading Schemes (Exclusion) Regulations 1997, regs 1–3 830
Foreign Package Holidays (Tour Operators and Travel Agents) Order
   1998, arts 1–5 832
Telecommunications (Data Protection and Privacy) (Direct Marketing)
   Regulations 1998, regs 1–15 834

PART III
EUROPEAN MATERIAL

Directive of the European Parliament and of the Council of 27
   January 1997 on cross-border credit transfers, Arts 1–14 847
Directive of the European Parliament of 20 May 1997 on the
   protection of consumers in respect of distance contracts,
   Arts 1–19, Annexes I, II 858
Commission Recommendation of 30 July 1997 concerning trans-
   actions by electronic payment instruments and in particular
   the relationship between issuer and holder, Arts 1–11 873
Directive of the European Parliament and of the Council of 16
   February 1998 on consumer protection in the indication of
   the prices of products offered to consumers, Arts 1–14 883
Directive of the European Parliament and of the Council of 19
   May 1998 on settlement finality in payment and securities
   settlement systems, Arts 1–14 890
Amended Proposal for a European Parliament and Council Directive
   combating late payment in commercial transactions, Arts 1–12,
   Annex 901
Common Position (EC) No 51/98 adopted by the Council on 24
   September 1998 with a view to adopting European Parliament
   and Council Directive on certain aspects of the sale of consumer
   goods and associated guarantees, Arts 1–12 912
Proposal for a Directive of the European Parliament and of the
   Council concerning the distance marketing of consumer
   financial services and amending Council Directive 90/619/EEC
   and Directives 97/7/EC and 98/27/EC, Arts 1–13, 17–19, Annex .. 923
Proposal for a European Parliament and Council Directive on certain
   legal aspects of electronic commerce in the internal market,
   Arts 1–3, 6–11, 22, 25–27, Annexes I, II....................................... 937

PART IV
INTERNATIONAL MATERIALS

United Nations Convention on the Carriage of Goods by Sea, 1978
(Hamburg Rules), Arts 1–26, 29–31 955

United Nations Convention on Contracts for the International Sale
of Goods, 1980 (Vienna Sales Convention), Arts 1–88, 90, 92–96,
98, 99 975

Uniform Customs and Practice for Documentary Credits (1993
Revision, ICC Publication No 500) 1007

PART V
CODES

The Banking Code (3rd Revised Edition) 1041

PART I

STATUTES

Statute of Frauds (1677)

(C 3)

An Act for prevention of Frauds and Perjuryes

[1 January 1677]

4   No action against executors, etc, upon a special promise, or upon any agreement, or contract for sale of lands, etc, unless agreement, etc, be in writing, and signed

... noe action shall be brought ... whereby to charge the defendant upon any speciall promise to answere for the debt default or miscarriages of another person ... unlesse the agreement upon which such action shall be brought or some memorandum or note thereof shall be in writeing and signed by the partie to be charged therewith or some other person thereunto by him lawfully authorized.

NOTES

Words omitted in the first place repealed by the Statute Law Revision Act 1883 and the Statute Law Revision Act 1948; words omitted in the second place repealed by the Law Reform (Enforcement of Contracts) Act 1954, s 1; words omitted in the third place repealed by the Law of Property Act 1925, s 207, Sch 7, and the Law Reform (Enforcement of Contracts) Act 1954, s 1.

Life Assurance Act 1774

(C 48)

An Act for regulating Insurances upon Lives, and for prohibiting all such Insurances except in cases where the Persons insuring shall have an Interest in the Life or Death of the Persons insured

Whereas it hath been found by experience that the making insurances on lives or other events wherein the assured shall have no interest hath introduced a mischievous kind of gaming:

1   No insurance to be made on lives, etc, by persons having no interest etc

From and after the passing of this Act no insurance shall be made by any person or persons, bodies politick or corporate, on the life or lives of any person or persons, or on any other event or events whatsoever, wherein the person or persons for whose use, benefit, or on whose account such policy or policies shall be made, shall have no interest, or by way of gaming or wagering; and that every assurance made contrary to the true intent and meaning hereof shall be null and void to all intents and purposes whatsoever.

2 No policies on lives without inserting the names of persons interested, etc

And ... it shall not be lawful to make any policy or policies on the life or lives of any person or persons, or other event or events, without inserting in such policy or policies the person or persons name or names interested therein, or for whose use, benefit, or on whose account such policy is so made or underwrote.

NOTES

Words omitted repealed by the Statute Law Revision Act 1888.

3 How much may be recovered where the insured hath interest in lives

And ... in all cases where the insured hath interest in such life or lives, event or events, no greater sum shall be recovered or received from the insurer or insurers than the amount of value of the interest of the insured in such life or lives, or other event or events.

NOTES

Words omitted repealed by the Statute Law Revision Act 1888.

4 Not to extend to insurances on ships, goods, etc

Provided, always, that nothing herein contained shall extend or be construed to extend to insurances bona fide made by any person or persons on ships, goods, or merchandises, but every such insurance shall be as valid and effectual in the law as if this Act had not been made.

Bills of Lading Act 1855

(C 111)

An Act to amend the Law relating to Bills of Lading

[14 August 1855]

NOTES

The short title was given to this Act by the Short Titles Act 1896.

WHEREAS by the custom of merchants, a bill of lading of goods being transferable by endorsement, the property in the goods may thereby pass to the endorsee, but nevertheless all rights in respect of the contract contained in the bill of lading continue in the original shipper or owner; and it is expedient that such rights should pass with the property: And whereas it frequently happens that the goods in respect of which bills of lading purport to be signed have not been laden on board, and it is proper that such bills of lading in the hands of a bonâ fide holder for value should not be questioned by the master or other person signing the same on the ground of the goods not having been laden as aforesaid:

1 Consignees and endorsees of bills of lading empowered to sue

Every consignee of goods named in a bill of lading, and every endorsee of a bill of lading, to whom the property in the goods therein mentioned shall pass upon or by reason of such consignment or endorsement, shall have transferred to and vested in him all rights of suit, and be subject to the same liabilities in respect of such goods as if the contract contained in the bill of lading had been made with himself.

NOTES

Repealed by the Carriage of Goods by Sea Act 1992, s 6.

2 Saving as to stoppage in transitu, and claims for freight, &c

Nothing herein contained shall prejudice or affect any right of stoppage in transitu, or any right to claim freight against the original shipper or owner, or any liability of the consignee or endorsee by reason or in consequence of his being such consignee or endorsee, or of his receipt of the goods by reason or in consequence of such consignment or endorsement.

3 Bill of lading in hands of consignee, &c conclusive evidence of shipment as against master, &c

Every bill of lading in the hands of a consignee or endorsee for valuable consideration, representing goods to have been shipped on board a vessel, shall be conclusive evidence of such shipment as against the master or other person signing the same, notwithstanding that such goods or some part thereof may not have been so shipped, unless such holder of the bill of lading shall have had actual notice at the time of receiving the same that the goods had not been in fact laden on board: Provided, that the master or other person so signing may exonerate himself in respect of such misrepresentation by showing that it was caused without any default on his part, and wholly by the fraud of the shipper, or of the holder, or some person under whom the holder claims.

Mercantile Law Amendment Act 1856

(C 97)

An Act to amend the Laws of England and Ireland affecting Trade and Commerce

[29 July 1856]

5 Surety who discharges the liability to be entitled to assignment of all securities held by the creditor, and to stand in the place of the creditor

Every person who, being surety for the debt or duty of another, or being liable with another for any debt or duty, shall pay such debt or perform such duty, shall be entitled to have assigned to him, or to a trustee for him, every judgment specialty, or other security which shall be held by the creditor in respect of such debt or duty, whether such judgment, specialty, or other security shall or shall not be deemed at law to have been satisfied by the payment of the debt or performance of the duty, and such person shall be entitled to stand in the place of the creditor, and to use all the remedies, and, if need be, and upon a proper indemnity, to use the name of the creditor, in any action or other proceeding, at law or in equity, in order to obtain from the principal debtor, or any co-surety, co-contractor, or co-debtor, as the case may be, indemnification for the advances made and loss sustained by the person who shall have so paid such debt or performed such duty, and such payment or performance so made by such surety shall not be pleadable in bar of any such action or other proceeding by him: Provided always, that no co-surety, co-contractor, or co-debtor shall be entitled to recover from any other co-surety, co-contractor, or co-debtor, by the means aforesaid, more than the just proportion to which, as between those parties themselves, such last-mentioned person shall be justly liable.

16   Short title

In citing this Act, it shall be sufficient to use the expression "The Mercantile Law Amendment Act 1856."

Bills of Sale Act 1878

(C 31)

An Act to consolidate and amend the Law for preventing Frauds upon Creditors by secret Bills of Sale of Personal Chattels

[22 July 1878]

1   Short title

This Act may be cited for all purposes as the Bills of Sale Act 1878.

3   Application

This Act shall apply to every bill of sale executed on or after the first day of January one thousand eight hundred and seventy-nine (whether the same be absolute, or subject or not subject to any trust) whereby the holder or grantee has power, either with or without notice, and either immediately or at any future time, to seize or take possession of any personal chattels comprised in or made subject to such bill of sale.

Application, construction and partial repeal of this Act: this Act remains in force so far as it relates to absolute bills of sale executed after its commencement; as to bills of sale given as security for the payment of money it is to be construed as one with the Bills of Sale Act (1878) Amendment Act 1882 (by virtue of s 3 of the 1882 Act), but as to such bills of sale only, this Act is repealed as far as it is inconsistent with the 1882 Act (by virtue of s 15 of the 1882 Act).

4 Interpretation of terms

In this Act the following words and expressions shall have the meanings in this section assigned to them respectively, unless there be something in the subject or context repugnant to such construction; (that is to say),

The expression "bill of sale" shall include bills of sale, assignments, transfers, declarations of trust without transfer, inventories of goods with receipt thereto attached, or receipts for purchase moneys of goods, and other assurances of personal chattels, and also powers of attorney, authorities, or licenses to take possession of personal chattels as security for any debt, and also any agreement, whether intended or not to be followed by the execution of any other instrument, by which a right in equity to any personal chattels, or to any charge or security thereon, shall be conferred, but shall not include the following documents; that is to say, assignments for the benefit of the creditors of the person making or giving the same, marriage settlements, transfers or assignments of any ship or vessel or any share thereof, transfers of goods in the ordinary course of business of any trade or calling, bills of sale of goods in foreign parts or at sea, bills of lading, India warrants, warehouse-keepers' certificates, warrants or orders for the delivery of goods, or any other documents used in the ordinary course of business as proof of the possession or control of goods, or authorising or purporting to authorise, either by indorsement or by delivery, the possessor of such document to transfer or receive goods thereby represented:

The expression "personal chattels" shall mean goods, furniture, and other articles capable of complete transfer by delivery, and (when separately assigned or charged) fixtures and growing crops, but shall not include chattel interests in real estate, nor fixtures (except trade machinery as herein-after defined), when assigned together with a freehold or leasehold interest in any land or building to which they are affixed, nor growing crops when assigned together with any interest in the land on which they grow, nor shares or interests in the stock, funds, or securities of any government, or in the capital or property of incorporated or joint stock companies, nor choses in action, nor any stock or produce upon any farm or lands which by virtue of any covenant or agreement or of the custom of the country ought not to be removed from any farm where the same are at the time of making or giving of such bill of sale:

Personal chattels shall be deemed to be in the "apparent possession" of the person making or giving a bill of sale, so long as they remain or are in or upon any house, mill, warehouse, building, works, yard, land, or other premises occupied by him, or are used and enjoyed by him in any place whatsoever, notwithstanding that formal possession thereof may have been taken by or given to any other person:

"Prescribed" means prescribed by rules made under the provisions of this Act.

5 Application of Act to trade machinery

From and after the commencement of this Act trade machinery shall, for the purposes of this Act, be deemed to be personal chattels, and any mode of disposition of trade machinery by the owner thereof which would be a bill of sale as to any other personal chattels shall be deemed to be a bill of sale within the meaning of this Act.

For the purposes of this Act—

"Trade machinery" means the machinery used in or attached to any factory or workshop;

1st Exclusive of the fixed motive-powers, such as the water-wheels and steam-engines, and the steam-boilers, donkey-engines, and other fixed appurtenances of the said motive-powers; and,

2nd Exclusive of the fixed power machinery, such as the shafts, wheels, drums, and their fixed appurtenances, which transmit the action of the motive-powers to the other machinery, fixed and loose, and

3rd Exclusive of the pipes for steam gas and water in the factory or workshop.

The machinery or effects excluded by this section from the definition of trade machinery shall not be deemed to be personal chattels within the meaning of this Act.

"Factory or workshop" means any premises on which any manual labour is exercised by way of trade, or for purposes of gain, in or incidental to the following purposes or any of them; that is to say,

    (a) In or incidental to the making any article or part of an article; or

    (b) In or incidental to the altering, repairing, ornamenting, finishing, of any article; or

    (c) In or incidental to the adapting for sale any article.

6 Certain instruments giving powers of distress to be subject to this Act

Every attornment instrument or agreement, not being a mining lease, whereby a power of distress is given or agreed to be given by any person to any other person by way of security for any present future or contingent debt or advance, and whereby any rent is reserved or made payable as a mode of providing for the payment of interest on such debt or advance, or otherwise for the purpose of such security only, shall be deemed to be a bill of sale, within the meaning of this Act, of any personal chattels which may be seized or taken under such power of distress.

Provided, that nothing in this section shall extend to any mortgage of any estate or interest in any land tenement or hereditament which the mortgagee, being in possession, shall have demised to the mortgagor as his tenant at a fair and reasonable rent.

7 Fixtures or growing crops not to be deemed separately assigned when the land passes by the same instrument

No fixtures or growing crops shall be deemed, under this Act, to be separately assigned or charged by reason only that they are assigned by separate words, or that power is given to sever them from the land or building to which they are affixed, or from the land on which they grow, without otherwise taking possession of or dealing with such land or building, or land, if by the same instrument any freehold or leasehold interest in the land or building to which such fixtures are affixed, or in the land on which such crops grow, is also conveyed or assigned to the same persons or person.

The same rule of construction shall be applied to all deeds or instruments, including fixtures or growing crops, executed before the commencement of this Act, and then subsisting and in force, in all questions arising under any bankruptcy liquidation assignment for the benefit of creditors, or execution of any process of any court, which shall take place or be issued after the commencement of this Act.

8 Avoidance of unregistered bill of sale in certain cases

Every bill of sale to which this Act applies shall be duly attested and shall be registered under this Act, within seven days after the making or giving thereof, and shall set forth the consideration for which such bill of sale was given, otherwise such bill of sale, as against all trustees or assignees of the estate of the person whose chattels, or any of them, are comprised in such bill of sale under the law relating to bankruptcy or liquidation, or under any assignment for the benefit of the creditors of such person, and also as against all sheriffs officers and other persons seizing any chattels comprised in such bill of sale, in the execution of any process of any court authorising the seizure of the chattels of the person by whom or of whose chattels such bill has been made, and also as against every person on whose behalf such process shall have been issued, shall be deemed fraudulent and void so far as regards the property in or right to the possession of any chattels comprised in such bill of sale which, at or after the time of filing the petition for bankruptcy or liquidation, or of the execution of such assignment, or of executing such process (as the case may be), and after the expiration of such seven days are in the possession or apparent possession of the person making such bill of sale (or of any person against whom the process has issued under or in the execution of which such bill has been made or given, as the case may be).

NOTES

This section is in terms repealed by the Bills of Sale Act (1878) Amendment Act 1882, s 15, but notwithstanding that, the effect of the Bills of Sale Act (1878) Amendment Act 1882, s 3, is that the repeal applies only to bills of sale to which the later Act applies, so that this section remains in force so far as regards bills of sale not given as a security for money (ie absolute bills of sale).

9 Avoidance of certain duplicate bills of sale

Where a subsequent bill of sale is executed within or on the expiration of seven days after the execution of a prior unregistered bill of sale, and comprises all or any part of the personal chattels comprised in such prior bill of sale, then, if such subsequent bill of sale is given as a security for the same debt as is secured by the prior bill of sale, or for any part of such debt, it shall, to the extent to which it is a security for the same debt or part thereof, and so far as respects the personal chattels or part thereof comprised in the prior bill, be absolutely void, unless it is proved to the satisfaction of the court having cognizance of the case that the subsequent bill of sale was bona fide given for the purpose of correcting some material error in the prior bill of sale, and not for the purpose of evading this Act.

10　Mode of registering bills of sale

A bill of sale shall be attested and registered under this Act in the following manner—

(1) The execution of every bill of sale shall be attested by a solicitor of the Supreme Court, and the attestation shall state that before the execution of the bill of sale the effect thereof has been explained to the grantor by the attesting solicitor.

(2) Such bill, with every schedule or inventory thereto annexed or therein referred to, and also a true copy of such bill and of every such schedule or inventory, and of every attestation of the execution of such bill of sale, together with an affidavit of the time of such bill of sale being made or given, and of its due execution and attestation, and a description of the residence and occupation of the person making or giving the same (or in case the same is made or given by any person under or in the execution of any process, then a description of the residence and occupation of the person against whom such process issued), and of every attesting witness to such bill of sale, shall be presented to and the said copy and affidavit shall be filed with the registrar within seven clear days after the making or giving of such bill of sale, in like manner as a warrant of attorney in any personal action given by a trader is now by law required to be filed:

(3) If the bill of sale is made or given subject to any defeasance or condition, or declaration of trust not contained in the body thereof such defeasance, condition, or declaration shall be deemed to be part of the bill, and shall be written on the same paper or parchment therewith before the registration, and shall be truly set forth in the copy filed under this Act therewith and as part thereof, otherwise the registration shall be void.

In case two or more bills of sale are given, comprising in whole or in part any of the same chattels, they shall have priority in the order of the date of their registration respectively as regards such chattels.

A transfer or assignment of a registered bill of sale need not be registered.

NOTES

Sub-s (1): as regards absolute bills of sale the requirements of sub-s (1) must still be observed, for, although repealed by the Bills of Sale Act (1878) Amendment

Act 1882, s 10, it remains in force so far as absolute bills are concerned by virtue of s 3 of that Act.

11 Renewal of registration

The registration of a bill of sale, whether executed before or after the commencement of this Act, must be renewed once at least every five years, and if a period of five years elapses from the registration or renewed registration of a bill of sale without a renewal or further renewal (as the case may be), the registration shall become void.

The renewal of a registration shall be effected by filing with the registrar an affidavit stating the date of the bill of sale and of the last registration thereof, and the names, residences, and occupations of the parties thereto as stated therein, and that the bill of sale is still a subsisting security.

Every such affidavit may be in the form set forth in the schedule (A) to this Act annexed.

A renewal of registration shall not become necessary by reason only of a transfer or assignment of a bill of sale.

12 Form of register

The registrar shall keep a book (in this Act called "the register") for the purposes of this Act, and shall, upon the filing of any bill of sale or copy under this Act, enter therein in the form set forth in the second schedule (B) to this Act annexed, or in any other prescribed form, the name, residence and occupation of the person by whom the bill was made or given (or in case the same was made or given by any person under or in the execution of process, then the name residence and occupation of the person against whom such process was issued, and also the name of the person or persons to whom or in whose favour the bill was given), and the other particulars shown in the said schedule or to be prescribed under this Act, and shall number all such bills registered in each year consecutively, according to the respective dates of their registration.

Upon the registration of any affidavit of renewal the like entry shall be made, with the addition of the date and number of the last previous entry relating to the same bill, and the bill of sale or copy originally filed shall be thereupon marked with the number affixed to such affidavit of renewal.

The registrar shall also keep an index of the names of the grantors of registered bills of sale with reference to entries in the register of the bills of sale given by each such grantor.

Such index shall be arranged in divisions corresponding with the letters of the alphabet, so that all grantors whose surnames begin with the same letter (and no others) shall be comprised in one division, but the arrangement within each such division need not be strictly alphabetical.

13   The registrar

The masters of the Supreme Court of Judicature attached to the Queen's Bench Division of the High Court of Justice, or such other officers as may for the time being be assigned for this purpose under the provisions of the Supreme Court of Judicature Acts, 1873 and 1875, shall be the registrar for the purposes of this Act, and any one of the said masters may perform all or any of the duties of the registrar.

14   Rectification of register

Any judge of the High Court of Justice on being satisfied that the omission to register a bill of sale or an affidavit or renewal thereof within the time prescribed by this Act, or the omission or mis-statement of the name residence or occupation of any person, was accidental or due to inadvertence, may in his discretion order such omission or mis-statement to be rectified by the insertion in the register of the true name residence or occupation, or by extending the time for such registration on such terms and conditions (if any) as to security, notice by advertisement or otherwise, or as to any other matter, as he thinks fit to direct.

15   Entry of satisfaction

Subject to and in accordance with any rules to be made under and for the purposes of this Act, the registrar may order a memorandum of satisfaction to be written upon any registered copy of a bill of sale, upon the prescribed evidence being given that the debt (if any) for which such bill of sale was made or given has been satisfied or discharged.

16   Copies may be taken, etc

Any person shall be entitled to have an office copy or extract of any registered bill of sale, and affidavit of execution filed therewith, or copy thereof, and of any affidavit filed therewith, if any, or registered affidavit of renewal, upon paying for the same at the like rate as for office copies of judgments of the High Court of Justice, and any copy of a registered bill of sale, and affidavit purporting to be an office copy thereof, shall in all courts and before all arbitrators or other persons, be admitted as prima facie evidence thereof, and of the fact and date of registration as shown thereon …

NOTES

Words omitted repealed by the Bills of Sale Act (1878) Amendment Act 1882, s 16.

20 Order and disposition

Chattels comprised in a bill of sale which has been and continues to be duly registered under this Act shall not be deemed to be in the possession, order, or disposition of the grantor of the bill of sale within the meaning of the Bankruptcy Act 1869.

NOTES

The provisions of this section are not affected, so far as regards absolute bills of sale, by the repeal of the section by the Bills of Sale Act (1878) Amendment Act 1882, s 15, the effect of s 3 of the 1882 Act being that the repeal only applies to bills which are within the latter Act (ie bills given as security for the payment of money).

21 Rules

Rules for the purposes of this Act may be made and altered from time to time by the like persons and in the like manner in which rules and regulations may be made under and for the purposes of the Supreme Court of Judicature Acts 1873 and 1875.

22 Time for registration

When the time for registering a bill of sale expires on a Sunday, or other day on which the registrar's office is closed, the registration shall be valid if made on the next following day on which the office is open.

24 Extent of Act

This Act shall not extend to Scotland or to Ireland.

SCHEDULES

SCHEDULE A

Section 11

I [*AB*] of do swear that a bill of sale, bearing date the day of 18 [*insert the date of the bill,*] and made between [*insert the names and descriptions of the parties in the original bill of sale*] and which said bill of sale [*or, and a copy of which said bill of sale, as the case may be*] was registered on the day of 18 [*insert date of registration*], is still a subsisting security.

Sworn, & *c*.

SCHEDULE B

Section 12

Satis-faction entered	No	By whom given (or against whom process issued)			To whom given	Nature of Instru-ment	Date	Date of Regist-ration	Date of Regis-tration of affidavit of renewal
		Name	Resi-dence	Occu-pation					

Bills of Sale Act (1878) Amendment Act 1882

(C 43)

An Act to amend the Bills of Sale Act 1878

[18 August 1882]

1 Short title

This Act may be cited for all purposes as the Bills of Sale Act (1878) Amendment Act 1882; and this Act and the Bills of Sale Act 1878 may be cited together as the Bills of Sale Acts 1878 and 1882.

3 Construction of Act

The Bills of Sale Act 1878 is herein-after referred to as "the principal Act," and this Act shall, so far as is consistent with the tenor thereof, be construed as one with the principal Act; but unless the context otherwise requires shall not apply to any bill of sale duly registered before the commencement of this Act so long as the registration thereof is not avoided by non-renewal or otherwise.

The expression "bill of sale," and other expressions in this Act, have the same meaning as in the principal Act, except as to bills of sale or other documents mentioned in section four of the principal Act, which may be given otherwise than by way of security for the payment of money, to which last-mentioned bills of sale and other documents this Act shall not apply.

4 Bill of sale to have schedule of property attached thereto

Every bill of sale shall have annexed thereto or written thereon a schedule containing an inventory of the personal chattels comprised in the bill of sale; and such bill of sale, save as herein-after mentioned, shall have effect only in respect of the personal chattels specifically described in the said schedule; and shall be void, except as against the grantor, in respect of any personal chattels not so specifically described.

5 Bill of sale not to affect after acquired property

Save as herein-after mentioned, a bill of sale shall be void, except as against the grantor, in respect of any personal chattels specifically described in the schedule thereto of which the grantor was not the true owner at the time of the execution of the bill of sale.

6 Exception as to certain things

Nothing contained in the foregoing sections of this Act shall render a bill of sale void in respect of any of the following things; (that is to say,)
  (1) Any growing crops separately assigned or charged where such crops were actually growing at the time when the bill of sale was executed.
  (2) Any fixtures separately assigned or charged, and any plant, or trade machinery where such fixtures, plant, or trade machinery are used in, attached to, or brought upon any land, farm, factory, workshop, shop, house, warehouse, or other place in substitution for any of the

like fixtures, plant, or trade machinery specifically described in the schedule to such bill of sale.

7 Bill of sale with power to seize except in certain events to be void

Personal chattels assigned under a bill of sale shall not be liable to be seized or taken possession of by the grantee for any other than the following causes:—

(1) If the grantor shall make default in payment of the sum or sums of money thereby secured at the time therein provided for payment, or in the performance of any covenant or agreement contained in the bill of sale and necessary for maintaining the security;

(2) If the grantor shall become a bankrupt, or suffer the said goods or any of them to be distrained for rent, rates, or taxes;

(3) If the grantor shall fraudulently either remove or suffer the said goods, or any of them, to be removed from the premises;

(4) If the grantor shall not, without reasonable excuse, upon demand in writing by the grantee, produce to him his last receipts for rent, rates, and taxes;

(5) If execution shall have been levied against the goods of the grantor under any judgment at law:

Provided that the grantor may within five days from the seizure or taking possession of any chattels on account of any of the above-mentioned causes, apply to the High Court, or to a judge thereof in chambers, and such court or judge, if satisfied that by payment of money or otherwise the said cause of seizure no longer exists, may restrain the grantee from removing or selling the said chattels, or may make such other order as may seem just.

[7A Defaults under consumer credit agreements

(1) Paragraph (1) of section 7 of this Act does not apply to a default relating to a bill of sale given by way of security for the payment of money under a regulated agreement to which section 87(1) of the Consumer Credit Act 1974 applies—

(a) unless the restriction imposed by section 88(2) of that Act has ceased to apply to the bill of sale; or

(b) if, by virtue of section 89 of that Act, the default is to be treated as not having occurred.

(2) Where paragraph (1) of section 7 of this Act does apply in relation to a bill of sale such as is mentioned in subsection (1) of this section, the

proviso to that section shall have effect with the substitution of "county court" for "High Court".]

8 Bill of sale to be void unless attested and registered

Every bill of sale shall be duly attested, and shall be registered under the principal Act within seven clear days after the execution thereof, or if it is executed in any place out of England then within seven clear days after the time at which it would in the ordinary course of post arrive in England if posted immediately after the execution thereof; and shall truly set forth the consideration for which it was given; otherwise such bill of sale shall be void in respect of the personal chattels comprised therein.

9 Form of bill of sale

A bill of sale made or given by way of security for the payment of money by the grantor thereof shall be void unless made in accordance with the form in the schedule to this Act annexed.

10 Attestation

The execution of every bill of sale by the grantor shall be attested by one or more credible witness or witnesses, not being a party or parties thereto ...

11 Local registration of contents of bills of sale

Where the affidavit (which under section ten of the principal Act is required to accompany a bill of sale when presented for registration) describes the residence of the person making or giving the same or of the person against whom the process is issued to be in some place outside [the London insolvency district] or where the bill of sale describes the chattels enumerated therein as being in some place outside [the London insolvency district], the registrar under the principal Act shall forthwith and within three clear days after registration in the principal registry, and in accordance with the prescribed directions, transmit an abstract in the prescribed form of the contents of such bill of sale to the county court registrar in whose district such places are situate, and if such places are in the districts of different registrars to each such registrar.

Every abstract so transmitted shall be filed, kept, and indexed by the registrar of the county court in the prescribed manner, and any person may search, inspect, make extracts from, and obtain copies of the abstract so registered in the like manner, and upon the like terms as to payment or otherwise as near as may be as in the case of bills of sale registered by the registrar under the principal Act.

NOTES

Words in square brackets substituted by the Insolvency Act 1985, s 235(1), Sch 8, para 1. For savings, see the Insolvency Act 1986, s 437, Sch 11, Pt II para 10.

12   Bill of sale under £30 to be void

Every bill of sale made or given in consideration of any sum under thirty pounds shall be void.

13   Chattels not to be removed or sold

All personal chattels seized or of which possession is taken … under or by virtue of any bill of sale (whether registered before or after the commencement of this Act), shall remain on the premises where they were so seized or so taken possession of, and shall not be removed or sold until after the expiration of five clear days from the day they were so seized or so taken possession of.

NOTES

Words omitted repealed by the Statute Law Revision Act 1898.

14   Bill of sale not to protect chattels against poor and parochial rates

A bill of sale to which this Act applies shall be no protection in respect of personal chattels included in such bill of sale which but for such bill of sale would have been liable to distress under a warrant for the recovery of taxes and poor and other parochial rates.

15 Repeal of part of Bills of Sale Act 1878

... all ... enactments contained in the principal Act which are inconsistent with this Act are repealed ...

NOTES

Words omitted repealed by the Statute Law Revision Act 1898.

16 Inspection of registered bills of sale

... any person shall be entitled at all reasonable times to search the register on payment of a fee of one shilling, or such other fee as may be prescribed, and subject to such regulations as may be prescribed, and shall be entitled at all reasonable times to inspect, examine, and make extracts from any and every registered bill of sale without being required to make a written application, or to specify any particulars in reference thereto, upon payment of one shilling for each bill of sale inspected, and such payment shall be made by a judicature stamp: Provided that the said extracts shall be limited to the dates of execution registration, renewal of registration, and satisfaction, to the names, addresses, and occupations of the parties, to the amount of the consideration, and to any further prescribed particulars.

NOTES

Words omitted repealed by the Statute Law Revision Act 1898.

17 Debentures to which Act not to apply

Nothing in this Act shall apply to any debentures issued by any mortgage, loan or other incorporated company, and secured upon the capital stock or goods, chattels, and effects of such company.

18 Extent of Act

This Act shall not extend to Scotland or Ireland.

SCHEDULE

Section 9

FORM OF BILL OF SALE

This Indenture made the day of, between *AB* of the one part, and *CD* of the other part, witnesseth that in consideration of the sum of £   now paid to *AB* by *CD* the receipt of which the said *AB* hereby acknowledges [*or whatever else the consideration may be*], he the said *AB* doth hereby assign unto *CD*, his executors, administrators, and assigns, all and singular the several chattels and things specifically described in the schedule hereto annexed by way of security for the payment of the sum of £   , and interest thereon at the rate of per cent per annum [*or whatever else may be the rate*]. And the said *AB* doth further agree and declare that he will duly pay to the said *CD* the principal sum aforesaid, together with the interest then due, by equal payments of £   on the day of [*or whatever else may be the stipulated times or time of payment*]. And the said *AB* doth also agree with the said *CD* that he will [*here insert terms as to insurance, payment of rent or otherwise, which the parties may agree to for the maintenance or defeasance of the security*].

Provided always, that the chattels hereby assigned shall not be liable to seizure or to be taken possession of by the said *CD* for any cause other than those specified in section seven of the Bills of Sale Act (1878) Amendment Act 1882.

In Witness, &c.

Signed and sealed by the said *AB* in the presence of me *EF* [*add witness' name, address, and description*].

Bills of Exchange Act 1882

(C 61)

An Act to codify the law relating to Bills of Exchange, Cheques, and Promissory Notes

[18 August 1882]

NOTES

By the Bills of Exchange (Time of Noting) Act 1917, s 2, the Bills of Exchange Act 1882 and the 1917 Act may be cited by the collective title of the Bills of Exchange Acts 1882 to 1917.

PART I
PRELIMINARY

1   Short title

This Act may be cited as the Bills of Exchange Act, 1882.

2   Interpretation of terms

In this Act, unless the context otherwise requires,—

"Acceptance" means an acceptance completed by delivery or notification.

"Action" includes counter claim and set off.

"Banker" includes a body of persons whether incorporated or not who carry on the business of banking.

"Bankrupt" includes any person whose estate is vested in a trustee or assignee under the law for the time being in force relating to bankruptcy.

"Bearer" means the person in possession of a bill or note which is payable to bearer.

"Bill" means bill of exchange, and "note" means promissory note.

"Delivery" means transfer of possession, actual or constructive, from one person to another.

"Holder" means the payee or indorsee of a bill or note who is in possession of it, or the bearer thereof.

"Indorsement" means an indorsement completed by delivery.

"Issue" means the first delivery of a bill or note, complete in form to a person who takes it as a holder.

"Person" includes a body of persons whether incorporated or not.

"Value" means valuable consideration.

"Written" includes printed, and "writing" includes print.

PART II
BILLS OF EXCHANGE

Form and interpretation

3   Bill of exchange defined

(1)   A bill of exchange is an unconditional order in writing, addressed by one person to another, signed by the person giving it, requiring the person

to whom it is addressed to pay on demand or at a fixed or determinable future time a sum certain in money to or to the order of a specified person, or to bearer.

(2)   An instrument which does not comply with these conditions, or which orders any act to be done in addition to the payment of money, is not a bill of exchange.

(3)   An order to pay out of a particular fund is not unconditional within the meaning of this section; but an unqualified order to pay, coupled with (a) an indication of a particular fund out of which the drawee is to re-imburse himself or a particular account to be debited with the amount, or (b) a statement of the transaction which gives rise to the bill, is unconditional.

(4)   A bill is not invalid by reason—
  (a) That it is not dated;
  (b) That it does not specify the value given, or that any value has been given therefor;
  (c) That it does not specify the place where it is drawn or the place where it is payable.

NOTES

   Bill of exchange: a bill of exchange drawn on or after 15 February 1971 is invalid if the sum payable is an amount of money wholly or partly in shillings or pence, see the Decimal Currency Act 1969, s 2(1).

4   Inland and foreign bills

(1)   An inland bill is a bill which is or on the face of it purports to be (a) both drawn and payable within the British Islands, or (b) drawn within the British Islands upon some person resident therein. Any other bill is a foreign bill.

For the purposes of this Act "British Islands" mean any part of the United Kingdom of Great Britain and Ireland the islands of Man, Guernsey, Jersey Alderney, and Sark, and the islands adjacent to any of them being part of the dominions of Her Majesty.

(2)   Unless the contrary appear on the face of the bill the holder may treat it as an inland bill.

5 Effect where different parties to bill are the same person

(1)   A bill may be drawn payable to, or to the order of, the drawer; or it may be drawn payable to, or to the order of, the drawee.

(2)   Where in a bill drawer and drawee are the same person, or where the drawee is a fictitious person or a person not having capacity to contract, the holder may treat the instrument, at his option, either as a bill of exchange or as a promissory note.

6 Address to drawee

(1)   The drawee must be named or otherwise indicated in a bill with reasonable certainty.

(2)   A bill may be addressed to two or more drawees whether they are partners or not, but an order addressed to two drawees in the alternative or to two or more drawees in succession is not a bill of exchange.

7 Certainty required as to payee

(1)   Where a bill is not payable to bearer, the payee must be named or otherwise indicated therein with reasonable certainty.

(2)   A bill may be made payable to two or more payees jointly, or it may be made payable in the alternative to one of two, or one or some of several payees. A bill may also be made payable to the holder of an office for the time being.

(3)   Where the payee is a fictitious or non-existing person the bill may be treated as payable to bearer.

8 What bills are negotiable

(1)   When a bill contains words prohibiting transfer, or indicating an intention that it should not be transferable, it is valid as between the parties thereto, but is not negotiable.

(2)   A negotiable bill may be payable either to order or to bearer.

(3)   A bill is payable to bearer which is expressed to be so payable, or on which the only or last indorsement is an indorsement in blank.

(4)   A bill is payable to order which is expressed to be so payable, or which is expressed to be payable to a particular person, and does not contain words prohibiting transfer or indicating an intention that it should not be transferable.

(5)   Where a bill, either originally or by indorsement, is expressed to be payable to the order of a specified person, and not to him or his order, it is nevertheless payable to him or his order at his option.

9   Sum payable

(1)   The sum payable by a bill is a sum certain within the meaning of this Act, although it was required to be paid—
   (a) With interest.
   (b) By stated instalments.
   (c) By stated instalments, with a provision that upon default in payment of any instalment the whole shall become due.
   (d) According to an indicated rate of exchange or according to a rate of exchange to be ascertained as directed by the bill.

(2)   Where the sum payable is expressed in words and also in figures, and there is a discrepancy between the two, the sum denoted by the words is the amount payable.

(3)   Where a bill is expressed to be payable with interest, unless the instrument otherwise provides, interest runs from the date of the bill, and if the bill is undated from the issue thereof.

10   Bill payable on demand

(1)   A bill is payable on demand—
   (a) Which is expressed to be payable on demand, or at sight, or on presentation; or
   (b) In which no time for payment was expressed.

(2)   Where a bill is accepted or indorsed when it is overdue, it shall, as regards the acceptor who so accepts, or any indorser who so indorses it, be deemed a bill payable on demand.

11   Bill payable at a future time

A bill is payable at a determinable future time within the meaning of this Act which is expressed to be payable—
(1) At a fixed period after date or sight.
(2) On or at a fixed period after the occurrence of a specified event which is certain to happen, though the time of happening may be uncertain.

An instrument expressed to be payable on a contingency is not a bill, and the happening of the event does not cure the defect.

12   Omission of date in bill payable after date

Where a bill expressed to be payable at a fixed period after date is issued undated, or where the acceptance of a bill payable at a fixed period after sight is undated, any holder may insert therein the true date of issue or acceptance, and the bill shall be payable accordingly.

Provided that (1) where the holder in good faith and by mistake inserts a wrong date, and (2) in every case where a wrong date is inserted, if the bill subsequently comes into the hands of a holder in due course the bill shall not be avoided thereby, but shall operate and be payable as if the date so inserted had been the true date.

13   Ante-dating and post-dating

(1)   Where a bill or an acceptance or any indorsement on a bill is dated, the date shall, unless the contrary be proved, be deemed to be the true date of the drawing, acceptance, or indorsement, as the case may be.

(2)   A bill is not invalid by reason only that it is ante-dated or post-dated, or that it bears date on a Sunday.

14   Computation of time of payment

Where a bill is not payable on demand the day on which it falls due is determined as follows—
[(1) The bill is due and payable in all cases on the last day of the time of payment as fixed by the bill or, if that is a non-business day, on the succeeding business day.]

(2) Where a bill is payable at a fixed period after date, after sight, or after the happening of a specified event, the time of payment is determined by excluding the day from which the time is to begin to run and by including the day of payment.

(3) Where a bill is payable at a fixed period after sight, the time begins to run from the date of the acceptance if the bill be accepted, and from the date of noting or protest if the bill be noted or protested for non-acceptance, or for non-delivery.

(4) The term "month" in a bill means calendar month.

NOTES

Sub-s (1): substituted by the Banking and Financial Dealings Act 1971, s 3(2).

15 Case of need

The drawer of a bill and any indorser may insert therein the name of a person to whom the holder may resort in case of need, that is to say, in case the bill is dishonoured by non-acceptance or non-payment. Such person is called the referee in case of need. It is in the option of the holder to resort to the referee in case of need or not as he may think fit.

16 Optional stipulations by drawer or indorser

The drawer of a bill, and any indorser, may insert therein an express stipulation—

(1) Negativing or limiting his own liability to the holder.

(2) Waiving as regards himself some or all of the holder's duties.

17 Definition and requisites of acceptance

(1) The acceptance of a bill is the signification by the drawee of his assent to the order of the drawer.

(2) An acceptance is invalid unless it complies with the following conditions, namely—

(a) It must be written on the bill and be signed by the drawee. The mere signature of the drawee without additional words is sufficient.

(b) It must not express that the drawee will perform his promise by any other means than the payment of money.

18 Time for acceptance

A bill may be accepted—
  (1) Before it has been signed by the drawer, or while otherwise incomplete:
  (2) When it is overdue, or after it has been dishonoured by a previous refusal to accept, or by non-payment:
  (3) When a bill payable after sight is dishonoured by non-acceptance, and the drawee subsequently accepts it, the holder, in the absence of any different agreement, is entitled to have the bill accepted as of the date of first presentment to the drawee for acceptance.

19 General and qualified acceptances

(1)   An acceptance is either (a) general or (b) qualified.

(2)   A general acceptance assents without qualification to the order of the drawer. A qualified acceptance in expressed terms varies the effect of the bill as drawn.

  In particular an acceptance is qualified which is—
  (a) conditional, that is to say, which makes payment by the acceptor dependent on the fulfilment of a condition therein stated:
  (b) partial, that is to say, an acceptance to pay part only of the amount for which the bill is drawn:
  (c) local, that is to say, an acceptance to pay only at a particular specified place:
      An acceptance to pay at a particular place is a general acceptance, unless it expressly states that the bill is to be paid there only and not elsewhere:
  (d) qualified as to time:
  (e) the acceptance of some one or more of the drawees, but not of all.

20 Inchoate instruments

(1)   Where a simple signature on a blank ... paper is delivered by the signer in order that it may be converted into a bill, it operates as a prima facie authority to fill it up as a complete bill for any amount ... , using the signature for that of the drawer, or the acceptor, or an indorser; and, in like manner, when a bill is wanting in any material particular, the person in possession of it has a prima facie authority to fill up the omission in any way he thinks fit.

(2)   In order that any such instrument when completed may be enforceable against any person who became a party thereto prior to its completion, it must be filled up within a reasonable time, and strictly in accordance with the authority given. Reasonable time for this purpose is a question of fact.

Provided that if any such instrument after completion is negotiated to a holder in due course it shall be valid and effectual for all purposes in his hands, and he may enforce it as if it had been filled up within a reasonable time and strictly in accordance with the authority given.

NOTES

Sub-s (1): words omitted repealed by the Finance Act 1970, s 36(8), Sch 8, Pt V, and the Finance Act (Northern Ireland) 1970, s 19, Sch 3, Pt III, in consequence of the abolition of stamp duties on bills of exchange and promissory notes.

21   Delivery

(1)   Every contract on a bill, whether it be the drawer's, the acceptor's, or an indorser's, is incomplete and revocable, until delivery of the instrument in order to give effect thereto.

Provided that where an acceptance is written on a bill, and the drawee gives notice to or according to the directions of the person entitled to the bill that he has accepted it, the acceptance then becomes complete and irrevocable.

(2)   As between immediate parties, and as regards a remote party other than a holder in due course, the delivery—
  (a) in order to be effectual must be made either by or under the authority of the party drawing, accepting, or indorsing, as the case may be:
  (b) may be shown to have been conditional or for a special purpose only, and not for the purpose of transferring the property in the bill.

But if the bill be in the hands of a holder in due course a valid delivery of the bill by all parties prior to him so as to make them liable to him is conclusively presumed.

(3)   Where a bill is no longer in the possession of a party who has signed it as drawer, acceptor, or indorser, a valid and unconditional delivery by him is presumed until the contrary is proved.

Capacity and authority of parties

22 Capacity of parties

(1)   Capacity to incur liability as a party to a bill is co-extensive with capacity to contract.

Provided that nothing in this section shall enable a corporation to make itself liable as drawer, acceptor, or indorser of a bill unless it is competent to it so to do under the law for the time being in force relating to corporations.

(2)   Where a bill is drawn or indorsed by an infant, minor, or corporation having no capacity or power to incur liability on a bill, the drawing or indorsement entitles the holder to receive payment of the bill, and to enforce it against any other party thereto.

23 Signature essential to liability

No person is liable as drawer, indorser, or acceptor of a bill who has not signed it as such: Provided that
  (1) Where a person signs a bill in a trade or assumed name, he is liable thereon as if he had signed it in his own name:
  (2) The signature of the name of a firm is equivalent to the signature by the person so signing of the names of all persons liable as partners in that firm.

24 Forged or unauthorised signature

Subject to the provisions of this Act, where a signature on a bill is forged or placed thereon without the authority of the person whose signature it purports to be, the forged or unauthorised signature is wholly inoperative, and no right to retain the bill or to give a discharge therefor or to enforce payment thereof against any party thereto can be acquired through or under that signature, unless the party against whom it is sought to retain or enforce payment of the bill is precluded from setting up the forgery or want of authority.

Provided that nothing in this section shall affect the ratification of an unauthorised signature not amounting to a forgery.

25 Procuration signatures

A signature by procuration operates as notice that the agent has but a limited authority to sign, and the principal is only bound by such signature if the agent in so signing was acting within the actual limits of his authority.

26 Person signing as agent or in representative capacity

(1)   Where a person signs a bill as drawer, indorser, or acceptor, and adds words to his signature, indicating that he signs for or on behalf of a principal, or in a representative character, he is not personally liable thereon; but the mere addition to his signature of words describing him as an agent, or as filling a representative character, does not exempt him from personal liability.

(2)   In determining whether a signature on a bill is that of the principal or that of the agent by whose hand it is written, the construction most favourable to the validity of the instrument shall be adopted.

The consideration for a bill

27 Value and holder for value

(1)   Valuable consideration for a bill may be constituted by,—
  (a) Any consideration sufficient to support a simple contract;
  (b) An antecedent debt or liability. Such a debt or liability is deemed valuable consideration whether the bill is payable on demand or at a future time.

(2)   Where value has at any time been given for a bill the holder is deemed to be a holder for value as regards the acceptor and all parties to the bill who became parties prior to such time.

(3)   Where the holder of a bill has a lien on it arising either from contract or by implication of law, he is deemed to be a holder for value to the extent of the sum for which he has a lien.

28   Accommodation bill or party

(1)   An accommodation party to a bill is a person who has signed a bill as drawer, acceptor, or indorser, without receiving value therefor, and for the purpose of lending his name to some other person.

(2)   An accommodation party is liable on the bill to a holder for value; and it is immaterial whether, when such holder took the bill, he knew such party to be an accommodation party or not.

29   Holder in due course

(1)   A holder in due course is a holder who has taken a bill, complete and regular on the face of it, under the following conditions; namely,
   (a) That he became the holder of it before it was overdue, and without notice that it had been previously dishonoured, if such was the fact:
   (b) That he took the bill in good faith and for value, and that at the time the bill was negotiated to him he had no notice of any defect in the title of the person who negotiated it.

(2)   In particular the title of a person who negotiates a bill is defective within the meaning of this Act when he obtained the bill, or the acceptance thereof, by fraud, duress, or force and fear, or other unlawful means, or an illegal consideration, or when he negotiates it in breach of faith, or under such circumstances as amount to a fraud.

(3)   A holder (whether for value or not), who derives his title to a bill through a holder in due course, and who is not himself a party to any fraud or illegality affecting it, has all the rights of that holder in due course as regards the acceptor and all parties to the bill prior to that holder.

30   Presumption of value and good faith

(1)   Every party whose signature appears on a bill is prima facie deemed to have become a party thereto for value.

(2)   Every holder of a bill is prima facie deemed to be a holder in due course; but if in an action on a bill it is admitted or proved that the acceptance, issue, or subsequent negotiation of the bill is affected with fraud, duress, or force and fear, or illegality, the burden of proof is shifted, unless and until the holder proves that, subsequent to the alleged fraud or illegality, value has in good faith been given for the bill.

Negotiation of bills

31 Negotiation of bill

(1)   A bill is negotiated when it is transferred from one person to another in such a manner as to constitute the transferee the holder of the bill.

(2)   A bill payable to bearer is negotiated by delivery.

(3)   A bill payable to order is negotiated by the indorsement of the holder completed by delivery.

(4)   Where the holder of a bill payable to his order transfers it for value without indorsing it, the transfer gives the transferee such title as the transferor had in the bill, and the transferee in addition acquires the right to have the indorsement of the transferor.

(5)   Where any person is under obligation to indorse a bill in a representative capacity, he may indorse the bill in such terms as to negative personal liability.

32 Requisites of a valid indorsement

An indorsement in order to operate as a negotiation must comply with the following conditions, namely,—
   (1) It must be written on the bill itself and be signed by the indorser. The simple signature of the indorser on the bill, without additional words, is sufficient.
       An indorsement written on an allonge, or on a "copy" of a bill issued or negotiated in a country where "copies" are recognised, is deemed to be written on the bill itself.
   (2) It must be an indorsement of the entire bill. A partial indorsement, that is to say, an indorsement which purports to transfer to the indorsee a part only of the amount payable, or which purports to transfer the bill to two or more indorsees severally, does not operate as a negotiation of the bill.
   (3) Where a bill is payable to the order of two or more payees or indorsees who are not partners all must indorse, unless the one indorsing has authority to indorse for the others.
   (4) Where, in a bill payable to order, the payee or indorsee is wrongly designated, or his name is mis-spelt, he may indorse the bill as therein described, adding, if he think fit, his proper signature.

(5) Where there are two or more indorsements on a bill, each indorsement is deemed to have been made in the order in which it appears on the bill, until the contrary is proved.

(6) An indorsement may be made in blank or special. It may also contain terms making it restrictive.

33 Conditional indorsement

Where a bill purports to be indorsed conditionally the condition may be disregarded by the payer, and payment to the indorsee is valid whether the condition has been fulfilled or not.

34 Indorsement in blank and special indorsement

(1) An indorsement in blank specifies no indorsee, and a bill so indorsed becomes payable to bearer.

(2) A special indorsement specifies the person to whom, or to whose order, the bill is to be payable.

(3) The provisions of this Act relating to a payee apply with the necessary modifications to an indorsee under a special indorsement.

(4) When a bill has been indorsed in blank, any holder may convert the blank indorsement into a special indorsement by writing above the indorser's signature a direction to pay the bill to or to the order of himself or some other person.

35 Restrictive indorsement

(1) An indorsement is restrictive which prohibits the further negotiation of the bill or which expresses that it is a mere authority to deal with the bill as thereby directed and not a transfer of the ownership thereof, as, for example, if a bill be indorsed "Pay D only," or "Pay D for the account of X," or "Pay D or order for collection."

(2) A restrictive indorsement gives the indorsee the right to receive payment of the bill and to sue any party thereto that his indorser could have sued, but gives him no power to transfer his rights as indorsee unless it expressly authorise him to do so.

(3)   Where a restrictive indorsement authorises further transfer, all subsequent indorsees take the bill with the same rights and subject to the same liabilities as the first indorsee under the restrictive indorsement.

36   Negotiation of overdue or dishonoured bill

(1)   Where a bill is negotiable in its origin it continues to be negotiable until it has been (a) restrictively indorsed or (b) discharged by payment or otherwise.

(2)   Where an overdue bill is negotiated, it can only be negotiated subject to any defect of title affecting it at its maturity, and thenceforward no person who takes it can acquire or give a better title than that which the person from whom he took it had.

(3)   A bill payable on demand is deemed to be overdue within the meaning and for the purposes of this section, when it appears on the face of it to have been in circulation for an unreasonable length of time. What is an unreasonable length of time for this purpose is a question of fact.

(4)   Except where an indorsement bears date after the maturity of the bill, every negotiation is prima facie deemed to have been effected before the bill was overdue.

(5)   Where a bill which is not overdue has been dishonoured any person who takes it with notice of the dishonour takes it subject to any defect of title attaching thereto at the time of dishonour, but nothing in this subsection shall affect the rights of a holder in due course.

37   Negotiation of bill to party already liable thereon

Where a bill is negotiated back to the drawer, or to a prior indorser or to the acceptor, such party may, subject to the provisions of this Act, re-issue and further negotiate the bill, but he is not entitled to enforce payment of the bill against any intervening party to whom he was previously liable.

38   Rights of the holder

The rights and powers of the holder of a bill are as follows—

(1) He may sue on the bill in his own name:

(2) Where he is a holder in due course, he holds the bill free from any defect of title of prior parties, as well as from mere personal defences available to prior parties among themselves, and may enforce payment against all parties liable on the bill:

(3) Where his title is defective (a) if he negotiates the bill to a holder in due course, that holder obtains a good and complete title to the bill, and (b) if he obtains payment of the bill the person who pays him in due course gets a valid discharge for the bill.

General duties of the holder

39   When presentment for acceptance is necessary

(1)   Where a bill is payable after sight, presentment for acceptance is necessary in order to fix the maturity of the instrument.

(2)   Where a bill expressly stipulates that it shall be presented for acceptance, or where a bill is drawn payable elsewhere than at the residence or place of business of the drawee, it must be presented for acceptance before it can be presented for payment.

(3)   In no other case is presentment for acceptance necessary in order to render liable any party to the bill.

(4)   Where the holder of a bill, drawn payable elsewhere than at the place of business or residence of the drawee, has not time, with the exercise of reasonable diligence, to present the bill for acceptance before presenting it for payment on the day that it falls due, the delay caused by presenting the bill for acceptance before presenting it for payment is excused, and does not discharge the drawer and indorsers.

40   Time for presenting bill payable after sight

(1)   Subject to the provisions of this Act, when a bill payable after sight is negotiated, the holder must either present it for acceptance or negotiate it within a reasonable time.

(2)   If he do not do so, the drawer and all indorsers prior to that holder are discharged.

(3)   In determining what is a reasonable time within the meaning of this section, regard shall be had to the nature of the bill, the usage of trade with respect to similar bills, and the facts of the particular case.

41 Rules as to presentment for acceptance, and excuses for non-presentment

(1)   A bill is duly presented for acceptance which is presented in accordance with the following rules—
  (a) The presentment must be made by or on behalf of the holder to the drawee or to some person authorised to accept or refuse acceptance on his behalf at a reasonable hour on a business day and before the bill is overdue:
  (b) Where a bill is addressed to two or more drawees, who are not partners, presentment must be made to them all, unless one has authority to accept for all, then presentment may be made to him only:
  (c) Where the drawee is dead presentment may be made to his personal representative:
  (d) Where the drawee is bankrupt, presentment may be made to him or to his trustee:
  (e) Where authorised by agreement or usage, a presentment through the post office is sufficient.

(2)   Presentment in accordance with these rules is excused, and a bill may be treated as dishonoured by non-acceptance—
  (a) Where the drawee is dead or bankrupt, or is a fictitious person or a person not having capacity to contract by bill:
  (b) Where, after the exercise of reasonable diligence, such presentment cannot be effected:
  (c) Where, although the presentment has been irregular, acceptance has been refused on some other ground.

(3)   The fact that the holder has reason to believe that the bill, on presentment, will be dishonoured does not excuse presentment.

42 Non-acceptance

When a bill is duly presented for acceptance and is not accepted within the customary time, the person presenting it must treat it as dishonoured by non-acceptance. If he do not, the holder shall lose his right of recourse against the drawer and indorsers.

43   Dishonour by non-acceptance and its consequences

(1)   A bill is dishonoured by non-acceptance—
   (a) when it is duly presented for acceptance, and such an acceptance as is prescribed by this Act is refused or cannot be obtained; or
   (b) when presentment for acceptance is excused and the bill is not accepted.

(2)   Subject to the provisions of this Act when a bill is dishonoured by non-acceptance, an immediate right of recourse against the drawer and indorsers accrues to the holder, and no presentment for payment is necessary.

44   Duties as to qualified acceptances

(1)   The holder of a bill may refuse to take a qualified acceptance, and if he does not obtain an unqualified acceptance may treat the bill as dishonoured by non-acceptance.

(2)   Where a qualified acceptance is taken, and the drawer or an indorser has not expressly or impliedly authorised the holder to take a qualified acceptance, or does not subsequently assent thereto, such drawer or indorser is discharged from his liability on the bill.

The provisions of this subsection do not apply to a partial acceptance, whereof due notice has been given. Where a foreign bill has been accepted as to part, it must be protested as to the balance.

(3)   When the drawer or indorser of a bill receives notice of a qualified acceptance, and does not within a reasonable time express his dissent to the holder he shall be deemed to have assented thereto.

45   Rules as to presentment for payment

Subject to the provisions of this Act a bill must be duly presented for payment. If it be not so presented the drawer and indorsers shall be discharged.

A bill is duly presented for payment which is presented in accordance with the following rules:—
   (1) Where the bill is not payable on demand, presentment must be made on the day it falls due.

(2) Where the bill is payable on demand, then, subject to the provisions of this Act, presentment must be made within a reasonable time after its issue in order to render the drawer liable, and within a reasonable time after its indorsement, in order to render the indorser liable.

In determining what is a reasonable time, regard shall be had to the nature of the bill, the usage of trade with regard to similar bills, and the facts of the particular case.

(3) Presentment must be made by the holder or by some person authorised to receive payment on his behalf at a reasonable hour on a business day, at the proper place as herein-after defined, either to the person designated by the bill as payer, or to some person authorised to pay or refuse payment on his behalf if with the exercise of reasonable diligence such person can there be found.

(4) A bill is presented at the proper place:—

   (a) Where a place of payment is specified in the bill and the bill is there presented.

   (b) Where no place of payment is specified, but the address of the drawee or acceptor is given in the bill, and the bill is there presented.

   (c) Where no place of payment is specified and no address given, and the bill is presented at the drawee's or acceptor's place of business if known, and if not, at his ordinary residence if known.

   (d) In any other case if presented to the drawee or acceptor wherever he can be found, or if presented at his last known place of business or residence.

(5) Where a bill is presented at the proper place, and after the exercise of reasonable diligence no person authorised to pay or refuse payment can be found there, no further presentment to the drawee or acceptor is required.

(6) Where a bill is drawn upon, or accepted by two or more persons who are not partners, and no place of payment is specified, presentment must be made to them all.

(7) Where the drawee or acceptor of a bill is dead, and no place of payment is specified, presentment must be made to a personal representative, if such there be, and with the exercise of reasonable diligence he can be found.

(8) Where authorised by agreement or usage a presentment through the post office is sufficient.

46 Excuses for delay or non-presentment for payment

(1) Delay in making presentment for payment is excused when the delay is caused by circumstances beyond the control of the holder, and not imputable

to his default, misconduct, or negligence. When the cause of delay ceases to operate presentment must be made with reasonable diligence.

(2)   Presentment for payment is dispensed with,—
   (a) Where, after the exercise of reasonable diligence presentment, as required by this Act, cannot be effected.
      The fact that the holder has reason to believe that the bill will, on presentment, be dishonoured, does not dispense with the necessity for presentment.
   (b) Where the drawee is a fictitious person.
   (c) As regards the drawer where the drawee or acceptor is not bound as between himself and the drawer, to accept or pay the bill, and the drawer has no reason to believe that the bill would be paid if presented.
   (d) As regards an indorser, where the bill was accepted or made for the accommodation of that indorser, and he has no reason to expect that the bill would be paid if presented.
   (e) By waiver of presentment, express or implied.

47   Dishonour by non-payment

(1)   A bill is dishonoured by non-payment (a) when it is duly presented for payment and payment is refused or cannot be obtained, or (b) when presentment is excused and the bill is overdue and unpaid.

(2)   Subject to the provisions of this Act, when a bill is dishonoured by non-payment, an immediate right of recourse against the drawer and indorsers accrues to the holder.

48   Notice of dishonour and effect of non-notice

Subject to the provisions of this Act, when a bill has been dishonoured by non-acceptance or by non-payment, notice of dishonour must be given to the drawer and each indorser, and any drawer or indorser to whom such notice is not given is discharged: Provided that—
   (1) Where a bill is dishonoured by non-acceptance, and notice of dishonour is not given, the rights of a holder in due course, subsequent to the omission, shall not be prejudiced by the omission.
   (2) Where a bill is dishonoured by non-acceptance, and due notice of dishonour is given, it shall not be necessary to give notice of a subsequent dishonour by non-payment unless the bill shall in the meantime have been accepted.

49 Rules as to notice of dishonour

Notice of dishonour in order to be valid and effectual must be given in accordance with the following rules:—

(1) The notice must be given by or on behalf of the holder, or by or on behalf of an indorser who, at the time of giving it, is himself liable on the bill.

(2) Notice of dishonour may be given by an agent either in his own name, or in the name of any party entitled to give notice whether that party be his principal or not.

(3) Where the notice is given by or on behalf of the holder, it enures for the benefit of all subsequent holders and all prior indorsers who have a right of recourse against the party to whom it is given.

(4) Where notice is given by or on behalf of an indorser entitled to give notice as herein-before provided, it enures for the benefit of the holder and all indorsers subsequent to the party to whom notice is given.

(5) The notice may be given in writing or by personal communication, and may be given in any terms which sufficiently identify the bill, and intimate that the bill has been dishonoured by non-acceptance or non-payment.

(6) The return of a dishonoured bill to the drawer or an indorser is, in point of form, deemed a sufficient notice of dishonour.

(7) A written notice need not be signed, and an insufficient written notice may be supplemented and validated by verbal communication. A misdescription of the bill shall not vitiate the notice unless the party to whom the notice is given is in fact misled thereby.

(8) Where notice of dishonour is required to be given to any person, it may be given either to the party himself, or to his agent in that behalf.

(9) Where the drawer or indorser is dead, and the party giving notice knows it, the notice must be given to a personal representative if such there be, and with the exercise of reasonable diligence he can be found.

(10) Where the drawer or indorser is bankrupt, notice may be given either to the party himself or to the trustee.

(11) Where there are two or more drawers or indorsers who are not partners, notice must be given to each of them, unless one of them has authority to receive such notice for the others.

(12) The notice may be given as soon as the bill is dishonoured and must be given within a reasonable time thereafter.

In the absence of special circumstances notice is not deemed to have been given within a reasonable time, unless—

    (a)   where the person giving and the person to receive notice reside in the same place, the notice is given or sent off in time to reach the latter on the day after the dishonour of the bill.

    (b)   where the person giving and the person to receive notice reside in different places, the notice is sent off on the day after the dishonour of the bill, if there be a post at a convenient hour on that day, and if there be no post on that day then by the next post thereafter.

(13)   Where a bill when dishonoured is in the hands of an agent, he may either himself give notice to the parties liable on the bill, or he may give notice to his principal. If he give notice to his principal, he must do so within the same time as if he were the holder, and the principal upon receipt of such notice has himself the same time for giving notice as if the agent had been an independent holder.

(14)   Where a party to a bill receives due notice of dishonour, he has after the receipt of such notice the same period of time for giving notice to antecedent parties that the holder has after the dishonour.

(15)   Where a notice of dishonour is duly addressed and posted, the sender is deemed to have given due notice of dishonour, notwithstanding any miscarriage by the post office.

50   Excuses for non-notice and delay

(1)   Delay in giving notice of dishonour is excused where the delay is caused by circumstances beyond the control of the party giving notice, and not imputable to his default, misconduct, or negligence. When the cause of delay ceases to operate the notice must be given with reasonable diligence.

(2)   Notice of dishonour is dispensed with—

  (a) When, after the exercise of reasonable diligence, notice as required by this Act cannot be given to or does not reach the drawer or indorser sought to be charged:

  (b) By waiver express or implied. Notice of dishonour may be waived before the time of giving notice has arrived, or after the omission to give due notice:

  (c) As regards the drawer in the following cases, namely, (1) where drawer and drawee are the same person, (2) where the drawee is a fictitious person or a person not having capacity to contract, (3) where the drawer is the person to whom the bill is presented for payment, (4) where the drawee or acceptor is as between himself and the drawer under no obligation to accept or pay the bill, (5) where the drawer has countermanded payment:

(d) As regards the indorser in the following cases, namely, (1) where the drawee is a fictitious person or a person not having capacity to contract, and the indorser was aware of the fact at the time he indorsed the bill, (2) where the indorser is the person to whom the bill is presented for payment, (3) where the bill was accepted or made for his accommodation.

51   Noting or protest of bill

(1)   Where an inland bill has been dishonoured it may, if the holder think fit, be noted for non-acceptance or non-payment, as the case may be; but it shall not be necessary to note or protest any such bill in order to preserve the recourse against the drawer or indorser.

(2)   Where a foreign bill, appearing on the face of it to be such, has been dishonoured by non-acceptance it must be duly protested for non-acceptance, and where such a bill, which has not been previously dishonoured by non-acceptance, is dishonoured by non-payment it must be duly protested for non-payment. If it be not so protested the drawer and indorsers are discharged. Where a bill does not appear on the face of it to be a foreign bill, protest thereof in case of dishonour is unnecessary.

(3)   A bill which has been protested for non-acceptance may be subsequently protested for non-payment.

(4)   Subject to the provisions of this Act, when a bill is noted or protested, [it may be noted on the day of its dishonour and must be noted not later than the next succeeding business day]. When a bill has been duly noted, the protest may be subsequently extended as of the date of the noting.

(5)   Where the acceptor of a bill becomes bankrupt or insolvent or suspends payment before it matures, the holder may cause the bill to be protested for better security against the drawer and indorsers.

(6)   A bill must be protested at the place where it is dishonoured: Provided that—
  (a) When a bill is presented through the post office, and returned by post dishonoured, it may be protested at the place to which it is returned and on the day of its return if received during business hours, and if not received during business hours, then not later than the next business day:
  (b) When a bill drawn payable at the place of business or residence of some person other than the drawee has been dishonoured by non-

acceptance, it must be protested for non-payment at the place where it is expressed to be payable, and no further presentment for payment to, or demand on, the drawee is necessary.

(7)   A protest must contain a copy of the bill, and must be signed by the notary making it, and must specify—
   (a) The person at whose request the bill is protested:
   (b) The place and date of protest, the cause or reason for protesting the bill, the demand made, and the answer given, if any, or the fact that the drawee or acceptor could not be found.

(8)   Where a bill is lost or destroyed, or is wrongly detained from the person entitled to hold it, protest may be made on a copy or written particulars thereof.

(9)   Protest is dispensed with by any circumstance which would dispense with notice of dishonour. Delay in noting or protesting is excused when the delay is caused by circumstances beyond the control of the holder, and not imputable to his default, misconduct, or negligence. When the cause of delay ceases to operate the bill must be noted or protested with reasonable diligence.

NOTES

Sub-s (4): words in square brackets substituted by the Bills of Exchange (Time of Noting) Act 1917, s 1.

52   Duties of holder as regards drawee or acceptor

(1)   When a bill is accepted generally presentment for payment is not necessary in order to render the acceptor liable.

(2)   When by the terms of a qualified acceptance presentment for payment is required, the acceptor, in the absence of an express stipulation to that effect, is not discharged by the omission to present the bill for payment on the day that it matures.

(3)   In order to render the acceptor of a bill liable it is not necessary to protest it, or that notice of dishonour should be given to him.

(4)   Where the holder of a bill presents it for payment, he shall exhibit the bill to the person from whom he demands payment, and when a bill is paid the holder shall forthwith deliver it up to the party paying it.

Liabilities of parties

53 Funds in hands of drawee

(1)   A bill, of itself, does not operate as an assignment of funds in the hands of the drawee available for the payment thereof, and the drawee of a bill who does not accept as required by this Act is not liable on the instrument. This sub-section shall not extend to Scotland.

(2)   ...

NOTES

Sub-s (2): applies to Scotland only.

54 Liability of acceptor

The acceptor of a bill, by accepting it—
   (1) Engages that he will pay it according to the tenor of his acceptance:
   (2) Is precluded from denying to a holder in due course—
     (a) The existence of the drawer, the genuineness of his signature, and his capacity and authority to draw the bill;
     (b) In the case of a bill payable to drawer's order, the then capacity of the drawer to indorse, but not the genuineness or validity of his indorsement;
     (c) In the case of a bill payable to the order of a third person, the existence of the payee and his then capacity to indorse, but not the genuineness or validity of his indorsement.

55 Liability of drawer or indorser

(1)   The drawer of a bill by drawing it—
   (a) Engages that on due presentment it shall be accepted and paid according to its tenor, and that if it be dishonoured he will compensate the holder or any indorser who is compelled to pay it, provided that the requisite proceedings on dishonour be duly taken;
   (b) Is precluded from denying to a holder in due course the existence of the payee and his then capacity to indorse.

(2)   The indorser of a bill by indorsing it—

(a) Engages that on due presentment it shall be accepted and paid according to its tenor, and that if it be dishonoured he will compensate the holder or a subsequent indorser who is compelled to pay it, provided that the requisite proceedings on dishonour be duly taken;
(b) Is precluded from denying to a holder in due course the genuineness and regularity in all respects of the drawer's signature and all previous indorsements;
(c) Is precluded from denying to his immediate or a subsequent indorsee that the bill was at the time of his indorsement a valid and subsisting bill, and that he had then a good title thereto.

56 Stranger signing bill liable as indorser

Where a person signs a bill otherwise than as drawer or acceptor, he thereby incurs the liabilities of an indorser to a holder in due course.

57 Measure of damages against parties to dishonoured bill

Where a bill is dishonoured, the measure of damages, which shall be deemed to be liquidated damages, shall be as follows—
(1) The holder may recover from any party liable on the bill, and the drawer who has been compelled to pay the bill may recover from the acceptor, and an indorser who has been compelled to pay the bill may recover from the acceptor or from the drawer, or from a prior indorser—
  (a) The amount of the bill:
  (b) Interest thereon from the time of presentment for payment if the bill is payable on demand, and from the maturity of the bill in any other case:
  (c) The expenses of noting, or, when protest is necessary, and the protest has been extended, the expenses of protest.
(2) ...
(3) Where by this Act interest may be recovered as damages, such interest may, if justice require it, be withheld wholly or in part, and where a bill is expressed to be payable with interest at a given rate, interest as damages may or may not be given at the same rate as interest proper.

NOTES

Sub-s (2): repealed by the Administration of Justice Act 1977, ss 4, 32(4), Sch 5, Pt I, except in relation to bills drawn before 29 August 1977.

58 Transferor by delivery and transferee

(1)   Where the holder of a bill payable to bearer negotiates it by delivery without indorsing it he is called a "transferor by delivery."

(2)   A transferor by delivery is not liable on the instrument.

(3)   A transferor by delivery who negotiates a bill thereby warrants to his immediate transferee being a holder for value that the bill is what it purports to be, that he has a right to transfer it, and that at the time of transfer he is not aware of any fact which renders it valueless.

Discharge of bill

59 Payment in due course

(1)   A bill is discharged by payment in due course by or on behalf of the drawee or acceptor.
  "Payment in due course" means payment made at or after the maturity of the bill to the holder thereof in good faith and without notice that his title to the bill is defective.

(2)   Subject to the provisions herein-after contained, when a bill is paid by the drawer or an indorser it is not discharged; but
  (a) Where a bill payable to, or to the order of, a third party is paid by the drawer, the drawer may enforce payment thereof against the acceptor, but may not re-issue the bill.
  (b) Where a bill is paid by an indorser, or where a bill payable to drawer's order is paid by the drawer, the party paying it is remitted to his former rights as regards the acceptor or antecedent parties and he may, if he thinks fit, strike out his own subsequent indorsements, and again negotiate the bill.

(3)   Where an accommodation bill is paid in due course by the party accommodated the bill is discharged.

60 Banker paying demand draft whereon indorsement is forged

When a bill payable to order on demand is drawn on a banker, and the banker on whom it is drawn pays the bill in good faith and in the ordinary course of business, it is not incumbent on the banker to show that the indorsement of the payee or any subsequent indorsement was made by or under the authority of the person whose indorsement it purports to be, and the banker is deemed to have paid the bill in due course, although such indorsement has been forged or made without authority.

61   Acceptor the holder at maturity

When the acceptor of a bill is or becomes the holder of it at or after its maturity, in his own right, the bill is discharged.

62   Express waiver

(1)   When the holder of a bill at or after its maturity absolutely and unconditionally renounces his rights against the acceptor the bill is discharged.

   The renunciation must be in writing, unless the bill is delivered up to the acceptor.

(2)   The liabilities of any party to a bill may in like manner be renounced by the holder before, at, or after its maturity; but nothing in this section shall affect the rights of a holder in due course without notice of the renunciation.

63   Cancellation

(1)   Where a bill is intentionally cancelled by the holder or his agent, and the cancellation is apparent thereon, the bill is discharged.

(2)   In like manner any party liable on a bill may be discharged by the intentional cancellation of his signature by the holder or his agent. In such case any indorser who would have had a right of recourse against the party whose signature is cancelled is also discharged.

(3)   A cancellation made unintentionally, or under a mistake, or without the authority of the holder is inoperative; but where a bill or any signature thereon appears to have been cancelled the burden of proof lies on the party who alleges that the cancellation was made unintentionally, or under a mistake, or without authority.

64 Alteration of bill

(1)   Where a bill or acceptance is materially altered without the assent of all parties liable on the bill, the bill is avoided except as against a party who has himself made, authorised, or assented to the alteration, and subsequent indorsers.

Provided that,
Where a bill has been materially altered, but the alteration is not apparent, and the bill is in the hands of a holder in due course, such holder may avail himself of the bill as if it had not been altered, and may enforce payment of it according to its original tenour.

(2)   In particular the following alterations are material, namely, any alteration of the date, the sum payable, the time of payment, the place of payment and, where a bill has been accepted generally, the addition of a place of payment without the acceptor's assent.

Acceptance and payment for honour

65 Acceptance for honour *suprà protest*

(1)   Where a bill of exchange has been protested for dishonour by non-acceptance, or protested for better security, and is not overdue, any person, not being a party already liable thereon, may, with the consent of the holder, intervene and accept the bill *suprà protest*, for the honour of any party liable thereon, or for the honour of the person for whose account the bill is drawn.

(2)   A bill may be accepted for honour for part only of the sum for which it is drawn.

(3)   An acceptance for honour *suprà protest* in order to be valid must—
  (a) be written on the bill, and indicate that it is an acceptance for honour:
  (b) be signed by the acceptor for honour.

(4)   Where an acceptance for honour does not expressly state for whose honour it is made, it is deemed to be an acceptance for the honour of the drawer.

(5)   Where a bill payable after sight is accepted for honour, its maturity is calculated from the date of the noting for non-acceptance, and not from the date of the acceptance for honour.

66 Liability of acceptor for honour

(1)   The acceptor for honour of a bill by accepting it engages that he will, on due presentment, pay the bill according to the tenor of his acceptance, if it is not paid by the drawee, provided it has been duly presented for payment, and protested for non-payment, and that he receives notice of these facts.

(2)   The acceptor for honour is liable to the holder and to all parties to the bill subsequent to the party for whose honour he has accepted.

67 Presentment to acceptor for honour

(1)   Where a dishonoured bill has been accepted for honour *suprà protest*, or contains a reference in case of need, it must be protested for non-payment before it is presented for payment to the acceptor for honour, or referee in case of need.

(2)   Where the address of the acceptor for honour is in the same place where the bill is protested for non-payment, the bill must be presented to him not later than the day following its maturity; and where the address of the acceptor for honour is in some place other than the place where it was protested for non-payment, the bill must be forwarded not later than the day following its maturity for presentment to him.

(3)   Delay in presentment or non-presentment is excused by any circumstance which would excuse delay in presentment for payment or non-presentment for payment.

(4)   When a bill of exchange is dishonoured by the acceptor for honour it must be protested for non-payment by him.

68 Payment for honour *suprà protest*

(1)   Where a bill has been protested for non-payment, any person may intervene and pay it *suprà protest* for the honour of any party liable thereon, or for the honour of the person for whose account the bill is drawn.

(2)   Where two or more persons offer to pay a bill for the honour of different parties, the person whose payment will discharge most parties to the bill shall have the preference.

(3)   Payment for honour *suprà protest*, in order to operate as such and not as a mere voluntary payment, must be attested by a notarial act of honour which may be appended to the protest or form an extension of it.

(4)   The notarial act of honour must be founded on a declaration made by the payer for honour, or his agent in that behalf, declaring his intention to pay the bill for honour, and for whose honour he pays.

(5)   Where a bill has been paid for honour, all parties subsequent to the party for whose honour it is paid are discharged, but the payer for honour is subrogated for, and succeeds to both the rights and duties of, the holder as regards the party for whose honour he pays, and all parties liable to that party.

(6)   The payer for honour on paying to the holder the amount of the bill and the notarial expenses incidental to its dishonour is entitled to receive both the bill itself and the protest. If the holder do not on demand deliver them up he shall be liable to the payer for honour in damages.

(7)   Where the holder of a bill refuses to receive payment *suprà protest* he shall lose his right of recourse against any party who would have been discharged by such payment.

Lost instruments

69   Holder's right to duplicate of lost bill

Where a bill has been lost before it is overdue the person who was the holder of it may apply to the drawer to give him another bill of the same tenor, giving security to the drawer if required to indemnify him against all persons whatever in case the bill alleged to have been lost shall be found again.

If the drawer on request as aforesaid refuses to give such duplicate bill he may be compelled to do so.

70   Action on lost bill

In any action or proceeding upon a bill, the court or a judge may order that the loss of the instrument shall not be set up, provided an indemnity be given to the satisfaction of the court or judge against the claims of any other person upon the instrument in question.

Bill in a set

71   Rules as to sets

(1)   Where a bill is drawn in a set, each part of the set being numbered, and containing a reference to the other parts the whole of the parts constitute one bill.

(2)   Where the holder of a set indorses two or more parts to different persons, he is liable on every such part, and every indorser subsequent to him is liable on the part he has himself indorsed as if the said parts were separate bills.

(3)   Where two or more parts of a set are negotiated to different holders in due course, the holder whose title first accrues is as between such holders deemed the true owner of the bill; but nothing in this sub-section shall affect the rights of a person who in due course accepts or pays the part first presented to him.

(4)   The acceptance may be written on any part, and it must be written on one part only.

   If the drawee accepts more than one part, and such accepted parts get into the hands of different holders in due course, he is liable on every such part as if it were a separate bill.

(5)   When the acceptor of a bill drawn in a set pays it without requiring the part bearing his acceptance to be delivered up to him, and that part at maturity is outstanding in the hands of a holder in due course, he is liable to the holder thereof.

(6)   Subject to the preceding rules, where any one part of a bill drawn in a set is discharged by payment or otherwise, the whole bill is discharged.

Conflict of laws

72   Rules where laws conflict

Where a bill drawn in one country is negotiated, accepted, or payable in another, the rights, duties, and liabilities of the parties thereto are determined as follows—

(1) The validity of a bill as regards requisites in form is determined by the law of the place of issue, and the validity as regards requisites in form of the supervening contracts, such as acceptance, or indorsement, or acceptance *suprà protest*, is determined by the law of the place where such contract was made.

Provided that—

(a)   Where a bill is issued out of the United Kingdom it is not invalid by reason only that it is not stamped in accordance with the law of the place of issue:

(b)   Where a bill, issued out of the United Kingdom, conforms, as regards requisites in form, to the law of the United Kingdom, it may, for the purpose of enforcing payment thereof, be treated as valid as between all persons who negotiate, hold, or become parties to it in the United Kingdom.

(2) Subject to the provisions of this Act, the interpretation of the drawing, indorsement, acceptance, or acceptance *suprà protest* of a bill, is determined by the law of the place where such contract is made.

Provided that where an inland bill is indorsed in a foreign country the indorsement shall as regards the payer be interpreted according to the law of the United Kingdom.

(3) The duties of the holder with respect to presentment for acceptance or payment and the necessity for or sufficiency of a protest or notice of dishonour, or otherwise, are determined by the law of the place where the act is done or the bill is dishonoured.

(4) ...

(5) Where a bill is drawn in one country and is payable in another, the due date thereof is determined according to the law of the place where it is payable.

NOTES

Sub-s (4): repealed by the Administration of Justice Act 1977, ss 4, 32(4), Sch 5, Pt I, except in relation to bills drawn before 29 August 1977.

PART III
CHEQUES ON A BANKER

73   Cheque defined

A cheque is a bill of exchange drawn on a banker payable on demand.

Except as otherwise provided in this Part, the provisions of this Act applicable to a bill of exchange payable on demand apply to a cheque.

74 Presentment of cheque for payment

Subject to the provisions of this Act—

(1) Where a cheque is not presented for payment within a reasonable time of its issue, and the drawer or the person on whose account it is drawn had the right at the time of such presentment as between him and the banker to have the cheque paid and suffers actual damage through the delay, he is discharged to the extent of such damage, that is to say, to the extent to which such drawer or person is a creditor of such banker to a larger amount than he would have been had such cheque been paid.

(2) In determining what is a reasonable time regard shall be had to the nature of the instrument, the usage of trade and of bankers, and the facts of the particular case.

(3) The holder of such cheque as to which such drawer or person is discharged shall be a creditor, in lieu of such drawer or person, of such banker to the extent of such discharge, and entitled to recover the amount from him.

[74A Presentment of cheque for payment: alternative place of presentment

Where the banker on whom a cheque is drawn—

(a) has by notice published in the London, Edinburgh and Belfast Gazettes specified an address at which cheques drawn on him may be presented, and

(b) has not by notice so published cancelled the specification of that address, the cheque is also presented at the proper place if it is presented there.]

NOTES

Commencement: 28 November 1996.
Inserted by the Deregulation (Bills of Exchange) Order 1996, SI 1996/2993, art 3.

[74B Presentment of cheque for payment: alternative means of presentment by banker

(1) A banker may present a cheque for payment to the banker on whom it is drawn by notifying him of its essential features by electronic means or otherwise, instead of by presenting the cheque itself.

(2)   If a cheque is presented for payment under this section, presentment need not be made at the proper place or at a reasonable hour on a business day.

(3)   If, before the close of business on the next business day following presentment of a cheque under this section, the banker on whom the cheque is drawn requests the banker by whom the cheque was presented to present the cheque itself—
  (a) the presentment under this section shall be disregarded, and
  (b) this section shall not apply in relation to the subsequent presentment of the cheque.

(4)   A request under subsection (3) above for the presentment of a cheque shall not constitute dishonour of the cheque by non-payment.

(5)   Where presentment of a cheque is made under this section, the banker who presented the cheque and the banker on whom it is drawn shall be subject to the same duties in relation to the collection and payment of the cheque as if the cheque itself had been presented for payment.

(6)   For the purposes of this section, the essential features of a cheque are—
  (a) the serial number of the cheque,
  (b) the code which identifies the banker on whom the cheque is drawn,
  (c) the account number of the drawer of the cheque, and
  (d) the amount of the cheque is entered by the drawer of the cheque.]

NOTES

Commencement: 28 November 1996.
Inserted by the Deregulation (Bills of Exchange) Order 1996, SI 1996/2993, art 4, in relation to cheques drawn on or after 28 November 1996.

[74C   Cheques presented for payment under section 74B: disapplication of section 52(4)

Section 52(4) above—
  (a) so far as relating to presenting a bill for payment, shall not apply to presenting a cheque for payment under section 74B above, and
  (b) so far as relating to a bill which is paid, shall not apply to a cheque which is paid following presentment under that section.]

NOTES

Commencement: 28 November 1996.
Inserted as noted to s 74B.

75 Revocation of banker's authority

The duty and authority of a banker to pay a cheque drawn on him by his customer are determined by—
  (1) Countermand of payment:
  (2) Notice of the customer's death.

Crossed cheques

76 General and special crossings defined

(1)   Where a cheque bears across its face an addition of—
  (a) The words "and company" or any abbreviation thereof between two parallel transverse lines, either with or without the words "not negotiable"; or
  (b) Two parallel transverse lines simply, either with or without the words "not negotiable";

that addition constitutes a crossing, and the cheque is crossed generally.

(2)   Where a cheque bears across its face an addition of the name of a banker, either with or without the words "not negotiable," that addition constitutes a crossing, and the cheque is crossed specially and to that banker.

77 Crossing by drawer or after issue

(1)   A cheque may be crossed generally or specially by the drawer.

(2)   Where a cheque is uncrossed, the holder may cross it generally or specially.

(3)   Where a cheque is crossed generally the holder may cross it specially.

(4)   Where a cheque is crossed generally or specially, the holder may add the words "not negotiable."

(5)   Where a cheque is crossed specially, the banker to whom it is crossed may again cross it specially to another banker for collection.

(6)   Where an uncrossed cheque, or a cheque crossed generally, is sent to a banker for collection, he may cross it specially to himself.

78   Crossing a material part of cheque

A crossing authorised by this Act is a material part of the cheque; it shall not be lawful for any person to obliterate or, except as authorised by this Act, to add to or alter the crossing.

79   Duties of banker as to crossed cheques

(1)   Where a cheque is crossed specially to more than one banker except when crossed to an agent for collection being a banker, the banker on whom it is drawn shall refuse payment thereof.

(2)   Where the banker on whom a cheque is drawn which is so crossed nevertheless pays the same, or pays a cheque crossed generally otherwise than to a banker, or if crossed specially otherwise than to the banker to whom it is crossed, or his agent for collection being a banker, he is liable to the true owner of the cheque for any loss he may sustain owing to the cheque having been so paid.

Provided that where a cheque is presented for payment which does not at the time of presentment appear to be crossed, or to have had a crossing which has been obliterated, or to have been added to or altered otherwise than as authorised by this Act, the banker paying the cheque in good faith and without negligence shall not be responsible or incur any liability, nor shall the payment be questioned by reason of the cheque having been crossed, or of the crossing having been obliterated or having been added to or altered otherwise than as authorised by this Act, and of payment having been made otherwise than to a banker or to the banker to whom the cheque is or was crossed, or to his agent for collection being a banker, as the case may be.

80   Protection to banker and drawer where cheque is crossed

Where the banker, on whom a crossed cheque [(including a cheque which under section 81A below or otherwise is not transferable)] is drawn, in good faith and without negligence pays it, if crossed generally, to a banker, and if crossed specially, to the banker to whom it is crossed, or his agent for collection being a banker, the banker paying the cheque, and, if the cheque has come into the hands of the payee, the drawer, shall respectively be entitled to the same rights and be placed in the same position as if payment of the cheque had been made to the true owner thereof.

NOTES

Words in square brackets inserted by the Cheques Act 1992, s 2.

81   Effect of crossing on holder

Where a person takes a crossed cheque which bears on it the words "not negotiable," he shall not have and shall not be capable of giving a better title to the cheque than that which the person from whom he took it had.

[81A   Non-transferable cheques

(1)   Where a cheque is crossed and bears across its face the words "account payee" or "a/c payee", either with or without the word "only", the cheque shall not be transferable, but shall only be valid as between the parties thereto.

(2)   A banker is not to be treated for the purposes of section 80 above as having been negligent by reason only of his failure to concern himself with any purported indorsement of a cheque which under subsection (1) above or otherwise is not transferable.]

NOTES

Inserted by the Cheques Act 1992, s 1.

PART IV
PROMISSORY NOTES

83   Promissory note defined

(1)   A promissory note is an unconditional promise in writing made by one person to another signed by the maker, engaging to pay, on demand or

at a fixed or determinable future time, a sum certain in money, to, or to the order of, a specified person or to bearer.

(2)   An instrument in the form of a note payable to maker's order is not a note within the meaning of this section unless and until it is indorsed by the maker.

(3)   A note is not invalid by reason only that it contains also a pledge of collateral security with authority to sell or dispose thereof.

(4)   A note which is, or on the face of it purports to be, both made and payable within the British Islands is an inland note. Any other note is a foreign note.

84   Delivery necessary

A promissory note is inchoate and incomplete until delivery thereof to the payee or bearer.

85   Joint and several notes

(1)   A promissory note may be made by two or more makers, and they may be liable thereon jointly, or jointly and severally according to its tenour.

(2)   Where a note runs "I promise to pay" and is signed by two or more persons it is deemed to be their joint and several note.

86   Note payable on demand

(1)   Where a note payable on demand has been indorsed, it must be presented for payment within a reasonable time of the indorsement. If it be not so presented the indorser is discharged.

(2)   In determining what is reasonable time, regard shall be had to the nature of the instrument, the usage of trade, and the facts of the particular case.

(3)   Where a note payable on demand is negotiated, it is not deemed to be overdue, for the purpose of affecting the holder with defects of title of which he had no notice, by reason that it appears that a reasonable time for presenting it for payment has elapsed since its issue.

87 Presentment of note for payment

(1)   Where a promissory note is in the body of it made payable at a particular place, it must be presented for payment at that place in order to render the maker liable. In any other case, presentment for payment is not necessary in order to render the maker liable.

(2)   Presentment for payment is necessary in order to render the indorser of a note liable.

(3)   Where a note is in the body of it made payable at a particular place, presentment at that place is necessary in order to render an indorser liable; but when a place of payment is indicated by way of memorandum only, presentment at that place is sufficient to render the indorser liable, but a presentment to the maker elsewhere, if sufficient in other respects, shall also suffice.

88 Liability of maker

The maker of a promissory note by making it—
  (1) Engages that he will pay it according to its tenour;
  (2) Is precluded from denying to a holder in due course the existence of the payee and his then capacity to indorse.

89 Application of Part II to notes

(1)   Subject to the provisions in this part, and except as by this section provided, the provisions of this Act relating to bills of exchange apply, with the necessary modifications, to promissory notes.

(2)   In applying those provisions the maker of a note shall be deemed to correspond with the acceptor of a bill, and the first indorser of a note shall be deemed to correspond with the drawer of an accepted bill payable to drawer's order.

(3)   The following provisions as to bills do not apply to notes; namely, provisions relating to—
  (a) Presentment for acceptance;
  (b) Acceptance;
  (c) Acceptance *suprà protest*;
  (d) Bills in a set.

(4)   Where a foreign note is dishonoured, protest thereof is unnecessary.

PART V
SUPPLEMENTARY

90 Good faith

A thing is deemed to be done in good faith, within the meaning of this Act where it is in fact done honestly, whether it is done negligently or not.

91 Signature

(1)   Where, by this Act, any instrument or writing is required to be signed by any person it is not necessary that he should sign it with his own hand, but it is sufficient if his signature is written thereon by some other person by or under his authority.

(2)   In the case of a corporation, where, by this Act, any instrument or writing is required to be signed, it is sufficient if the instrument or writing be sealed with the corporate seal.

But nothing in this section shall be construed as requiring the bill or note of a corporation to be under seal.

92 Computation of time

Where, by this Act, the time limited for doing any act or thing is less than three days, in reckoning time, non-business days are excluded.
  "Non-business days" for the purposes of this Act mean—
    (a) [Saturday] Sunday, Good Friday, Christmas Day:
    (b) A bank holiday under [the Banking and Financial Dealings Act 1971]:
    (c) A day appointed by Royal proclamation as a public fast or thanks-giving day:
    [(d) a day declared by an order under section 2 of the Banking and Financial Dealings Act 1971 to be a non-business day].

Any other day is a business day.

NOTES

Words in square brackets in para (a) inserted by the Banking and Financial Dealings Act 1971, s 3(1), (3); words in square brackets in para (b) substituted by the Banking and Financial Dealings Act 1971, s 4(4); para (d) inserted by the Banking and Financial Dealings Act 1971, s 4(4).

93 When noting equivalent to protest

For the purposes of this Act, where a bill or note is required to be protested within a specified time or before some further proceeding is taken, it is sufficient that the bill has been noted for protest before the expiration of the specified time or the taking of the proceeding; and the formal protest may be extended at any time thereafter as of the date of the noting.

94 Protest when notary not accessible

Where a dishonoured bill or note is authorised or required to be protested, and the services of a notary cannot be obtained at the place where the bill is dishonoured, any householder or substantial resident of the place may, in the presence of two witnesses, give a certificate, signed by them, attesting the dishonour of the bill, and the certificate shall in all respects operate as if it were a formal protest of the bill.

The form given in Schedule 1 to this Act may be used with necessary modifications, and if used shall be sufficient.

95 Dividend warrants may be crossed

The provisions of this Act as to crossed cheques shall apply to a warrant for payment of dividend.

97 Savings

(1)   The rules in bankruptcy relating to bills of exchange, promissory notes, and cheques, shall continue to apply thereto notwithstanding anything in this Act contained.

(2)   The rules of common law including the law merchant, save in so far as they are inconsistent with the express provisions of this Act, shall continue to apply to bills of exchange, promissory notes, and cheques.

(3)   Nothing in this Act or in any repeal effected thereby shall affect—

(a) ... any law or enactment for the time being in force relating to the revenue:

(b) The provisions of the Companies Act, 1862, or Acts amending it, or any Act relating to joint stock banks or companies:

(c) The provisions of any Act relating to or confirming the privileges of the Bank of England or the Bank of Ireland respectively:

(d) The validity of any usage relating to dividend warrants, or the indorsements thereof.

NOTES

Sub-s (3): words omitted in para (a) repealed by the Statute Law Revision Act 1898.

99 Construction with other Acts, etc

Where any Act or document refers to any enactment repealed by this Act, the Act or document shall be construed, and shall operate, as if it referred to the corresponding provisions of this Act.

SCHEDULES

FIRST SCHEDULE

Section 94

Form of protest which may be used when the services of a notary cannot be obtained.

Know all men that I, *AB* [householder], of in the county of in the United Kingdom, at the request of *CD*, there being no notary public available, did on the day of 188 at demand payment [*or* acceptance] of the bill of exchange hereunder written, from *EF*, to which demand he made answer [state answer, if any] wherefore I now, in the presence of *GH* and *JK* do protest the said bill of exchange.

(Signed) *AB*

GH Witnesses.

JK

NB—The bill itself should be annexed, or a copy of the bill and all that is written thereon should be underwritten.

Factors Act 1889

(C 45)

An Act to amend and consolidate the Factors Acts

[26 August 1889]

Preliminary

1 Definitions

For the purposes of this Act—

(1) The expression "mercantile agent" shall mean a mercantile agent having in the customary course of his business as such agent authority either to sell goods, or to consign goods for the purpose of sale, or to buy goods, or to raise money on the security of goods:

(2) A person shall be deemed to be in possession of goods or of the documents of title to goods, where the goods or documents are in his actual custody or are held by any other person subject to his control or for him or on his behalf:

(3) The expression "goods" shall include wares and merchandise:

(4) The expression "document of title" shall include any bill of lading, dock warrant, warehouse-keeper's certificate, and warrant or order for the delivery of goods, and any other document used in the ordinary course of business as proof of the possession or control of goods, or authorising or purporting to authorise, either by endorsement or by delivery, the possessor of the document to transfer or receive goods thereby represented:

(5) The expression "pledge" shall include any contract pledging, or giving a lien or security on, goods, whether in consideration of an original advance or of any further or continuing advance or of any pecuniary liability:

(6) The expression "person" shall include any body of persons corporate or unincorporate.

Dispositions by mercantile agents

2 Powers of mercantile agent with respect to disposition of goods

(1) Where a mercantile agent is, with the consent of the owner, in possession of goods or of the documents of title to goods, any sale, pledge,

or other disposition of the goods, made by him when acting in the ordinary course of business of a mercantile agent, shall, subject to the provisions of this Act, be as valid as if he were expressly authorised by the owner of the goods to make the same, provided that the person taking under the disposition acts in good faith, and has not at the time of the disposition notice that the person making the disposition has not authority to make the same.

(2)   Where a mercantile agent has, with the consent of the owner, been in possession of goods or of the documents of title to goods, any sale, pledge, or other disposition, which would have been valid if the consent had continued shall be valid notwithstanding the determination of the consent: provided that the person taking under the disposition has not at the time thereof notice that the consent has been determined.

(3)   Where a mercantile agent has obtained possession of any documents of title to goods by reason of his being or having been, with the consent of the owner, in possession of the goods represented thereby, or of any other documents of title to the goods, his possession of the first-mentioned documents shall, for the purposes of this Act, be deemed to be with the consent of the owner.

(4)   For the purposes of this Act the consent of the owner shall be presumed in the absence of evidence to the contrary.

3   Effect of pledges of documents of title

A pledge of the documents of title to goods shall be deemed to be a pledge of the goods.

4   Pledge for antecedent debt

Where a mercantile agent pledges goods as security for a debt or liability due from the pledgor to the pledgee before the time of the pledge, the pledgee shall acquire no further right to the goods than could have been enforced by the pledgor at the time of the pledge.

5   Rights acquired by exchange of goods or documents

The consideration necessary for the validity of a sale, pledge, or other disposition, of goods, in pursuance of this Act, may be either a payment in cash, or the delivery or transfer of other goods, or of a document of title to goods, or of a negotiable security, or any other valuable consideration; but where goods are pledged by a mercantile agent in consideration of the delivery or transfer of other goods, or of a document of title to goods, or of a negotiable security, the pledgee shall acquire no right or interest in the goods so pledged in excess of the value of the goods, documents, or security when so delivered or transferred in exchange.

6 Agreements through clerks, etc

For the purposes of this Act an agreement made with a mercantile agent through a clerk or other person authorised in the ordinary course of business to make contracts of sale or pledge on his behalf shall be deemed to be an agreement with the agent.

7 Provisions as to consignors and consignees

(1)   Where the owner of goods has given possession of the goods to another person for the purpose of consignment or sale, or has shipped the goods in the name of another person, and the consignee of the goods has not had notice that such person is not the owner of the goods, the consignee shall, in respect of advances made to or for the use of such person, have the same lien on the goods as if such person were the owner of the goods, and may transfer any such lien to another person.

(2)   Nothing in this section shall limit or effect the validity of any sale, pledge, or disposition, by a mercantile agent.

Dispositions by sellers and buyers of goods

8 Disposition by seller remaining in possession

Where a person, having sold goods, continues, or is, in possession of the goods or of the documents of title to the goods, the delivery or transfer by that person, or by a mercantile agent acting for him, of the goods or documents of title under any sale, pledge, or other disposition thereof, or under any agreement for sale pledge, or other disposition thereof, to any person receiving the same in good faith and without notice of the previous sale, shall have the same effect as if the person making the delivery or transfer were expressly authorised by the owner of the goods to make the same.

9 Disposition by buyer obtaining possession

Where a person, having bought or agreed to buy goods, obtains with the consent of the seller possession of the goods or the documents of title to the goods, the delivery or transfer, by that person or by a mercantile agent acting for him, of the goods or documents of title, under any sale, pledge, or other disposition thereof, or under any agreement for sale, pledge, or other disposition thereof, to any person receiving the same in good faith and without notice of any lien or other right of the original seller in respect of the goods, shall have the same effect as if the person making the delivery or transfer were a mercantile agent in possession of the goods or documents of title with the consent of the owner.

[For the purposes of this section—
  (i)   the buyer under a conditional sale agreement shall be deemed not to be a person who has bought or agreed to buy goods, and
  (ii) "conditional sale agreement" means an agreement for the sale of goods which is a consumer credit agreement within the meaning of the Consumer Credit Act 1974 under which the purchase price or part of it is payable by instalments, and the property in the goods is to remain in the seller (notwithstanding that the buyer is to be in possession of the goods) until such conditions as to the payment of instalments or otherwise as may be specified in the agreement are fulfilled.]

NOTES

Words in square brackets added by the Consumer Credit Act 1974, s 192(3)(a), Sch 4, Pt I, para 2.

10 Effect of transfer of documents on vendor's lien or right of stoppage in transitu

Where a document of title to goods has been lawfully transferred to a person as a buyer or owner of the goods, and that person transfers the document to a person who takes the document in good faith and for valuable consideration, the last-mentioned transfer shall have the same effect for defeating any vendor's lien or right of stoppage in transitu as the transfer of a bill of lading has for defeating the right of stoppage in transitu.

Supplemental

11 Mode of transferring documents

For the purposes of this Act, the transfer of a document may be by endorsement, or, where the document is by custom or by its express terms transferable by delivery, or makes the goods deliverable to the bearer, then by delivery.

12 Saving for rights of true owner

(1)   Nothing in this Act shall authorise an agent to exceed or depart from his authority as between himself and his principal, or exempt him from any liability, civil or criminal, for so doing.

(2)   Nothing in this Act shall prevent the owner of goods from recovering the goods from an agent or his trustee in bankruptcy at any time before the sale or pledge thereof, or shall prevent the owner of goods pledged by an agent from having the right to redeem the goods at any time before the sale thereof, on satisfying the claim for which the goods were pledged, and paying to the agent, if by him required, any money in respect of which the agent would by law be entitled to retain the goods or the documents of title thereto, or any of them, by way of lien as against the owner, or from recovering from any person with whom the goods have been pledged any balance of money remaining in his hands as the produce of the sale of the goods after deducting the amount of his lien.

(3)   Nothing in this Act shall prevent the owner of goods sold by an agent from recovering from the buyer the price agreed to be paid for the same, or any part of that price, subject to any right of set off on the part of the buyer against the agent.

13 Saving for common law powers of agent

The provisions of this Act shall be construed in amplification and not in derogation of the powers exercisable by an agent independently of this Act.

16   Extent of Act

This Act shall not extend to Scotland.

17   Short title

This Act may be cited as the Factors Act 1889.

Bills of Sale Act 1890

(C 53)

An Act to exempt certain letters of hypothecation from the operation of the Bills of Sale Act 1882

[10 August 1890]

1   Exemption of letters of hypothecation of imported goods from 41 & 42 Vict c 31, and 45 & 46 Vict c 43, s 9

[An instrument charging or creating any security on or declaring trusts of imported goods given or executed at any time prior to their deposit in a warehouse, factory, or store, or to their being reshipped for export, or delivered to a purchaser not being the person giving or executing such instrument shall not be deemed a bill of sale within the meaning of the Bills of Sale Acts 1878 and 1882.]

NOTES

Substituted by the Bills of Sale Act 1891, s 1.

2   Savings of 46 & 47 Vict c 52, s 44

Nothing in this Act shall affect the operation of section forty-four of the Bankruptcy Act 1883 in respect of any goods comprised in any such instrument as is herein-before described, if such goods would but for this Act be goods within the meaning of sub-section three of that section.

3   Short title

This Act may be cited as the Bills of Sale Act 1890.

Partnership Act 1890

(C 39)

An Act to declare and amend the Law of Partnership

[14 August 1890]

Relations of Partners to persons dealing with them

5 Power of partner to bind the firm

Every partner is an agent of the firm and his other partners for the purpose of the business of the partnership; and the acts of every partner who does any act for carrying on in the usual way business of the kind carried on by the firm of which he is a member bind the firm and his partners, unless the partner so acting has in fact no authority to act for the firm in the particular matter, and the person with whom he is dealing either knows that he has no authority, or does not know or believe him to be a partner.

14 Persons liable by "holding out"

(1)   Every one who by words spoken or written or by conduct represents himself, or who knowingly suffers himself to be represented, as a partner in a particular firm, is liable as a partner to any one who has on the faith of any such representation given credit to the firm, whether the representation has or has not been made or communicated to the person so giving credit by or with the knowledge of the apparent partner making the representation or suffering it to be made.

(2)   Provided that where after a partner's death the partnership business is continued in the old firm's name, the continued use of that name or of the deceased partner's name as part thereof shall not of itself make his executors or administrators estate or effects liable for any partnership debts contracted after his death.

Supplemental

50 Short title

This Act may be cited as the Partnership Act 1890.

Marine Insurance Act 1906

(C 41)

An Act to codify the Law relating to Marine Insurance

[21 December 1906]

NOTES

Marine Insurance Acts 1906 and 1909. By the Marine Insurance (Gambling Policies) Act 1909, s 2, that Act and this Act may be cited together by this collective title.

Marine Insurance

1 Marine insurance defined

A contract of marine insurance is a contract whereby the insurer undertakes to indemnify the assured, in manner and to the extent thereby agreed, against marine losses, that is to say, the losses incident to marine adventure.

2 Mixed sea and land risks

(1)   A contract of marine insurance may, by its express terms, or by usage of trade, be extended so as to protect the assured against losses on inland waters or on any land risk which may be incidental to any sea voyage.

(2)   Where a ship in course of building, or the launch of a ship, or any adventure analogous to a marine adventure, is covered by a policy in the form of a marine policy, the provisions of this Act, in so far as applicable, shall apply thereto; but, except as by this section provided, nothing in this Act shall alter or affect any rule of law applicable to any contract of insurance other than a contract of marine insurance as by this Act defined.

3 Marine adventure and maritime perils defined

(1)   Subject to the provisions of this Act, every lawful marine adventure may be the subject of a contract of marine insurance.

(2)   In particular there is a marine adventure where—
  (a) Any ship goods or other moveables are exposed to maritime perils. Such property is in this Act referred to as "insurable property";
  (b) The earning or acquisition of any freight, passage money, commission, profit, or other pecuniary benefit, or the security for any advances, loan, or disbursements, is endangered by the exposure of insurable property to maritime perils;
  (c) Any liability to a third party may be incurred by the owner of, or other person interested in or responsible for, insurable property, by reason of maritime perils.

"Maritime perils" means the perils consequent on, or incidental to, the navigation of the sea, that is to say, perils of the seas, fire, war perils, pirates, rovers, thieves, captures, seisures, restraints, and detainments of princes and peoples, jettisons, barratry, and any other perils, either of the like kind or which may be designated by the policy.

Insurable Interest

4   Avoidance of wagering or gaming contracts

(1)   Every contract of marine insurance by way of gaming or wagering is void.

(2)   A contract of marine insurance is deemed to be a gaming or wagering contract—
  (a) Where the assured has not an insurable interest as defined by this Act, and the contract is entered into with no expectation of acquiring such an interest; or
  (b) Where the policy is made "interest or no interest," or "without further proof of interest than the policy itself," or "without benefit of salvage to the insurer," or subject to any other like term:

Provided that, where there is no possibility of salvage, a policy may be effected without benefit of salvage to the insurer.

5   Insurable interest defined

(1)   Subject to the provisions of this Act, every person has an insurable interest who is interested in a marine adventure.

(2)   In particular a person is interested in a marine adventure where he stands in any legal or equitable relation to the adventure or to any insurable property at risk therein, in consequence of which he may benefit by the safety or due arrival of insurable property, or may be prejudiced by its loss, or damage thereto, or by the detention thereof, or may incur liability in respect thereof.

6   When interest must attach

(1)   The assured must be interested in the subject-matter insured at the time of the loss though he need not be interested when the insurance is effected:

Provided that where the subject-matter is insured "lost or not lost," the assured may recover although he may not have acquired his interest until after the loss, unless at the time of effecting the contract of insurance the assured was aware of the loss, and the insurer was not.

(2)   Where the assured has no interest at the time of the loss, he cannot acquire interest by any act or election after he is aware of the loss.

7   Defeasible or contingent interest

(1)   A defeasible interest is insurable, as also is a contingent interest.

(2)   In particular, where the buyer of goods has insured them, he has an insurable interest, notwithstanding that he might, at his election, have rejected the goods, or have treated them as at the seller's risk, by reason of the latter's delay in making delivery or otherwise.

8   Partial interest

A partial interest of any nature is insurable.

9   Re-insurance

(1)   The insurer under a contract of marine insurance has an insurable interest in his risk, and may re-insure in respect of it.

(2)   Unless the policy otherwise provides, the original assured has no right or interest in respect of such re-insurance.

10 Bottomry

The lender of money on bottomry or respondentia has an insurable interest in respect of the loan.

11 Master's and seamen's wages

The master or any member of the crew of a ship has an insurable interest in respect of his wages.

12 Advance freight

In the case of advance freight, the person advancing the freight has an insurable interest, in so far as such freight is not repayable in case of loss.

13 Charges of insurance

The assured has an insurable interest in the charges of any insurance which he may effect.

14 Quantum of interest

(1)   Where the subject-matter insured is mortgaged, the mortgagor has an insurable interest in the full value thereof, and the mortgagee has an insurable interest in respect of any sum due or to become due under the mortgage.

(2)   A mortgagee, consignee, or other person having an interest in the subject-matter insured may insure on behalf and for the benefit of other persons interested as well as for his own benefit.

(3)   The owner of insurable property has an insurable interest in respect of the full value thereof, notwithstanding that some third person may have agreed, or be liable, to indemnify him in case of loss.

15 Assignment of interest

Where the assured assigns or otherwise parts with his interest in the subject-matter insured, he does not thereby transfer to the assignee his rights under the contract of insurance, unless there be an express or implied agreement with the assignee to that effect.

But the provisions of this section do not affect a transmission of interest by operation of law.

Insurable Value

16 Measure of insurable value

Subject to any express provision or valuation in the policy, the insurable value of the subject-matter insured must be ascertained as follows:—

(1) In insurance on ship, the insurable value is the value, at the commencement of the risk, of the ship, including her outfit, provisions and stores for the officers and crew, money advanced for seamen's wages, and other disbursements (if any) incurred to make the ship fit for the voyage or adventure contemplated by the policy, plus the charges of insurance upon the whole:

> The insurable value, in the case of a steamship, includes also the machinery, boilers, and coals and engine stores if owned by the assured, and, in the case of a ship engaged in a special trade, the ordinary fittings requisite for that trade:

(2) In insurance on freight, whether paid in advance or otherwise, the insurable value is the gross amount of the freight at the risk of the assured, plus the charges of insurance:

(3) In insurance on goods or merchandise, the insurable value is the prime cost of the property insured, plus the expenses of and incidental to shipping and the charges of insurance upon the whole:

(4) In insurance on any other subject-matter, the insurable value is the amount at the risk of the assured when the policy attaches, plus the charges of insurance.

Disclosure and Representations

17 Insurance is uberrimae fidei

A contract of marine insurance is a contract based upon the utmost good faith, and, if the utmost good faith be not observed by either party, the contract may be avoided by the other party.

18 Disclosure by assured

(1) Subject to the provisions of this section, the assured must disclose to the insurer, before the contract is concluded, every material circumstance which is known to the assured, and the assured is deemed to know every circumstance which, in the ordinary course of business, ought to be known by him. If the assured fails to make such disclosure, the insurer may avoid the contract.

(2) Every circumstance is material which would influence the judgment of a prudent insurer in fixing the premium, or determining whether he will take the risk.

(3) In the absence of inquiry the following circumstances need not be disclosed, namely:—
  (a) Any circumstance which diminishes the risk;
  (b) Any circumstance which is known or presumed to be known to the insurer. The insurer is presumed to know matters of common notoriety or knowledge, and matters which an insurer in the ordinary course of his business, as such, ought to know;
  (c) Any circumstance as to which information is waived by the insurer;
  (d) Any circumstance which it is superfluous to disclose by reason of any express or implied warranty.

(4) Whether any particular circumstance, which is not disclosed, be material or not is, in each case, a question of fact.

(5) The term "circumstance" includes any communication made to, or information received by, the assured.

19 Disclosure by agent effecting insurance

Subject to the provisions of the preceding section as to circumstances which need not be disclosed, where an insurance is effected for the assured by an agent, the agent must disclose to the insurer—
  (a) Every material circumstance which is known to himself, and an agent to insure is deemed to know every circumstance which in the ordinary course of business ought to be known by, or to have been communicated to, him; and
  (b) Every material circumstance which the assured is bound to disclose, unless it come to his knowledge too late to communicate it to the agent.

20   Representations pending negotiation of contract

(1)   Every material representation made by the assured or his agent to the insurer during the negotiations for the contract, and before the contract is concluded, must be true. If it be untrue the insurer may avoid the contract.

(2)   A representation is material which would influence the judgment of a prudent insurer in fixing the premium, or determining whether he will take the risk.

(3)   A representation may be either a representation as to a matter of fact, or as to a matter of expectation or belief.

(4)   A representation as to matter of fact is true, if it be substantially correct, that is to say, if the difference between what is represented and what is actually correct would not be considered material by a prudent insurer.

(5)   A representation as to a matter of expectation or belief is true if it be made in good faith.

(6)   A representation may be withdrawn or corrected before the contract is concluded.

(7)   Whether a particular representation be material or not is, in each case, a question of fact.

21   When contract is deemed to be concluded

A contract of marine insurance is deemed to be concluded when the proposal of the assured is accepted by the insurer, whether the policy be then issued or not; and, for the purpose of showing when the proposal was accepted, reference may be made to the slip or covering note or other customary memorandum of the contract, ...

NOTES

    Words omitted repealed by the Finance Act 1959, s 37(5), Sch 8, Pt II, and the Finance Act (Northern Ireland) 1959, s 17(2), Sch 3, Pt II.

The Policy

22   Contract must be embodied in policy

Subject to the provisions of any statute, a contract of marine insurance is inadmissible in evidence unless it is embodied in a marine policy in accordance with this Act. The policy may be executed and issued either at the time when the contract is concluded, or afterwards.

23   What policy must specify

A marine policy must specify—
   (1)   The name of the assured, or of some person who effects the insurance on his behalf:
   (2)–(5) ...

NOTES

Sub-ss (2)–(5): repealed by the Finance Act 1959, ss 30(5), (7), 37(5), Sch 8, Pt II, and the Finance Act (Northern Ireland) 1959, ss 5(5), (7), 17(2), Sch 3, Pt II.

24   Signature of insurer

(1)   A marine policy must be signed by or on behalf of the insurer, provided that in the case of a corporation the corporate seal may be sufficient, but nothing in this section shall be construed as requiring the subscription of a corporation to be under seal.

(2)   Where a policy is subscribed by or on behalf of two or more insurers, each subscription, unless the contrary be expressed, constitutes a distinct contract with the assured.

25   Voyage and time policies

(1)   Where the contract is to insure the subject-matter "at and from", or from one place to another or others, the policy is called a "voyage policy", and where the contract is to insure the subject-matter for a definite period

of time the policy is called a "time policy". A contract for both voyage and time may be included in the same policy.

(2)   ...

26   Designation of subject-matter

(1)   The subject-matter insured must be designated in a marine policy with reasonable certainty.

(2)   The nature and extent of the interest of the assured in the subject-matter insured need not be specified in the policy.

(3)   Where the policy designates the subject-matter insured in general terms, it shall be construed to apply to the interest intended by the assured to be covered.

(4)   In the application of this section regard shall be had to any usage regulating the designation of the subject-matter insured.

27   Valued policy

(1)   A policy may be either valued or unvalued.

(2)   A valued policy is a policy which specifies the agreed value of the subject-matter insured.

(3)   Subject to the provisions of this Act, and in the absence of fraud, the value fixed by the policy is, as between the insurer and assured, conclusive of the insurable value of the subject intended to be insured, whether the loss be total or partial.

(4)   Unless the policy otherwise provides, the value fixed by the policy is not conclusive for the purpose of determining whether there has been a constructive total loss.

28 Unvalued policy

An unvalued policy is a policy which does not specify the value of the subject-matter insured, but, subject to the limit of the sum insured, leaves the insurable value to be subsequently ascertained, in the manner hereinbefore specified.

29 Floating policy by ship or ships

(1)   A floating policy is a policy which describes the insurance in general terms, and leaves the name of the ship or ships and other particulars to be defined by subsequent declaration.

(2)   The subsequent declaration or declarations may be made by indorsement on the policy, or in other customary manner.

(3)   Unless the policy otherwise provides, the declarations must be made in the order of dispatch or shipment. They must, in the case of goods, comprise all consignments within the terms of the policy, and the value of the goods or other property must be honestly stated, but an omission or erroneous declaration may be rectified even after loss or arrival, provided the omission or declaration was made in good faith.

(4)   Unless the policy otherwise provides, where a declaration of value is not made until after notice of loss or arrival, the policy must be treated as an unvalued policy as regards the subject-matter of that declaration.

30 Construction of terms in policy

(1)   A policy may be in the form in the First Schedule to this Act.

(2)   Subject to the provisions of this Act, and unless the context of the policy otherwise requires, the terms and expressions mentioned in the First Schedule to this Act shall be construed as having the scope and meaning in that schedule assigned to them.

31 Premium to be arranged

(1)   Where an insurance is effected at a premium to be arranged, and no arrangement is made, a reasonable premium is payable.

(2)   Where an insurance is effected on the terms that an additional premium is to be arranged in a given event, and that event happens but no arrangement is made, then a reasonable additional premium is payable.

Double Insurance

32   Double insurance

(1)   Where two or more policies are effected by or on behalf of the assured on the same adventure and interest or any part thereof, and the sums insured exceed the indemnity allowed by this Act, the assured is said to be over-insured by double insurance.

(2)   Where the assured is over-insured by double insurance—
 (a) The assured, unless the policy otherwise provides, may claim payment from the insurers in such order as he may think fit, provided that he is not entitled to receive any sum in excess of the indemnity allowed by this Act;
 (b) Where the policy under which the assured claims is a valued policy, the assured must give credit as against the valuation for any sum received by him under any other policy without regard to the actual value of the subject-matter insured;
 (c) Where the policy under which the assured claims is an unvalued policy he must give credit, as against the full insurable value, for any sum received by him under any other policy;
 (d) Where the assured receives any sum in excess of the indemnity allowed by this Act, he is deemed to hold such sum in trust for the insurers, according to their right of contribution among themselves.

Warranties, etc

33   Nature of warranty

(1)   A warranty, in the following sections relating to warranties, means a promissory warranty, that is to say, a warranty by which the assured undertakes that some particular thing shall or shall not be done, or that some condition shall be fulfilled, or whereby he affirms or negatives the existence of a particular state of facts.

(2)   A warranty may be express or implied.

(3)   A warranty, as above defined, is a condition which must be exactly complied with, whether it be material to the risk or not. If it be not so complied with, then, subject to any express provision in the policy, the insurer is discharged from liability as from the date of the breach of warranty, but without prejudice to any liability incurred by him before that date.

34   When breach of warranty excused

(1)   Non-compliance with a warranty is excused when, by reason of a change of circumstances, the warranty ceases to be applicable to the circumstances of the contract, or when compliance with the warranty is rendered unlawful by any subsequent law.

(2)   Where a warranty is broken, the assured cannot avail himself of the defence that the breach has been remedied, and the warranty complied with, before loss.

(3)   A breach of warranty may be waived by the insurer.

3   Express warranties

(1)   An express warranty may be in any form of words from which the intention to warrant is to be inferred.

(2)   An express warranty must be included in, or written upon, the policy, or must be contained in some document incorporated by reference into the policy.

(3)   An express warranty does not exclude an implied warranty, unless it be inconsistent therewith.

36   Warranty of neutrality

(1)   Where insurable property, whether ship or goods, is expressly warranted neutral, there is an implied condition that the property shall have a neutral character at the commencement of the risk, and that, so far as the assured can control the matter, its neutral character shall be preserved during the risk.

(2)   Where a ship is expressly warranted "neutral" there is also an implied condition that, so far as the assured can control the matter, she shall be

properly documented, that is to say, that she shall carry the necessary papers to establish her neutrality, and that she shall not falsify or suppress her papers, or use simulated papers. If any loss occurs through breach of this condition, the insurer may avoid the contract.

37   No implied warranty of nationality

There is no implied warranty as to the nationality of a ship, or that her nationality shall not be changed during the risk.

38   Warranty of good safety

Where the subject-matter insured is warranted "well" or "in good safety" on a particular day, it is sufficient if it be safe at any time during that day.

39   Warranty of seaworthiness of ship

(1)   In a voyage policy there is an implied warranty that at the commencement of the voyage the ship shall be seaworthy for the purpose of the particular adventure insured.

(2)   Where the policy attaches while the ship is in port, there is also an implied warranty that she shall, at the commencement of the risk, be reasonably fit to encounter the ordinary perils of the port.

(3)   Where the policy relates to a voyage which is performed in different stages, during which the ship requires different kinds of or further preparation or equipment, there is an implied warranty that at the commencement of each stage the ship is seaworthy in respect of such preparation or equipment for the purposes of that stage.

(4)   A ship is deemed to be seaworthy when she is reasonably fit in all respects to encounter the ordinary perils of the seas of the adventure insured.

(5)   In a time policy there is no implied warranty that the ship shall be seaworthy at any stage of the adventure, but where, with the privity of the assured, the ship is sent to sea in an unseaworthy state, the insurer is not liable for any loss attributable to unseaworthiness.

40   No implied warranty that goods are seaworthy

(1)   In a policy on goods or other moveables there is no implied warranty that the goods or moveables are seaworthy.

(2)   In a voyage policy on goods or other moveables there is an implied warranty that at the commencement of the voyage the ship is not only seaworthy as a ship, but also that she is reasonably fit to carry the goods or other moveables to the destination contemplated by the policy.

41   Warranty of legality

There is an implied warranty that the adventure insured is a lawful one, and that, so far as the assured can control the matter, the adventure shall be carried out in a lawful manner.

The Voyage

42   Implied condition as to commencement of risk

(1)   Where the subject-matter is insured by a voyage policy "at and from" or "from" a particular place, it is not necessary that the ship should be at that place when the contract is concluded, but there is an implied condition that the adventure shall be commenced within a reasonable time, and that if the adventure be not so commenced the insurer may avoid the contract.

(2)   The implied condition may be negatived by showing that the delay was caused by circumstances known to the insurer before the contract was concluded, or by showing that he waived the condition.

43   Alteration of port of departure

Where the place of departure is specified by the policy, and the ship instead of sailing from that place sails from any other place, the risk does not attach.

44   Sailing for different destination

Where the destination is specified in the policy, and the ship, instead of sailing for that destination, sails for any other destination, the risk does not attach.

45 Change of voyage

(1)   Where, after the commencement of the risk, the destination of the ship is voluntarily changed from the destination contemplated by the policy, there is said to be a change of voyage.

(2)   Unless the policy otherwise provides, where there is a change of voyage, the insurer is discharged from liability as from the time of change, that is to say, as from the time when the determination to change it is manifested; and it is immaterial that the ship may not in fact have left the course of voyage contemplated by the policy when the loss occurs.

46 Deviation

(1)   Where a ship, without lawful excuse, deviates from the voyage contemplated by the policy, the insurer is discharged from liability as from the time of deviation, and it is immaterial that the ship may have regained her route before any loss occurs.

(2)   There is a deviation from the voyage contemplated by the policy—
   (a) Where the course of the voyage is specifically designated by the policy, and that course is departed from; or
   (b) Where the course of the voyage is not specifically designated by the policy, but the usual and customary course is departed from.

(3)   The intention to deviate is immaterial; there must be a deviation in fact to discharge the insurer from his liability under the contract.

47 Several ports of discharge

(1)   Where several ports of discharge are specified by the policy, the ship may proceed to all or any of them, but, in the absence of any usage or sufficient cause to the contrary, she must proceed to them, or such of them as she goes to, in the order designated by the policy. If she does not there is a deviation.

(2)   Where the policy is to "ports of discharge", within a given area, which are not named, the ship must, in the absence of any usage or sufficient cause to the contrary, proceed to them, or such of them as she goes to, in their geographical order. If she does not there is a deviation.

48   Delay in voyage

In the case of a voyage policy, the adventure insured must be prosecuted throughout its course with reasonable dispatch, and, if without lawful excuse it is not so prosecuted, the insurer is discharged from liability as from the time when the delay became unreasonable.

49   Excuses for deviation or delay

(1)   Deviation or delay in prosecuting the voyage contemplated by the policy is excused—
  (a) Where authorised by any special term in the policy; or
  (b) Where caused by circumstances beyond the control of the master and his employer; or
  (c) Where reasonably necessary in order to comply with an express or implied warranty; or
  (d) Where reasonably necessary for the safety of the ship or subject-matter insured; or
  (e) For the purpose of saving human life, or aiding a ship in distress where human life may be in danger; or
  (f) Where reasonably necessary for the purpose of obtaining medical or surgical aid for any person on board the ship; or
  (g) Where caused by the barratrous conduct of the master or crew, if barratry be one of the perils insured against.

(2)   When the cause excusing the deviation or delay ceases to operate, the ship must resume her course, and prosecute her voyage, with reasonable dispatch.

Assignment of Policy

50   When and how policy is assignable

(1)   A marine policy is assignable unless it contains terms expressly prohibiting assignment. It may be assigned either before or after loss.

(2)   Where a marine policy has been assigned so as to pass the beneficial interest in such policy, the assignee of the policy is entitled to sue thereon in his own name; and the defendant is entitled to make any defence arising out of the contract which he would have been entitled to make if the action

had been brought in the name of the person by or on behalf of whom the policy was effected.

(3)   A marine policy may be assigned by indorsement thereon or in other customary manner.

51   Assured who has no interest cannot assign

Where the assured has parted with or lost his interest in the subject-matter insured, and has not, before or at the time of so doing, expressly or impliedly agreed to assign the policy, any subsequent assignment of the policy is inoperative:

Provided that nothing in this section affects the assignment of a policy after loss.

The Premium

52   When premium payable

Unless otherwise agreed, the duty of the assured or his agent to pay the premium, and the duty of the insurer to issue the policy to the assured or his agent, are concurrent conditions, and the insurer is not bound to issue the policy until payment or tender of the premium.

53   Policy effected through broker

(1)   Unless otherwise agreed, where a marine policy is effected on behalf of the assured by a broker, the broker is directly responsible to the insurer for the premium, and the insurer is directly responsible to the assured for the amount which may be payable in respect of losses, or in respect of returnable premium.

(2)   Unless otherwise agreed, the broker has, as against the assured, a lien upon the policy for the amount of the premium and his charges in respect of effecting the policy; and, where he has dealt with the person who employs him as a principal, he has also a lien on the policy in respect of any balance on any insurance account which may be due to him from such person, unless when the debt was incurred he had reason to believe that such person was only an agent.

54 Effect of receipt on policy

Where a marine policy effected on behalf of the assured by a broker acknowledges the receipt of the premium, such acknowledgment is, in the absence of fraud, conclusive as between the insurer and the assured, but not as between the insurer and broker.

Loss and Abandonment

55 Included and excluded losses

(1)   Subject to the provisions of this Act, and unless the policy otherwise provides, the insurer is liable for any loss proximately caused by a peril insured against, but, subject as aforesaid, he is not liable for any loss which is not proximately caused by a peril insured against.

(2)   In particular,—
  (a) The insurer is not liable for any loss attributable to the wilful misconduct of the assured, but, unless the policy otherwise provides, he is liable for any loss proximately caused by a peril insured against, even though the loss would not have happened but for the misconduct or negligence of the master or crew;
  (b) Unless the policy otherwise provides, the insurer on ship or goods is not liable for any loss proximately caused by delay, although the delay be caused by a peril insured against;
  (c) Unless the policy otherwise provides, the insurer is not liable for ordinary wear and tear, ordinary leakage and breakage, inherent vice or nature of the subject-matter insured, or for any loss proximately caused by rats or vermin, or for any injury to machinery not proximately caused by maritime perils.

56 Partial and total loss

(1)   A loss may be either total or partial. Any loss other than a total loss, as herein-after defined, is a partial loss.

(2)   A total loss may be either an actual total loss, or a constructive total loss.

(3)   Unless a different intention appears from the terms of the policy, an insurance against total loss includes a constructive, as well as an actual, total loss.

(4)   Where the assured brings an action for a total loss and the evidence proves only a partial loss, he may, unless the policy otherwise provides, recover for a partial loss.

(5)   Where goods reach their destination in specie, but by reason of obliteration of marks, or otherwise, they are incapable of identification, the loss, if any, is partial, and not total.

57   Actual total loss

(1)   Where the subject-matter insured is destroyed, or so damaged as to cease to be a thing of the kind insured, or where the assured is irretrievably deprived thereof, there is an actual total loss.

(2)   In the case of an actual total loss no notice of abandonment need be given.

58   Missing ship

Where the ship concerned in the adventure is missing, and after the lapse of a reasonable time no news of her has been received, an actual total loss may be presumed.

59   Effect of transhipment, etc

Where, by a peril insured against, the voyage is interrupted at an intermediate port or place, under such circumstances as, apart from any special stipulation in the contract of affreightment, to justify the master in landing and re-shipping the goods or other moveables, or in transhipping them, and sending them on to their destination, the liability of the insurer continues, notwithstanding the landing or transhipment.

60   Constructive total loss defined

(1)   Subject to any express provision in the policy, there is a constructive total loss where the subject-matter insured is reasonably abandoned on account of its actual total loss appearing to be unavoidable, or because it could not be preserved from actual total loss without an expenditure which would exceed its value when the expenditure had been incurred.

(2) In particular, there is a constructive total loss—
 (i) Where the assured is deprived of the possession of his ship or goods by a peril insured against, and (a) it is unlikely that he can recover the ship or goods, as the case may be, or (b) the cost of recovering the ship or goods, as the case may be, would exceed their value when recovered; or
 (ii) In the case of damage to a ship, where she is so damaged by a peril insured against that the cost of repairing the damage would exceed the value of the ship when repaired.

In estimating the cost of repairs, no deduction is to be made in respect of general average contributions to those repairs payable by other interests, but account is to be taken of the expense of future salvage operations and of any future general average contributions to which the ship would be liable if repaired; or
 (iii) In the case of damage to goods, where the cost of repairing the damage and forwarding the goods to their destination would exceed their value on arrival.

61 Effect of constructive total loss

Where there is a constructive total loss the assured may either treat the loss as a partial loss, or abandon the subject-matter insured to the insurer and treat the loss as if it were an actual total loss.

62 Notice of abandonment

(1) Subject to the provisions of this section, where the assured elects to abandon the subject-matter insured to the insurer, he must give notice of abandonment. If he fails to do so the loss can only be treated as a partial loss.

(2) Notice of abandonment may be given in writing, or by word of mouth, or partly in writing and partly by word of mouth, and may be given in terms which indicate the intention of the assured to abandon his insured interest in the subject-matter insured unconditionally to the insurer.

(3) Notice of abandonment must be given with reasonable diligence after the receipt of reliable information of the loss, but where the information is of a doubtful character the assured is entitled to a reasonable time to make inquiry.

(4)   Where notice of abandonment is properly given, the rights of the assured are not prejudiced by the fact that the insurer refuses to accept the abandonment.

(5)   The acceptance of an abandonment may be either express or implied from the conduct of the insurer. The mere silence of the insurer after notice is not an acceptance.

(6)   Where a notice of abandonment is accepted the abandonment is irrevocable. The acceptance of the notice conclusively admits liability for the loss and the sufficiency of the notice.

(7)   Notice of abandonment is unnecessary where, at the time when the assured receives information of the loss, there would be no possibility of benefit to the insurer if notice were given to him.

(8)   Notice of abandonment may be waived by the insurer.

(9)   Where an insurer has re-insured his risk, no notice of abandonment need be given by him.

63   Effect of abandonment

(1)   Where there is a valid abandonment the insurer is entitled to take over the interest of the assured in whatever may remain of the subject-matter insured, and all proprietary rights incidental thereto.

(2)   Upon the abandonment of a ship, the insurer thereof is entitled to any freight in course of being earned, and which is earned by her subsequent to the casualty causing the loss, less the expenses of earning it incurred after the casualty; and, where the ship is carrying the owner's goods, the insurer is entitled to a reasonable remuneration for the carriage of them subsequent to the casualty causing the loss.

Partial Losses (including Salvage and General Average and Particular Charges)

64   Particular average loss

(1)   A particular average loss is a partial loss of the subject-matter insured, caused by a peril insured against, and which is not a general average loss.

(2)   Expenses incurred by or on behalf of the assured for the safety or preservation of the subject-matter insured, other than general average and salvage charges, are called particular charges. Particular charges are not included in particular average.

65   Salvage charges

(1)   Subject to any express provision in the policy, salvage charges incurred in preventing a loss by perils insured against may be recovered as a loss by those perils.

(2)   "Salvage charges" means the charges recoverable under maritime law by a salvor independently of contract. They do not include the expenses of services in the nature of salvage rendered by the assured or his agents, or any person employed for hire by them, for the purpose of averting a peril insured against. Such expenses, where properly incurred, may be recovered as particular charges or as a general average loss, according to the circumstances under which they were incurred.

66   General average loss

(1)   A general average loss is a loss caused by or directly consequential on a general average act. It includes a general average expenditure as well as a general average sacrifice.

(2)   There is a general average act where any extraordinary sacrifice or expenditure is voluntarily and reasonably made or incurred in time of peril for the purpose of preserving the property imperilled in the common adventure.

(3)   Where there is a general average loss, the party on whom it falls is entitled, subject to the conditions imposed by maritime law, to a rateable contribution from the other parties interested, and such contribution is called a general average contribution.

(4)   Subject to any express provision in the policy, where the assured has incurred a general average expenditure, he may recover from the insurer in respect of the proportion of the loss which falls upon him; and, in the case of a general average sacrifice, he may recover from the insurer in respect of the whole loss without having enforced his right of contribution from the other parties liable to contribute.

(5)   Subject to any express provision in the policy, where the assured has paid, or is liable to pay, a general average contribution in respect of the subject insured, he may recover therefor from the insurer.

(6)   In the absence of express stipulation, the insurer is not liable for any general average loss or contribution where the loss was not incurred for the purpose of avoiding, or in connexion with the avoidance of, a peril insured against.

(7)   Where ship, freight, and cargo, or any two of those interests, are owned by the same assured, the liability of the insurer in respect of general average losses or contributions is to be determined as if those subjects were owned by different persons.

Measure of Indemnity

67   Extent of liability of insurer for loss

(1)   The sum which the assured can recover in respect of a loss on a policy by which he is insured, in the case of an unvalued policy to the full extent of the insurable value, or, in the case of a valued policy to the full extent of the value fixed by the policy, is called the measure of indemnity.

(2)   Where there is a loss recoverable under the policy, the insurer, or each insurer if there be more than one, is liable for such proportion of the measure of indemnity as the amount of his subscription bears to the value fixed by the policy in the case of a valued policy, or to the insurable value in the case of an unvalued policy.

68   Total loss

Subject to the provisions of this Act and to any express provision in the policy, where there is a total loss of the subject-matter insured,—
  (1) If the policy be a valued policy, the measure of indemnity is the sum fixed by the policy:
  (2) If the policy be an unvalued policy, the measure of indemnity is the insurable value of the subject-matter insured.

69　Partial loss of ship

Where a ship is damaged, but is not totally lost, the measure of indemnity, subject to any express provision in the policy, is as follows:—

   (1)　Where the ship has been repaired, the assured is entitled to the reasonable cost of the repairs, less the customary deductions, but not exceeding the sum insured in respect of any one casualty:

   (2)　Where the ship has been only partially repaired, the assured is entitled to the reasonable cost of such repairs, computed as above, and also to be indemnified for the reasonable depreciation, if any, arising from the unrepaired damage, provided that the aggregate amount shall not exceed the cost of repairing the whole damage, computed as above:

   (3)　Where the ship has not been repaired, and has not been sold in her damaged state during the risk, the assured is entitled to be indemnified for the reasonable depreciation arising from the unrepaired damage, but not exceeding the reasonable cost of repairing such damage, computed as above.

70　Partial loss of freight

Subject to any express provision in the policy, where there is a partial loss of freight, the measure of indemnity is such proportion of the sum fixed by the policy in the case of a valued policy, or of the insurable value in the case of an unvalued policy, as the proportion of freight lost by the assured bears to the whole freight at the risk of the assured under the policy.

71　Partial loss of goods, merchandise, etc

Where there is a partial loss of goods, merchandise, or other moveables, the measure of indemnity, subject to any express provision in the policy, is as follows:—

   (1)　Where part of the goods, merchandise or other moveables insured by a valued policy is totally lost, the measure of indemnity is such proportion of the sum fixed by the policy as the insurable value of the part lost bears to the insurable value of the whole, ascertained as in the case of an unvalued policy:

   (2)　Where part of the goods, merchandise, or other moveables insured by an unvalued policy is totally lost, the measure of indemnity is the insurable value of the part lost, ascertained as in case of total loss:

(3) Where the whole or any part of the goods or merchandise insured has been delivered damaged at its destination, the measure of indemnity is such proportion of the sum fixed by the policy in the case of a valued policy, or of the insurable value in the case of an unvalued policy, as the difference between the gross sound and damaged values at the place of arrival bears to the gross sound value:

(4) "Gross value" means the wholesale price or, if there be no such price, the estimated value, with, in either case, freight, landing charges, and duty paid beforehand; provided that, in the case of goods or merchandise customarily sold in bond, the bonded price is deemed to be the gross value. "Gross proceeds" means the actual price obtained at a sale where all charges on sale are paid by the sellers.

72 Apportionment of valuation

(1)   Where different species of property are insured under a single valuation, the valuation must be apportioned over the different species in proportion to their respective insurable values, as in the case of an unvalued policy. The insured value of any part of a species is such proportion of the total insured value of the same as the insurable value of the part bears to the insurable value of the whole, ascertained in both cases as provided by this Act.

(2)   Where a valuation has to be apportioned, and particulars of the prime cost of each separate species, quality, or description of goods cannot be ascertained, the division of the valuation may be made over the net arrived sound values of the different species, qualities, or descriptions of goods.

73 General average contributions and salvage charges

(1)   Subject to any express provision in the policy, where the assured has paid, or is liable for, any general average contribution, the measure of indemnity is the full amount of such contribution, if the subject-matter liable to contribution is insured for its full contributory value; but, if such subject-matter be not insured for its full contributory value, or if only part of it be insured, the indemnity payable by the insurer must be reduced in proportion to the under insurance, and where there has been a particular average loss which constitutes a deduction from the contributory value, and for which the insurer is liable, that amount must be deducted from the insured value in order to ascertain what the insurer is liable to contribute.

(2)   Where the insurer is liable for salvage charges the extent of his liability must be determined on the like principle.

74   Liabilities to third parties

Where the assured has effected an insurance in express terms against any liability to a third party, the measure of indemnity, subject to any express provision in the policy, is the amount paid or payable by him to such third party in respect of such liability.

75   General provisions as to measure of indemnity

(1)   Where there has been a loss in respect of any subject-matter not expressly provided for in the foregoing provisions of this Act, the measure of indemnity shall be ascertained, as nearly as may be, in accordance with those provisions, in so far as applicable to the particular case.

(2)   Nothing in the provisions of this Act relating to the measure of indemnity shall affect the rules relating to double insurance, or prohibit the insurer from disproving interest wholly or in part, or from showing that at the time of the loss the whole or any part of the subject-matter insured was not at risk under the policy.

76   Particular average warranties

(1)   Where the subject-matter insured is warranted free from particular average, the assured cannot recover for a loss of part, other than a loss incurred by a general average sacrifice unless the contract contained in the policy be apportionable; but, if the contract be apportionable, the assured may recover for a total loss of any apportionable part.

(2)   Where the subject-matter insured is warranted free from particular average, either wholly or under a certain percentage, the insurer is nevertheless liable for salvage charges, and for particular charges and other expenses properly incurred pursuant to the provisions of the suing and labouring clause in order to avert a loss insured against.

(3)   Unless the policy otherwise provides, where the subject-matter insured is warranted free from particular average under a specified

percentage, a general average loss cannot be added to a particular average loss to make up the specified percentage.

(4)   For the purpose of ascertaining whether the specified percentage has been reached, regard shall be had only to the actual loss suffered by the subject-matter insured. Particular charges and the expenses of and incidental to ascertaining and proving the loss must be excluded.

77 Successive losses

(1)   Unless the policy otherwise provides, and subject to the provisions of this Act, the insurer is liable for successive losses, even though the total amount of such losses may exceed the sum insured.

(2)   Where, under the same policy, a partial loss, which has not been repaired or otherwise made good, is followed by a total loss, the assured can only recover in respect of the total loss:

Provided that nothing in this section shall affect the liability of the insurer under the suing and labouring clause.

78 Suing and labouring clause

(1)   Where the policy contains a suing and labouring clause, the engagement thereby entered into is deemed to be supplementary to the contract of insurance, and the assured may recover from the insurer any expenses properly incurred pursuant to the clause, notwithstanding that the insurer may have paid for a total loss, or that the subject-matter may have been warranted free from particular average, either wholly or under a certain percentage.

(2)   General average losses and contributions and salvage charges, as defined by this Act, are not recoverable under the suing and labouring clause.

(3)   Expenses incurred for the purpose of averting or diminishing any loss not covered by the policy are not recoverable under the suing and labouring clause.

(4)   It is the duty of the assured and his agents, in all cases, to take such measures as may be reasonable for the purpose of averting or minimising a loss.

Rights of Insurer on Payment

79   Right of subrogation

(1)   Where the insurer pays for a total loss, either of the whole, or in the case of goods of any apportionable part, of the subject-matter insured, he thereupon becomes entitled to take over the interest of the assured in whatever may remain of the subject-matter so paid for, and he is thereby subrogated to all the rights and remedies of the assured in and in respect of that subject-matter as from the time of the casualty causing the loss.

(2)   Subject to the foregoing provisions, where the insurer pays for a partial loss, he acquires no title to the subject-matter insured, or such part of it as may remain, but he is thereupon subrogated to all rights and remedies of the assured in and in respect of the subject-matter insured as from the time of the casualty causing the loss, in so far as the assured has been indemnified, according to this Act, by such payment for the loss.

80   Right of contribution

(1)   Where the assured is over-insured by double insurance, each insurer is bound, as between himself and the other insurers, to contribute rateably to the loss in proportion to the amount for which he is liable under his contract.

(2)   If any insurer pays more than his proportion of the loss, he is entitled to maintain an action for contribution against the other insurers, and is entitled to the like remedies as a surety who has paid more than his proportion of the debt.

81   Effect of under insurance

Where the assured is insured for an amount less than the insurable value or, in the case of a valued policy, for an amount less than the policy valuation, he is deemed to be his own insurer in respect of the uninsured balance.

Return of Premium

82 Enforcement of return

Where the premium or a proportionate part thereof is, by this Act, declared to be returnable,—
   (a) If already paid, it may be recovered by the assured from the insurer; and
   (b) If unpaid, it may be retained by the assured or his agent.

83 Return by agreement

Where the policy contains a stipulation for the return of the premium, or a proportionate part thereof, on the happening of a certain event, and that event happens, the premium, or, as the case may be, the proportionate part thereof, is thereupon returnable to the assured.

84 Return for failure of consideration

(1)   Where the consideration for the payment of the premium totally fails, and there has been no fraud or illegality on the part of the assured or his agents, the premium is thereupon returnable to the assured.

(2)   Where the consideration for the payment of the premium is apportionable and there is a total failure of any apportionable part of the consideration, a proportionate part of the premium is, under the like conditions, thereupon returnable to the assured.

(3)   In particular—
   (a) Where the policy is void, or is avoided by the insurer as from the commencement of the risk, the premium is returnable, provided that there has been no fraud or illegality on the part of the assured; but if the risk is not apportionable, and has once attached, the premium is not returnable;
   (b) Where the subject-matter insured, or part thereof, has never been imperilled, the premium, or, as the case may be, a proportionate part thereof, is returnable:
      Provided that where the subject-matter has been insured "lost or not lost" and has arrived in safety at the time when the contract is concluded, the premium is not returnable unless, at such time, the insurer knew of the safe arrival.

(c) Where the assured has no insurable interest throughout the currency of the risk, the premium is returnable, provided that this rule does not apply to a policy effected by way of gaming or wagering;

(d) Where the assured has a defeasible interest which is terminated during the currency of the risk, the premium is not returnable;

(e) Where the assured has over-insured under an unvalued policy, a proportionate part of the premium is returnable;

(f) Subject to the foregoing provisions, where the assured has over-insured by double insurance, a proportionate part of the several premiums is returnable:

> Provided that, if the policies are effected at different times, and any earlier policy has at any time borne the entire risk, or if a claim has been paid on the policy in respect of the full sum insured thereby, no premium is returnable in respect of that policy, and when the double insurance is effected knowingly by the assured no premium is returnable.

Mutual Insurance

85 Modification of Act in case of mutual insurance

(1) Where two or more persons mutually agree to insure each other against marine losses there is said to be a mutual insurance.

(2) The provisions of this Act relating to the premium do not apply to mutual insurance, but a guarantee, or such other arrangement as may be agreed upon, may be substituted for the premium.

(3) The provisions of this Act, in so far as they may be modified by the agreement of the parties, may in the case of mutual insurance be modified by the terms of the policies issued by the association, or by the rules and regulations of the association.

(4) Subject to the exceptions mentioned in this section, the provisions of this Act apply to a mutual insurance.

Supplemental

86 Ratification by assured

Where a contract of marine insurance is in good faith effected by one person on behalf of another, the person on whose behalf it is effected may ratify the contract even after he is aware of a loss.

87 Implied obligations varied by agreement or usage

(1)   Where any right, duty, or liability would arise under a contract of marine insurance by implication of law, it may be negatived or varied by express agreement, or by usage, if the usage be such as to bind both parties to the contract.

(2)   The provisions of this section extend to any right, duty, or liability declared by this Act which may be lawfully modified by agreement.

88 Reasonable time, etc, a question of fact

Where by this Act any reference is made to reasonable time, reasonable premium, or reasonable diligence, the question what is reasonable is a question of fact.

89 Slip as evidence

Where there is a duly stamped policy, reference may be made, as heretofore, to the slip or covering note, in any legal proceeding.

90 Interpretation of terms

In this Act, unless the context or subject-matter otherwise requires,—
    "Action" includes counter-claim and set off:
    "Freight" includes the profit derivable by a shipowner from the employment of his ship to carry his own goods or moveables, as well as freight payable by a third party, but does not include passage money:
    "Moveables" means any moveable tangible property, other than the ship, and includes money, valuable securities, and other documents:
    "Policy" means a marine policy.

91 Savings

(1)   Nothing in this Act, or in any repeal effected thereby, shall affect—
    (a) The provisions of the Stamp Act 1891, or any enactment for the time being in force relating to the revenue;
    (b) The provisions of the Companies Act 1862, or any enactment amending or substituted for the same;

(c) The provisions of any statute not expressly repealed by this Act.

(2)   The rules of the common law including the law merchant, save in so far as they are inconsistent with the express provisions of this Act, shall continue to apply to contracts of marine insurance.

94   Short title

This Act may be cited as the Marine Insurance Act 1906.

FIRST SCHEDULE

Section 30

FORM OF POLICY

BE IT KNOWN THAT   as well in   own name as for and in the name and names of all and every other person or persons to whom the same doth, may, or shall appertain, in part or in all doth make assurance and cause   and them, and every of them, to be insured lost or not lost, at and from

Upon any kind of goods and merchandise, and also upon the body, tackle, apparel, ordnance, munition, artillery, boat, and other furniture, of and in the good ship or vessel called the   whereof is master under God, for this present voyage,   or whosoever else shall go for master in the said ship, or by whatsoever other name or names the said ship, or the master thereof, is or shall be named or called; beginning the adventure upon the said goods and merchandises from the loading thereof aboard the said ship.

upon the said ship, etc

and so shall continue and endure, during her abode there, upon the said ship, etc

And further, until the said ship, with all her ordnance, tackle, apparel, etc, and goods and merchandises whatsoever shall be arrived at

upon the said ship, etc, until she hath moored at anchor twenty-four hours in good safety; and upon the goods and merchandises, until the same be there discharged and safely landed. And it shall be lawful for the said ship, etc, in this voyage to proceed and sail to and touch and stay at any ports or places whatsoever

without prejudice to this insurance. The said ship, etc, goods and merchandises, etc, for so much as concerns the assured by agreement between the assured and assurers in this policy, are and shall be valued at

Touching the adventures and perils which we the assurers are contented to bear and do take upon us in this voyage: they are of the seas, men of war, fire, enemies, pirates, rovers, thieves, jettisons, letters of mart and countermart, surprisals, takings at sea, arrests, restraints, and detainments of all kings, princes, and people, of what nation, condition, or quality soever, barratry of the master and mariners, and of all other perils, losses, and misfortunes, that have or shall come to the hurt, detriment, or damage of the said goods and merchandises, and ship, etc, or any part thereof. And in case of any loss or misfortune it shall be lawful to the assured, their factors, servants and assigns, to sue, labour, and travel for, in and about the defence, safeguards, and recovery of the said goods and merchandises, and ship, etc, or any part thereof, without prejudice to this insurance; to the charges whereof we, the assurers, will contribute each one according to the rate and quantity of his sum herein assured. And it is especially declared and agreed that no acts of the insurer or insured in recovering, saving, or preserving the property insured shall be considered as a waiver, or acceptance of abandonment. And it is agreed by us, the insurers, that this writing or policy of assurance shall be of as much force and effect as the surest writing or policy of assurance heretofore made in Lombard Street, or in the Royal Exchange, or elsewhere in London. And so we, the assurers, are contented, and do hereby promise and bind ourselves, each one for his own part, our heirs, executors, and goods to the assured, their executors, administrators, and assigns, for the true performance of the premises, confessing ourselves paid the consideration due unto us for this assurance by the assured, at and after the rate of

IN WITNESS whereof we, the assurers, have subscribed our names and sums assured in London.

N.B.—Corn, fish, salt, fruit, flour, and seed are warranted free from average, unless general, or the ship be stranded—sugar, tobacco, hemp, flax, hides and skins are warranted free from average, under five pounds per cent., and all other goods, also the ship and freight, are warranted free from average, under three pounds per cent. unless general, or the ship be stranded.

RULES FOR CONSTRUCTION OF POLICY

The following are the rules referred to by this Act for the construction of a policy in the above or other like form, where the context does not otherwise require:—

1.   Where the subject-matter is insured "lost or not lost," and the loss has occurred before the contract is concluded, the risk attaches, unless at such time the assured was aware of the loss, and the insurer was not.

2.   Where the subject-matter is insured "from" a particular place, the risk does not attach until the ship starts on the voyage insured.

3.— (a) Where a ship is insured "at and from" a particular place, and she is at that place in good safety when the contract is concluded, the risk attaches immediately.
  (b) If she be not at that place when the contract is concluded, the risk attaches as soon as she arrives there in good safety, and, unless the policy otherwise provides, it is immaterial that she is covered by another policy for a specified time after arrival.
  (c) Where chartered freight is insured "at and from" a particular place, and the ship is at that place in good safety when the contract is concluded the risk attaches immediately. If she be not there when the contract is concluded, the risk attaches as soon as she arrives there in good safety.
  (d) Where freight, other than chartered freight, is payable without special conditions and is insured "at and from" a particular place, the risk attaches pro rata as the goods or merchandise are shipped; provided that if there be cargo in readiness which belongs to the shipowner, or which some other person has contracted with him to ship, the risk attaches as soon as the ship is ready to receive such cargo.

4.   Where goods or other moveables are insured "from the loading thereof," the risk does not attach until such goods or moveables are actually on board, and the insurer is not liable for them while in transit from the shore to ship.

5.   Where the risk on goods or other moveables continues until they are "safely landed," they must be landed in the customary manner and within a reasonable time after arrival at the port of discharge, and if they are not so landed the risk ceases.

6.   In the absence of any further license or usage, the liberty to touch and stay "at any port or place whatsoever" does not authorise the ship to

depart from the course of her voyage from the port of departure to the port of destination.

7.   The term "perils of the seas" refers only to fortuitous accidents or casualties of the seas. It does not include the ordinary action of the winds and waves.

8.   The term "pirates" includes passengers who mutiny and rioters who attack the ship from the shore.

9.   The term "thieves" does not cover clandestine theft or a theft committed by any one of the ship's company, whether crew or passengers.

10.   The term "arrests, etc, of kings, princes, and people" refers to political or executive acts, and does not include a loss caused by riot or by ordinary judicial process.

11.   The term "barratry" includes every wrongful act wilfully committed by the master or crew to the prejudice of the owner, or, as the case may be, the charterer.

12.   The term "all other perils" includes only perils similar in kind to the perils specifically mentioned in the policy.

13.   The term "average unless general" means a partial loss of the subject-matter insured other than a general average loss, and does not include "particular charges."

14.   Where the ship has stranded, the insurer is liable for the excepted losses, although the loss is not attributable to the stranding, provided that when the stranding takes place the risk has attached and, if the policy be on goods, that the damaged goods are on board.

15.   The term "ship" includes the hull, materials and outfit, stores and provisions for the officers and crew, and, in the case of vessels engaged in a special trade, the ordinary fittings requisite for the trade, and also, in the case of a steamship, the machinery, boilers, and coals and engine stores, if owned by the assured.

16.   The term "freight" includes the profit derivable by a shipowner from the employment of his ship to carry his own goods or moveables, as well as freight payable by a third party, but does not include passage money.

17.   The term "goods" means goods in the nature of merchandise, and does not include personal effects or provisions and stores for use on board.

In the absence of any usage to the contrary, deck cargo and living animals must be insured specifically, and not under the general denomination of goods.

Marine Insurance (Gambling Policies) Act 1909

(C 12)

An Act to prohibit Gambling on Loss by Maritime Perils

[20 October 1909]

1 Prohibition of gambling on loss by maritime perils

(1)   If—

(a) any person effects a contract of marine insurance without having any bona fide interest, direct or indirect, either in the safe arrival of the ship in relation to which the contract is made or in the safety or preservation of the subject-matter insured, or a bona fide expectation of acquiring such an interest; or

(b) any person in the employment of the owner of a ship, not being a part owner of the ship, effects a contract of marine insurance in relation to the ship, and the contract is made "interest or no interest," or "without further proof of interest than the policy itself," or "without benefit of salvage to the insurer," or subject to any other like term,

the contract shall be deemed to be a contract by way of gambling on loss by maritime perils, and the person effecting it shall be guilty of an offence, and shall be liable, on summary conviction, to imprisonment, with or without hard labour, for a term not exceeding six months or to a fine not exceeding [level 3 on the standard scale], and in either case to forfeit to the Crown any money he may receive under the contract.

(2)   Any broker or other person through whom, and any insurer with whom, any such contract is effected shall be guilty of an offence and liable on summary conviction to the like penalties if he acted knowing that the contract was by way of gambling on loss by maritime perils within the meaning of this Act.

(3)   Proceedings under this Act shall not be instituted without the consent in England of the Attorney-General, in Scotland of the Lord Advocate, and in Ireland of the Attorney-General for Ireland.

(4)   Proceedings shall not be instituted under this Act against a person (other than a person in the employment of the owner of the ship in relation to which the contract was made) alleged to have effected a contract by way of gambling on loss by maritime perils until an opportunity has been afforded him of showing that the contract was not such a contract as aforesaid, and any information given by that person for that purpose shall not be admissible in evidence against him in any prosecution under this Act.

(5)   If proceedings under this Act are taken against any person (other than a person in the employment of the owner of the ship in relation to which the contract was made) for effecting such a contract, and the contract was made "interest or no interest," or "without further proof of interest than the policy itself," or "without benefit of salvage to the insurer," or subject to any other like term, the contract shall be deemed to be a contract by way of gambling on loss by maritime perils unless the contrary is proved.

(6)   For the purpose of giving jurisdiction under this Act, every offence shall be deemed to have been committed either in the place in which the same actually was committed or in any place in which the offender may be.

(7)   Any person aggrieved by an order or decision of a court of summary jurisdiction under this Act, may appeal to [the Crown Court].

(8)   For the purposes of this Act the expression "owner" includes charterer.

(9)   Subsection (7) of this section shall not apply to Scotland.

NOTES

Sub-s (1): words in square brackets substituted by virtue of the Criminal Justice Act 1982, ss 38, 46.

Sub-s (7): words in square brackets substituted by the Courts Act 1971, s 56, Sch 9, Pt I.

2   Short title

This Act may be cited as the Marine Insurance (Gambling Policies) Act 1909, and the Marine Insurance Act 1906 and this Act may be cited together as the Marine Insurance Acts 1906 and 1909.

Law of Property Act 1925

(C 20)

An Act to consolidate the enactments relating to Conveyancing and the Law of Property in England and Wales

[9 April 1925]

NOTES

Modification: this Act (with the exception of s 75) has been modified by the Solicitors' Incorporated Practices Order 1991, SI 1991/2684, arts 2–5, Sch 1, so that any reference to a solicitor is to be construed as including a reference to a recognised body within the meaning of the Administration of Justice Act 1985, s 9.

PART II
CONTRACTS, CONVEYANCES AND OTHER INSTRUMENTS

Conveyances and other Instruments

53 Instruments required to be in writing

(1)   Subject to the provisions hereinafter contained with respect to the creation of interests in land by parol—

    (a)   no interest in land can be created or disposed of except by writing signed by the person creating or conveying the same, or by his agent thereunto lawfully authorised in writing, or by will, or by operation of law;

    (b)   a declaration of trust respecting any land or any interest therein must be manifested and proved by some writing signed by some person who is able to declare such trust or by his will;

    (c)   a disposition of an equitable interest or trust subsisting at the time of the disposition, must be in writing signed by the person disposing of the same, or by his agent thereunto lawfully authorised in writing or by will.

(2)   This section does not affect the creation or operation of resulting, implied or constructive trusts.

PART IV
EQUITABLE INTERESTS AND THINGS IN ACTION

136   Legal assignments of things in action

(1)   Any absolute assignment by writing under the hand of the assignor (not purporting to be by way of charge only) of any debt or other legal thing in action, of which express notice in writing has been given to the debtor, trustee or other person from whom the assignor would have been entitled to claim such debt or thing in action, is effectual in law (subject to equities having priority over the right of the assignee) to pass and transfer from the date of such notice—

(a) the legal right to such debt or thing in action;

(b) all legal and other remedies for the same; and

(c) the power to give a good discharge for the same without the concurrence of the assignor:

Provided that, if the debtor, trustee or other person liable in respect of such debt or thing in action has notice—

(a) that the assignment is disputed by the assignor or any person claiming under him; or

(b) of any other opposing or conflicting claims to such debt or thing in action;

he may, if he thinks fit, either call upon the persons making claim thereto to interplead concerning the same, or pay the debt or other thing in action into court under the provisions of the Trustee Act 1925.

(2)   This section does not affect the provisions of the Policies of Assurance Act 1867.

[(3)   The county court has jurisdiction (including power to receive payment of money or securities into court) under the proviso to subsection (1) of this section where the amount or value of the debt or thing in action does not exceed [£30,000].]

NOTES

Sub-s (3): added by the County Courts Act 1984, s 148(1), Sch 2, Pt II, para 4; sum in square brackets substituted by the High Court and County Courts Jurisdiction Order 1991, SI 1991/724, art 2(8), Schedule, Pt I.

137 Dealings with life interests, reversions and other equitable interests

(1)   The law applicable to dealings with equitable things in action which regulates the priority of competing interests therein, shall, as respects dealings with equitable interests in land, capital money, and securities representing capital money effected after the commencement of this Act, apply to and regulate the priority of competing interests therein.

This subsection applies whether or not the money or securities are in court.

(2)(i) In the case of a dealing with an equitable interest in settled land, capital money or securities representing capital money, the persons to be served with notice of the dealing shall be the trustees of the settlement; and where the equitable interest is created by a derivative or subsidiary settlement, the persons to be served with notice shall be the trustees of that settlement.

(ii)   In the case of a dealing with an equitable interest in [land subject to a trust of land, or the proceeds of the sale of such land, the persons to be served with notice shall be the trustees].

(iii) In any other case the person to be served with notice of a dealing with an equitable interest in land shall be the estate owner of the land affected.

The persons on whom notice is served pursuant to this subsection shall be affected thereby in the same manner as if they had been trustees of personal property out of which the equitable interest was created or arose.

This subsection does not apply where the money or securities are in court.

(3)   A notice, otherwise than in writing, given to, or received by, a trustee after the commencement of this Act as respects any dealing with an equitable interest in real or personal property, shall not affect the priority of competing claims of purchasers in that equitable interest.

(4)   Where, as respects any dealing with an equitable interest in real or personal property—

(a) the trustees are not persons to whom a valid notice of the dealing can be given; or

(b) there are no trustees to whom a notice can be given; or

(c) for any other reason a valid notice cannot be served, or cannot be served without unreasonable cost or delay;

a purchaser may at his own cost require that—

   (i)   a memorandum of the dealing be endorsed, written on or permanently annexed to the instrument creating the trust;

  (ii) the instrument be produced to him by the person having the possession or custody thereof to prove that a sufficient memorandum has been placed thereon or annexed thereto.

Such memorandum shall, as respects priorities, operate in like manner as if notice in writing of the dealing had been given to trustees duly qualified to receive the notice at the time when the memorandum is placed on or annexed to the instrument creating the trust.

(5)   Where the property affected is settled land, the memorandum shall be placed on or annexed to the trust instrument and not the vesting instrument.

Where the property affected is land [subject to a trust of land], the memorandum shall be placed on or annexed to the instrument whereby the equitable interest is created.

(6)   Where the trust is created by statute or by operation of law, or in any other case where there is no instrument whereby the trusts are declared, the instrument under which the equitable interest is acquired or which is evidence of the devolution thereof shall, for the purposes of this section, be deemed the instrument creating the trust.

In particular, where the trust arises by reason of an intestacy, the letters of administration or probate in force when the dealing was effected shall be deemed such instrument.

(7)   Nothing in this section affects any priority acquired before the commencement of this Act.

(8)   Where a notice in writing of a dealing with an equitable interest in real or personal property has been served on a trustee under this section, the trustees from time to time of the property affected shall be entitled to the custody of the notice, and the notice shall be delivered to them by any person who for the time being may have the custody thereof; and subject to the payment of costs, any person interested in the equitable interest may require production of the notice.

(9)   The liability of the estate owner of the legal estate affected to produce documents and furnish information to persons entitled to equitable interests therein shall correspond to the liability of a trustee for sale to produce documents and furnish information to persons entitled to equitable interests in the proceeds of sale of the land.

(10)   This section does not apply until a trust has been created, and in this section "dealing" includes a disposition by operation of law.

   Sub-ss (2), (5): words in square brackets substituted by the Trusts of Land and Appointment of Trustees Act 1996, s 25(1), Sch 3, para 4(1), (15)(a), (b), subject to savings contained in ss 3, 18(3), 25(5) of the 1996 Act.

PART XII
CONSTRUCTION, JURISDICTION, AND GENERAL PROVISIONS

209   Short title, commencement, extent

(1)   This Act may be cited as the Law of Property Act 1925.

(2)   …

(3)   This Act extends to England and Wales only.

NOTES
   Sub-s (2): repealed by the Statute Law Revision Act 1950.

Third Parties (Rights Against Insurers) Act 1930

(C 25)

An Act to confer on third parties rights against insurers of third-party risks in the event of the insured becoming insolvent, and in certain other events
[10 July 1930]

1   Rights of third parties against insurers on bankruptcy, etc, of the insured

(1)   Where under any contract of insurance a person (hereinafter referred to as the insured) is insured against liabilities to third parties which he may incur, then—

(a) in the event of the insured becoming bankrupt or making a composition or arrangement with his creditors; or

(b) in the case of the insured being a company, in the event of a winding-up order [or an administration order] being made, or a resolution for a voluntary winding-up being passed, with respect to the company, or of a receiver or manager of the company's business or undertaking being duly appointed, or of possession being taken, by or on behalf of the holders of any debentures secured by a floating charge, of any property comprised in or subject to the charge [or of [a voluntary arrangement proposed for the purposes of Part I of the Insolvency Act 1986 being approved under that Part]];

if, either before or after that event, any such liability as aforesaid is incurred by the insured, his rights against the insurer under the contract in respect of the liability shall, notwithstanding anything in any Act or rule of law to the contrary, be transferred to and vest in the third party to whom the liability was so incurred.

(2)   Where [the estate of any person falls to be administered in accordance with an order under section [421 of the Insolvency Act 1986]], then, if any debt provable in bankruptcy [in Scotland, any claim accepted in the sequestration] is owing by the deceased in respect of a liability against which he was insured under a contract of insurance as being a liability to a third party, the deceased debtor's rights against the insurer under the contract in respect of that liability shall, notwithstanding anything in [any such order], be transferred to and vest in the person to whom the debt is owing.

(3)   In so far as any contract of insurance made after the commencement of this Act in respect of any liability of the insured to third parties purports, whether directly or indirectly, to avoid the contract or to alter the rights of the parties thereunder upon the happening to the insured of any of the events specified in paragraph (a) or paragraph (b) of subsection (1) of this section or upon the [estate of any person falling to be administered in accordance with an order under section [421 of the Insolvency Act 1986]], the contract shall be of no effect.

(4)   Upon a transfer under subsection (1) or subsection (2) of this section, the insurer shall, subject to the provisions of section three of this Act, be under the same liability to the third party as he would have been under to the insured, but—

(a) if the liability of the insurer to the insured exceeds the liability of the insured to the third party, nothing in this Act shall affect the rights of the insured against the insurer in respect of the excess; and

(b) if the liability of the insurer to the insured is less than the liability of the insured to the third party, nothing in this Act shall affect the rights of the third party against the insured in respect of the balance.

(5) For the purposes of this Act, the expression "liabilities to third parties", in relation to a person insured under any contract of insurance, shall not include any liability of that person in the capacity of insurer under some other contract of insurance.

(6) This Act shall not apply—
(a) where a company is wound up voluntarily merely for the purposes of reconstruction or of amalgamation with another company; or
(b) to any case to which subsections (1) and (2) of section seven of the Workmen's Compensation Act 1925 applies.

NOTES

Sub-s (1): words in first pair of square brackets and words in second (outer) pair of square brackets added by the Insolvency Act 1985, s 235, Sch 8, para 7; words in third pair of square brackets substituted by s 439(2) of, and Sch 14 to, the 1986 Act.

Sub-s (2): words in first and final pairs of square brackets substituted by the Insolvency Act 1985, s 235, Sch 8, para 7 and the Insolvency Act 1986, s 437, Sch 11; words in third pair of square brackets inserted by the Bankruptcy (Scotland) Act 1985, s 75(1), Sch 7, Pt I, para 6(1); words in fourth pair of square brackets substituted by s 439(2) of, and Sch 14 to, the 1986 Act.

Sub-s (3): words in first pair of square brackets substituted by the Insolvency Act 1985, s 235, Sch 8, para 7 and the Insolvency Act 1986, s 437, Sch 11; further amended by s 439(2) of, and Sch 14 to, the 1986 Act.

2 Duty to give necessary information to third parties

(1) In the event of any person becoming bankrupt or making a composition or arrangement with his creditors, or in the event of [the estate of any person falling to be administered in accordance with an order under section [421 of the Insolvency Act 1986]], or in the event of a winding-up order [or an administration order] being made, or a resolution for a voluntary winding-up being passed, with respect to any company or of a receiver or manager of the company's business or undertaking being duly appointed or of possession being taken by or on behalf of the holders of any debentures secured by a floating charge of any property comprised in or subject to the charge it shall be the duty of the bankrupt, debtor, personal representative of the deceased debtor or company, and, as the case may be, of the trustee in bankruptcy, trustee, liquidator, [administrator,] receiver, or manager, or

person in possession of the property to give at the request of any person claiming that the bankrupt, debtor, deceased debtor, or company is under a liability to him such information as may reasonably be required by him for the purpose of ascertaining whether any rights have been transferred to and vested in him by this Act and for the purpose of enforcing such rights, if any, and any contract of insurance, in so far as it purports, whether directly or indirectly, to avoid the contract or to alter the rights of the parties thereunder upon the giving of any such information in the events aforesaid or otherwise to prohibit or prevent the giving thereof in the said events shall be of no effect.

[(1A) The reference in subsection (1) of this section to a trustee includes a reference to the supervisor of a [voluntary arrangement proposed for the purposes of, and approved under, Part I or Part VIII of the Insolvency Act 1986].]

(2)   If the information given to any person in pursuance of subsection (1) of this section discloses reasonable ground for supposing that there have or may have been transferred to him under this Act rights against any particular insurer, that insurer shall be subject to the same duty as is imposed by the said subsection on the persons therein mentioned.

(3)   The duty to give information imposed by this section shall include a duty to allow all contracts of insurance, receipts for premiums, and other relevant documents in the possession or power of the person on whom the duty is so imposed to be inspected and copies thereof to be taken.

NOTES

Sub-s (1): words in first (outer), third and final pairs of square brackets substituted or added by the Insolvency Act 1985, s 235, Sch 8, para 7; words in second (inner) pair of square brackets substituted by s 439(2) of, and Sch 14 to, the 1986 Act.

Sub-s (1A): inserted by the Insolvency Act 1985, s 235, Sch 8, para 7, words in square brackets therein substituted by s 439(2) of, and Sch 14 to, the 1986 Act.

3   Settlement between insurers and insured persons

Where the insured has become bankrupt or where in the case of the insured being a company, a winding-up order [or an administration order] has been made or a resolution for a voluntary winding-up has been passed, with respect to the company, no agreement made between the insurer and the insured after liability has been incurred to a third party and after the commencement of the bankruptcy or winding-up [or the day of the making of the administration order], as the case may be, nor any waiver, assignment, or other disposition made by, or payment made to the insured after the commencement [or day] aforesaid shall be effective to defeat or affect the rights transferred to the third party under this Act, but those rights shall be the same as if no such agreement, waiver, assignment, disposition or payment had been made.

NOTES

Words in square brackets inserted by the Insolvency Act 1985, s 235, Sch 8, para 7.

5   Short title

This Act may be cited as the Third Parties (Rights Against Insurers) Act 1930.

Law Reform (Frustrated Contracts) Act 1943

(C 40)

An Act to amend the law relating to the frustration of contracts
                                                              [5 August 1943]

1   Adjustment of rights and liabilities of parties to frustrated contracts

(1)   Where a contract governed by English law has become impossible of performance or been otherwise frustrated, and the parties thereto have for that reason been discharged from the further performance of the contract, the following provisions of this section shall, subject to the provisions of section two of this Act, have effect in relation thereto.

(2)   All sums paid or payable to any party in pursuance of the contract before the time when the parties were so discharged (in this Act referred to as "the time of discharge") shall, in the case of sums so paid, be recoverable from him as money received by him for the use of the party by whom the sums were paid, and, in the case of sums so payable, cease to be so payable:

Provided that, if the party to whom the sums were so paid or payable incurred expenses before the time of discharge in, or for the purpose of, the performance of the contract, the court may, if it considers it just to do so having regard to all the circumstances of the case, allow him to retain or, as the case may be, recover the whole or any part of the sums so paid or payable, not being an amount in excess of the expenses so incurred.

(3) Where any party to the contract has, by reason of anything done by any other party thereto in, or for the purpose of, the performance of the contract, obtained a valuable benefit (other than a payment of money to which the last foregoing subsection applies) before the time of discharge, there shall be recoverable from him by the said other party such sum (if any), not exceeding the value of the said benefit to the party obtaining it, as the court considers just, having regard to all the circumstances of the case and, in particular,—

    (a) the amount of any expenses incurred before the time of discharge by the benefited party in, or for the purpose of, the performance of the contract, including any sums paid or payable by him to any other party in pursuance of the contract and retained or recoverable by that party under the last foregoing subsection, and

    (b) the effect, in relation to the said benefit, of the circumstances giving rise to the frustration of the contract.

(4) In estimating, for the purposes of the foregoing provisions of this section, the amount of any expenses incurred by any party to the contract, the court may, without prejudice to the generality of the said provisions, include such sum as appears to be reasonable in respect of overhead expenses and in respect of any work or services performed personally by the said party.

(5) In considering whether any sum ought to be recovered or retained under the foregoing provisions of this section by any party to the contract, the court shall not take into account any sums which have, by reason of the circumstances giving rise to the frustration of the contract, become payable to that party under any contract of insurance unless there was an obligation to insure imposed by an express term of the frustrated contract or by or under any enactment.

(6) Where any person has assumed obligations under the contract in consideration of the conferring of a benefit by any other party to the contract upon any other person, whether a party to the contract or not, the court may, if in all the circumstances of the case it considers it just to do so, treat for the purposes of subsection (3) of this section any benefit so conferred as a benefit obtained by the person who has assumed the obligations as aforesaid.

2 Provision as to application of this Act

(1)   This Act shall apply to contracts, whether made before or after the commencement of this Act, as respects which the time of discharge is on or after the first day of July, nineteen hundred and forty-three, but not to contracts as respects which the time of discharge is before the said date.

(2)   This Act shall apply to contracts to which the Crown is a party in like manner as to contracts between subjects.

(3)   Where any contract to which this Act applies contains any provision which, upon the true construction of the contract, is intended to have effect in the event of circumstances arising which operate, or would but for the said provision operate, to frustrate the contract, or is intended to have effect whether such circumstances arise or not, the court shall give effect to the said provision and shall only give effect to the foregoing section of this Act to such extent, if any, as appears to the court to be consistent with the said provision.

(4)   Where it appears to the court that a part of any contract to which this Act applies can properly be severed from the remainder of the contract, being a part wholly performed before the time of discharge, or so performed except for the payment in respect of that part of the contract of sums which are or can be ascertained under the contract, the court shall treat that part of the contract as if it were a separate contract and had not been frustrated and shall treat the foregoing section of this Act as only applicable to the remainder of that contract.

(5)   This Act shall not apply—
   (a) to any charterparty, except a time charterparty or a charterparty by way of demise, or to any contract (other than a charterparty) for the carriage of goods by sea; or
   (b) to any contract of insurance, save as is provided by subsection (5) of the foregoing section; or
   (c) to any contract to which [section 7 of the Sale of Goods Act 1979] (which avoids contracts for the sale of specific goods which perish before the risk has passed to the buyer) applies, or to any other contract for the sale, or for the sale and delivery, of specific goods, where the contract is frustrated by reason of the fact that the goods have perished.

NOTES

Sub-s (5): words in square brackets in para (c) substituted by the Sale of Goods Act 1979, s 63, Sch 2, para 2.

3 Short title and interpretation

(1)   This Act may be cited as the Law Reform (Frustrated Contracts) Act 1943.

(2)   In this Act the expression "court" means, in relation to any matter, the court or arbitrator by or before whom the matter falls to be determined.

Cheques Act 1957

(C 36)

An Act to amend the law relating to cheques and certain other instruments
[17 July 1957]

1 Protection of bankers paying unindorsed or irregularly indorsed cheques, etc

(1)   Where a banker in good faith and in the ordinary course of business pays a cheque drawn on him which is not indorsed or is irregularly indorsed, he does not, in doing so, incur any liability by reason only of the absence of, or irregularity in, indorsement, and he is deemed to have paid it in due course.

(2)   Where a banker in good faith and in the ordinary course of business pays any such instrument as the following, namely,—
   (a) a document issued by a customer of his which, though not a bill of exchange, is intended to enable a person to obtain payment from him of the sum mentioned in the document;
   (b) a draft payable on demand drawn by him upon himself, whether payable at the head office or some other office of his bank;

he does not, in doing so, incur any liability by reason only of the absence of, or irregularity in, indorsement, and the payment discharges the instrument.

2 Rights of bankers collecting cheques not indorsed by holders

A banker who gives value for, or has a lien on, a cheque payable to order which the holder delivers to him for collection without indorsing it, has such (if any) rights as he would have had if, upon delivery, the holder had indorsed it in blank.

3   Unindorsed cheques as evidence of payment

[(1)]   An unindorsed cheque which appears to have been paid by the banker
on whom it is drawn is evidence of the receipt by the payee of the sum
payable by the cheque.

[(2)   For the purposes of subsection (1) above, a copy of a cheque to which
that subsection applies is evidence of the cheque if—
   (a) the copy is made by the banker in whose possession the cheque is
        after presentment and,
   (b) it is certified by him to be a true copy of the original.]

NOTES

   Sub-s (1): numbered as such by the Deregulation (Bills of Exchange) Order 1996,
SI 1996/2993, art 5.
   Sub-s (2): added by SI 1996/2993, art 5.

4   Protection of bankers collecting payment of cheques, etc

(1)   Where a banker, in good faith and without negligence,—
   (a) receives payment for a customer of an instrument to which this
        section applies; or
   (b) having credited a customer's account with the amount of such an
        instrument, receives payment thereof for himself;

and the customer has no title, or a defective title, to the instrument, the
banker does not incur any liability to the true owner of the instrument by
reason only of having received payment thereof.

(2)   This section applies to the following instruments, namely,—
   (a) cheques [(including cheques which under section 81A(1) of the Bills
        of Exchange Act 1882 or otherwise are not transferable)];
   (b) any document issued by a customer of a banker which, though not a
        bill of exchange, is intended to enable a person to obtain payment
        from that banker of the sum mentioned in the document;
   (c) any document issued by a public officer which is intended to enable
        a person to obtain payment from the Paymaster General or the
        Queen's and Lord Treasurer's Remembrancer of the sum mentioned
        in the document but is not a bill of exchange;
   (d) any draft payable on demand drawn by a banker upon himself,
        whether payable at the head office or some other office of his
        bank.

(3) A banker is not to be treated for the purposes of this section as having been negligent by reason only of his failure to concern himself with absence of, or irregularity in, indorsement of an instrument.

NOTES

Sub-s (2): words in square brackets in para (a) inserted by the Cheques Act 1992, s 3.

5   Application of certain provisions of Bills of Exchange Act, 1882, to instruments not being bills of exchange

The provisions of the Bills of Exchange Act, 1882, relating to crossed cheques shall, so far as applicable, have effect in relation to instruments (other than cheques) to which the last foregoing section applies as they have effect in relation to cheques.

6   Construction, saving and repeal

(1) This Act shall be construed as one with the Bills of Exchange Act, 1882.

(2) The foregoing provisions of this Act do not make negotiable any instrument which, apart from them, is not negotiable.

(3) ...

NOTES

Sub-s (3): repealed by the Statute Law (Repeals) Act 1974.

7   Provisions as to Northern Ireland

This Act extends to Northern Ireland, ...

NOTES

Words omitted repealed by the Northern Ireland Constitution Act 1973, s 41(1), Sch 6, Pt I.

8 Short title and commencement

(1)   This Act may be cited as the Cheques Act 1957.

(2)   This Act shall come into operation at the expiration of a period of three months beginning with the day on which it is passed.

Hire-Purchase Act 1964

(C 53)

An Act to amend the law relating to hire-purchase and credit-sale, and, in relation thereto, to amend the enactments relating to the sale of goods; to make provision with respect to dispositions of motor vehicles which have been let or agreed to be sold by way of hire-purchase or conditional sale; to amend the Advertisements (Hire-Purchase) Act 1957; and for purposes connected with the matters aforesaid

[16 July 1964]

[PART III
TITLE TO MOTOR VEHICLES ON HIRE-PURCHASE OR CONDITIONAL SALE

27 Protection of purchasers of motor vehicles

(1)   This section applies where a motor vehicle has been bailed or (in Scotland) hired under a hire-purchase agreement, or has been agreed to be sold under a conditional sale agreement, and, before the property in the vehicle has become vested in the debtor, he disposes of the vehicle to another person.

(2)   Where the disposition referred to in subsection (1) above is to a private purchaser, and he is a purchaser of the motor vehicle in good faith, without notice of the hire-purchase or conditional sale agreement (the "relevant agreement") that disposition shall have effect as if the creditor's title to the vehicle has been vested in the debtor immediately before that disposition.

(3)   Where the person to whom the disposition referred to in subsection (1) above is made (the "original purchaser") is a trade or finance purchaser, then if the person who is the first private purchaser of the motor vehicle after that disposition (the "first private purchaser") is a purchaser of the vehicle in good faith without notice of the relevant agreement, the

disposition of the vehicle to the first private purchaser shall have effect as if the title of the creditor to the vehicle had been vested in the debtor immediately before he disposed of it to the original purchaser.

(4)   Where, in a case within subsection (3) above—
   (a) the disposition by which the first private purchaser becomes a purchaser of the motor vehicle in good faith without notice of the relevant agreement is itself a bailment or hiring under a hire-purchase agreement, and
   (b) the person who is the creditor in relation to that agreement disposes of the vehicle to the first private purchaser, or a person claiming under him, by transferring to him the property in the vehicle in pursuance of a provision in the agreement in that behalf,

the disposition referred to in paragraph (b) above (whether or not the person to whom it is made is a purchaser in good faith without notice of the relevant agreement) shall as well as the disposition referred to in paragraph (a) above, have effect as mentioned in subsection (3) above.

(5)   The preceding provisions of this section apply—
   (a) notwithstanding anything in [section 21 of the Sale of Goods Act 1979] (sale of goods by a person not the owner), but
   (b) without prejudice to the provisions of the Factors Act (as defined by [section 61(1) of the said Act of 1979]) or of any other enactment enabling the apparent owner of goods to dispose of them as if he were the true owner.

(6)   Nothing in this section shall exonerate the debtor from any liability (whether criminal or civil) to which he would be subject apart from this section; and, in a case where the debtor disposes of the motor vehicle to a trade or finance purchaser, nothing in this section shall exonerate—
   (a) that trade or finance purchaser; or
   (b) any other trade or finance purchaser who becomes a purchaser of the vehicle and is not a person claiming under the first private purchaser,

from any liability (whether criminal or civil) to which he would be subject apart from this section.]

NOTES

This Part of this Act (ie this section and ss 28, 29 post) was substituted by the Consumer Credit Act 1974, s 192(3)(a), Sch 4, para 22.

Sub-s (5): words in square brackets substituted by the Sale of Goods Act 1979, s 63, Sch 2, para 4.

[28 Presumptions relating to dealings with motor vehicles

(1) Where in any proceedings (whether criminal or civil) relating to a motor vehicle it is proved—
- (a) that the vehicle was bailed or (in Scotland) hired under a hire-purchase agreement, or was agreed to be sold under a conditional sale agreement and
- (b) that a person (whether a party to the proceedings or not) became a private purchaser of the vehicle in good faith without notice of the hire-purchase or conditional sale agreement (the "relevant agreement"),

this section shall have effect for the purposes of the operation of section 27 of this Act in relation to those proceedings.

(2) It shall be presumed for those purposes, unless the contrary is proved, that the disposition of the vehicle to the person referred to in subsection (1)(b) above (the "relevant purchaser") was made by the debtor.

(3) If it is proved that that disposition was not made by the debtor, then it shall be presumed for those purposes, unless the contrary is proved—
- (a) that the debtor disposed of the vehicle to a private purchaser purchasing in good faith without notice of the relevant agreement, and
- (b) that the relevant purchaser is or was a person claiming under the person to whom the debtor so disposed of the vehicle.

(4) If it is proved that the disposition of the vehicle to the relevant purchaser was not made by the debtor, and that the person to whom the debtor disposed of the vehicle (the "original purchaser") was a trade or finance purchaser, then it shall be presumed for those purposes, unless the contrary is proved,—
- (a) that the person who, after the disposition of the vehicle to the original purchaser, first became a private purchaser of the vehicle was a purchaser in good faith without notice of the relevant agreement, and
- (b) that the relevant purchaser is or was a person claiming under the original purchaser.

(5) Without prejudice to any other method of proof, where in any proceedings a party thereto admits a fact, that fact shall, for the purposes of this section, be taken as against him to be proved in relation to those proceedings.]

Substituted as noted to s 27.

[29 Interpretation of Part III

(1)   In this Part of this Act—

"conditional sale agreement" means an agreement for the sale of goods under which the purchase price or part of it is payable by instalments, and the property in the goods is to remain in the seller (notwithstanding that the buyer is to be in possession of the goods) until such conditions as to the payment of instalments or otherwise as may be specified in the agreement are fulfilled;

"creditor" means the person by whom goods are bailed or (in Scotland) hired under a hire-purchase agreement or as the case may be, the seller under a conditional sale agreement, or the person to whom his rights and duties have passed by assignment or operation of law;

"disposition" means any sale or contract of sale (including a conditional sale agreement), any bailment or (in Scotland) hiring under a hire-purchase agreement and any transfer of the property in goods in pursuance of a provision in that behalf contained in a hire-purchase agreement, and includes, any transaction purporting to be a disposition (as so defined), and 'dispose of' shall be construed accordingly.

"hire-purchase agreement" means an agreement, other than a conditional sale agreement, under which—

(a) goods are bailed or (in Scotland) hired in return for periodical payments by the person to whom they are bailed or hired, and

(b) the property in the goods will pass to that person if the terms of the agreement are complied with and one or more of the following occurs—

(i)   the exercise of an option to purchase by that person,

(ii)   the doing of any other specified act by any party to the agreement,

(iii) the happening of any other specified events; and

"motor vehicle" means a mechanically propelled vehicle intended or adapted for use on roads to which the public has access.

(2)   In this Part of this Act, "trade or finance purchaser" means a purchaser who, at the time of the disposition made to him, carries on a business which consists, wholly or partly,—

(a) of purchasing motor vehicles for the purpose of offering or exposing them for sale, or

(b) of providing finance by purchasing motor vehicles for the purpose of bailing or (in Scotland) hiring them under hire-purchase agreements or agreeing to sell them under conditional sale agreements,

and "private purchaser" means a purchaser who, at the time of the disposition made to him, does not carry on any such business.

(3)   For the purposes of this Part of this Act a person becomes a purchaser of a motor vehicle if, and at the time when, a disposition of the vehicle is made to him; and a person shall be taken to be a purchaser of a motor vehicle without notice of a hire-purchase agreement or conditional sale agreement if, at the time of the disposition made to him, he has no actual notice that the vehicle is or was the subject of any such agreement.

(4)   In this Part of this Act the "debtor" in relation to a motor vehicle which has been bailed or hired under a hire-purchase agreement, or, as the case may be, agreed to be sold under a conditional sale agreement, means the person who at the material time (whether the agreement has before that time been terminated or not) is either—
(a) the person to whom the vehicle is bailed or hired under that agreement, or
(b) is, in relation to the agreement, the buyer, including a person who at the time is, by virtue of section 130(4) of the Consumer Credit Act 1974 treated as a bailee or (in Scotland) a custodier of the vehicle.

(5)   In this Part of this Act any reference to the title of the creditor to a motor vehicle which has been bailed or (in Scotland) hired under a hire-purchase agreement, or agreed to be sold under a conditional sale agreement, and is disposed of by the debtor, is a reference to such title (if any) to the vehicle as, immediately before that disposition, was vested in the person who then was the creditor in relation to the agreement.]

NOTES

Substituted as noted to s 27.

PART V
SUPPLEMENTARY PROVISIONS

37 Short title, citation and extent

(1)   This Act may be cited as the Hire-Purchase Act 1964.

(2)–(4) ...

(5)   This Act shall not extend to Northern Ireland.

Sub-ss (2), (4): repealed by the Hire-Purchase Act 1965, s 59, Sch 6.
Sub-ss (3), (4): repealed by the Hire-Purchase (Scotland) Act 1965, s 55, Sch 6.

Carriage of Goods by Road Act 1965

(C 37)

An Act to give effect to the Convention on the Contract for the International Carriage of Goods by Road signed at Geneva on 19th May 1956; and for purposes connected therewith

[5 August 1965]

1 Convention to have force of law

Subject to the following provisions of this Act, the provisions of the Convention on the Contract for the International Carriage of Goods by Road (in this Act referred to as "the Convention"), as set out in the Schedule to this Act, shall have the force of law in the United Kingdom so far as they relate to the rights and liabilities of persons concerned in the carriage of goods by road under a contract to which the Convention applies.

2 Designation of High Contracting Parties

(1)   Her Majesty may by Order in Council from time to time certify who are the High Contracting Parties to the Convention and in respect of what territories they are respectively parties.

(2)   An Order in Council under this section shall, except so far as it has been superseded by a subsequent Order, be conclusive evidence of the matters so certified.

3 Power of court to take account of other proceedings

(1)   A court before which proceedings are brought to enforce a liability which is limited by article 23 in the Schedule to this Act may at any stage

of the proceedings make any such order as appears to the court to be just and equitable in view of the provisions of the said article 23 and of any other proceedings which have been, or are likely to be, commenced in the United Kingdom or elsewhere to enforce the liability in whole or in part.

(2)   Without prejudice to the preceding subsection, a court before which proceedings are brought to enforce a liability which is limited by the said article 23 shall, where the liability is, or may be, partly enforceable in other proceedings in the United Kingdom or elsewhere, have jurisdiction to award an amount less than the court would have awarded if the limitation applied solely to the proceedings before the court, or to make any part of its award conditional on the result of any other proceedings.

4   Registration of foreign judgments

(1)   Subject to the next following subsection, Part I of the Foreign Judgments (Reciprocal Enforcement) Act 1933 (in this section referred to as "the Act of 1933") shall apply, whether or not it would otherwise have so applied, to any judgment which—
  (a) has been given in any such action as is referred to in paragraph 1 of article 31 in the Schedule to this Act, and
  (b) has been so given by any court or tribunal of a territory in respect of which one of the High Contracting Parties, other than the United Kingdom, is a party to the Convention, and
  (c) has become enforceable in that territory.

(2)   In the application of Part I of the Act of 1933 in relation to any such judgment as is referred to in the preceding subsection, section 4 of that Act shall have effect with the omission of subsections (2) and (3).

(3)   The registration, in accordance with Part I of the Act of 1933, of any such judgment as is referred to in subsection (1) of this section shall constitute, in relation to that judgment, compliance with the formalities for the purposes of paragraph 3 of article 31 in the Schedule to this Act.

5   Contribution between carriers

(1)   Where a carrier under a contract to which the Convention applies is liable in respect of any loss or damage for which compensation is payable under the Convention, nothing in [section 1 of the Civil Liability (Contribution) Act 1978], or section 3(2) of the Law Reform (Miscellaneous Provisions) (Scotland) Act 1940 shall confer on him any right to recover

contribution in respect of that loss or damage from any other carrier who, in accordance with article 34 in the Schedule to this Act, is a party to the contract of carriage.

(2)   The preceding subsection shall be without prejudice to the operation of article 37 in the Schedule to this Act.

Sub-s (1): words in square brackets substituted by the Civil Liability (Contribution) Act 1978, s 9(1), Sch 1, para 7.

6   Actions against High Contracting Parties

Every High Contracting Party to the Convention shall, for the purposes of any proceedings brought in a court in the United Kingdom in accordance with the provisions of article 31 in the Schedule to this Act to enforce a claim in respect of carriage undertaken by that Party, be deemed to have submitted to the jurisdiction of that court, and accordingly rules of court may provide for the manner in which any such action is to be commenced and carried on; but nothing in this section shall authorise the issue of execution, or in Scotland the execution of diligence, against the property of any High Contracting Party.

7   Arbitrations

(1)   Any reference in the preceding provisions of this Act to a court includes a reference to an arbitration tribunal acting by virtue of article 33 in the Schedule to this Act.

(2)   For the purposes of article 32 in the Schedule to this Act, as it has effect (by virtue of the said article 33) in relation to arbitrations,—
  [(a) as respects England and Wales and Northern Ireland, the provisions of section 14(3) to (5) of the Arbitration Act 1996 (which determine the time at which an arbitration is commenced) apply;]
  (c) ...

Sub-s (2): para (a) substituted, for paras (a), (b) as originally enacted, and para (b) repealed, by the Arbitration Act 1996, s 107, Sch 3, para 21, Sch 4; para (c) applies to Scotland only.

8 Resolution of conflicts between Conventions on carriage of goods

(1)   If it appears to Her Majesty in Council that there is any conflict between the provisions of this Act (including the provisions of the Convention as set out in the Schedule to this Act) and any provisions relating to the carriage of goods for reward by land, sea or air contained in—
   (a) any other Convention which has been signed or ratified by or on behalf of Her Majesty's Government in the United Kingdom before the passing of this Act, or
   (b) any enactment of the Parliament of the United Kingdom giving effect to such a Convention,

Her Majesty may by Order in Council make such provision as may seem to Her to be appropriate for resolving that conflict by amending or modifying this Act or any such enactment.

(2)   Any statutory instrument made by virtue of this section shall be subject to annulment in pursuance of a resolution of either House of Parliament.

[8A Amendments consequential on revision of Convention

(1)   If at any time it appears to Her Majesty in Council that Her Majesty's Government in the United Kingdom have agreed to any revision of the Convention, Her Majesty may by Order in Council make such amendment of—
   [(a) this Act; and]
   (c) section 5(1) of the Carriage by Air and Road Act 1979,

as appear to Her to be appropriate in consequence of the revision.

(2)   In the preceding subsection "revision" means an omission from, addition to or alteration of the Convention and includes replacement of the Convention or part of it by another convention.

(3)   An Order in Council under this section shall not be made unless a draft of the Order has been laid before Parliament and approved by a resolution of each House of Parliament.]

NOTES

Inserted by the Carriage by Air and Road Act 1979, s 3(3).

Sub-s (1): para (a) substituted for original paras (a), (b) by the International Transport Conventions Act 1983, s 9, Sch 2, para 2.

13 Application to Crown

This Act shall bind the Crown.

14 Short title, interpretation and commencement

(1)   This Act may be cited as the Carriage of Goods by Road Act 1965.

(2)   The persons who, for the purposes of this Act, are persons concerned in the carriage of goods by road under a contract to which the Convention applies are—
  (a) the sender,
  (b) the consignee,
  (c) any carrier who, in accordance with article 34 in the Schedule to this Act or otherwise, is a party to the contract of carriage,
  (d) any person for whom such a carrier is responsible by virtue of article 3 in the Schedule to this Act,
  (e) any person to whom the rights and liabilities of any of the persons referred to in paragraphs (a) to (d) to this subsection have passed (whether by assignment or assignation or by operation of law).

(3)   Except in so far as the context otherwise requires, any reference in this Act to an enactment shall be construed as a reference to that enactment as amended or extended by or under any other enactment.

(4)   This Act shall come into operation on such day as Her Majesty may by Order in Council appoint; but nothing in this Act shall apply in relation to any contract or the carriage of goods by road made before the day so appointed.

SCHEDULE

Section 1

CONVENTION ON THE CONTRACT FOR THE INTERNATIONAL CARRIAGE
OF GOODS BY ROAD

CHAPTER I
SCOPE OF APPLICATION

Article 1

1.   This Convention shall apply to every contract for the carriage of goods by road in vehicles for reward, when the place of taking over of the goods and the place designated for delivery, as specified in the contract, are situated in two different countries, of which at least one is a Contracting country, irrespective of the place of residence and the nationality of the parties.

2.   For the purposes of this Convention, "vehicles" means motor vehicles, articulated vehicles, trailers and semi-trailers as defined in article 4 of the Convention on Road Traffic dated 19th September 1949.

3.   This Convention shall apply also where carriage coming within its scope is carried out by States or by governmental institutions or organizations.

4.   This Convention shall not apply—
   (a) to carriage performed under the terms of any international postal convention;
   (b) to funeral consignments;
   (c) to furniture removal.

5.   The Contracting Parties agree not to vary any of the provisions of this Convention by special agreements between two or more of them, except to make it inapplicable to their frontier traffic or to authorise the use in transport operations entirely confined to their territory of consignment notes representing a title to the goods.

Article 2

1.   Where the vehicle containing the goods is carried over part of the journey by sea, rail, inland waterways or air, and, except where the

provisions of article 14 are applicable, the goods are not unloaded from the vehicle, this Convention shall nevertheless apply to the whole of the carriage. Provided that to the extent that it is proved that any loss, damage or delay in delivery of the goods which occurs during the carriage by the other means of transport was not caused by an act or omission of the carrier by road, but by some event which could only have occurred in the course of and by reason of the carriage by that other means of transport, the liability of the carrier by road shall be determined not by this Convention but in the manner in which the liability of the carrier by the other means of transport would have been determined if a contract for the carriage of the goods alone had been made by the sender with the carrier by the other means of transport in accordance with the conditions prescribed by law for the carriage of goods by that means of transport. If, however, there are no such prescribed conditions, the liability of the carrier by road shall be determined by this Convention.

2. If the carrier by road is also himself the carrier by the other means of transport, his liability shall also be determined in accordance with the provisions of paragraph 1 of this article, but as if, in his capacities as carrier by road and as carrier by the other means of transport, he were two separate persons.

CHAPTER II
PERSONS FOR WHOM THE CARRIER IS RESPONSIBLE

Article 3

For the purposes of this Convention the carrier shall be responsible for the acts and omissions of his agents and servants and of any other persons of whose services he makes use for the performance of the carriage, when such agents, servants or other persons are acting within the scope of their employment, as if such acts or omissions were his own.

CHAPTER III
CONCLUSION AND PERFORMANCE OF THE CONTRACT OF CARRIAGE

Article 4

The contract of carriage shall be confirmed by the making out of a consignment note. The absence, irregularity or loss of the consignment note shall not affect the existence or the validity of the contract of carriage which shall remain subject to the provisions of this Convention.

Article 5

1.   The consignment note shall be made out in three original copies signed by the sender and by the carrier. These signatures may be printed or replaced by the stamps of the sender and the carrier if the law of the country in which the consignment note has been made out so permits. The first copy shall be handed to the sender, the second shall accompany the goods and the third shall be retained by the carrier.

2.   When the goods which are to be carried have to be loaded in different vehicles, or are of different kinds or are divided into different lots, the sender or the carrier shall have the right to require a separate consignment note to be made out for each vehicle used, or for each kind or lot of goods.

Article 6

1.   The consignment note shall contain the following particulars—
   (a) the date of the consignment note and the place at which it is made out;
   (b) the name and address of the sender;
   (c) the name and address of the carrier;
   (d) the place and the date of taking over of the goods and the place designated for delivery;
   (e) the name and address of the consignee;
   (f) the description in common use of the nature of the goods and the method of packing, and, in the case of dangerous goods, their generally recognised description;
   (g) the number of packages and their special marks and numbers;
   (h) the gross weight of the goods or their quantity otherwise expressed;
   (i) charges relating to the carriage (carriage charges, supplementary charges, customs duties and other charges incurred from the making of the contract to the time of delivery);
   (j) the requisite instructions for Customs and other formalities;
   (k) a statement that the carriage is subject, notwithstanding any clause to the contrary, to the provisions of this Convention.

2.   Where applicable, the consignment note shall also contain the following particulars—
   (a) a statement that transhipment is not allowed;
   (b) the charges which the sender undertakes to pay;

(c) the amount of "cash on delivery" charges;

(d) a declaration of the value of the goods and the amount representing special interest in delivery;

(e) the sender's instructions to the carrier regarding insurance of the goods;

(f) the agreed time-limit within which the carriage is to be carried out;

(g) a list of the documents handed to the carrier.

3.   The parties may enter in the consignment note any other particulars which they may deem useful.

Article 7

1.   The sender shall be responsible for all expenses, loss and damage sustained by the carrier by reason of the inaccuracy or inadequacy of—

(a) the particulars specified in article 6, paragraph 1(b), (d), (e), (f), (g), (h) and (j);

(b) the particulars specified in article 6, paragraph 2;

(c) any other particulars or instructions given by him to enable the consignment note to be made out or for the purpose of their being entered therein.

2.   If, at the request of the sender, the carrier enters in the consignment note the particulars referred to in paragraph 1 of this article, he shall be deemed, unless the contrary is proved, to have done so on behalf of the sender.

3.   If the consignment note does not contain the statement specified in article 6, paragraph 1(k), the carrier shall be liable for all expenses, loss and damage sustained through such omission by the person entitled to dispose of the goods.

Article 8

1.   On taking over the goods, the carrier shall check—

(a) the accuracy of the statements in the consignment note as to the number of packages and their marks and numbers, and

(b) the apparent condition of the goods and their packaging.

2.   Where the carrier has no reasonable means of checking the accuracy of the statements referred to in paragraph 1(a) of this article, he shall enter his reservations in the consignment note together with the grounds on which

they are based. He shall likewise specify the grounds for any reservations which he makes with regard to the apparent condition of the goods and their packaging. Such reservations shall not bind the sender unless he has expressly agreed to be bound by them in the consignment note.

3.   The sender shall be entitled to require the carrier to check the gross weight of the goods or their quantity otherwise expressed. He may also require the contents of the packages to be checked. The carrier shall be entitled to claim the cost of such checking. The result of the checks shall be entered in the consignment note.

Article 9

1.   The consignment note shall be *prima facie* evidence of the making of the contract of carriage, the conditions of the contract and the receipt of the goods by the carrier.

2.   If the consignment note contains no specific reservations by the carrier, it shall be presumed, unless the contrary is proved, that the goods and their packaging appeared to be in good condition when the carrier took them over and that the number of packages, their marks and numbers corresponded with the statements in the consignment note.

Article 10

The sender shall be liable to the carrier for damage to persons, equipment or other goods, and for any expenses due to defective packing of the goods, unless the defect was apparent or known to the carrier at the time when he took over the goods and he made no reservations concerning it.

Article 11

1.   For the purposes of the Customs or other formalities which have to be completed before delivery of the goods, the sender shall attach the necessary documents to the consignment note or place them at the disposal of the carrier and shall furnish him with all the information which he requires.

2.   The carrier shall not be under any duty to enquire into either the accuracy or the adequacy of such documents and information. The sender shall be liable to the carrier for any damage caused by the absence,

inadequacy or irregularity of such documents and information, except in the case of some wrongful act or neglect on the part of the carrier.

3.   The liability of the carrier for the consequences arising from the loss or incorrect use of the documents specified in and accompanying the consignment note or deposited with the carrier shall be that of an agent, provided that the compensation payable by the carrier shall not exceed that payable in the event of loss of the goods.

Article 12

1.   The sender has the right to dispose of the goods, in particular by asking the carrier to stop the goods in transit, to change the place at which delivery is to take place or to deliver the goods to a consignee other than the consignee indicated in the consignment note.

2.   This right shall cease to exist when the second copy of the consignment note is handed to the consignee or when the consignee exercises his right under article 13, paragraph 1; from that time onwards the carrier shall obey the orders of the consignee.

3.   The consignee shall, however, have the right of disposal from the time when the consignment note is drawn up, if the sender makes an entry to that effect in the consignment note.

4.   If in exercising his right of disposal the consignee has ordered the delivery of the goods to another person, that other person shall not be entitled to name other consignees.

5.   The exercise of the right of disposal shall be subject to the following conditions:
   (a) that the sender or, in the case referred to in paragraph 3 of this article, the consignee who wishes to exercise the right produces the first copy of the consignment note on which the new instructions to the carrier have been entered and indemnifies the carrier against all expenses, loss and damage involved in carrying out such instructions;
   (b) that the carrying out of such instructions is possible at the time when the instructions reach the person who is to carry them out and does not either interfere with the normal working of the carrier's undertaking or prejudice the senders or consignees of other consignments;
   (c) that the instructions do not result in a division of the consignment.

6.   When, by reason of the provisions of paragraph 5(b) of this article, the carrier cannot carry out the instructions which he receives, he shall immediately notify the person who gave him such instructions.

7.   A carrier who has not carried out the instructions given under the conditions provided for in this article, or who has carried them out without requiring the first copy of the consignment note to be produced, shall be liable to the person entitled to make a claim for any loss or damage caused thereby.

Article 13

1.   After arrival of the goods at the place designated for delivery, the consignee shall be entitled to require the carrier to deliver to him, against a receipt, the second copy of the consignment note and the goods. If the loss of the goods is established or if the goods have not arrived after the expiry of the period provided for in article 19, the consignee shall be entitled to enforce in his own name against the carrier any rights arising from the contract of carriage.

2.   The consignee who avails himself of the rights granted to him under paragraph 1 of this article shall pay the charges shown to be due on the consignment note, but in the event of dispute on this matter the carrier shall not be required to deliver the goods unless security has been furnished by the consignee.

Article 14

1.   If for any reason it is or becomes impossible to carry out the contract in accordance with the terms laid down in the consignment note before the goods reach the place designated for delivery, the carrier shall ask for instructions from the person entitled to dispose of the goods in accordance with the provisions of article 12.

2.   Nevertheless, if circumstances are such as to allow the carriage to be carried out under conditions differing from those laid down in the consignment note and if the carrier has been unable to obtain instructions in reasonable time from the person entitled to dispose of the goods in accordance with the provisions of article 12, he shall take such steps as seem to him to be in the best interests of the person entitled to dispose of the goods.

Article 15

1.   Where circumstances prevent delivery of the goods after their arrival at the place designated for delivery, the carrier shall ask the sender for his instructions. If the consignee refuses the goods the sender shall be entitled to dispose of them without being obliged to produce the first copy of the consignment note.

2.   Even if he has refused the goods, the consignee may nevertheless require delivery so long as the carrier has not received instructions to the contrary from the sender.

3.   When circumstances preventing delivery of the goods arise after the consignee, in exercise of his rights under article 12, paragraph 3, has given an order for the goods to be delivered to another person, paragraphs 1 and 2 of this article shall apply as if the consignee were the sender and that other person were the consignee.

Article 16

1.   The carrier shall be entitled to recover the cost of his request for instructions, and any expenses entailed in carrying out such instructions, unless such expenses were caused by the wrongful act or neglect of the carrier.

2.   In the cases referred to in article 14, paragraph 1, and in article 15, the carrier may immediately unload the goods for account of the person entitled to dispose of them and thereupon the carriage shall be deemed to be at an end. The carrier shall then hold the goods on behalf of the person so entitled. He may however entrust them to a third party, and in that case he shall not be under any liability except for the exercise of reasonable care in the choice of such third party. The charges due under the consignment note and all other expenses shall remain chargeable against the goods.

3.   The carrier may sell the goods, without awaiting instructions from the person entitled to dispose of them, if the goods are perishable or their condition warrants such a course, or when the storage expenses would be out of proportion to the value of the goods. He may also proceed to the sale of the goods in other cases if after the expiry of a reasonable period he has not received from the person entitled to dispose of the goods instructions to the contrary which he may reasonably be required to carry out.

4.   If the goods have been sold pursuant to this article, the proceeds of sale, after deduction of the expenses chargeable against the goods, shall be placed at the disposal of the person entitled to dispose of the goods. If these charges exceed the proceeds of sale, the carrier shall be entitled to the difference.

5.   The procedure in the case of sale shall be determined by the law or custom of the place where the goods are situated.

CHAPTER IV
LIABILITY OF THE CARRIER

Article 17

1.   The carrier shall be liable for the total or partial loss of the goods and for damage thereto occurring between the time when he takes over the goods and the time of delivery, as well as for any delay in delivery.

2.   The carrier shall however be relieved of liability if the loss, damage or delay was caused by the wrongful act or neglect of the claimant, by the instructions of the claimant given otherwise than as the result of a wrongful act or neglect on the part of the carrier, by inherent vice of the goods or through circumstances which the carrier could not avoid and the consequences of which he was unable to prevent.

3.   The carrier shall not be relieved of liability by reason of the defective condition of the vehicle used by him in order to perform the carriage, or by reason of the wrongful act or neglect of the person from whom he may have hired the vehicle or of the agents or servants of the latter.

4.   Subject to article 18, paragraphs 2 to 5, the carrier shall be relieved of liability when the loss or damage arises from the special risks inherent in one or more of the following circumstances—
  (a) use of open unsheeted vehicles, when their use has been expressly agreed and specified in the consignment note;
  (b) the lack of, or defective condition of packing in the case of goods which, by their nature, are liable to wastage or to be damaged when not packed or when not properly packed;
  (c) handling, loading, stowage or unloading of the goods by the sender, the consignee or persons acting on behalf of the sender or the consignee;

   (d) the nature of certain kinds of goods which particularly exposes them to total or partial loss or to damage, especially through breakage, rust, decay, desiccation, leakage, normal wastage, or the action of moth or vermin;

   (e) insufficiency or inadequacy of marks or numbers on the packages;

   (f) the carriage of livestock.

5.   Where under this article the carrier is not under any liability in respect of some of the factors causing the loss, damage or delay, he shall only be liable to the extent that those factors for which he is liable under this article have contributed to the loss, damage or delay.

Article 18

1.   The burden of proving that loss, damage or delay was due to one of the causes specified in article 17, paragraph 2, shall rest upon the carrier.

2.   When the carrier establishes that in the circumstances of the case, the loss or damage could be attributed to one or more of the special risks referred to in article 17, paragraph 4, it shall be presumed that it was so caused. The claimant shall however be entitled to prove that the loss or damage was not, in fact, attributable either wholly or partly to one of these risks.

3.   This presumption shall not apply in the circumstances set out in article 17, paragraph 4(a), if there has been an abnormal shortage, or a loss of any package.

4.   If the carriage is performed in vehicles specially equipped to protect the goods from the effects of heat, cold, variations in temperature or the humidity of the air, the carrier shall not be entitled to claim the benefit of article 17, paragraph 4(d), unless he proves that all steps incumbent on him in the circumstances with respect to the choice, maintenance and use of such equipment were taken and that he complied with any special instructions issued to him.

5.   The carrier shall not be entitled to claim the benefit of article 17, paragraph 4(f), unless he proves that all steps normally incumbent on him in the circumstances were taken and that he complied with any special instructions issued to him.

Article 19

Delay in delivery shall be said to occur when the goods have not been delivered within the agreed time-limit or when, failing an agreed time-limit, the actual duration of the carriage having regard to the circumstances of the case, and in particular, in the case of partial loads, the time required for making up a complete load in the normal way, exceeds the time it would be reasonable to allow a diligent carrier.

Article 20

1.   The fact that goods have not been delivered within thirty days following the expiry of the agreed time-limit, or, if there is no agreed time-limit, within sixty days from the time when the carrier took over the goods, shall be conclusive evidence of the loss of the goods, and the person entitled to make a claim may thereupon treat them as lost.

2.   The person so entitled may, on receipt of compensation for the missing goods, request in writing that he shall be notified immediately should the goods be recovered in the course of the year following the payment of compensation. He shall be given a written acknowledgment of such request.

3.   Within the thirty days following receipt of such notification, the person entitled as aforesaid may require the goods to be delivered to him against payment of the charges shown to be due on the consignment note and also against refund of the compensation he received less any charges included therein but without prejudice to any claims to compensation for delay in delivery under article 23 and, where applicable, article 26.

4.   In the absence of the request mentioned in paragraph 2 or of any instructions given within the period of thirty days specified in paragraph 3, or if the goods are not recovered until more than one year after the payment of compensation, the carrier shall be entitled to deal with them in accordance with the law of the place where the goods are situated.

Article 21

Should the goods have been delivered to the consignee without collection of the "cash on delivery" charge which should have been collected by the carrier under the terms of the contract of carriage, the carrier shall be liable to the sender for compensation not exceeding the amount of such charge without prejudice to his right of action against the consignee.

Article 22

1.   When the sender hands goods of a dangerous nature to the carrier, he shall inform the carrier of the exact nature of the danger and indicate,

if necessary, the precautions to be taken. If this information has not been entered in the consignment note, the burden of proving, by some other means, that the carrier knew the exact nature of the danger constituted by the carriage of the said goods shall rest upon the sender or the consignee.

2.   Goods of a dangerous nature which, in the circumstances referred to in paragraph 1 of this article, the carrier did not know were dangerous, may, at any time or place, be unloaded, destroyed or rendered harmless by the carrier without compensation; further, the sender shall be liable for all expenses, loss or damage arising out of their handing over for carriage or of their carriage.

Article 23

1.   When, under the provisions of this Convention, a carrier is liable for compensation in respect of total or partial loss of goods, such compensation shall be calculated by reference to the value of the goods at the place and time at which they were accepted for carriage.

2.   The value of the goods shall be fixed according to the commodity exchange price or, if there is no such price, according to the current market price or, if there is no commodity exchange price or current market price, by reference to the normal value of goods of the same kind and quality.

[3.   Compensation shall not, however, exceed 8.33 units of account per kilogram of gross weight short.]

4.   In addition, the carriage charges, Customs duties and other charges incurred in respect of the carriage of the goods shall be refunded in full in case of total loss and in proportion to the loss sustained in case of partial loss, but no further damages shall be payable.

5.   In the case of delay, if the claimant proves that damage has resulted therefrom the carrier shall pay compensation for such damage not exceeding the carriage charges.

6.   Higher compensation may only be claimed where the value of the goods or a special interest in delivery has been declared in accordance with articles 24 and 26.

[7.   The unit of account mentioned in this Convention is the Special Drawing Right as defined by the International Monetary Fund. The amount mentioned in paragraph 3 of this article shall be converted into the national currency of the State of the Court seised of the case on the basis of the value of that currency on the date of the judgment or the date agreed upon by the Parties.]

Article 24

The sender may, against payment of a surcharge to be agreed upon, declare in the consignment note a value for the goods exceeding the limit laid down in article 23, paragraph 3, and in that case the amount of the declared value shall be substituted for that limit.

Article 25

1.   In case of damage, the carrier shall be liable for the amount by which the goods have diminished in value, calculated by reference to the value of the goods fixed in accordance with article 23, paragraphs 1, 2 and 4.

2.   The compensation may not, however, exceed—
    (a) if the whole consignment has been damaged, the amount payable in the case of total loss;
    (b) if part only of the consignment has been damaged, the amount payable in the case of loss of the part affected.

Article 26

1.   The sender may, against payment of a surcharge to be agreed upon, fix the amount of a special interest in delivery in the case of loss or damage or of the agreed time-limit being exceeded, by entering such amount in the consignment note.

2.   If a declaration of a special interest in delivery has been made, compensation for the additional loss or damage proved may be claimed, up to the total amount of the interest declared, independently of the compensation provided for in articles 23, 24 and 25.

Article 27

1.   The claimant shall be entitled to claim interest on compensation payable. Such interest, calculated at five per centum per annum, shall

accrue from the date on which the claim was sent in writing to the carrier or, if no such claim has been made, from the date on which legal proceedings were instituted.

2. When the amounts on which the calculation of the compensation is based are not expressed in the currency of the country in which payment is claimed, conversion shall be at the rate of exchange applicable on the day and at the place of payment of compensation.

Article 28

1. In cases where, under the law applicable, loss, damage or delay arising out of carriage under this Convention gives rise to an extra-contractual claim, the carrier may avail himself of the provisions of this Convention which exclude his liability or which fix or limit the compensation due.

2. In cases where the extra-contractual liability for loss, damage or delay of one of the persons for whom the carrier is responsible under the terms of article 3 is in issue, such person may also avail himself of the provisions of this Convention which exclude the liability of the carrier or which fix or limit the compensation due.

Article 29

1. The carrier shall not be entitled to avail himself of the provisions of this chapter which exclude or limit his liability or which shift the burden of proof if the damage was caused by his wilful misconduct or by such default on his part as, in accordance with the law of the court or tribunal seised of the case, is considered as equivalent to wilful misconduct.

2. The same provision shall apply if the wilful misconduct or default is committed by the agents or servants of the carrier or by any other persons of whose services he makes use for the performance of the carriage, when such agents, servants or other persons are acting within the scope of their employment. Furthermore, in such a case such agents, servants or other persons shall not be entitled to avail themselves, with regard to their personal liability, of the provisions of this chapter referred to in paragraph 1.

CHAPTER V
CLAIMS AND ACTIONS

Article 30

1.   If the consignee takes delivery of the goods without duly checking their condition with the carrier or without sending him reservations giving a general indication of the loss or damage, not later than the time of delivery in the case of apparent loss or damage and within seven days of delivery, Sundays and public holidays excepted, in the case of loss or damage which is not apparent, the fact of his taking delivery shall be *prima facie* evidence that he has received the goods in the condition described in the consignment note. In the case of loss or damage which is not apparent the reservations referred to shall be made in writing.

2.   When the condition of the goods has been duly checked by the consignee and the carrier, evidence contradicting the result of this checking shall only be admissible in the case of loss or damage which is not apparent and provided that the consignee has duly sent reservations in writing to the carrier within seven days, Sundays and public holidays excepted, from the date of checking.

3.   No compensation shall be payable for delay in delivery unless a reservation has been sent in writing to the carrier, within twenty-one days from the time that the goods were placed at the disposal of the consignee.

4.   In calculating the time-limits provided for in this Article the date of delivery, or the date of checking, or the date when the goods were placed at the disposal of the consignee, as the case may be, shall not be included.

5.   The carrier and the consignee shall give each other every reasonable facility for making the requisite investigations and checks.

Article 31

1.   In legal proceedings arising out of carriage under this Convention, the plaintiff may bring an action in any court or tribunal of a contracting country designated by agreement between the parties and, in addition, in the courts or tribunals of a country within whose territory
   (a) the defendant is ordinarily resident, or has his principal place of business, or the branch or agency through which the contract of carriage was made, or

(b) the place where the goods were taken over by the carrier or the place designated for delivery is situated,

and in no other courts or tribunals.

2.   Where in respect of a claim referred to in paragraph 1 of this article an action is pending before a court or tribunal competent under that paragraph, or where in respect of such a claim a judgment has been entered by such a court or tribunal no new action shall be started between the same parties on the same grounds unless the judgment of the court or tribunal before which the first action was brought is not enforceable in the country in which the fresh proceedings are brought.

3.   When a judgment entered by a court or tribunal of a contracting country in any such action as is referred to in paragraph 1 of this article has become enforceable in that country, it shall also become enforceable in each of the other contracting States, as soon as the formalities required in the country concerned have been complied with. The formalities shall not permit the merits of the case to be re-opened.

4.   The provisions of paragraph 3 of this article shall apply to judgments after trial, judgments by default and settlements confirmed by an order of the court, but shall not apply to interim judgments or to awards of damages, in addition to costs against a plaintiff who wholly or partly fails in his action.

5.   Security for costs shall not be required in proceedings arising out of carriage under this Convention from nationals of contracting countries resident or having their place of business in one of those countries.

Article 32

1.   The period of limitation for an action arising out of carriage under this Convention shall be one year. Nevertheless, in the case of wilful misconduct, or such default as in accordance with the law of the court or tribunal seised of the case, is considered as equivalent to wilful misconduct, the period of limitation shall be three years. The period of limitation shall begin to run:
(a) in the case of partial loss, damage or delay in delivery, from the date of delivery;
(b) in the case of total loss, from the thirtieth day after the expiry of the agreed time-limit or where there is no agreed time-limit from the sixtieth day from the date on which the goods were taken over by the carrier;

(c) in all other cases, on the expiry of a period of three months after the making of the contract of carriage.

The day on which the period of limitation begins to run shall not be included in the period.

2.   A written claim shall suspend the period of limitation until such date as the carrier rejects the claim by notification in writing and returns the documents attached thereto. If a part of the claim is admitted the period of limitation shall start to run again only in respect of that part of the claim still in dispute. The burden of proof of the receipt of the claim, or of the reply and of the return of the documents, shall rest with the party relying upon these facts. The running of the period of limitation shall not be suspended by further claims having the same object.

3.   Subject to the provisions of paragraph 2 above, the extension of the period of limitation shall be governed by the law of the court or tribunal seised of the case. That law shall also govern the fresh accrual of rights of action.

4.   A right of action which has become barred by lapse of time may not be exercised by way of counter-claim or set-off.

Article 33

The contract of carriage may contain a clause conferring competence on an arbitration tribunal if the clause conferring competence on the tribunal provides that the tribunal shall apply this Convention.

CHAPTER VI
PROVISIONS RELATING TO CARRIAGE PERFORMED BY SUCCESSIVE CARRIERS

Article 34

If carriage governed by a single contract is performed by successive road carriers, each of them shall be responsible for the performance of the whole operation, the second carrier and each succeeding carrier becoming a party to the contract of carriage, under the terms of the consignment note, by reason of his acceptance of the goods and the consignment note.

Article 35

1.   A carrier accepting the goods from a previous carrier shall give the latter a dated and signed receipt. He shall enter his name and address on the second copy of the consignment note. Where applicable, he shall enter on the second copy of the consignment note and on the receipt reservations of the kind provided for in article 8, paragraph 2.

2.   The provisions of article 9 shall apply to the relations between successive carriers.

Article 36

Except in the case of a counter-claim or a set-off raised in an action concerning a claim based on the same contract of carriage, legal proceedings in respect of liability for loss, damage or delay may only be brought against the first carrier, the last carrier or the carrier who was performing that portion of the carriage during which the event causing the loss, damage or delay occurred; an action may be brought at the same time against several of these carriers.

Article 37

A carrier who has paid compensation in compliance with the provisions of this Convention, shall be entitled to recover such compensation, together with interest thereon and all costs and expenses incurred by reason of the claim, from the other carriers who have taken part in the carriage, subject to the following provisions:
 (a) the carrier responsible for the loss or damage shall be solely liable for the compensation whether paid by himself or by another carrier;
 (b) when the loss or damage has been caused by the action of two or more carriers, each of them shall pay an amount proportionate to his share of liability; should it be impossible to apportion the liability, each carrier shall be liable in proportion to the share of the payment for the carriage which is due to him;
 (c) if it cannot be ascertained to which carriers liability is attributable for the loss or damage, the amount of the compensation shall be apportioned between all the carriers as laid down in (b) above.

Article 38

If one of the carriers is insolvent, the share of the compensation due from him and unpaid by him shall be divided among the other carriers in proportion to the share of the payment for the carriage due to them.

Article 39

1.   No carrier against whom a claim is made under articles 37 and 38 shall be entitled to dispute the validity of the payment made by the carrier making the claim if the amount of the compensation was determined by judicial authority after the first mentioned carrier had been given due notice of the proceedings and afforded an opportunity of entering an appearance.

2.   A carrier wishing to take proceedings to enforce his right of recovery may make his claim before the competent court or tribunal of the country in which one of the carriers concerned is ordinarily resident, or has his principal place of business or the branch or agency through which the contract of carriage was made. All the carriers concerned may be made defendants in the same action.

3.   The provisions of article 31, paragraphs 3 and 4, shall apply to judgments entered in the proceedings referred to in articles 37 and 38.

4.   The provisions of article 32 shall apply to claims between carriers. The period of limitation shall, however, begin to run either on the date of the final judicial decision fixing the amount of compensation payable under the provisions of this Convention, or, if there is no such judicial decision, from the actual date of payment.

Article 40

Carriers shall be free to agree among themselves on provisions other than those laid down in articles 37 and 38.

CHAPTER VII
NULLITY OF STIPULATIONS CONTRARY TO THE CONVENTION

Article 41

1.   Subject to the provisions of Article 40, any stipulation which would directly or indirectly derogate from the provisions of this Convention shall be null and void. The nullity of such a stipulation shall not involve the nullity of the other provisions of the contract.

2.   In particular, a benefit of insurance in favour of the carrier or any other similar clause, or any clause shifting the burden of proof shall be null and void.

[*Chapter VIII of the Convention is not reproduced. This deals with the coming into force of the Convention, the settlement of disputes between the High Contracting Parties and related matters.*]

PROTOCOL OF SIGNATURE

1.   This Convention shall not apply to traffic between the United Kingdom of Great Britain and Northern Ireland and the Republic of Ireland.

NOTES

Article 23: para 3 substituted and para 7 added by the Carriage by Air and Road Act 1979, s 4(2).

Misrepresentation Act 1967

(C 7)

An Act to amend the law relating to innocent misrepresentations and to amend sections 11 and 35 of the Sale of Goods Act 1893

[22 March 1967]

1   Removal of certain bars to rescission for innocent misrepresentation

Where a person has entered into a contract after a misrepresentation has been made to him, and—
    (a) the misrepresentation has become a term of the contract; or
    (b) the contract has been performed;

or both, then, if otherwise he would be entitled to rescind the contract without alleging fraud, he shall be so entitled, subject to the provisions of this Act, notwithstanding the matters mentioned in paragraphs (a) and (b) of this section.

2   Damages for misrepresentation

(1)   Where a person has entered into a contract after a misrepresentation has been made to him by another party thereto and as a result thereof he

has suffered loss, then, if the person making the misrepresentation would be liable to damages in respect thereof had the misrepresentation been made fraudulently, that person shall be so liable notwithstanding that the misrepresentation was not made fraudulently, unless he proves that he had reasonable ground to believe and did believe up to the time the contract was made that the facts represented were true.

(2)   Where a person has entered into a contract after a misrepresentation has been made to him otherwise than fraudulently, and he would be entitled, by reason of the misrepresentation, to rescind the contract, then, if it is claimed, in any proceedings arising out of the contract, that the contract ought to be or has been rescinded, the court or arbitrator may declare the contract subsisting and award damages in lieu of rescission, if of opinion that it would be equitable to do so, having regard to the nature of the misrepresentation and the loss that would be caused by it if the contract were upheld, as well as to the loss that rescission would cause to the other party.

(3)   Damages may be awarded against a person under subsection (2) of this section whether or not he is liable to damages under subsection (1) thereof, but where he is so liable any award under the said subsection (2) shall be taken into account in assessing his liability under the said subsection (1).

[3 Avoidance of provision excluding liability for misrepresentation

If a contract contains a term which would exclude or restrict—
  (a) any liability to which a party to a contract may be subject by reason of any misrepresentation made by him before the contract was made; or
  (b) any remedy available to another party to the contract by reason of such a misrepresentation,

that term shall be of no effect except in so far as it satisfies the requirement of reasonableness as stated in section 11(1) of the Unfair Contract Terms Act 1977; and it is for those claiming that the term satisfies that requirement to show that it does.]

NOTES

Substituted by the Unfair Contract Terms Act 1977, s 8(1).

5 Saving for past transactions

Nothing in this Act shall apply in relation to any misrepresentation or contract of sale which is made before the commencement of this Act.

6 Short title, commencement and extent

(1)   This Act may be cited as the Misrepresentation Act 1967.

(2)   This Act shall come into operation at the expiration of the period of one month beginning with the date on which it is passed.

(3)   ...

(4)   This Act does not extend to Northern Ireland.

NOTES

Sub-s (3): applies to Scotland only.

Trade Descriptions Act 1968

(C 29)

An Act to replace the Merchandise Marks Acts 1887 to 1953 by fresh provisions prohibiting misdescriptions of goods, services, accommodation and facilities provided in the course of trade; to prohibit false or misleading indications as to the price of goods; to confer power to require information or instructions relating to goods to be marked on or to accompany the goods or to be included in advertisements; to prohibit the unauthorised use of devices or emblems signifying royal awards; to enable the Parliament of Northern Ireland to make laws relating to merchandise marks; and for purposes connected with those matters

[30 May 1968]

Prohibition of false trade descriptions

I Prohibition of false trade descriptions

(1)   Any person who, in the course of a trade or business,—
  (a) applies a false trade description to any goods; or

(b) supplies or offers to supply any goods to which a false trade description is applied;

shall, subject to the provisions of this Act, be guilty of an offence.

(2)   Sections 2 to 6 of this Act shall have effect for the purposes of this section and for the interpretation of expressions used in this section, wherever they occur in this Act.

2   Trade description

(1)   A trade description is an indication, direct or indirect, and by whatever means given of any of the following matters with respect to any goods or parts of goods, that is to say—
 (a) quantity, size or gauge;
 (b) method of manufacture, production, processing or reconditioning;
 (c) composition;
 (d) fitness for purpose, strength, performance, behaviour or accuracy;
 (e) any physical characteristics not included in the preceding paragraphs;
 (f) testing by any person and results thereof;
 (g) approval by any person or conformity with a type approved by any person;
 (h) place or date of manufacture, production, processing or reconditioning;
 (i) person by whom manufactured, produced, processed or reconditioned;
 (j) other history, including previous ownership or use.

(2)   The matters specified in subsection (1) of this section shall be taken—
 (a) in relation to any animal, to include sex, breed or cross, fertility and soundness;
 (b) in relation to any semen, to include the identity and characteristics of the animal from which it was taken and measure of dilution.

(3)   In this section "quantity" includes length, width, height, area, volume, capacity, weight and number.

(4)   Notwithstanding anything in the preceding provisions of this section, the following shall be deemed not to be trade descriptions, that is to say, any description or mark applied in pursuance of—
 (a) ... ;
 (b) section 2 of the Agricultural Produce (Grading and Marking) Act 1928 (as amended by the Agricultural Produce (Grading and Marking) Amendment Act 1931) or any corresponding enactment of the Parliament of Northern Ireland;

    (c) the Plant Varieties and Seeds Act 1964;

    (d) the Agriculture and Horticulture Act 1964 [or any Community grading rules within the meaning of Part III of that Act];

    (e) the Seeds Act (Northern Ireland) 1965;

    (f) the Horticulture Act (Northern Ireland) 1966;

  [(g) the Consumer Protection Act 1987;]

  [(h) the Plant Varieties Act 1997;]

[any statement made in respect of, or mark applied to, any material in pursuance of Part IV of the Agriculture Act 1970, any name or expression to which a meaning has been assigned under section 70 of that Act when applied to any material in the circumstances specified in that section] ... any mark prescribed by a system of classification compiled under section 5 of the Agriculture Act 1967 [and any designation, mark or description applied in pursuance of a scheme brought into force under section 6(1) or an order made under section 25(1) of the Agriculture Act 1970].

(5)   Notwithstanding anything in the preceding provisions of this section,

    [(a)]where provision is made under [the Food Safety Act 1990] or the [Food Safety (Northern Ireland) Order 1991] [or the Consumer Protection Act 1987] prohibiting the application of a description except to goods in the case of which the requirements specified in that provision are complied with, that description, when applied to such goods, shall be deemed not to be a trade description.

    [(b) where by virtue of any provision made under Part V of the Medicines Act 1968 (or made under any provisions of the said Part V as applied by an order made under section 104 or section 105 of that Act) anything which, in accordance with this Act, constitutes the application of a trade description to goods is subject to any requirements or restrictions imposed by that provision, any particular description specified in that provision, when applied to goods in circumstances to which those requirements or restrictions are applicable, shall be deemed not to be a trade description].

NOTES

Sub-s (4): para (a) repealed, and words in square brackets in para (d) added, by the European Communities Act 1972, s 4, Sch 3, Pt II, Sch 4, para 4(2); para (g) inserted by the Consumer Safety Act 1978, s 7(8), substituted by the Consumer Protection Act 1987, s 48(1), Sch 4, para 2(1); para (h) inserted by the Plant Varieties Act 1997, s 51(4); other words in square brackets substituted or added and second words omitted repealed by the Agriculture Act 1970, ss 6(4), 87(3), 113(3), Sch 5, Pt V.

Sub-s (5): the letter "(a)" and para (b) added by the Medicines Act 1968, s 135(1), Sch 5, para 16; in para (a), words in first pair of square brackets substituted by the

Food Safety Act 1990, s 59(1), Sch 3, para 6, words in second pair of square brackets substituted by the Food Safety (Northern Ireland) Order 1991, SI 1991/762, art 51(1), Sch 2, para 8, words in final pair of square brackets substituted by the Consumer Protection Act 1987, s 48(1), Sch 4, para 2(1).

3　False trade description

(1)　A false trade description is a trade description which is false to a material degree.

(2)　A trade description which, though not false, is misleading, that is to say, likely to be taken for such an indication of any of the matters specified in section 2 of this Act as would be false to a material degree, shall be deemed to be a false trade description.

(3)　Anything which, though not a trade description, is likely to be taken for an indication of any of those matters and, as such an indication, would be false to a material degree, shall be deemed to be a false trade description.

(4)　A false indication, or anything likely to be taken as an indication which would be false, that any goods comply with a standard specified or recognised by any person or implied by the approval of any person shall be deemed to be a false trade description, if there is no such person or no standard so specified, recognised or implied.

4　Applying a trade description to goods

(1)　A person applies a trade description to goods if he—
  (a)　affixes or annexes it to or in any manner marks it on or incorporates it with—
    (i)　the goods themselves, or
    (ii)　anything in, on or with which the goods are supplied; or
  (b)　places the goods in, on or with anything which the trade description has been affixed or annexed to, marked on or incorporated with, or places any such thing with the goods; or
  (c)　uses the trade description in any manner likely to be taken as referring to the goods.

(2)　An oral statement may amount to the use of a trade description.

(3)　Where goods are supplied in pursuance of a request in which a trade description is used and circumstances are such as to make it reasonable to infer that the goods are supplied as goods corresponding to that trade description, the person supplying the goods shall be deemed to have applied that trade description to the goods.

5 Trade descriptions used in advertisements

(1)   The following provisions of this section shall have effect where in an advertisement a trade description is used in relation to any class of goods.

(2)   The trade description shall be taken as referring to all goods of the class, whether or not in existence at the time the advertisement is published—
 (a) for the purpose of determining whether an offence has been committed under paragraph (a) of section 1(1) of this Act; and
 (b) where goods of the class are supplied or offered to be supplied by a person publishing or displaying the advertisement, also for the purpose of determining whether an offence has been committed under paragraph (b) of the said section 1(1).

(3)   In determining for the purposes of this section whether any goods are of a class to which a trade description used in an advertisement relates regard shall be had not only to the form and content of the advertisement but also to the time, place, manner and frequency of its publication and all other matters making it likely or unlikely that a person to whom the goods are supplied would think of the goods as belonging to the class in relation to which the trade description is used in the advertisement.

6 Offer to supply

A person exposing goods for supply or having goods in his possession for supply shall be deemed to offer to supply them.

Power to define terms and to require display, etc of information

7 Definition orders

Where it appears to the Board of Trade—

(a) that it would be in the interest of persons to whom any goods are supplied; or

(b) that it would be in the interest of persons by whom any goods are exported and would not be contrary to the interest of persons to whom such goods are supplied in the United Kingdom;

that any expressions used in relation to the goods should be understood as having definite meanings, the Board may by order assign such meanings either—

(i) to those expressions when used in the course of a trade or business as, or as part of, a trade description applied to the goods; or

(ii) to those expressions when so used in such circumstances as may be specified in the order;

and where such a meaning is so assigned to an expression it shall be deemed for the purposes of this Act to have that meaning when used as mentioned in paragraph (i) or, as the case may be, paragraph (ii) of this section.

8 Marking orders

(1)   Where it appears to the Board of Trade necessary or expedient in the interest of persons to whom any goods are supplied that the goods should be marked with or accompanied by any information (whether or not amounting to or including a trade description) or instruction relating to the goods, the Board may, subject to the provisions of this Act, by order impose requirements for securing that the goods are so marked or accompanied, and regulate or prohibit the supply of goods with respect to which the requirements are not complied with; and the requirements may extend to the form and manner in which the information or instruction is to be given.

(2)   Where an order under this section is in force with respect to goods of any description, any person who, in the course of any trade or business, supplies or offers to supply goods of that description in contravention of the order shall, subject to the provisions of this Act, be guilty of an offence.

(3)   An order under this section may make different provision for different circumstances and may, in the case of goods supplied in circumstances where the information or instruction required by the order would not be conveyed until after delivery, require the whole or part thereof to be also displayed near the goods.

9 Information, etc to be given in advertisements

(1)   Where it appears to the Board of Trade necessary or expedient in the interest of persons to whom any goods are to be supplied that any description of advertisements of the goods should contain or refer to any information (whether or not amounting to or including a trade description) relating to the goods the Board may, subject to the provisions of this Act, by order impose requirements as to the inclusion of that information, or of an indication of the means by which it may be obtained, in such description of advertisements of the goods as may be specified in the order.

(2)   An order under this section may specify the form and manner in which any such information or indication is to be included in advertisements of any description and may make different provision for different circumstances.

(3)   Where an advertisement of any goods to be supplied in the course of any trade or business fails to comply with any requirement imposed under this section, any person who publishes the advertisement shall, subject to the provisions of this Act, be guilty of an offence.

10   Provisions supplementary to sections 8 and 9

(1)   A requirement imposed by an order under section 8 or section 9 of this Act in relation to any goods shall not be confined to goods manufactured or produced in any one country or any one of a number of countries or to goods manufactured or produced outside any one or more countries, unless—
   (a) it is imposed with respect to a description of goods in the case of which the Board of Trade are satisfied that the interest of persons in the United Kingdom to whom goods of that description are supplied will be sufficiently protected if the requirement is so confined; and
   (b) the Board of Trade are satisfied that the order is compatible with the international obligations of the United Kingdom.

(2)   Where any requirements with respect to any goods are for the time being imposed by such an order and the Board of Trade are satisfied, on the representation of persons appearing to the Board to have a substantial interest in the matter, that greater hardship would be caused to such persons if the requirements continued to apply than is justified by the interest of persons to whom such goods are supplied, the power of the Board to relax or discontinue the requirements by a further order may be exercised without the consultation and notice required by section 38(3) of this Act.

Misstatements other than false trade descriptions

12 False representations as to royal approval or award, etc

(1) If any person, in the course of any trade or business, gives, by whatever means, any false indication, direct or indirect, that any goods or services supplied by him or any methods adopted by him are or are of a kind supplied to or approved by Her Majesty or any member of the Royal Family, he shall, subject to the provisions of this Act, be guilty of an offence.

(2) If any person, in the course of any trade or business, uses, without the authority of Her Majesty, any device or emblem signifying the Queen's Award to Industry or anything so nearly resembling such a device or emblem as to be likely to deceive, he shall, subject to the provisions of this Act, be guilty of an offence.

13 False representations as to supply of goods or services

If any person, in the course of any trade or business, gives, by whatever means, any false indication, direct or indirect, that any goods or services supplied by him are of a kind supplied to any person he shall, subject to the provisions of this Act, be guilty of an offence.

14 False or misleading statements as to services, etc

(1) It shall be an offence for any person in the course of any trade or business—
    (a) to make a statement which he knows to be false; or
    (b) recklessly to make a statement which is false;

as to any of the following matters, that is to say,—
    (i) the provision in the course of any trade or business of any services, accommodation or facilities;
    (ii) the nature of any services, accommodation or facilities provided in the course of any trade or business;
    (iii) the time at which, manner in which or persons by whom any services, accommodation or facilities are so provided;
    (iv) the examination, approval or evaluation by any person of any services, accommodation or facilities so provided; or
    (v) the location or amenities of any accommodation so provided.

(2)   For the purposes of this section—

(a) anything (whether or not a statement as to any of the matters specified in the preceding subsection) likely to be taken for such a statement as to any of those matters as would be false shall be deemed to be a false statement as to that matter; and

(b) a statement made regardless of whether it is true or false shall be deemed to be made recklessly, whether or not the person making it had reasons for believing that it might be false.

(3)   In relation to any services consisting of or including the application of any treatment or process or the carrying out of any repair, the matters specified in subsection (1) of this section shall be taken to include the effect of the treatment, process or repair.

(4)   In this section "false" means false to a material degree and "services" does not include anything done under a contract of service.

15   Orders defining terms for purposes of section 14

Where it appears to the Board of Trade that it would be in the interest of persons for whom any services, accommodation or facilities are provided in the course of any trade or business that any expressions used with respect thereto should be understood as having definite meanings, the Board may by order assign such meanings to those expressions when used as, or as part of, such statements as are mentioned in section 14 of this Act with respect to those services, accommodation or facilities; and where such a meaning is so assigned to an expression it shall be deemed for the purposes of this Act to have that meaning when so used.

Prohibition of importation of certain goods

16   Prohibition of importation of goods bearing false indication of origin

Where a false trade description is applied to any goods outside the United Kingdom and the false indication, or one of the false indications, given, or likely to be taken as given, thereby is an indication of the place of manufacture, production, processing or reconditioning of the goods or any part thereof, the goods shall not be imported into the United Kingdom.

Provisions as to offences

18 Penalty for offences

A person guilty of an offence under this Act for which no other penalty is specified shall be liable—
  (a) on summary conviction, to a fine not exceeding [the prescribed sum] and
  (b) on conviction on indictment, to a fine or imprisonment for a term not exceeding two years or both.

NOTES

Reference to "the prescribed sum" in para (a) substituted by virtue of the Magistrates' Courts Act 1980, s 32(2).

19 Time limit for prosecutions

(1)   No prosecution for an offence under this Act shall be commenced after the expiration of three years from the commission of the offence or one year from its discovery by the prosecutor, whichever is the earlier.

(2)   Notwithstanding anything in [section 127(1) of the Magistrates' Courts Act 1980], a magistrates' court may try an information for an offence under this Act if the information was laid at any time within twelve months from the commission of the offence.

(3)   ...

(4)   Subsections (2) and (3) of this section do not apply where—
  (a) the offence was committed by the making of an oral statement; or
  (b) the offence was one of supplying goods to which a false trade description is applied, and the trade description was applied by an oral statement; or
  (c) the offence was one where a false trade description is deemed to have been applied to goods by virtue of section 4(3) of this Act and the goods were supplied in pursuance of an oral request.

NOTES

Sub-s (2): words in square brackets substituted by the Magistrates' Courts Act 1980, s 154, Sch 7, para 75.
Sub-s (3): applies to Scotland only.

20 Offences by corporations

(1) Where an offence under this Act which has been committed by a body corporate is proved to have been committed with the consent and connivance of, or to be attributable to any neglect on the part of, any director, manager, secretary or other similar officer of the body corporate, or any person who was purporting to act in any such capacity, he as well as the body corporate shall be guilty of that offence and shall be liable to be proceeded against and punished accordingly.

(2) In this section "director", in relation to any body corporate established by or under any enactment for the purpose of carrying on under national ownership any industry or part of an industry or undertaking, being a body corporate whose affairs are managed by the members thereof, means a member of that body corporate.

21 Accessories to offences committed abroad

(1) Any person who, in the United Kingdom, assists in or induces the commission in any other country of an act in respect of goods which, if the act were committed in the United Kingdom, would be an offence under section 1 of this Act shall be guilty of an offence, except as provided by subsection (2) of this section, but only if either—
  (a) the false trade description concerned is an indication (or anything likely to be taken as an indication) that the goods or any part thereof were manufactured, produced, processed or reconditioned in the United Kingdom; or
  (b) the false description concerned—
    (i) consists of or comprises an expression (or anything likely to be taken as an expression) to which a meaning is assigned by an order made by virtue of section 7(b) of this Act, and
    (ii) where that meaning is so assigned only in circumstances specified in the order, the trade description is used in those circumstances.

(2) A person shall not be guilty of an offence under subsection (1) of this section if, by virtue of section 32 of this Act, the act, though committed in the United Kingdom, would not be an offence under section 1 of this Act had the goods been intended for despatch to the other country.

(3) Any person who, in the United Kingdom, assists in or induces the commission outside the United Kingdom of an act which, if committed in

the United Kingdom, would be an offence under section 12 of this Act shall be guilty of an offence.

22 Restrictions on institution of proceedings and admission of evidence

(1)   Where any act or omission constitutes both an offence under this Act and an offence under any provision contained in or having effect by virtue of Part IV of the [Weights and Measures Act 1985] or [Part V of the Weights and Measures (Northern Ireland) Order 1981]—
   (a) proceedings for the offence shall not be instituted under this Act, except by virtue of section 23 thereof, without the service of such a notice as is required by [subsection (3) of section 83 of the said Act of 1985] or, as the case may be, [paragraph (3) of Article 46 of the said Order of 1981], nor after the expiration of the period mentioned in paragraph (c) of that subsection [or, as the case may be, that paragraph]; and
   (b) [sections 35, 36 and 37(1) and (2) of the said Act of 1985] or, as the case may be, [of Article 24 of the said Order of 1981], shall, with the necessary modifications, apply as if the offence under this Act were an offence under Part IV of that Act [or, as the case may be, Part V of that Order,] or any instrument made thereunder.

(2)   Where any act or omission constitutes both an offence under this Act and an offence under the food and drugs laws, evidence on behalf of the prosecution concerning any sample procured for analysis shall not be admissible in proceedings for the offence under this Act unless the relevant provisions of those laws have been complied with.

...

[(2A) In subsection (2) of this section—
   "the food and drugs laws" means the Food Safety Act 1990, the Medicines Act 1968 and the [Food Safety (Northern Ireland) Order 1991] and any instrument made thereunder;
   "the relevant provisions" means—
      (i)   in relation to the said Act of 1990, section 31 and regulations made thereunder;
      (ii) in relation to the said Act of 1968, so much of Schedule 3 to that Act as is applicable to the circumstances in which the sample was procured; and
      (iii) in relation to the said Order, [Article 31 and regulations made thereunder],

or any provisions replacing any of those provisions by virtue of section 17 of the said Act of 1990, paragraph 27 of Schedule 3 to the said Act of 1968 or [Article 16] of the said Order.]

(3)   The Board of Trade may by order provide that in proceedings for an offence under this Act in relation to such goods as may be specified in the order (other than proceedings for an offence falling within the preceding provisions of this section) evidence on behalf of the prosecution concerning any sample procured for analysis shall not be admissible unless the sample has been dealt with in such manner as may be specified in the order.

23   Offences due to fault of other person

Where the commission by any person of an offence under this Act is due to the act or default of some other person that other person shall be guilty of the offence, and a person may be charged with and convicted of the offence by virtue of this section whether or not proceedings are taken against the first-mentioned person.

Defences

24   Defence of mistake, accident, etc

(1)   In any proceedings for an offence under this Act it shall, subject to subsection (2) of this section, be a defence for the person charged to prove—
   (a) that the commission of the offence was due to a mistake or to reliance on information supplied to him or to the act or default of another person, an accident or some other cause beyond his control; and

(b) that he took all reasonable precautions and exercised all due diligence to avoid the commission of such an offence by himself or any person under his control.

(2)   If in any case the defence provided by the last foregoing subsection involves the allegation that the commission of the offence was due to the act or default of another person or to reliance on information supplied by another person, the person charged shall not, without leave of the court, be entitled to rely on that defence unless, within a period ending seven clear days before the hearing, he has served on the prosecutor a notice in writing giving such information identifying or assisting in the identification of that other person as was then in his possession.

(3)   In any proceedings for an offence under this Act of supplying or offering to supply goods to which a false trade description is applied it shall be a defence for the person charged to prove that he did not know, and could not with reasonable diligence have ascertained, that the goods did not conform to the description or that the description had been applied to the goods.

25   Innocent publication of advertisement

In proceedings for an offence under this Act committed by the publication of an advertisement it shall be a defence for the person charged to prove that he is a person whose business it is to publish or arrange for the publication of advertisements and that he received the advertisement for publication in the ordinary course of business and did not know and had no reason to suspect that its publication would amount to an offence under this Act.

Enforcement

26   Enforcing authorities

(1)   It shall be the duty of every local weights and measures authority to enforce within their area the provisions of this Act and of any order made under this Act; . . .

(2)   Every local weights and measures authority shall, whenever the Board of Trade so direct, make to the Board a report on the exercise of

their functions under this Act in such form and containing such particulars as the Board may direct.

(3)–(5) ...

NOTES

Sub-s (1): words omitted repealed by the Weights and Measures Act 1985, ss 96(1), 98(1), Sch 11, para 18(2), Sch 13, Pt I.

Sub-ss (3), (4): repealed by the Local Government, Planning and Land Act 1980, ss 1(4), 194, Sch 4, para 10(a), Sch 34, Pt IV.

Sub-s (5): applies to Scotland only.

27   Power to make test purchases

A local weights and measures authority shall have power to make, or to authorise any of their officers to make on their behalf, such purchases of goods, and to authorise any of their officers to secure the provision of such services, accommodation or facilities, as may appear expedient for the purpose of determining whether or not the provisions of this Act and any order made thereunder are being complied with.

28   Power to enter premises and inspect and seize goods and documents

(1)   A duly authorised officer of a local weights and measures authority or of a Government department may, at all reasonable hours and on production, if required, of his credentials, exercise the following powers, that is to say,—

(a) he may, for the purpose of ascertaining whether any offence under this Act has been committed, inspect any goods and enter any premises other than premises used only as a dwelling;

(b) if he has reasonable cause to suspect that an offence under this Act has been committed, he may, for the purpose of ascertaining whether it has been committed, require any person carrying on a trade or business or employed in connection with a trade or business to produce any books or documents relating to the trade or business and may take copies of, or of any entry in, any such book or document;

(c) if he has reasonable cause to believe that an offence under this Act has been committed, he may seize and detain any goods for the purpose of ascertaining, by testing or otherwise, whether the offence has been committed;

(d) he may seize and detain any goods or documents which he has reason to believe may be required as evidence in proceedings for an offence under this Act;

(e) he may, for the purpose of exercising his powers under this subsection to seize goods, but only if and to the extent that it is reasonably necessary in order to secure that the provisions of this Act and of any order made thereunder are duly observed, require any person having authority to do so to break open any container or open any vending machine and, if that person does not comply with the requirement, he may do so himself.

(2)   An officer seizing any goods or documents in the exercise of his powers under this section shall inform the person from whom they are seized and, in the case of goods seized from a vending machine, the person whose name and address are stated on the machine as being the proprietor's or, if no name and address are so stated, the occupier of the premises on which the machine stands or to which it is affixed.

(3)   If a justice of the peace, on sworn information in writing—
   (a) is satisfied that there is reasonable ground to believe either—
      (i) that any goods, books or documents which a duly authorised officer has power under this section to inspect are on any premises and that their inspection is likely to disclose evidence of the commission of an offence under this Act; or
      (ii) that any offence under this Act has been, is being or is about to be committed on any premises; and
   (b) is also satisfied either—
      (i) that admission to the premises has been or is likely to be refused and that notice of intention to apply for a warrant under this subsection has been given to the occupier; or
      (ii) that an application for admission, or the giving of such a notice, would defeat the object of the entry or that the premises are unoccupied or that the occupier is temporarily absent and it might defeat the object of the entry to await his return,

the justice may by warrant under his hand, which shall continue in force for a period of one month, authorise an officer of a local weights and measures authority or of a Government department to enter the premises, if need be by force.

...

(4)   An officer entering any premises by virtue of this section may take with him such other persons and such equipment as may appear to him

necessary; and on leaving any premises which he has entered by virtue of a warrant under the preceding subsection he shall, if the premises are unoccupied or the occupier is temporarily absent, leave them as effectively secured against trespassers as he found them.

(5)   If any person discloses to any person—
    (a) any information with respect to any manufacturing process or trade secret obtained by him in premises which he has entered by virtue of this section; or
    (b) any information obtained by him in pursuance of this Act;

he shall be guilty of an offence unless the disclosure was made in or for the purpose of the performance by him or any other person of functions under this Act.

[(5A)   Subsection (5) of this section does not apply to disclosure for a purpose specified in [section 38(2)(a), (b) or (c) of the Consumer Protection Act 1987.]]

(6)   If any person who is not a duly authorised officer of a local weights and measures authority or of a Government department purports to act as such under this section he shall be guilty of an offence.

(7)   Nothing in this section shall be taken to compel the production by a solicitor of a document containing a privileged communication made by or to him in that capacity or to authorise the taking of possession of any such document which is in his possession.

NOTES

Sub-s (3): words omitted apply to Scotland only.
Sub-s (5A): inserted by the Consumer Credit Act 1974, s 192(3)(a), Sch 4, Pt I, para 28, words in square brackets therein substituted by the Consumer Protection Act 1987, s 48(1), Sch 4, para 2(2).

29   Obstruction of authorised officers

(1)   Any person who—
    (a) wilfully obstructs an officer of a local weights and measures authority or of a Government department acting in pursuance of this Act; or
    (b) wilfully fails to comply with any requirement properly made to him by such an officer under section 28 of this Act; or
    (c) without reasonable cause fails to give such an officer so acting any other assistance or information which he may reasonably require of him for the purpose of the performance of his functions under this Act,

shall be guilty of an offence and liable, on summary conviction, to a fine not exceeding [level 3 on the standard scale].

(2)　If any person, in giving any such information as is mentioned in the preceding subsection, makes any statement which he knows to be false, he shall be guilty of an offence.

(3)　Nothing in this section shall be construed as requiring a person to answer any question or give any information if to do so might incriminate him.

NOTES

Sub-s (1): reference to a level on the standard scale substituted by virtue of the Criminal Justice Act 1982, ss 38, 46.

30　Notice of test and intended prosecution

(1)　Where any goods seized or purchased by an officer in pursuance of this Act are submitted to a test, then—
  (a) if the goods were seized, the officer shall inform the person mentioned in section 28(2) of this Act of the result of the test;
  (b) if the goods were purchased and the test leads to the institution of proceedings for an offence under this Act, the officer shall inform the person from whom the goods were purchased, or, in the case of goods sold through a vending machine, the person mentioned in section 28(2) of this Act, of the result of the test;

and shall, where as a result of the test proceedings for an offence under this Act are instituted against any person, allow him to have the goods tested on his behalf if it is reasonably practicable to do so.

(2)–(4)…

NOTES

Sub-ss (2)–(4): repealed by the Fair Trading Act 1973, s 139, Sch 13.

31　Evidence by certificate

(1)　The Board of Trade may by regulations provide that certificates issued by such persons as may be specified by the regulations in relation to such matters as may be so specified shall, subject to the provisions of this section, be received in evidence of those matters in any proceedings under this Act.

(2) Such a certificate shall not be received in evidence—
   (a) unless the party against whom it is to be given in evidence has been served with a copy thereof not less than seven days before the hearing; or
   (b) if that party has, not less than three days before the hearing, served on the other party a notice requiring the attendance of the person issuing the certificate.

(3) ...

(4) For the purposes of this section any document purporting to be such a certificate as is mentioned in this section shall be deemed to be such a certificate unless the contrary is shown.

(5) Regulations under this section shall be made by statutory instrument which shall be subject to annulment in pursuance of a resolution of either House of Parliament.

NOTES

Sub-s (3): applies to Scotland only.

Miscellaneous and supplemental

32 Power to exempt goods sold for export, etc

[(1)] In relation to goods which are intended—
   (a) for despatch to a destination outside the United Kingdom and any designated country within the meaning of [section 24(2)(b) of the Weights and Measures Act 1985] or section 15(5)(b) of the Weights and Measures Act (Northern Ireland) 1967; or
   (b) for use as stores within the meaning of the [Customs and Excise Management Act 1979] in a ship or aircraft on a voyage or flight to an eventual destination outside the United Kingdom; or
   (c) for use by Her Majesty's forces or by a visiting force within the meaning of any of the provisions of Part I of the Visiting Forces Act 1952; or
   [(d) for industrial use within the meaning of the Weights and Measures Act 1985 or for constructional use;]

section 1 of this Act shall apply as if there were omitted from the matters included in section 2(1) of this Act those specified in paragraph (a) thereof;

and, if the Board of Trade by order specify any other of those matters for the purposes of this section with respect to any description of goods, the said section 1 shall apply, in relation to goods of that description which are intended for despatch to a destination outside the United Kingdom and such country (if any) as may be specified in the order, as if the matters so specified were also omitted from those included in the said section 2(1).

[(2)   In this section "constructional use", in relation to any goods, means the use of those goods in constructional work (or, if the goods are explosives within the meaning of the Explosives Acts 1875 and 1923, in mining, quarrying or demolition work) in the course of the carrying on of a business.]

NOTES

Sub-s (1): numbered as such, words in square brackets in para (a) substituted, para (d) substituted, by the Weights and Measures Act 1985, s 97, Sch 12, para 4(1), (2); words in square brackets in para (b) substituted by the Customs and Excise Management Act 1979, s 177(1), Sch 4, para 12, Table, Pt I.

Sub-s (2): added by the Weights and Measures Act 1985, s 97, Sch 12, para 4(2).

33   Compensation for loss, etc of goods seized under s 28

(1)   Where, in the exercise of his powers under section 28 of this Act, an officer of a local weights and measures authority or of a Government department seizes and detains any goods and their owner suffers loss by reason thereof or by reason that the goods, during the detention, are lost or damaged or deteriorate, then, unless the owner is convicted of an offence under this Act committed in relation to the goods, the authority or department shall be liable to compensate him for the loss so suffered.

(2)   Any disputed question as to the right to or the amount of any compensation payable under this section shall be determined by arbitration and, in Scotland, by a single arbiter appointed, failing agreement between the parties, by the sheriff.

34   Trade marks containing trade descriptions

The fact that a trade description is a trade mark, or part of a trade mark, … does not prevent it from being a false trade description when applied to any goods, except where the following conditions are satisfied, that is to say—
    (a)   that it could have been lawfully applied to the goods if this Act had not been passed; and

(b) that on the day this Act is passed the trade mark either is registered under the Trade Marks Act 1938 or is in use to indicate a connection in the course of trade between such goods and the proprietor of the trade mark; and

(c) that the trade mark as applied is used to indicate such a connection between the goods and the proprietor of the trade mark or[, in the case of a registered trade mark, a person licensed to use it]; and

(d) that the person who is the proprietor of the trade mark is the same person as, or a successor in title of, the proprietor on the day this Act is passed.

Words omitted repealed, and words in square brackets substituted, by the Trade Marks Act 1994, s 106(1), Sch 4, para 4.

Modification: references to trade marks or registered trade marks within the meaning of the Trade Marks Act 1938 shall, unless the context otherwise requires, be construed as references to trade marks or registered trade marks within the meaning of the Trade Marks Act 1994; see the Trade Marks Act 1994, Sch 4, para 1.

35 Saving for civil rights

A contract for the supply of any goods shall not be void or unenforceable by reason only of a contravention of any provision of this Act.

36 Country of origin

(1) For the purposes of this Act goods shall be deemed to have been manufactured or produced in the country in which they last underwent a treatment or process resulting in a substantial change.

(2) The Board of Trade may by order specify—

(a) in relation to any description of goods, what treatment or process is to be regarded for the purposes of this section as resulting or not resulting in a substantial change;

(b) in relation to any description of goods different parts of which were manufactured or produced in different countries, or of goods assembled in a country different from that in which their parts were manufactured or produced, in which of those countries the goods are to be regarded for the purposes of this Act as having been manufactured or produced.

37 Market research experiments

(1) In this section "market research experiment" means any activities conducted for the purpose of ascertaining the opinion of persons (in this section referred to as "participants") of—
  (a) any goods; or
  (b) anything in, on or with which the goods are supplied; or
  (c) the appearance or any other characteristic of the goods or of any such thing; or
  (d) the name or description under which the goods are supplied.

(2) This section applies to any market research experiment with respect to which the following conditions are satisfied, that is to say—
  (a) that any participant to whom any goods are supplied in the course of the experiment is informed, at or before the time at which they are supplied to him, that they are supplied for such a purpose as is mentioned in subsection (1) of this section, and
  (b) that no consideration in money or money's worth is given by a participant for the goods or any goods supplied to him for comparison.

(3) Neither section 1 nor section 8 of this Act shall apply in relation to goods supplied or offered to be supplied, whether to a participant or any other person, in the course of a market research experiment to which this section applies.

38 Orders

(1) Any power to make an order under the preceding provisions of this Act shall be exercisable by statutory instrument, which shall be subject to annulment in pursuance of a resolution of either House of Parliament, and includes power to vary or revoke such an order by a subsequent order.

(2) Any order under the preceding provisions of this Act which relates to any agricultural, horticultural or fishery produce, whether processed or not, food, feeding stuffs or ingredients of food or feeding stuffs, fertilisers or any goods used as pesticides or for similar purposes shall be made by the Board of Trade acting jointly with the following Ministers, that is to say, if the order extends to England and Wales, the Minister of Agriculture, Fisheries and Food, and if it extends to Scotland or Northern Ireland, the Secretary of State concerned.

(3)　The following provisions shall apply to the making of an order under section 7, 8, 9, 15 or 36 of this Act, except in the case mentioned in section 10(2) thereof, that is to say—

　(a)　before making the order the Board of Trade shall consult with such organisations as appear to them to be representative of interests substantially affected by it and shall publish, in such manner as the Board think appropriate, notice of their intention to make the order and of the place where copies of the proposed order may be obtained; and

　(b)　the order shall not be made until the expiration of a period of twenty-eight days from the publication of the notice and may then be made with such modifications (if any) as the Board of Trade think appropriate having regard to any representations received by them.

39　Interpretation

(1)　The following provisions shall have effect, in addition to sections 2 to 6 of this Act, for the interpretation in this Act of expressions used therein, that is to say,—

"advertisement" includes a catalogue, a circular and a price list;

"goods" includes ships and aircraft, things attached to land and growing crops;

"premises" includes any place and any stall, vehicle, ship or aircraft; and

"ship" includes any boat and any other description of vessel used in navigation.

(2)　For the purposes of this Act, a trade description or statement published in any newspaper, book or periodical or in any film or sound or television broadcast [or in any programme included in any programme service (within the meaning of the Broadcasting Act 1990) other than a sound or television broadcasting service] shall not be deemed to be a trade description applied or statement made in the course of a trade or business unless it is or forms part of an advertisement.

NOTES

Sub-s (2): words in square brackets originally inserted by the Cable and Broadcasting Act 1984, s 57(1), Sch 5, para 19, substituted by the Broadcasting Act 1990, s 203(1), Sch 20, para 11.

43　Short title and commencement

(1)　This Act may be cited as the Trade Descriptions Act 1968.

(2)   This Act shall come into force on the expiration of the period of six months beginning with the day on which it is passed.

Carriage of Goods by Sea Act 1971

(C 19)

An Act to amend the law with respect to the carriage of goods by sea
[8 April 1971]

NOTES

Hovercraft: by the Hovercraft (Civil Liability) Order 1986, SI 1986/1305, arts 4, 5, 10, Sch 2, Sch 4, this Act applies, in relation to the carriage of goods by hovercraft (other than passengers' baggage) as it applies in relation to goods on board or carried by ship.

1   Application of Hague Rules as amended

(1)   In this Act, "the Rules" means the International Convention for the unification of certain rules of law relating to bills of lading signed at Brussels on 25th August 1924, as amended by the Protocol signed at Brussels on 23rd February 1968 [and by the Protocol signed at Brussels on 21st December 1979].

(2)   The provisions of the Rules, as set out in the Schedule to this Act, shall have the force of law.

(3)   Without prejudice to subsection (2) above, the said provisions shall have effect (and have the force of law) in relation to and in connection with the carriage of goods by sea in ships where the port of shipment is a port in the United Kingdom, whether or not the carriage is between ports in two different States within the meaning of Article X of the Rules.

(4)   Subject to subsection (6) below, nothing in this section shall be taken as applying anything in the Rules to any contract for the carriage of goods by sea, unless the contract expressly or by implication provides for the issue of a bill of lading or any similar document of title.

(5)   ...

(6)   Without prejudice to Article X(c) of the Rules, the Rules shall have the force of law in relation to—

(a) any bill of lading if the contract contained in or evidenced by it expressly provides that the Rules shall govern the contract, and

(b) any receipt which is a non-negotiable document marked as such if the contract contained in or evidenced by it is a contract for the carriage of goods by sea which expressly provides that the Rules are to govern the contract as if the receipt were a bill of lading,

but subject, where paragraph (b) applies, to any necessary modifications and in particular with the omission in Article III of the Rules of the second sentence of paragraph 4 and of paragraph 7.

(7)   If and so far as the contract contained in or evidenced by a bill of lading or receipt within paragraph (a) or (b) of subsection (6) above applies to deck cargo or live animals, the Rules as given the force of law by that subsection shall have effect as if Article I(c) did not exclude deck cargo and live animals.

In this subsection "deck cargo" means cargo which by the contract of carriage is stated as being carried on deck and is so carried.

NOTES

Sub-s (1): words in square brackets inserted by the Merchant Shipping Act 1981, s 2(1) and by virtue of the Merchant Shipping Act 1995, s 314(2), Sch 13, para 45(1), (2).

Sub-s (5): repealed by the Merchant Shipping Act 1981, s 5(3), Schedule.

[1A   Conversion of special drawing rights into sterling

(1)   For the purposes of Article IV of the Rules the value on a particular day of one special drawing right shall be treated as equal to such a sum in sterling as the International Monetary Fund have fixed as being the equivalent of one special drawing right—

(a) for that day; or

(b) if no sum has been so fixed for that day, for the last day before that day for which a sum has been so fixed.

(2)   A certificate given by or on behalf of the Treasury stating—

(a) that a particular sum in sterling has been fixed as aforesaid for a particular day; or

(b) that no sum has been so fixed for a particular day and that a particular sum in sterling has been so fixed for a day which is the last day for which a sum has been so fixed before the particular day, shall be conclusive evidence of those matters for the purposes of subsection (1) above; and a document purporting to be such a certificate shall in any proceedings be received in evidence and, unless the contrary is proved, be deemed to be such a certificate.

(3)   The Treasury may charge a reasonable fee for any certificate given in pursuance of subsection (2) above, and any fee received by the Treasury by virtue of this subsection shall be paid into the Consolidated Fund.]

NOTES

Commencement: 1 January 1996.
Inserted by the Merchant Shipping Act 1995, s 314(2), Sch 13, para 45(1), (3).

2   Contracting States, etc

(1)   If Her Majesty by Order in Council certifies to the following effect, that is to say, that for the purposes of the Rules—
  (a) a State specified in the Order is a contracting State, or is a contracting State in respect of any place or territory so specified; or
  (b) any place or territory specified in the Order forms part of a State so specified (whether a contracting State or not),

the Order shall, except so far as it has been superseded by a subsequent Order, be conclusive evidence of the matters so certified.

(2)   An Order in Council under this section may be varied or revoked by a subsequent Order in Council.

3   Absolute warranty of seaworthiness not to be implied in contracts to which Rules apply

There shall not be implied in any contract for the carriage of goods by sea to which the Rules apply by virtue of this Act any absolute undertaking by the carrier of the goods to provide a seaworthy ship.

4   Application of Act to British possessions, etc

(1)   Her Majesty may by Order in Council direct that this Act shall extend, subject to such exceptions, adaptations and modifications as may be specified in the Order, to all or any of the following territories, that is—

(a) any colony (not being a colony for whose external relations a country other than the United Kingdom is responsible),

(b) any country outside Her Majesty's dominions in which Her Majesty has jurisdiction in right of Her Majesty's Government of the United Kingdom.

(2)   An Order in Council under this section may contain such transitional and other consequential and incidental provisions as appear to Her Majesty to be expedient, including provisions amending or repealing any legislation about the carriage of goods by sea forming part of the law of any of the territories mentioned in paragraphs (a) and (b) above.

(3)   An Order in Council under this section may be varied or revoked by a subsequent Order in Council.

5   Extension of application of Rules to carriage from ports in British possessions, etc

(1)   Her Majesty may by Order in Council provide that section 1(3) of this Act shall have effect as if the reference therein to the United Kingdom included a reference to all or any of the following territories, that is—

(a) the Isle of Man;

(b) any of the Channel Islands specified in the Order;

(c) any colony specified in the Order (not being a colony for whose external relations a country other than the United Kingdom is responsible);

(d)   ...

(e) any country specified in the Order, being a country outside Her Majesty's dominions in which Her Majesty has jurisdiction in right of Her Majesty's Government of the United Kingdom.

(2)   An Order in Council under this section may be varied or revoked by a subsequent Order in Council.

NOTES

Sub-s (1): para (d) repealed by the Statute Law (Repeals) Act 1989.

6   Supplemental

(1)   This Act may be cited as the Carriage of Goods by Sea Act 1971.

(2)   It is hereby declared that this Act extends to Northern Ireland.

(3)   The following enactments shall be repealed, that is—
   (a) the Carriage of Goods by Sea Act 1924,
   (b) section 12(4)(a) of the Nuclear Installations Act 1965,

and without prejudice to section 38(1) of the Interpretation Act 1889, the reference to the said Act of 1924 in section 1(1)(i)(ii) of the Hovercraft Act 1968 shall include a reference to this Act.

[(4)   It is hereby declared that for the purposes of Article VIII of the Rules section 186 of the Merchant Shipping Act 1995 (which entirely exempts shipowners and others in certain circumstances or loss of, or damage to, goods) is a provision relating to limitation of liability.]

(5)   This Act shall come into force on such day as Her Majesty may by Order in Council appoint, and, for the purposes of the transition from the law in force immediately before the day appointed under this subsection to the provisions of this Act, the Order appointing the day may provide that those provisions shall have effect subject to such transitional provisions as may be contained in the Order.

NOTES

   Sub-s (4): substituted by the Merchant Shipping Act 1995, s 314(3), Sch 13, para 45(1), (4).

SCHEDULE

Section 3

THE HAGUE RULES AS AMENDED BY THE BRUSSELS PROTOCOL 1968

ARTICLE I

In these Rules the following words are employed, with the meanings set out below:—
   (a) "Carrier" includes the owner or the charterer who enters into a contract of carriage with a shipper.
   (b) "Contract of carriage" applies only to contracts of carriage covered by a bill of lading or any similar document of title, in so far as such document relates to the carriage of goods by sea, including any bill of lading or any similar document as aforesaid issued under or pursuant to a charter party from the moment at which such bill of

lading or similar document of title regulates the relations between a carrier and a holder of the same.

(c) "Goods" includes goods, wares, merchandise, and articles of every kind whatsoever except live animals and cargo which by the contract of carriage is stated as being carried on deck and is so carried.

(d) "Ship" means any vessel used for the carriage of goods by sea.

(e) "Carriage of goods" covers the period from the time when the goods are loaded on to the time they are discharged from the ship.

ARTICLE II

Subject to the provisions of Article VI, under every contract of carriage of goods by sea the carrier, in relation to the loading, handling, stowage, carriage, custody, care and discharge of such goods, shall be subject to the responsibilities and liabilities, and entitled to the rights and immunities hereinafter set forth.

ARTICLE III

1. The carrier shall be bound before and at the beginning of the voyage to exercise due diligence to—
   (a) Make the ship seaworthy.
   (b) Properly man, equip and supply the ship.
   (c) Make the holds, refrigerating and cool chambers, and all other parts of the ship in which goods are carried, fit and safe for their reception, carriage and preservation.

2. Subject to the provisions of Article IV, the carrier shall properly and carefully load, handle, stow, carry, keep, care for, and discharge the goods carried.

3. After receiving the goods into his charge the carrier or the master or agent of the carrier shall, on demand of the shipper, issue to the shipper a bill of lading showing among other things—
   (a) The leading marks necessary for identification of the goods as the same are furnished in writing by the shipper before the loading of such goods starts, provided such marks are stamped or otherwise shown clearly upon the goods if uncovered, or on the cases or coverings in which such goods are contained, in such a manner as should ordinarily remain legible until the end of the voyage.
   (b) Either the number of packages or pieces, or the quantity, or weight, as the case may be, as furnished in writing by the shipper.
   (c) The apparent order and condition of the goods.

Provided that no carrier, master or agent of the carrier shall be bound to state or show in the bill of lading any marks, number, quantity, or weight which he has reasonable ground for suspecting not accurately to represent the goods actually received, or which he has had no reasonable means of checking.

4.   Such a bill of lading shall be prima facie evidence of the receipt by the carrier of the goods as therein described in accordance with paragraph 3(a), (b) and (c). However, proof to the contrary shall not be admissible when the bill of lading has been transferred to a third party acting in good faith.

5.   The shipper shall be deemed to have guaranteed to the carrier the accuracy at the time of shipment of the marks, number, quantity and weight, as furnished by him, and the shipper shall indemnify the carrier against all loss, damages and expenses arising or resulting from inaccuracies in such particulars. The right of the carrier to such indemnity shall in no way limit his responsibility and liability under the contract of carriage to any person other than the shipper.

6.   Unless notice of loss or damage and the general nature of such loss or damage be given in writing to the carrier or his agent at the port of discharge before or at the time of the removal of the goods into the custody of the person entitled to delivery thereof under the contract of carriage, or, if the loss or damage be not apparent, within three days, such removal shall be prima facie evidence of the delivery by the carrier of the goods as described in the bill of lading.

The notice in writing need not be given if the state of the goods has, at the time of their receipt, been the subject of joint survey or inspection.

Subject to paragraph 6bis the carrier and the ship shall in any event be discharged from all liability whatsoever in respect of the goods, unless suit is brought within one year of their delivery or of the date when they should have been delivered. This period may, however, be extended if the parties so agree after the cause of action has arisen.

In the case of any actual or apprehended loss or damage the carrier and the receiver shall give all reasonable facilities to each other for inspecting and tallying the goods.

6bis. An action for indemnity against a third person may be brought even after the expiration of the year provided for in the preceding paragraph if brought within the time allowed by the law of the Court seized of the case.

However, the time allowed shall be not less than three months, commencing from the day when the person bringing such action for indemnity has settled the claim or has been served with process in the action against himself.

7.   After the goods are loaded the bill of lading to be issued by the carrier, master, or agent of the carrier, to the shipper shall, if the shipper so demands, be a "shipped" bill of lading, provided that if the shipper shall have previously taken up any document of title to such goods, he shall surrender the same as against the issue of the "shipped" bill of lading, but at the option of the carrier such document of title may be noted at the port of shipment by the carrier, master, or agent with the name or names of the ship or ships upon which the goods have been shipped and the date or dates of shipment, and when so noted, if it shows the particulars mentioned in paragraph 3 of Article III, shall for the purpose of this article be deemed to constitute a "shipped" bill of lading.

8.   Any clause, covenant, or agreement in a contract of carriage relieving the carrier or the ship from liability for loss or damage to, or in connection with, goods arising from negligence, fault, or failure in the duties and obligations provided in this article or lessening such liability otherwise than as provided in these Rules, shall be null and void and of no effect. A benefit of insurance in favour of the carrier or similar clause shall be deemed to be a clause relieving the carrier from liability.

ARTICLE IV

1.   Neither the carrier nor the ship shall be liable for loss or damage arising or resulting from unseaworthiness unless caused by want of due diligence on the part of the carrier to make the ship seaworthy, and to secure that the ship is properly manned, equipped and supplied, and to make the holds, refrigerating and cool chambers and all other parts of the ship in which goods are carried fit and safe for their reception, carriage and preservation in accordance with the provisions of paragraph 1 of Article III. Whenever loss or damage has resulted from unseaworthiness the burden of proving the exercise of due diligence shall be on the carrier or other person claiming exemption under this article.

2.   Neither the carrier nor the ship shall be responsible for loss or damage arising or resulting from—
  (a) Act, neglect, or default of the master, mariner, pilot, or the servants of the carrier in the navigation or in the management of the ship.
  (b) Fire, unless caused by the actual fault or privity of the carrier.
  (c) Perils, dangers and accidents of the sea or other navigable waters.

(d) Act of God.

(e) Act of war.

(f) Act of public enemies.

(g) Arrest or restraint of princes, rulers or people, or seizure under legal process.

(h) Quarantine restrictions.

(i) Act or omission of the shipper or owner of the goods, his agent or representative.

(j) Strikes or lockouts or stoppage or restraint of labour from whatever cause, whether partial or general.

(k) Riots and civil commotions.

(l) Saving or attempting to save life or property at sea.

(m) Wastage in bulk or weight or any other loss or damage arising from inherent defect, quality or vice of the goods.

(n) Insufficiency of packing.

(o) Insufficiency or inadequacy of marks.

(p) Latent defects not discoverable by due diligence.

(q) Any other cause arising without the actual fault or privity of the carrier, or without the fault or neglect of the agents or servants of the carrier, but the burden of proof shall be on the person claiming the benefit of this exception to show that neither the actual fault or privity of the carrier nor the fault or neglect of the agents or servants of the carrier contributed to the loss or damage.

3.   The shipper shall not be responsible for loss or damage sustained by the carrier or the ship arising or resulting from any cause without the act, fault or neglect of the shipper, his agents or his servants.

4.   Any deviation in saving or attempting to save life or property at sea or any reasonable deviation shall not be deemed to be an infringement or breach of these Rules or of the contract of carriage, and the carrier shall not be liable for any loss or damage resulting therefrom.

5. (a) Unless the nature and value of such goods have been declared by the shipper before shipment and inserted in the bill of lading, neither the carrier nor the ship shall in any event be or become liable for any loss or damage to or in connection with the goods in an amount exceeding [666.67 units of account] per package or unit or [2 units of account per kilogramme] of gross weight of the goods lost or damaged, whichever is the higher.

   (b) The total amount recoverable shall be calculated by reference to the value of such goods at the place and time at which the goods are discharged from the ship in accordance with the contract or should have been so discharged.

The value of the goods shall be fixed according to the commodity exchange price, or, if there be no such price, according to the current market price, or, if there be no commodity exchange price or current market price, by reference to the normal value of goods of the same kind and quality.

(c) Where a container, pallet or similar article of transport is used to consolidate goods, the number of packages or units enumerated in the bill of lading as packed in such article of transport shall be deemed the number of packages or units for the purpose of this paragraph as far as these packages or units are concerned. Except as aforesaid such article of transport shall be considered the package or unit.

[(d) The unit of account mentioned in this Article is the special drawing right as defined by the International Monetary Fund. The amounts mentioned in sub-paragraph (a) of this paragraph shall be converted into national currency on the basis of the value of that currency on a date to be determined by the law of the Court seized of the case.]

(e) Neither the carrier nor the ship shall be entitled to the benefit of the limitation of liability provided for in this paragraph if it is proved that the damage resulted from an act or omission of the carrier done with intent to cause damage, or recklessly and with knowledge that damage would probably result.

(f) The declaration mentioned in sub-paragraph (a) of this paragraph, if embodied in the bill of lading, shall be prima facie evidence, but shall not be binding or conclusive on the carrier.

(g) By agreement between the carrier, master or agent of the carrier and the shipper other maximum amounts than those mentioned in sub-paragraph (a) of this paragraph may be fixed, provided that no maximum amount so fixed shall be less than the appropriate maximum mentioned in that sub-paragraph.

(h) Neither the carrier nor the ship shall be responsible in any event for loss or damage to, or in connection with, goods if the nature or value thereof has been knowingly mis-stated by the shipper in the bill of lading.

6. Goods of an inflammable, explosive or dangerous nature to the shipment whereof the carrier, master or agent of the carrier has not consented with knowledge of their nature and character, may at any time before discharge be landed at any place, or destroyed or rendered innocuous by the carrier without compensation and the shipper of such goods shall be liable for all damages and expenses directly or indirectly arising out of or resulting from such shipment. If any such goods shipped with such knowledge and consent shall become a danger to the ship or cargo, they may in like manner be landed at any place, or destroyed or rendered innocuous by the carrier without liability on the part of the carrier except to general average, if any.

ARTICLE IV BIS

1.   The defences and limits of liability provided for in these Rules shall apply in any action against the carrier in respect of loss or damage to goods covered by a contract of carriage whether the action be founded in contract or in tort.

2.   If such an action is brought against a servant or agent of the carrier (such servant or agent not being an independent contractor), such servant or agent shall be entitled to avail himself of the defences and limits of liability which the carrier is entitled to invoke under these Rules.

3.   The aggregate of the amounts recoverable from the carrier, and such servants and agents, shall in no case exceed the limit provided for in these Rules.

4.   Nevertheless, a servant or agent of the carrier shall not be entitled to avail himself of the provisions of this article, if it is proved that the damage resulted from an act or omission of the servant or agent done with intent to cause damage or recklessly and with knowledge that damage would probably result.

ARTICLE V

A carrier shall be at liberty to surrender in whole or in part all or any of his rights and immunities or to increase any of his responsibilities and obligations under these Rules, provided such surrender or increase shall be embodied in the bill of lading issued to the shipper. The provisions of these Rules shall not be applicable to charter parties, but if bills of lading are issued in the case of a ship under a charter party they shall comply with the terms of these Rules.

Nothing in these Rules shall be held to prevent the insertion in a bill of lading of any lawful provision regarding general average.

ARTICLE VI

Notwithstanding the provisions of the preceding articles, a carrier, master or agent of the carrier and a shipper shall in regard to any particular goods be at liberty to enter into any agreement in any terms as to the responsibility and liability of the carrier for such goods, and as to the rights and immunities of the carrier in respect of such goods, or his obligation as to seaworthiness, so far as this stipulation is not contrary to public policy, or the care or diligence of his servants or agents in regard to the loading, handling, stowage, carriage, custody, care and discharge of the goods carried by sea, provided that in this case no bill of lading has been or shall be issued and that the terms agreed shall be embodied in a receipt which shall be a non-negotiable document and shall be marked as such.

Any agreement so entered into shall have full legal effect.

Provided that this article shall not apply to ordinary commercial shipments made in the ordinary course of trade, but only to other shipments where the character or condition of the property to be carried or the circumstances, terms and conditions under which the carriage is to be performed are such as reasonably to justify a special agreement.

ARTICLE VII

Nothing herein contained shall prevent a carrier or a shipper from entering into any agreement, stipulation, condition, reservation or exemption as to the responsibility and liability of the carrier or the ship for the loss or damage to, or in connection with, the custody and care and handling of goods prior to the loading on, and subsequent to the discharge from, the ship on which the goods are carried by sea.

ARTICLE VIII

The provisions of these Rules shall not affect the rights and obligations of the carrier under any statute for the time being in force relating to the limitation of the liability of owners of sea-going vessels.

ARTICLE IX

These Rules shall not affect the provisions of any international Convention or national law governing liability for nuclear damage.

ARTICLE X

The provisions of these Rules shall apply to every bill of lading relating to the carriage of goods between ports in two different States if:
  (a) the bill of lading is issued in a contracting State, or

(b) the carriage is from a port in a contracting State, or

(c) the contract contained in or evidenced by the bill of lading provides that these Rules or legislation of any State giving effect to them are to govern the contract,

whatever may be the nationality of the ship, the carrier, the shipper, the consignee, or any other interested person.

[The last two paragraphs of this article are not reproduced. They require contracting States to apply the Rules to bills of lading mentioned in the article and authorise them to apply the Rules to other bills of lading.]

[Articles 11 to 16 of the International Convention for the unification of certain rules of law relating to bills of lading signed at Brussels on 25th August 1924 are not reproduced. They deal with the coming into force of the Convention, procedure for ratification, accession and denunciation, and the right to call for a fresh conference to consider amendments to the Rules contained in the Convention.]

NOTES

Article IV: words in square brackets in para 5 substituted by the Merchant Shipping Act 1981, s 2(3), (4) and by virtue of the Merchant Shipping Act 1995, s 314(2), Sch 13, para 45(2).

Words in italics in square brackets at the end of the Schedule are as set out in the Queen's Printer's version of the Act.

Powers of Attorney Act 1971

(C 27)

An Act to make new provision in relation to powers of attorney and the delegation by trustees of their trusts, powers and discretions

[12 May 1971]

4 Powers of attorney given as security

(1) Where a power of attorney is expressed to be irrevocable and is given to secure—

(a) a proprietary interest of the donee of the power; or

(b) the performance of an obligation owed to the donee,

then, so long as the donee has that interest or the obligation remains undischarged, the power shall not be revoked—

    (i)   by the donor without the consent of the donee; or

    (ii) by the death, incapacity or bankruptcy of the donor or, if the donor is a body corporate, by its winding up or dissolution.

(2)   A power of attorney given to secure a proprietary interest may be given to the person entitled to the interest and persons deriving title under him to that interest, and those persons shall be duly constituted donees of the power for all purposes of the power but without prejudice to any right to appoint substitutes given by the power.

(3)   This section applies to powers of attorney whenever created.

5   Protection of donee and third persons where power of attorney is revoked

(1)   A donee of a power of attorney who acts in pursuance of the power at a time when it has been revoked shall not, by reason of the revocation, incur any liability (either to the donor or to any other person) if at that time he did not know that the power had been revoked.

(2)   Where a power of attorney has been revoked and a person, without knowledge of the revocation, deals with the donee of the power, the transaction between them shall, in favour of that person, be as valid as if the power had then been in existence.

(3)   Where the power is expressed in the instrument creating it to be irrevocable and to be given by way of security then, unless the person dealing with the donee knows that it was not in fact given by way of security, he shall be entitled to assume that the power is incapable of revocation except by the donor acting with the consent of the donee and shall accordingly be treated for the purposes of subsection (2) of this section as having knowledge of the revocation only if he knows that it has been revoked in that manner.

(4)   Where the interest of a purchaser depends on whether a transaction between the donee of a power of attorney and another person was valid by virtue of subsection (2) of this section, it shall be conclusively presumed in favour of the purchaser that that person did not at the material time know of the revocation of the power if—

    (a)   the transaction between that person and the donee was completed within twelve months of the date on which the power came into operation; or

    (b)   that person makes a statutory declaration, before or within three months after the completion of the purchase, that he did not at the material time know of the revocation of the power.

(5)   Without prejudice to subsection (3) of this section, for the purposes of this section knowledge of the revocation of a power of attorney includes knowledge of the occurrence of any event (such as the death of the donor) which has the effect of revoking the power.

(6)   In this section "purchaser" and "purchase" have the meanings specified in section 205(1) of the Law of Property Act 1925.

(7)   This section applies whenever the power of attorney was created but only to acts and transactions after the commencement of this Act.

7 Execution of instruments etc by donee of power of attorney

[(1)   If the donee of a power of attorney is an individual, he may, if he thinks fit—
   (a)   execute any instrument with his own signature, and]
   (b)   do any other thing in his own name,

by the authority of the donor of the power; and any document executed or thing done in that manner shall be as effective as if executed or done by the donee with the signature ... , or, as the case may be, in the name, of the donor of the power.

(2)   For the avoidance of doubt it is hereby declared that an instrument to which subsection (3) ... of section 74 of the Law of Property Act 1925 applies may be executed either as provided in [that subsection] or as provided in this section.

(3)   This section is without prejudice to any statutory direction requiring an instrument to be executed in the name of an estate owner within the meaning of the said Act of 1925.

(4)   This section applies whenever the power of attorney was created.

NOTES

Sub-ss (1), (2): words in square brackets substituted, and words omitted repealed, by the Law of Property (Miscellaneous Provisions) Act 1989, ss 1(8), 4, Sch 1, para 7, Sch 2.

11   Short title, repeals, consequential amendments, commencement and extent

(1)   This Act may be cited as the Powers of Attorney Act 1971.

(2)   The enactments specified in Schedule 2 to this Act are hereby repealed to the extent specified in the third column of that Schedule.

(3)   ...

(4)   This Act shall come into force on 1st October 1971.

(5)   Section 3 of this Act extends to Scotland and Northern Ireland but, save as aforesaid, this Act extends to England and Wales only.

NOTES

Sub-s (3): in part amends the Law of Property Act 1925, s 125(2); remainder repealed by the Supreme Court Act 1981, s 152(4), Sch 7.

Unsolicited Goods and Services Act 1971

(C 30)

An Act to make provision for the greater protection of persons receiving unsolicited goods, and to amend the law with respect to charges for entries in directories

[12 May 1971]

NOTES

Unsolicited Goods and Services Acts 1971 and 1975: by the Unsolicited Goods and Services (Amendment) Act 1975, s 4(1), that Act and this Act may be cited together by this collective title.

1   Rights of recipient of unsolicited goods

(1)   In the circumstances specified in the following subsection, a person who after the commencement of this Act receives unsolicited goods, may as between himself and the sender, use, deal with or dispose of them as if

they were an unconditional gift to him, and any right of the sender to the goods shall be extinguished.

(2)   The circumstances referred to in the preceding subsection are that the goods were sent to the recipient with a view to his acquiring them, that the recipient has no reasonable cause to believe that they were sent with a view to their being acquired for the purposes of a trade or business and has neither agreed to acquire nor agreed to return them, and either—

(a) that during the period of six months beginning with the day on which the recipient received the goods the sender did not take possession of them and the recipient did not unreasonably refuse to permit the sender to do so; or

(b) that not less than thirty days before the expiration of the period aforesaid the recipient gave notice to the sender in accordance with the following subsection, and that during the period of thirty days beginning with the day on which the notice was given the sender did not take possession of the goods and the recipient did not unreasonably refuse to permit the sender to do so.

(3)   A notice in pursuance of the preceding subsection shall be in writing and shall—

(a) state the recipient's name and address and, if possession of the goods in question may not be taken by the sender at that address, the address at which it may be so taken;

(b) contain a statement, however expressed, that the goods are unsolicited,

and may be sent by post.

(4)   In this section "sender", in relation to any goods, includes any person on whose behalf or with whose consent the goods are sent, and any other person claiming through or under the sender or any such person.

2   Demands and threats regarding payment

(1)   A person who, not having reasonable cause to believe there is a right to payment, in the course of any trade or business makes a demand for payment, or asserts a present or prospective right to payment, for what he knows are unsolicited goods sent (after the commencement of this Act) to another person with a view to his acquiring them, shall be guilty of an offence and on summary conviction shall be liable to a fine not exceeding [level 4 on the standard scale].

(2)   A person who, not having reasonable cause to believe there is a right to payment, in the course of any trade or business and with a view to obtaining any payment for what he knows are unsolicited goods sent as aforesaid—

    (a)   threatens to bring any legal proceedings; or

    (b)   places or causes to be placed the name of any person on a list of defaulters or debtors or threatens to do so; or

    (c)   invokes or causes to be invoked any other collection procedure or threatens to do so,

shall be guilty of an offence and shall be liable on summary conviction to a fine not exceeding [level 5 on the standard scale].

NOTES

    References to levels on the standard scale substituted by virtue of the Criminal Justice Act 1982, ss 38, 46.

3   Directory entries

(1)   A person shall not be liable to make any payment, and shall be entitled to recover any payment made by him, by way of charge for including or arranging for the inclusion in a directory of an entry relating to that person or his trade or business, unless there has been signed by him or on his behalf an order complying with this section or a note complying with this section of his agreement to the charge and, in the case of a note of agreement to the charge, before the note was signed, a copy of it was supplied, for retention by him, to him or to a person acting on his behalf.

(2)   A person shall be guilty of an offence punishable on summary conviction with a fine not exceeding [the prescribed sum] if, in a case where a payment in respect of a charge would, in the absence of an order or note of agreement to the charge complying with this section, be recoverable from him in accordance with the terms of subsection (1) above, he demands payment, or asserts a present or prospective right to payment, of the charge or any part of it, without knowing or having reasonable cause to believe that the entry to which the charge relates was ordered in accordance with this section or a proper note of agreement has been duly signed.

(3)   For the purposes of subsection (1) above, an order for an entry in a directory must be made by means of an order form or other stationery belonging to the person to whom, or to whose trade or business, the entry

is to relate and bearing, in print, the name and address (or one or more of the addresses) of that person; and the note required by this section of a person's agreement to a charge must state the amount of the charge immediately above the place for signature, and—

(a) must identify the directory or proposed directory, and give the following particulars of it—
  (i) the proposed date of publication of the directory or of the issue in which the entry is to be included and the name and address of the person producing it;
  (ii) if the directory or that issue is to be put on sale, the price at which it is to be offered for sale and the minimum number of copies which are to be available for sale;
  (iii) if the directory or that issue is to be distributed free of charge (whether or not it is also to be put on sale), the minimum number of copies which are to be so distributed; and
(b) must set out or give reasonable particulars of the entry in respect of which the charge would be payable.

(4)   Nothing in this section shall apply to a payment due under a contract entered into before the commencement of this Act, or entered into by the acceptance of an offer made before that commencement.

NOTES

Sub-s (2): words in square brackets substituted by virtue of the Magistrates' Courts Act 1980, s 32(2).

Sub-s (3): for the words in italics there are substituted the words "shall comply with the requirements of regulations under section 3A of this Act applicable thereto" by the Unsolicited Goods and Services (Amendment) Act 1975, s 2(1), as from a day to be appointed.

[3A   Contents and form of notes of agreement, invoices and similar documents

(1)   For the purposes of this Act, the Secretary of State may make regulations as to the contents and form of notes of agreement, invoices and similar documents; and, without prejudice to the generality of the foregoing, any such regulations may—

(a) require specified information to be included,
(b) prescribe the manner in which specified information is to be included,
(c) prescribe such other requirements (whether as to presentation, type, size, colour or disposition of lettering, quality or colour of paper or

otherwise) as the Secretary of State may consider appropriate for securing that specified information is clearly brought to the attention of the recipient of any note of agreement, invoice or similar document,

(d) make different provision for different classes or descriptions of notes of agreement, invoices and similar documents or for the same class or description in different circumstances,

(e) contain such supplementary and incidental provisions as the Secretary of State may consider appropriate.

(2)   Any reference in this section to a note of agreement includes any such copy as is mentioned in section 3(1) of this Act.

(3)   Regulations under this section shall be made by statutory instrument and shall be subject to annulment in pursuance of a resolution of either House of Parliament.]

NOTES

Inserted by the Unsolicited Goods and Services (Amendment) Act 1975, s 1.
Regulations: the Unsolicited Goods and Services (Invoices etc) Regulations 1975, SI 1975/732.

4   Unsolicited publications

(1)   A person shall be guilty of an offence if he sends or causes to be sent to another person any book, magazine or leaflet (or advertising material for any such publication) which he knows or ought reasonably to know is unsolicited and which describes or illustrates human sexual techniques.

(2)   A person found guilty of an offence under this section shall be liable on summary conviction to a fine not exceeding [level 5 on the standard scale].

(3)   A prosecution for an offence under this section shall not in England and Wales be instituted except by, or with the consent of, the Director of Public Prosecutions.

NOTES

Sub-s (2): reference to a level on the standard scale substituted by virtue of the Criminal Justice Act 1982, ss 38, 46.

5 Offences by corporations

(1)   Where an offence under this Act which has been committed by a body corporate is proved to have been committed with the consent or connivance of, or to be attributable to any neglect on the part of, any director, manager, secretary, or other similar officer of the body corporate, or of any person who was purporting to act in any such capacity, he as well as the body corporate shall be guilty of that offence and shall be liable to be proceeded against and punished accordingly.

(2)   Where the affairs of a body corporate are managed by its members, this section shall apply in relation to the acts or defaults of a member in connection with his functions of management as if he were a director of the body corporate.

6 Interpretation

(1)   In this Act, unless the context or subject matter otherwise requires,—
"acquire" includes hire;
"send" includes deliver, and "sender" shall be construed accordingly;
"unsolicited" means, in relation to goods sent to any person, that they are sent without any prior request made by him or on his behalf.

[(2)   For the purposes of this Act any invoice or similar document stating the amount of any payment and not complying with the requirements of regulations under section 3A of this Act applicable thereto shall be regarded as asserting a right to the payment.]

NOTES

Sub-s (2): substituted with savings by the Unsolicited Goods and Services (Amendment) Act 1975, s 2(2).

7 Citation, commencement and extent

(1)   This Act may be cited as the Unsolicited Goods and Services Act 1971.

(2)   This Act shall come into force at the expiration of three months beginning with the day on which it is passed.

(3)   This Act does not extend to Northern Ireland.

Supply of Goods (Implied Terms) Act 1973

(C 13)

An Act to amend the law with respect to the terms to be implied in contracts of sale of goods and hire-purchase agreements and on the exchange of goods for trading stamps, and with respect to the terms of conditional sale agreements: and for connected purposes

[18 April 1973]

Hire-purchase agreements

[8 Implied terms as to title

(1)   In every hire-purchase agreement, other than one to which subsection (2) below applies, there is—
   (a) an implied [term] on the part of the creditor that he will have a right to sell the goods at the time when the property is to pass; and
   (b) an implied [term] that—
      (i) the goods are free, and will remain free until the time when the property is to pass, from any charge or encumbrance not disclosed or known to the person to whom the goods are bailed or (in Scotland) hired before the agreement is made, and
      (ii) that person will enjoy quiet possession of the goods except so far as it may be disturbed by any person entitled to the benefit of any charge or encumbrance so disclosed or known.

(2)   In a hire-purchase agreement, in the case of which there appears from the agreement or is to be inferred from the circumstances of the agreement an intention that the creditor should transfer only such title as he or a third person may have, there is—
   (a) an implied [term] that all charges or encumbrances known to the creditor and not known to the person to whom the goods are bailed or hired have been disclosed to that person before the agreement is made; and
   (b) an implied [term] that neither—
      (i) the creditor; nor
      (ii) in a case where the parties to the agreement intend that any title which may be transferred shall be only such title as a third

person may have, that person; nor

(iii)   anyone claiming through or under the creditor or that third person otherwise than under a charge or encumbrance disclosed or known to the person to whom the goods are bailed or hired, before the agreement is made;

will disturb the quiet possession of the person to whom the goods are bailed or hired.

[(3)   As regards England and Wales and Northern Ireland, the term implied by subsection (1)(a) above is a condition and the terms implied by subsections (1)(b), (2)(a) and (2)(b) above are warranties.]]

NOTES

Commencement: 3 January 1995 (sub-s (3)); 19 May 1985 (sub-ss (1), (2)).

Substituted by the Consumer Credit Act 1974, s 192(3)(a), Sch 4, para 35.

Sub-ss (1), (2): words in square brackets substituted by the Sale and Supply of Goods Act 1994, s 7(1), Sch 2 para 4(1), (2)(a).

Sub-s (3): inserted by the Sale and Supply of Goods Act 1994, s 7(1), Sch 2, para 4(1), (2)(b).

[9   Bailing or hiring by description

(1)   Where under a hire-purchase agreement goods are bailed or (in Scotland) hired by description, there is an implied [term] that the goods will correspond with the description, and if under the agreement the goods are bailed or hired by reference to a sample as well as a description, it is not sufficient that the bulk of the goods corresponds with the sample if the goods do not also correspond with the description.

[(1A) As regards England and Wales and Northern Ireland, the term implied by subsection (1) above is a condition.]

(2)   Goods shall not be prevented from being bailed or hired by description by reason only that, being exposed for sale, bailment or hire, they are selected by the person to whom they are bailed or hired.]

NOTES

Commencement: 3 January 1995 (sub-s (1A)); 19 May 1985 (sub-ss (1), (2)).

Substituted by the Consumer Credit Act 1974, s 192(3)(a), Sch 4, para 35.

Sub-s (1): word in square brackets substituted by the Sale and Supply of Goods Act 1994, s 7(1), Sch 2, para 4(1), (3)(a).

Sub-s (1A): inserted by the Sale and Supply of Goods Act 1994, s 7(1), Sch 2, para 4(1), (3)(b).

[10 Implied undertakings as to quality or fitness

(1)   Except as provided by this section and section 11 below and subject to the provisions of any other enactment, including any enactment of the Parliament of Northern Ireland, or the Northern Ireland Assembly, there is no implied [term] as to the quality or fitness for any particular purpose of goods bailed or (in Scotland) hired under a hire-purchase agreement.

[(2)   Where the creditor bails or hires goods under a hire purchase agreement in the course of a business, there is an implied term that the goods supplied under the agreement are of satisfactory quality.

(2A)   For the purposes of this Act, goods are of satisfactory quality if they meet the standard that a reasonable person would regard as satisfactory, taking account of any description of the goods, the price (if relevant) and all the other relevant circumstances.

(2B)   For the purposes of this Act, the quality of goods includes their state and condition and the following (among others) are in appropriate cases aspects of the quality of goods—
  (a) fitness for all the purposes for which goods of the kind in question are commonly supplied,
  (b) appearance and finish,
  (c) freedom from minor defects,
  (d) safety, and
  (e) durability.

(2C)   The term implied by subsection (2) above does not extend to any matter making the quality of goods unsatisfactory—
  (a) which is specifically drawn to the attention of the person to whom the goods are bailed or hired before the agreement is made,
  (b) where that person examines the goods before the agreement is made, which that examination ought to reveal, or
  (c) where the goods are bailed or hired by reference to a sample, which would have been apparent on a reasonable examination of the sample]

(3)   Where the creditor bails or hires goods under a hire-purchase agreement in the course of a business and the person to whom the goods are bailed or hired, expressly or by implication, makes known—
   (a) to the creditor in the course of negotiations conducted by the creditor in relation to the making of the hire-purchase agreement, or
   (b) to a credit-broker in the course of negotiations conducted by that broker in relation to goods sold by him to the creditor before forming the subject matter of the hire-purchase agreement,

any particular purpose for which the goods are being bailed or hired, there is an implied [term] that the goods supplied under the agreement are reasonably fit for that purpose, whether or not that is a purpose for which such goods are commonly supplied, except where the circumstances show that the person to whom the goods are bailed or hired does not rely, or that it is unreasonable for him to rely, on the skill or judgment of the creditor or credit-broker.

(4)   An implied [term] as to quality or fitness for a particular purpose may be annexed to a hire-purchase agreement by usage.

(5)   The preceding provisions of this section apply to a hire-purchase agreement made by a person who in the course of a business is acting as agent for the creditor as they apply to an agreement made by the creditor in the course of a business, except where the creditor is not bailing or hiring in the course of a business and either the person to whom the goods are bailed or hired knows that fact or reasonable steps are taken to bring it to the notice of that person before the agreement is made.

(6)   In subsection (3) above and this subsection—
   (a) "credit-broker" means a person acting in the course of a business of credit brokerage.
   (b) "credit brokerage" means the effecting of introductions of individuals desiring to obtain credit—
      (i)   to persons carrying on any business so far as it relates to the provision of credit, or
      (ii) to other persons engaged in credit brokerage.]

[(7)   As regards England and Wales and Northern Ireland, the terms implied by subsections (2) and (3) above are conditions.]

NOTES

Commencement: 3 January 1995 (sub-ss (2), (2A), (2B), (2C), (7)); 19 May 1985 (sub-ss (1), (3)–(6)).

Substituted by the Consumer Credit Act 1974, s 192(3)(a), Sch 4, para 35.

Sub-ss (1), (3), (4): words in square brackets substituted by the Sale and Supply of Goods Act 1994, s 7(1), Sch 2, para 4(1), (4)(b).

Sub-ss (2), (2A), (2B), (2C): substituted, for sub-s (2) as originally enacted, by the Sale and Supply of Goods Act 1994, s 7(1), Sch 2, para 4(1), (4)(a).

Sub-s (7): added by the Sale and Supply of Goods Act 1994, s 7(1), Sch 2, para 4(1), (4)(c).

[11 Samples

[(1)] Where under a hire-purchase agreement goods are bailed or (in Scotland) hired by reference to a sample, there is an implied [term]—

   (a) that the bulk will correspond with the sample in quality; and

   (b) that the person to whom the goods are bailed or hired will have a reasonable opportunity of comparing the bulk with the sample; and

   (c) that the goods will be free from any defect, [making their quality unsatisfactory], which would not be apparent on reasonable examination of the sample.]

[(2) As regards England and Wales and Northern Ireland, the term implied by subsection (1) above is a condition.]

NOTES

Commencement: 3 January 1995 (sub-s (2)); 19 May 1985 (sub-s (1)).

Substituted by the Consumer Credit Act 1974, s 192(3)(a), Sch 4, para 35.

Sub-s (1): numbered as such and words in square brackets substituted by the Sale and Supply of Goods Act 1994, s 7(1), Sch 2, para 4(1), (5)(a)–(c).

Sub-s (2): inserted by the Sale and Supply of Goods Act 1994, s 7(1), Sch 2, para 4(1), (5)(d).

[11A Modification of remedies for breach of statutory condition in non-consumer cases

(1) Where in the case of a hire purchase agreement—

   (a) the person to whom goods are bailed would, apart from this subsection, have the right to reject them by reason of a breach on the part of the creditor of a term implied by section 9, 10 or 11(1)(a) or (c) above, but

   (b) the breach is so slight that it would be unreasonable for him to reject them, then, if the person to whom the goods are bailed does not deal as consumer, the breach is not to be treated as a breach of condition but may be treated as a breach of warranty.

(2)   This section applies unless a contrary intention appears in, or is to be implied from, the agreement.

(3)   It is for the creditor to show—
   (a) that a breach fell within subsection (1)(b) above, and
   (b) that the person to whom the goods were bailed did not deal as consumer.

(4)   The references in this section to dealing as consumer are to be construed in accordance with Part I of the Unfair Contract Terms Act 1977.

(5)   This section does not apply to Scotland.]

NOTES
   Commencement: 3 January 1995.
   Inserted by the Sale and Supply of Goods Act 1994, s 7(1), Sch 2, **para 4(1)**, (6).

[12   Exclusion of implied terms

An express term does not negative a term implied by this Act unless inconsistent with it.]

NOTES
   Commencement: 3 January 1995.
   Substituted by the Sale and Supply of Goods Act 1994, s 7(1), Sch 2, para 4(1), (7).

[14   Special provisions as to conditional sale agreements

(1)   [Section 11(4) of the Sale of Goods Act 1979] (whereby in certain circumstances a breach of a condition in a contract of sale is treated only as a breach of warranty) shall not apply to [a conditional sale agreement where the buyer deals as consumer within Part I of the Unfair Contract Terms Act 1977 ...].

(2)   In England and Wales and Northern Ireland a breach of a condition (whether express or implied) to be fulfilled by the seller under any such agreement shall be treated as a breach of warranty, and not as grounds for rejecting the goods and treating the agreement as repudiated, if (but only if) it would have fallen to be so treated had the condition been contained or implied in a corresponding hire-purchase agreement as a condition to be fulfilled by the creditor.]

NOTES

Substituted by the Consumer Credit Act 1974, s 192(3)(a), Sch 4, para 36.

Sub-s (1): words in first pair of square brackets substituted by the Sale of Goods Act 1979, s 63, Sch 2, para 16; words in second pair of square brackets substituted by the Unfair Contract Terms Act 1977, s 31(3), Sch 3; words omitted repealed by the Statute Law (Repeals) Act 1981.

[15   Supplementary

(1)   In sections 8 to 14 above and this section—

"business" includes a profession and the activities of any government department (including a Northern Ireland department), [or local or public authority];

"buyer" and "seller" includes a person to whom rights and duties under a conditional sale agreement have passed by assignment or operation of law;

. . . . .

"conditional sale agreement" means an agreement for the sale of goods under which the purchase price or part of it is payable by instalments, and the property in the goods is to remain in the seller (notwithstanding that the buyer is to be in possession of the goods) until such conditions as to the payment of instalments or otherwise as may be specified in the agreement are fulfilled;

. . . . .

"creditor" means the person by whom the goods are bailed or (in Scotland) hired under a hire-purchase agreement or the person to whom his rights and duties under the agreement have passed by assignment or operation of law; and

"hire-purchase agreement" means an agreement, other than a conditional sale agreement, under which—

(a) goods are bailed or (in Scotland) hired in return for periodical payments by the person to whom they are bailed or hired, and

(b) the property in the goods will pass to that person if the terms of the agreements are complied with and one or more of the following occurs—

    (i) the exercise of an option to purchase by that person,

    (ii) the doing of any other specified act by any party to the agreement,

    (iii) the happening of any other specified event.

(2) ...

(3) In section 14(2) above "corresponding hire-purchase agreement" means, in relation to a conditional sale agreement, a hire-purchase agreement relating to the same goods as the conditional sale agreement and made between the same parties and at the same time and in the same circumstances and, as nearly as may be, in the same terms as the conditional sale agreement.

(4) Nothing in sections 8 to 13 above shall prejudice the operation of any other enactment including any enactment of the Parliament of Northern Ireland or the Northern Ireland Assembly or any rule of law whereby any [term], other than one relating to quality or fitness, is to be implied in any hire-purchase agreement.]

NOTES

Substituted by the Consumer Credit Act 1974, s 192(3)(a), Sch 4, para 36.

Sub-s (1): in definition "business" words in square brackets substituted, and second definition omitted repealed, by the Unfair Contract Terms Act 1977, s 31(3), (4), Schs 3, 4; first definition omitted repealed by the Sale and Supply of Goods Act 1994, s 7(1), Sch 2 para 4(1), (9)(a), Sch 3.

Sub-s (2): repealed by the Sale and Supply of Goods Act 1994, s 7(1), Sch 2, para 4(1), (9)(b), Sch 3.

Sub-s (4): word in square brackets substituted by the Sale and Supply of Goods Act 1994, s 7(1), Sch 2 para 4(1), (9)(c).

Miscellaneous

18 Short title, citation, interpretation, commencement, repeal and saving

(1) This Act may be cited as the Supply of Goods (Implied Terms) Act 1973.

(2) ...

(3)   This Act shall come into operation at the expiration of a period of one month beginning with the date on which it is passed.

(4)   ...

(5)   This Act does not apply to contracts of sale or hire-purchase agreements made before its commencement.

NOTES

Sub-s (2): repealed by the Sale of Goods Act 1979, s 63(2), Sch 3.
Sub-s (4): repeals the Hire-Purchase Act 1965, ss 17–20, 29(3)(c), the Hire-Purchase (Scotland) Act 1965, ss 17–20, 29(3)(c) and the Hire-Purchase Act (Northern Ireland) 1966, ss 17–20, 29(3)(c).

Fair Trading Act 1973

(C 41)

An Act to provide for the appointment of a Director General of Fair Trading and of a Consumer Protection Advisory Committee, and to confer on the Director General and the Committee so appointed, on the Secretary of State, on the Restrictive Practices Court and on certain other courts new functions for the protection of consumers; to make provisions, in substitution for the Monopolies and Restrictive Practices (Inquiry and Control) Act 1948 and the Monopolies and Mergers Act 1965, for the matters dealt with in those Acts and related matters, including restrictive labour practices; to amend the Restrictive Trade Practices Act 1956 and the Restrictive Trade Practices Act 1968, to make provision for extending the said Act of 1956 to agreements relating to services, and to transfer to the Director General of Fair Trading the functions of the Registrar of Restrictive Trading Agreements; to make provision with respect to pyramid selling and similar trading schemes; to make new provision in place of section 30(2) to (4) of the Trade Descriptions Act 1968; and for purposes connected with those matters

[25 July 1973]

PART I
INTRODUCTORY

1   Director General of Fair Trading

(1)   The Secretary of State shall appoint an officer to be known as the Director General of Fair Trading (in this Act referred to as "the Director")

for the purpose of performing the functions assigned or transferred to the Director by or under this Act.

(2)   An appointment of a person to hold office as the Director shall not be for a term exceeding five years; but previous appointment to that office shall not affect eligibility for re-appointment.

(3)   The Director may at any time resign his office as the Director by notice in writing addressed to the Secretary of State; and the Secretary of State may remove any person from that office on the ground of incapacity or misbehaviour.

(4)   Subject to subsections (2) and (3) of this section, the Director shall hold and vacate office as such in accordance with the terms of his appointment.

(5)   The Director may appoint such staff as he may think fit, subject to the approval of the Minister for the Civil Service as to numbers and as to terms and conditions of service.

(6)   The provisions of Schedule 1 to this Act shall have effect with respect to the Director.

2   General functions of Director

(1)   Without prejudice to any other functions assigned or transferred to him by or under this Act, it shall be the duty of the Director, so far as appears to him to be practicable from time to time,—

(a) to keep under review the carrying on of commercial activities in the United Kingdom which relate to goods supplied to consumers in the United Kingdom or produced with a view to their being so supplied, or which relate to services supplied for consumers in the United Kingdom, and to collect information with respect to such activities, and the persons by whom they are carried on, with a view to his becoming aware of, and ascertaining the circumstances relating to, practices which may adversely affect the economic interests of consumers in the United Kingdom, and

(b) to receive and collate evidence becoming available to him with respect to such activities as are mentioned in the preceding paragraph and which appears to him to be evidence of practices which may adversely affect the interests (whether they are economic interests or interests with respect to health, safety or other matters) of consumers in the United Kingdom.

(2)   It shall also be the duty of the Director, so far as appears to him to be practicable from time to time, to keep under review the carrying on of commercial activities in the United Kingdom, and to collect information with respect to those activities, and the persons by whom they are carried on, with a view to his becoming aware of, and ascertaining the circumstances relating to, monopoly situations or uncompetitive practices.

(3)   It shall be the duty of the Director, where either he considers it expedient or he is requested by the Secretary of State to do so,—

    (a) to give information and assistance to the Secretary of State with respect to any of the matters in respect of which the Director has any duties under subsections (1) and (2) of this section, or

    (b) subject to the provisions of Part II of this Act in relation to recommendations under that Part of this Act, to make recommendations to the Secretary of State as to any action which in the opinion of the Director it would be expedient for the Secretary of State or any other Minister to take in relation to any of the matters in respect of which the Director has any such duties.

(4)   It shall also be the duty of the Director to have regard to evidence becoming available to him with respect to any course of conduct on the part of a person carrying on a business which appears to be conduct detrimental to the interests of consumers in the United Kingdom and (in accordance with the provisions of Part III of this Act) to be regarded as unfair to them, with a view to considering what action (if any) he should take under Part III of this Act.

(5)   It shall be the duty of the Director to have regard to the needs of regional development and to the desirability of dispersing administrative offices from London in making decisions on the location of offices for his staff.

3 Consumer Protection Advisory Committee

(1)   There shall be established an advisory committee to be called the Consumer Protection Advisory Committee (in this Act referred to as "the Advisory Committee") for the purpose of performing the functions assigned to that Committee by Part II of this Act.

(2)   Subject to subsection (6) of this section, the Advisory Committee shall consist of not less than ten and not more than fifteen members, who shall be appointed by the Secretary of State.

(3)   The Secretary of State may appoint persons to the Advisory Committee either as full-time members or as part-time members.

(4)   Of the members of the Advisory Committee, the Secretary of State shall appoint one to be chairman and one to be deputy chairman of the Advisory Committee.

(5)   In appointing persons to be members of the Advisory Committee, the Secretary of State shall have regard to the need for securing that the Advisory Committee will include—

(a) one or more persons appearing to him to be qualified to advise on practices relating to goods supplied to consumers in the United Kingdom or produced with a view to their being so supplied, or relating to services supplied for consumers in the United Kingdom, by virtue of their knowledge of or experience in the supply (whether to consumers or not) of such goods or by virtue of their knowledge of or experience in the supply of such services;

(b) one or more persons appearing to him to be qualified to advise on such practices as are mentioned in the preceding paragraph by virtue of their knowledge of or experience in the enforcement of the [Weights and Measures Act 1985] or the Trade Descriptions Act 1968 or other similar enactments; and

(c) one or more persons appearing to him to be qualified to advise on such practices by virtue of their knowledge of or experience in organisations established, or activities carried on for the protection of consumers.

(6)   The Secretary of State may by order made by statutory instrument increase the maximum number of members of the Advisory Committee to such number as he may think fit.

(7)   The provisions of Schedule 2 to this Act shall have effect with respect to the Advisory Committee.

NOTES

Sub-s (5): words in square brackets substituted by the Weights and Measures Act 1985, s 97, Sch 12, para 6.

12   Powers of Secretary of State in relation to functions of Director

(1)   The Secretary of State may give general directions indicating considerations to which the Director should have particular regard in determining the order of priority in which—

(a) matters are to be brought under review in the performance of his duty under section 2(1) of this Act, or

(b) classes of goods or services are to be brought under review by him for the purpose of considering whether a monopoly situation exists or may exist in relation to them.

(2)   The Secretary of State may also give general directions indicating—

(a) considerations to which in cases where it appears to the Director that a practice may adversely affect the interests of consumers in the United Kingdom, he should have particular regard in determining whether to make a recommendation to the Secretary of State under section 2(3)(b) of this Act, or

(b) considerations to which, in cases where it appears to the Director that a consumer trade practice may adversely affect the economic interests of consumers in the United Kingdom, he should have particular regard in determining whether to make a reference to the Advisory Committee under Part II of this Act, or

(c) considerations to which, in cases where it appears to the Director that a monopoly situation exists or may exist, he should have particular regard in determining whether to make a monopoly reference to the Commission under Part IV of this Act.

(3)   The Secretary of State, on giving any directions under this section, shall arrange for those directions to be published in such manner as the Secretary of State thinks most suitable in the circumstances.

PART II
REFERENCES TO CONSUMER PROTECTION ADVISORY COMMITTEE

General provisions

13   Meaning of "consumer trade practice"

In this Act "consumer trade practice" means any practice which is for the time being carried on in connection with the supply of goods (whether by way of sale or otherwise) to consumers or in connection with the supply of services for consumers and which relates—

(a) to the terms or conditions (whether as to price or otherwise) on or subject to which goods or services are or are sought to be supplied, or

(b) to the manner in which those terms or conditions are communicated to persons to whom goods are or are sought to be supplied or for whom services are or are sought to be supplied, or,

(c) to promotion (by advertising, labelling or marking of goods, canvassing or otherwise) of the supply of goods or of the supply of services, or

(d) to methods of salesmanship employed in dealing with consumers, or

(e) to the way in which goods are packed or otherwise got up for the purpose of being supplied, or

(f) to methods of demanding or securing payment for goods or services supplied.

14   General provisions as to references to Advisory Committee

(1)   Subject to sections 15 and 16 of this Act, the Secretary of State or any other Minister or the Director may refer to the Advisory Committee the question whether a consumer trade practice specified in the reference adversely affects the economic interests of consumers in the United Kingdom.

(2)   The Secretary of State or any other Minister by whom a reference is made under this section shall transmit a copy of the reference to the Director.

(3)   On any reference made to the Advisory Committee under this section the Advisory Committee shall consider the question so referred to them and shall prepare a report on that question and (except as otherwise provided by section 21(3) of this Act) submit that report to the person by whom the reference was made.

(4)   Subject to the provisions of section 133 of this Act, it shall be the duty of the Director, where he is requested by the Advisory Committee to do so for the purpose of assisting the Committee in carrying out an investigation on a reference made to them under this section, to give to the Committee—

(a) any information which is in his possession and which relates to matters falling within the scope of the investigation, and

(b) any other assistance which the Committee may require, and which it is within his power to give, in relation to any such matters.

(5)   The Advisory Committee shall transmit to the Secretary of State a copy of every report which is made by them under this section to a person

other than the Secretary of State, and shall transmit to the Director a copy of every report which is made by them under this section to a person other than the Director.

15 Exclusion from s 14 in respect of certain services

No reference under section 14 of this Act shall be made to the Advisory Committee by the Secretary of State or by any other Minister or by the Director if it appears to him—

(a) that the consumer trade practice in question is carried on in connection only with the supply of services of a description specified in Schedule 4 to this Act, and

(b) that a monopoly situation exists or may exist in relation to the supply of services of that description.

16 Restriction on references under s 14 in respect of certain goods and services

(1) No reference under section 14 of this Act shall be made to the Advisory Committee by the Director except with the consent of the appropriate Minister, if it appears to the Director that the consumer trade practice in question—

(a) is carried on in connection only with the supply, by a body corporate to which this section applies, of goods or services of a description specified in Part I of Schedule 5 to this Act, ...

(b) ... [or

(c) is carried on in connection only with the supply of electricity by a licence holder within the meaning of Part I of the Electricity Act 1989] [or Part II of the Electricity (Northern Ireland) Order 1992] [or

(d) is carried on in connection only with the conveyance or supply of gas by a licence holder within the meaning of Part II of the Gas (Northern Ireland) Order 1996;].

(2) This section applies to any body corporate which fulfils the following conditions, that is to say—

(a) that the affairs of the body corporate are managed by its members, and

(b) that by virtue of an enactment those members are appointed by a Minister;

and in this section "Minister" includes a Minister of the Government of Northern Ireland, and "the appropriate Minister", in relation to a body corporate, means the Minister by whom members of that body corporate are appointed.

[(2A) In this section "the appropriate Minister", in relation to a licence holder within the meaning of Part I of the Electricity Act 1989, means the Secretary of State responsible for matters relating to energy.]

[(2B) In this section "the appropriate Minister" in relation to a licence holder within the meaning of Part II of the Electricity (Northern Ireland) Order 1992 [or Part II of the Gas (Northern Ireland) Order 1996], means the Department of Economic Development.]

(3)   The Secretary of State may by order made by statutory instrument vary any of the provisions of Schedule 5 to this Act, either by adding one or more further entries or by altering or deleting any entry for the time being contained in it; and any reference in this Act to that Schedule shall be construed as a reference to that Schedule as for the time being in force.

NOTES

Sub-s (1): para (b) and word "or" immediately preceding it repealed by the Telecommunications Act 1984, s 109(6), Sch 7, Pt I, para (c) and word "or" immediately preceding it added by the Electricity Act 1989, s 112(1), Sch 16, para 16(1), (2) and words in square brackets in para (c) inserted by the Electricity (Northern Ireland) Order 1992, SI 1992/231, art 95(1), Sch 12, para 9(a), para (d) and word "or" immediately preceding it added by the Gas (Northern Ireland) Order 1996, SI 1996/275, art 71(1), Sch 6.

Sub-s (2A): inserted by the Electricity Act 1989, s 112(1), Sch 16, para 16(1), (3).

Sub-s (2B): inserted by the Electricity (Northern Ireland) Order 1992, SI 1992/231, art 95(1), Sch 12, para 9(b), and words in square brackets therein inserted by the Gas (Northern Ireland) Order 1996, SI 1996/275, art 71(1).

17   Reference to Advisory Committee proposing recommendation to Secretary of State to make an order

(1)   This section applies to any reference made to the Advisory Committee by the Director under section 14 of this Act which includes proposals in accordance with the following provisions of this section.

(2)   Where it appears to the Director that a consumer trade practice has the effect, or is likely to have the effect,—
   (a) of misleading consumers as to, or withholding from them adequate information as to, or an adequate record of, their rights and obligations under relevant consumer transactions, or

(b) of otherwise misleading or confusing consumers with respect to any matter in connection with relevant consumer transactions, or

(c) of subjecting consumers to undue pressure to enter into relevant consumer transactions, or

(d) of causing the terms or conditions, on or subject to which consumers enter into relevant consumer transactions, to be so adverse to them as to be inequitable,

any reference made by the Director under section 14 of this Act with respect to that consumer trade practice may, if the Director thinks fit, include proposals for recommending to the Secretary of State that he should exercise his powers under the following provisions of this Part of this Act with respect to that consumer trade practice.

(3)   A reference to which this section applies shall state which of the effects specified in subsection (2) of this section it appears to the Director that the consumer trade practice in question has or is likely to have.

(4)   Where the Director makes a reference to which this section applies, he shall arrange for it to be published in full in the London, Edinburgh and Belfast Gazettes.

(5)   In this Part of this Act "relevant consumer transaction", in relation to a consumer trade practice, means any transaction to which a person is, or may be invited to become, a party in his capacity as a consumer in relation to that practice.

18   No such recommendation to be made except in pursuance of reference to which s 17 applies

The Director shall not make any recommendation to the Secretary of State to exercise his powers under the following provisions of this Part of this Act except by way of making a reference to the Advisory Committee to which section 17 of this Act applies.

19   Scope of recommendation proposed in reference to which s 17 applies

(1)   In formulating any proposals which, in accordance with the provisions of section 17 of this Act, are included in a reference to which that section applies, the Director shall have regard—

(a) to the particular respects in which it appears to him that the consumer trade practice specified in the reference may adversely affect the economic interests of consumers in the United Kingdom, and

(b) to the class of relevant consumer transactions, or the classes (whether being some or all classes) of such transactions in relation to which it appears to him that the practice may so affect those consumers;

and the proposed recommendation shall be for an order making, in relation to relevant consumer transactions of that class or of those classes, as the case may be, such provision specified in the proposals as the Director may consider requisite for the purpose of preventing the continuance of that practice, or causing it to be modified, in so far as it may so affect those consumers in those respects.

(2) Without prejudice to the generality of the preceding subsection, for the purpose mentioned in that subsection any such proposals may in particular recommend the imposition by such an order of prohibitions or requirements of any description specified in Schedule 6 to this Act.

(3) In that Schedule, in its application to any such proposals, "the specified consumer trade practice" means the consumer trade practice specified in the reference in which the proposals are made, "specified consumer transactions" means transactions which are relevant consumer transactions in relation to that consumer trade practice and are of a description specified in the proposals, and "specified" (elsewhere than in those expressions) means specified in the proposals.

20 Time-limit and quorum for report on reference to which s 17 applies

(1) A report of the Advisory Committee on a reference to which section 17 of this Act applies shall not have effect, and no action shall be taken in relation to it under the following provisions of this Part of this Act, unless the report is made before the end of the period of three months beginning with the date of the reference or of such further period or periods (if any) as may be allowed by the Secretary of State.

(2) The Secretary of State shall not allow any further period for such a report except after consulting the Advisory Committee and considering

any representations made by them with respect to the proposal to allow a further period.

(3)   No such further period shall be longer than three months; but (subject to subsection (2) of this section) two or more further periods may be allowed in respect of the same reference.

(4)   The quorum necessary for a meeting of the Advisory Committee held for the final settling of a report of the Committee on a reference to which section 17 of this Act applies shall be not less than two-thirds of the members of the Committee.

21   Report of Advisory Committee on reference to which s 17 applies

(1)   A report of the Advisory Committee on a reference to which section 17 of this Act applies shall state the conclusions of the Committee on the questions—
  (a) whether the consumer trade practice specified in the reference adversely affects the economic interests of consumers in the United Kingdom, and
  (b) if so, whether it does so by reason, or partly by reason, that it has or is likely to have such one or more of the effects specified in section 17(2) of this Act as are specified in the report.

(2)   If, in their conclusions set out in such a report, the Advisory Committee find that the consumer trade practice specified in the reference does adversely affect the economic interests of consumers in the United Kingdom, and does so wholly or partly for the reason mentioned in subsection (1)(b) of this section, the report shall state whether the Committee—
  (a) agree with the proposals set out in the reference, or
  (b) would agree with those proposals if they were modified in a manner specified in the report, or
  (c) disagree with the proposals and do not desire to suggest any such modifications.

(3)   Every report of the Advisory Committee on a reference to which section 17 of this Act applies shall be made to the Secretary of State, and shall set out in full the reference on which it is made.

Order in pursuance of report of Advisory Committee

22 Order of Secretary of State in pursuance of report on reference to which s 17 applies

(1)   The provisions of this section shall have effect where a report of the Advisory Committee on a reference to which section 17 of this Act applies has been laid before Parliament in accordance with the provisions of Part VII of this Act, and the report states that the Committee—
  (a) agree with the proposals set out in the reference, or
  (b) would agree with those proposals if they were modified in a manner specified in the report.

(2)   In the circumstances mentioned in the preceding subsection, the Secretary of State may, if he thinks fit, by an order made by statutory instrument make such provision as—
  (a) in a case falling within paragraph (a) of the preceding subsection, is in his opinion appropriate for giving effect to the proposals set out in the reference, or
  (b) in a case falling within paragraph (b) of that subsection, is in his opinion appropriate for giving effect either to the proposals as set out in the reference or to those proposals as modified in the manner specified in the report, as the Secretary of State may in his discretion determine.

(3)   Any such order may contain such supplementary or incidental provisions as the Secretary of State may consider appropriate in the circumstances; and (without prejudice to the generality of this subsection) any such order may restrict the prosecution of offences under the next following section in respect of contraventions of the order where those contraventions also constitute offences under another enactment.

(4)   No such order, and no order varying or revoking any such order, shall be made under this section unless a draft of the order has been laid before Parliament and approved by a resolution of each House of Parliament.

23 Penalties for contravention of order under s 22

Subject to the following provisions of this Part of this Act, any person who contravenes a prohibition imposed by an order under section 22 of this Act, or who does not comply with a requirement imposed by such an order which applies to him, shall be guilty of an offence and shall be liable—

(a) on summary conviction, to a fine not exceeding [the prescribed sum];

(b) on conviction on indictment, to a fine or to imprisonment for a term not exceeding two years or both.

Reference to "the prescribed sum" substituted by virtue of the Magistrates' Courts Act 1980, s 32(2).

24 Offences due to default of other person

Where the commission by any person of an offence under section 23 of this Act is due to the act or default of some other person, that other person shall be guilty of the offence, and a person may be charged with and convicted of the offence by virtue of this section whether or not proceedings are taken against the first-mentioned person.

25 Defences in proceedings under s 23

(1) In any proceedings for an offence under section 23 of this Act it shall, subject to subsection (2) of this section, be a defence for the person charged to prove—

(a) that the commission of the offence was due to a mistake, or to reliance on information supplied to him, or to the act or default of another person, an accident or some other cause beyond his control, and

(b) that he took all reasonable precautions and exercised all due diligence to avoid the commission of such an offence by himself or any person under his control.

(2) If in any case the defence provided by the preceding subsection involves the allegation that the commission of the offence was due to the act or default of another person or to reliance on information supplied by another person, the person charged shall not, without leave of the court, be entitled to rely on that defence unless, within a period ending seven clear days before the hearing, he has served on the prosecutor a notice in writing, giving such information identifying or assisting in the identification of that other person as was then in his possession.

(3) In proceedings for an offence under section 23 of this Act committed by the publication of an advertisement, it shall be a defence for the person charged to prove that he is a person whose business it is to publish or arrange for the publication of advertisements, and that he received the

advertisement for publication in the ordinary course of business and did not know and had no reason to suspect that its publication would amount to an offence under section 23 of this Act.

26 Limitation of effect of orders under s 22

A contract for the supply of goods or services shall not be void or unenforceable by reason only of a contravention of an order made under section 22 of this Act; and, subject to the provisions of section 33 of the Interpretation Act 1889 (which relates to offences under two or more laws), the provisions of this Part of this Act shall not be construed as—

(a) conferring a right of action in any civil proceedings (other than proceedings for the recovery of a fine) in respect of any contravention of such an order, or

(b) affecting any restriction imposed by or under any other enactment, whether public, local or private, or

(c) derogating from any right of action or other remedy (whether civil or criminal) in proceedings instituted otherwise than under this Part of this Act.

Enforcement of orders

27 Enforcing authorities

(1) It shall be the duty of every local weights and measures authority to enforce within their area the provisions of any order made under section 22 of this Act ...

(2) ...

NOTES

Sub-s (1): words omitted repealed by the Weights and Measures Act 1985, s 98(1), Sch 13, Pt I.

Sub-s (2): applies to Scotland only.

28 Power to make test purchases

A local weights and measures authority may make, or may authorise any of their officers to make on their behalf, such purchases of goods, and may authorise any of their officers to obtain such services, as may be expedient for the purpose of determining whether or not the provisions of any order made under section 22 of this Act are being complied with.

29 Power to enter premises and inspect and seize goods and documents

(1)   A duly authorised officer of a local weights and measures authority, or a person duly authorised in writing by the Secretary of State, may at all reasonable hours, and on production, if required, of his credentials, exercise the following powers, that is to say—

   (a) he may, for the purpose of ascertaining whether any offence under section 23 of this Act has been committed, inspect any goods and enter any premises other than premises used only as a dwelling;

   (b) if he has reasonable cause to suspect that an offence under that section has been committed, he may, for the purpose of ascertaining whether it has been committed, require any person carrying on a business or employed in connection with a business to produce any books or documents relating to the business and may take copies of, or of any entry in, any such book or document;

   (c) if he has reasonable cause to believe that such an offence has been committed, he may seize and detain any goods for the purpose of ascertaining, by testing or otherwise, whether the offence has been committed;

   (d) he may seize and detain any goods or documents which he has reason to believe may be required as evidence in proceedings for such an offence;

   (e) he may, for the purpose of exercising his powers under this subsection to seize goods, but only if and to the extent that it is reasonably necessary in order to secure that the provisions of an order made under section 22 of this Act are duly observed, require any person having authority to do so to break open any container or open any vending machine and, if that person does not comply with the requirement, he may do so himself.

(2)   A person seizing any goods or documents in the exercise of his powers under this section shall inform the person from whom they are seized and, in the case of goods seized from a vending machine, the person whose name and address are stated on the machine as being the proprietor's or, if no name and address are so stated, the occupier of the premises on which the machine stands or to which it is affixed.

(3)   If a justice of the peace, on sworn information in writing,—

   (a) is satisfied that there is reasonable ground to believe either—

      (i)   that any goods, books or documents which a person has power under this section to inspect are on any premises and that their inspection is likely to disclose evidence of the commission of an offence under section 23 of this Act, or

      (ii) that any offence under section 23 has been, is being or is about to be committed on any premises, and

   (b) is also satisfied either—

      (i)   that admission to the premises has been or is likely to be refused and that notice of intention to apply for a warrant under this subsection has been given to the occupier, or

      (ii) that an application for admission, or the giving of such a notice, would defeat the object of the entry or that the premises are unoccupied or that the occupier is temporarily absent, and it might defeat the object of the entry to await his return,

the justice may by warrant under his hand, which shall continue in force for a period of one month, authorise any such officer or other person as is mentioned in subsection (1) of this section to enter the premises, if need be by force.

...

(4)   A person entering any premises by virtue of this section may take with him such other persons and such equipment as may appear to him necessary; and on leaving any premises which he has entered by virtue of a warrant under subsection (3) of this section he shall, if the premises are unoccupied or the occupier is temporarily absent, leave them as effectively secured against trespassers as he found them.

(5)   Nothing in this section shall be taken to compel the production by a barrister, advocate or solicitor of a document containing a privileged communication made by or to him in that capacity or to authorise the taking of possession of any such document which is in his possession.

NOTES

Sub-s (3): words omitted apply to Scotland only.

30   Offences in connection with exercise of powers under s 29

(1)   Subject to subsection (6) of this section, any person who—

(a) wilfully obstructs any such officer or person as is mentioned in subsection (1) of section 29 of this Act acting in the exercise of any powers conferred on him by or under that section, or

(b) wilfully fails to comply with any requirement properly made to him by such an officer or person under that section, or

(c) without reasonable cause fails to give such an officer or person so acting any other assistance or information which he may reasonably require of him for the purpose of the performance of his functions under this Part of this Act,

shall be guilty of an offence.

(2) If any person, in giving any such information as is mentioned in subsection (1)(c) of this section, makes any statement which he knows to be false, he shall be guilty of an offence.

(3) If any person discloses to any other person—
(a) any information with respect to any manufacturing process or trade secret obtained by him in premises which he has entered by virtue of section 29 of this Act, or

(b) any information obtained by him under that section or by virtue of subsection (1) of this section,

he shall, unless the disclosure was made in the performance of his duty, be guilty of an offence.

(4) If any person who is neither a duly authorised officer of a weights and measures authority nor a person duly authorised in that behalf by the Secretary of State purports to act as such under section 29 of this Act or under this section, he shall be guilty of an offence.

(5) Any person guilty of an offence under subsection (1) of this section shall be liable on summary conviction to a fine not exceeding [level 3 on the standard scale]; and any person guilty of an offence under subsection (2), subsection (3) or subsection (4) of this section shall be liable—
(a) on summary conviction, to a fine not exceeding [the prescribed sum];

(b) on conviction on indictment, to a fine or to imprisonment for a term not exceeding two years or to both.

(6) Nothing in this section shall be construed as requiring a person to answer any question or give any information if to do so might incriminate that person or (where that person is married) the husband or wife of that person.

NOTES

Sub-s (5): reference to "level 3 on the standard scale" substituted by virtue of the Criminal Justice Act 1982, ss 38, 46, and reference to "the prescribed sum" in para (a) substituted by virtue of the Magistrates' Courts Act 1980, s 32(2).

31 Notice of test

Where any goods seized or purchased by a person in pursuance of this Part of this Act are submitted to a test, then—

(a) if the goods were seized, he shall inform any such person as is mentioned in section 29(2) of this Act of the result of the test;

(b) if the goods were purchased and the test leads to the institution of proceedings for an offence under section 23 of this Act, he shall inform the person from whom the goods were purchased, or, in the case of goods sold through a vending machine, the person mentioned in relation to such goods in section 29(2) of this Act, of the result of the test;

and where, as a result of the test, proceedings for an offence under section 23 of this Act are instituted against any person, he shall allow that person to have the goods tested on his behalf if it is reasonably practicable to do so.

32 Compensation for loss in respect of goods seized under s 29

(1)   Where in the exercise of his powers under section 29 of this Act a person seizes and detains any goods, and their owner suffers loss by reason of their being seized or by reason that the goods, during the detention, are lost or damaged or deteriorate, unless the owner is convicted of an offence under section 23 of this Act committed in relation to the goods, the appropriate authority shall be liable to compensate him for the loss so suffered.

(2)   Any disputed question as to the right to or the amount of any compensation payable under this section shall be determined by arbitration and, in Scotland, by a single arbiter appointed, failing agreement between the parties, by the sheriff.

(3)   In this section "the appropriate authority"—

(a) in relation to goods seized by an officer of a local weights and measures authority, means that authority, and

(b) in any other case, means the Secretary of State.

PART III
ADDITIONAL FUNCTIONS OF DIRECTOR FOR PROTECTION OF
CONSUMERS

34 Action by Director with respect to course of conduct detrimental to interests of consumers

(1)   Where it appears to the Director that the person carrying on a business has in the course of that business persisted in a course of conduct which—

    (a) is detrimental to the interests of consumers in the United Kingdom, whether those interests are economic interests or interests in respect of health, safety or other matters, and

    (b) in accordance with the following provisions of this section is to be regarded as unfair to consumers,

the Director shall use his best endeavours, by communication with that person or otherwise, to obtain from him a satisfactory written assurance that he will refrain from continuing that course of conduct and from carrying on any similar course of conduct in the course of that business.

(2)   For the purposes of subsection (1)(b) of this section a course of conduct shall be regarded as unfair to consumers if it consists of contraventions of one or more enactments which impose duties, prohibitions or restrictions enforceable by criminal proceedings, whether any such duty, prohibition or restriction is imposed in relation to consumers as such or not and whether the person carrying on the business has or has not been convicted of any offence in respect of any such contravention.

(3)   A course of conduct on the part of the person carrying on a business shall also be regarded for those purposes as unfair to consumers if it consists of things done, or omitted to be done, in the course of that business in breach of contract or in breach of a duty (other than a contractual duty) owed to any person by virtue of any enactment or rule of law and enforceable by civil proceedings, whether (in any such case) civil proceedings in respect of the breach of contract or breach of duty have been brought or not.

(4)   For the purpose of determining whether it appears to him that a person has persisted in such a course of conduct as is mentioned in subsection (1) of this section, the Director shall have regard to either or both of the following, that is to say—

    (a) complaints received by him, whether from consumers or from other persons;

(b) any other information collected by or furnished to him, whether by virtue of this Act or otherwise.

35 Proceedings before Restrictive Practices Court

If, in the circumstances specified in subsection (1) of section 34 of this Act,—
  (a) the Director is unable to obtain from the person in question such an assurance as is mentioned in that subsection, or
  (b) that person has given such an assurance and it appears to the Director that he has failed to observe it,

the Director may bring proceedings against him before the *Restrictive Practices Court*.

NOTES

Words in italics repealed, subject to a saving, by the Competition Act 1998, s 74(1), (2), Sch 12, para 1(1), (5), Sch 13, Pt V, as from a day to be appointed under s 76(3) thereof.

36 Evidence in proceedings under s 35

(1)   For the purposes of section 11 of the Civil Evidence Act 1968, section 10 of the Law Reform (Miscellaneous Provisions) (Scotland) Act 1968 or section 7 of the Civil Evidence Act (Northern Ireland) 1971 (each of which relates to convictions as evidence in civil proceedings), proceedings under section 35 of this Act shall (without prejudice to the generality of the relevant definition) be taken to be civil proceedings within the meaning of the Act in question.

(2)   Where in any proceedings under section 35 of this Act the Director alleges such a breach of contract or breach of duty as is mentioned in section 34(3) of this Act, a judgment of any court given in civil proceedings, which includes a finding that the breach of contract or breach of duty in question was committed,—
  (a) shall be admissible in evidence for the purpose of proving the breach of contract or breach of duty, and
  (b) shall, unless the contrary is proved, be taken to be sufficient evidence that the breach of contract or breach of duty was committed.

(3)   For the purposes of subsection (2) of this section no account shall be taken of a judgment given in any civil proceedings if it has subsequently

been reversed on appeal, or has been varied on appeal so as to negative the finding referred to in that subsection.

(4)   In subsection (1) of this section "the relevant definition" means section 18(1) of the Civil Evidence Act 1968, section 17(1) of the Law Reform (Miscellaneous Provisions) (Scotland) Act 1968 or section 14(1) of the Civil Evidence Act (Northern Ireland) 1971, as the case may be.

37   Order of, or undertaking given to, Court in proceedings under s 35

(1)   Where in any proceedings before the *Restrictive Practices Court* under section 35 of this Act—
  (a) the Court finds that the person against whom the proceedings are brought (in this section referred to as "the respondent") has in the course of a business carried on by him persisted in such a course of conduct as is mentioned in section 34(1) of this Act, and
  (b) the respondent does not give an undertaking to the Court under subsection (3) of this section which is accepted by the Court, and
  (c) it appears to the Court that, unless an order is made against the respondent under this section, he is likely to continue that course of conduct or to carry on a similar course of conduct,

the Court may make an order against the respondent under this section.

(2)   An order of the Court under this section shall (with such degree of particularity as appears to the Court to be sufficient for the purposes of the order) indicate the nature of the course of conduct to which the finding of the Court under subsection (1)(a) of this section relates, and shall direct the respondent—
  (a) to refrain from continuing that course of conduct, and
  (b) to refrain from carrying on any similar course of conduct in the course of his business.

(3)   Where in any proceedings under section 35 of this Act the Court makes such a finding as is mentioned in subsection (1)(a) of this section, and the respondent offers to give to the Court an undertaking either—
  (a) to refrain as mentioned in paragraphs (a) and (b) of subsection (2) of this section, or
  (b) to take particular steps which, in the opinion of the Court, would suffice to prevent a continuance of the course of conduct to which the complaint relates and to prevent the carrying on by the

respondent of any similar course of conduct in the course of his business,

the Court may, if it thinks fit, accept that undertaking instead of making an order under this section.

NOTES

Sub-s (1): words in italics repealed, subject to a saving, by the Competition Act 1998, s 74(1), (2), Sch 12, para 1(1), (5), Sch 13, Pt V, as from a day to be appointed under s 76(3) thereof.

38 Provisions as to persons consenting to or conniving at courses of conduct detrimental to interests of consumers

(1)   The provisions of this section shall have effect where it appears to the Director—
(a) that a body corporate has in the course of a business carried on by that body persisted in such a course of conduct as is mentioned in section 34(1) of this Act, and
(b) that the course of conduct in question has been so persisted in with the consent or connivance of a person (in this and the next following section referred to as "the accessory") who at a material time fulfilled the relevant conditions in relation to that body.

(2)   For the purposes of this section a person shall be taken to fulfil the relevant conditions in relation to a body corporate at any time if that person either—
(a) is at that time a director, manager, secretary or other similar officer of the body corporate or a person purporting to act in any such capacity, or
(b) whether being an individual or a body of persons, corporate or unincorporate, has at that time a controlling interest in that body corporate.

(3)   If, in the circumstances specified in subsection (1) of this section,—
(a) the Director has used his best endeavours to obtain from the accessory such an assurance as is mentioned in the next following subsection and has been unable to obtain such an assurance from him, or
(b) the accessory has given such an assurance to the Director and it appears to the Director that he has failed to observe it,

the Director may bring proceedings against the accessory before the *Restrictive Practices Court.*

(4)   The assurance referred to in subsection (3) of this section is a satisfactory written assurance given by the accessory that he will refrain—
  (a) from continuing to consent to or connive at the course of conduct in question;
  (b) from carrying on any similar course of conduct in the course of any business which may at any time be carried on by him; and
  (c) from consenting to or conniving at the carrying on of any such course of conduct by any other body corporate in relation to which, at any time when that course of conduct is carried on, he fulfils the relevant conditions.

(5)   Proceedings may be brought against the accessory under this section whether or not any proceedings are brought under section 35 of this Act against the body corporate referred to in subsection (1) of this section.

(6)   Section 36 of this Act shall have effect in relation to proceedings under this section as it has effect in relation to proceedings under section 35 of this Act.

(7)   For the purposes of this section a person (whether being an individual or a body of persons, corporate or unincorporate) has a controlling interest in a body corporate if (but only if) that person can, directly or indirectly, determine the manner in which one-half of the votes which could be cast at a general meeting of the body corporate are to be cast on matters, and in circumstances, not of such a description as to bring into play any special voting rights or restrictions on voting rights.

NOTES

Sub-s (3): words in italics repealed, subject to a saving, by the Competition Act 1998, s 74(1), (2), Sch 12, para 1(1), (5), Sch 13, Pt V, as from a day to be appointed under s 76(3) thereof.

39 Order of, or undertaking given to, Court in proceedings under s 38

(1)   Where in any proceedings brought against the accessory before the *Restrictive Practices Court* under section 38 of this Act—
  (a) the Court finds that the conditions specified in paragraphs (a) and (b) of subsection (1) of that section are fulfilled in the case of the accessory, and

(b) the accessory does not give an undertaking to the Court under subsection (3) of this section which is accepted by the Court, and

(c) it appears to the Court that, unless an order is made against the accessory under this section, it is likely that he will not refrain from acting in one or more of the ways mentioned in paragraphs (a) to (c) of subsection (4) of that section,

the Court may make an order against the accessory under this section.

(2)   An order of the Court under this section shall (with such degree of particularity as appears to the Court to be sufficient for the purposes of the order) indicate the nature of the course of conduct to which the finding of the Court under subsection (1)(a) of this section relates, and shall direct the accessory, in relation to the course of conduct so indicated, to refrain from acting in any of the ways mentioned in paragraphs (a) to (c) of subsection (4) of section 38 of this Act.

(3)   Where in any proceedings under section 38 of this Act the Court makes such a finding as is mentioned in subsection (1)(a) of this section, and the accessory offers to give to the Court an undertaking either—

(a) to refrain from acting in any of the ways mentioned in paragraphs (a) to (c) of subsection (4) of that section, or

(b) to take particular steps which, in the opinion of the Court, would suffice to prevent him from acting in any of those ways,

the Court may, if it thinks fit, accept that undertaking instead of making an order under this section.

NOTES

Sub-s (1): words in italics repealed, subject to a saving, by the Competition Act 1998, s 74(1), (2), Sch 12, para 1(1), (5), Sch 13, Pt V, as from a day to be appointed under s 76(3) thereof.

40   Provisions as to interconnected bodies corporate

(1)   This section applies to any order made under section 37 or section 39 of this Act.

(2)   Where an order to which this section applies is made against a body corporate which is a member of a group of interconnected bodies corporate, the *Restrictive Practices Court*, on making the order, may direct that it shall be binding upon all members of the group as if each of them were the body corporate against which the order is made.

(3) Where an order to which this section applies has been made against a body corporate, and at a time when that order is in force—

(a) the body corporate becomes a member of a group of interconnected bodies corporate, or

(b) a group of interconnected bodies corporate of which it is a member is increased by the addition of one or more further members,

the *Restrictive Practices Court*, on the application of the Director, may direct that the order shall thereafter be binding upon each member of the group as if it were the body corporate against which the order was made.

(4) The power conferred by subsection (3) of this section shall be exercisable—

(a) whether, at the time when the original order was made, the body corporate against which it was made was a member of a group of interconnected bodies corporate or not, and

(b) if it was such a member, whether a direction under subsection (2) of this section was given or not.

NOTES

Sub-ss (2), (3): words in italics repealed, subject to a saving, by the Competition Act 1998, s 74(1), (2), Sch 12, para 1(1), (5), Sch 13, Pt V, as from a day to be appointed under s 76(3) thereof.

41 Concurrent jurisdiction of other courts in certain cases

(1) In any case where—

(a) the Director could bring proceedings against a person before the *Restrictive Practices Court* under section 35 or section 38 of this Act, and

(b) it appears to the Director that the conditions specified in the next following subsection are fulfilled,

the Director may, if he thinks fit, bring those proceedings in an appropriate alternative court instead of bringing them before the *Restrictive Practices Court*; and, in relation to any proceedings brought by virtue of this section, the appropriate alternative court in which they are brought shall have the like jurisdiction as the *Restrictive Practices Court* would have had if they had been brought in that Court.

(2) The conditions referred to in the preceding subsection are—

(a) that neither the person against whom the proceedings are to be brought nor the person against whom any associated proceedings have been or are intended to be brought is a body corporate having a share capital, paid up or credited as paid up, of an amount exceeding £10,000, and

(b) that neither those proceedings nor any associated proceedings involve or are likely to involve the determination of a question (whether of law or of fact) of such general application as to justify its being reserved for determination by the *Restrictive Practices Court.*

(3) For the purposes of this section, the following shall be appropriate alternative courts in relation to proceedings in respect of a course of conduct maintained in the course of a business, that is to say, the county court for any district (or, in Northern Ireland, any division) in which, or, in Scotland, any sheriff court within whose jurisdiction, that business is carried on.

(4) In relation to any proceedings brought in an appropriate alternative court by virtue of this section, or to any order made in any such proceedings, any reference in section 37, in section 39 or section 40 of this Act to the *Restrictive Practices Court* shall be construed as a reference to the appropriate alternative court in which the proceedings are brought.

(5) In this section "associated proceedings"—
(a) in relation to proceedings under section 35 of this Act, means proceedings under section 38 of this Act against a person as being a person consenting to or conniving at the course of conduct in question, and
(b) in relation to proceedings under section 38 of this Act, means proceedings under section 35 of this Act against a person as being the person by whom the course of conduct in question has been maintained.

NOTES

Sub-ss (1), (2), (4): words in italics repealed, subject to a saving, by the Competition Act 1998, s 74(1), (2), Sch 12, para 1(1), (5), Sch 13, Pt V, as from a day to be appointed under s 76(3) thereof.

[41A   Meaning of "relevant Court"

In this Part of this Act, "relevant Court", in relation to proceedings in respect of a course of conduct maintained in the course of a business, means any of the following courts in whose jurisdiction that business is carried on—
(a) in England and Wales or Northern Ireland, the High Court;
(b) ...]

Commencement: to be appointed.

Inserted by the Competition Act 1998, s 74(1), Sch 12, para 1(1), (6), as from a day to be appointed under s 76(3) thereof.

Para (b): applies to Scotland only.

42 Appeals from decisions or orders of courts under Part III

(1) Notwithstanding anything in any other enactment, an appeal, whether on a question of fact or on a question of law, shall lie from any decision or order of any court in proceedings under Part III of this Act[; but this subsection is subject to subsection (3) of this section].

(2) Any such appeal shall lie—
  (a) in the case of proceedings in England and Wales, to the Court of Appeal;
  (b) ...
  (c) in the case of proceedings in Northern Ireland, to the Court of Appeal in Northern Ireland.

[(3) ...]

Commencement: to be appointed (sub-s (1)); 1 November 1973 (sub-ss (1), (2)).

Sub-s (1): words in square brackets added by the Competition Act 1998, s 74(1), Sch 12, para 1(1), (7)(a), as from a day to be appointed under s 76(3) thereof.

Sub-s (2): para (b) applies to Scotland only.

Sub-s (3): added by the Competition Act 1998, s 74(1), Sch 12, para 1(1), (7)(c), as from a day to be appointed under s 76(3) thereof; applies to Scotland only.

PART XI
PYRAMID SELLING AND SIMILAR TRADING SCHEMES

[118 Trading schemes to which Part XI applies

(1) This Part of this Act applies to any trading scheme if—
  (a) the prospect is held out to participants of receiving payments or other benefits in respect of any of the matters specified in subsection (2) of this section; and
  (b) (subject to subsection (7) of this section) either or both of the conditions in subsections (3) and (4) of this section are fulfilled in relation to the scheme.

(2)   The matters referred to in paragraph (a) of subsection (1) of this section are—
   (a) the introduction by any person of other persons who become participants in a trading scheme;
   (b) the continued participation of participants in a trading scheme;
   (c) the promotion, transfer or other change of status of participants within a trading scheme;
   (d) the supply of goods or services by any person to or for other persons;
   (e) the acquisition of goods or services by any person.

(3)   The condition in this subsection is that—
   (a) goods or services, or both, are to be provided by the person promoting the scheme (in this Part of this Act referred to as "the promoter") or, in the case of a scheme promoted by two or more persons acting in concert (in this Part of this Act referred to as "the promoters"), by one or more of those persons; and
   (b) the goods or services so provided—
      (i) are to be supplied to or for other persons under transactions effected by participants (whether in the capacity of agents of the promoter or of one of the promoters or in any other capacity), or
      (ii) are to be used for the purposes of the supply of goods or services to or for other persons under such transactions.

(4)   The condition in this subsection is that goods or services, or both, are to be supplied by the promoter or any of the promoters to or for persons introduced to him or any of the other promoters (or an employee or agent of his or theirs) by participants.

(5)   For the purposes of this Part of this Act a prospect of a kind mentioned in paragraph (a) of subsection (1) of this section shall be treated as being held out to a participant whether it is held out so as to confer on him a legally enforceable right or not.

(6)   This Part of this Act does not apply to any trading scheme—
   (a) under which the promoter or any of the promoters or participants is to carry on, or to purport to carry on, investment business in the United Kingdom (within the meaning of section 1 of the Financial Services Act 1986); or
   (b) which otherwise falls within a description prescribed by regulations made by the Secretary of State by statutory instrument.

(7)   The Secretary of State may by order made by statutory instrument—
   (a) disapply paragraph (b) of Subsection (1) of this section in relation to a trading scheme of a kind specified in the order; or

(b) amend or repeal paragraph (a) of subsection (6) of this section;

and no such order, and no order varying or revoking any such order, shall be made under this subsection unless a draft of the order has been laid before Parliament and approved by a resolution of each House of Parliament.

(8)   In this Part of this Act—

"goods" includes property of any description and a right to, or interest in, property;

"participant" means, in relation to a trading scheme, a person (other than the promoter or any of the promoters) participating in the scheme;

"trading scheme" includes any arrangements made in connection with the carrying on of a business, whether those arrangements are made or recorded wholly or partly in writing or not;

and any reference to the provision or supply of goods shall be construed as including a reference to the grant or transfer of a right or interest.

(9)   In this section any reference to the provision or supply of goods or services by a person shall be construed as including a reference to the provision or supply of goods or services under arrangements to which that person is a party.]

NOTES

Commencement: 6 February 1997.
Substituted by the Trading Schemes Act 1996, s 1.

119 Regulations relating to such trading schemes

(1)   Regulations made by the Secretary of State by statutory instrument may make provision with respect to the issue, circulation or distribution of [any form of advertisement, prospectus, circular or notice which contains any information] calculated to lead directly or indirectly to persons becoming participants in such a trading scheme, and may prohibit any such [advertisement, prospectus, circular or notice] from being issued, circulated or distributed unless it complies with such requirements as to the matters to be included or not included in it as may be prescribed by the regulations.

(2)   Regulations made by the Secretary of State by statutory instrument may prohibit the promoter or any of the promoters of, or any participant in, a trading scheme to which this Part of this Act applies from—

(a) supplying any goods to a participant in the trading scheme, or

(b) supplying any training facilities or other services for such a participant, or

(c) providing any goods or services under a transaction effected by such a participant, or

(d) being a party to any arrangements under which goods or services are supplied or provided as mentioned in any of the preceding paragraphs, or

(e) accepting from any such participant any payment, or any undertaking to make a payment, in respect of any goods or services supplied or provided as mentioned in any of paragraphs (a) to (d) of this subsection or in respect of any goods or services to be so supplied or provided,

unless (in any such case) such requirements as are prescribed by the regulations are complied with.

(3)   Any requirements prescribed by regulations under subsection (2) of this section shall be such as the Secretary of State considers necessary or expedient for the purpose of preventing participants in trading schemes to which this Part of this Act applies from being unfairly treated; and, without prejudice to the generality of this subsection, any such requirements may include provisions—

(a) requiring the rights and obligations of every participant under such a trading scheme to be set out in full in an agreement in writing made between the participant and the promoter or (if more than one) each of the promoters;

(b) specifying rights required to be conferred on every such participant, and obligations required to be assumed by the promoter or promoters, under any such trading scheme; or

(c) imposing restrictions on the liabilities to be incurred by such a participant in respect of any of the matters mentioned in paragraphs (a) to (e) of subsection (2) of this section.

(4)   Regulations made under subsection (2) of this section—

(a) may include provision for enabling a person who has made a payment as a participant in a trading scheme to which this Part of this Act applies, in circumstances where any of the requirements prescribed by the regulations were not complied with, to recover the whole or part of that payment from any person to whom or for whose benefit it was paid, and

(b) subject to any provision made in accordance with the preceding paragraph, may prescribe the degree to which anything done in contravention of the regulations is to be treated as valid or invalid for the purposes of any civil proceedings.

(5)   The power to make regulations under this section may be exercised so as to make different provision—

(a) in relation to different descriptions of trading schemes to which this Part of this Act applies, or

(b) in relation to trading schemes which are or were in operation on a date specified in the regulations and trading schemes which are or were not in operation on that date,

or in relation to different descriptions of participants in such trading schemes.

120 Offences under Part XI

(1)   Subject to the next following section, any person who issues, circulates or distributes, or causes another person to issue, circulate or distribute, a [advertisement, prospectus, circular or notice] in contravention of any regulations made under subsection (1) of section 119 of this Act shall be guilty of an offence.

(2)   Any person who contravenes any regulations made under subsection (2) of that section shall be guilty of an offence.

(3)   If any person who is a participant in a trading scheme to which this Part of this Act applies, or has applied or been invited to become a participant in such a trading scheme,—

(a) makes any payment to or for the benefit of the promoter or (if there is more than one) any of the promoters, or to or for the benefit of a participant in the trading scheme, and

(b) is induced to make that payment by reason that the prospect is held out to him of receiving payments or other benefits in respect of the introduction of other persons who become participants in the trading scheme,

any person to whom or for whose benefit that payment is made shall be guilty of an offence.

(4)   If the promoter or any of the promoters of a trading scheme to which this Part of this Act applies, or any other person acting in accordance with

such a trading scheme, by holding out to any person such a prospect as is mentioned in subsection (3)(b) of this section, attempts to induce him—

(a) if he is already a participant in the trading scheme, to make any payment to or for the benefit of the promoter or any of the promoters or to or for the benefit of a participant in the trading scheme, or

(b) if he is not already a participant in the trading scheme, to become such a participant and to make any such payment as is mentioned in the preceding paragraph,

the person attempting to induce him to make that payment shall be guilty of an offence.

(5) In determining, for the purposes of subsection (3) or subsection (4) of this section, whether an inducement or attempt to induce is made by holding out such a prospect as is therein mentioned, it shall be sufficient if such a prospect constitutes or would constitute a substantial part of the inducement.

(6) Where the person by whom an offence is committed under subsection (3) or subsection (4) of this section is not the sole promoter of the trading scheme in question, any other person who is the promoter or (as the case may be) one of the promoters of the trading scheme shall, subject to the next following section, also be guilty of that offence.

(7) Nothing in subsections (3) to (6) of this section shall be construed as limiting the circumstances in which the commission of any act may constitute an offence under subsection (1) or subsection (2) of this section.

(8) In this section any reference to the making of a payment to or for the benefit of a person shall be construed as including the making of a payment partly to or for the benefit of that person and partly to or for the benefit of one or more other persons.

NOTES

Sub-s (1): words in square brackets substituted by the Trading Schemes Act 1996, s 2(2).

121 Defences in certain proceedings under Part XI

(1) Where a person is charged with an offence under subsection (1) of section 120 of this Act in respect of an advertisement, it shall be a defence for him to prove that he is a person whose business it is to publish or

arrange for the publication of advertisements, and that he received the advertisement for publication in the ordinary course of business and did not know, and had no reason to suspect, that its publication would amount to an offence under that subsection.

(2)   Where a person is charged with an offence by virtue of subsection (6) of section 120 of this Act, it shall be a defence for him to prove—
  (a) that the trading scheme to which the charge relates was in operation before the commencement of this Act, and
  (b) that the act constituting the offence was committed without his consent or connivance.

122   Penalties for offences under Part XI

A person guilty of an offence under this Part of this Act shall be liable—
  (a) on summary conviction, to a fine not exceeding [the prescribed sum] or to imprisonment for a term not exceeding three months or to both;
  (b) on conviction on indictment, to a fine or to imprisonment for a term not exceeding two years or to both.

NOTES

Reference to "the prescribed sum" substituted by virtue of the Magistrates' Courts Act 1980, s 32(2).

123   Enforcement provisions

(1)   The provisions of sections 29 to 32 of this Act shall have effect for the purposes of this Part of this Act as if in those provisions—
  (a) references to a weights and measures authority or a duly authorised officer of such an authority were omitted, and
  (b) any reference to an offence under section 23 of this Act were a reference to an offence under this Part of this Act.

(2)   For the purposes of the application to Northern Ireland of those provisions as applied by the preceding subsection—
  (a) any reference to the Secretary of State shall be construed as a reference to the Ministry of Commerce for Northern Ireland, and
  (b) paragraphs (c) and (d) of section 33(2) of this Act shall have effect for the purposes of the application of Part II of this Act to Northern Ireland.

PART XII
MISCELLANEOUS AND SUPPLEMENTARY PROVISIONS

124 Publication of information and advice

(1)   With respect to any matter in respect of which the Director has any duties under section 2(1) of this Act, he may arrange for the publication, in such form and in such manner as he may consider appropriate, of such information and advice as it may appear to him to be expedient to give to consumers in the United Kingdom.

(2)   In arranging for the publication of any such information or advice, the Director shall have regard to the need for excluding, so far as that is practicable,—
   (a) any matter which relates to the private affairs of an individual, where the publication of that matter would or might, in the opinion of the Director, seriously and prejudicially affect the interests of that individual, and
   (b) any matter which relates specifically to the affairs of a particular body of persons, whether corporate or unincorporate, where publication of that matter would or might, in the opinion of the Director, seriously and prejudicially affect the interests of that body.

(3)   Without prejudice to the exercise of his powers under subsection (1) of this section, it shall be the duty of the Director to encourage relevant associations to prepare, and to disseminate to their members, codes of practice for guidance in safeguarding and promoting the interests of consumers in the United Kingdom.

(4)   In this section "relevant association" means any association (whether incorporated or not) whose membership consists wholly or mainly of persons engaged in the production or supply of goods or in the supply of services or of persons employed by or representing persons so engaged and whose objects or activities include the promotion of the interests of persons so engaged.

125 Annual and other reports of Director

(1)   The Director shall, as soon as practicable after the end of the year 1974 and of each subsequent calendar year, make to the Secretary of State a report on his activities, and the activities of the Advisory Committee and of the Commission, during that year.

(2)   Every such report shall include a general survey of developments, during the year to which it relates, in respect of matters falling within the scope of the Director's duties under any enactment (including any enactment contained in this Act, other than this section) [and shall set out any directions given to the Director under section 2(2) of the Consumer Credit Act 1974 during that year].

(3)   The Secretary of State shall lay a copy of every report made by the Director under subsection (1) of this section before each House of Parliament, and shall arrange for every such report to be published in such manner as he may consider appropriate.

(4)   The Director may also prepare such other reports as appear to him to be expedient with respect to such matters as are mentioned in subsection (2) of this section, and may arrange for any such report to be published in such manner as he may consider appropriate.

(5)   In making any report under this Act the Director shall have regard to the need for excluding, so far as that is practicable, any such matter as is specified in paragraph (a) or paragraph (b) of section 124(2) of this Act.

(6)   For the purposes of this section any period between the commencement of this Act and the end of the year 1973 shall be treated as included in the year 1974.

NOTES

Sub-s (2): words in square brackets added by the Consumer Credit Act 1974, s 5.

129   Time-limit for prosecutions

(1)   No prosecution for an offence under this Act shall be commenced after the expiration of three years from the commission of the offence or one year from its discovery by the prosecutor, whichever is the earlier.

(2)   Notwithstanding anything in [section 127(1) of the Magistrates' Courts Act 1980], a magistrates' court may try an information for an offence under this Act if the information was laid within twelve months from the commission of the offence.

(3)   ...

(4)   In the application of this section to Northern Ireland, for the references in subsection (2) to [section 127(1) of the Magistrates' Courts Act 1980], and to the trial and laying of an information there shall be substituted respectively references to [Article 19(1) of the Magistrates' Courts (Northern Ireland) Order 1981] and to the hearing and determination and making of a complaint [and as if in that subsection for the words "an offence under this Act" there were substituted the words "an offence under section 30(1) or 46(2) of this Act"].

NOTES

Sub-s (2): words in square brackets substituted by the Magistrates' Courts Act 1980, s 154, Sch 7, para 118.

Sub-s (3): applies to Scotland only.

Sub-s (4): words in first pair of square brackets substituted by the Magistrates' Courts Act 1980, s 154, Sch 7, para 118; words in second pair of square brackets substituted by the Magistrates' Courts (Northern Ireland) Order 1981, SI 1981/1675, art 170(2), Sch 6, Pt I; words in final pair of square brackets added by the Criminal Justice (Northern Ireland) Order 1980, SI 1980/704, art 12, Sch 1, Pt II.

130   Notice to Director of intended prosecution

(1)   Where a local weights and measures authority in England or Wales proposes to institute proceedings for an offence under section 23 of this Act, or for an offence under the Trade Descriptions Act 1968, other than an offence under section 28(5) or section 29 of that Act, [or for an offence under any provision made by or under Part III of the Consumer Protection Act 1987,] [or for an offence under section 1 of, or paragraph 6 of the Schedule to, the Property Misdescriptions Act 1991,] [or for an offence under [any of sections 1A to 2 or 5B of the Timeshare Act 1992]] it shall, as between the authority and the Director, be the duty of the authority to give to the Director notice of the intended proceedings, together with a summary of the facts on which the charges are to be founded, and to postpone institution of the proceedings until either—

   (a) twenty-eight days have elapsed since the giving of that notice, or
   (b) the Director has notified the authority that he has received the notice and the summary of the facts.

(2)   In relation to offences under the Trade Descriptions Act 1968, the preceding subsection shall have effect subject to the transitional provisions having effect by virtue of section 139 of this Act.

NOTES

Sub-s (1): words in first pair of square brackets inserted by the Consumer Protection Act 1987, s 48, Sch 4, para 3; words in second pair of square brackets inserted by the Property Misdescriptions Act 1991, s 3, Schedule, para 2(1); words in third (outer) pair of square brackets inserted by the Timeshare Act 1992, s 10, Schedule, para 2(1); words in final (inner) pair of square brackets substituted by the Timeshare Regulations 1997, SI 1997/1081, reg 13(5).

131 Notification of convictions and judgments to Director

(1) Where in any criminal proceedings a person is convicted of an offence by or before a court in the United Kingdom, or a judgment is given against a person in civil proceedings in any such court, and it appears to the court—

    (a) having regard to the functions of the Director under Part III of this Act [or under the Estate Agents Act 1979], that it would be expedient for the conviction or judgment to be brought to his attention, and

    (b) that it may not be brought to his attention unless arrangements for the purpose are made by the court,

the court may make arrangements for that purpose notwithstanding that the proceedings have been finally disposed of by the court.

(2) In this section "judgment" includes any order or decree, and any reference to the giving of a judgment shall be construed accordingly.

NOTES

Sub-s (1): words in square brackets in para (a) inserted by the Estate Agents Act 1979, s 9(5).

132 Offences by bodies corporate

(1) Where an offence under section 23, section 46, section 85(6) [section 93B] or Part XI of this Act, which has been committed by a body corporate, is proved to have been committed with the consent or connivance of, or to be attributable to any neglect on the part of, any director, manager, secretary or other similar officer of the body corporate, or any person who was purporting to act in any such capacity, he as well as the body corporate shall be guilty of that offence and be liable to be proceeded against and punished accordingly.

(2) Where the affairs of a body corporate are managed by its members, subsection (1) of this section shall apply in relation to the acts and defaults

of a member in connection with his functions of management as if he were a director of the body corporate.

NOTES

Sub-s (1): words in square brackets inserted by the Companies Act 1989, s 153, Sch 20, para 17.

134   Provisions as to orders

(1)   Any statutory instrument whereby any order is made under any of the preceding provisions of this Act, other than a provision which requires a draft of the order to be laid before Parliament before making the order, or whereby any regulations are made under this Act, shall be subject to annulment in pursuance of a resolution of either House of Parliament.

(2)   Any power conferred by any provision of this Act to make an order by statutory instrument shall include power to revoke or vary the order by a subsequent order made under that provision.

137   General interpretation provisions

(1)   In this Act—
   "the Act of 1948" means the Monopolies and Restrictive Practices (Inquiry and Control) Act 1948;
   ["the Act of 1976" means the Restrictive Trade Practices Act 1976];

. . . . .

   "the Act of 1965" means the Monopolies and Mergers Act 1965;

. . . . .

   "contract of employment" means a contract of service or of apprenticeship, whether it is express or implied, and (if it is express) whether it is oral or in writing;
   "scale" (where the reference is to the scale on which any services are, or are to be, made available, supplied or obtained) means scale measured in terms of money or money's worth or in any other manner.

(2)   Except in so far as the context otherwise requires, in this Act, ... , the following expressions have the meanings hereby assigned to them respectively, that is to say—

"the Advisory Committee" means the Consumer Protection Advisory Committee;

"agreement" means any agreement or arrangement, in whatever way and in whatever form it is made, and whether it is, or is intended to be, legally enforceable or not;

"business" includes a professional practice and includes any other undertaking which is carried on for gain or reward or which is an undertaking in the course of which goods or services are supplied otherwise than free of charge;

"commercial activities in the United Kingdom" means any of the following, that is to say, the production and supply of goods in the United Kingdom, the supply of services in the United Kingdom and the export of goods from the United Kingdom;

"the Commission" means the [Competition] Commission;

"complex monopoly situation" has the meaning assigned to it by section 11 of this Act;

"consumer" (subject to subsection (6) of this section) means any person who is either—

(a) a person to whom goods are or are sought to be supplied (whether by way of sale or otherwise) in the course of a business carried on by the person supplying or seeking to supply them, or

(b) a person for whom services are or are sought to be supplied in the course of a business carried on by the person supplying or seeking to supply them,

and who does not receive or seek to receive the goods or services in the course of a business carried on by him;

"the Director" means the Director General of Fair Trading;

"enactment" includes an enactment of the Parliament of Northern Ireland;

"goods" includes buildings and other structures, and also includes ships, aircraft and hovercraft, ... ;

"group" (where the reference is to a group of persons fulfilling specified conditions, other than the condition of being interconnected bodies corporate) means any two or more persons fulfilling those conditions, whether apart from fulfilling them they would be regarded as constituting a group or not;

"merger reference" has the meaning assigned to it by section 5(3) of this Act;

"merger situation qualifying for investigation" has the meaning assigned to it by section 64(8) of this Act;

"Minister" includes a government department but shall not by virtue of this provision be taken to include the establishment consisting of the Director and his staff, and, except where the contrary is expressly provided, does not include any Minister or department of the Government of Northern Ireland;

"monopoly reference" and "monopoly situation" have the meanings assigned to them by section 5(3) of this Act;

"newspaper merger reference" has the meaning assigned to it by section 59(3) of this Act;

"practice" means any practice, whether adopted in pursuance of an agreement or otherwise;

"price" includes any charge or fee, by whatever name called;

"produce", in relation to the production of minerals or other substances, includes getting them, and, in relation to the production of animals or fish, includes taking them;

"supply", in relation to the supply of goods, includes supply by way of sale, lease, hire or hire-purchase, and, in relation to buildings or other structures, includes the construction of them by a person for another person;

"uncompetitive practices" means practices having the effect of preventing, restricting or distorting competition in connection with any commercial activities in the United Kingdom;

"worker" (subject to subsection (7) of this section) has the meaning assigned to it by section 167 of the Industrial Relations Act 1971.

(3) In the provisions of this Act ... "the supply of services" does not include the rendering of any services under a contract of employment but, ...

(a) includes the undertaking and performance for gain or reward of engagements (whether professional or other) for any matter other than the supply of goods, and

(b) includes both the rendering of services to order and the provision of services by making them available to potential users; [and

(c) includes the making of arrangements for a person to put or keep on land a caravan (within the meaning of Part I of the Caravan Sites and Control of Development Act 1960) other than arrangements by virtue of which the person may occupy the caravan as his only or main residence] [and

(d) includes the making of arrangements for the use by public service vehicles (within the meaning of the Public Passenger Vehicles Act 1981) of a parking place which is used as a point at which passengers on services provided by means of such vehicles may be taken up or set down] [and

(e) includes the making of arrangements permitting use of the tunnel system (within the meaning of the Channel Tunnel Act 1987) by a person operating services for the carriage of passengers or goods by rail] [and

(f) includes the making of arrangements, by means of such an agreement as is mentioned in section 189(2) of the Broadcasting Act 1990, for the sharing of the use of any telecommunication apparatus (within the meaning of Schedule 2 to the Telecommunications Act 1984)] [and

(g) includes the supply of network services and station services, within the meaning of Part I of the Railways Act 1993;]

and any reference in those provisions to services supplied or to be supplied, or to services provided or to be provided, shall be construed accordingly.

[(3A) The Secretary of State may by order made by statutory instrument—

(a) provide that "the supply of services" in the provisions of this Act is to include, or to cease to include, any activity specified in the order which consists in, or in making arrangements in connection with, permitting the use of land; and

(b) for that purpose, amend or repeal any of paragraphs (c), (d), (e) or (g) of subsection (3) above.

(3B)   No order under subsection (3A) above is to be made unless a draft of the order has been laid before Parliament and approved by a resolution of each House of Parliament.

(3C)   The provisions of Schedule 9 to this Act apply in the case of a draft of any such order as they apply in the case of a draft of an order to which section 91(1) above applies.]

(4)   ...

(5)   For the purposes of the provisions of this Act ... , any two bodies corporate are to be treated as interconnected if one of them is a body corporate of which the other is a subsidiary (within the meaning of [section 736 of the Companies Act 1985]) or if both of them are subsidiaries (within the meaning of that section) of one and the same body corporate; and in those provisions "interconnected bodies corporate" shall be construed accordingly, and "group of interconnected bodies corporate" means a group consisting of two or more bodies corporate all of whom are interconnected with each other.

(6)   For the purposes of the application of any provision of this Act in relation to goods or services of a particular description or to which a particular practice applies, "consumers" means persons who are consumers

(as defined by subsection (2) of this section) in relation to goods or services of that description or in relation to goods or services to which that practice applies.

(7) For the purposes of the application of this Act to Northern Ireland, the definition of "worker" in subsection (2) of this section shall apply as if the Industrial Relations Act 1971 extended to Northern Ireland but, in section 167(2)(a) of that Act, references to general medical services, pharmaceutical services, general dental services or general ophthalmic services provided under the enactments mentioned in that subsection were references to the corresponding services provided in Northern Ireland under the corresponding enactments there in force.

(8) Except in so far as the context otherwise requires, any reference in this Act to an enactment shall be construed as a reference to that enactment as amended or extended by or under any other enactment, including this Act.

NOTES

Commencement: 1 April 1999 (sub-ss (3A)–(3C)); 14 September 1973 (sub-ss (1)–(3), (4)–(8)).

Sub-s (1): words in square brackets substituted and words omitted in the second place repealed by the Restrictive Trade Practices Act 1976, s 44, Schs 5, 6; words omitted in the first place repealed by the Resale Prices Act 1976, s 29, Sch 3, Pt I.

Sub-s (2): words omitted in the first place repealed by the Restrictive Trade Practices Act 1976, s 44, Sch 6; in definition "the Commission" word in square brackets substituted by the Competition Act 1998 (Competition Commission) Transitional, Consequential and Supplemental Provisions Order 1999, SI 1999/506, art 14; words omitted in the second place repealed by the Electricity Act 1989, s 112(4), Sch 18.

Sub-s (3): words omitted repealed by the Restrictive Trade Practices Act 1976, s 44, Sch 6; para (c) and word immediately preceding it inserted by the Competition Act 1980, s 23; para (d) and word immediately preceding it inserted by the Transport Act 1985, s 116(1); para (e) and word immediately preceding it inserted by the Channel Tunnel Act 1987, s 33(10); para (f) and word immediately preceding it inserted by the Broadcasting Act 1990, s 192(1); para (g) and word immediately preceding it inserted by the Railways Act 1993, s 66(4).

Sub-ss (3A)–(3C): inserted by the Competition Act 1998, s 68, as from a day to be appointed under s 76(3) thereof.

Sub-s (4): repealed by the Electricity Act 1989, s 112(4), Sch 18.

Sub-s (5): words omitted repealed by the Restrictive Trade Practices Act 1976, s 44, Sch 6; words in square brackets substituted by the Companies Consolidation (Consequential Provisions) Act 1985, s 30, Sch 2.

138 Supplementary interpretation provisions

(1)   This section applies to the following provisions of this Act, that is to say, section 2(4), Parts II and III, section 137(6), and the definition of "consumer" contained in section 137(2).

(2)   For the purposes of any provisions to which this section applies it is immaterial whether any person supplying goods or services has a place of business in the United Kingdom or not.

(3)   For the purposes of any provisions to which this section applies any goods or services supplied wholly or partly outside the United Kingdom, if they are supplied in accordance with arrangements made in the United Kingdom, whether made orally or by one or more documents delivered in the United Kingdom or by correspondence posted from and to addresses in the United Kingdom, shall be treated as goods supplied to, or services supplied for, persons in the United Kingdom.

(4)   In relation to the supply of goods under a hire-purchase agreement, a credit-sale agreement or a conditional sale agreement, the person conducting any antecedent negotiations, as well as the owner or seller, shall for the purposes of any provisions to which this section applies be treated as a person supplying or seeking to supply the goods.

[(5)   In subsection (4) of this section, the following expressions have the meanings given by, or referred to in, section 189 of the Consumer Credit Act 1974—
    "antecedent negotiations",
    "conditional sale agreement",
    "credit-sale agreement",
    "hire-purchase agreement".]

(6)   In any provisions to which this section applies—
    (a) any reference to a person to or for whom goods or services are supplied shall be construed as including a reference to any guarantor of such a person, and
    (b) any reference to the terms or conditions on or subject to which goods or services are supplied shall be construed as including a reference to the terms or conditions on or subject to which any person undertakes to act as such a guarantor;

and in this subsection "guarantor", in relation to a person to or for whom goods or services are supplied, includes a person who undertakes to indemnify the supplier of the goods or services against any loss which he may incur in respect of the supply of the goods or services to or for that person.

(7)   For the purposes of any provisions to which this section applies goods or services supplied by a person carrying on a business shall be taken to be supplied in the course of that business if payment for the supply of the goods or services is made or (whether under a contract or by virtue of an enactment or otherwise) is required to be made.

NOTES

Sub-s (5): substituted by the Consumer Credit Act 1974, s 192(3)(a), Sch 4, Pt I, para 37.

140   Short title, citation, commencement and extent

(1)   This Act may be cited as the Fair Trading Act 1973.

(2)   ...

(3)   This Act shall come into operation on such day as the Secretary of State may by order made by statutory instrument appoint: and different dates may be so appointed for, or for different purposes of, any one or more of the provisions of this Act (including, in the case of section 139 of this Act, the amendment or repeal of different enactments specified in Schedule 12 or Schedule 13 to this Act or of different provisions of any enactment so specified).

(4)   Where any provision of this Act, other than a provision contained in Schedule 11, refers to the commencement of this Act, it shall be construed as referring to the day appointed under this section for the coming into operation of that provision.

(5)   This Act extends to Northern Ireland.

NOTES

Sub-s (2): repealed by the Restrictive Trade Practices Act 1976, s 44, Sch 6.

SCHEDULES

SCHEDULE I

Section 1

DIRECTOR GENERAL OF FAIR TRADING

1.   There shall be paid to the Director such remuneration, and such travelling and other allowances, as the Secretary of State with the approval of the Minister for the Civil Service may determine.

2.   In the case of any such holder of the office of the Director as may be determined by the Secretary of State with the approval of the Minister for the Civil Service, there shall be paid such pension, allowance or gratuity to or in respect of him on his retirement or death, or such contributions or payments towards provision for such a pension, allowance or gratuity, as may be so determined.

3.   If, when any person ceases to hold office as the Director, it appears to the Secretary of State with the approval of the Minister for the Civil Service that there are special circumstances which make it right that he should receive compensation, there may be paid to him a sum by way of compensation of such amount as may be so determined.

4.   ...

5.   The Director shall have an official seal for the authentication of documents required for the purposes of his functions.

6.   The Documentary Evidence Act 1868 shall have effect as if the Director were included in the first column of the Schedule to that Act, as if the Director and any person authorised to act on behalf of the Director were mentioned in the second column of that Schedule, and as if the regulations referred to in that Act included any document issued by the Director or by any such person.

7.   Anything authorised or required by or under this Act or any other enactment to be done by the Director, other than the making of a statutory instrument, may be done by any member of the staff of the Director who is authorised generally or specially in that behalf in writing by the Director.

NOTES

Para 4: repealed by the House of Commons Disqualification Act 1975, s 10(2), Sch 3 and the Northern Ireland Assembly Disqualification Act 1975, s 5(2), Sch 3, Pt I.

Transfer of functions: By the Transfer of Functions (Treasury and Minister for the Civil Service) Order 1995, SI 1995/269, art 3, Schedule, para 9, the Minister for the Civil Service's functions under paras 1–3 of this Schedule (which were transferred to the Treasury by the Transfer of Functions (Minister for the Civil Service and Treasury) Order 1981, SI 1981/1670), became functions of the Minister with effect from 1 April 1995.

SCHEDULE 4

Sections 14 and 109

SERVICES EXCLUDED FROM SECTIONS 14 AND 109

1.   Legal services (that is to say, the services of barristers, advocates or solicitors in their capacity as such).

2.   Medical services (that is to say, the provision of medical or surgical advice or attendance and the performance of surgical operations).

3.   Dental services (that is to say, any services falling within the practice of dentistry within the meaning of the Dentists Act [1984]).

4.   Ophthalmic services (that is to say, the testing of sight).

5.   Veterinary services (that is to say, any services which constitute veterinary surgery within the meaning of the Veterinary Surgeons Act 1966).

6.   Nursing services (that is to say, any services which constitute nursing within the meaning of the Nurses Act 1957, the Nurses (Scotland) Act 1951 or the Nurses and Midwives Act (Northern Ireland) 1970).

7.   The services of midwives, physiotherapists or chiropodists in their capacity as such.

8.   The services of architects in their capacity as such.

9.   Accounting and auditing services (that is to say, the making or preparation of accounts or accounting records and the examination, verification and auditing of financial statements).

[10. The services of registered patent agents (within the meaning of Part V of the Copyright, Designs and Patents Act 1988) in their capacity as such.]

[10A. The services of persons carrying on for gain in the United Kingdom the business of acting as agents or other representatives of other persons for the purpose of applying for or obtaining European patents or for the purpose of conducting proceedings [in relation to applications for or otherwise] in connection with such patents before the European Patent Office or the comptroller and whose names appear on the European list (within the meaning of [Part V of the Copyright, Designs and Patents Act 1988]) in their capacity as such persons.]

11. The services of parliamentary agents entered in the register in either House of Parliament as agents entitled to practise both in promoting and in opposing Bills, in their capacity as such parliamentary agents.

12. The services of surveyors (that is to say, of surveyors of land, of quantity surveyors, of surveyors of buildings or other structures and of surveyors of ships) in their capacity as such surveyors.

13. The services of professional engineers or technologists (that is to say, of persons practising or employed as consultants in the field of—
    (a) civil engineering;
    (b) mechanical, aeronautical, marine, electrical or electronic engineering;
    (c) mining, quarrying, soil analysis or other forms of minerology or geology;
    (d) agronomy, forestry, livestock rearing or ecology;
    (e) metallurgy, chemistry, biochemistry or physics; or
    (f) any other form of engineering or technology analogous to those mentioned in the preceding sub-paragraphs),

in their capacity as such engineers or technologists.

14. Services consisting of the provision—
    (a) of primary, secondary or further education within the meaning of [the Education Act 1996,] the Education (Scotland) Acts 1939 to 1971 or the Education and Libraries (Northern Ireland) Order 1972, or
    (b) of university or other higher education not falling within the preceding sub-paragraph.

15.   The services of ministers of religion in their capacity as such ministers.

Para 3: date in square brackets substituted by the Dentists Act 1984, s 54(1), Sch 5, para 6.

Para 10: substituted by the Copyright, Designs and Patents Act 1988, s 303(1), Sch 7, para 15.

Para 10A: inserted by the Patents Act 1977, s 132(6), Sch 5, para 7; words in first pair of square brackets inserted by the Administration of Justice Act 1985, s 60(2)(a), (6); words in second pair of square brackets substituted by the Copyright, Designs and Patents Act 1988, s 303(1), Sch 7, para 15.

Para 14: words in square brackets substituted by the Education Act 1996, s 582(1), Sch 37, Pt I, para 26.

Modification: the reference to solicitors in para 1 includes a reference to bodies recognised under the Administration of Justice Act 1985, s 9, by virtue of the Solicitors' Incorporated Practices Order 1991, SI 1991/2684, arts 2–5, Sch 1.

SCHEDULE 5

Sections 16, 50, 51

GOODS AND SERVICES REFERRED TO IN SECTION 16

PART I
GENERAL RESTRICTION

[1], 3 ...

[4.   The carriage of passengers by road in Northern Ireland.]

[5.   Services for the carriage of passengers, or of goods, by railway, network services and station services, within the meaning of Part I of the Railways Act 1993, but excluding the carriage of passengers or goods on shuttle services (within the meaning of the Channel Tunnel Act 1987).]

6.   The services of conveying, receiving, collecting, despatching and delivering letters.

7.   The running of any system for the conveyance, through the agency of electric, magnetic, electro-magnetic, electro-chemical or electro-

mechanical energy, of any of the matters specified [in paragraphs (a) to (d) of section 4(1) of the Telecommunications Act 1984].

NOTES

Para 1: (as substituted for original paras 1, 2 by the Gas Act 1986, s 67(1), Sch 7, para 15(4)) repealed by the Gas Act 1995, s 17(5), Sch 6.

Para 3: repealed by the Electricity Act 1989, s 112(1), (4), Sch 16, para 16(5), Sch 18 and the Electricity (Northern Ireland) Order 1992, SI 1992/231, art 95(1), (4), Sch 12, para 11, Sch 14.

Para 4: substituted by the Transport Act 1985, s 114(2)(a).

Para 5: substituted by the Railways Act 1993, s 66(5).

Para 7: words in square brackets substituted by the Telecommunications Act 1984, s 109, Sch 4, para 57(4).

SCHEDULE 6

Section 19

MATTERS FALLING WITHIN SCOPE OF PROPOSALS UNDER SECTION 17

1. Prohibition of the specified consumer trade practice either generally or in relation to specified consumer transactions.

2. Prohibition of specified consumer transactions unless carried out at specified times or at a place of a specified description.

3. Prohibition of the inclusion in specified consumer transactions of terms or conditions purporting to exclude or limit the liability of a party to such a transaction in respect of specified matters.

4. A requirement that contracts relating to specified consumer transactions shall include specified terms or conditions.

5. A requirement that contracts or other documents relating to specified consumer transactions shall comply with specified provisions as to lettering (whether as to size, type, colouring or otherwise).

6. A requirement that specified information shall be given to parties to specified consumer transactions.

Consumer Credit Act 1974

(C 39)

An Act to establish for the protection of consumers a new system, administered by the Director General of Fair Trading, of licensing and other control of traders concerned with the provision of credit, or the supply of goods on hire or hire-purchase, and their transactions, in place of the present enactments regulating moneylenders, pawnbrokers and hire-purchase traders and their transactions, and for related matters
[31 July 1974]

PART I
DIRECTOR GENERAL OF FAIR TRADING

1 General functions of Director

(1)   It is the duty of the Director General of Fair Trading ("the Director")—
- (a) to administer the licensing system set up by this Act,
- (b) to exercise the adjudicating functions conferred on him by this Act in relation to the issue, renewal, variation, suspension and revocation of licences, and other matters,
- (c) generally to superintend the working and enforcement of this Act, and regulations made under it, and
- (d) where necessary or expedient, himself to take steps to enforce this Act, and regulations so made.

(2)   It is the duty of the Director, so far as appears to him to be practicable and having regard both to the national interest and the interests of persons carrying on businesses to which this Act applies and their customers, to keep under review and from time to time advise the Secretary of State about—
- (a) social and commercial developments in the United Kingdom and elsewhere relating to the provision of credit or bailment or (in Scotland) hiring of goods to individuals, and related activities; and
- (b) the working and enforcement of this Act and orders and regulations made under it.

2 Powers of Secretary of State

(1)   The Secretary of State may by order—

(a) confer on the Director additional functions concerning the provision of credit or bailment or (in Scotland) hiring of goods to individuals, and related activities, and

(b) regulate the carrying out by the Director of his functions under this Act.

(2) The Secretary of State may give general directions indicating considerations to which the Director should have particular regard in carrying out his functions under this Act, and may give specific directions on any matter connected with the carrying out by the Director of those functions.

(3) The Secretary of State, on giving any directions under subsection (2), shall arrange for them to be published in such manner as he thinks most suitable for drawing them to the attention of interested persons.

(4) With the approval of the Secretary of State and the Treasury, the Director may charge, for any service or facility provided by him under this Act, a fee of an amount specified by general notice (the "specified fee").

(5) Provision may be made under subsection (4) for reduced fees, or no fees at all, to be paid for certain services or facilities by persons of a specified description, and references in this Act to the specified fee shall, in such cases, be construed accordingly.

(6) An order under subsection (1)(a) shall be made by statutory instrument and shall be of no effect unless a draft of the order has been laid before and approved by each House of Parliament.

(7) References in subsection (2) to the functions of the Director under this Act do not include the making of a determination to which section 41 or 150 (appeals from Director to Secretary of State) applies.

4 Dissemination of information and advice

The Director shall arrange for the dissemination, in such form and manner as he considers appropriate, of such information and advice as it may appear to him expedient to give to the public in the United Kingdom about the operation of this Act, the credit facilities available to them, and other matters within the scope of his functions under this Act.

6 Form etc of application

(1)   An application to the Director under this Act is of no effect unless the requirements of this section are satisfied.

(2)   The application must be in writing, and in such form, and accompanied by such particulars, as the Director may specify by general notice, and must be accompanied by the specified fee.

(3)   After giving preliminary consideration to an application, the Director may by notice require the applicant to furnish him with such further information relevant to the application as may be described in the notice, and may require any information furnished by the applicant (whether at the time of the application or subsequently) to be verified in such manner as the Director may stipulate.

(4)   The Director may by notice require the applicant to publish details of his application at a time or times and in a manner specified in the notice.

7 Penalty for false information

A person who, in connection with any application or request to the Director under this Act, or in response to any invitation or requirement of the Director under this Act, knowingly or recklessly gives information to the Director which, in a material particular, is false or misleading, commits an offence.

PART II
CREDIT AGREEMENTS, HIRE AGREEMENTS AND LINKED TRANSACTIONS

8 Consumer credit agreements

(1)   A personal credit agreement is an agreement between an individual ("the debtor") and any other person ("the creditor") by which the creditor provides the debtor with credit of any amount.

(2)   A consumer credit agreement is a personal credit agreement by which the creditor provides the debtor with credit not exceeding [£25,000].

(3)   A consumer credit agreement is a regulated agreement within the meaning of this Act if it is not an agreement (an "exempt agreement") specified in or under section 16.

NOTES

Sub-s (2): sum in square brackets substituted by the Consumer Credit (Increase of Monetary Limits) Order 1983, SI 1983/1878, art 4, Schedule, Pt II, as amended by SI 1998/996.

9 Meaning of credit

(1)   In this Act "credit" includes a cash loan, and any other form of financial accommodation.

(2)   Where credit is provided otherwise than in sterling, it shall be treated for the purposes of this Act as provided in sterling of an equivalent amount.

(3)   Without prejudice to the generality of subsection (1), the person by whom goods are bailed or (in Scotland) hired to an individual under a hire-purchase agreement shall be taken to provide him with fixed-sum credit to finance the transaction of an amount equal to the total price of the goods less the aggregate of the deposit (if any) and the total charge for credit.

(4)   For the purposes of this Act, an item entering into the total charge for credit shall not be treated as credit even though time is allowed for its payment.

10 Running-account credit and fixed-sum credit

(1)   For the purposes of this Act—
   (a) running-account credit is a facility under a personal credit agreement whereby the debtor is enabled to receive from time to time (whether in his own person, or by another person) from the creditor or a third party cash, goods and services (or any of them) to an amount or value such that, taking into account payments made by or to the credit of the debtor, the credit limit (if any) is not at any time exceeded; and
   (b) fixed-sum credit is any other facility under a personal credit agreement whereby the debtor is enabled to receive credit (whether in one amount or by instalments).

(2)   In relation to running-account credit, "credit limit" means, as respects any period, the maximum debit balance which, under the credit agreement, is allowed to stand on the account during that period, disregarding any term of the agreement allowing that maximum to be exceeded merely temporarily.

(3)   For the purposes of section 8(2), running-account credit shall be taken not to exceed the amount specified in that subsection ("the specified amount") if—

   (a) the credit limit does not exceed the specified amount; or

   (b) whether or not there is a credit limit, and if there is, notwithstanding that it exceeds the specified amount,—

      (i) the debtor is not enabled to draw at any one time an amount which, so far as (having regard to section 9(4)) it represents credit, exceeds the specified amount, or

      (ii) the agreement provides that, if the debit balance rises above a given amount (not exceeding the specified amount), the rate of the total charge for credit increases or any other condition favouring the creditor or his associate comes into operation, or

      (iii) at the time the agreement is made it is probable, having regard to the terms of the agreement and any other relevant considerations, that the debit balance will not at any time rise above the specified amount.

11   Restricted-use credit and unrestricted-use credit

(1)   A restricted-use credit agreement is a regulated consumer credit agreement—

   (a) to finance a transaction between the debtor and the creditor, whether forming part of that agreement or not, or

   (b) to finance a transaction between the debtor and a person (the "supplier") other than the creditor, or

   (c) to refinance any existing indebtedness of the debtor's, whether to the creditor or another person,

and "restricted-use credit" shall be construed accordingly.

(2)   An unrestricted-use credit agreement is a regulated consumer credit agreement not falling within subsection (1), and "unrestricted-use credit" shall be construed accordingly.

(3)   An agreement does not fall within subsection (1) if the credit is in fact provided in such a way as to leave the debtor free to use it as he chooses, even though certain uses would contravene that or any other agreement.

(4)   An agreement may fall within subsection (1)(b) although the identity of the supplier is unknown at the time the agreement is made.

12 Debtor-creditor-supplier agreements

A debtor-creditor-supplier agreement is a regulated consumer credit agreement being—

(a) a restricted-use credit agreement which falls within section 11(1)(a), or

(b) a restricted-use credit agreement which falls within section 11(1)(b) and is made by the creditor under pre-existing arrangements, or in contemplation of future arrangements, between himself and the supplier, or

(c) an unrestricted-use credit agreement which is made by the creditor under pre-existing arrangements between himself and a person (the "supplier") other than the debtor in the knowledge that the credit is to be used to finance a transaction between the debtor and the supplier.

13 Debtor-creditor agreements

A debtor-creditor agreement is a regulated consumer credit agreement being—

(a) a restricted-use credit agreement which falls within section 11(1)(b) but is not made by the creditor under pre-existing arrangements, or in contemplation of future arrangements, between himself and the supplier, or

(b) a restricted-use credit agreement which falls within section 11(1)(c), or

(c) an unrestricted-use credit agreement which is not made by the creditor under pre-existing arrangements between himself and a person (the "supplier") other than the debtor in the knowledge that the credit is to be used to finance a transaction between the debtor and the supplier.

14 Credit-token agreements

(1) A credit-token is a card, check, voucher, coupon, stamp, form, booklet or other document or thing given to an individual by a person carrying on a consumer credit business, who undertakes—

(a) that on the production of it (whether or not some other action is also required) he will supply cash, goods and services (or any of them) on credit, or

(b) that where, on the production of it to a third party (whether or not any other action is also required), the third party supplies cash, goods and services (or any of them), he will pay the third party for them (whether or not deducting any discount or commission), in return for payment to him by the individual.

(2)   A credit-token agreement is a regulated agreement for the provision of credit in connection with the use of a credit-token.

(3)   Without prejudice to the generality of section 9(1), the person who gives to an individual an undertaking falling within subsection (1)(b) shall be taken to provide him with credit drawn on whenever a third party supplies him with cash, goods or services.

(4)   For the purposes of subsection (1), use of an object to operate a machine provided by the person giving the object or a third party shall be treated as the production of the object to him.

15   Consumer hire agreements

(1)   A consumer hire agreement is an agreement made by a person with an individual (the "hirer") for the bailment or (in Scotland) the hiring of goods to the hirer, being an agreement which—
   (a)   is not a hire-purchase agreement, and
   (b)   is capable of subsisting for more than three months, and
   (c)   does not require the hirer to make payments exceeding [£25,000].

(2)   A consumer hire agreement is a regulated agreement if it is not an exempt agreement.

NOTES

Sub-s (1): sum in square brackets substituted by the Consumer Credit (Increase of Monetary Limits) Order 1983, SI 1983/1878, art 4, Schedule, Pt II, as amended by SI 1998/996.

16   Exempt agreements

(1)   This Act does not regulate a consumer credit agreement where the creditor is a local authority ... , or a body specified, or of a description specified, in an order made by the Secretary of State, being—
   (a)   an insurance company,
   (b)   a friendly society,
   (c)   an organisation of employers or organisation of workers,

   (d) a charity,

   (e) a land improvement company, ...

   (f) a body corporate named or specifically referred to in any public general Act

[(ff) a body corporate named or specifically referred to in an order made under—

      section 156(4), *444(1)* or 447(2)(a) of the Housing Act 1985,

      [section 156(4) of that Act as it has effect by virtue of section 17 of the Housing Act 1996 (the right to acquire),]

      section 2 of the Home Purchase Assistance and Housing Corporation Guarantee Act 1978 or section 31 of the Tenants' Rights, &c (Scotland) Act 1980, or

      Article 154(1)(a) or 156AA of the Housing (Northern Ireland) Order 1981 or Article 10(6A) of the Housing (Northern Ireland) Order 1983; or]

  [(g) a building society][, or

  (h) an authorised institution or wholly-owned subsidiary (within the meaning of the Companies Act 1985) of such an institution].

(2)   Subsection (1) applies only where the agreement is—

  (a) a debtor-creditor-supplier agreement financing—

     (i) the purchase of land, or

     (ii) the provision of dwellings on any land,

    and secured by a land mortgage on that land, or

  (b) a debtor-creditor agreement secured by any land mortgage; or

  (c) a debtor-creditor-supplier agreement financing a transaction which is a linked transaction in relation to—

     (i) an agreement falling within paragraph (a), or

     (ii) an agreement falling within paragraph (b) financing—

       (aa) the purchase of any land, or

       (bb) the provision of dwellings on any land,

         and secured by a land mortgage on the land referred to in paragraph (a) or, as the case may be, the land referred to in sub-paragraph (ii).

(3)   The Secretary of State shall not make, vary or revoke an order—

  (a) under subsection (1)(a) without consulting the Minister of the Crown responsible for insurance companies,

  (b) under subsection (1)(b) ... without consulting the Chief Registrar of Friendly Societies,

  (c) under subsection (1)(d) without consulting the Charity Commissioners, ...

  (d) under subsection (1)(e)[, (f) or (ff)] without consulting any Minister of the Crown with responsibilities concerning the body in question [or

(e)   under subsection (1)(g) without consulting the Building Societies
      Commission and the Treasury] [or

(f)   under subsection (1)(h) without consulting the Treasury and the
      [Financial Services Authority]].

(4)   An order under subsection (1) relating to a body may be limited so as to
apply only to agreements by that body of a description specified in the order.

(5)   The Secretary of State may by order provide that this Act shall not
regulate other consumer credit agreements where—
  (a)   the number of payments to be made by the debtor does not exceed
        the number specified for that purpose in the order, or
  (b)   the rate of the total charge for credit does not exceed the rate so
        specified, or
  (c)   an agreement has a connection with a country outside the United
        Kingdom.

(6)   The Secretary of State may by order provide that this Act shall not
regulate consumer hire agreements of a description specified in the order
where—
  (a)   the owner is a body corporate authorised by or under any enactment
        to supply electricity, gas or water, and
  (b)   the subject of the agreement is a meter or metering equipment,

[or where the owner is a public telecommunications operator specified in
the order].

[(6A)   This Act does not regulate a consumer credit agreement where the
creditor is a housing authority and the agreement is secured by a land
mortgage of a dwelling.

(6B)   In subsection (6A) "housing authority" means—
  (a)   as regards England and Wales, [the Housing Corporation ... and] an
        authority or body within section 80(1) of the Housing Act 1985 (the
        landlord condition for secure tenancies), other than a housing
        association or a housing trust which is a charity;
  (b)   ...
  (c)   as regards Northern Ireland, the Northern Ireland Housing
        Executive.]

(7)   Nothing in this section affects the application of sections 137 to 140
(extortionate credit bargains).

(8)   ...

(9)   In the application of this section to Northern Ireland subsection (3) shall have effect as if any reference to a Minister of the Crown were a reference to a Northern Ireland department, any reference to the Chief Registrar of Friendly Societies were a reference to the Registrar of Friendly Societies for Northern Ireland, and any reference to the Charity Commissioners were a reference to the Department of Finance for Northern Ireland.

NOTES

Sub-s (1): words omitted repealed and para (g) inserted by the Building Societies Act 1986, s 120, Sch 18, Pt I, para 10(2), Sch 19, Pt I; para (ff) inserted by the Housing and Planning Act 1986, s 22(2), (4), with respect to agreements made after 7 January 1987, number in italics therein repealed by the Housing Act 1996, s 227, Sch 19, Pt XIV, as from a day to be appointed, and words in square brackets therein inserted by the Housing Act 1996 (Consequential Amendments) (No 2) Order 1997, SI 1997/627, art 2, Schedule, para 2; para (h) added by the Banking Act 1987, s 88.

Sub-s (3): words omitted from para (b) repealed by the Employment Protection Act 1975, s 125(3), Sch 18; word omitted from para (c) repealed and para (e) inserted by the Building Societies Act 1986, s 120, Sch 18, Pt I, para 10(3), Sch 19, Pt I; in para (d) words in square brackets substituted with respect to agreements made after 7 January 1987 by the Housing and Planning Act 1986, s 22(2), (4); para (f) added by the Banking Act 1987, s 88(3), words in square brackets therein substituted by the Bank of England Act 1998, s 23(1), Sch 5, para 36.

Sub-s (6): words in square brackets substituted by the Telecommunications Act 1984, s 109, Sch 4, para 60(1).

Sub-s (6A): inserted with respect to agreements made after 7 January 1987 by the Housing and Planning Act 1986, s 22(3), (4).

Sub-s (6B): inserted with respect to agreements made after 7 January 1987 by the Housing and Planning Act 1986, s 22(3), (4), words in square brackets therein inserted by the Housing Act 1988, s 140(1), Sch 17, Pt I, para 20; words omitted from para (a) repealed, subject to transitional provisions, by the Government of Wales Act 1998, ss 141, 152, Sch 18, Pt VI; para (b) applies to Scotland only.

Sub-s (8): applies to Scotland only.

Modification: sub-s (1) modified, in relation to the reference to an institution authorised under the Banking Act, by the Banking Coordination (Second Council Directive) Regulations 1992, SI 1992/3218, reg 82(1), Sch 10, para 7.

17   Small agreements

(1)   A small agreement is—
   (a) a regulated consumer credit agreement for credit not exceeding [£50], other than a hire-purchase or conditional sale agreement; or
   (b) a regulated consumer hire agreement which does not require the hirer to make payments exceeding [£50],

being an agreement which is either unsecured or secured by a guarantee or indemnity only (whether or not the guarantee or indemnity is itself secured).

(2)   Section 10(3)(a) applies for the purposes of subsection (1) as it applies for the purposes of section 8(2).

(3)   Where—
   (a) two or more small agreements are made at or about the same time between the same parties, and
   (b) it appears probable that they would instead have been made as a single agreement but for the desire to avoid the operation of provisions of this Act which would have applied to that single agreement but, apart from this subsection, are not applicable to the small agreements,

this Act applies to the small agreements as if they were regulated agreements other than small agreements.

(4)   If, apart from this subsection, subsection (3) does not apply to any agreements but would apply if, for any party or parties to any of the agreements, there were substituted an associate of that party, or associates of each of those parties, as the case may be, then subsection (3) shall apply to the agreements.

NOTES

   Sub-s (1): sums in square brackets substituted by the Consumer Credit (Increase of Monetary Limits) Order 1983, SI 1983/1878, art 3, Schedule, Pt I.

18 Multiple agreements

(1)   This section applies to an agreement (a "multiple agreement") if its terms are such as—
   (a) to place a part of it within one category of agreement mentioned in this Act, and another part of it within a different category of agreements so mentioned, or within a category of agreement not so mentioned, or
   (b) to place it, or a part of it, within two or more categories of agreement so mentioned.

(2)   Where a part of an agreement falls within subsection (1), that part shall be treated for the purposes of this Act as a separate agreement.

(3)   Where an agreement falls within subsection (1)(b), it shall be treated as an agreement in each of the categories in question, and this Act shall apply to it accordingly.

(4)   Where under subsection (2) a part of a multiple agreement is to be treated as a separate agreement, the multiple agreement shall (with any necessary modifications) be construed accordingly; and any sum payable under the multiple agreement, if not apportioned by the parties, shall for the purposes of proceedings in any court relating to the multiple agreement be apportioned by the court as may be requisite.

(5)   In the case of an agreement for running-account credit, a term of the agreement allowing the credit limit to be exceeded merely temporarily shall not be treated as a separate agreement or as providing fixed-sum credit in respect of the excess.

(6)   This Act does not apply to a multiple agreement so far as the agreement relates to goods if under the agreement payments are to be made in respect of the goods in the form of rent (other than a rent-charge) issuing out of land.

19   Linked transactions

(1)   A transaction entered into by the debtor or hirer, or a relative of his, with any other person ("the other party"), except one for the provision of security, is a linked transaction in relation to an actual or prospective regulated agreement (the "principal agreement") of which it does not form part if—
 (a) the transaction is entered into in compliance with a term of the principal agreement; or
 (b) the principal agreement is a debtor-creditor-supplier agreement and the transaction is financed, or to be financed, by the principal agreement; or
 (c) the other party is a person mentioned in subsection (2), and a person so mentioned initiated the transaction by suggesting it to the debtor or hirer, or his relative, who enters into it—
  (i) to induce the creditor or owner to enter into the principal agreement, or
  (ii) for another purpose related to the principal agreement, or
  (iii) where the principal agreement is a restricted-use credit agreement, for a purpose related to a transaction financed, or to be financed, by the principal agreement.

(2)   The persons referred to in subsection (1)(c) are—
(a) the creditor or owner, or his associate;
(b) a person who, in the negotiation of the transaction, is represented by a credit-broker who is also a negotiator in antecedent negotiations for the principal agreement;
(c) a person who, at the time the transaction is initiated, knows that the principal agreement has been made or contemplates that it might be made.

(3)   A linked transaction entered into before the making of the principal agreement has no effect until such time (if any) as that agreement is made.

(4)   Regulations may exclude linked transactions of the prescribed description from the operation of subsection (3).

20   Total charge for credit

(1)   The Secretary of State shall make regulations containing such provisions as appear to him appropriate for determining the true cost to the debtor of the credit provided or to be provided under an actual or prospective consumer credit agreement (the "total charge for credit"), and regulations so made shall prescribe—
(a) what items are to be treated as entering into the total charge for credit, and how their amount is to be ascertained;
(b) the method of calculating the rate of the total charge for credit.

(2)   Regulations under subsection (1) may provide for the whole or part of the amount payable by the debtor or his relative under any linked transaction to be included in the total charge for credit, whether or not the creditor is a party to the transaction or derives benefit from it.

PART III
LICENSING OF CREDIT AND HIRE BUSINESSES

Licensing principles

21   Businesses needing a licence

(1)   Subject to this section, a licence is required to carry on a consumer credit business or consumer hire business.

(2)   A local authority does not need a licence to carry on a business.

(3)   A body corporate empowered by a public general Act naming it to carry on a business does not need a licence to do so.

22   Standard and group licences

(1)   A licence may be—
    (a) a standard licence, that is a licence, issued by the Director to a person named in the licence on an application made by him, which, during the prescribed period, covers such activities as are described in the licence, or
    (b) a group licence, that is a licence, issued by the Director (whether on the application of any person or of his own motion), which, during such period as the Director thinks fit or, if he thinks fit, indefinitely, covers such persons and activities as are described in the licence.

(2)   A licence is not assignable or, subject to section 37, transmissible on death or in any other way.

(3)   Except in the case of a partnership or an unincorporated body of persons, a standard licence shall not be issued to more than one person.

(4)   A standard licence issued to a partnership or an unincorporated body of persons shall be issued in the name of the partnership or body.

(5)   The Director may issue a group licence only if it appears to him that the public interest is better served by doing so than by obliging the persons concerned to apply separately for standard licences.

(6)   The persons covered by a group licence may be described by general words, whether or not coupled with the exclusion of named persons, or in any other way the Director thinks fit.

(7)   The fact that a person is covered by a group licence in respect of certain activities does not prevent a standard licence being issued to him in respect of those activities or any of them.

(8)   A group licence issued on the application of any person shall be issued to that person, and general notice shall be given of the issue of any group licence (whether on application or not).

Modified, in relation to standard licences held by a European institution or quasi-European authorised institution, by the Banking Coordination (Second Council Directive) Regulations 1992, SI 1992/3218, reg 57.

23 Authorisation of specific activities

(1)   Subject to this section, a licence to carry on a business covers all lawful activities done in the course of that business, whether by the licensee or other persons on his behalf.

(2)   A licence may limit the activities it covers, whether by authorising the licensee to enter into certain types of agreement only, or in any other way.

(3)   A licence covers the canvassing off trade premises of debtor-creditor-supplier agreements or regulated consumer hire agreements only if, and to the extent that, the licence specifically so provides; and such provision shall not be included in a group licence.

(4)   Regulations may be made specifying other activities which, if engaged in by or on behalf of the person carrying on a business, require to be covered by an express term in his licence.

24 Control of name of business

A standard licence authorises the licensee to carry on a business under the name or names specified in the licence, but not under any other name.

25 Licensee to be a fit person

(1)   A standard licence shall be granted on the application of any person if he satisfies the Director that—
  (a) he is a fit person to engage in activities covered by the licence, and
  (b) the name or names under which he applies to be licensed is or are not misleading or otherwise undesirable.

(2)   In determining whether an applicant for a standard licence is a fit person to engage in any activities, the Director shall have regard to any circumstances appearing to him to be relevant, and in particular any evidence

tending to show that the applicant, or any of the applicant's employees, agents or associates (whether past or present) or, where the applicant is a body corporate, any person appearing to the Director to be a controller of the body corporate or an associate of any such person, has—

  (a) committed any offence involving fraud or other dishonesty, or violence,

  (b) contravened any provision made by or under this Act, or by or under any other enactment regulating the provision of credit to individuals or other transactions with individuals,

  (c) practised discrimination on grounds of sex, colour, race or ethnic or national origins in, or in connection with, the carrying on of any business, or

  (d) engaged in business practices appearing to the Director to be deceitful or oppressive, or otherwise unfair or improper (whether unlawful or not).

(3)   In subsection (2), "associate", in addition to the persons specified in section 184, includes a business associate.

NOTES

Modified by the Banking Coordination (Second Council Directive) Regulations 1992, SI 1992/3218, reg 58.

26 Conduct of business

Regulations may be made as to the conduct by a licensee of his business, and may in particular specify—

  (a) the books and other records to be kept by him, and

  (b) the information to be furnished by him to persons with whom he does business or seeks to do business, and the way it is to be furnished.

NOTES

Modified, in relation to European institutions, by the Banking Coordination (Second Council Directive) Regulations 1992, SI 1992/3218, reg 59(1).

Issue of licences

27 Determination of applications

(1)   Unless the Director determines to issue a licence in accordance with an application he shall, before determining the application, by notice—

(a) inform the applicant, giving his reasons, that, as the case may be, he is minded to refuse the application, or to grant it in terms different from those applied for, describing them, and

(b) invite the applicant to submit to the Director representations in support of his application in accordance with section 34.

(2)   If the Director grants the application in terms different from those applied for then, whether or not the applicant appeals, the Director shall issue the licence in the terms approved by him unless the applicant by notice informs him that he does not desire a licence in those terms.

28   Exclusion from group licence

Where the Director is minded to issue a group licence (whether on the application of any person or not), and in doing so to exclude any person from the group by name, he shall, before determining the matter,—

(a) give notice of that fact to the person proposed to be excluded, giving his reasons, and

(b) invite that person to submit to the Director representations against his exclusion in accordance with section 34.

Renewal, variation, suspension and revocation of licences

29   Renewal

(1)   If the licensee under a standard licence, or the original applicant for, or any licensee under, a group licence of limited duration, wishes the Director to renew the licence, whether on the same terms (except as to expiry) or on varied terms, he must, during the period specified by the Director by general notice or such longer period as the Director may allow, make an application to the Director for its renewal.

(2)   The Director may of his own motion renew any group licence.

(3)   The preceding provisions of this Part apply to the renewal of a licence as they apply to the issue of a licence, except that section 28 does not apply to a person who was already excluded in the licence up for renewal.

(4)   Until the determination of an application under subsection (1) and, where an appeal lies from the determination, until the end of the appeal

period, the licence shall continue in force, notwithstanding that apart from this subsection it would expire earlier.

(5)   On the refusal of an application under this section, the Director may give directions authorising a licensee to carry into effect agreements made by him before the expiry of the licence.

(6)   General notice shall be given of the renewal of a group licence.

30   Variation by request

(1)   On an application made by the licensee, the Director may if he thinks fit by notice to the licensee vary a standard licence in accordance with the application.

(2)   In the case of a group licence issued on the application of any person, the Director, on an application made by that person, may if he thinks fit by notice to that person vary the terms of the licence in accordance with the application; but the Director shall not vary a group licence under this subsection by excluding a named person, other than the person making the request, unless that named person consents in writing to his exclusion.

(3)   In the case of a group licence from which (whether by name or description) a person is excluded, the Director, on an application made by that person, may if he thinks fit, by notice to that person, vary the terms of the licence so as to remove the exclusion.

(4)   Unless the Director determines to vary a licence in accordance with an application he shall, before determining the application, by notice—
  (a) inform the applicant, giving his reasons, that he is minded to refuse the application, and
  (b) invite the applicant to submit to the Director representations in support of his application in accordance with section 34.

(5)   General notice shall be given that a variation of a group licence has been made under this section.

31   Compulsory variation

(1)   Where at a time during the currency of a licence the Director is of the opinion that, if the licence had expired at that time, he would, on an application for its renewal or further renewal on the same terms (except as to expiry),

have been minded to grant the application but on different terms, and that therefore the licence should be varied, he shall proceed as follows.

(2)   In the case of a standard licence the Director shall, by notice—
  (a) inform the licensee of the variations the Director is minded to make in the terms of the licence, stating his reasons, and
  (b) invite him to submit to the Director representations as to the proposed variations in accordance with section 34.

(3)   In the case of a group licence the Director shall—
  (a) give general notice of the variations he is minded to make in the terms of the licence, stating his reasons, and
  (b) in the notice invite any licensee to submit to him representations as to the proposed variations in accordance with section 34.

(4)   In the case of a group licence issued on application the Director shall also—
  (a) inform the original applicant of the variations the Director is minded to make in the terms of the licence, stating his reasons, and
  (b) invite him to submit to the Director representations as to the proposed variations in accordance with section 34.

(5)   If the Director is minded to vary a group licence by excluding any person (other than the original applicant) from the group by name the Director shall, in addition, take the like steps under section 28 as are required in the case mentioned in that section.

(6)   General notice shall be given that a variation of any group licence has been made under this section.

(7)   A variation under this section shall not take effect before the end of the appeal period.

32 Suspension and revocation

(1)   Where at a time during the currency of a licence the Director is of the opinion that if the licence had expired at that time he would have been minded not to renew it, and that therefore it should be revoked or suspended, he shall proceed as follows.

(2)   In the case of a standard licence the Director shall, by notice—
  (a) inform the licensee that, as the case may be, the Director is minded to revoke the licence, or suspend it until a specified date or indefinitely, stating his reasons, and

(b) invite him to submit representations as to the proposed revocation or suspension in accordance with section 34.

(3) In the case of a group licence the Director shall—
  (a) give general notice that, as the case may be, he is minded to revoke the licence, or suspend it until a specified date or indefinitely, stating his reasons, and
  (b) in the notice invite any licensee to submit to him representations as to the proposed revocation or suspension in accordance with section 34.

(4) In the case of a group licence issued on application the Director shall also—
  (a) inform the original applicant that, as the case may be, the Director is minded to revoke the licence, or suspend it until a specified date or indefinitely, stating his reasons, and
  (b) invite him to submit representations as to the proposed revocation or suspension in accordance with section 34.

(5) If he revokes or suspends the licence, the Director may give directions authorising a licensee to carry into effect agreements made by him before the revocation or suspension.

(6) General notice shall be given of the revocation or suspension of a group licence.

(7) A revocation or suspension under this section shall not take effect before the end of the appeal period.

(8) Except for the purposes of section 29, a licensee under a suspended licence shall be treated, in respect of the period of suspension, as if the licence had not been issued; and where the suspension is not expressed to end on a specified date it may, if the Director thinks fit, be ended by notice given by him to the licensee or, in the case of a group licence, by general notice.

33 Application to end suspension

(1) On an application made by a licensee the Director may, if he thinks fit, by notice to the licensee end the suspension of a licence, whether the suspension was for a fixed or indefinite period.

(2) Unless the Director determines to end the suspension in accordance with the application he shall, before determining the application, by notice—

(a) inform the applicant, giving his reasons, that he is minded to refuse the application, and

(b) invite the applicant to submit to the Director representations in support of his application in accordance with section 34.

(3)   General notice shall be given that a suspension of a group licence has been ended under this section.

(4)   In the case of a group licence issued on application—

(a) the references in subsection (1) to a licensee include the original applicant;

(b) the Director shall inform the original applicant that a suspension of a group licence has been ended under this section.

Miscellaneous

34 Representations to Director

(1)   Where this section applies to an invitation by the Director to any person to submit representations, the Director shall invite that person, within 21 days after the notice containing the invitation is given to him or published, or such longer period as the Director may allow,—

(a) to submit his representations in writing to the Director, and

(b) to give notice to the Director, if he thinks fit, that he wishes to make representations orally,

and where notice is given under paragraph (b) the Director shall arrange for the oral representations to be heard.

(2)   In reaching his determination the Director shall take into account any representations submitted or made under this section.

(3)   The Director shall give notice of his determination to the persons who were required to be invited to submit representations about it or, where the invitation to submit representations was required to be given by general notice, shall give general notice of the determination.

35 The register

(1)   The Director shall establish and maintain a register, in which he shall cause to be kept particulars of—

(a) applications not yet determined for the issue, variation or renewal of licences, or for ending the suspension of a licence;

(b) licences which are in force, or have at any time been suspended or revoked, with details of any variation of the terms of a licence;

(c) decisions given by him under this Act, and any appeal from those decisions; and

(d) such other matters (if any) as he thinks fit.

(2)   The Director shall give general notice of the various matters required to be entered in the register, and of any change in them made under subsection (1)(d).

(3)   Any person shall be entitled, on payment of the specified fee—
   (a) to inspect the register during ordinary office hours and take copies of any entry, or
   (b) to obtain from the Director a copy, certified by the Director to be correct, of any entry in the register.

(4)   The Director may, if he thinks fit, determine that the right conferred by subsection (3)(a) shall be exercisable in relation to a copy of the register instead of, or in addition to, the original.

(5)   The Director shall give general notice of the place or places where, and times when, the register or a copy of it may be inspected.

36   Duty to notify changes

(1)   Within 21 working days after a change takes place in any particulars entered in the register in respect of a standard licence or the licensee under section 35(1)(d) (not being a change resulting from action taken by the Director), the licensee shall give the Director notice of the change; and the Director shall cause any necessary amendment to be made in the register.

(2)   Within 21 working days after—
   (a) any change takes place in the officers of—
      (i)   a body corporate, or an unincorporated body of persons, which is the licensee under a standard licence, or
      (ii) a body corporate which is a controller of a body corporate which is such a licensee, or
   (b) a body corporate which is such a licensee becomes aware that a person has become or ceased to be a controller of the body corporate, or

(c) any change takes place in the members of a partnership which is such a licensee (including a change on the amalgamation of the partnership with another firm, or a change whereby the number of partners is reduced to one),

the licensee shall give the Director notice of the change.

(3) Within 14 working days after any change takes place in the officers of a body corporate which is a controller of another body corporate which is a licensee under a standard licence, the controller shall give the licensee notice of the change.

(4) Within 14 working days after a person becomes or ceases to be a controller of a body corporate which is a licensee under a standard licence, that person shall give the licensee notice of the fact.

(5) Where a change in a partnership has the result that the business ceases be carried on under the name, or any of the names, specified in a standard licence the licence shall cease to have effect.

(6) Where the Director is given notice under subsection (1) or (2) of any change, and subsection (5) does not apply, the Director may by notice require the licensee to furnish him with such information, verified in such manner, as the Director may stipulate.

37 Death, bankruptcy etc of licensee

(1) A licence held by one individual terminates if he—
   (a) dies, or
   (b) is adjudged bankrupt, or
   (c) becomes a patient within the meaning of Part VIII of the Mental Health Act 1959.

(2) In relation to a licence held by one individual, or a partnership or other unincorporated body of persons, or a body corporate, regulations may specify other events relating to the licensee on the occurrence of which the licence is to terminate.

(3) Regulations may—
   (a) provide for the termination of a licence by subsection (1), or under subsection (2), to be deferred for a period not exceeding 12 months, and

(b) authorise the business of the licensee to be carried on under the licence by some other person during the period of deferment, subject to such conditions as may be prescribed.

(4)   This section does not apply to group licences.

39   Offences against Part III

(1)   A person who engages in any activities for which a licence is required when he is not a licensee under a licence covering those activities commits an offence.

(2)   A licensee under a standard licence who carries on business under a name not specified in the licence commits an offence.

(3)   A person who fails to give the Director or a licensee notice under section 36 within the period required commits an offence.

40   Enforcement of agreements made by unlicensed trader

(1)   A regulated agreement, other than a non-commercial agreement, if made when the creditor or owner was unlicensed, is enforceable against the debtor or hirer only where the Director has made an order under this section which applies to the agreement.

(2)   Where during any period an unlicensed person (the "trader") was carrying on a consumer credit business or consumer hire business, he or his successor in title may apply to the Director for an order that regulated agreements made by the trader during that period are to be treated as if he had been licensed.

(3)   Unless the Director determines to make an order under subsection (2) in accordance with the application, he shall, before determining the application, by notice—
    (a) inform the applicant, giving his reasons, that, as the case may be, he is minded to refuse the application, or to grant it in terms different from those applied for, describing them, and
    (b) invite the applicant to submit to the Director representations in support of his application in accordance with section 34.

(4)   In determining whether or not to make an order under subsection (2) in respect of any period the Director shall consider, in addition to any other relevant factors—
   (a) how far, if at all, debtors or hirers under regulated agreements made by the trader during that period were prejudiced by the trader's conduct,
   (b) whether or not the Director would have been likely to grant a licence covering that period on an application by the trader, and
   (c) the degree of culpability for the failure to obtain a licence.

(5)   If the Director thinks fit, he may in an order under subsection (2)—
   (a) limit the order to specified agreements, or agreements of a specified description or made at a specified time;
   (b) make the order conditional on the doing of specified acts by the applicant.

41   Appeals to Secretary of State under Part III

(1)   If, in the case of a determination by the Director such as is mentioned in column 1 of the table set out at the end of this section, a person mentioned in relation to that determination in column 2 of the table is aggrieved by the determination he may, within the prescribed period, and in the prescribed manner, appeal to the Secretary of State.

(2)   Regulations may make provision as to the persons by whom (on behalf of the Secretary of State) appeals under this section are to be heard, the manner in which they are to be conducted, and any other matter connected with such appeals.

(3)   On an appeal under this section, the Secretary of State may give such directions for disposing of the appeal as he thinks just, including a direction for the payment of costs by any party to the appeal.

(4)   A direction under subsection (3) for payment of costs may be made a rule of the High Court on the application of the party in whose favour it is given.

(5)   ...

TABLE

Determination	Appellant
Refusal to issue, renew or vary licence in accordance with terms of application.	The applicant.
Exclusion of person from group licence.	The person excluded.
Refusal to give directions in respect of a licensee under section 29(5) or 32(5).	The licensee.
Compulsory variation, or suspension or revocation, of standard licence.	The licensee.
Compulsory variation, or suspension or revocation, of group licence.	The original applicant or any licensee.
Refusal to end suspension of licence in accordance with terms of application.	The applicant.
Refusal to make order under section 40(2) in accordance with terms of application.	The applicant.

NOTES

Sub-s (5): applies to Scotland only.

Modified by the Tribunals and Inquiries Act 1992, s 11(6), and by the Banking Coordination (Second Council Directive) Regulations 1992, SI 1992/3218, reg 18(6), Sch 5, para 5.

PART IV
SEEKING BUSINESS

Advertising

43 Advertisements to which Part IV applies

(1)   This Part applies to any advertisement, published for the purposes of a business carried on by the advertiser, indicating that he is willing—
  (a) to provide credit, or
  (b) to enter into an agreement for the bailment or (in Scotland) the hiring of goods by him.

(2)   An advertisement does not fall within subsection (1) if the advertiser does not carry on—
  (a) a consumer credit business or consumer hire business, or
  (b) a business in the course of which he provides credit to individuals secured on land, or
  (c) a business which comprises or relates to unregulated agreements where—
      (i)   the [law applicable to] the agreement is the law of a country outside the United Kingdom, and
      (ii) if the [law applicable to] the agreement were the law of a part of the United Kingdom it would be a regulated agreement.

(3)   An advertisement does not fall within subsection (1)(a) if it indicates—
  (a) that the credit must exceed [£25,000], and that no security is required, or the security is to consist of property other than land, or
  (b) that the credit is available only to a body corporate.

(4)   An advertisement does not fall within subsection (1)(b) if it indicates that the advertiser is not willing to enter into a consumer hire agreement.

(5)   The Secretary of State may by order provide that this Part shall not apply to other advertisements of a description specified in the order.

NOTES

Sub-s (2): words in square brackets substituted by the Contracts (Applicable Law) Act 1990, s 5, Sch 4, para 2.
Sub-s (3): sum in square brackets substituted by the Consumer Credit (Increase of Monetary Limits) Order 1983, SI 1983/1878, art 4, Schedule, Pt II, as amended by SI 1998/996.

44   Form and content of advertisements

(1)   The Secretary of State shall make regulations as to the form and content of advertisements to which this Part applies, and the regulations shall contain such provisions as appear to him appropriate with a view to ensuring that, having regard to its subject-matter and the amount of detail included in it, an advertisement conveys a fair and reasonably comprehensive indication of the nature of the credit or hire facilities offered by the advertiser and of their true cost to persons using them.

(2)   Regulations under subsection (1) may in particular—
  (a) require specified information to be included in the prescribed manner in advertisements, and other specified material to be excluded;

(b) contain requirements to ensure that specified information is clearly brought to the attention of persons to whom advertisements are directed, and that one part of an advertisement is not given insufficient or excessive prominence compared with another.

45 Prohibition of advertisement where goods etc not sold for cash

If an advertisement to which this Part applies indicates that the advertiser is willing to provide credit under a restricted-use credit agreement relating to goods or services to be supplied by any person, but at the time when the advertisement is published that person is not holding himself out as prepared to sell the goods or provide the services (as the case may be) for cash, the advertiser commits an offence.

46 False or misleading advertisements

(1)   If an advertisement to which this Part applies conveys information which in a material respect is false or misleading the advertiser commits an offence.

(2)   Information stating or implying an intention on the advertiser's part which he has not got is false.

47 Advertising infringements

(1)   Where an advertiser commits an offence against regulations made under section 44 or against section 45 or 46, or would be taken to commit such an offence but for the defence provided by section 168, a like offence is committed by—
   (a) the publisher of the advertisement, and
   (b) any person who, in the course of a business carried on by him, devised the advertisement, or a part of it relevant to the first-mentioned offence, and
   (c) where the advertiser did not procure the publication of the advertisement, the person who did procure it.

(2)   In proceedings for an offence under subsection (1)(a) it is a defence for the person charged to prove that—
   (a) the advertisement was published in the course of a business carried on by him, and

(b) he received the advertisement in the course of that business, and did not know and had no reason to suspect that its publication would be an offence under this Part.

Canvassing etc

48 Definition of canvassing off trade premises (regulated agreements)

(1) An individual (the "canvasser") canvasses a regulated agreement off trade premises if he solicits the entry (as debtor or hirer) of another individual (the "consumer") into the agreement by making oral representations to the consumer, or any other individual, during a visit by the canvasser to any place (not excluded by subsection (2)) where the consumer, or that other individual, as the case may be, is, being a visit—
   (a) carried out for the purpose of making such oral representations to individuals who are at that place, but
   (b) not carried out in response to a request made on a previous occasion.

(2) A place is excluded from subsection (1) if it is a place where a business is carried on (whether on a permanent or temporary basis) by—
   (a) the creditor or owner, or
   (b) a supplier, or
   (c) the canvasser, or the person whose employee or agent the canvasser is, or
   (d) the consumer.

49 Prohibition of canvassing debtor-creditor agreements off trade premises

(1) It is an offence to canvass debtor-creditor agreements off trade premises.

(2) It is also an offence to solicit the entry of an individual (as debtor) into a debtor-creditor agreement during a visit carried out in response to a request made on a previous occasion, where—
   (a) the request was not in writing signed by or on behalf of the person making it, and
   (b) if no request for the visit had been made, the soliciting would have constituted the canvassing of a debtor-creditor agreement off trade premises.

(3)   Subsections (1) and (2) do not apply to any soliciting for an agreement enabling the debtor to overdraw on a current account of any description kept with the creditor, where—

(a) the Director has determined that current accounts of that description kept with the creditor are excluded from subsections (1) and (2), and

(b) the debtor already keeps an account with the creditor (whether a current account or not).

(4)   A determination under subsection (3)(a)—

(a) may be made subject to such conditions as the Director thinks fit, and

(b) shall be made only where the Director is of opinion that it is not against the interests of debtors.

(5)   If soliciting is done in breach of a condition imposed under subsection (4)(a), the determination under subsection (3)(a) does not apply to it.

50   Circulars to minors

(1)   A person commits an offence who, with a view to financial gain, sends to a minor any document inviting him to—

(a) borrow money, or

(b) obtain goods on credit or hire, or

(c) obtain services on credit, or

(d) apply for information or advice on borrowing money or otherwise obtaining credit, or hiring goods.

(2)   In proceedings under subsection (1) in respect of the sending of a document to a minor, it is a defence for the person charged to prove that he did not know, and had no reasonable cause to suspect, that he was a minor.

(3)   Where a document is received by a minor at any school or educational establishment for minors, a person sending it to him at that establishment knowing or suspecting it to be such an establishment shall be taken to have reasonable cause to suspect that he is a minor.

51   Prohibition of unsolicited credit-tokens

(1)   It is an offence to give a person a credit-token if he has not asked for it.

(2)   To comply with subsection (1) a request must be contained in a document signed by the person making the request, unless the credit-token agreement is a small debtor-creditor-supplier agreement.

(3)   Subsection (1) does not apply to the giving of a credit-token to a person—
(a) for use under a credit-token agreement already made, or
(b) in renewal or replacement of a credit-token previously accepted by him under a credit-token agreement which continues in force, whether or not varied.

Miscellaneous

52 Quotations

(1)   Regulations may be made—
(a) as to the form and content of any document (a "quotation") by which a person who carries on a consumer credit business or consumer hire business, or a business in the course of which he provides credit to individuals secured on land, gives prospective customers information about the terms on which he is prepared to do business;
(b) requiring a person carrying on such a business to provide quotations to such persons and in such circumstances as are prescribed.

(2)   Regulations under subsection (1)(a) may in particular contain provisions relating to quotations such as are set out in relation to advertisements in section 44.

53 Duty to display information

Regulations may require a person who carries on a consumer credit business or consumer hire business, or a business in the course of which he provides credit to individuals secured on land, to display in the prescribed manner, at any premises where the business is carried on to which the public have access, prescribed information about the business.

54 Conduct of business regulations

Without prejudice to the generality of section 26, regulations under that section may include provisions further regulating the seeking of business by a licensee who carries on a consumer credit business or a consumer hire business.

NOTES

Modified, in relation to European institutions, by the Banking Coordination (Second Council Directive) Regulations 1992, SI 1992/3218, reg 59(2).

PART V
ENTRY INTO CREDIT OR HIRE AGREEMENTS

Preliminary matters

55 Disclosure of information

(1)   Regulations may require specified information to be disclosed in the prescribed manner to the debtor or hirer before a regulated agreement is made.

(2)   A regulated agreement is not properly executed unless regulations under subsection (1) were complied with before the making of the agreement.

56 Antecedent negotiations

(1)   In this Act "antecedent negotiations" means any negotiations with the debtor or hirer—
   (a) conducted by the creditor or owner in relation to the making of any regulated agreement, or
   (b) conducted by a credit-broker in relation to goods sold or proposed to be sold by the credit-broker to the creditor before forming the subject-matter of a debtor-creditor-supplier agreement within section 12(a), or
   (c) conducted by the supplier in relation to a transaction financed or proposed to be financed by a debtor-creditor-supplier agreement within section 12(b) or (c),

and "negotiator" means the person by whom negotiations are so conducted with the debtor or hirer.

(2)   Negotiations with the debtor in a case falling within subsection (1)(b) or (c) shall be deemed to be conducted by the negotiator in the capacity of agent of the creditor as well as in his actual capacity.

(3)   An agreement is void if, and to the extent that, it purports in relation to an actual or prospective regulated agreement—
- (a) to provide that a person acting as, or on behalf of, a negotiator is to be treated as the agent of the debtor or hirer, or
- (b) to relieve a person from liability for acts or omissions of any person acting as, or on behalf of, a negotiator.

(4)   For the purposes of this Act, antecedent negotiations shall be taken to begin when the negotiator and the debtor or hirer first enter into communication (including communication by advertisement), and to include any representations made by the negotiator to the debtor or hirer and any other dealings between them.

57   Withdrawal from prospective agreement

(1)   The withdrawal of a party from a prospective regulated agreement shall operate to apply this Part to the agreement, any linked transaction and any other thing done in anticipation of the making of the agreement as it would apply if the agreement were made and then cancelled under section 69.

(2)   The giving to a party of a written or oral notice which, however expressed, indicates the intention of the other party to withdraw from a prospective regulated agreement operates as a withdrawal from it.

(3)   Each of the following shall be deemed to be the agent of the creditor or owner for the purpose of receiving a notice under subsection (2)—
- (a) a credit-broker or supplier who is the negotiator in antecedent negotiations, and
- (b) any person who, in the course of a business carried on by him, acts on behalf of the debtor or hirer in any negotiations for the agreement.

(4)   Where the agreement, if made, would not be a cancellable agreement, subsection (1) shall nevertheless apply as if the contrary were the case.

58   Opportunity for withdrawal from prospective land mortgage

(1)   Before sending to the debtor or hirer, for his signature, an unexecuted agreement in a case where the prospective regulated agreement is to be secured on land (the "mortgaged land"), the creditor or owner shall give the debtor or hirer a copy of the unexecuted agreement which contains a notice in the prescribed form indicating the right of the debtor or hirer to

withdraw from the prospective agreement, and how and when the right is exercisable, together with a copy of any other document referred to in the unexecuted agreement.

(2)   Subsection (1) does not apply to—
   (a) a restricted-use credit agreement to finance the purchase of the mortgaged land, or
   (b) an agreement for a bridging loan in connection with the purchase of the mortgaged land or other land.

59   Agreement to enter future agreement void

(1)   An agreement is void if, and to the extent that, it purports to bind a person to enter as debtor or hirer into a prospective regulated agreement.

(2)   Regulations may exclude from the operation of subsection (1) agreements such as are described in the regulations.

Making the agreement

60   Form and content of agreements

(1)   The Secretary of State shall make regulations as to the form and content of documents embodying regulated agreements, and the regulations shall contain such provisions as appear to him appropriate with a view to ensuring that the debtor or hirer is made aware of—
   (a) the rights and duties conferred or imposed on him by the agreement,
   (b) the amount and rate of the total charge for credit (in the case of a consumer credit agreement),
   (c) the protection and remedies available to him under this Act, and
   (d) any other matters which, in the opinion of the Secretary of State, it is desirable for him to know about in connection with the agreement.

(2)   Regulations under subsection (1) may in particular—
   (a) require specified information to be included in the prescribed manner in documents, and other specified material to be excluded;
   (b) contain requirements to ensure that specified information is clearly brought to the attention of the debtor or hirer, and that one part of a document is not given insufficient or excessive prominence compared with another.

(3)   If, on an application made to the Director by a person carrying on a consumer credit business or a consumer hire business, it appears to the Director impracticable for the applicant to comply with any requirement of regulations under subsection (1) in a particular case, he may, by notice to the applicant, direct that the requirement be waived or varied in relation to such agreements, and subject to such conditions (if any), as he may specify, and this Act and the regulations shall have effect accordingly.

(4)   The Director shall give a notice under subsection (3) only if he is satisfied that to do so would not prejudice the interests of debtors or hirers.

61   Signing of agreement

(1)   A regulated agreement is not properly executed unless—
  (a) a document in the prescribed form itself containing all the prescribed terms and conforming to regulations under section 60(1) is signed in the prescribed manner both by the debtor or hirer and by or on behalf of the creditor or owner, and
  (b) the document embodies all the terms of the agreement, other than implied terms, and
  (c) the document is, when presented or sent to the debtor or hirer for signature, in such a state that all its terms are readily legible.

(2)   In addition, where the agreement is one to which section 58(1) applies, it is not properly executed unless—
  (a) the requirements of section 58(1) were complied with, and
  (b) the unexecuted agreement was sent, for his signature, to the debtor or hirer by post not less than seven days after a copy of it was given to him under section 58(1), and
  (c) during the consideration period, the creditor or owner refrained from approaching the debtor or hirer (whether in person, by telephone or letter, or in any other way) except in response to a specific request made by the debtor or hirer after the beginning of the consideration period, and
  (d) no notice of withdrawal by the debtor or hirer was received by the creditor or owner before the sending of the unexecuted agreement.

(3)   In subsection (2)(c), "the consideration period" means the period beginning with the giving of the copy under section 58(1) and ending—
  (a) at the expiry of seven days after the day on which the unexecuted agreement is sent, for his signature, to the debtor or hirer, or

(b) on its return by the debtor or hirer after signature by him,

whichever first occurs.

(4)　　Where the debtor or hirer is a partnership or an unincorporated body of persons, subsection (1)(a) shall apply with the substitution for "by the debtor or hirer" of "by or on behalf of the debtor or hirer".

62　Duty to supply copy of unexecuted agreement

(1)　　If the unexecuted agreement is presented personally to the debtor or hirer for his signature, but on the occasion when he signs it the document does not become an executed agreement, a copy of it, and of any other document referred to in it, must be there and then delivered to him.

(2)　　If the unexecuted agreement is sent to the debtor or hirer for his signature, a copy of it, and of any other document referred to in it, must be sent to him at the same time.

(3)　　A regulated agreement is not properly executed if the requirements of this section are not observed.

63　Duty to supply copy of executed agreement

(1)　　If the unexecuted agreement is presented personally to the debtor or hirer for his signature, and on the occasion when he signs it the document becomes an executed agreement, a copy of the executed agreement, and of any other document referred to in it, must be there and then delivered to him.

(2)　　A copy of the executed agreement, and of any other document referred to in it, must be given to the debtor or hirer within the seven days following the making of the agreement unless—
　　(a) subsection (1) applies, or
　　(b) the unexecuted agreement was sent to the debtor or hirer for his signature and, on the occasion of his signing it, the document became an executed agreement.

(3)　　In the case of a cancellable agreement, a copy under subsection (2) must be sent by post.

(4)　　In the case of a credit-token agreement, a copy under subsection (2) need not be given within the seven days following the making of the agreement if it is given before or at the time when the credit-token is given to the debtor.

(5)   A regulated agreement is not properly executed if the requirements of this section are not observed.

64   Duty to give notice of cancellation rights

(1)   In the case of a cancellable agreement, a notice in the prescribed form indicating the right of the debtor or hirer to cancel the agreement, how and when that right is exercisable, and the name and address of a person to whom notice of cancellation may be given,—
(a) must be included in every copy given to the debtor or hirer under section 62 or 63, and
(b) except where section 63(2) applied, must also be sent by post to the debtor or hirer within the seven days following the making of the agreement.

(2)   In the case of a credit-token agreement, a notice under subsection (1)(b) need not be sent by post within the seven days following the making of the agreement if either—
(a) it is sent by post to the debtor or hirer before the credit-token is given to him, or
(b) it is sent by post to him together with the credit-token.

(3)   Regulations may provide that except where section 63(2) applied a notice sent under subsection (1)(b) shall be accompanied by a further copy of the executed agreement, and of any other document referred to in it.

(4)   Regulations may provide that subsection (1)(b) is not to apply in the case of agreements such as are described in the regulations, being agreements made by a particular person, if—
(a) on an application by that person to the Director, the Director has determined that, having regard to—
(i)   the manner in which antecedent negotiations for agreements with the applicant of that description are conducted, and
(ii) the information provided to debtors or hirers before such agreements are made,
the requirement imposed by subsection (1)(b) can be dispensed with without prejudicing the interests of debtors or hirers; and
(b) any conditions imposed by the Director in making the determination are complied with.

(5)   A cancellable agreement is not properly executed if the requirements of this section are not observed.

65   Consequences of improper execution

(1)   An improperly-executed regulated agreement is enforceable against the debtor or hirer on an order of the court only.

(2)   A retaking of goods or land to which a regulated agreement relates is an enforcement of the agreement.

66   Acceptance of credit-tokens

(1)   The debtor shall not be liable under a credit-token agreement for use made of the credit-token by any person unless the debtor had previously accepted the credit-token, or the use constituted an acceptance of it by him.

(2)   The debtor accepts a credit-token when—
  (a) it is signed, or
  (b) a receipt for it is signed, or
  (c) it is first used,

either by the debtor himself or by a person who, pursuant to the agreement, is authorised by him to use it.

Cancellation of certain agreements within cooling-off period

67   Cancellable agreements

A regulated agreement may be cancelled by the debtor or hirer in accordance with this Part if the antecedent negotiations included oral representations made when in the presence of the debtor or hirer by an individual acting as, or on behalf of, the negotiator, unless—
  (a) the agreement is secured on land, or is a restricted-use credit agreement to finance the purchase of land or is an agreement for a bridging loan in connection with the purchase of land, or
  (b) the unexecuted agreement is signed by the debtor or hirer at premises at which any of the following is carrying on any business (whether on a permanent or temporary basis)—

   (i)   the creditor or owner;
  (ii)   any party to a linked transaction (other than the debtor or hirer or a relative of his);
 (iii)   the negotiator in any antecedent negotiations.

68 Cooling-off period

The debtor or hirer may serve notice of cancellation of a cancellable agreement between his signing of the unexecuted agreement and—
(a) the end of the fifth day following the day on which he received a copy under section 63(2) or a notice under section 64(1)(b), or
(b) if (by virtue of regulations made under section 64(4)) section 64(1)(b) does not apply, the end of the fourteenth day following the day on which he signed the unexecuted agreement.

69 Notice of cancellation

(1)   If within the period specified in section 68 the debtor or hirer under a cancellable agreement serves on—
(a) the creditor or owner, or
(b) the person specified in the notice under section 64(1), or
(c) a person who (whether by virtue of subsection (6) or otherwise) is the agent of the creditor or owner,

a notice (a "notice of cancellation") which, however expressed and whether or not conforming to the notice given under section 64(1), indicates the intention of the debtor or hirer to withdraw from the agreement, the notice shall operate—
  (i)   to cancel the agreement, and any linked transaction, and
 (ii)   to withdraw any offer by the debtor or hirer, or his relative, to enter into a linked transaction.

(2)   In the case of a debtor-creditor-supplier agreement for restricted-use credit financing—
(a) the doing of work or supply of goods to meet an emergency, or
(b) the supply of goods which, before service of the notice of cancellation, had by the act of the debtor or his relative become incorporated in any land or thing not comprised in the agreement or any linked transaction,

subsection (1) shall apply with the substitution of the following for paragraph (i)—

"(i) to cancel only such provisions of the agreement and any linked transaction as—
   (aa) relate to the provision of credit, or
   (bb) require the debtor to pay an item in the total charge for credit, or
   (cc) subject the debtor to any obligation other than to pay for the doing of the said work, or the supply of the said goods".

(3)   Except so far as is otherwise provided, references in this Act to the cancellation of an agreement or transaction do not include a case within subsection (2).

(4)   Except as otherwise provided by or under this Act, an agreement or transaction cancelled under subsection (1) shall be treated as if it had never been entered into.

(5)   Regulations may exclude linked transactions of the prescribed description from subsection (1)(i) or (ii).

(6)   Each of the following shall be deemed to be the agent of the creditor or owner for the purpose of receiving a notice of cancellation—
   (a) a credit-broker or supplier who is the negotiator in antecedent negotiations, and
   (b) any person who, in the course of a business carried on by him, acts on behalf of the debtor or hirer in any negotiations for the agreement.

(7)   Whether or not it is actually received by him, a notice of cancellation sent by post to a person shall be deemed to be served on him at the time of posting.

70 Cancellation: recovery of money paid by debtor or hirer

(1)   On the cancellation of a regulated agreement, and of any linked transaction,—
   (a) any sum paid by the debtor or hirer, or his relative, under or in contemplation of the agreement or transaction, including any item in the total charge for credit, shall become repayable, and
   (b) any sum, including any item in the total charge for credit, which but for the cancellation is, or would or might become, payable by the debtor or hirer, or his relative, under the agreement or transaction shall cease to be, or shall not become, so payable, and
   (c) in the case of a debtor-creditor-supplier agreement falling within section 12(b) any sum paid on the debtor's behalf by the creditor to the supplier shall become repayable to the creditor.

(2)   If, under the terms of a cancelled agreement or transaction, the debtor or hirer, or his relative, is in possession of any goods, he shall have a lien on them for any sum repayable to him under subsection (1) in respect of that agreement or transaction, or any other linked transaction.

(3)   A sum repayable under subsection (1) is repayable by the person to whom it was originally paid, but in the case of a debtor-creditor-supplier agreement falling within section 12(b) the creditor and the supplier shall be under a joint and several liability to repay sums paid by the debtor, or his relative, under the agreement or under a linked transaction falling within section 19(1)(b) and accordingly, in such a case, the creditor shall be entitled, in accordance with rules of court, to have the supplier made a party to any proceedings brought against the creditor to recover any such sums.

(4)   Subject to any agreement between them, the creditor shall be entitled to be indemnified by the supplier for loss suffered by the creditor in satisfying his liability under subsection (3), including costs reasonably incurred by him in defending proceedings instituted by the debtor.

(5)   Subsection (1) does not apply to any sum which, if not paid by a debtor, would be payable by virtue of section 71, and applies to a sum paid or payable by a debtor for the issue of a credit-token only where the credit-token has been returned to the creditor or surrendered to a supplier.

(6)   If the total charge for credit includes an item in respect of a fee or commission charged by a credit-broker, the amount repayable under subsection (1) in respect of that item shall be the excess over [£5] of the fee or commission.

(7)   If the total charge for credit includes any sum payable or paid by the debtor to a credit-broker otherwise than in respect of a fee or commission charged by him, that sum shall for the purposes of subsection (6) be treated as if it were such a fee or commission.

(8)   So far only as is necessary to give effect to section 69(2), this section applies to an agreement or transaction within that subsection as it applies to a cancelled agreement or transaction.

NOTES

Sub-s (6): sum in square brackets substituted by the Consumer Credit (Further Increase of Monetary Amounts) Order 1998, SI 1998/997, art 3, Schedule.

71 Cancellation: repayment of credit

(1)   Notwithstanding the cancellation of a regulated consumer credit agreement, other than a debtor-creditor-supplier agreement for restricted-use credit, the agreement shall continue in force so far as it relates to repayment of credit and payment of interest.

(2)   If, following the cancellation of a regulated consumer credit agreement, the debtor repays the whole or a portion of a credit—
  (a) before the expiry of one month following service of the notice of cancellation, or
  (b) in the case of a credit repayable by instalments, before the date on which the first instalment is due,

no interest shall be payable on the amount repaid.

(3)   If the whole of a credit repayable by instalments is not repaid on or before the date specified in subsection (2)(b), the debtor shall not be liable to repay any of the credit except on receipt of a request in writing in the prescribed form, signed by or on behalf of the creditor, stating the amounts of the remaining instalments (recalculated by the creditor as nearly as may be in accordance with the agreement and without extending the repayment period), but excluding any sum other than principal and interest.

(4)   Repayment of a credit, or payment of interest, under a cancelled agreement shall be treated as duly made if it is made to any person on whom, under section 69, a notice of cancellation could have been served, other than a person referred to in section 69(6)(b).

72 Cancellation: return of goods

(1)   This section applies where any agreement or transaction relating to goods, being—
  (a) a restricted-use debtor-creditor-supplier agreement, a consumer hire agreement, or a linked transaction to which the debtor or hirer under any regulated agreement is a party, or
  (b) a linked transaction to which a relative of the debtor or hirer under any regulated agreement is a party,

is cancelled after the debtor or hirer (in a case within paragraph (a)) or the relative (in a case within paragraph (b)) has acquired possession of the goods by virtue of the agreement or transaction.

(2)   In this section—
  (a) "the possessor" means the person who has acquired possession of the goods as mentioned in subsection (1),
  (b) "the other party" means the person from whom the possessor acquired possession, and
  (c) "the pre-cancellation period" means the period beginning when the possessor acquired possession and ending with the cancellation.

(3)   The possessor shall be treated as having been under a duty throughout the pre-cancellation period—
  (a) to retain possession of the goods, and
  (b) to take reasonable care of them.

(4)   On the cancellation, the possessor shall be under a duty, subject to any lien, to restore the goods to the other party in accordance with this section, and meanwhile to retain possession of the goods and take reasonable care of them.

(5)   The possessor shall not be under any duty to deliver the goods except at his own premises and in pursuance of a request in writing signed by or on behalf of the other party and served on the possessor either before, or at the time when, the goods are collected from those premises.

(6)   If the possessor—
  (a) delivers the goods (whether at his own premises or elsewhere) to any person on whom, under section 69, a notice of cancellation could have been served (other than a person referred to in section 69(6)(b)), or
  (b) sends the goods at his own expense to such a person,

he shall be discharged from any duty to retain the goods or deliver them to any person.

(7)   Where the possessor delivers the goods as mentioned in subsection (6)(a) his obligation to take care of the goods shall cease; and if he sends the goods as mentioned in subsection (6)(b), he shall be under a duty to take reasonable care to see that they are received by the other party and not damaged in transit, but in other respects his duty to take care of the goods shall cease.

(8)   Where, at any time during the period of 21 days following the cancellation, the possessor receives such a request as is mentioned in subsection (5), and unreasonably refuses or unreasonably fails to comply with it, his duty to take reasonable care of the goods shall continue until

he delivers or sends the goods as mentioned in subsection (6), but if within that period he does not receive such a request his duty to take reasonable care of the goods shall cease at the end of that period.

(9)   The preceding provisions of this section do not apply to—
  (a) perishable goods, or
  (b) goods which by their nature are consumed by use and which, before the cancellation, were so consumed, or
  (c) goods supplied to meet an emergency, or
  (d) goods which, before the cancellation, had become incorporated in any land or thing not comprised in the cancelled agreement or a linked transaction.

(10)   Where the address of the possessor is specified in the executed agreement, references in this section to his own premises are to that address and no other.

(11)   Breach of a duty imposed by this section is actionable as a breach of statutory duty.

73   Cancellation: goods given in part-exchange

(1)   This section applies on the cancellation of a regulated agreement where, in antecedent negotiations, the negotiator agreed to take goods in part-exchange (the "part-exchange goods") and those goods have been delivered to him.

(2)   Unless, before the end of the period of ten days beginning with the date of cancellation, the part-exchange goods are returned to the debtor or hirer in a condition substantially as good as when they were delivered to the negotiator, the debtor or hirer shall be entitled to recover from the negotiator a sum equal to the part-exchange allowance (as defined in subsection (7)(b)).

(3)   In the case of a debtor-creditor-supplier agreement within section 12(b), the negotiator and the creditor shall be under a joint and several liability to pay to the debtor a sum recoverable under subsection (2).

(4)   Subject to any agreement between them, the creditor shall be entitled to be indemnified by the negotiator for loss suffered by the creditor in satisfying his liability under subsection (3), including costs reasonably incurred by him in defending proceedings instituted by the debtor.

(5)   During the period of ten days beginning with the date of cancellation, the debtor or hirer, if he is in possession of goods to which the cancelled agreement relates, shall have a lien on them for—
  (a) delivery of the part-exchange goods, in a condition substantially as good as when they were delivered to the negotiator, or
  (b) a sum equal to the part-exchange allowance;

and if the lien continues to the end of that period it shall thereafter subsist only as a lien for a sum equal to the part-exchange allowance.

(6)   Where the debtor or hirer recovers from the negotiator or creditor, or both of them jointly, a sum equal to the part-exchange allowance, then, if the title of the debtor or hirer to the part-exchange goods has not vested in the negotiator, it shall so vest on the recovery of that sum.

(7)   For the purposes of this section—
  (a) the negotiator shall be treated as having agreed to take goods in part-exchange if, in pursuance of the antecedent negotiations, he either purchased or agreed to purchase those goods or accepted or agreed to accept them as part of the consideration for the cancelled agreement, and
  (b) the part-exchange allowance shall be the sum agreed as such in the antecedent negotiations or, if no such agreement was arrived at, such sum as it would have been reasonable to allow in respect of the part-exchange goods if no notice of cancellation had been served.

(8)   In an action brought against the creditor for a sum recoverable under subsection (2), he shall be entitled, in accordance with rules of court, to have the negotiator made a party to the proceedings.

Exclusion of certain agreements from Part V

74 Exclusion of certain agreements from Part V

(1)   This Part (except section 56) does not apply to—
  (a) a non-commercial agreement, or
  (b) a debtor-creditor agreement enabling the debtor to overdraw on a current account, or
  (c) a debtor-creditor agreement to finance the making of such payments arising on, or connected with, the death of a person as may be prescribed.

(2)   This Part (except sections 55 and 56) does not apply to a small debtor-creditor-supplier agreement for restricted-use credit.

[(2A) In the case of an agreement to which the Consumer Protection (Cancellation of Contracts Concluded away from Business Premises) Regulations 1987 apply the reference in subsection (2) to a small agreement shall be construed as if in section 17(1)(a) and (b) "£35" were substituted for "£50".]

(3)   Subsection (1)(b) or (c) applies only where the Director so determines, and such a determination—
   (a) may be made subject to such conditions as the Director thinks fit, and
   (b) shall be made only if the Director is of opinion that it is not against the interests of debtors.

[(3A) Notwithstanding anything in subsection (3)(b) above, in relation to a debtor-creditor agreement under which the creditor is the Bank of England or a bank within the meaning of the Bankers' Books Evidence Act 1879, the Director shall make a determination that subsection (1)(b) above applies unless he considers that it would be against the public interest to do so.]

(4)   If any term of an agreement falling within subsection [(1)(c)] or (2) is expressed in writing, regulations under section 60(1) shall apply to that term (subject to section 60(3)) as if the agreement was a regulated agreement not falling within subsection [(1)(c)] or (2).

NOTES

Sub-s (2A): inserted by the Consumer Protection (Cancellation of Contracts Concluded away from Business Premises) Regulations 1987, SI 1987/2117, reg 9.

Sub-s (3A): added by the Banking Act 1979, s 38(1).

Sub-s (4): words in square brackets substituted by the Banking Act 1979, s 38(1).

PART VI
MATTERS ARISING DURING CURRENCY OF CREDIT OR HIRE AGREEMENTS

75 Liability of creditor for breaches by supplier

(1)   If the debtor under a debtor-creditor-supplier agreement falling within section 12(b) or (c) has, in relation to a transaction financed by the

agreement, any claim against the supplier in respect of a misrepresentation or breach of contract, he shall have a like claim against the creditor, who, with the supplier, shall accordingly be jointly and severally liable to the debtor.

(2) Subject to any agreement between them, the creditor shall be entitled to be indemnified by the supplier for loss suffered by the creditor in satisfying his liability under subsection (1), including costs reasonably incurred by him in defending proceedings instituted by the debtor.

(3) Subsection (1) does not apply to a claim—
    (a) under a non-commercial agreement, or
    (b) so far as the claim relates to any single item to which the supplier has attached a cash price not exceeding [£100] or more than [£30,000].

(4) This section applies notwithstanding that the debtor, in entering into the transaction, exceeded the credit limit or otherwise contravened any term of the agreement.

(5) In an action brought against the creditor under subsection (1) he shall be entitled, in accordance with rules of court, to have the supplier made a party to the proceedings.

NOTES

Sub-s (3): sums in square brackets substituted by the Consumer Credit (Increase of Monetary Limits) Order 1983, SI 1983/1878, arts 3, 4, Schedule, Pts I, II.

76 Duty to give notice before taking certain action

(1) The creditor or owner is not entitled to enforce a term of a regulated agreement by—
    (a) demanding earlier payment of any sum, or
    (b) recovering possession of any goods or land, or
    (c) treating any right conferred on the debtor or hirer by the agreement as terminated, restricted or deferred,

except by or after giving the debtor or hirer not less than seven days' notice of intention to do so.

(2) Subsection (1) applies only where—
    (a) a period for the duration of the agreement is specified in the agreement, and

    (b) that period has not ended when the creditor or owner does an act mentioned in subsection (1),

but so applies notwithstanding that, under the agreement, any party is entitled to terminate it before the end of the period so specified.

(3)   A notice under subsection (1) is ineffective if not in the prescribed form.

(4)   Subsection (1) does not prevent a creditor from treating the right to draw on any credit as restricted or deferred and taking such steps as may be necessary to make the restriction or deferment effective.

(5)   Regulations may provide that subsection (1) is not to apply to agreements described by the regulations.

(6)   Subsection (1) does not apply to a right of enforcement arising by reason of any breach by the debtor or hirer of the regulated agreement.

77 Duty to give information to debtor under fixed-sum credit agreement

(1)   The creditor under a regulated agreement for fixed-sum credit, within the prescribed period after receiving a request in writing to that effect from the debtor and payment of a fee of [£1], shall give the debtor a copy of the executed agreement (if any) and of any other document referred to in it, together with a statement signed by or on behalf of the creditor showing, according to the information to which it is practicable for him to refer,—
    (a) the total sum paid under the agreement by the debtor;
    (b) the total sum which has become payable under the agreement by the debtor but remains unpaid, and the various amounts comprised in that total sum, with the date when each became due; and
    (c) the total sum which is to become payable under the agreement by the debtor, and the various amounts comprised in that total sum, with the date, or mode of determining the date, when each becomes due.

(2)   If the creditor possesses insufficient information to enable him to ascertain the amounts and dates mentioned in subsection (1)(c), he shall be taken to comply with that paragraph if his statement under subsection (1) gives the basis on which, under the regulated agreement, they would fall to be ascertained.

(3)   Subsection (1) does not apply to—

   (a) an agreement under which no sum is, or will or may become, payable by the debtor, or

   (b) a request made less than one month after a previous request under that subsection relating to the same agreement was complied with.

(4)   If the creditor under an agreement fails to comply with subsection (1)—

   (a) he is not entitled, while the default continues, to enforce the agreement; and

   (b) if the default continues for one month he commits an offence.

(5)   This section does not apply to a non-commercial agreement.

NOTES

Sub-s (1): sum in square brackets substituted by the Consumer Credit (Further Increase in Monetary Amounts) Order 1998, SI 1998/997, art 3, Schedule.

78 Duty to give information to debtor under running-account credit agreement

(1)   The creditor under a regulated agreement for running-account credit, within the prescribed period after receiving a request in writing to that effect from the debtor and payment of a fee of [£1], shall give the debtor a copy of the executed agreement (if any) and of any other document referred to in it, together with a statement signed by or on behalf of the creditor showing, according to the information to which it is practicable for him to refer,—

   (a) the state of the account, and

   (b) the amount, if any, currently payable under the agreement by the debtor to the creditor, and

   (c) the amounts and due dates of any payments which, if the debtor does not draw further on the account, will later become payable under the agreement by the debtor to the creditor.

(2)   If the creditor possesses insufficient information to enable him to ascertain the amounts and dates mentioned in subsection (1)(c), he shall be taken to comply with that paragraph if his statement under subsection (1) gives the basis on which, under the regulated agreement, they would fall to be ascertained.

(3)   Subsection (1) does not apply to—

(a) an agreement under which no sum is, or will or may become, payable by the debtor, or

(b) a request made less than one month after a previous request under that subsection relating to the same agreement was complied with.

(4) Where running-account credit is provided under a regulated agreement, the creditor shall give the debtor statements in the prescribed form, and with the prescribed contents—

(a) showing according to the information to which it is practicable for him to refer, the state of the account at regular intervals of not more than twelve months, and

(b) where the agreement provides, in relation to specified periods, for the making of payments by the debtor, or the charging against him of interest or any other sum, showing according to the information to which it is practicable for him to refer the state of the account at the end of each of those periods during which there is any movement in the account.

(5) A statement under subsection (4) shall be given within the prescribed period after the end of the period to which the statement relates.

(6) If the creditor under an agreement fails to comply with subsection (1)—

(a) he is not entitled, while the default continues, to enforce the agreement; and

(b) if the default continues for one month he commits an offence.

(7) This section does not apply to a non-commercial agreement, and subsections (4) and (5) do not apply to a small agreement.

NOTES

Sub-s (1): sum in square brackets substituted by the Consumer Credit (Further Increase in Monetary Amounts) Order 1998, SI 1998/997, art 3, Schedule.

79 Duty to give hirer information

(1) The owner under a regulated consumer hire agreement, within the prescribed period after receiving a request in writing to that effect from the hirer and payment of a fee of [£1], shall give to the hirer a copy of the executed agreement and of any other document referred to in it, together with a statement signed by or on behalf of the owner showing, according to the information to which it is practicable for him to refer, the total sum

which has become payable under the agreement by the hirer but remains unpaid and the various amounts comprised in that total sum, with the date when each became due.

(2)   Subsection (1) does not apply to—
    (a) an agreement under which no sum is, or will or may become, payable by the hirer, or
    (b) a request made less than one month after a previous request under that subsection relating to the same agreement was complied with.

(3)   If the owner under an agreement fails to comply with subsection (1)—
    (a) he is not entitled, while the default continues, to enforce the agreement; and
    (b) if the default continues for one month he commits an offence.

(4)   This section does not apply to a non-commercial agreement.

NOTES

Sub-s (1): sum in square brackets substituted by the Consumer Credit (Further Increase in Monetary Amounts) Order 1998, SI 1998/997, art 3, Schedule.

80 Debtor or hirer to give information about goods

(1)   Where a regulated agreement, other than a non-commercial agreement, requires the debtor or hirer to keep goods to which the agreement relates in his possession or control, he shall, within seven working days after he has received a request in writing to that effect from the creditor or owner, tell the creditor or owner where the goods are.

(2)   If the debtor or hirer fails to comply with subsection (1), and the default continues for 14 days, he commits an offence.

81 Appropriation of payments

(1)   Where a debtor or hirer is liable to make to the same person payments in respect of two or more regulated agreements, he shall be entitled, on making any payment in respect of the agreements which is not sufficient to discharge the total amount then due under all the agreements, to appropriate the sum so paid by him—
    (a) in or towards the satisfaction of the sum due under any one of the agreements, or

(b) in or towards the satisfaction of the sums due under any two or more of the agreements in such proportions as he thinks fit.

(2)   If the debtor or hirer fails to make any such appropriation where one or more of the agreements is—

(a) a hire-purchase agreement or conditional sale agreement, or

(b) a consumer hire agreement, or

(c) an agreement in relation to which any security is provided,

the payment shall be appropriated towards the satisfaction of the sums due under the several agreements respectively in the proportions which those sums bear to one another.

82   Variation of agreements

(1)   Where, under a power contained in a regulated agreement, the creditor or owner varies the agreement, the variation shall not take effect before notice of it is given to the debtor or hirer in the prescribed manner.

(2)   Where an agreement (a "modifying agreement") varies or supplements an earlier agreement, the modifying agreement shall for the purposes of this Act be treated as—

(a) revoking the earlier agreement, and

(b) containing provisions reproducing the combined effect of the two agreements,

and obligations outstanding in relation to the earlier agreement shall accordingly be treated as outstanding instead in relation to the modifying agreement.

(3)   If the earlier agreement is a regulated agreement but (apart from this subsection) the modifying agreement is not then, unless the modifying agreement is for running account credit, it shall be treated as a regulated agreement.

(4)   If the earlier agreement is a regulated agreement for running-account credit, and by the modifying agreement the creditor allows the credit limit to be exceeded but intends the excess to be merely temporary, Part V (except section 56) shall not apply to the modifying agreement.

(5)   If—

(a) the earlier agreement is a cancellable agreement, and

(b) the modifying agreement is made within the period applicable under section 68 to the earlier agreement,

then, whether or not the modifying agreement would, apart from this subsection, be a cancellable agreement, it shall be treated as a cancellable agreement in respect of which a notice may be served under section 68 not later than the end of the period applicable under that section to the earlier agreement.

(6)   Except under subsection (5), a modifying agreement shall not be treated as a cancellable agreement.

(7)   This section does not apply to a non-commercial agreement.

83   Liability for misuse of credit facilities

(1)   The debtor under a regulated consumer credit agreement shall not be liable to the creditor for any loss arising from use of the credit facility by another person not acting, or to be treated as acting, as the debtor's agent.

(2)   This section does not apply to a non-commercial agreement, or to any loss in so far as it arises from misuse of an instrument to which section 4 of the Cheques Act 1957 applies.

84   Misuse of credit-tokens

(1)   Section 83 does not prevent the debtor under a credit-token agreement from being made liable to the extent of [£50] (or the credit limit if lower) for loss to the creditor arising from use of the credit-token by other persons during a period beginning when the credit-token ceases to be in the possession of any authorised person and ending when the credit-token is once more in the possession of an authorised person.

(2)   Section 83 does not prevent the debtor under a credit-token agreement from being made liable to any extent for loss to the creditor from use of the credit-token by a person who acquired possession of it with the debtor's consent.

(3)   Subsections (1) and (2) shall not apply to any use of the credit-token after the creditor has been given oral or written notice that it is lost or stolen, or is for any other reason liable to misuse.

(4)   Subsections (1) and (2) shall not apply unless there are contained in the credit-token agreement in the prescribed manner particulars of the name, address and telephone number of a person stated to be the person to whom notice is to be given under subsection (3).

(5)   Notice under subsection (3) takes effect when received, but where it is given orally, and the agreement so requires, it shall be treated as not taking effect if not confirmed in writing within seven days.

(6)   Any sum paid by the debtor for the issue of the credit-token, to the extent (if any) that it has not been previously offset by use made of the credit token, shall be treated as paid towards satisfaction of any liability under subsection (1) or (2).

(7)   The debtor, the creditor, and any person authorised by the debtor to use the credit-token, shall be authorised persons for the purposes of subsection (1).

(8)   Where two or more credit-tokens are given under one credit-token agreement, the preceding provisions of this section apply to each credit-token separately.

NOTES

Sub-s (1): figure in square brackets substituted by the Consumer Credit (Further Increase in Monetary Amounts) Order 1998, SI 1998/997, art 3, Schedule.

85   Duty on issue of new credit-tokens

(1)   Whenever, in connection with a credit-token agreement, a credit-token (other than the first) is given by the creditor to the debtor, the creditor shall give the debtor a copy of the executed agreement (if any) and of any other document referred to in it.

(2)   If the creditor fails to comply with this section—
   (a) he is not entitled, while the default continues, to enforce the agreement; and
   (b) if the default continues for one month he commits an offence.

(3)   This section does not apply to a small agreement.

86   Death of debtor or hirer

(1)   The creditor or owner under a regulated agreement is not entitled, by reason of the death of the debtor or hirer, to do an act specified in paragraphs (a) to (e) of section 87(1) if at the death the agreement is fully secured.

(2)   If at the death of the debtor or hirer a regulated agreement is only partly secured or is unsecured, the creditor or owner is entitled, by reason of the death of the debtor or hirer, to do an act specified in paragraphs (a) to (e) of section 87(1) on an order of the court only.

(3)   This section applies in relation to the termination of an agreement only where—
   (a) a period for its duration is specified in the agreement, and
   (b) that period has not ended when the creditor or owner purports to terminate the agreement,

but so applies notwithstanding that, under the agreement, any party is entitled to terminate it before the end of the period so specified.

(4)   This section does not prevent the creditor from treating the right to draw on any credit as restricted or deferred, and taking such steps as may be necessary to make the restriction or deferment effective.

(5)   This section does not affect the operation of any agreement providing for payment of sums—
   (a) due under the regulated agreement, or
   (b) becoming due under it on the death of the debtor or hirer,

out of the proceeds of a policy of assurance on his life.

(6)   For the purposes of this section an act is done by reason of the death of the debtor or hirer if it is done under a power conferred by the agreement which is—
   (a) exercisable on his death, or
   (b) exercisable at will and exercised at any time after his death.

PART VII
DEFAULT AND TERMINATION

Default notices

87 Need for default notice

(1)   Service of a notice on the debtor or hirer in accordance with section 88 (a "default notice") is necessary before the creditor or owner can become entitled, by reason of any breach by the debtor or hirer of a regulated agreement,—
   (a) to terminate the agreement, or
   (b) to demand earlier payment of any sum, or
   (c) to recover possession of any goods or land, or
   (d) to treat any right conferred on the debtor or hirer by the agreement as terminated, restricted or deferred, or
   (e) to enforce any security.

(2)   Subsection (1) does not prevent the creditor from treating the right to draw upon any credit as restricted or deferred, and taking such steps as may be necessary to make the restriction or deferment effective.

(3)   The doing of an act by which a floating charge becomes fixed is not enforcement of a security.

(4)   Regulations may provide that subsection (1) is not to apply to agreements described by the regulations.

88 Contents and effect of default notice

(1)   The default notice must be in the prescribed form and specify—
   (a) the nature of the alleged breach;
   (b) if the breach is capable of remedy, what action is required to remedy it and the date before which that action is to be taken;
   (c) if the breach is not capable of remedy, the sum (if any) required to be paid as compensation for the breach, and the date before which it is to be paid.

(2)   A date specified under subsection (1) must not be less than seven days after the date of service of the default notice, and the creditor or

owner shall not take action such as is mentioned in section 87(1) before the date so specified or (if no requirement is made under subsection (1)) before those seven days have elapsed.

(3)   The default notice must not treat as a breach failure to comply with a provision of the agreement which becomes operative only on breach of some other provision, but if the breach of that other provision is not duly remedied or compensation demanded under subsection (1) is not duly paid, or (where no requirement is made under subsection (1)) if the seven days mentioned in subsection (2) have elapsed, the creditor or owner may treat the failure as a breach and section 87(1) shall not apply to it.

(4)   The default notice must contain information in the prescribed terms about the consequences of failure to comply with it.

(5)   A default notice making a requirement under subsection (1) may include a provision for the taking of action such as is mentioned in section 87(1) at any time after the restriction imposed by subsection (2) will cease, together with a statement that the provision will be ineffective if the breach is duly remedied or the compensation duly paid.

89   Compliance with default notice

If before the date specified for that purpose in the default notice the debtor or hirer takes the action specified under section 88(1)(b) or (c) the breach shall be treated as not having occurred.

Further restriction of remedies for default

90   Retaking of protected hire-purchase etc goods

(1)   At any time when—
   (a) the debtor is in breach of a regulated hire-purchase or a regulated conditional sale agreement relating to goods, and
   (b) the debtor has paid to the creditor one-third or more of the total price of the goods, and
   (c) the property in the goods remains in the creditor,

the creditor is not entitled to recover possession of the goods from the debtor except on an order of the court.

(2)   Where under a hire-purchase or conditional sale agreement the creditor is required to carry out any installation and the agreement specifies, as part of the total price, the amount to be paid in respect of the installation (the "installation charge") the reference in subsection (1)(b) to one third of the total price shall be construed as a reference to the aggregate of the installation charge and one third of the remainder of the total price.

(3)   In a case where—
   (a) subsection (1)(a) is satisfied, but not subsection (1)(b), and
   (b) subsection (1)(b) was satisfied on a previous occasion in relation to an earlier agreement, being a regulated hire-purchase or regulated conditional sale agreement, between the same parties, and relating to any of the goods comprised in the later agreement (whether or not other goods were also included),

subsection (1) shall apply to the later agreement with the omission of paragraph (b).

(4)   If the later agreement is a modifying agreement, subsection (3) shall apply with the substitution, for the second reference to the later agreement, of a reference to the modifying agreement.

(5)   Subsection (1) shall not apply, or shall cease to apply, to an agreement if the debtor has terminated, or terminates, the agreement.

(6)   Where subsection (1) applies to an agreement at the death of the debtor, it shall continue to apply (in relation to the possessor of the goods) until the grant of probate or administration, or (in Scotland) confirmation (on which the personal representative would fall to be treated as the debtor).

(7)   Goods falling within this section are in this Act referred to as "protected goods".

91   Consequences of breach of s 90

If goods are recovered by the creditor in contravention of section 90—
   (a) the regulated agreement, if not previously terminated, shall terminate, and
   (b) the debtor shall be released from all liability under the agreement, and shall be entitled to recover from the creditor all sums paid by the debtor under the agreement.

92   Recovery of possession of goods or land

(1)   Except under an order of the court, the creditor or owner shall not be entitled to enter any premises to take possession of goods subject to a regulated hire-purchase agreement, regulated conditional sale agreement or regulated consumer hire agreement.

(2)   At any time when the debtor is in breach of a regulated conditional sale agreement relating to land, the creditor is entitled to recover possession of the land from the debtor, or any person claiming under him, on an order of the court only.

(3)   An entry in contravention of subsection (1) or (2) is actionable as a breach of statutory duty.

93   Interest not to be increased on default

The debtor under a regulated consumer credit agreement shall not be obliged to pay interest on sums which, in breach of the agreement, are unpaid by him at a rate—
  (a) where the total charge for credit includes an item in respect of interest, exceeding the rate of that interest, or
  (b) in any other case, exceeding what would be the rate of the total charge for credit if any items included in the total charge for credit by virtue of section 20(2) were disregarded.

Early payment by debtor

94   Right to complete payments ahead of time

(1)   The debtor under a regulated consumer credit agreement is entitled at any time, by notice to the creditor and the payment to the creditor of all amounts payable by the debtor to him under the agreement (less any rebate allowable under section 95), to discharge the debtor's indebtedness under the agreement.

(2)   A notice under subsection (1) may embody the exercise by the debtor of any option to purchase goods conferred on him by the agreement, and deal with any other matter arising on, or in relation to, the termination of the agreement.

95 Rebate on early settlement

(1)   Regulations may provide for the allowance of a rebate of charges for credit to the debtor under a regulated consumer credit agreement where, under section 94, on refinancing, on breach of the agreement, or for any other reason, his indebtedness is discharged or becomes payable before the time fixed by the agreement, or any sum becomes payable by him before the time so fixed.

(2)   Regulations under subsection (1) may provide for calculation of the rebate by reference to any sums paid or payable by the debtor or his relative under or in connection with the agreement (whether to the creditor or some other person), including sums under linked transactions and other items in the total charge for credit.

96 Effect on linked transactions

(1)   Where for any reason the indebtedness of the debtor under a regulated consumer credit agreement is discharged before the time fixed by the agreement, he, and any relative of his, shall at the same time be discharged from any liability under a linked transaction, other than a debt which has already become payable.

(2)   Subsection (1) does not apply to a linked transaction which is itself an agreement providing the debtor or his relative with credit.

(3)   Regulations may exclude linked transactions of the prescribed description from the operation of subsection (1).

97 Duty to give information

(1)   The creditor under a regulated consumer credit agreement, within the prescribed period after he has received a request in writing to that effect from the debtor, shall give the debtor a statement in the prescribed form indicating, according to the information to which it is practicable for him to refer, the amount of the payment required to discharge the debtor's indebtedness under the agreement, together with the prescribed particulars showing how the amount is arrived at.

(2)   Subsection (1) does not apply to a request made less than one month after a previous request under that subsection relating to the same agreement was complied with.

(3)   If the creditor fails to comply with subsection (1)—

  (a) he is not entitled, while the default continues, to enforce the agreement; and

  (b) if the default continues for one month he commits an offence.

Termination of agreements

98   Duty to give notice of termination (non-default cases)

(1)   The creditor or owner is not entitled to terminate a regulated agreement except by or after giving the debtor or hirer not less than seven days' notice of the termination.

(2)   Subsection (1) applies only where—

  (a) a period for the duration of the agreement is specified in the agreement, and

  (b) that period has not ended when the creditor or owner does an act mentioned in subsection (1),

but so applies notwithstanding that, under the agreement, any party is entitled to terminate it before the end of the period so specified.

(3)   A notice under subsection (1) is ineffective if not in the prescribed form.

(4)   Subsection (1) does not prevent a creditor from treating the right to draw on any credit as restricted or deferred and taking such steps as may be necessary to make the restriction or deferment effective.

(5)   Regulations may provide that subsection (1) is not to apply to agreements described by the regulations.

(6)   Subsection (1) does not apply to the termination of a regulated agreement by reason of any breach by the debtor or hirer of the agreement.

99   Right to terminate hire-purchase etc agreements

(1)   At any time before the final payment by the debtor under a regulated hire-purchase or regulated conditional sale agreement falls due, the debtor shall be entitled to terminate the agreement by giving notice to any person entitled or authorised to receive the sums payable under the agreement.

(2)   Termination of an agreement under subsection (1) does not affect any liability under the agreement which has accrued before the termination.

(3)   Subsection (1) does not apply to a conditional sale agreement relating to land after the title to the land has passed to the debtor.

(4)   In the case of a conditional sale agreement relating to goods, where the property in the goods, having become vested in the debtor, is transferred to a person who does not become the debtor under the agreement, the debtor shall not thereafter be entitled to terminate the agreement under subsection (1).

(5)   Subject to subsection (4), where a debtor under a conditional sale agreement relating to goods, terminates the agreement under this section after the property in the goods has become vested in him, the property in the goods shall thereupon vest in the person (the "previous owner") in whom it was vested immediately before it became vested in the debtor:

Provided that if the previous owner has died, or any other event has occurred whereby that property, if vested in him immediately before that event, would thereupon have vested in some other person, the property shall be treated as having devolved as if it had been vested in the previous owner immediately before his death or immediately before that event, as the case may be.

100   Liability of debtor on termination of hire-purchase etc agreement

(1)   Where a regulated hire-purchase or regulated conditional sale agreement is terminated under section 99 the debtor shall be liable, unless the agreement provides for a smaller payment, or does not provide for any payment, to pay to the creditor the amount (if any) by which one-half of the total price exceeds the aggregate of the sums paid and the sums due in respect of the total price immediately before the termination.

(2)   Where under a hire-purchase or conditional sale agreement the creditor is required to carry out any installation and the agreement specifies, as part of the total price, the amount to be paid in respect of the installation (the "installation charge") the reference in subsection (1) to one-half of the total price shall be construed as a reference to the aggregate of the installation charge and one-half of the remainder of the total price.

(3)   If in any action the court is satisfied that a sum less than the amount specified in subsection (1) would be equal to the loss sustained by the

creditor in consequence of the termination of the agreement by the debtor, the court may make an order for the payment of that sum in lieu of the amount specified in subsection (1).

(4)   If the debtor has contravened an obligation to take reasonable care of the goods or land, the amount arrived at under subsection (1) shall be increased by the sum required to recompense the creditor for that contravention, and subsection (2) shall have effect accordingly.

(5)   Where the debtor, on the termination of the agreement, wrongfully retains possession of goods to which the agreement relates, then, in any action brought by the creditor to recover possession of the goods from the debtor, the court, unless it is satisfied that having regard to the circumstances it would not be just to do so, shall order the goods to be delivered to the creditor without giving the debtor an option to pay the value of the goods.

101   Right to terminate hire agreement

(1)   The hirer under a regulated consumer hire agreement is entitled to terminate the agreement by giving notice to any person entitled or authorised to receive the sums payable under the agreement.

(2)   Termination of an agreement under subsection (1) does not affect any liability under the agreement which has accrued before the termination.

(3)   A notice under subsection (1) shall not expire earlier than eighteen months after the making of the agreement, but apart from that the minimum period of notice to be given under subsection (1), unless the agreement provides for a shorter period, is as follows.

(4)   If the agreement provides for the making of payments by the hirer to the owner at equal intervals, the minimum period of notice is the length of one interval or three months, whichever is less.

(5)   If the agreement provides for the making of such payments at differing intervals, the minimum period of notice is the length of the shortest interval or three months, whichever is less.

(6)   In any other case, the minimum period of notice is three months.

(7)   This section does not apply to—

   (a) any agreement which provides for the making by the hirer of payments which in total (and without breach of the agreement) exceed [£1,500] in any year, or

   (b) any agreement where—

      (i) goods are bailed or (in Scotland) hired to the hirer for the purposes of a business carried on by him, or the hirer holds himself out as requiring the goods for those purposes, and

     (ii) the goods are selected by the hirer, and acquired by the owner for the purposes of the agreement at the request of the hirer from any person other than the owner's associate, or

   (c) any agreement where the hirer requires, or holds himself out as requiring, the goods for the purpose of bailing or hiring them to other persons in the course of a business carried on by him.

(8)   If, on an application made to the Director by a person carrying on a consumer hire business, it appears to the Director that it would be in the interest of hirers to do so, he may by notice to the applicant direct that this section shall not apply to consumer hire agreements made by the applicant, and subject to such conditions (if any) as the Director may specify, this Act shall have effect accordingly.

(9)   In the case of a modifying agreement subsection (3) shall apply with the substitution, for "the making of the agreement" of "the making of the original agreement".

NOTES

   Sub-s (7): sum in square brackets substituted by the Consumer Credit (Further Increase in Monetary Amounts) Order 1998, SI 1998/997, art 3, Schedule.

102 Agency for receiving notice of rescission

(1)   Where the debtor or hirer under a regulated agreement claims to have a right to rescind the agreement, each of the following shall be deemed to be the agent of the creditor or owner for the purpose of receiving any notice rescinding the agreement which is served by the debtor or hirer—

   (a) a credit-broker or supplier who was the negotiator in antecedent negotiations, and

   (b) any person who, in the course of a business carried on by him, acted on behalf of the debtor or hirer in any negotiations for the agreement.

(2)   In subsection (1) "rescind" does not include—
  (a) service of a notice of cancellation, or
  (b) termination of an agreement under section 99 or 101, or by the exercise of a right or power in that behalf expressly conferred by the agreement.

103   Termination statements

(1)   If an individual (the "customer") serves on any person (the "trader") a notice—
  (a) stating that—
    (i)   the customer was the debtor or hirer under a regulated agreement described in the notice, and the trader was the creditor or owner under the agreement, and
    (ii) the customer has discharged his indebtedness to the trader under the agreement, and
    (iii) the agreement has ceased to have any operation; and
  (b) requiring the trader to give the customer a notice, signed by or on behalf of the trader, confirming that those statements are correct,

the trader shall, within the prescribed period after receiving the notice, either comply with it or serve on the customer a counter-notice stating that, as the case may be, he disputes the correctness of the notice or asserts that the customer is not indebted to him under the agreement.

(2)   Where the trader disputes the correctness of the notice he shall give particulars of the way in which he alleges it to be wrong.

(3)   Subsection (1) does not apply in relation to any agreement if the trader has previously complied with that subsection on the service of a notice under it with respect to that agreement.

(4)   Subsection (1) does not apply to a non-commercial agreement.

(5)   If the trader fails to comply with subsection (1), and the default continues for one month, he commits an offence.

PART VIII
SECURITY

General

105 Form and content of securities

(1)   Any security provided in relation to a regulated agreement shall be expressed in writing.

(2)   Regulations may prescribe the form and content of documents ("security instruments") to be made in compliance with subsection (1).

(3)   Regulations under subsection (2) may in particular—
  (a) require specified information to be included in the prescribed manner in documents, and other specified material to be excluded;
  (b) contain requirements to ensure that specified information is clearly brought to the attention of the surety, and that one part of a document is not given insufficient or excessive prominence compared with another.

(4)   A security instrument is not properly executed unless—
  (a) a document in the prescribed form, itself containing all the prescribed terms and conforming to regulations under subsection (2), is signed in the prescribed manner by or on behalf of the surety, and
  (b) the document embodies all the terms of the security, other than implied terms, and
  (c) the document, when presented or sent for the purpose of being signed by or on behalf of the surety, is in such a state that its terms are readily legible, and
  (d) when the document is presented or sent for the purpose of being signed by or on behalf of the surety there is also presented or sent a copy of the document.

(5)   A security instrument is not properly executed unless—
  (a) where the security is provided after, or at the time when, the regulated agreement is made, a copy of the executed agreement, together with a copy of any other document referred to in it, is given to the surety at the time the security is provided, or
  (b) where the security is provided before the regulated agreement is made, a copy of the executed agreement, together with a copy of any other document referred to in it, is given to the surety within seven days after the regulated agreement is made.

(6)   Subsection (1) does not apply to a security provided by the debtor or hirer.

(7)   If—
  (a) in contravention of subsection (1) a security is not expressed in writing, or
  (b) a security instrument is improperly executed,

the security (so far as provided in relation to a regulated agreement) is enforceable against the surety on an order of the court only.

(8)   If an application for an order under subsection (7) is dismissed (except on technical grounds only) section 106 (ineffective securities) shall apply to the security.

(9)   Regulations under section 60(1) shall include provision requiring documents embodying regulated agreements also to embody any security provided in relation to a regulated agreement by the debtor or hirer.

10   Ineffective securities

Where, under any provision of this Act, this section is applied to any security provided in relation to a regulated agreement, then, subject to section 177 (saving for registered charges),—
  (a) the security, so far as it is so provided, shall be treated as never having effect;
  (b) any property lodged with the creditor or owner solely for the purposes of the security as so provided shall be returned by him forthwith;
  (c) the creditor or owner shall take any necessary action to remove or cancel an entry in any register, so far as the entry relates to the security as so provided; and
  (d) any amount received by the creditor or owner on realisation of the security shall, so far as it is referable to the agreement, be repaid to the surety.

107   Duty to give information to surety under fixed-sum credit agreement

(1)   The creditor under a regulated agreement for fixed-sum credit in relation to which security is provided, within the prescribed period after receiving a request in writing to that effect from the surety and payment of a fee of [£1], shall give to the surety (if a different person from the debtor)—

(a) a copy of the executed agreement (if any) and of any other document referred to in it;

(b) a copy of the security instrument (if any); and

(c) a statement signed by or on behalf of the creditor showing, according to the information to which it is practicable for him to refer,—

    (i) the total sum paid under the agreement by the debtor,

    (ii) the total sum which has become payable under the agreement by the debtor but remains unpaid, and the various amounts comprised in that total sum, with the date when each became due, and

    (iii) the total sum which is to become payable under the agreement by the debtor, and the various amounts comprised in that total sum, with the date, or mode of determining the date, when each becomes due.

(2) If the creditor possesses insufficient information to enable him to ascertain the amount and dates mentioned in subsection (1)(c)(iii), he shall be taken to comply with that sub-paragraph if his statement under subsection (1)(c) gives the basis on which, under the regulated agreement, they would fall to be ascertained.

(3) Subsection (1) does not apply to—

(a) an agreement under which no sum is, or will or may become, payable by the debtor, or

(b) a request made less than one month after a previous request under that subsection relating to the same agreement was complied with.

(4) If the creditor under an agreement fails to comply with subsection (1)—

(a) he is not entitled, while the default continues, to enforce the security, so far as provided in relation to the agreement; and

(b) if the default continues for one month he commits an offence.

(5) This section does not apply to a non-commercial agreement.

NOTES

Sub-s (1): sum in square brackets substituted by the Consumer Credit (Further Increase in Monetary Amounts) Order 1998, SI 1998/997, art 3, Schedule.

108 Duty to give information to surety under running-account credit agreement

(1) The creditor under a regulated agreement for running-account credit in relation to which security is provided, within the prescribed period after

receiving a request in writing to that effect from the surety and payment of a fee of [£1], shall give to the surety (if a different person from the debtor)—

  (a) a copy of the executed agreement (if any) and of any other document referred to in it;

  (b) a copy of the security instrument (if any); and

  (c) a statement signed by or on behalf of the creditor showing, according to the information to which it is practicable for him to refer,—

   (i) the state of the account, and

   (ii) the amount, if any, currently payable under the agreement by the debtor to the creditor, and

   (iii) the amounts and due dates of any payments which, if the debtor does not draw further on the account, will later become payable under the agreement by the debtor to the creditor.

(2)   If the creditor possesses insufficient information to enable him to ascertain the amounts and dates mentioned in subsection (1)(c)(iii), he shall be taken to comply with that sub-paragraph if his statement under subsection (1)(c) gives basis on which, under the regulated agreement, they would fall to be ascertained.

(3)   Subsection (1) does not apply to—

  (a) an agreement under which no sum is, or will or may become, payable by the debtor, or

  (b) a request made less than one month after a previous request under that subsection relating to the same agreement was complied with.

(4)   If the creditor under an agreement fails to comply with subsection (1)—

  (a) he is not entitled, while the default continues, to enforce the security, so far as provided in relation to the agreement; and

  (b) if the default continues for one month he commits an offence.

(5)   This section does not apply to a non-commercial agreement.

NOTES

Sub-s (1): sum in square brackets substituted by the Consumer Credit (Further Increase in Monetary Amounts) Order 1998, SI 1998/997, art 3, Schedule.

109   Duty to give information to surety under consumer hire agreement

(1)   The owner under a regulated consumer hire agreement in relation to which security is provided, within the prescribed period after receiving

a request in writing to that effect from the surety and payment of a fee of [£1], shall give to the surety (if a different person from the hirer)—

   (a) a copy of the executed agreement and of any other document referred to in it;

   (b) a copy of the security instrument (if any); and

   (c) a statement signed by or on behalf of the owner showing, according to the information to which it is practicable for him to refer, the total sum which has become payable under the agreement by the hirer but remains unpaid and the various amounts comprised in that total sum, with the date when each became due.

(2)   Subsection (1) does not apply to—

   (a) an agreement under which no sum is, or will or may become, payable by the hirer, or

   (b) a request made less than one month after a previous request under that subsection relating to the same agreement was complied with.

(3)   If the owner under an agreement fails to comply with subsection (1)—

   (a) he is not entitled, while the default continues, to enforce the security, so far as provided in relation to the agreement; and

   (b) if the default continues for one month he commits an offence.

(4)   This section does not apply to a non-commercial agreement.

NOTES

Sub-s (1): sum in square brackets substituted by the Consumer Credit (Further Increase in Monetary Amounts) Order 1998, SI 1998/997, art 3, Schedule.

110 Duty to give information to debtor or hirer

(1)   The creditor or owner under a regulated agreement, within the prescribed period after receiving a request in writing to that effect from the debtor or hirer and payment of a fee of [£1], shall give the debtor or hirer a copy of any security instrument executed in relation to the agreement after the making of the agreement.

(2)   Subsection (1) does not apply to—

   (a) a non-commercial agreement, or

   (b) an agreement under which no sum is, or will or may become, payable by the debtor or hirer, or

   (c) a request made less than one month after a previous request under subsection (1) relating to the same agreement was complied with.

(3)   If the creditor or owner under an agreement fails to comply with subsection (1)—

   (a) he is not entitled, while the default continues, to enforce the security (so far as provided in relation to the agreement); and
   (b) if the default continues for one month he commits an offence.

NOTES

   Sub-s (1): sum in square brackets substituted by the Consumer Credit (Further Increase in Monetary Amounts) Order 1998, SI 1998/997, art 3, Schedule.

111   Duty to give surety copy of default etc notice

(1)   When a default notice or a notice under section 76(1) or 98(1) is served on a debtor or hirer, a copy of the notice shall be served by the creditor or owner on any surety (if a different person from the debtor or hirer).

(2)   If the creditor or owner fails to comply with subsection (1) in the case of any surety, the security is enforceable against the surety (in respect of the breach or other matter to which the notice relates) on an order of the court only.

112   Realisation of securities

Subject to section 121, regulations may provide for any matters relating to the sale or other realisation, by the creditor or owner, of property over which any right has been provided by way of security in relation to an actual or prospective regulated agreement, other than a non-commercial agreement.

113   Act not to be evaded by use of security

(1)   Where a security is provided in relation to an actual or prospective regulated agreement, the security shall not be enforced so as to benefit the creditor or owner, directly or indirectly, to an extent greater (whether as respects the amount of any payment or the time or manner of its being made) than would be the case if the security were not provided and any obligations of the debtor or hirer, or his relative, under or in relation to the agreement were carried out to the extent (if any) to which they would be enforced under this Act.

(2)   In accordance with subsection (1), where a regulated agreement is enforceable on an order of the court or the Director only, any security provided in relation to the agreement is enforceable (so far as provided in relation to the agreement) where such an order has been made in relation to the agreement, but not otherwise.

(3)   Where—
  (a) a regulated agreement is cancelled under section 69(1) or becomes subject to section 69(2), or
  (b) a regulated agreement is terminated under section 91, or
  (c) in relation to any agreement an application for an order under section 40(2), 65(1), 124(1) or 149(2) is dismissed (except on technical grounds only), or
  (d) a declaration is made by the court under section 142(1) (refusal of enforcement order) as respects any regulated agreement,

section 106 shall apply to any security provided in relation to the agreement.

(4)   Where subsection (3)(d) applies and the declaration relates to a part only of the regulated agreement, section 106 shall apply to the security only so far as it concerns that part.

(5)   In the case of a cancelled agreement, the duty imposed on the debtor or hirer by section 71 or 72 shall not be enforceable before the creditor or owner has discharged any duty imposed on him by section 106 (as applied by subsection (3)(a)).

(6)   If the security is provided in relation to a prospective agreement or transaction, the security shall be enforceable in relation to the agreement or transaction only after the time (if any) when the agreement is made; and until that time the person providing the security shall be entitled, by notice to the creditor or owner, to require that section 106 shall thereupon apply to the security.

(7)   Where an indemnity [or guarantee] is given in a case where the debtor or hirer is a minor, or [an indemnity is given in a case where he] is otherwise not of full capacity, the reference in subsection (1) to the extent to which his obligations would be enforced shall be read in relation to the indemnity [or guarantee] as a reference to the extent to which [those obligations] would be enforced if he were of full capacity.

(8)   Subsections (1) to (3) also apply where a security is provided in relation to an actual or prospective linked transaction, and in that case—

(a) references to the agreement shall be read as references to the linked transaction, and

(b) references to the creditor or owner shall be read as references to any person (other than the debtor or hirer, or his relative) who is a party, or prospective party, to the linked transaction.

NOTES

Sub-s (7): words in square brackets inserted or substituted by the Minors' Contracts Act 1987, s 4(1).

Pledges

114 Pawn-receipts

(1) At the time he receives the article, a person who takes any article in pawn under a regulated agreement shall give to the person from whom he receives it a receipt in the prescribed form (a "pawn-receipt").

(2) A person who takes any article in pawn from an individual whom he knows to be, or who appears to be and is, a minor commits an offence.

(3) This section and sections 115 to 122 do not apply to—
   (a) a pledge of documents of title [or of bearer bonds], or
   (b) a non-commercial agreement.

NOTES

Sub-s (3): words in square brackets inserted by the Banking Act 1979, s 38(2).

115 Penalty for failure to supply copies of pledge agreement, etc

If the creditor under a regulated agreement to take any article in pawn fails to observe the requirements of sections 62 to 64 or 114(1) in relation to the agreement he commits an offence.

116 Redemption period

(1) A pawn is redeemable at any time within six months after it was taken.

(2)   Subject to subsection (1), the period within which a pawn is redeemable shall be the same as the period fixed by the parties for the duration of the credit secured by the pledge, or such longer period as they may agree.

(3)   If the pawn is not redeemed by the end of the period laid down by subsections (1) and (2) (the "redemption period"), it nevertheless remains redeemable until it is realised by the pawnee under section 121, except where under section 120(1)(a) the property in it passes to the pawnee.

(4)   No special charge shall be made for redemption of a pawn after the end of the redemption period, and charges in respect of the safe keeping of the pawn shall not be at a higher rate after the end of the redemption period than before.

117   Redemption procedure

(1)   On surrender of the pawn-receipt, and payment of the amount owing, at any time when the pawn is redeemable, the pawnee shall deliver the pawn to the bearer of the pawn-receipt.

(2)   Subsection (1) does not apply if the pawnee knows or has reasonable cause to suspect that the bearer of the pawn-receipt is neither the owner of the pawn nor authorised by the owner to redeem it.

(3)   The pawnee is not liable to any person in tort or delict for delivering the pawn where subsection (1) applies, or refusing to deliver it where the person demanding delivery does not comply with subsection (1) or, by reason of subsection (2), subsection (1) does not apply.

118   Loss etc of pawn-receipt

(1)   A person (the "claimant") who is not in possession of the pawn-receipt but claims to be the owner of the pawn, or to be otherwise entitled or authorised to redeem it, may do so at any time when it is redeemable by tendering to the pawnee in place of the pawn-receipt—
  (a) a statutory declaration made by the claimant in the prescribed form, and with the prescribed contents, or
  (b) where the pawn is security for fixed-sum credit not exceeding [£75] or running-account credit on which the credit limit does not exceed [£75], and the pawnee agrees, a statement in writing in the prescribed form, and with the prescribed contents, signed by the claimant.

(2)   On compliance by the claimant with subsection (1), section 117 shall apply as if the declaration or statement were the pawn-receipt, and the pawn- receipt itself shall become inoperative for the purposes of section 117.

NOTES

Sub-s (1): sums in square brackets substituted by the Consumer Credit (Further Increase in Monetary Amounts) Order 1998, SI 1998/997, art 3, Schedule.

119   Unreasonable refusal to deliver pawn

(1)   If a person who has taken a pawn under a regulated agreement refuses without reasonable cause to allow the pawn to be redeemed, he commits an offence.

(2)   On the conviction in England or Wales of a pawnee under subsection (1) where the offence does not amount to theft, section 28 (orders for restitution) of the Theft Act 1968, and any provision of the Theft Act 1968 relating to that section, shall apply as if the pawnee had been convicted of stealing the pawn.

(3)   On the conviction in Northern Ireland of a pawnee under subsection (1) where the offence does not amount to theft, section 27 (orders for restitution) of the Theft Act (Northern Ireland) 1969, and any provision of the Theft Act (Northern Ireland) 1969 relating to that section, shall apply as if the pawnee had been convicted of stealing the pawn.

120   Consequence of failure to redeem

(1)   If at the end of the redemption period the pawn has not been redeemed—
  (a) notwithstanding anything in section 113, the property in the pawn passes to the pawnee where the redemption period is six months and the pawn is security for fixed-sum credit not exceeding [£75] or running- account credit on which the credit limit does not exceed [£75]; or
  (b) in any other case the pawn becomes realisable by the pawnee.

(2)   Where the debtor or hirer is entitled to apply to the court for a time order under section 129, subsection (1) shall apply with the substitution, for "at the end of the redemption period" of "after the expiry of five days following the end of the redemption period".

NOTES

Sub-s (1): sums in square brackets substituted by the Consumer Credit (Further Increase in Monetary Amounts) Order 1998, SI 1998/997, art 3, Schedule.

121 Realisation of pawn

(1)   When a pawn has become realisable by him, the pawnee may sell it, after giving to the pawnor (except in such cases as may be prescribed) not less than the prescribed period of notice of the intention to sell, indicating in the notice the asking price and such other particulars as may be prescribed.

(2)   Within the prescribed period after the sale takes place, the pawnee shall give the pawnor the prescribed information in writing as to the sale, its proceeds and expenses.

(3)   Where the net proceeds of sale are not less than the sum which, if the pawn had been redeemed on the date of the sale, would have been payable for its redemption, the debt secured by the pawn is discharged and any surplus shall be paid by the pawnee to the pawnor.

(4)   Where subsection (3) does not apply, the debt shall be treated as from the date of sale as equal to the amount by which the net proceeds of sale fall short of the sum which would have been payable for the redemption of the pawn on that date.

(5)   In this section the "net proceeds of sale" is the amount realised (the "gross amount") less the expenses (if any) of the sale.

(6)   If the pawnor alleges that the gross amount is less than the true market value of the pawn on the date of sale, it is for the pawnee to prove that he and any agents employed by him in the sale used reasonable care to ensure that the true market value was obtained, and if he fails to do so subsections (3) and (4) shall have effect as if the reference in subsection (5) to the gross amount were a reference to the true market value.

(7)   If the pawnor alleges that the expenses of the sale were unreasonably high, it is for the pawnee to prove that they were reasonable, and if he fails to do so subsections (3) and (4) shall have effect as if the reference in subsection (5) to expenses were a reference to reasonable expenses.

Negotiable instruments

123 Restrictions on taking and negotiating instruments

(1)   A creditor or owner shall not take a negotiable instrument, other than a bank note or cheque, in discharge of any sum payable—
  (a) by the debtor or hirer under a regulated agreement, or
  (b) by any person as surety in relation to the agreement.

(2)   The creditor or owner shall not negotiate a cheque taken by him in discharge of a sum payable as mentioned in subsection (1), except to a banker (within the meaning of the Bills of Exchange Act 1882).

(3)   The creditor or owner shall not take a negotiable instrument as security for the discharge of any sum payable as mentioned in subsection (1).

(4)   A person takes a negotiable instrument as security for the discharge of a sum if the sum is intended to be paid in some other way, and the negotiable instrument is to be presented for payment only if the sum is not paid in that way.

(5)   This section does not apply where the regulated agreement is a non-commercial agreement.

(6)   The Secretary of State may by order provide that this section shall not apply where the regulated agreement has a connection with a country outside the United Kingdom.

124 Consequences of breach of s 123

(1)   After any contravention of section 123 has occurred in relation to a sum payable as mentioned in section 123(1)(a), the agreement under which the sum is payable is enforceable against the debtor or hirer on an order of the court only.

(2)   After any contravention of section 123 has occurred in relation to a sum payable by any surety, the security is enforceable on an order of the court only.

(3)   Where an application for an order under subsection (2) is dismissed (except on technical grounds only) section 106 shall apply to the security.

125 Holders in due course

(1)   A person who takes a negotiable instrument in contravention of section 123(1) or (3) is not a holder in due course, and is not entitled to enforce the instrument.

(2)   Where a person negotiates a cheque in contravention of section 123(2), his doing so constitutes a defect in his title within the meaning of the Bills of Exchange Act 1882.

(3)   If a person mentioned in section 123(1)(a) or (b) ("the protected person") becomes liable to a holder in due course of an instrument taken from the protected person in contravention of section 123(1) or (3), or taken from the protected person and negotiated in contravention of section 123(2), the creditor or owner shall indemnify the protected person in respect of that liability.

(4)   Nothing in this Act affects the rights of the holder in due course of any negotiable instrument.

Land mortgages

126 Enforcement of land mortgages

A land mortgage securing a regulated agreement is enforceable (so far as provided in relation to the agreement) on an order of the court only.

PART IX
JUDICIAL CONTROL

Enforcement of certain regulated agreements and securities

127 Enforcement orders in cases of infringement

(1)   In the case of an application for an enforcement order under—
   (a) section 65(1) (improperly executed agreements), or
   (b) section 105(7)(a) or (b) (improperly executed security instruments), or

(c) section 111(2) (failure to serve copy of notice on surety), or

(d) section 124(1) or (2) (taking of negotiable instrument in contravention of section 123),

the court shall dismiss the application if, but (subject to subsections (3) and (4)) only if, it considers it just to do so having regard to—

(i) prejudice caused to any person by the contravention in question, and the degree of culpability for it; and

(ii) the powers conferred on the court by subsection (2) and sections 135 and 136.

(2) If it appears to the court just to do so, it may in an enforcement order reduce or discharge any sum payable by the debtor or hirer, or any surety, so as to compensate him for prejudice suffered as a result of the contravention in question.

(3) The court shall not make an enforcement order under section 65(1) if section 61(1)(a) (signing of agreements) was not complied with unless a document (whether or not in the prescribed form and complying with regulations under section 60(1)) itself containing all the prescribed terms of the agreement was signed by the debtor or hirer (whether or not in the prescribed manner).

(4) The court shall not make an enforcement order under section 65(1) in the case of a cancellable agreement if—

(a) a provision of section 62 or 63 was not complied with, and the creditor or owner did not give a copy of the executed agreement, and of any other document referred to in it, to the debtor or hirer before the commencement of the proceedings in which the order is sought, or

(b) section 64(1) was not complied with.

(5) Where an enforcement order is made in a case to which subsection (3) applies, the order may direct that the regulated agreement is to have effect as if it did not include a term omitted from the document signed by the debtor or hirer.

128 Enforcement orders on death of debtor or hirer

The court shall make an order under section 86(2) if, but only if, the creditor or owner proves that he has been unable to satisfy himself that the present and future obligations of the debtor or hirer under the agreement are likely to be discharged.

Extension of time

129 Time orders

(1)   [Subject to subsection (3) below] if it appears to the court just to do so—
  (a) on an application for an enforcement order; or
  (b) on an application made by a debtor or hirer under this paragraph after service on him of—
     (i)   a default notice, or
     (ii) a notice under section 76(1) or 98(1); or
  (c) in an action brought by a creditor or owner to enforce a regulated agreement or any security, or recover possession of any goods or land to which a regulated agreement relates,

the court may make an order under this section (a "time order").

(2)   A time order shall provide for one or both of the following, as the court considers just—
  (a) the payment by the debtor or hirer or any surety of any sum owed under a regulated agreement or a security by such instalments, payable at such times, as the court, having regard to the means of the debtor or hirer and any surety, considers reasonable;
  (b) the remedying by the debtor or hirer of any breach of a regulated agreement (other than the non-payment of money) within such period as the court may specify.

[(3)   ...]

130 Supplemental provisions about time orders

(1)   Where in accordance with rules of court an offer to pay any sum by instalments is made by the debtor or hirer and accepted by the creditor or owner, the court may in accordance with rules of court make a time order under section 129(2)(a) giving effect to the offer without hearing evidence of means.

(2) In the case of a hire-purchase or conditional sale agreement only, a time order under section 129(2)(a) may deal with sums which, although not payable by the debtor at the time the order is made, would if the agreement continued in force become payable under it subsequently.

(3) A time order under section 129(2)(a) shall not be made where the regulated agreement is secured by a pledge if, by virtue of regulations made under section 76(5), 87(4) or 98(5), service of a notice is not necessary for enforcement of the pledge.

(4) Where, following the making of a time order in relation to a regulated hire-purchase or conditional sale agreement or a regulated consumer hire agreement, the debtor or hirer is in possession of the goods, he shall be treated (except in the case of a debtor to whom the creditor's title has passed) as a bailee or (in Scotland) a custodier of the goods under the terms of the agreement, notwithstanding that the agreement has been terminated.

(5) Without prejudice to anything done by the creditor or owner before the commencement of the period specified in a time order made under section 129(2)(b) ("the relevant period"),—
   (a) he shall not while the relevant period subsists take in relation to the agreement any action such as is mentioned in section 87(1);
   (b) where—
      (i) a provision of the agreement ("the secondary provision") becomes operative only on breach of another provision of the agreement ("the primary provision"), and
      (ii) the time order provides for the remedying of such a breach of the primary provision within the relevant period,
   he shall not treat the secondary provision as operative before the end of that period;
   (c) if while the relevant period subsists the breach to which the order relates is remedied it shall be treated as not having occurred.

(6) On the application of any person affected by a time order, the court may vary or revoke the order.

Protection of property pending proceedings

131 Protection orders

The court, on the application of the creditor or owner under a regulated agreement, may make such orders as it thinks just for protecting any property of the creditor or owner, or property subject to any security, from damage or depreciation pending the determination of any proceedings under this Act, including orders restricting or prohibiting use of the property or giving directions as to its custody.

Hire and hire-purchase etc agreements

132 Financial relief for hirer

(1)   Where the owner under a regulated consumer hire agreement recovers possession of goods to which the agreement relates otherwise than by action, the hirer may apply to the court for an order that—
  (a) the whole or part of any sum paid by the hirer to the owner in respect of the goods shall be repaid, and
  (b) the obligation to pay the whole or part of any sum owed by the hirer to the owner in respect of the goods shall cease,

and if it appears to the court just to do so, having regard to the extent of the enjoyment of the goods by the hirer, the court shall grant the application in full or in part.

(2)   Where in proceedings relating to a regulated consumer hire agreement the court makes an order for the delivery to the owner of goods to which the agreement relates the court may include in the order the like provision as may be made in an order under subsection (1).

133 Hire-purchase etc agreements: special powers of court

(1)   If, in relation to a regulated hire-purchase or conditional sale agreement, it appears to the court just to do so—
  (a) on an application for an enforcement order or time order; or
  (b) in an action brought by the creditor to recover possession of goods to which the agreement relates,

the court may—
  (i) make an order (a "return order") for the return to the creditor of goods to which the agreement relates,
  (ii) make an order (a "transfer order") for the transfer to the debtor of the creditor's title to certain goods to which the agreement relates ("the transferred goods"), and the return to the creditor of the remainder of the goods.

(2)   In determining for the purposes of this section how much of the total price has been paid ("the paid-up sum"), the court may—

(a) treat any sum paid by the debtor, or owed by the creditor, in relation to the goods as part of the paid-up sum;

(b) deduct any sum owed by the debtor in relation to the goods (otherwise than as part of the total price) from the paid-up sum,

and make corresponding reductions in amounts so owed.

(3)   Where a transfer order is made, the transferred goods shall be such of the goods to which the agreement relates as the court thinks just; but a transfer order shall be made only where the paid-up sum exceeds the part of the total price referable to the transferred goods by an amount equal to at least one-third of the unpaid balance of the total price.

(4)   Notwithstanding the making of a return order or transfer order, the debtor may at any time before the goods enter the possession of the creditor, on payment of the balance of the total price and the fulfilment of any other necessary conditions, claim the goods ordered to be returned to the creditor.

(5)   When, in pursuance of a time order or under this section, the total price of goods under a regulated hire-purchase agreement or regulated conditional sale agreement is paid and any other necessary conditions are fulfilled, the creditor's title to the goods vests in the debtor.

(6)   If, in contravention of a return order or transfer order, any goods to which the order relates are not returned to the creditor, the court, on the application of the creditor, may—

(a) revoke so much of the order as relates to those goods, and

(b) order the debtor to pay the creditor the unpaid portion of so much of the total price as is referable to those goods.

(7)   For the purposes of this section, the part of the total price referable to any goods is the part assigned to those goods by the agreement or (if no such assignment is made) the part determined by the court to be reasonable.

134   Evidence of adverse detention in hire-purchase etc cases

(1)   Where goods are comprised in a regulated hire-purchase agreement, regulated conditional sale agreement or regulated consumer hire agreement, and the creditor or owner—

(a) brings an action or makes an application to enforce a right to recover possession of the goods from the debtor or hirer, and

(b) proves that a demand for the delivery of the goods was included in the default notice under section 88(5), or that, after the right to recover possession of the goods accrued but before the action was begun or the application was made, he made a request in writing to the debtor or hirer to surrender the goods,

then, for the purposes of the claim of the creditor or owner to recover possession of the goods, the possession of them by the debtor or hirer shall be deemed to be adverse to the creditor or owner.

(2)   In subsection (1) "the debtor or hirer" includes a person in possession of the goods at any time between the debtor's or hirer's death and the grant of probate or administration, or (in Scotland) confirmation.

(3)   Nothing in this section affects a claim for damages for conversion or (in Scotland) for delict.

Supplemental provisions as to orders

135   Power to impose conditions, or suspend operation of order

(1)   If it considers it just to do so, the court may in an order made by it in relation to a regulated agreement include provisions—
  (a) making the operation of any term of the order conditional on the doing of specified acts by any party to the proceedings;
  (b) suspending the operation of any term of the order either—
    (i)   until such time as the court subsequently directs, or
    (ii) until the occurrence of a specified act or omission.

(2)   The court shall not suspend the operation of a term requiring the delivery up of goods by any person unless satisfied that the goods are in his possession or control.

(3)   In the case of a consumer hire agreement, the court shall not so use its powers under subsection (1)(b) as to extend the period for which, under the terms of the agreement, the hirer is entitled to possession of the goods to which the agreement relates.

(4)   On the application of any person affected by a provision included under subsection (1), the court may vary the provision.

136 Power to vary agreements and securities

The court may in an order made by it under this Act include such provision as it considers just for amending any agreement or security in consequence of a term of the order.

Extortionate credit bargains

137 Extortionate credit bargains

(1)   If the court finds a credit bargain extortionate it may reopen the credit agreement so as to do justice between the parties.

(2)   In this section and sections 138 to 140—
   (a) "credit agreement" means any agreement between an individual (the "debtor") and any other person (the "creditor") by which the creditor provides the debtor with credit of any amount, and
   (b) "credit bargain"—
     (i)   where no transaction other than the credit agreement is to be taken into account in computing the total charge for credit, means the credit agreement, or
     (ii) where one or more other transactions are to be so taken into account, means the credit agreement and those other transactions, taken together.

138 When bargains are extortionate

(1)   A credit bargain is extortionate if it—
   (a) requires the debtor or a relative of his to make payments (whether unconditionally, or on certain contingencies) which are grossly exorbitant, or
   (b) otherwise grossly contravenes ordinary principles of fair dealing.

(2)   In determining whether a credit bargain is extortionate, regard shall be had to such evidence as is adduced concerning—
   (a) interest rates prevailing at the time it was made,
   (b) the factors mentioned in subsections (3) to (5), and
   (c) any other relevant considerations.

(3)   Factors applicable under subsection (2) in relation to the debtor include—

(a) his age, experience, business capacity and state of health; and

(b) the degree to which, at the time of making the credit bargain, he was under financial pressure, and the nature of that pressure.

(4) Factors applicable under subsection (2) in relation to the creditor include—

(a) the degree of risk accepted by him, having regard to the value of any security provided;

(b) his relationship to the debtor; and

(c) whether or not a colourable cash price was quoted for any goods or services included in the credit bargain.

(5) Factors applicable under subsection (2) in relation to a linked transaction include the question how far the transaction was reasonably required for the protection of debtor or creditor, or was in the interest of the debtor.

139 Reopening of extortionate agreements

(1) A credit agreement may, if the court thinks just, be reopened on the ground that the credit bargain is extortionate—

(a) on an application for the purpose made by the debtor or any surety to the High Court, county court or sheriff court; or

(b) at the instance of the debtor or a surety in any proceedings to which the debtor and creditor are parties, being proceedings to enforce the agreement, any security relating to it, or any linked transaction; or

(c) at the instance of the debtor or a surety in other proceedings in any court where the amount paid or payable under the credit agreement is relevant.

(2) In reopening the agreement, the court may, for the purpose of relieving the debtor or a surety from payment of any sum in excess of that fairly due and reasonable, by order—

(a) direct accounts to be taken, or (in Scotland) an accounting to be made, between any persons,

(b) set aside the whole or part of any obligation imposed on the debtor or surety by the credit bargain or any related agreement,

(c) require the creditor to repay the whole or part of any sum paid under the credit bargain or any related agreement by the debtor or a surety, whether paid to the creditor or any other person,

(d) direct the return to the surety of any property provided for the purposes of the security, or

(e) alter the terms of the credit agreement or any security instrument.

(3)   An order may be made under subsection (2) notwithstanding that its effect is to place a burden on the creditor in respect of an advantage unfairly enjoyed by another person who is a party to a linked transaction.

(4)   An order under subsection (2) shall not alter the effect of any judgment.

(5)   In England and Wales, an application under subsection (1)(a) shall be brought only in the county court in the case of—
  (a) a regulated agreement, or
  (b) an agreement (not being a regulated agreement) under which the creditor provides the debtor with fixed-sum credit ... or running-account credit ...

[(5A) ...]

(6)   ...

(7)   In Northern Ireland an application under subsection (1)(a) may be brought in the county court in the case of—
  (a) a regulated agreement, or
  (b) an agreement (not being a regulated agreement) under which the creditor provides the debtor with fixed-sum credit not exceeding [£15,000] or running-account credit on which the credit limit does not exceed [£15,000].

NOTES

Sub-s (5): in sub-para (b) words omitted revoked by the High Court and County Courts Jurisdiction Order 1991, SI 1991/724, art 2(1), (8), Schedule, Pt I.

Sub-s (5A): inserted by the Administration of Justice Act 1982, s 37, Sch 3, Pt II, para 4; repealed by SI 1991/724, art 2(8), Schedule, Pt I.

Sub-s (6): applies to Scotland only.

Sub-s (7): sums in square brackets substituted by virtue of the County Courts (Financial Limits) Order (Northern Ireland) 1993, SR 1993/282, art 2, Schedule.

140   Interpretation of sections 137 to 139

Where the credit agreement is not a regulated agreement, expressions used in sections 137 to 139 which, apart from this section, apply only to regulated agreements, shall be construed as nearly as may be as if the credit agreement were a regulated agreement.

141 Jurisdiction and parties

(1)   In England and Wales, the county court shall have jurisdiction to hear and determine—
  (a) any action by the creditor or owner to enforce a regulated agreement or any security relating to it;
  (b) any action to enforce any linked transaction against the debtor or hirer or his relative;

and such an action shall not be brought in any other court.

(2)   Where an action or application is brought in the High Court which, by virtue of this Act, ought to have been brought in the county court it shall not be treated as improperly brought, but shall be transferred to the county court.

(3)–(3B) ...

(4)   In Northern Ireland the county court shall have jurisdiction to hear and determine any action or application falling within subsection (1).

(5)   Except as may be provided by rules of court, all the parties to a regulated agreement, and any surety, shall be made parties to any proceedings relating to the agreement.

NOTES

   Sub-ss (3)–(3B): apply to Scotland only.

142 Power to declare rights of parties

(1)   Where under any provision of this Act a thing can be done by a creditor or owner on an enforcement order only, and either—
  (a) the court dismisses (except on technical grounds only) an application for an enforcement order, or
  (b) where no such application has been made or such an application has been dismissed on technical grounds only, an interested party applies to the court for a declaration under this subsection,

the court may if it thinks just make a declaration that the creditor or owner is not entitled to do that thing, and thereafter no application for an enforcement order in respect of it shall be entertained.

(2)   Where—
    (a) a regulated agreement or linked transaction is cancelled under section 69(1), or becomes subject to section 69(2), or
    (b) a regulated agreement is terminated under section 91, and an interested party applies to the court for a declaration under this subsection, the court may make a declaration to that effect.

PART X
ANCILLARY CREDIT BUSINESS

Definitions

145   Types of ancillary credit business

(1)   An ancillary credit business is any business so far as it comprises or relates to—
    (a) credit brokerage,
    (b) debt-adjusting,
    (c) debt-counselling,
    (d) debt-collecting, or
    (e) the operation of a credit reference agency.

(2)   Subject to section 146(5), credit brokerage is the effecting of introductions—
    (a) of individuals desiring to obtain credit—
        (i)   to persons carrying on businesses to which this sub-paragraph applies, or
        (ii) in the case of an individual desiring to obtain credit to finance the acquisition or provision of a dwelling occupied or to be occupied by himself or his relative, to any person carrying on a business in the course of which he provides credit secured on land, or
    (b) of individuals desiring to obtain goods on hire to persons carrying on businesses to which this paragraph applies, or
    (c) of individuals desiring to obtain credit, or to obtain goods on hire, to other credit-brokers.

(3)   Subsection (2)(a)(i) applies to—
   (a) a consumer credit business;
   (b) a business which comprises or relates to consumer credit agreements being, otherwise than by virtue of section 16(5)(a), exempt agreements;
   (c) a business which comprises or relates to unregulated agreements where—
      (i)   the [law applicable to] the agreement is the law of a country outside the United Kingdom, and
      (ii) if the [law applicable to] the agreement were the law of a part of the United Kingdom it would be a regulated consumer credit agreement.

(4)   Subsection (2)(b) applies to—
   (a) a consumer hire business;
   (b) a business which comprises or relates to unregulated agreements where—
      (i)   the [law applicable to] the agreement is the law of a country outside the United Kingdom, and
      (ii) if the [law applicable to] the agreement were the law of a part of the United Kingdom it would be a regulated consumer hire agreement.

(5)   Subject to section 146(6), debt-adjusting is, in relation to debts due under consumer credit agreements or consumer hire agreements,—
   (a) negotiating with the creditor or owner, on behalf of the debtor or hirer, terms for the discharge of a debt, or
   (b) taking over, in return for payments by the debtor or hirer, his obligation to discharge a debt, or
   (c) any similar activity concerned with the liquidation of a debt.

(6)   Subject to section 146(6), debt-counselling is the giving of advice to debtors or hirers about the liquidation of debts due under consumer credit agreements or consumer hire agreements.

(7)   Subject to section 146(6), debt-collecting is the taking of steps to procure payment of debts due under consumer credit agreements or consumer hire agreements.

(8)   A credit reference agency is a person carrying on a business comprising the furnishing of persons with information relevant to the financial standing of individuals, being information collected by the agency for that purpose.

Sub-ss (3), (4): words in square brackets substituted by the Contracts (Applicable Law) Act 1990, s 5, Sch 4, para 2.

146 Exceptions from section 145

(1)   A barrister or advocate acting in that capacity is not to be treated as doing so in the course of any ancillary credit business.

(2)   A solicitor engaging in contentious business (as defined in [section 87(1) of the Solicitors Act 1974]) is not to be treated as doing so in the course of any ancillary credit business.

(3)   . . .

(4)   A solicitor in Northern Ireland engaging in [contentious business (as defined in Article 3(2) of the Solicitors (Northern Ireland) Order 1976)], is not to be treated as doing so in the course of any ancillary credit business.

(5)   For the purposes of section 145(2), introductions effected by an individual by canvassing off trade premises either debtor-creditor-supplier agreements falling within section 12(a) or regulated consumer hire agreements shall be disregarded if—
(a) the introductions are not effected by him in the capacity of an employee, and
(b) he does not by any other method effect introductions falling within section 145(2).

(6)   It is not debt-adjusting, debt-counselling or debt-collecting for a person to do anything in relation to a debt arising under an agreement if—
(a) he is the creditor or owner under the agreement, otherwise than by virtue of an assignment, or
(b) he is the creditor or owner under the agreement by virtue of an assignment made in connection with the transfer to the assignee of any business other than a debt-collecting business, or
(c) he is the supplier in relation to the agreement, or
(d) he is a credit-broker who has acquired the business of the person who was the supplier in relation to the agreement, or
(e) he is a person prevented by subsection (5) from being treated as a credit-broker, and the agreement was made in consequence of an introduction (whether made by him or another person) which, under subsection (5), is to be disregarded.

Sub-ss (2), (4): words in square brackets substituted by the Arbitration Act 1996, s 107(1), Sch 3, para 28.

Sub-s (3): applies to Scotland only.

Solicitors: any reference to solicitor(s) etc modified to include references to bodies recognised under the Administration of Justice Act 1985, s 9, by the Solicitors' Incorporated Practices Order 1991, SI 1991/2684, arts 4, 5, Sch 1.

Licensing

147 Application of Part III

(1)   The provisions of Part III (except section 40) apply to an ancillary credit business as they apply to a consumer credit business.

(2)   Without prejudice to the generality of section 26, regulations under that section (as applied by subsection (1)) may include provisions regulating the collection and dissemination of information by credit reference agencies.

148 Agreement for services of unlicensed trader

(1)   An agreement for the services of a person carrying on an ancillary credit business (the "trader"), if made when the trader was unlicensed, is enforceable against the other party (the "customer") only where the Director has made an order under subsection (2) which applies to the agreement.

(2)   The trader or his successor in title may apply to the Director for an order that agreements within subsection (1) are to be treated as if made when the trader was licensed.

(3)   Unless the Director determines to make an order under subsection (2) in accordance with the application, he shall, before determining the application, by notice—
   (a) inform the trader, giving his reasons, that, as the case may be, he is minded to refuse the application, or to grant it in terms different from those applied for, describing them, and
   (b) invite the trader to submit to the Director representations in support of his application in accordance with section 34.

(4)   In determining whether or not to make an order under subsection (2) in respect of any period the Director shall consider, in addition to any other relevant factors,—

(a) how far, if at all, customers under agreements made by the trader during that period were prejudiced by the trader's conduct,

(b) whether or not the Director would have been likely to grant a licence covering that period on an application by the trader, and

(c) the degree of culpability for the failure to obtain a licence.

(5)   If the Director thinks fit, he may in an order under subsection (2)—

(a) limit the order to specified agreements, or agreements of a specified description or made at a specified time;

(b) make the order conditional on the doing of specified acts by the trader.

149 Regulated agreements made on introductions by unlicensed credit-broker

(1)   A regulated agreement made by a debtor or hirer who, for the purpose of making that agreement, was introduced to the creditor or owner by an unlicensed credit-broker is enforceable against the debtor or hirer only where—

(a) on the application of the credit-broker, the Director has made an order under section 148(2) in respect of a period including the time when the introduction was made, and the order does not (whether in general terms or specifically) exclude the application of this paragraph to the regulated agreement, or

(b) the Director has made an order under subsection (2) which applies to the agreement.

(2)   Where during any period individuals were introduced to a person carrying on a consumer credit business or consumer hire business by an unlicensed credit-broker for the purpose of making regulated agreements with the person carrying on that business, that person or his successor in title may apply to the Director for an order that regulated agreements so made are to be treated as if the credit-broker had been licensed at the time of the introduction.

(3)   Unless the Director determines to make an order under subsection (2) in accordance with the application, he shall, before determining the application, by notice—

(a) inform the applicant, giving his reasons, that, as the case may be, he is minded to refuse the application, or to grant it in terms different from those applied for, describing them, and

(b) invite the applicant to submit to the Director representations in support of his application in accordance with section 34.

(4)   In determining whether or not to make an order under subsection (2) the Director shall consider, in addition to any other relevant factors—

(a) how far, if at all, debtors or hirers under regulated agreements to which the application relates were prejudiced by the credit-broker's conduct, and

(b) the degree of culpability of the applicant in facilitating the carrying on by the credit-broker of his business when unlicensed.

(5)   If the Director thinks fit, he may in an order under subsection (2)—

(a) limit the order to specified agreements, or agreements of a specified description or made at a specified time;

(b) make the order conditional on the doing of specified acts by the applicant.

150 Appeals to Secretary of State against licensing decisions

Section 41 (as applied by section 147(1)) shall have effect as if the following entry were included in the table set out at the end—

Determination	*Appellant*
Refusal to make order under section 148(2) or 149(2) in accordance with terms of the application	The applicant

Seeking business

151 Advertisements

(1)   Sections 44 to 47 apply to an advertisement published for the purposes of a business of credit brokerage carried on by any person, whether it advertises the services of that person or the services of persons to whom he effects introductions, as they apply to an advertisement to which Part IV applies.

(2)   Sections 44, 46 and 47 apply to an advertisement, published for the purposes of a business carried on by the advertiser, indicating that he is willing to advise on debts, or engage in transactions concerned with the liquidation of debts, as they apply to an advertisement to which Part IV applies.

(3)   The Secretary of State may by order provide that an advertisement published for the purposes of a business of credit brokerage, debt-adjusting or debt-counselling shall not fall within subsection (1) or (2) if it is of a description specified in the order.

(4)   An advertisement does not fall within subsection (2) if it indicates that the advertiser is not willing to act in relation to consumer credit agreements and consumer hire agreements.

(5)   In subsections (1) and (3) "credit brokerage" includes the effecting of introductions of individuals desiring to obtain credit to any person carrying on a business in the course of which he provides credit secured on land.

152   Application of sections 52 to 54 to credit brokerage etc

(1)   Sections 52 to 54 apply to a business of credit brokerage, debt-adjusting or debt-counselling as they apply to a consumer credit business.

(2)   In their application to a business of credit brokerage, sections 52 and 53 shall apply to the giving of quotations and information about the business of any person to whom the credit-broker effects introductions as well as to the giving of quotations and information about his own business.

153   Definition of canvassing off trade premises (agreements for ancillary credit services)

(1)   An individual (the "canvasser") canvasses off trade premises the services of a person carrying on an ancillary credit business if he solicits the entry of another individual (the "consumer") into an agreement for the provision to the consumer of those services by making oral representations to the consumer, or any other individual, during a visit by the canvasser to any place (not excluded by subsection (2)) where the consumer, or that other individual, as the case may be, is, being a visit—
   (a) carried out for the purpose of making such oral representations to individuals who are at that place, but
   (b) not carried out in response to a request made on a previous occasion.

(2)   A place is excluded from subsection (1) if it is a place where (whether on a permanent or temporary basis)—
   (a) the ancillary credit business is carried on, or

(b) any business is carried on by the canvasser or the person whose employee or agent the canvasser is, or by the consumer.

154 Prohibition of canvassing certain ancillary credit services off trade premises

It is an offence to canvass off trade premises the services of a person carrying on a business of credit brokerage, debt-adjusting or debt-counselling.

155 Right to recover brokerage fees

(1) The excess over [£5] of a fee or commission for his services charged by a credit-broker to an individual to whom this subsection applies shall cease to be payable or, as the case may be, shall be recoverable by the individual if the introduction does not result in his entering into a relevant agreement within the six months following the introduction (disregarding any agreement which is cancelled under section 69(1) or becomes subject to section 69(2)).

(2) Subsection (1) applies to an individual who sought an introduction for a purpose which would have been fulfilled by his entry into—
  (a) a regulated agreement, or
  (b) in the case of an individual such as is referred to in section 145(2)(a)(ii), an agreement for credit secured on land, or
  (c) an agreement such as is referred to in section 145(3)(b) or (c) or (4)(b).

(3) An agreement is a relevant agreement for the purposes of subsection (1) in relation to an individual if it is an agreement such as is referred to in subsection (2) in relation to that individual.

(4) In the case of an individual desiring to obtain credit under a consumer credit agreement, any sum payable or paid by him to a credit-broker otherwise than as a fee or commission for the credit-broker's services shall for the purposes of subsection (1) be treated as such a fee or commission if it enters, or would enter, into the total charge for credit.

NOTES

Sub-s (1): sum in square brackets substituted by the Consumer Credit (Further Increase in Monetary Amounts) Order 1998, SI 1998/997, art 3, Schedule.

Entry into agreements

156　Entry into agreements

Regulations may make provision, in relation to agreements entered into in the course of a business of credit brokerage, debt-adjusting or debt-counselling, corresponding, with such modifications as the Secretary of State thinks fit, to the provision which is or may be made by or under sections 55, 60, 61, 62, 63, 65, 127, 179 or 180 in relation to agreements to which those sections apply.

Credit reference agencies

157　Duty to disclose name etc of agency

(1)　A creditor, owner or negotiator, within the prescribed period after receiving a request in writing to that effect from the debtor or hirer, shall give him notice of the name and address of any credit reference agency from which the creditor, owner or negotiator has, during the antecedent negotiations, applied for information about his financial standing.

(2)　Subsection (1) does not apply to a request received more than 28 days after the termination of the antecedent negotiations, whether on the making of the regulated agreement or otherwise.

(3)　If the creditor, owner or negotiator fails to comply with subsection (1) he commits an offence.

158　Duty of agency to disclose filed information

(1)　A credit reference agency, within the prescribed period after receiving,
  (a) a request in writing to that effect from any *individual* (the "consumer") and
  (b) such particulars as the agency may reasonably require to enable them to identify the file, and
  (c) a fee of [£2],

shall give the consumer a copy of the file relating to *him* kept by the agency.

(2)   When giving a copy of the file under subsection (1), the agency shall also give the consumer a statement in the prescribed form of *his* rights under section 159.

(3)   If the agency does not keep a file relating to the consumer it shall give *him* notice of that fact, but need not return any money paid.

(4)   If the agency contravenes any provision of this section it commits an offence.

(5)   In this Act "file", in relation to an individual, means all the information about him kept by a credit reference agency, regardless of how the information is stored and "copy of the file", as respects information not in plain English, means a transcript reduced into plain English.

NOTES

Sub-s (1): for the first word in italics there are substituted with savings the words "partnership or other unincorporated body of persons not consisting entirely of bodies corporate", and for the second word in italics there is substituted the word "it" by the Data Protection Act 1998, s 62(1), Sch 14, para 20, as from a day to be appointed, for savings see Sch 14, para 20 thereto; sum in square brackets substituted by the Consumer Credit (Further Increase in Monetary Amounts) Order 1998, SI 1998/997, art 3, Schedule.

Sub-s (2): for the word in italics there are substituted with savings the words "the consumer's" by the Data Protection Act 1998, s 62(1), Sch 14, para 20, as from a day to be appointed; for savings see Sch 14, para 20 thereto.

Sub-s (3): for the word in italics there are substituted with savings the words "the consumer" by the Data Protection Act 1998, s 62(1), Sch 14, para 20, as from a day to be appointed; for savings see Sch 14, para 20 thereto.

159 Correction of wrong information

(1)   A consumer given information under section 158 who considers that an entry in his file is incorrect, and that if it is not corrected he is likely to be prejudiced, may give notice to the agency requiring it either to remove the entry from the file or amend it.

(2)   Within 28 days after receiving a notice under subsection (1), the agency shall by notice inform the *consumer* that it has—
  (a) removed the entry from the file, or
  (b) amended the entry, or
  (c) taken no action,

and if the notice states that the agency has amended the entry it shall include a copy of the file so far as it comprises the amended entry.

(3)   Within 28 days after receiving a notice under subsection (2) or, where no such notice was given, within 28 days after the expiry of the period mentioned in subsection (2), the *consumer* may, unless he has been informed by the agency that it has removed the entry from his file, serve a further notice on the agency requiring it to add to the file an accompanying notice of correction (not exceeding 200 words) drawn up by the *consumer* and include a copy of it when furnishing information included in or based on that entry.

(4)   Within 28 days after receiving a notice under subsection (3), the agency, unless it intends to apply to the *Director* under subsection (5), shall by notice inform the *consumer* that it has received the notice under subsection (3) and intends to comply with it.

(5)   If—
   (a) the *consumer* has not received a notice under subsection (4) within the time required, or
   (b) it appears to the agency that it would be improper for it to publish a notice of correction because it is incorrect, or unjustly defames any person, or is frivolous or scandalous, or is for any other reason unsuitable,

the *consumer* or, as the case may be, the agency may, in the prescribed manner and on payment of the specified fee, apply to the *Director*, who may make such order on the application as he thinks fit.

(6)   If a person to whom an order under this section is directed fails to comply with it within the period specified in the order he commits an offence.

[(7)   The Data Protection Commissioner may vary or revoke any order made by him under this section.

(8)   In this section "the relevant authority" means—
   (a) where the objector is a partnership or other unincorporated body of persons, the Director, and
   (b) in any other case, the Data Protection Commissioner.]

NOTES

Commencement: to be appointed (sub-ss (7), (8)); 31 July 1974 (sub-ss (1)–(6)).
   Sub-s (1): substituted with savings by the Data Protection Act 1998, s 62(1), (2), as from a day to be appointed, as follows (for savings see Sch 14, para 20 thereto)—

"(1)   Any individual (the "objector") given—
   (a) information under section 7 of the Data Protection Act 1998 by a credit reference agency, or
   (b) information under section 158,

who considers that an entry in his file is incorrect, and that if it is not corrected he is likely to be prejudiced, may give notice to the agency requiring it either to remove the entry from the file or amend it."

Sub-ss (2), (3): for the words in italics there are substituted with savings the word "objector" by the Data Protection Act 1998, s 62(1), (3)(a), as from a day to be appointed; for savings see Sch 14, para 20 thereto.

Sub-s (4): for the first word in italics there are substituted with saving the words "the relevant authority" and for the second word in italics there is substituted with savings the word "objector" by the Data Protection Act 1998, s 62(1), (3)(a), (b), as from a day to be appointed; for savings see Sch 14, para 20 thereto.

Sub-s (5): for the first and second words in italics there are substituted with savings the word "objector", and for the third word in italics there are substituted with saving the words "the relevant authority" by the Data Protection Act 1998, s 62(1), (3)(a), (b), as from a day to be appointed; for savings see Sch 14, para 20 thereto.

Sub-ss (7), (8): added with savings by the Data Protection Act 1998, s 62(1), (4), as from a day to be appointed; for savings see Sch 14, para 20 thereto.

160   Alternative procedure for business consumers

(1)   The Director, on an application made by a credit reference agency, may direct that this section shall apply to the agency if he is satisfied—
   (a) that compliance with section 158 in the case of consumers who carry on a business would adversely affect the service provided to its customers by the agency, and
   (b) that, having regard to the methods employed by the agency and to any other relevant factors, it is probable that consumers carrying on a business would not be prejudiced by the making of the direction.

(2)   Where an agency to which this section applies receives a request, particulars and a fee under section 158(1) from a consumer who carries on a business, and section 158(3) does not apply, the agency, instead of complying with section 158, may elect to deal with the matter under the following subsections.

(3)   Instead of giving the consumer a copy of the file, the agency shall within the prescribed period give notice to the consumer that it is proceeding under this section, and by notice give the consumer such information included in or based on entries in the file as the Director may direct, together with a statement in the prescribed form of the consumer's rights under subsections (4) and (5).

(4)   If within 28 days after receiving the information given *him* under subsection (3), or such longer period as the Director may allow, the consumer—

(a) gives notice to the Director that *he* is dissatisfied with the information, and

(b) satisfies the Director that *he* has taken such steps in relation to the agency as may be reasonable with a view to removing the cause of *his* dissatisfaction, and

(c) pays the Director the specified fee,

the Director may direct the agency to give the Director a copy of the file, and the Director may disclose to the consumer such of the information on the file as the Director thinks fit.

(5)   Section 159 applies with any necessary modifications to information given to the consumer under this section as it applies to information given under section 158.

(6)   If an agency making an election under subsection (2) fails to comply with subsection (3) or (4) it commits an offence.

(7)   In this section "consumer" has the same meaning as in section 158.]

NOTES

Commencement: to be appointed (sub-s (7)); 31 July 1974 (sub-ss (1)–(6)).

Sub-s (4): for the first word in italics there are substituted the words "to the consumer", for the second and third words in italics there are substituted the words "the consumer", and for the fourth word in italics there are substituted the words "the consumer's" by the Data Protection Act 1998, s 62(5)(a), as from a day to be appointed.

Sub-s (7): added by the Data Protection Act 1998, s 62(5)(b), as from a day to be appointed.

PART XI
ENFORCEMENT OF ACT

161   Enforcement authorities

(1)   The following authorities ("enforcement authorities") have a duty to enforce this Act and regulations made under it—

(a) the Director,

(b) in Great Britain, the local weights and measures authority,

(c) in Northern Ireland, the Department of Commerce for Northern Ireland.

(2)   Where a local weights and measures authority in England or Wales propose to institute proceedings for an offence under this Act (other than an offence under section 162(6), 165(1) or (2) or 174(5)) it shall, as between the authority and the Director, be the duty of the authority to give the Director notice of the intended proceedings, together with a summary of the facts on which the charges are to be founded, and postpone institution of the proceedings until either—

(a) 28 days have expired since that notice was given, or

(b) the Director has notified them of receipt of the notice and summary.

(3)   Every local weights and measures authority shall, whenever the Director requires, report to him in such form and with such particulars as he requires on the exercise of their functions under this Act.

(4)–(6) . . .

NOTES

Sub-ss (4)–(6): repealed by the Local Government, Planning and Land Act 1980, ss 1(4), 194, Sch 4, para 10, Sch 34, Pt IV.

162   Powers of entry and inspection

(1)   A duly authorised officer of an enforcement authority, at all reasonable hours and on production, if required, of his credentials, may—

(a) in order to ascertain whether a breach of any provision of or under this Act has been committed, inspect any goods and enter any premises (other than premises used only as a dwelling);

(b) if he has reasonable cause to suspect that a breach of any provision of or under this Act has been committed, in order to ascertain whether it has been committed, require any person—

(i) carrying on, or employed in connection with, a business to produce any books or documents relating to it; or

(ii) having control of any information relating to a business recorded otherwise than in a legible form to provide a document containing a legible reproduction of the whole or any part of the information, and take copies of, or of any entry in, the books or documents;

(c) if he has reasonable cause to believe that a breach of any provision of or under this Act has been committed, seize and detain any goods in order to ascertain (by testing or otherwise) whether such a breach has been committed;

(d) seize and detain any goods, books or documents which he has reason to believe may be required as evidence in proceedings for an offence under this Act;

(e) for the purpose of exercising his powers under this subsection to seize goods, books or documents, but only if and to the extent that it is reasonably necessary for securing that the provisions of this Act and of any regulations made under it are duly observed, require any person having authority to do so to break open any container and, if that person does not comply, break it open himself.

(2) An officer seizing goods, books or documents in exercise of his powers under this section shall not do so without informing the person he seizes them from.

(3) If a justice of the peace, on sworn information in writing, or, in Scotland, a sheriff or a magistrate or justice of the peace, on evidence on oath,—

(a) is satisfied that there is reasonable ground to believe either—

(i) that any goods, books or documents which a duly authorised officer has power to inspect under this section are on any premises and their inspection is likely to disclose evidence of a breach of any provision of or under this Act; or

(ii) that a breach of any provision of or under this Act has been, is being or is about to be committed on any premises; and

(b) is also satisfied either—

(i) that admission to the premises has been or is likely to be refused and that notice of intention to apply for a warrant under this subsection has been given to the occupier; or

(ii) that an application for admission, or the giving of such a notice, would defeat the object of the entry or that the premises are unoccupied or that the occupier is temporarily absent and it might defeat the object of the entry to wait for his return,

the justice or, as the case may be, the sheriff or magistrate may by warrant under his hand, which shall continue in force for a period of one month, authorise an officer of an enforcement authority to enter the premises (by force if need be).

(4) An officer entering premises by virtue of this section may take such other persons and equipment with him as he thinks necessary; and on leaving premises entered by virtue of a warrant under subsection (3) shall, if they are unoccupied or the occupier is temporarily absent, leave them as effectively secured against trespassers as he found them.

(5)   Regulations may provide that, in cases described by the regulations, an officer of a local weights and measures authority is not to be taken to be duly authorised for the purposes of this section unless he is authorised by the Director.

(6)   A person who is not a duly authorised officer of an enforcement authority, but purports to act as such under this section, commits an offence.

(7)   Nothing in this section compels a barrister, advocate or solicitor to produce a document containing a privileged communication made by or to him in that capacity or authorises the seizing of any such document in his possession.

163   Compensation for loss

(1)   Where, in exercising his powers under section 162, an officer of an enforcement authority seizes and detains goods and their owner suffers loss by reason of—
   (a) that seizure, or
   (b) the loss, damage or deterioration of the goods during detention,

then, unless the owner is convicted of an offence under this Act committed in relation to the goods, the authority shall compensate him for the loss so suffered.

(2)   Any dispute as to the right to or amount of any compensation under subsection (1) shall be determined by arbitration.

164   Power to make test purchases etc

(1)   An enforcement authority may—
   (a) make, or authorise any of their officers to make on their behalf, such purchases of goods; and
   (b) authorise any of their officers to procure the provision of such services or facilities or to enter into such agreements or other transactions,

as may appear to them expedient for determining whether any provisions made by or under this Act are being complied with.

(2)   Any act done by an officer authorised to do it under subsection (1) shall be treated for the purposes of this Act as done by him as an individual on his own behalf.

(3)   Any goods seized by an officer under this Act may be tested, and in the event of such a test he shall inform the person mentioned in section 162(2) of the test results.

(4)   Where any test leads to proceedings under this Act, the enforcement authority shall—
   (a) if the goods were purchased, inform the person they were purchased from of the test results, and
   (b) allow any person against whom the proceedings are taken to have the goods tested on his behalf if it is reasonably practicable to do so.

165   Obstruction of authorised officers

(1)   Any person who—
   (a) wilfully obstructs an officer of an enforcement authority acting in pursuance of this Act; or
   (b) wilfully fails to comply with any requirement properly made to him by such an officer under section 162; or
   (c) without reasonable cause fails to give such an officer (so acting) other assistance or information he may reasonably require in performing his functions under this Act,

commits an offence.

(2)   If any person, in giving such information as is mentioned in subsection (1)(c), makes any statement which he knows to be false, he commits an offence.

(3)   Nothing in this section requires a person to answer any question or give any information if to do so might incriminate that person or (where that person is married) the husband or wife of that person.

166   Notification of convictions and judgments to Director

Where a person is convicted of an offence or has a judgment given against him by or before any court in the United Kingdom and it appears to the court—
   (a) having regard to the functions of the Director under this Act, that the conviction or judgment should be brought to the Director's attention, and
   (b) that it may not be brought to his attention unless arrangements for that purpose are made by the court,

the court may make such arrangements notwithstanding that the proceedings have been finally disposed of.

167 Penalties

(1)   An offence under a provision of this Act specified in column 1 of Schedule 1 is triable in the mode or modes indicated in column 3, and on conviction is punishable as indicated in column 4 (where a period of time indicates the maximum term of imprisonment, and a monetary amount indicates the maximum fine, for the offence in question).

(2)   A person who contravenes any regulations made under section 44, 52, 53, or 112, or made under section 26 by virtue of section 54, commits an offence.

168 Defences

(1)   In any proceedings for an offence under this Act it is a defence for the person charged to prove—
   (a) that his act or omission was due to a mistake, or to reliance on information supplied to him, or to an act or omission by another person, or to an accident or some other cause beyond his control, and
   (b) that he took all reasonable precautions and exercised all due diligence to avoid such an act or omission by himself or any person under his control.

(2)   If in any case the defence provided by subsection (1) involves the allegation that the act or omission was due to an act or omission by another person or to reliance on information supplied by another person, the person charged shall not, without leave of the court, be entitled to rely on that defence unless, within a period ending seven clear days before the hearing, he has served on the prosecutor a notice giving such information identifying or assisting in the identification of that other person as was then in his possession.

169 Offences by bodies corporate

Where at any time a body corporate commits an offence under this Act with the consent or connivance of, or because of neglect by, any individual, the individual commits the like offence if at that time—

(a) he is a director, manager, secretary or similar officer of the body corporate, or

(b) he is purporting to act as such an officer, or

(c) the body corporate is managed by its members, of whom he is one.

170 No further sanctions for breach of Act

(1)   A breach of any requirement made (otherwise than by any court) by or under this Act shall incur no civil or criminal sanction as being such a breach, except to the extent (if any) expressly provided by or under this Act.

(2)   In exercising his functions under this Act the Director may take account of any matter appearing to him to constitute a breach of a requirement made by or under this Act, whether or not any sanction for that breach is provided by or under this Act and, if it is so provided, whether or not proceedings have been brought in respect of the breach.

(3)   Subsection (1) does not prevent the grant of an injunction, or the making of an order of certiorari, mandamus or prohibition or as respects Scotland the grant of an interdict or of an order under section 91 of the Court of Session Act 1868 (order for specific performance of statutory duty).

171 Onus of proof in various proceedings

(1)   If an agreement contains a term signifying that in the opinion of the parties section 10(3)(b)(iii) does not apply to the agreement, it shall be taken not to apply unless the contrary is proved.

(2)   It shall be assumed in any proceedings, unless the contrary is proved, that when a person initiated a transaction as mentioned in section 19(1)(c) he knew the principal agreement had been made, or contemplated that it might be made.

(3)   Regulations under section 44 or 52 may make provision as to the onus of proof in any proceedings to enforce the regulations.

(4)   In proceedings brought by the creditor under a credit-token agreement—

(a) it is for the creditor to prove that the credit-token was lawfully supplied to the debtor, and was accepted by him, and

(b) if the debtor alleges that any use made of the credit-token was not authorised by him, it is for the creditor to prove either—

    (i) that the use was so authorised, or

    (ii) that the use occurred before the creditor had been given notice under section 84(3).

(5)   In proceedings under section 50(1) in respect of a document received by a minor at any school or other educational establishment for minors, it is for the person sending it to him at that establishment to prove that he did not know or suspect it to be such an establishment.

(6)   In proceedings under section 119(1) it is for the pawnee to prove that he had reasonable cause to refuse to allow the pawn to be redeemed.

(7)   If, in proceedings referred to in section 139(1), the debtor or any surety alleges that the credit bargain is extortionate it is for the creditor to prove the contrary.

172   Statements by creditor or owner to be binding

(1)   A statement by a creditor or owner is binding on him if given under—
section 77(1),
section 78(1),
section 79(1),
section 97(1),
section 107(1)(c),
section 108(1)(c), or
section 109(1)(c).

(2)   Where a trader—
    (a) gives a customer a notice in compliance with section 103(1)(b), or
    (b) gives a customer a notice under section 103(1) asserting that the customer is not indebted to him under an agreement,

the notice is binding on the trader.

(3)   Where in proceedings before any court—
    (a) it is sought to rely on a statement or notice given as mentioned in subsection (1) or (2), and
    (b) the statement or notice is shown to be incorrect,

the court may direct such relief (if any) to be given to the creditor or owner from the operation of subsection (1) or (2) as appears to the court to be just.

173   Contracting-out forbidden

(1)   A term contained in a regulated agreement or linked transaction, or in any other agreement relating to an actual or prospective regulated agreement or linked transaction, is void if, and to the extent that, it is inconsistent with a provision for the protection of the debtor or hirer or his relative or any surety contained in this Act or in any regulation made under this Act.

(2)   Where a provision specifies the duty or liability of the debtor or hirer or his relative or any surety in certain circumstances, a term is inconsistent with that provision if it purports to impose, directly or indirectly, an additional duty or liability on him in those circumstances.

(3)   Notwithstanding subsection (1), a provision of this Act under which a thing may be done in relation to any person on an order of the court or the Director only shall not be taken to prevent its being done at any time with that person's consent given at that time, but the refusal of such consent shall not give rise to any liability.

PART XII
SUPPLEMENTAL

General

174   Restrictions on disclosure of information

(1)   No information obtained under or by virtue of this Act about any individual shall be disclosed without his consent.

(2)   No information obtained under or by virtue of this Act about any business shall be disclosed except, so long as the business continues to be carried on, with the consent of the person for the time being carrying it on.

(3)   Subsections (1) and (2) do not apply to any disclosure of information made—
   (a) for the purpose of facilitating the performance of any functions, under this Act, the Trade Descriptions Act 1968 or Part II or III or section 125 (annual and other reports of Director) of the Fair Trading Act 1973 [or the Estate Agents Act 1979] [or the Competition Act 1980]

[or the Telecommunications Act 1984] [or the Gas Act 1986] [or the Airports Act 1986] [or the Consumer Protection Act 1987] [or Part II of the Consumer Protection (Northern Ireland) Order 1987] [or the Control of Misleading Advertisements Regulations 1988] [or the Courts and Legal Services Act 1990] [or the Railways Act 1993] [or the Coal Industry Act 1994] [or the Water Act 1989] [the Water Act 1991 or any of the other consolidation Acts (within the meaning of section 206 of that Act of 1991)] [or the Electricity Act 1989] [or the Electricity (Northern Ireland) Order 1992] [or the Gas (Northern Ireland) Order 1996] [or Part IV of the Airports (Northern Ireland) Order 1994] of the Secretary of State, any other Minister, [the Director General of Telecommunications,] [the Director General of Gas Supply,] [the Civil Aviation Authority] [the Director General of Water Services,] [the Director General of Electricity Supply] [or the Director General of Electricity Supply for Northern Ireland] [the Rail Regulator] [or the Director General of Gas for Northern Ireland] [the Rail Regulator] [the Authorised Conveyancing Practitioners board, the Coal Authority] any enforcement authority or any Northern Ireland department, or

(b) in connection with the investigation of any criminal offence or for the purposes of any criminal proceedings, or

(c) for the purposes of any civil proceedings brought under or by virtue of this Act or under Part III of the Fair Trading Act 1973 [or under the Control of Misleading Advertisements Regulations 1988].

[(3A) Subsections (1) and (2) do not apply to any disclosure of information by the Director to the [Financial Services Authority] for the purpose of enabling or assisting the [Authority] to discharge its functions under the Banking Act 1987 or the Director to discharge his functions under this Act.]

(4) Nothing in subsections (1) and (2) shall be construed—
(a) as limiting the particulars which may be entered in the register; or
(b) as applying to any information which has been made public as part of the register.

(5) Any person who discloses information in contravention of this section commits an offence.

NOTES

Sub-s (3): amended by the Estate Agents Act 1979, s 10(4)(b), the Competition Act 1980, s 19(4)(d), the Telecommunications Act 1984, s 109,

Sch 4, para 60(2), the Gas Act 1986, s 67(1), Sch 7, para 19, the Airports Act 1986, s 83(1), Sch 4, para 4, the Consumer Protection Act 1987, s 48(1), Sch 4, para 4, the Control of Misleading Advertisements Regulations 1988, SI 1988/915, reg 7(6)(b), the Water Act 1989, s 190(1), Sch 25, para 47, the Electricity Act 1989, s 112(1), Sch 16, para 17(1), (2), the Water Consolidation (Consequential Provisions) Act 1991, s 2(1), Sch 1, para 26, the Electricity (Northern Ireland) Order 1992, SI 1992/231, art 95(1), Sch 12, para 14, the Railways Act 1993, s 152(1), Sch 12, para 8, the Coal Industry Act 1994, s 67, Sch 9, para 15, the Airports (Northern Ireland) Order 1994, SI 1994/426, art 71(2), Sch 9, para 4, and the Gas (Northern Ireland) Order 1996, SI 1996/275, art 71(1), Sch 6; and amended by the Courts and Legal Services Act 1990, s 125(3), Sch 18, para 6, as from a day to be appointed.

Sub-s (3A): inserted by the Banking Act 1987, s 87, words in square brackets therein substituted by the Bank of England Act 1998, s 23(1), Sch 5, para 60.

Modification: sub-s (3A) modified, in relation to references to the Bank's functions under the Banking Act and the Director's functions under the Consumer Credit Act, by the Banking Coordination (Second Council Directive) Regulations 1992, SI 1992/3218, reg 62.

175   Duty of persons deemed to be agents

Where under this Act a person is deemed to receive a notice or payment as agent of the creditor or owner under a regulated agreement, he shall be deemed to be under a contractual duty to the creditor or owner to transmit the notice, or remit the payment, to him forthwith.

177   Saving for registered charges

(1)   Nothing in this Act affects the rights of a proprietor of a registered charge (within the meaning of the Land Registration Act 1925), who—
  (a) became the proprietor under a transfer for valuable consideration without notice of any defect in the title arising (apart from this section) by virtue of this Act, or
  (b) derives title from such a proprietor.

(2)   Nothing in this Act affects the operation of section 104 of the Law of Property Act 1925 (protection of purchaser where mortgagee exercises power of sale).

(3)   Subsection (1) does not apply to a proprietor carrying on a business of debt-collecting.

(4)   Where, by virtue of subsection (1), a land mortgage is enforced which apart from this section would be treated as never having effect, the original creditor or owner shall be liable to indemnify the debtor or hirer against any loss thereby suffered by him.

(5)   . . .

(6)   In the application of this section to Northern Ireland—
- (a) any reference to the proprietor of a registered charge (within the meaning of the Land Registration Act 1925) shall be construed as a reference to the registered owner of a charge under the Local Registration of Title (Ireland) Act 1891 or Part IV of the Land Registration Act (Northern Ireland) 1970, and
- (b) for the reference to section 104 of the Law of Property Act 1925 there shall be substituted a reference to section 21 of the Conveyancing and Law of Property Act 1881 and section 5 of the Conveyancing Act 1911.

NOTES

Sub-s (5): applies to Scotland only.

Regulations, orders, etc

179   Power to prescribe form etc of secondary documents

(1)   Regulations may be made as to the form and content of credit-cards, trading-checks, receipts, vouchers and other documents or things issued by creditors, owners or suppliers under or in connection with regulated agreements or by other persons in connection with linked transactions, and may in particular—
- (a) require specified information to be included in the prescribed manner in documents, and other specified material to be excluded;
- (b) contain requirements to ensure that specified information is clearly brought to the attention of the debtor or hirer, or his relative, and that one part of a document is not given insufficient or excessive prominence compared with another.

(2)   If a person issues any document or thing in contravention of regulations under subsection (1) then, as from the time of the contravention but without prejudice to anything done before it, this Act shall apply as if

the regulated agreement had been improperly executed by reason of a contravention of regulations under section 60(1).

180 Power to prescribe form etc of copies

(1)   Regulations may be made as to the form and content of documents to be issued as copies of any executed agreement, security instrument or other document referred to in this Act, and may in particular—
  (a) require specified information to be included in the prescribed manner in any copy, and contain requirements to ensure that such information is clearly brought to the attention of a reader of the copy;
  (b) authorise the omission from a copy of certain material contained in the original, or the inclusion of such material in condensed form.

(2)   A duty imposed by any provision of this Act (except section 35) to supply a copy of any document—
  (a) is not satisfied unless the copy supplied is in the prescribed form and conforms to the prescribed requirements;
  (b) is not infringed by the omission of any material, or its inclusion in condensed form, if that is authorised by regulations;

and references in this Act to copies shall be construed accordingly.

(3)   Regulations may provide that a duty imposed by this Act to supply a copy of a document referred to in an unexecuted agreement or an executed agreement shall not apply to documents of a kind specified in the regulations.

181 Power to alter monetary limits etc

(1)   The Secretary of State may by order made by statutory instrument amend, or further amend, any of the following provisions of this Act so as to reduce or increase a sum mentioned in that provision, namely, sections 8(2), 15(1)(c), 17(1), 43(3)(a), 70(6), 75(3)(b), 77(1), 78(1), 79(1), 84(1), 101(7)(a), 107(1), 108(1), 109(1), 110(1), 118(1)(b), 120(1)(a), 139(5) and (7), 155(1) and 158(1).

(2)   An order under subsection (1) amending section 8(2), 15(1)(c), 17(1), 43(3)(a), 75(3)(b) or 139(5) or (7) shall be of no effect unless a draft of the order has been laid before and approved by each House of Parliament.

182   Regulations and orders

(1)   Any power of the Secretary of State to make regulations or orders under this Act, except the power conferred by sections 2(1)(a), 181 and 192, shall be exercisable by statutory instrument subject to annulment in pursuance of a resolution of either House of Parliament.

(2)   Where a power to make regulations or orders is exercisable by the Secretary of State by virtue of this Act, regulations or orders made in the exercise of that power may—
  (a) make different provision in relation to different cases or classes of case, and
  (b) exclude certain cases or classes of case, and
  (c) contain such transitional provisions as the Secretary of State thinks fit.

(3)   Regulations may provide that specified expressions, when used as described by the regulations, are to be given the prescribed meaning, notwithstanding that another meaning is intended by the person using them.

(4)   Any power conferred on the Secretary of State by this Act to make orders includes power to vary or revoke an order so made.

183   Determinations etc by Director

The Director may vary or revoke any determination or direction made or given by him under this Act (other than Part III, or Part III as applied by section 147).

Interpretation

184   Associates

(1)   A person is an associate of an individual if that person is the individual's husband or wife, or is a relative, or the husband or wife of a relative, of the individual or of the individual's husband or wife.

(2)   A person is an associate of any person with whom he is in partnership, and of the husband or wife or a relative of any individual with whom he is in partnership.

(3)   A body corporate is an associate of another body corporate—

(a) if the same person is a controller of both, or a person is a controller of one and persons who are his associates, or he and persons who are his associates, are the controllers of the other; or

(b) if a group of two or more persons is a controller of each company, and the groups either consist of the same persons or could be regarded as consisting of the same persons by treating (in one or more cases) a member of either group as replaced by a person of whom he is an associate.

(4)   A body corporate is an associate of another person if that person is a controller of it or if that person and persons who are his associates together are controllers of it.

(5)   In this section "relative" means brother, sister, uncle, aunt, nephew, niece, lineal ancestor or lineal descendant, and references to a husband or wife include a former husband or wife and a reputed husband or wife; and for the purposes of this subsection a relationship shall be established as if any illegitimate child, step-child or adopted child of a person had been a child born to him in wedlock.

185   Agreement with more than one debtor or hirer

(1)   Where an actual or prospective regulated agreement has two or more debtors or hirers (not being a partnership or an unincorporated body of persons)—

(a) anything required by or under this Act to be done to or in relation to the debtor or hirer shall be done to or in relation to each of them; and

(b) anything done under this Act by or on behalf of one of them shall have effect as if done by or on behalf of all of them.

(2)   Notwithstanding subsection (1)(a), where running-account credit is provided to two or more debtors jointly, any of them may by a notice signed by him (a "dispensing notice") authorise the creditor not to comply in his case with section 78(4) (giving of periodical statement of account); and the dispensing notice shall have effect accordingly until revoked by a further notice given by the debtor to the creditor:

Provided that—

(a) a dispensing notice shall not take effect if previous dispensing notices are operative in the case of the other debtor, or each of the other debtors, as the case may be;

(b) any dispensing notices operative in relation to an agreement shall cease to have effect if any of the debtors dies.

[(c) a dispensing notice which is operative in relation to an agreement shall be operative also in relation to any subsequent agreement which, in relation to the earlier agreement, is a modifying agreement]

(3)   Subsection (1)(b) does not apply for the purposes of section 61(1)(a) or 127(3).

(4)   Where a regulated agreement has two or more debtors or hirers (not being a partnership or an unincorporated body of persons), section 86 applies to the death of any of them.

(5)   An agreement for the provision of credit, or the bailment or (in Scotland) the hiring of goods, to two or more persons jointly where—
   (a) one or more of those persons is an individual, and
   (b) one or more of them is a body corporate,

is a consumer credit agreement or consumer hire agreement if it would have been one had they all been individuals; and the body corporate or bodies corporate shall accordingly be included among the debtors or hirers under the agreement.

(6)   Where subsection (5) applies, references in this Act to the signing of any document by the debtor or hirer shall be construed in relation to a body corporate as referring to a signing on behalf of the body corporate.

NOTES

Sub-s (2): para (c) added by the Banking Act 1979, s 38(3).

186   Agreement with more than one creditor or owner

Where an actual or prospective regulated agreement has two or more creditors or owners, anything required by or under this Act to be done to, or in relation to, or by, the creditor or owner shall be effective if done to, or in relation to, or by, any one of them.

187   Arrangements between creditor and supplier

(1)   A consumer credit agreement shall be treated as entered into under pre-existing arrangements between a creditor and a supplier if it is entered

into in accordance with, or in furtherance of, arrangements previously made between persons mentioned in subsection (4)(a), (b) or (c).

(2)   A consumer credit agreement shall be treated as entered into in contemplation of future arrangements between a creditor and a supplier if it is entered into in the expectation that arrangements will subsequently be made between persons mentioned in subsection (4)(a), (b) or (c) for the supply of cash, goods and services (or any of them) to be financed by the consumer credit agreement.

(3)   Arrangements shall be disregarded for the purposes of subsection (1) or (2) if—
  (a) they are arrangements for the making, in specified circumstances, of payments to the supplier by the creditor, and
  (b) the creditor holds himself out as willing to make, in such circumstances, payments of the kind to suppliers generally.

[(3A) Arrangements shall also be disregarded for the purposes of subsections (1) and (2) if they are arrangements for the electronic transfer of funds from a current account at a bank within the meaning of the Bankers' Books Evidence Act 1879.]

(4)   The persons referred to in subsections (1) and (2) are—
  (a) the creditor and the supplier;
  (b) one of them and an associate of the other's;
  (c) an associate of one and an associate of the other's.

(5)   Where the creditor is an associate of the supplier's, the consumer credit agreement shall be treated, unless the contrary is proved, as entered into under pre-existing arrangements between the creditor and the supplier.

NOTES

Sub-s (3A): inserted by the Banking Act 1987, s 89.

188   Examples of use of new terminology

(1)   Schedule 2 shall have effect for illustrating the use of terminology employed in this Act.

(2)   The examples given in Schedule 2 are not exhaustive.

(3)   In the case of conflict between Schedule 2 and any other provision of this Act, that other provision shall prevail.

(4)   The Secretary of State may by order amend Schedule 2 by adding further examples or in any other way.

189   Definitions

(1)   In this Act, unless the context otherwise requires—
   "advertisement" includes every form of advertising, whether in a
      publication, by television or radio, by display of notices, signs, labels,
      showcards or goods, by distribution of samples, circulars, catalogues,
      price lists or other material, by exhibition of pictures, models or
      films, or in any other way, and references to the publishing of
      advertisements shall be construed accordingly;
   "advertiser" in relation to an advertisement, means any person indicated
      by the advertisement as willing to enter into transactions to which
      the advertisement relates;
   "ancillary credit business" has the meaning given by section 145(1);
   "antecedent negotiations" has the meaning given by section 56;
   "appeal period" means the period beginning on the first day on which an
      appeal to the Secretary of State may be brought and ending on the
      last day on which it may be brought or, if it is brought, ending on its
      final determination, or abandonment;

. . . . .

   "associate" shall be construed in accordance with section 184;
   ["authorised institution" means an institution authorised under the
      Banking Act 1987;]
   "bill of sale" has the meaning given by section 4 of the Bills of Sale
      Act 1878 or, for Northern Ireland, by section 4 of the Bills of Sale
      (Ireland) Act 1879;
   ["building society" means a building society within the meaning of the
      Building Societies Act 1986;]
   "business" includes profession or trade, and references to a business
      apply subject to subsection (2);
   "cancellable agreement" means a regulated agreement which, by virtue
      of section 67, may be cancelled by the debtor or hirer;
   "canvass" shall be construed in accordance with sections 48 and 153;
   "cash" includes money in any form;
   "charity" means as respects England and Wales a charity registered under
      [the Charities Act 1993] or an exempt charity (within the meaning
      of that Act), and as respects Scotland and Northern Ireland an
      institution or other organisation established for charitable purposes
      only ("organisation" including any persons administering a trust and

"charitable" being construed in the same way as if it were contained in the Income Tax Acts);

"conditional sale agreement" means an agreement for the sale of goods or land under which the purchase price or part of it is payable by instalments, and the property in the goods or land is to remain in the seller (notwithstanding that the buyer is to be in possession of the goods or land) until such conditions as to the payment of instalments or otherwise as may be specified in the agreement are fulfilled;

"consumer credit agreement" has the meaning given by section 8, and includes a consumer credit agreement which is cancelled under section 69(1), or becomes subject to section 69(2), so far as the agreement remains in force;

"consumer credit business" means any business so far as it comprises or relates to the provision of credit under regulated consumer credit agreements;

"consumer hire agreement" has the meaning given by section 15;

"consumer hire business" means any business so far as it comprises or relates to the bailment or (in Scotland) the hiring of goods under regulated consumer hire agreements;

"controller", in relation to a body corporate, means a person—

(a) in accordance with whose directions or instructions the directors of the body corporate or of another body corporate which is its controller (or any of them) are accustomed to act, or

(b) who, either alone or with any associate or associates, is entitled to exercise or control the exercise of, one third or more of the voting power at any general meeting of the body corporate or of another body corporate which is its controller;

"copy" shall be construed in accordance with section 180;

. . . . .

"court" means in relation to England and Wales the county court, in relation to Scotland the sheriff court and in relation to Northern Ireland the High Court or the county court;

"credit" shall be construed in accordance with section 9;

"credit-broker" means a person carrying on a business of credit brokerage;

"credit brokerage" has the meaning given by section 145(2);

"credit limit" has the meaning given by section 10(2);

"creditor" means the person providing credit under a consumer credit agreement or the person to whom his rights and duties under the agreement have passed by assignment or operation of law, and in relation to a prospective consumer credit agreement, includes the prospective creditor;

"credit reference agency" has the meaning given by section 145(8);

"credit-sale agreement" means an agreement for the sale of goods, under which the purchase price or part of it is payable by instalments, but which is not a conditional sale agreement;

"credit-token" has the meaning given by section 14(1);

"credit-token agreement" means a regulated agreement for the provision of credit in connection with the use of a credit-token;

"debt-adjusting" has the meaning given by section 145(5);

"debt-collecting" has the meaning given by section 145(7);

"debt-counselling" has the meaning given by section 145(6);

"debtor" means the individual receiving credit under a consumer credit agreement or the person to whom his rights and duties under the agreement have passed by assignment or operation of law, and in relation to a prospective consumer credit agreement includes the prospective debtor;

"debtor-creditor agreement" has the meaning given by section 13;

"debtor-creditor-supplier agreement" has the meaning given by section 12;

"default notice" has the meaning given by section 87(1);

"deposit" means any sum payable by a debtor or hirer by way of deposit or down-payment, or credited or to be credited to him on account of any deposit or down-payment, whether the sum is to be or has been paid to the creditor or owner or any other person, or is to be or has been discharged by a payment of money or a transfer or delivery of goods or by any other means;

"Director" means the Director General of Fair Trading;

"electric line" has the meaning given by [the Electricity Act 1989] or, for Northern Ireland, [the Electricity (Northern Ireland) Order 1992];

"embodies" and related words shall be construed in accordance with subsection (4);

"enforcement authority" has the meaning given by section 161(1);

"enforcement order" means an order under section 65(1), 105(7)(a) or (b), 111(2) or 124(1) or (2);

"executed agreement" means a document, signed by or on behalf of the parties, embodying the terms of a regulated agreement, or such of them as have been reduced to writing;

"exempt agreement" means an agreement specified in or under section 16;

"finance" means to finance wholly or partly and "financed" and "refinanced" shall be construed accordingly;

"file" and "copy of the file" have the meanings given by section 158(5);

"fixed-sum credit" has the meaning given by section 10(1)(b);

"friendly society" means a society registered under the Friendly Societies Acts 1896 to 1971 … ;

"future arrangements" shall be construed in accordance with section 187;

"general notice" means a notice published by the Director at a time and in a manner appearing to him suitable for securing that the notice is seen within a reasonable time by persons likely to be affected by it;

"give", means, deliver or send by post to;

"goods" has the meaning given by [section 61(1) of the Sale of Goods Act 1979];

"group licence" has the meaning given by section 22(1)(b);

"High Court" means Her Majesty's High Court of Justice, or the Court of Session in Scotland or the High Court of Justice in Northern Ireland;

"hire-purchase agreement" means an agreement, other than a conditional sale agreement, under which—

  (a) goods are bailed or (in Scotland) hired in return for periodical payments by the person to whom they are bailed or hired, and

  (b) the property in the goods will pass to that person if the terms of the agreement are complied with and one or more of the following occurs—

   (i)   the exercise of an option to purchase by that person,

   (ii) the doing of any other specified act by any party to the agreement,

   (iii) the happening of any other specified event;

"hirer" means the individual to whom goods are bailed or (in Scotland) hired under a consumer hire agreement, or the person to whom his rights and duties under the agreement have passed by assignment or operation of law, and in relation to a prospective consumer hire agreement includes the prospective hirer;

"individual" includes a partnership or other unincorporated body of persons not consisting entirely of bodies corporate;

"installation" means—

  (a) the installing of any electric line or any gas or water pipe,

  (b) the fixing of goods to the premises where they are to be used, and the alteration of premises to enable goods to be used on them,

  (c) where it is reasonably necessary that goods should be constructed or erected on the premises where they are to be used, any work carried out for the purpose of constructing or erecting them on those premises;

"insurance company" has the meaning given by [section 96(1) of the Insurance Companies Act 1982], but does not include a friendly society or an organisation of workers or organisation of employers;

"judgment" includes an order or decree made by any court;

"land", includes an interest in land, and in relation to Scotland includes heritable subjects of whatever description;

"land improvement company" means an improvement company as defined by section 7 of the Improvement of Land Act 1899;

"land mortgage" includes any security charged on land;

"licence" means a licence under Part III (including that Part as applied to ancillary credit businesses by section 147);

"licensed", in relation to any act, means authorised by a licence to do the act or cause or permit another person to do it;

"licensee", in the case of a group licence, includes any person covered by the licence;

"linked transaction" has the meaning given by section 19(1);

"local authority", in relation to England ... , means ... a county council, a London borough council, a district council, the Common Council of the City of London, or the Council of Isles of Scilly, [in relation to Wales means a county council or a county borough council,] and in relation to Scotland, means a [council constituted under section 2 of the Local Government etc (Scotland) Act 1994], and, in relation to Northern Ireland, means a district council;

. . . . .

"modifying agreement" has the meaning given by section 82(2);

. . . . .

"multiple agreement" has the meaning given by section 18(1);

"negotiator" has the meaning given by section 56(1);

"non-commercial agreement" means a consumer credit agreement or a consumer hire agreement not made by the creditor or owner in the course of a business carried on by him;

"notice" means notice in writing;

"notice of cancellation" has the meaning given by section 69(1);

"owner" means a person who bails or (in Scotland) hires out goods under a consumer hire agreement or the person to whom his rights and duties under the agreement have passed by assignment or operation of law, and in relation to a prospective consumer hire agreement, includes the prospective bailor or persons from whom the goods are to be hired;

"pawn" means any article subject to a pledge;

"pawn-receipt" has the meaning given by section 114;

"pawnee" and "pawnor" include any person to whom the rights and duties of the original pawnee or the original pawnor, as the case may be, have passed by assignment or operation of law;

"payment" includes tender;

"personal credit agreement" has the meaning given by section 8(1);

"pledge" means the pawnee's rights over an article taken in pawn;

"prescribed" means prescribed by regulations made by the Secretary of State;

"pre-existing arrangements" shall be construed in accordance with section 187;

"principal agreement" has the meaning given by section 19(1);

"protected goods" has the meaning given by section 90(7);

"quotation" has the meaning given by section 52(1)(a);

"redemption period" has the meaning given by section 116(3);

"register" means the register kept by the Director under section 35;

"regulated agreement" means a consumer credit agreement, or consumer hire agreement, other than an exempt agreement, and "regulated" and "unregulated" shall be construed accordingly;

"regulations" means regulations made by the Secretary of State;

"relative", except in section 184, means a person who is an associate by virtue of section 184(1);

"representation" includes any condition or warranty, and any other statement or undertaking, whether oral or in writing;

"restricted-use credit agreement" and "restricted-use credit" have the meanings given by section 11(1);

"rules of court", in relation to Northern Ireland means, in relation to the High Court, rules made under section 7 of the Northern Ireland Act 1962, and, in relation to any other court, rules made by the authority having for the time being power to make rules regulating the practice and procedure in that court;

"running-account credit" shall be construed in accordance with section 10;

"security", in relation to an actual or prospective consumer credit agreement or consumer hire agreement, or any linked transaction, means a mortgage, charge, pledge, bond, debenture, indemnity, guarantee, bill, note or other right provided by the debtor or hirer, or at his request (express or implied), to secure the carrying out of the obligations of the debtor or hirer under the agreement;

"security instrument" has the meaning given by section 105(2);

"serve on" means deliver or send by post to;

"signed" shall be construed in accordance with subsection (3);

"small agreement" has the meaning given by section 17(1), and "small" in relation to an agreement within any category shall be construed accordingly;

"specified fee" shall be construed in accordance with section 2(4) and (5);

"standard licence" has the meaning given by section 22(1)(a);

"supplier" has the meaning given by section 11(1)(b) or 12(c) or 13(c) or, in relation to an agreement falling within section 11(1)(a), means the creditor, and includes a person to whom the rights and duties of a supplier (as so defined) have passed by assignment or operation of law, or (in relation to a prospective agreement) the prospective supplier;

"surety" means the person by whom any security is provided, or the person to whom his rights and duties in relation to the security have passed by assignment or operation of law;

"technical grounds" shall be construed in accordance with subsection (5);

"time order" has the meaning given by section 129(1);

"total charge for credit" means a sum calculated in accordance with regulations under section 20(1);

"total price" means the total sum payable by the debtor under a hire-purchase agreement or a conditional sale agreement, including any sum payable on the exercise of an option to purchase, but excluding any sum payable as a penalty or as compensation or damages for a breach of the agreement;

"unexecuted agreement" means a document embodying the terms of a prospective regulated agreement, or such of them as it is intended to reduce to writing;

"unlicensed" means without a licence but applies only in relation to acts for which a licence is required;

"unrestricted-use credit agreement" and "unrestricted-use credit" have the meanings given by section 11(2);

"working day" means any day other than—

(a) Saturday or Sunday,

(b) Christmas Day or Good Friday,

(c) a bank holiday within the meaning given by section 1 of the Banking and Financial Dealings Act 1971.

(2)   A person is not to be treated as carrying on a particular type of business merely because occasionally he enters into transactions belonging to a business of that type.

(3)   Any provision of this Act requiring a document to be signed is complied with by a body corporate if the document is sealed by that body.

This subsection does not apply to Scotland.

(4)   A document embodies a provision if the provision is set out either in the document itself or in another document referred to in it.

(5)   An application dismissed by the court or the Director shall, if the court or the Director (as the case may be) so certifies, be taken to be dismissed on technical grounds only.

(6)   Except in so far as the context otherwise requires, any reference in this Act to an enactment shall be construed as a reference to that enactment as amended by or under any other enactment, including this Act.

(7)   In this Act, except where otherwise indicated—
   (a) a reference to a numbered Part, section or Schedule is a reference to the Part or section of, or the Schedule to, this Act so numbered, and
   (b) a reference in a section to a numbered subsection is a reference to the subsection of that section so numbered, and
   (c) a reference in a section, subsection or Schedule to a numbered paragraph is a reference to the paragraph of that section, subsection or Schedule so numbered.

NOTES

Sub-s (1): words omitted apply to Scotland only, repealed in part by the Age of Legal Capacity (Scotland) Act 1991, s 10, Sch 2; definition "authorised institution" inserted by the Banking Act 1987, s 88; definition "building society" substituted by the Building Societies Act 1986, s 120, Sch 18, Pt I, para 10(4); in definition "charity" words in square brackets substituted by the Charities Act 1993, s 98(1), Sch 6, para 30; in definition "electric line" words in first pair of square brackets substituted by the Electricity Act 1989, s 112(1), Sch 16, para 17(1), (3), words in second pair of square brackets substituted by the Electricity (Northern Ireland) Order 1992, SI 1992/231, art 95(1), Sch 12, para 15; in definition "friendly society" words omitted repealed by the Friendly Societies Act 1992, s 120, Sch 22, Pt I; in definition "goods" words in square brackets substituted by the Sale of Goods Act 1979, s 63, Sch 2, para 18; in definition "insurance company" words in square brackets substituted by the Insurance Companies Act 1982, s 99(2), Sch 5, para 14; in definition "local authority" words omitted in the first place repealed and words in first pair of square brackets inserted, by the Local Government (Wales) Act 1994, s 66(6), (8), Sch 16, para 45, Sch 18, words omitted in the second place repealed by the Local Government Act 1985, s 102, Sch 17, and words in second pair of square brackets substituted by the Local Government etc (Scotland) Act 1994, s 180(1), Sch 13, para 94.

Miscellaneous

192   Transitional and commencement provisions, amendments and repeals

(1)   The provisions of Schedule 3 shall have effect for the purposes of this Act.

(2)   The appointment of a day for the purposes of any provision of Schedule 3 shall be effected by an order of the Secretary of State made by statutory instrument; and any such order shall include a provision amending Schedule 3 so as to insert an express reference to the day appointed.

(3)   Subject to subsection (4)—
   (a) the enactments specified in Schedule 4 shall have effect subject to the amendments specified in that Schedule (being minor amendments or amendments consequential on the preceding provisions of this Act), and
   (b) the enactments specified in Schedule 5 are hereby repealed to the extent shown in column 3 of that Schedule.

(4)   The Secretary of State shall by order made by statutory instrument provide for the coming into operation of the amendments contained in Schedule 4 and the repeals contained in Schedule 5, and those amendments and repeals shall have effect only as provided by an order so made.

193   Short title and extent

(1)   This Act may be cited as the Consumer Credit Act 1974.

(2)   This Act extends to Northern Ireland.

SCHEDULES

SCHEDULE I

Section 167

PROSECUTION AND PUNISHMENT OF OFFENCES

1 Section	2 Offence	3 Mode of prosecution	4 Imprisonment or fine
7 ...	Knowingly or recklessly giving false information to Director.	(a) Summarily. (b) On indictment.	The prescribed sum. 2 years or a fine or both.
9(1) ...	Engaging in activities requiring a licence when not a licensee.	(a) Summarily. (b) On indictment.	The prescribed sum. 2 years or a fine or both.

1 Section	2 Offence	3 Mode of prosecution	4 Imprisonment or fine
9(2) ...	Carrying on business under a name not specified in licence.	(a) Summarily. (b) On indictment.	The prescribed sum. 2 years or a fine or both.
39(3) ...	Failure to notify changes in registered particulars.	(a) Summarily. (b) On indictment.	The prescribed sum. 2 years or a fine or both.
45 ...	Advertising credit where goods etc not available for cash.	(a) Summarily. (b) On indictment.	2 years or a fine or both.
46(1) ...	False or misleading advertisements.	(a) Summarily. (b) On indictment.	The prescribed sum. 2 years or a fine or both.
47(1) ...	Advertising infringements.	(a) Summarily. (b) On indictment.	The prescribed sum. 2 years or a fine or both.
49(1) ...	Canvassing debtor-creditor agreements off trade premises.	(a) Summarily. (b) On indictment.	The prescribed sum. 1 year or a fine or both.
49(2) ...	Soliciting debtor-creditor agreements during visits made in response to previous oral requests.	(a) Summarily. (b) On indictment.	The prescribed sum. 1 year or a fine or both.
50(1) ...	Sending circulars to minors.	(a) Summarily. (b) On indictment.	The prescribed sum. 1 year or a fine or both.
51(1) ...	Supplying unsolicited credit-tokens.	(a) Summarily. (b) On indictment.	The prescribed sum. 2 years or a fine or both.
77(4) ...	Failure of creditor under fixed-sum credit agreement to supply copies of documents etc.	Summarily.	Level 4 on the standard scale.
78(6) ...	Failure of creditor under running-account credit agreement to supply copies of documents etc.	Summarily.	Level 4 on the standard scale.
79(3) ...	Failure of owner under consumer hire agreement to supply copies of documents etc.	Summarily.	Level 4 on the standard scale.
80(2) ...	Failure to tell creditor or owner whereabouts of goods.	Summarily.	Level 3 on the standard scale.
85(2) ...	Failure of creditor to supply copy of credit-token agreement.	Summarily.	Level 4 on the standard scale.
97(3) ...	Failure to supply debtor with statement of amount required to discharge agreement.	Summarily.	Level 3 on the standard scale.
103(5) ...	Failure to deliver notice relating to discharge of agreements.	Summarily.	Level 3 on the standard scale.

1 *Section*	2 *Offence*	3 *Mode of prosecution*	4 *Imprisonment or fine*
107(4) ...	Failure of creditor to give information to surety under fixed-sum credit agreement.	Summarily.	Level 4 on the standard scale.
108(4) ...	Failure of creditor to give information to surety under running-account credit agreement.	Summarily.	Level 4 on the standard scale.
109(3) ...	Failure of owner to give information to surety under consumer hire agreement.	Summarily.	Level 4 on the standard scale.
110(3) ...	Failure of creditor or owner to supply a copy of any security instrument to debtor or hirer.	Summarily.	Level 4 on the standard scale.
114(2) ...	Taking pledges from minors.	(a) Summarily. (b) On indictment.	The prescribed sum. 1 year or a fine or both.
115 ...	Failure to supply copies of a pledge agreement or pawn-receipt.	Summarily.	Level 4 on the standard scale.
119(1) ...	Unreasonable refusal to allow pawn to be redeemed.	Summarily.	Level 4 on the standard scale.
154 ...	Canvassing ancillary credit services off trade premises.	(a) Summarily. (b) On indictment.	The prescribed sum. 1 year or a fine or both.
157(3) ...	Refusal to give name etc of credit reference agency.	Summarily.	Level 4 on the standard scale.
158(4) ...	Failure of credit reference agency to disclose filed information.	Summarily.	Level 4 on the standard scale.
159(6) ...	Failure of credit reference agency to correct information.	Summarily.	Level 4 on the standard scale.
160(6) ...	Failure of credit reference agency to comply with section 160(3) or (4).	Summarily.	Level 4 on the standard scale.
162(6) ...	Impersonation of enforcement authority officers.	(a) Summarily. (b) On indictment.	The prescribed sum. 2 years or a fine or both.
165(1) ...	Obstruction of enforcement authority officers.	Summarily.	Level 4 on the standard scale.
165(2) ...	Giving false information to enforcement authority officers.	(a) Summarily. (b) On indictment.	The prescribed sum. 2 years or a fine or both.
167(2) ...	Contravention of regulations under section 44, 52, 53, 54 or 112.	(a) Summarily. (b) On indictment.	The prescribed sum. 2 years or a fine or both.
174(5) ...	Wrongful disclosure of information.	(a) Summarily. (b) On indictment.	The prescribed sum. 2 years or a fine or both.

NOTES

The references to the "prescribed sum" in this Schedule are substituted by virtue of the Magistrates' Courts Act 1980, s 32(2).

The references to levels 3 and 4 on the standard scale are substituted by virtue of the Criminal Justice Act 1982, ss 38, 46.

SCHEDULE 2

Section 188(1)

EXAMPLES OF USE OF NEW TERMINOLOGY

PART I
LIST OF TERMS

Term	Defined in section	Illustrated by example(s)
Advertisement	189(1)	2
Advertiser	189(1)	2
Antecedent negotiations	56	1, 2, 3, 4
Cancellable agreement	67	4
Consumer credit agreement	8	5, 6, 7, 15, 19, 21
Consumer hire agreement	15	20, 24
Credit	9	16, 19, 21
Credit-broker	189(1)	2
Credit limit	10(2)	6, 7, 19, 22, 23
Creditor	189(1)	1, 2, 3, 4
Credit-sale agreement	189(1)	5
Credit-token	14	3, 14, 16
Credit-token agreement	14	3, 14, 16, 22
Debtor-creditor agreement	13	8, 16, 17, 18
Debtor-creditor-supplier agreement	12	8, 16
Fixed-sum credit	10	9, 10, 17, 23

Term	Defined in section	Illustrated by example(s)
Hire-purchase agreement	189(1)	10
Individual	189(1)	19, 24
Linked transaction	19	11
Modifying agreement	82(2)	24
Multiple agreement	18	16, 18
Negotiator	56(1)	1, 2, 3, 4
Personal credit agreement	8(1)	19
Pre-existing arrangements	187	8, 21
Restricted-use credit	11	10, 12, 13, 14, 16
Running-account credit	10	15, 16, 18, 23
Small agreement	17	16, 17, 22
Supplier	189(1)	3, 14
Total charge for credit	20	5, 10
Total price	189(1)	10
Unrestricted-use credit	11	8, 12, 16, 17, 18.

PART II
EXAMPLES

Example 1

Facts. Correspondence passes between an employee of a moneylending company (writing on behalf of the company) and an individual about the terms on which the company would grant him a loan under a regulated agreement.

Analysis. The correspondence constitutes antecedent negotiations falling within section 56(1)(a), the moneylending company being both creditor and negotiator.

Example 2

Facts. Representations are made about goods in a poster displayed by a shopkeeper near the goods, the goods being selected by a customer who has read the poster and then sold by the shopkeeper to a finance company introduced by him (with whom he has a business relationship). The goods are disposed of by the finance company to the customer under a regulated hire-purchase agreement.

Analysis. The representations in the poster constitute antecedent negotiations falling within section 56(1)(b), the shopkeeper being the credit-broker and negotiator and the finance company being the creditor. The poster is an advertisement and the shopkeeper is the advertiser.

Example 3

Facts. Discussions take place between a shopkeeper and a customer about goods the customer wishes to buy using a credit-card issued by the D Bank under a regulated agreement.

Analysis. The discussions constitute antecedent negotiations falling within section 56(1)(c), the shopkeeper being the supplier and negotiator and the D Bank the creditor. The credit-card is a credit-token as defined in section 14(1), and the regulated agreement under which it was issued is a credit-token agreement as defined in section 14(2).

Example 4

Facts. Discussions take place and correspondence passes between a secondhand car dealer and a customer about a car, which is then sold by the dealer to the customer under a regulated conditional sale agreement. Subsequently, on a revocation of that agreement by consent, the car is resold by the dealer to a finance company introduced by him (with whom he has a business relationship), who in turn dispose of it to the same customer under a regulated hire-purchase agreement.

Analysis. The discussions and correspondence constitute antecedent negotiations in relation both to the conditional sale agreement and the hire-purchase agreement. They fall under section 56(1)(a) in relation to the conditional sale agreement, the dealer being the creditor and the negotiator. In relation to the hire-purchase agreement they fall within section 56(1)(b), the dealer continuing to be treated as the negotiator but the finance company now being the creditor. Both agreements are cancellable if the discussions took place when the individual conducting the negotiations (whether the "negotiator" or his employee or agent) was in the presence of the debtor, unless the unexecuted agreement was signed by the debtor at trade premises (as defined in section 67(b)). If the discussions all took place by telephone however, or the unexecuted agreement was signed by the debtor on trade premises (as so defined) the agreements are not cancellable.

Example 5

Facts. E agrees to sell to F (an individual) an item of furniture in return for 24 monthly instalments of £10 payable in arrears. The property in the goods passes to F immediately.

Analysis. This is a credit-sale agreement (see definition of "credit-sale agreement" in section 189(1)). The credit provided amounts to £240 less the amount which, according to regulations made under section 20(1), constitutes the total charge for credit. (This amount is required to be deducted by section 9(4)). Accordingly the agreement falls within section 8(2) and is a consumer credit agreement.

Example 6

Facts. The G Bank grants H (an individual) an unlimited overdraft, with an increased rate of interest on so much of any debit balance as exceeds £2,000.

Analysis. Although the overdraft purports to be unlimited, the stipulation for increased interest above £2,000 brings the agreement within section 10(3)(b)(ii) and it is a consumer credit agreement.

Example 7

Facts. J is an individual who owns a small shop which usually carries a stock worth about £1,000. K makes a stocking agreement under which he undertakes to provide on short-term credit the stock needed from time to time by J without any specified limit.

Analysis. Although the agreement appears to provide unlimited credit, it is probable, having regard to the stock usually carried by J, that his indebtedness to K will not at any time rise above £5,000. Accordingly the agreement falls within section 10(3)(b)(iii) and is a consumer credit agreement.

Example 8

Facts. U, a moneylender, lends £500 to V (an individual) knowing he intends to use it to buy office equipment from W. W introduced V to U, it being his practice to introduce customers needing finance to him. Sometimes U gives W a commission for this and sometimes not. U pays the £500 direct to V.

Analysis. Although this appears to fall under section 11(1)(b), it is excluded by section 11(3) and is therefore (by section 11(2)) an unrestricted-use credit agreement. Whether it is a debtor-creditor agreement (by section 13(c)) or a debtor-creditor-supplier agreement (by section 12(c)) depends on whether the previous dealings between U and W amount to "pre-existing arrangements", that is whether the agreement can be taken to have been entered into "in accordance with, or in furtherance of" arrangements previously made between U and W, as laid down in section 187(1).

Example 9

Facts. A agrees to lend B (an individual) £4,500 in nine monthly instalments of £500.

Analysis. This is a cash loan and is a form of credit (see section 9 and definition of "cash" in section 189(1)). Accordingly it falls within section 10(1)(b) and is fixed-sum credit amounting to £4,500.

Example 10

Facts. C (in England) agrees to bail goods to D (an individual) in return for periodical payments. The agreement provides for the property in the goods to pass to D on payment of a total of £7,500 and the exercise by D of an option to purchase. The sum of £7,500 includes a down-payment of £1,000. It also includes an amount which, according to regulations made under section 20(1), constitutes a total charge for credit of £1,500.

Analysis. This is a hire-purchase agreement with a deposit of £1,000 and a total price of £7,500 (see definitions of "hire-purchase agreement", "deposit" and "total price" in section 189(1)). By section 9(3), it is taken to provide credit amounting to £7,500 - (£1,500 + £1,000), which equals £5,000. Under section 8(2), the agreement is therefore a consumer credit agreement, and under sections 9(3) and 11(1) it is a restricted-use credit agreement for fixed-sum credit. A similar result would follow if the agreement by C had been a hiring agreement in Scotland.

Example 11

Facts. X (an individual) borrows £500 from Y (Finance). As a condition of the granting of the loan X is required—
  (a) to execute a second mortgage on his house in favour of Y (Finance), and
  (b) to take out a policy of insurance on his life with Y (Insurances).

In accordance with the loan agreement, the policy is charged to Y (Finance) as collateral security for the loan. The two companies are associates within the meaning of section 184(3).

Analysis. The second mortgage is a transaction for the provision of security and accordingly does not fall within section 19(1), but the taking out of the insurance policy is a linked transaction falling within section 19(1)(a). The charging of the policy is a separate transaction (made between different parties) for the provision of security and again is excluded from section 19(1). The only linked transaction is therefore the taking out of the insurance policy. If X had not been required by the loan agreement to take out the policy, but it had been done at the suggestion of Y (Finance) to induce them to enter into the loan agreement, it would have been a linked transaction under section 19(1)(c)(i) by virtue of section 19(2)(a).

Example 12

Facts. The N Bank agrees to lend O (an individual) £2,000 to buy a car from P. To make sure the loan is used as intended, the N Bank stipulates that the money must be paid by it direct to P.

Analysis. The agreement is a consumer credit agreement by virtue of section 8(2). Since it falls within section 11(1)(b), it is a restricted-use credit agreement, P being the supplier. If the N Bank had not stipulated for direct payment to the supplier, section 11(3) would have operated and made the agreement into one for unrestricted-use credit.

Example 13

Facts. Q, a debt-adjuster, agrees to pay off debts owed by R (an individual) to various moneylenders. For this purpose the agreement provides for the making of a loan by Q to R in return for R's agreeing to repay the loan by instalments with interest. The loan money is not paid over to R but retained by Q and used to pay off the moneylenders.

Analysis. This is an agreement to refinance existing indebtedness of the debtor's, and if the loan by Q does not exceed £5,000 is a restricted-use credit agreement falling within section 11(1)(c).

Example 14

Facts. On payment of £1, S issues to T (an individual) a trading check under which T can spend up to £20 at any shop which has agreed, or in future agrees, to accept S's trading checks.

Analysis. The trading check is a credit-token falling within section 14(1)(b). The credit-token agreement is a restricted-use credit agreement within section 11(1)(b), any shop in which the credit-token is used being the "supplier". The fact that further shops may be added after the issue of the credit-token is irrelevant in view of section 11(4).

Example 15

Facts. A retailer, L, agrees with M (an individual) to open an account in M's name and, in return for M's promise to pay a specified minimum sum into the account each month and to pay a monthly charge for credit, agrees to allow to be debited to the account, in respect of purchases made by M from L, such sums as will not increase the debit balance at any time beyond the credit limit, defined in the agreement as a given multiple of the specified minimum sum.

Analysis. This arrangement provides credit falling within the definition of running-account credit in section 10(1)(a). Provided the credit limit is not over £5,000, the agreement falls within section 8(2) and is a consumer credit agreement for running-account credit.

Example 16

Facts. Under an unsecured agreement, A (Credit), an associate of the A Bank, issues to B (an individual) a credit-card for use in obtaining cash on credit from A (Credit), to be paid by branches of the A Bank (acting as agent of A (Credit)), or goods or cash from suppliers or banks who have agreed to honour credit-cards issued by A (Credit). The credit limit is £30.

Analysis. This is a credit-token agreement falling within section 14(1)(a) and (b). It is a regulated consumer credit agreement for running-account credit. Since the credit limit does not exceed £30, the agreement is a small agreement. So far as the agreement relates to goods it is a debtor-creditor-supplier agreement within section 12(b), since it provides restricted-use credit under section 11(1)(b). So far as it relates to cash it is a debtor-creditor agreement within section 13(c) and the credit it provides is unrestricted-use credit. This is therefore a multiple agreement. In that the whole agreement falls within several of the categories of agreement mentioned in this Act, it is, by section 18(3), to be treated as an agreement in each of those categories. So far as it is a debtor-creditor-supplier agreement providing restricted-use credit it is, by section 18(2), to be treated as a separate agreement; and similarly so far as it is a debtor-creditor agreement providing unrestricted-used credit. (See also Example 22.)

Example 17

Facts. The manager of the C Bank agrees orally with D (an individual) to open a current account in D's name. Nothing is said about overdraft facilities. After maintaining the account in credit for some weeks, D draws a cheque in favour of E for an amount exceeding D's credit balance by £20. E presents the cheque and the Bank pay it.

Analysis. In drawing the cheque D, by implication, requests the Bank to grant him an overdraft of £20 on its usual terms as to interest and other charges. In deciding to honour the cheque, the Bank by implication accepts the offer. This constitutes a regulated small consumer credit agreement for unrestricted-use, fixed-sum credit. It is a debtor-creditor agreement, and falls within section 74(1)(b) if covered by a determination under section 74(3). (Compare Example 18.)

Example 18

Facts. F (an individual) has had a current account with the G Bank for many years. Although usually in credit, the account has been allowed by the Bank to become overdrawn from time to time. The maximum such overdraft has been is about £1,000. No explicit agreement has ever been made about overdraft facilities. Now, with a credit balance of £500, F draws a cheque for £1,300.

Analysis. It might well be held that the agreement with F (express or implied) under which the Bank operate his account includes an implied term giving him the right to overdraft facilities up to say £1,000. If so, the agreement is a regulated consumer credit agreement for unrestricted-use, running-account credit. It is a debtor-creditor agreement, and falls within section 74(1)(b) if covered by a direction under section 74(3). It is also a multiple agreement, part of which (i.e. the part not dealing with the overdraft), as referred to in section 18(1)(a), falls within a category of agreement not mentioned in this Act. (Compare Example 17.)

Example 19

Facts. H (a finance house) agrees with J (a partnership of individuals) to open an unsecured loan account in J's name on which the debit balance is not to exceed £7,000 (having regard to payments into the account made from time to time by J). Interest is to be payable in advance on this sum, with provision for yearly adjustments. H is entitled to debit the account with interest, a "setting-up" charge, and other charges. Before J has an opportunity to draw on the account it is initially debited with £2,250 for advance interest and other charges.

Analysis. This is a personal running-account credit agreement (see sections 8(1) and 10(1)(a), and definition of "individual" in section 189(1)). By section 10(2) the credit limit is £7,000. By section 9(4) however the initial debit of £2,250, and any other charges later debited to the account by H, are not to be treated as credit even though time is allowed for their payment. Effect is given to this by section 10(3). Although the credit limit of £7,000 exceeds the amount (£5,000) specified in section 8(2) as the maximum for a consumer credit agreement, so that the agreement is not within section 10(3)(a), it is caught by section 10(3)(b)(i). At the beginning J can effectively draw (as credit) no more than £4,750, so the agreement is a consumer credit agreement.

Example 20

Facts. K (in England) agrees with L (an individual) to bail goods to L for a period of three years certain at £2,000 a year, payable quarterly. The agreement contains no provision for the passing of the property in the goods to L.

Analysis. This is not a hire-purchase agreement (see paragraph (b) of the definition of that term in section 189(1)) and is capable of subsisting for more than three months. Paragraphs (a) and (b) of section 15(1) are therefore satisfied, but paragraph (c) is not. The payments by L must exceed £5,000 if he conforms to the agreement. It is true that under section 101 L has a right to terminate the agreement on giving K three months' notice expiring not earlier than eighteen months after the making of the agreement, but that section applies only where the agreement is a regulated consumer hire agreement apart from the section (see subsection (1)). So the agreement is not a consumer hire agreement, though it would be if the hire charge were say £1,500 a year, or there were a "break" clause in it operable by either party before the hire charges exceeded £5,000. A similar result would follow if the agreement by K had been a hiring agreement in Scotland.

Example 21

Facts. The P Bank decides to issue cheque cards to its customers under a scheme whereby the Bank undertakes to honour cheques of up to £30 in every case where the payee has taken the cheque in reliance on the cheque card, whether the customer has funds in his account or not. The P Bank writes to the major retailers advising them of this scheme and also publicises it by advertising. The Bank issues a cheque card to Q (an individual), who uses it to pay by cheque for goods costing £20 bought by Q from R, a major retailer. At the time, Q has £500 in his account at the P Bank.

Analysis. The agreement under which the cheque card is issued to Q is a consumer credit agreement even though at all relevant times Q has more than £30 in his account. This is because Q is free to draw out his whole balance and then use the cheque card, in which case the Bank has bound itself to honour the cheque. In other words the cheque card agreement provides Q with credit, whether he avails himself of it or not. Since the amount of the credit is not subject to any express limit, the cheque card can be used any number of times. It may be presumed however that section 10(3)(b)(iii) will apply. The agreement is an unrestricted-use debtor- creditor agreement (by section 13(c)). Although the P Bank wrote to R informing R of the P Bank's willingness to honour any cheque taken by R in reliance on a cheque card, this does not constitute pre-existing arrangements as mentioned in section 13(c) because section 187(3) operates to prevent it. The agreement is not a credit-token agreement within section 14(1)(b) because payment by the P Bank to R, would be a payment of the cheque and not a payment for the goods.

Example 22

Facts. The facts are as in Example 16. On one occasion B uses the credit-card in a way which increases his debit balance with A (Credit) to £40. A

(Credit) writes to B agreeing to allow the excess on that occasion only, but stating that it must be paid off within one month.

Analysis. In exceeding his credit limit B, by implication, requests A (Credit) to allow him a temporary excess (compare Example 17). A (Credit) is thus faced by B's action with the choice of treating it as a breach of contract or granting his implied request. He does the latter. If he had done the former, B would be treated as taking credit to which he was not entitled (section 14(3)) and, subject to the terms of his contract with A (Credit), would be liable to damages for breach of contract. As it is, the agreement to allow the excess varies the original credit-token agreement by adding a new term. Under section 10(2), the new term is to be disregarded in arriving at the credit limit, so that the credit-token agreement at no time ceases to be a small agreement. By section 82(2) the later agreement is deemed to revoke the original agreement and contain provisions reproducing the combined effect of the two agreements. By section 82(4), this later agreement is exempted from Part V (except section 56).

Example 23

Facts. Under an oral agreement made on 10th January, X (an individual) has an overdraft on his current account at the Y Bank with a credit limit of £100. On 15th February, when his overdraft standards at £90, X draws a cheque for £25. It is the first time that X has exceeded his credit limit, and on 16th February the bank honours the cheque.

Analysis. The agreement of 10th January is a consumer credit agreement for running-account credit. The agreement of 15th–16th February varies the earlier agreement by adding a term allowing the credit limit to be exceeded merely temporarily. By section 82(2) the later agreement is deemed to revoke the earlier agreement and reproduce the combined effect of the two agreements. By section 82(4), Part V of this Act (except section 56) does not apply to the later agreement. By section 18(5), a term allowing a merely temporary excess over the credit limit is not to be treated as a separate agreement, or as providing fixed-sum credit. The whole of the £115 owed to the Bank by X on 16th February is therefore running-account credit.

Example 24

Facts. On 1st March 1975 Z (in England) enters into an agreement with A (an unincorporated body of persons) to bail to A equipment consisting of two components (component P and component Q). The agreement is not a hire-purchase agreement and is for a fixed term of 3 years, so paragraphs (a) and (b) of section 15(1) are both satisfied. The rental is payable monthly at a rate of £2,400 a year, but the agreement provides that this is to be reduced to £1,200 a year for the remainder of the agreement if at any time during its currency A returns component Q to the owner Z. On 5th May 1976 A is incorporated as A Ltd., taking over A's assets and liabilities. On 1st March 1977, A Ltd. returns component Q. On 1st January 1978, Z and A Ltd. agree to extend the earlier agreement by one year, increasing the rental for the final year by £250 to £1,450.

Analysis. When entered into on 1st March 1975, the agreement is a consumer hire agreement. A falls within the definition of "individual" in section 189(1) and if A returns component Q before 1st May 1976 the total rental will not exceed £5,000 (see section 15(1)(c)). When this date is passed without component Q having been returned it is obvious that the total rental must now exceed £5,000. Does this mean that the agreement then ceases to be a consumer hire agreement? The answer is no, because there has been no change in the terms of the agreement, and without such a change the agreement cannot move from one category to the other. Similarly, the fact that A's rights and duties under the agreement pass to a body corporate on 5th May 1976 does not cause the agreement to cease to be a consumer hire agreement (see the definition of "hirer" in section 189(1)).

The effect of the modifying agreement of 1st January 1978 is governed by section 82(2), which requires it to be treated as containing provisions reproducing the combined effect of the two actual agreements, that is to say as providing that—

(a) obligations outstanding on 1st January 1978 are to be treated as outstanding under the modifying agreement;

(b) the modifying agreement applies at the old rate of hire for the months of January and February 1978, and

(c) for the year beginning 1st March 1978 A Ltd. will be the bailee of component P at a rental of £1,450.

The total rental under the modifying agreement is £1,850. Accordingly the modifying agreement is a regulated agreement. Even if the total rental under the modifying agreement exceeded £5,000 it would still be regulated because of the provisions of section 82(3).

Unsolicited Goods and Services (Amendment) Act 1975

(C 13)

An Act to amend the Unsolicited Goods and Services Act 1971, to enable the Secretary of State to make regulations with respect to the contents and form of notes of agreement, invoices and similar documents and to provide for conviction on indictment in relation to an offence under s 3(2) of the said Act; and for connected matters

[20 March 1975]

NOTES

Unsolicited Goods and Services Acts 1971 and 1975: by s 4(1) of this Act, the 1971 Act and this Act may be cited together by this collective title.

3 Provision for offence under section 3(2) of the Act of 1971 to be prosecuted on indictment

(1)   An offence under section 3 (2) of the Act of 1971 may be prosecuted on indictment; and a person convicted on indictment of an offence under that section shall be liable to a fine.

(2)   This section applies only to offences committed after the coming into operation of this section.

4 Short title, citation, commencement, transitional provisions and extent

(1)   This Act may be cited as the Unsolicited Goods and Services (Amendment) Act 1975 and the Unsolicited Goods and Services Act 1971 and this Act may be cited together as the Unsolicited Goods and Services Acts 1971 and 1975.

(2)   Sections 1 and 3 of this Act and this section shall come into operation on the passing of this Act but any regulations made by virtue of the said section 1 shall not come into operation before the date appointed by order under subsection (3) below for the coming into operation of section 2 of this Act.

(3)   Section 2 of this Act shall come into operation on such date as the Secretary of State may by order made by statutory instrument appoint; and different dates may be appointed by order under this subsection for different provisions of that section.

(4)   The amendments made to sections 3(3) and 6(2) of the Act of 1971 by section 2 of this Act and any regulations made by virtue of section 1 of this Act shall not apply to any note of agreement signed, or invoice or similar document sent before the date appointed by order under subsection (3) above for the coming into operation of the said section 2.

(5)   This Act shall not extend to Northern Ireland.

Torts (Interference With Goods) Act 1977

(C 32)

An Act to amend the law concerning conversion and other torts affecting goods

[22 July 1977]

Preliminary

1   Definition of "wrongful interference with goods"

In this Act "wrongful interference", or "wrongful interference with goods", means—
   (a) conversion of goods (also called trover),
   (b) trespass to goods,
   (c) negligence so far as it results in damage to goods or to an interest in goods,
   (d) subject to section 2, any other tort so far as it results in damage to goods or to an interest in goods

[and references in this Act (however worded) to proceedings for wrongful interference or to a claim or right to claim for wrongful interference shall include references to proceedings by virtue of Part I of the Consumer Protection Act 1987 [or Part II of the Consumer Protection (Northern Ireland) Order 1987] (product liability) in respect of any damage to goods or to an interest in goods or, as the case may be, to a claim or right to claim by virtue of that Part in respect of any such damage].

NOTES

Words in first (outer) pair of square brackets added by the Consumer Protection Act 1987, s 48, Sch 4, and words in second (inner) pair of square brackets inserted by the Consumer Protection (Northern Ireland) Order 1987, SI 1987/2049, art 35(1), Sch 3, para 3.

Detention of goods

2 Abolition of detinue

(1)   Detinue is abolished.

(2)   An action lies in conversion for loss or destruction of goods which a bailee has allowed to happen in breach of his duty to his bailor (that is to say it lies in a case which is not otherwise conversion, but would have been detinue before detinue was abolished).

3 Form of judgment where goods are detained

(1)   In proceedings for wrongful interference against a person who is in possession or in control of the goods relief may be given in accordance with this section, so far as appropriate.

(2)   The relief is—
   (a) an order for delivery of the goods, and for payment of any consequential damages, or
   (b) an order for delivery of the goods, but giving the defendant the alternative of paying damages by reference to the value of the goods, together in either alternative with payment of any consequential damages, or
   (c) damages.

(3)   Subject to rules of court—
   (a) relief shall be given under only one of paragraphs (a), (b) and (c) of subsection (2),
   (b) relief under paragraph (a) of subjection (2) is at the discretion of the court, and the claimant may choose between the others.

(4)   If it is shown to the satisfaction of the court that an order under subsection (2)(a) has not been complied with, the court may—

    (a) revoke the order, or the relevant part of it, and

    (b) make an order for payment of damages by reference to the value of the goods.

(5)   Where an order is made under subsection (2)(b) the defendant may satisfy the order by returning the goods at any time before execution of judgment, but without prejudice to liability to pay any consequential damages.

(6)   An order for delivery of the goods under subsection (2)(a) or (b) may impose such conditions as may be determined by the court, or pursuant to rules of court, and in particular, where damages by reference to the value of the goods would not be the whole of the value of the goods, may require an allowance to be made by the claimant to reflect the difference.

For example, a bailor's action against the bailee may be one in which the measure of damages is not the full value of the goods, and then the court may order delivery of the goods, but require the bailor to pay the bailee a sum reflecting the difference.

(7)   Where under subjection (1) or subsection (2) of section 6 an allowance is to be made in respect of an improvement of the goods, and an order is made under subsection (2)(a) or (b), the court may assess the allowance to be made in respect of the improvement, and by the order require, as a condition for delivery of the goods, that allowance to be made by the claimant.

(8)   This section is without prejudice—

    (a) to the remedies afforded by section 133 of the Consumer Credit Act 1974, or

    (b) to the remedies afforded by sections 35, 42 and 44 of the Hire-Purchase Act 1965, or to those sections of the Hire-Purchase Act (Northern Ireland) 1966 (so long as those sections respectively remain in force), or

    (c) to any jurisdiction to afford ancillary or incidental relief.

4   Interlocutory relief where goods are detained

(1)   In this section "proceedings" means proceedings for wrongful interference.

(2)   On the application of any person in accordance with rules of court, the High Court shall, in such circumstances as may be specified in the rules, have power to make an order providing for the delivery up of any goods which are or may become the subject matter of subsequent proceedings in the court, or as to which any question may arise in proceedings.

(3)   Delivery shall be, as the order may provide, to the claimant or to a person appointed by the court for the purpose, and shall be on such terms and conditions as may be specified in the order.

(4)   The power to make rules of court under section [84 of the Supreme Court Act 1981] or under section 7 of the Northern Ireland Act 1962 shall include power to make rules of court as to the manner in which an application for such an order can be made, and as to the circumstances in which such an order can be made; and any such rules may include such incidental, supplementary and consequential provisions as the authority making the rules may consider necessary or expedient.

(5)   The preceding provisions of this section shall have effect in relation to county courts as they have effect in relation to the High Court, and as if in those provisions references to rules of court and to section [84] of the said Act of [1981] or section 7 of the Northern Ireland Act 1962 included references to county court rules and to [section 75 of the County Courts Act 1984] or [Article 47 of the County Courts (Northern Ireland) Order 1980].

NOTES

Sub-s (4): words in square brackets substituted by the Supreme Court Act 1981, s 152(1), Sch 5.

Sub-s (5): words in first and second pairs of square brackets substituted by the Supreme Court Act 1981, s 152(1), Sch 5; words in third pair of square brackets substituted by the County Courts Act 1984, s 148(1), Sch 2, Pt V, para 64; words in final pair of square brackets substituted by the County Courts (Northern Ireland) Order 1980, SI 1980/397, art 68(2), Sch 1, Pt II.

Damages

5   Extinction of title on satisfaction of claim for damages

(1)   Where damages for wrongful interference are, or would fall to be, assessed on the footing that the claimant is being compensated—

    (a) for the whole of his interest in the goods, or

    (b) for the whole of his interest in the goods subject to a reduction for contributory negligence,

payment of the assessed damages (under all heads), or as the case may be settlement of a claim for damages for the wrong (under all heads), extinguishes the claimant's title to that interest.

(2)   In subsection (1) the reference to the settlement of the claim includes—

    (a) where the claim is made in court proceedings, and the defendant has paid a sum into court to meet the whole claim, the taking of that sum by the claimant, and

    (b) where the claim is made in court proceedings, and the proceedings are settled or compromised, the payment of what is due in accordance with the settlement or compromise, and

    (c) where the claim is made out of court and is settled or compromised, the payment of what is due in accordance with the settlement or compromise.

(3)   It is hereby declared that subsection (1) does not apply where damages are assessed on the footing that the claimant is being compensated for the whole of his interest in the goods, but the damages paid are limited to some lesser amount by virtue of any enactment or rule of law.

(4)   Where under section 7(3) the claimant accounts over to another person (the "third party") so as to compensate (under all heads) the third party for the whole of his interest in the goods, the third party's title to that interest is extinguished.

(5)   This section has effect subject to any agreement varying the respective rights of the parties to the agreement, and where the claim is made in court proceedings has effect subject to any order of the court.

6   Allowance for improvement of the goods

(1)   If in proceedings for wrongful interference against a person (the "improver") who has improved the goods, it is shown that the improver acted in the mistaken but honest belief that he had a good title to them, an allowance shall be made for the extent to which, at the time as at which the goods fall to be valued in assessing damages, the value of the goods is attributable to the improvement.

(2)   If, in proceedings for wrongful interference against a person ("the purchaser") who has purported to purchase the goods—
  (a) from the improver, or
  (b) where after such a purported sale the goods passed by a further purported sale on one or more occasions, on any such occasion,

it is shown that the purchaser acted in good faith, an allowance shall be made on the principle set out in subsection (1).

   For example, where a person in good faith buys a stolen car from the improver and is sued in conversion by the true owner the damages may be reduced to reflect the improvement, but if the person who bought the stolen car from the improver sues the improver for failure of consideration, and the improver acted in good faith, subsection (3) below will ordinarily make a comparable reduction in the damages he recovers from the improver.

(3)   If in a case within subsection (2) the person purporting to sell the goods acted in good faith, then in proceedings by the purchaser for recovery of the purchase price because of failure of consideration, or in any other proceedings founded on that failure of consideration, an allowance shall, where appropriate, be made on the principle set out in subsection (1).

(4)   This section applies, with the necessary modifications, to a purported bailment or other disposition of goods as it applies to a purported sale of goods.

Liability to two or more claimants

7   Double liability

(1)   In this section "double liability" means the double liability of the wrongdoer which can arise—
  (a) where one of two or more rights of action for wrongful interference is founded on a possessory title, or
  (b) where the measure of damages in an action for wrongful interference founded on a proprietary title is or includes the entire value of the goods, although the interest is one of two or more interests in the goods.

(2)   In proceedings to which any two or more claimants are parties, the relief shall be such as to avoid double liability of the wrongdoer as between those claimants.

(3)   On satisfaction, in whole or in part, of any claim for an amount exceeding that recoverable if subsection (2) applied, the claimant is liable to account over to the other person having a right to claim to such extent as will avoid double liability.

(4)   Where, as the result of enforcement of a double liability, any claimant is unjustly enriched to any extent, he shall be liable to reimburse the wrongdoer to that extent.

For example, if a converter of goods pays damages first to a finder of the goods, and then to the true owner, the finder is unjustly enriched unless he accounts over to the true owner under subsection (3); and then the true owner is unjustly enriched and becomes liable to reimburse the converter of the goods.

8   Competing rights to the goods

(1)   The defendant in an action for wrongful interference shall be entitled to show, in accordance with rules of court, that a third party has a better right than the plaintiff as respects all or any part of the interest claimed by the plaintiff, or in right of which he sues, and any rule of law (sometimes called jus tertii) to the contrary is abolished.

(2)   Rules of court relating to proceedings for wrongful interference may—
  (a) require the plaintiff to give particulars of his title,
  (b) require the plaintiff to identify any person who, to his knowledge, has or claims any interest in the goods,
  (c) authorise the defendant to apply for directions as to whether any person should be joined with a view to establishing whether he has a better right than the plaintiff, or has a claim as a result of which the defendant might be doubly liable,
  (d) where a party fails to appear on an application within paragraph (c), or to comply with any direction given by the court on such an application, authorise the court to deprive him of any right of action against the defendant for the wrong either unconditionally, or subject to such terms or conditions as may be specified.

(3)   Subsection (2) is without prejudice to any other power of making rules of court.

Conversion and trespass to goods

10 Co-owners

(1)   Co-ownership is no defence to an action founded on conversion or trespass to goods where the defendant without the authority of the other co-owner—
  (a) destroys the goods, or disposes of the goods in a way giving a good title to the entire property in the goods, or otherwise does anything equivalent to the destruction of the other's interest in the goods, or
  (b) purports to dispose of the goods in a way which would give a good title to the entire property in the goods if he was acting with the authority of all co-owners of the goods.

(2)   Subsection (1) shall not effect the law concerning execution or enforcement of judgments, or concerning any form of distress.

(3)   Subsection (1)(a) is by way of restatement of existing law so far as it relates to conversion.

11   Minor amendments

(1)   Contributory negligence is no good defence in proceedings founded on conversion, or on intentional trespass to goods.

(2)   Receipt of goods by way of pledge is conversion if the delivery of the goods is conversion.

(3)   Denial of title is not of itself conversion.

Uncollected goods

12   Bailee's power of sale

(1)   This section applies to goods in the possession or under the control of a bailee where—
  (a) the bailor is in breach of an obligation to take delivery of the goods or, if the terms of the bailment so provide, to give directions as to their delivery, or
  (b) the bailee could impose such an obligation by giving notice to the bailor, but is unable to trace or communicate with the bailor, or

(c) the bailee can reasonably expect to be relieved of any duty to safeguard the goods on giving notice to the bailor, but is unable to trace or communicate with the bailor.

(2)   In the cases in Part I of Schedule 1 to this Act a bailee may, for the purposes of subsection (1), impose an obligation on the bailor to take delivery of the goods, or as the case may be to give directions as to their delivery, and in those cases the said Part I sets out the methods of notification.

(3)   If the bailee—
   (a) has in accordance with Part II of Schedule 1 to this Act given notice to the bailor of his intention to sell the goods under this subsection, or
   (b) has failed to trace or communicate with the bailor with a view to giving him such a notice, after having taken reasonable steps for the purpose,

and is reasonably satisfied that the bailor owns the goods, he shall be entitled, as against the bailor, to sell the goods.

(4)   Where subsection (3) applies but the bailor did not in fact own the goods, a sale under this section, or under section 13, shall not give a good title as against the owner, or as against a person claiming under the owner.

(5)   A bailee exercising his powers under subsection (3) shall be liable to account to the bailor for the proceeds of sale, less any costs of sale, and—
   (a) the account shall be taken on the footing that the bailee should have adopted the best method of sale reasonably available in the circumstances, and
   (b) where subsection (3)(a) applies, any sum payable in respect of the goods by the bailor to the bailee which accrued due before the bailee gave notice of intention to sell the goods shall be deductible from the proceeds of sale.

(6)   A sale duly made under this section gives a good title to the purchaser as against the bailor.

(7)   In this section, section 13, and Schedule 1 to this Act,
   (a) "bailor" and "bailee" include their respective successors in title, and
   (b) references to what is payable, paid or due to the bailee in respect of the goods include references to what would be payable by the bailor to the bailee as a condition of delivery of the goods at the relevant time.

(8)   This section, and Schedule 1 to this Act, have effect subject to the terms of the bailment.

(9)   This section shall not apply where the goods were bailed before the commencement of this Act.

13   Sale authorised by the court

(1)   If a bailee of the goods to which section 12 applies satisfies the court that he is entitled to sell the goods under section 12, or that he would be so entitled if he had given any notice required in accordance with Schedule 1 to this Act, the court—
  (a) may authorise the sale of the goods subject to such terms and conditions, if any, as may be specified in the order, and
  (b) may authorise the bailee to deduct from the proceeds of sale any costs of sale and any amount due from the bailor to the bailee in respect of the goods, and
  (c) may direct the payment into court of the net proceeds of sale, less any amount deducted under paragraph (b), to be held to the credit of the bailor.

(2)   A decision of the court authorising a sale under this section shall, subject to any right of appeal, be conclusive, as against the bailor, of the bailee's entitlement to sell the goods, and gives a good title to the purchaser as against the bailor.

(3)   In this section "the court" means the High Court or a county court, [and a county court shall have jurisdiction in the proceedings save that, in Northern Ireland, a county court shall only have jurisdiction in proceedings if the value of the goods does not exceed the county court limit mentioned in Article 10(1) of the County Courts (Northern Ireland) Order 1980].

NOTES

Sub-s (3): words in square brackets substituted by the High Court and County Courts Jurisdiction Order 1991, SI 1991/724, art 2(1), (8), Schedule, Pt I.

Supplemental

14   Interpretation

(1)   In this Act, unless the context otherwise requires—

. . . . .

"enactment" includes an enactment contained in an Act of the Parliament of Northern Ireland or an Order in Council made under the Northern Ireland (Temporary Provisions) Act 1972, or in a Measure of the Northern Ireland Assembly,

"goods" includes all chattels personal other than things in action and money,

"High Court" includes the High Court of Justice in Northern Ireland.

(2)   References in this Act to any enactment include references to that enactment as amended, extended or applied by or under that or any other enactment.

NOTES

Sub-s (1): definition omitted repealed by the High Court and County Courts Jurisdiction Order 1991, SI 1991/724, art 2(8), Schedule, Pt I.

16   Extent and application to the Crown

(1)   . . .

(2)   This Act, except section 15, extends to Northern Ireland.

(3)   This Act shall bind the Crown, but as regards the Crown's liability in tort shall not bind the Crown further than the Crown is made liable in tort by the Crown Proceedings Act 1947.

NOTES

Sub-s (1): applies to Scotland only.

17   Short title, etc

(1)   This Act may be cited as the Torts (Interference with Goods) Act 1977.

(2)   This Act shall come into force on such day as the Lord Chancellor may by order contained in a statutory instrument appoint, and such an order may appoint different dates for different provisions or for different purposes.

(3)   Schedule 2 to this Act contains transitional provisions.

SCHEDULES

SCHEDULE 1

Section 12

UNCOLLECTED GOODS

PART 1
POWER TO IMPOSE OBLIGATION TO COLLECT GOODS

1.—(1) For the purposes of section 12(1) a bailee may, in the circumstances specified in this Part of this Schedule, by notice given to the bailor impose on him an obligation to take delivery of the goods.

(2)   The notice shall be in writing, and may be given either—
  (a) by delivering it to the bailor, or
  (b) by leaving it at his proper address, or
  (c) by post.

(3)   The notice shall—
  (a) specify the name and address of the bailee, and give sufficient particulars of the goods and the address or place where they are held, and
  (b) state that the goods are ready for delivery to the bailor, or where combined with a notice terminating the contract of bailment, will be ready for delivery when the contract is terminated, and
  (c) specify the amount, if any, which is payable by the bailor to the bailee in respect of the goods and which became due before the giving of the notice.

(4)   Where the notice is sent by post it may be combined with a notice under Part II of this Schedule if the notice is sent by post in a way complying with paragraph 6(4).

(5)   References in this Part of this Schedule to taking delivery of the goods include, where the terms of the bailment admit, references to giving directions as to their delivery.

(6)   This Part of this Schedule is without prejudice to the provisions of any contract requiring the bailor to take delivery of the goods.

Goods accepted for repair or other treatment

2.   If a bailee has accepted goods for repair or other treatment on the terms (expressed or implied) that they will be re-delivered to the bailor when the repair or other treatment has been carried out, the notice may be given at any time after the repair or other treatment has been carried out.

Goods accepted for valuation or appraisal

3.   If a bailee has accepted goods in order to value or appraise them, the notice may be given at any time after the bailee has carried out the valuation or appraisal.

Storage, warehousing, etc

4.—(1) If a bailee is in possession of goods which he has held as custodian, and his obligation as custodian has come to an end, the notice may be given at any time after the ending of the obligation, or may be combined with any notice terminating his obligation as custodian.

(2)   This paragraph shall not apply to goods held by a person as mercantile agent, that is to say by a person having in the customary course of his business as a mercantile agent authority either to sell goods or to consign goods for the purpose of sale, or to buy goods, or to raise money on the security of goods.

Supplemental

5.   Paragraphs 2, 3 and 4 apply whether or not the bailor has paid any amount due to the bailee in respect of the goods, and whether or not the bailment is for reward, or in the course of business, or gratuitous.

PART II
NOTICE OF INTENTION TO SELL GOODS

6.—(1) A notice under section 12(3) shall—
  (a) specify the name and address of the bailee, and give sufficient particulars of the goods and the address or place where they are held, and
  (b) specify the date on or after which the bailee proposes to sell the goods, and

(c) specify the amount, if any, which is payable by the bailor to the bailee in respect of the goods, and which became due before the giving of the notice.

(2)   The period between giving of the notice and the date specified in the notice as that on or after which the bailee proposes to exercise the power of sale shall be such as will afford the bailor a reasonable opportunity of taking delivery of the goods.

(3)   If any amount is payable in respect of the goods by the bailor to the bailee, and became due before giving of the notice, the said period shall be not less than three months.

(4)   The notice shall be in writing and shall be sent by post in a registered letter, or by the recorded delivery service.

7.—(1) The bailee shall not give a notice under section 12(3), or exercise his right to sell the goods pursuant to such a notice, at a time when he has notice that, because of a dispute concerning the goods, the bailor is questioning or refusing to pay all or any part of what the bailee claims to be due to him in respect of the goods.

(2)   This paragraph shall be left out of account in determining under section 13(1) whether a bailee of goods is entitled to sell the goods under section 12, or would be so entitled if he had given any notice required in accordance with this Schedule.

Supplemental

8.   For the purposes of this Schedule, and of section 26 of the Interpretation Act 1889 in its application to this Schedule, the proper address of the person to whom a notice is to be given shall be—
   (a) in the case of a body corporate, a registered or principal office of the body corporate, and
   (b) in any other case, the last known address of the person.

Unfair Contract Terms Act 1977

(C 50)

An Act to impose further limits on the extent to which under the law of England and Wales and Northern Ireland civil liability for breach of contract, or for negligence or other breach of duty, can be avoided by means of contract terms and otherwise, and under the law of Scotland civil liability can be avoided by means of contract terms

[26 October 1977]

PART I
AMENDMENT OF LAW FOR ENGLAND AND WALES AND NORTHERN IRELAND

Introductory

I Scope of Part I

(1)   For the purposes of this Part of this Act, "negligence" means the breach—

    (a) of any obligation, arising from the express or implied terms of a contract, to take reasonable care or exercise reasonable skill in the performance of the contract;

    (b) of any common law duty to take reasonable care or exercise reasonable skill (but not any stricter duty);

    (c) of the common duty of care imposed by the Occupiers' Liability Act 1957 or the Occupiers' Liability Act (Northern Ireland) 1957.

(2)   This Part of this Act is subject to Part III; and in relation to contracts, the operation of sections 2 to 4 and 7 is subject to the exceptions made by Schedule 1.

(3)   In the case of both contract and tort, sections 2 to 7 apply (except where the contrary is stated in section 6(4)) only to business liability, that is liability for breach of obligations or duties arising—

    (a) from things done or to be done by a person in the course of a business (whether his own business or another's); or

    (b) from the occupation of premises used for business purposes of the occupier;

and references to liability are to be read accordingly [but liability of an occupier of premises for breach of an obligation or duty towards a person obtaining access to the premises for recreational or educational purposes, being liability for loss or damage suffered by reason of the dangerous state of the premises, is not a business liability of the occupier unless granting that person such access for the purposes concerned falls within the business purposes of the occupier].

(4)   In relation to any breach of duty or obligation, it is immaterial for any purpose of this Part of this Act whether the breach was inadvertent or intentional, or whether liability for it arises directly or vicariously.

NOTES

Sub-s (3): words in square brackets added by the Occupiers' Liability Act 1984, s 2.

Avoidance of liability for negligence, breach of contract, etc

2 Negligence liability

(1)   A person cannot by reference to any contract term or to a notice given to persons generally or to particular persons exclude or restrict his liability for death or personal injury resulting from negligence.

(2)   In the case of other loss or damage, a person cannot so exclude or restrict his liability for negligence except in so far as the term or notice satisfies the requirement of reasonableness.

(3)   Where a contract term or notice purports to exclude or restrict liability for negligence a person's agreement to or awareness of it is not of itself to be taken as indicating his voluntary acceptance of any risk.

3 Liability arising in contract

(1)   This section applies as between contracting parties where one of them deals as consumer or on the other's written standard terms of business.

(2)   As against that party, the other cannot by reference to any contract term—
   (a) when himself in breach of contract, exclude or restrict any liability of his in respect of the breach; or
   (b) claim to be entitled—
      (i)   to render a contractual performance substantially different from that which was reasonably expected of him, or
      (ii) in respect of the whole or any part of his contractual obligation, to render no performance at all,

except in so far as (in any of the cases mentioned above in this subsection) the contract term satisfies the requirement of reasonableness.

4 Unreasonable indemnity clauses

(1)   A person dealing as consumer cannot by reference to any contract term be made to indemnify another person (whether a party to the contract or not) in respect of liability that may be incurred by the other for negligence or breach of contract, except in so far as the contract term satisfies the requirement of reasonableness.

(2)   This section applies whether the liability in question—
  (a) is directly that of the person to be indemnified or is incurred by him vicariously;
  (b) is to the person dealing as consumer or to someone else.

Liability arising from sale or supply of goods

5 "Guarantee" of consumer goods

(1)   In the case of goods of a type ordinarily supplied for private use or consumption, where loss or damage—
  (a) arises from the goods proving defective while in consumer use; and
  (b) results from the negligence of a person concerned in the manufacture or distribution of the goods,

liability for the loss or damage cannot be excluded or restricted by reference to any contract term or notice contained in or operating by reference to a guarantee of the goods.

(2)   For these purposes—
  (a) goods are to be regarded as "in consumer use" when a person is using them, or has them in his possession for use, otherwise than exclusively for the purposes of a business; and
  (b) anything in writing is a guarantee if it contains or purports to contain some promise or assurance (however worded or presented) that defects will be made good by complete or partial replacement, or by repair, monetary compensation or otherwise.

(3)   This section does not apply as between the parties to a contract under or in pursuance of which possession or ownership of the goods passed.

6 Sale and hire-purchase

(1)   Liability for breach of the obligations arising from—

(a) section 12 of the Sale of Goods Act 1979] (seller's implied undertakings as to title, etc.);

(b) section 8 of the Supply of Goods (Implied Terms) Act 1973 (the corresponding thing in relation to hire-purchase),

cannot be excluded or restricted by reference to any contract term.

(2)   As against a person dealing as consumer, liability for breach of the obligations arising from—

(a) [section 13, 14 or 15 of the 1979 Act] (seller's implied undertakings as to conformity of goods with description or sample, or as to their quality or fitness for a particular purpose);

(b) section 9, 10 or 11 of the 1973 Act (the corresponding things in relation to hire-purchase),

cannot be excluded or restricted by reference to any contract term.

(3)   As against a person dealing otherwise than as consumer, the liability specified in subsection (2) above can be excluded or restricted by reference to a contract term, but only in so far as the term satisfies the requirement of reasonableness.

(4)   The liabilities referred to in this section are not only the business liabilities defined by section 1(3), but include those arising under any contract of sale of goods or hire-purchase agreement.

NOTES

Sub-ss (1), (2): words in square brackets substituted by the Sale of Goods Act 1979, s 63, Sch 2, para 19.

7   Miscellaneous contracts under which goods pass

(1)   Where the possession or ownership of goods passes under or in pursuance of a contract not governed by the law of sale of goods or hire-purchase, subsections (2) to (4) below apply as regards the effect (if any) to be given to contract terms excluding or restricting liability for breach of obligation arising by implication of law from the nature of the contract.

(2)   As against a person dealing as consumer, liability in respect of the goods' correspondence with description or sample, or their quality or fitness for any particular purpose, cannot be excluded or restricted by reference to any such term.

(3)   As against a person dealing otherwise than as consumer, that liability can be excluded or restricted by reference to such a term, but only in so far as the term satisfies the requirement of reasonableness.

[(3A) Liability for breach of the obligations arising under section 2 of the Supply of Goods and Services Act 1982 (implied terms about title etc in certain contracts for the transfer of the property in goods) cannot be excluded or restricted by references to any such term.]

(4)   Liability in respect of—
  (a) the right to transfer ownership of the goods, or give possession; or
  (b) the assurance of quiet possession to a person taking goods in pursuance of the contract,

cannot [(in a case to which subsection (3A) above does not apply)] be excluded or restricted by reference to any such term except in so far as the term satisfies the requirement of reasonableness.

(5)   This section does not apply in the case of goods passing on a redemption of trading stamps within the Trading Stamps Act 1964 or the Trading Stamps Act (Northern Ireland) 1965.

NOTES

Sub-s (3A): inserted by the Supply of Goods and Services Act 1982, s 17(2).
Sub-s (4): words in square brackets inserted by the Supply of Goods and Services Act 1982, s 17(3).

Other provisions about contracts

9   Effect of breach

(1)   Where for reliance upon it a contract term has to satisfy the requirement of reasonableness, it may be found to do so and be given effect accordingly notwithstanding that the contract has been terminated either by breach or by a party electing to treat it as repudiated.

(2)   Where on a breach the contract is nevertheless affirmed by a party entitled to treat it as repudiated, this does not of itself exclude the requirement of reasonableness in relation to any contract term.

10 Evasion by means of secondary contract

A person is not bound by any contract term prejudicing or taking away rights of his which arise under, or in connection with the performance of, another contract, so far as those rights extend to the enforcement of another's liability which this Part of this Act prevents that other from excluding or restricting.

Explanatory provisions

11 The "reasonableness" test

(1)　In relation to a contract term, the requirement of reasonableness for the purposes of this Part of this Act, section 3 of the Misrepresentation Act 1967 and section 3 of the Misrepresentation Act (Northern Ireland) 1967 is that the term shall have been a fair and reasonable one to be included having regard to the circumstances which were, or ought reasonably to have been, known to or in the contemplation of the parties when the contract was made.

(2)　In determining for the purposes of section 6 or 7 above whether a contract term satisfies the requirement of reasonableness, regard shall be had in particular to the matters specified in Schedule 2 to this Act; but this subsection does not prevent the court or arbitrator from holding, in accordance with any rule of law, that a term which purports to exclude or restrict any relevant liability is not a term of the contract.

(3)　In relation to a notice (not being a notice having contractual effect), the requirement of reasonableness under this Act is that it should be fair and reasonable to allow reliance on it, having regard to all the circumstances obtaining when the liability arose or (but for the notice) would have arisen.

(4)　Where by reference to a contract term or notice a person seeks to restrict liability to a specified sum of money, and the question arises (under this or any other Act) whether the term or notice satisfies the requirement of reasonableness, regard shall be had in particular (but without prejudice to subsection (2) above in the case of contract terms) to—
  (a)　the resources which he could expect to be available to him for the purpose of meeting the liability should it arise; and
  (b)　how far it was open to him to cover himself by insurance.

(5)   It is for those claiming that a contract term or notice satisfies the requirement of reasonableness to show that it does.

12   "Dealing as consumer"

(1)   A party to a contract "deals as consumer" in relation to another party if—

    (a) he neither makes the contract in the course of a business nor holds himself out as doing so; and

    (b) the other party does make the contract in the course of a business; and

    (c) in the case of a contract governed by the law of sale of goods or hire-purchase, or by section 7 of this Act, the goods passing under or in pursuance of the contract are of a type ordinarily supplied for private use or consumption.

(2)   But on a sale by auction or by competitive tender the buyer is not in any circumstances to be regarded as dealing as consumer.

(3)   Subject to this, it is for those claiming that a party does not deal as consumer to show that he does not.

13   Varieties of exemption clause

(1)   To the extent that this Part of this Act prevents the exclusion or restriction of any liability it also prevents—

    (a) making the liability or its enforcement subject to restrictive or onerous conditions;

    (b) excluding or restricting any right or remedy in respect of the liability, or subjecting a person to any prejudice in consequence of his pursuing any such right or remedy;

    (c) excluding or restricting rules of evidence or procedure;

and (to that extent) sections 2 and 5 to 7 also prevent excluding or restricting liability by reference to terms and notices which exclude or restrict the relevant obligation or duty.

(2)   But an agreement in writing to submit present or future differences to arbitration is not to be treated under this Part of this Act as excluding or restricting any liability.

14   Interpretation of Part I

In this Part of this Act—
"business" includes a profession and the activities of any government
department or local or public authority;
"goods" has the same meaning as in [the Sale of Goods Act 1979];
"hire-purchase agreement" has the same meaning as in the Consumer
Credit Act 1974;
"negligence" has the meaning given by section 1(1);
"notice" includes an announcement, whether or not in writing, and any
other communication or pretended communication; and
"personal injury" includes any disease and any impairment of physical
or mental condition.

NOTES

Words in square brackets in definition "goods" substituted by the Sale of Goods
Act 1979, s 63, Sch 2, para 20.

PART III
PROVISIONS APPLYING TO WHOLE OF UNITED KINGDOM

Miscellaneous

26   International supply contracts

(1)   The limits imposed by this Act on the extent to which a person may
exclude or restrict liability by reference to a contract term do not apply to
liability arising under such a contract as is described in subsection (3) below.

(2)   The terms of such a contract are not subject to any requirement of
reasonableness under section 3 or 4 . . .

(3)   Subject to subsection (4), that description of contract is one whose
characteristics are the following—
   (a) either it is a contract of sale of goods or it is one under or in pursuance
      of which the possession or ownership of goods passes; and
   (b) it is made by parties whose places of business (or, if they have none,
      habitual residences) are in the territories of different States (the

Channel Islands and the Isle of Man being treated for this purpose as different States from the United Kingdom).

(4)   A contract falls within subsection (3) above only if either—
  (a) the goods in question are, at the time of the conclusion of the contract, in the course of carriage, or will be carried, from the territory of one State to the territory of another; or
  (b) the acts constituting the offer and acceptance have been done in the territories of different States; or
  (c) the contract provides for the goods to be delivered to the territory of a State other than that within whose territory those acts were done.

NOTES

Sub-s (2): words omitted apply to Scotland only.

27   Choice of law clauses

(1)   Where the [law applicable to] a contract is the law of any part of the United Kingdom only by choice of the parties (and apart from that choice would be the law of some country outside the United Kingdom) sections 2 to 7 and 16 to 21 of this Act do not operate as part [of the law applicable to the contract].

(2)   This Act has effect notwithstanding any contract term which applies or purports to apply the law of some country outside the United Kingdom, where (either or both)—
  (a) the term appears to the court, or arbitrator or arbiter to have been imposed wholly or mainly for the purpose of enabling the party imposing it to evade the operation of this Act; or
  (b) in the making of the contract one of the parties dealt as consumer, and he was then habitually resident in the United Kingdom, and the essential steps necessary for the making of the contract were taken there, whether by him or by others on his behalf.

(3)   . . .

NOTES

Sub-s (1): words in square brackets substituted by the Contracts (Applicable Law) Act 1990, s 5, Sch 4, para 4.
Sub-s (3): applies to Scotland only.

29 Saving for other relevant legislation

(1)   Nothing in this Act removes or restricts the effect of, or prevents reliance upon, any contractual provision which—
   (a) is authorised or required by the express terms or necessary implication of an enactment; or
   (b) being made with a view to compliance with an international agreement to which the United Kingdom is a party, does not operate more restrictively than is contemplated by the agreement.

(2)   A contract term is to be taken—
   (a) for the purposes of Part I of this Act, as satisfying the requirement of reasonableness; and
   (b) . . .

if it is incorporated or approved by, or incorporated pursuant to a decision or ruling of, a competent authority acting in the exercise of any statutory jurisdiction or function and is not a term in a contract to which the competent authority is itself a party.

(3)   In this section—
   "competent authority" means any court, arbitrator or arbiter, government department or public authority;
   "enactment" means any legislation (including subordinate legislation) of the United Kingdom or Northern Ireland and any instrument having effect by virtue of such legislation; and
   "statutory" means conferred by an enactment.

NOTES
Sub-s (2): para (b) applies to Scotland only.

General

31 Commencement; amendments; repeals

(1)   This Act comes into force on 1st February 1978.

(2)   Nothing in this Act applies to contracts made before the date on which it comes into force, but subject to this, it applies to liability for any loss or damage which is suffered on or after that date.

(3)   The enactments specified in Schedule 3 to this Act are amended as there shown.

(4)   The enactments specified in Schedule 4 to this Act are repealed to the extent specified in column 3 of that Schedule.

32   Citation and extent

(1)   This Act may be cited as the Unfair Contract Terms Act 1977.

(2)   Part I of this Act extends to England and Wales and to Northern Ireland; but it does not extend to Scotland.

(3)   . . .

(4)   This Part of this Act extends to the whole of the United Kingdom.

NOTES

Sub-s (3): applies to Scotland only.

SCHEDULES

SCHEDULE I

Section 1(2)

SCOPE OF SECTIONS 2 TO 4 AND 7

1.   Sections 2 to 4 of this Act do not extend to—
   (a) any contract of insurance (including a contract to pay an annuity on human life);
   (b) any contract so far as it relates to the creation or transfer of an interest in land, or to the termination of such an interest, whether by extinction, merger, surrender, forfeiture or otherwise;
   (c) any contract so far as it relates to the creation or transfer of a right or interest in any patent, trade mark, copyright [or design right], registered design, technical or commercial information or other intellectual property, or relates to the termination of any such right or interest;
   (d) any contract so far as it relates—

    (i)   to the formation or dissolution of a company (which means any body corporate or unincorporated association and includes a partnership), or

   (ii)   to its constitution or the rights or obligations of its corporators or members;

  (e)   any contract so far as it relates to the creation or transfer of securities or of any right or interest in securities.

2.   Section 2(1) extends to—

  (a)   any contract of marine salvage or towage;

  (b)   any charterparty of a ship or hovercraft; and

  (c)   any contract for the carriage of goods by ship or hovercraft;

but subject to this sections 2 to 4 and 7 do not extend to any such contract except in favour of a person dealing as consumer.

3.   Where goods are carried by ship or hovercraft in pursuance of a contract which either—

  (a)   specifies that as the means of carriage over part of the journey to be covered, or

  (b)   makes no provision as to the means of carriage and does not exclude that means,

then sections 2(2), 3 and 4 do not, except in favour of a person dealing as consumer, extend to the contract as it operates for and in relation to the carriage of the goods by that means.

4.   Section 2(1) and (2) do not extend to a contract of employment, except in favour of the employee.

5.   Section 2(1) does not affect the validity of any discharge and indemnity given by a person, on or in connection with an award to him of compensation for pneumoconiosis attributable to employment in the coal industry, in respect of any further claim arising from his contracting that disease.

NOTES

Para 1: words in square brackets in sub-para (c) inserted by the Copyright, Designs and Patents Act 1988, s 303(1), Sch 7, para 24.

SCHEDULE 2

Sections 11(2), 24(2)

"GUIDELINES" FOR APPLICATION OF REASONABLENESS TEST

The matters to which regard is to be had in particular for the purposes of sections 6(3), 7(3) and (4), 20 and 21 are any of the following which appear to be relevant—

(a) the strength of the bargaining positions of the parties relative to each other, taking into account (among other things) alternative means by which the customer's requirements could have been met;

(b) whether the customer received an inducement to agree to the term, or in accepting it had an opportunity of entering into a similar contract with other persons, but without having to accept a similar term;

(c) whether the customer knew or ought reasonably to have known of the existence and extent of the term (having regard, among other things, to any custom of the trade and any previous course of dealing between the parties);

(d) where the term excludes or restricts any relevant liability if some condition is not complied with, whether it was reasonable at the time of the contract to expect that compliance with that condition would be practicable;

(e) whether the goods were manufactured, processed or adapted to the special order of the customer.

Civil Liability (Contribution) Act 1978

(C 47)

An Act to make new provision for contribution between persons who are jointly or severally, or both jointly and severally, liable for the same damage and in certain other similar cases where two or more persons have paid or may be required to pay compensation for the same damage; and to amend the law relating to proceedings against persons jointly liable for the same debt or jointly or severally, or both jointly and severally, liable for the same damage

[31 July 1978]

Proceedings for contribution

1 Entitlement to contribution

(1)   Subject to the following provisions of this section, any person liable in respect of any damage suffered by another person may recover contribution from any other person liable in respect of the same damage (whether jointly with him or otherwise).

(2)   A person shall be entitled to recover contribution by virtue of subsection (1) above notwithstanding that he has ceased to be liable in respect of the damage in question since the time when the damage occurred, provided that he was so liable immediately before he made or was ordered or agreed to make the payment in respect of which the contribution is sought.

(3)   A person shall be liable to make contribution by virtue of subsection (1) above notwithstanding that he has ceased to be liable in respect of the damage in question since the time when the damage occurred, unless he ceased to be liable by virtue of the expiry of a period of limitation or prescription which extinguished the right on which the claim against him in respect of the damage was based.

(4)   A person who has made or agreed to make any payment in bona fide settlement or compromise of any claim made against him in respect of any damage (including a payment into court which has been accepted) shall be entitled to recover contribution in accordance with this section without regard to whether or not he himself is or ever was liable in respect of the damage, provided, however, that he would have been liable assuming that the factual basis of the claim against him could be established.

(5)   A judgment given in any action brought in any part of the United Kingdom by or on behalf of the person who suffered the damage in question against any person from whom contribution is sought under this section shall be conclusive in the proceedings for contribution as to any issue determined by that judgment in favour of the person from whom the contribution is sought.

(6)   References in this section to a person's liability in respect of any damage are references to any such liability which has been or could be established in an action brought against him in England and Wales by or on behalf of the person who suffered the damage; but it is immaterial

whether any issue arising in any such action was or would be determined (in accordance with the rules of private international law) by reference to the law of a country outside England and Wales.

2 Assessment of contribution

(1) Subject to subsection (3) below, in any proceedings for contribution under section 1 above the amount of the contribution recoverable from any person shall be such as may be found by the court to be just and equitable having regard to the extent of that person's responsibility for the damage in question.

(2) Subject to subsection (3) below, the court shall have power in any such proceedings to exempt any person from liability to make contribution, or to direct that the contribution to be recovered from any person shall amount to a complete indemnity.

(3) Where the amount of the damages which have or might have been awarded in respect of the damage in question in any action brought in England and Wales by or on behalf of the person who suffered it against the person from whom the contribution is sought was or would have been subject to—
    (a) any limit imposed by or under any enactment or by any agreement made before the damage occurred;
    (b) any reduction by virtue of section 1 of the Law Reform (Contributory Negligence) Act 1945 or section 5 of the Fatal Accidents Act 1976; or
    (c) any corresponding limit or reduction under the law of a country outside England and Wales;

the person from whom the contribution is sought shall not by virtue of any contribution awarded under section 1 above be required to pay in respect of the damage a greater amount than the amount of those damages as so limited or reduced.

Proceedings for the same debt or damage

3 Proceedings against persons jointly liable for the same debt or damage

Judgment recovered against any person liable in respect of any debt or damage shall not be a bar to an action, or to the continuance of an action, against any other person who is (apart from any such bar) jointly liable with him in respect of the same debt or damage.

4 Successive actions against persons liable (jointly or otherwise) for the same damage

If more than one action is brought in respect of any damage by or on behalf of the person by whom it was suffered against persons liable in respect of the damage (whether jointly or otherwise) the plaintiff shall not be entitled to costs in any of those actions, other than that in which judgment is first given, unless the court is of the opinion that there was reasonable ground for bringing the action.

Supplemental

5 Application to the Crown

Without prejudice to section 4(1) of the Crown Proceedings Act 1947 (indemnity and contribution), this Act shall bind the Crown, but nothing in this Act shall be construed as in any way affecting Her Majesty in Her private capacity (including in right of Her Duchy of Lancaster) or the Duchy of Cornwall.

6 Interpretation

(1)   A person is liable in respect of any damage for the purposes of this Act if the person who suffered it (or anyone representing his estate or dependants) is entitled to recover compensation from him in respect of that damage (whatever the legal basis of his liability, whether tort, breach of contract, breach of trust or otherwise).

(2)   References in this Act to an action brought by or on behalf of the person who suffered any damage include references to an action brought for the benefit of his estate or dependants.

(3)   In this Act "dependants" has the same meaning as in the Fatal Accidents Act 1976.

(4)   In this Act, except in section 1(5) above, "action" means an action brought in England and Wales.

7 Savings

(1)   Nothing in this Act shall affect any case where the debt in question became due or (as the case may be) the damage in question occurred before the date on which it comes into force.

(2)   A person shall not be entitled to recover contribution or liable to make contribution in accordance with section 1 above by reference to any liability based on breach of any obligation assumed by him before the date on which this Act comes into force.

(3)   The right to recover contribution in accordance with section 1 above supersedes any right, other than an express contractual right, to recover contribution (as distinct from indemnity) otherwise than under this Act in corresponding circumstances; but nothing in this Act shall affect—
   (a) any express or implied contractual or other right to indemnity; or
   (b) any express contractual provision regulating or excluding contribution;

which would be enforceable apart from this Act (or render enforceable any agreement for indemnity or contribution which would not be enforceable apart from this Act).

10   Short title, commencement and extent

(1)   This Act may be cited as the Civil Liability (Contribution) Act 1978.

(2)   This Act shall come into force on 1st January next following the date on which it is passed.

(3)   . . .

NOTES
   Sub-s (3): applies to Scotland only.

Banking Act 1979

(C 37)

An Act to regulate the acceptance of deposits in the course of a business; to confer functions on the Bank of England with respect to the control of institutions carrying on deposit-taking businesses; to give further protection to persons who are depositors with such institutions; to make provision with respect to advertisements inviting the making of deposits; to restrict the use of names and descriptions associated with banks and banking; to prohibit fraudulent inducement to make a deposit; to amend the Consumer Credit Act 1974 and the law with respect to instruments to which section 4 of the Cheques Act 1957 applies; to repeal certain enactments relating to banks and banking; and for purposes connected therewith

[4 April 1979]

PART IV
MISCELLANEOUS AND GENERAL

47   Defence of contributory negligence

In any circumstances in which proof of absence of negligence on the part of a banker would be a defence in proceedings by reason of section 4 of the Cheques Act 1957, a defence of contributory negligence shall also be available to the banker notwithstanding the provisions of section 11(1) of the Torts (Interference with Goods) Act 1977.

52   Short title, commencement and extent

(1)   This Act may be cited as the Banking Act 1979.

(2)   This Act extends to Northern Ireland.

(3)   This Act shall come into operation on such day as the Treasury may appoint by order made by statutory instrument; and different days may be so appointed for different provisions of this Act and for such different purposes of the same provision as may be specified in the order.

(4)   Any reference in any provision of this Act to "the appointed day" shall be construed as a reference to the day appointed for the purposes of that provision; and any reference in this Act to the day appointed for the purposes of any provision of this Act—
    (a)   shall be construed as a reference to the day appointed under this section for the coming into operation of that provision; and
    (b)   where different days are appointed for different purposes of that provision, shall be construed, unless an order under this section otherwise provides, as a reference to the first day so appointed.

Sale of Goods Act 1979

(C 54)

An Act to consolidate the law relating to the sale of goods
[6 December 1979]

PART I
CONTRACTS TO WHICH ACT APPLIES

1 Contracts to which Act applies

(1)   This Act applies to contracts of sale of goods made on or after (but not to those made before) 1 January 1894.

(2)   In relation to contracts made on certain dates, this Act applies subject to the modification of certain of its sections as mentioned in Schedule 1 below.

(3)   Any such modification is indicated in the section concerned by a reference to Schedule 1 below.

(4)   Accordingly, where a section does not contain such a reference, this Act applies in relation to the contract concerned without such modification of the section.

PART II
FORMATION OF THE CONTRACT

Contract of sale

2 Contract of sale

(1)   A contract of sale of goods is a contract by which the seller transfers or agrees to transfer the property in goods to the buyer for a money consideration, called the price.

(2)   There may be a contract of sale between one part owner and another.

(3)   A contract of sale may be absolute or conditional.

(4)   Where under a contract of sale the property in the goods is transferred from the seller to the buyer the contract is called a sale.

(5)   Where under a contract of sale the transfer of the property in the goods is to take place at a future time or subject to some condition later to be fulfilled the contract is called an agreement to sell.

(6)   An agreement to sell becomes a sale when the time elapses or the conditions are fulfilled subject to which the property in the goods is to be transferred.

3   Capacity to buy and sell

(1)   Capacity to buy and sell is regulated by the general law concerning capacity to contract and to transfer and acquire property.

(2)   Where necessaries are sold and delivered to a minor or to a person who by reason of mental incapacity or drunkenness is incompetent to contract, he must pay a reasonable price for them.

(3)   In subsection (2) above "necessaries" means goods suitable to the condition in life of the minor or other person concerned and to his actual requirements at the time of the sale and delivery.

Formalities of contract

4   How contract of sale is made

(1)   Subject to this and any other Act, a contract of sale may be made in writing (either with or without seal), or by word of mouth, or partly in writing and partly by word of mouth, or may be implied from the conduct of the parties.

(2)   Nothing in this section affects the law relating to corporations.

Subject matter of contract

5 Existing or future goods

(1)   The goods which form the subject of a contract of sale may be either existing goods, owned or possessed by the seller, or goods to be manufactured or acquired by him after the making of the contract of sale, in this Act called future goods.

(2)   There may be a contract for the sale of goods the acquisition of which by the seller depends on a contingency which may or may not happen.

(3)   Where by a contract of sale the seller purports to effect a present sale of future goods, the contract operates as an agreement to sell the goods.

6 Goods which have perished

Where there is a contract for the sale of specific goods, and the goods without the knowledge of the seller have perished at the time when the contract is made, the contract is void.

7 Goods perishing before sale but after agreement to sell

Where there is an agreement to sell specific goods and subsequently the goods, without any fault on the part of the seller or buyer, perish before the risk passes to the buyer, the agreement is avoided.

The price

8 Ascertainment of price

(1)   The price in a contract of sale may be fixed by the contract, or may be left to be fixed in a manner agreed by the contract, or may be determined by the course of dealing between the parties.

(2)   Where the price is not determined as mentioned in subsection (1) above the buyer must pay a reasonable price.

(3)   What is a reasonable price is a question of fact dependent on the circumstances of each particular case.

9   Agreement to sell at valuation

(1)   Where there is an agreement to sell goods on the terms that the price is to be fixed by the valuation of a third party, and he cannot or does not make the valuation, the agreement is avoided; but if the goods or any part of them have been delivered to and appropriated by the buyer he must pay a reasonable price for them.

(2)   Where the third party is prevented from making the valuation by the fault of the seller or buyer, the party not at fault may maintain an action for damages against the party at fault.

[Implied terms etc]

10   Stipulations about time

(1)   Unless a different intention appears from the terms of the contract, stipulations as to time of payment are not of the essence of a contract of sale.

(2)   Whether any other stipulation as to time is or is not of the essence of the contract depends on the terms of the contract.

(3)   In a contract of sale "month" prima facie means calendar month.

NOTES

    Cross-heading preceding this section substituted by the Sale and Supply of Goods Act 1994, s 7(1), Sch 2, para 5(1), (10).

11   When condition to be treated as warranty

[(1)   This section does not apply to Scotland.]

(2)   Where a contract of sale is subject to a condition to be fulfilled by the seller, the buyer may waive the condition, or may elect to treat the breach of the condition as a breach of warranty and not as a ground for treating the contract as repudiated.

(3)   Whether a stipulation in a contract of sale is a condition, the breach of which may give rise to a right to treat the contract as repudiated, or a warranty, the breach of which may give rise to a claim for damages but not to a right to reject the goods and treat the contract as repudiated, depends in each case on the construction of the contract; and a stipulation may be a condition, though called a warranty in the contract.

(4)   [Subject to section 35A below] Where a contract of sale is not severable and the buyer has accepted the goods or part of them, the breach of a condition to be fulfilled by the seller can only be treated as a breach of warranty, and not as a ground for rejecting the goods and treating the contract as repudiated, unless there is an express or implied term of the contract to that effect.

(5)   . . .

(6)   Nothing in this section affects a condition or warranty whose fulfilment is excused by law by reason of impossibility or otherwise.

(7)   Paragraph 2 of Schedule 1 below applies in relation to a contract made before 22 April 1967 or (in the application of this Act to Northern Ireland) 28 July 1967.

NOTES

Commencement: 3 January 1995 (sub-s (1)); 1 January 1980 (remainder).
Sub-s (1): substituted by the Sale and Supply of Goods Act 1994, s 7(1), Sch 2, para 5(1), (2).
Sub-s (4): words in square brackets inserted by the Sale and Supply of Goods Act 1994, s 3(2).
Sub-s (5): repealed by the Sale and Supply of Goods Act 1994, s 7, Sch 2, para 5(1), (2), Sch 3.

12   Implied terms about title, etc

(1)   In a contract of sale, other than one to which subsection (3) below applies, there is an implied [term] on the part of the seller that in the case of a sale he has a right to sell the goods, and in the case of an agreement to sell he will have such a right at the time when the property is to pass.

(2)   In a contract of sale, other than one to which subsection (3) below applies, there is also an implied [term] that—

(a) the goods are free, and will remain free until the time when the property is to pass, from any charge or encumbrance not disclosed or known to the buyer before the contract is made, and

(b) the buyer will enjoy quiet possession of the goods except so far as it may be disturbed by the owner or other person entitled to the benefit of any charge or encumbrance so disclosed or known.

(3) This subsection applies to a contract of sale in the case of which there appears from the contract or is to be inferred from its circumstances an intention that the seller should transfer only such title as he or a third person may have.

(4) In a contract to which subsection (3) above applies there is an implied [term] that all charges or encumbrances known to the seller and not known to the buyer have been disclosed to the buyer before the contract is made.

(5) In a contract to which subsection (3) above applies there is also an implied [term] that none of the following will disturb the buyer's quiet possession of the goods, namely—

(a) the seller;

(b) in a case where the parties to the contract intend that the seller should transfer only such title as a third person may have, that person;

(c) anyone claiming through or under the seller or that third person otherwise than under a charge or encumbrance disclosed or known to the buyer before the contract is made.

[(5A) As regards England and Wales and Northern Ireland, the term implied by subsection (1) above is a condition and the terms implied by subsections (2), (4) and (5) above are warranties.]

(6) Paragraph 3 of Schedule 1 below applies in relation to a contract made before 18 May 1973.

NOTES

Commencement: 3 January 1995 (sub-s (5A)); 1 January 1980 (remainder).

Sub-ss (1), (2), (4), (5): words in square brackets substituted by the Sale and Supply of Goods Act 1994, s 7(1), Sch 2, para 5(1), (3)(a).

Sub-s (5A): inserted by the Sale and Supply of Goods Act 1994, s 7(1), Sch 2, para 5(1), (3)(b).

13   Sale by description

(1)   Where there is a contract for the sale of goods by description, there is an implied [term] that the goods will correspond with the description.

[(1A) As regards England and Wales and Northern Ireland, the term implied by subsection (1) above is a condition.]

(2)   If the sale is by sample as well as by description it is not sufficient that the bulk of the goods corresponds with the sample if the goods do not also correspond with the description.

(3)   A sale of goods is not prevented from being a sale by description by reason only that, being exposed for sale or hire, they are selected by the buyer.

(4)   Paragraph 4 of Schedule 1 below applies in relation to a contract made before 18th May 1973.

NOTES

   Commencement: 3 January 1995 (sub-s (1A)); 1 January 1980 (remainder).
   Sub-s (1): word in square brackets substituted by the Sale and Supply of Goods Act 1994, s 7(1), Sch 2, para 5(1), (4)(a).
   Sub-s (1A): inserted by the Sale and Supply of Goods Act 1994, s 7(1), Sch 2, para 5(1), (4)(b).

14   Implied terms about quality or fitness

(1)   Except as provided by this section and section 15 below and subject to any other enactment, there is no implied [term] about the quality or fitness for any particular purpose of goods supplied under a contract of sale.

[(2)   Where the seller sells goods in the course of a business, there is an implied term that the goods supplied under the contract are of satisfactory quality.

(2A)   For the purposes of this Act, goods are of satisfactory quality if they meet the standard that a reasonable person would regard as satisfactory, taking account of any description of the goods, the price (if relevant) and all the other relevant circumstances.

(2B)   For the purposes of this Act, the quality of goods includes their state and condition and the following (among others) are in appropriate cases aspects of the quality of goods—
  (a) fitness for all the purposes for which goods of the kind in question are commonly supplied,
  (b) appearance and finish,
  (c) freedom from minor defects,
  (d) safety, and
  (e) durability.

(2C)   The term implied by subsection (2) above does not extend to any matter making the quality of goods unsatisfactory—
  (a) which is specifically drawn to the buyer's attention before the contract is made,
  (b) where the buyer examines the goods before the contract is made, which that examination ought to reveal, or
  (c) in the case of a contract for sale by sample, which would have been apparent on a reasonable examination of the sample.]

(3)   Where the seller sells goods in the course of a business and the buyer, expressly or by implication, makes known—
  (a) to the seller, or
  (b) where the purchase price or part of it is payable by instalments and the goods were previously sold by a credit-broker to the seller, to that credit-broker,

any particular purpose for which the goods are being bought, there is an implied [term] that the goods supplied under the contract are reasonably fit for that purpose, whether or not that is a purpose for which such goods are commonly supplied, except where the circumstances show that the buyer does not rely, or that it is unreasonable for him to rely, on the skill or judgment of the seller or credit-broker.

(4)   An implied [term] about quality or fitness for a particular purpose may be annexed to a contract of sale by usage.

(5)   The preceding provisions of this section apply to a sale by a person who in the course of a business is acting as agent for another as they apply to a sale by a principal in the course of a business, except where that other is not selling in the course of a business and either the buyer knows that fact or reasonable steps are taken to bring it to the notice of the buyer before the contract is made.

[(6)   As regard England and Wales and Northern Ireland, the terms implied by subsections (2) and (3) above are conditions.]

(7)   Paragraph 5 of Schedule 1 below applies in relation to a contract made on or after 18 May 1973 and before the appointed day, and paragraph 6 in relation to one made before 18th May 1973.

(8)   In subsection (7) above and paragraph 5 of Schedule 1 below references to the appointed day are to the day appointed for the purposes of those provisions by an order of the Secretary of State made by statutory instrument.

NOTES

Commencement: 3 January 1995 (sub-ss (2)–(2C), (6)); 1 January 1980 (remainder).

Sub-ss (1), (3), (4): words in square brackets substituted by the Sale and Supply of Goods Act 1994, s 7(1), Sch 2, para 5(1). (5)(a).

Sub-ss (2)–(2C): substituted, for sub-s (2) as originally enacted, by the Sale and Supply of Goods Act 1994, s 1(1).

Sub-s (6): substituted by the Sale and Supply of Goods Act 1994, s 7(1), Sch 2, para (1), 5(5)(b).

Sale by sample

15 Sale by sample

(1)   A contract of sale is a contract for sale by sample where there is an express or implied term to that effect in the contract.

(2)   In the case of a contract for sale by sample there is an implied [term]—
  (a) that the bulk will correspond with the sample in quality;
  (b) . . .
  (c) that the goods will be free from any defect, [making their quality unsatisfactory], which would not be apparent on reasonable examination of the sample.

[(3)   As regards England and Wales and Northern Ireland, the term implied by subsection (2) above is a condition.]

(4)   Paragraph 7 of Schedule 1 below applies in relation to a contract made before 18 May 1973.

NOTES

Commencement: 3 January 1995 (sub-s (3)); 1 January 1980 (remainder).

Sub-s (2): words in square brackets substituted and para (b) repealed, by the Sale and Supply of Goods Act 1994, ss 1(2), 7, Sch 2, para 5(1), (6)(a), Sch 3.

Sub-s (3): substituted by the Sale and Supply of Goods Act 1994, s 7, Sch 2, para 5(1), (6)(b).

[*Miscellaneous*

15A Modifications of remedies for breach of condition in non-consumer cases

(1)   Where in the case of a contract of sale—
    (a) the buyer would, apart from this subsection, have the right to reject goods by reason of a breach on the part of the seller of a term implied by section 13, 14 or 15 above, but
    (b) the breach is so slight that it would be unreasonable for him to reject them,

then, if the buyer does not deal as consumer, the breach is not to be treated as a breach of condition but may be treated as a breach of warranty.

(2)   This section applies unless a contrary intention appears in, or is to be implied from, the contract.

(3)   It is for the seller to show that a breach fell within subsection (1)(b) above.

(4)   This section does not apply to Scotland.]

NOTES

Commencement: 3 January 1995.
Inserted, together with preceding cross-heading, by the Sale and Supply of Goods Act 1994, s 4(1).
Sub-s (4): applies to Scotland only.

PART III
EFFECTS OF THE CONTRACT

Transfer of property as between seller and buyer

16 Goods must be ascertained

[Subject to section 20A below] Where there is a contract for the sale of unascertained goods no property in the goods is transferred to the buyer unless and until the goods are ascertained.

. square brackets inserted by the Sale of Goods (Amendment) Act 1995,

17   Property passes when intended to pass

(1)   Where there is a contract for the sale of specific or ascertained goods the property in them is transferred to the buyer at such time as the parties to the contract intend it to be transferred.

(2)   For the purpose of ascertaining the intention of the parties regard shall be had to the terms of the contract, the conduct of the parties and the circumstances of the case.

18   Rules for ascertaining intention

Unless a different intention appears, the following are rules for ascertaining the intention of the parties as to the time at which the property in the goods is to pass to the buyer.

Rule 1.—Where there is an unconditional contract for the sale of specific goods in a deliverable state the property in the goods passes to the buyer when the contract is made, and it is immaterial whether the time of payment or the time of delivery, or both, be postponed.

Rule 2.—Where there is a contract for the sale of specific goods and the seller is bound to do something to the goods for the purpose of putting them into a deliverable state, the property does not pass until the thing is done and the buyer has notice that it has been done.

Rule 3.—Where there is a contract for the sale of specific goods in a deliverable state but the seller is bound to weigh, measure, test, or do some other act or thing with reference to the goods for the purpose of ascertaining the price, the property does not pass until the act or thing is done and the buyer has notice that it has been done.

Rule 4.—When goods are delivered to the buyer on approval or on sale or return or other similar terms the property in the goods passes to the buyer:—

(a) when he signifies his approval or acceptance to the seller or does any other act adopting the transaction;

(b) if he does not signify his approval or acceptance to the seller but retains the goods without giving notice of rejection, then, if a time has been fixed for the return of the goods, on the expiration of that time, and, if no time has been fixed, on the expiration of a reasonable time.

Rule 5.—(1) Where there is a contract for the sale of unascertained or future goods by description, and goods of that description and in a deliverable state are unconditionally appropriated to the contract, either by the seller with the assent of the buyer or by the buyer with the assent of the seller, the property in the goods then passes to the buyer; and the assent may be express or implied, and may be given either before or after the appropriation is made.

(2) Where, in pursuance of the contract, the seller delivers the goods to the buyer or to a carrier or other bailee or custodier (whether named by the buyer or not) for the purpose of transmission to the buyer, and does not reserve the right of disposal, he is to be taken to have unconditionally appropriated the goods to the contract.

[(3) Where there is a contract for the sale of a specified quantity of unascertained goods in a deliverable state forming part of a bulk which is identified either in the contract or by subsequent agreement between the parties and the bulk is reduced to (or to less than) that quantity, then, if the buyer under that contract is the only buyer to whom goods are then due out of the bulk—

(a) the remaining goods are to be taken as appropriated to that contract at the time when the bulk is so reduced; and

(b) the property in those goods then passes to that buyer.

(4) Paragraph (3) above applies also (with the necessary modifications) where a bulk is reduced to (or to less than) the aggregate of the quantities due to a single buyer under separate contracts relating to that bulk and he is the only buyer to whom goods are then due out of that bulk.]

NOTES

Rule 5: paras (3), (4) added by the Sale of Goods (Amendment) Act 1995, s 1(2).

19 Reservation of right of disposal

(1)   Where there is a contract for the sale of specific goods or where goods are subsequently appropriated to the contract, the seller may, by the terms of the contract or appropriation, reserve the right of disposal of the goods until certain conditions are fulfilled; and in such a case, notwithstanding the delivery of the goods to the buyer, or to a carrier or other bailee or custodier for the purpose of transmission to the buyer, the property in the goods does not pass to the buyer until the conditions imposed by the seller are fulfilled.

goods are shipped, and by the bill of lading the goods are
o the order of the seller or his agent, the seller is prima facie
o reserve the right of disposal.

(3)   Where the seller of goods draws on the buyer for the price, and
transmits the bill of exchange and bill of lading to the buyer together to
secure acceptance or payment of the bill of exchange, the buyer is bound to
return the bill of lading if he does not honour the bill of exchange, and if he
wrongfully retains the bill of lading the property in the goods does not pass
to him.

20   Risk prima facie passes with property

(1)   Unless otherwise agreed, the goods remain at the seller's risk until
the property in them is transferred to the buyer, but when the property in
them is transferred to the buyer the goods are at the buyer's risk whether
delivery has been made or not.

(2)   But where delivery has been delayed through the fault of either
buyer or seller the goods are at the risk of the party at fault as regards any
loss which might not have occurred but for such fault.

(3)   Nothing in this section affects the duties or liabilities of either seller
or buyer as a bailee or custodier of the goods of the other party.

[20A   Undivided shares in goods forming part of a bulk

(1)   This section applies to a contract for the sale of a specified quantity
of unascertained goods if the following conditions are met—
  (a) the goods or some of them form part of a bulk which is identified
      either in the contract or by subsequent agreement between the
      parties; and
  (b) the buyer has paid the price for some or all of the goods which are
      the subject of the contract and which form part of the bulk.

(2)   Where this section applies, then (unless the parties agree otherwise),
as soon as the conditions specified in paragraphs (a) and (b) of subsection (1)
above are met or at such later time as the parties may agree—
  (a) property in an undivided share in the bulk is transferred to the
      buyer, and
  (b) the buyer becomes an owner in common of the bulk.

(3)   Subject to subsection (4) below, for the purposes of this section, the undivided share of a buyer in a bulk at any time shall be such share as the quantity of goods paid for and due to the buyer out of the bulk bears to the quantity of goods in the bulk at that time.

(4)   Where the aggregate of the undivided shares of buyers in a bulk determined under subsection (3) above would at any time exceed the whole of the bulk at that time, the undivided share in the bulk of each buyer shall be reduced proportionately so that the aggregate of the undivided shares is equal to the whole bulk.

(5)   Where a buyer has paid the price for only some of the goods due to him out of a bulk, any delivery to the buyer out of the bulk shall, for the purposes of this section, be ascribed in the first place to the goods in respect of which payment has been made.

(6)   For the purposes of this section payment of part of the price for any goods shall be treated as payment for a corresponding part of the goods.]

NOTES

Commencement: 19 September 1995.
Inserted, together with s 20B, by the Sale of Goods (Amendment) Act 1995, s 1(3).

[20B   Deemed consent by co-owner to dealings in bulk goods

(1)   A person who has become an owner in common of a bulk by virtue of section 20A above shall be deemed to have consented to—
  (a) any delivery of goods out of the bulk to any other owner in common of the bulk, being goods which are due to him under his contract;
  (b) any dealing with or removal, delivery or disposal of goods in the bulk by any other person who is an owner in common of the bulk in so far as the goods fall within that co-owner's undivided share in the bulk at the time of the dealing, removal, delivery or disposal.

(2)   No cause of action shall accrue to anyone against a person by reason of that person having acted in accordance with paragraph (a) or (b) of subsection (1) above in reliance on any consent deemed to have been given under that subsection.

(3)   Nothing in this section or section 20A above shall—

(a) impose an obligation on a buyer of goods out of a bulk to compensate any other buyer of goods out of that bulk for any shortfall in the goods received by that other buyer;

(b) affect any contractual arrangement between buyers of goods out of a bulk for adjustments between themselves; or

(c) affect the rights of any buyer under his contract.]

NOTES

Commencement: 19 September 1995.
Inserted as noted to s 20A.

Transfer of title

21 Sale by person not the owner

(1) Subject to this Act, where goods are sold by a person who is not their owner, and who does not sell them under the authority or with the consent of the owner, the buyer acquires no better title to the goods than the seller had, unless the owner of the goods is by his conduct precluded from denying the seller's authority to sell.

(2) Nothing in this Act affects—

(a) the provisions of the Factors Acts or any enactment enabling the apparent owner of goods to dispose of them as if he were their true owner;

(b) the validity of any contract of sale under any special common law or statutory power of sale or under the order of a court of competent jurisdiction.

22 Market overt

(1) . . .

(2) This section does not apply to Scotland.

(3) Paragraph 8 of Schedule 1 below applies in relation to a contract under which goods were sold before 1st January 1968 or (in the application of this Act to Northern Ireland) 29th August 1967.

NOTES

Sub-s (1): repealed in relation to any contract for sale of goods made after 3 January 1995 by the Sale of Goods (Amendment) Act 1994, ss 1, 3(2).

23   Sale under voidable title

When the seller of goods has a voidable title to them, but his title has not been avoided at the time of the sale, the buyer acquires a good title to the goods, provided he buys them in good faith and without notice of the seller's defect of title.

24   Seller in possession after sale

Where a person having sold goods continues or is in possession of the goods, or of the documents of title to the goods, the delivery or transfer by that person, or by a mercantile agent acting for him, of the goods or documents of title under any sale, pledge, or other disposition thereof, to any person receiving the same in good faith and without notice of the previous sale, has the same effect as if the person making the delivery or transfer were expressly authorised by the owner of the goods to make the same.

25   Buyer in possession after sale

(1)   Where a person having bought or agreed to buy goods obtains, with the consent of the seller, possession of the goods or the documents of title to the goods, the delivery or transfer by that person, or by a mercantile agent acting for him, of the goods or documents of title, under any sale, pledge, or other disposition thereof, to any person receiving the same in good faith and without notice of any lien or other right of the original seller in respect of the goods, has the same effect as if the person making the delivery or transfer were a mercantile agent in possession of the goods or documents of title with the consent of the owner.

(2)   For the purposes of subsection (1) above—
   (a) the buyer under a conditional sale agreement is to be taken not to be a person who has bought or agreed to buy goods, and
   (b) "conditional sale agreement" means an agreement for the sale of goods which is a consumer credit agreement within the meaning of the

Consumer Credit Act 1974 under which the purchase price or part of it is payable by instalments, and the property in the goods is to remain in the seller (notwithstanding that the buyer is to be in possession of the goods) until such conditions as to the payment of instalments or otherwise as may be specified in the agreement are fulfilled.

(3)   Paragraph 9 of Schedule 1 below applies in relation to a contract under which a person buys or agrees to buy goods and which is made before the appointed day.

(4)   In subsection (3) above and paragraph 9 of Schedule 1 below references to the appointed day are to the day appointed for the purposes of those provisions by an order of the Secretary of State made by statutory instrument.

26   Supplementary to sections 24 and 25

In sections 24 and 25 above "mercantile agent" means a mercantile agent having in the customary course of his business as such agent authority either—
  (a) to sell goods, or
  (b) to consign goods for the purpose of sale, or
  (c) to buy goods, or
  (d) to raise money on the security of goods.

PART IV
PERFORMANCE OF THE CONTRACT

27   Duties of seller and buyer

It is the duty of the seller to deliver the goods, and of the buyer to accept and pay for them, in accordance with the terms of the contract of sale.

28   Payment and delivery are concurrent conditions

Unless otherwise agreed, delivery of the goods and payment of the price are concurrent conditions, that is to say, the seller must be ready and willing to give possession of the goods to the buyer in exchange for the price and the buyer must be ready and willing to pay the price in exchange for possession of the goods.

29 Rules about delivery

(1)   Whether it is for the buyer to take possession of the goods or for the seller to send them to the buyer is a question depending in each case on the contract, express or implied, between the parties.

(2)   Apart from any such contract, express or implied, the place of delivery is the seller's place of business if he has one, and if not, his residence; except that, if the contract is for the sale of specific goods, which to the knowledge of the parties when the contract is made are in some other place, then that place is the place of delivery.

(3)   Where under the contract of sale the seller is bound to send the goods to the buyer, but no time for sending them is fixed, the seller is bound to send them within a reasonable time.

(4)   Where the goods at the time of sale are in the possession of a third person, there is no delivery by seller to buyer unless and until the third person acknowledges to the buyer that he holds the goods on his behalf; but nothing in this section affects the operation of the issue or transfer of any document of title to goods.

(5)   Demand or tender of delivery may be treated as ineffectual unless made at a reasonable hour; and what is a reasonable hour is a question of fact.

(6)   Unless otherwise agreed, the expenses of and incidental to putting the goods into a deliverable state must be borne by the seller.

30 Delivery of wrong quantity

(1)   Where the seller delivers to the buyer a quantity of goods less than he contracted to sell, the buyer may reject them, but if the buyer accepts the goods so delivered he must pay for them at the contract rate.

(2)   Where the seller delivers to the buyer a quantity of goods larger than he contracted to sell, the buyer may accept the goods included in the contract and reject the rest, or he may reject the whole.

[(2A) A buyer who does not deal as consumer may not—
  (a) where the seller delivers a quantity of goods less than he contracted to sell, reject the goods under subsection (1) above, or
  (b) where the seller delivers a quantity of goods larger than he contracted to sell, reject the whole under subsection (2) above,

if the shortfall or, as the case may be, excess is so slight that it would be unreasonable for him to do so.

(2B)   It is for the seller to show that a shortfall or excess fell within subsection (2A) above.

(2C)   Subsections (2A) and (2B) above do not apply to Scotland.

(2D), (2E)   . . .]

(3)   Where the seller delivers to the buyer a quantity of goods larger than he contracted to sell and the buyer accepts the whole of the goods so delivered he must pay for them at the contract rate.

(4)   . . .

(5)   This section is subject to any usage of trade, special agreement, or course of dealing between the parties.

NOTES

Commencement: 3 January 1995 (sub-ss (2A)–(2E)); 1 January 1980 (remainder).

Sub-ss (2A)–(2C): inserted, together with sub-ss (2D), (2E), by the Sale and Supply of Goods Act 1994, ss 4(2), 5(2).

Sub-ss (2D), (2E): inserted, together with sub-ss (2A)–(2C), by the Sale and Supply of Goods Act 1994, ss 4(2), 5(2); apply to Scotland only.

Sub-s (4): repealed by the Sale and Supply of Goods Act 1994, ss 3(3), 7(2), Sch 3.

31   Instalment deliveries

(1)   Unless otherwise agreed, the buyer of goods is not bound to accept delivery of them by instalments.

(2)   Where there is a contract for the sale of goods to be delivered by stated instalments, which are to be separately paid for, and the seller makes defective deliveries in respect of one or more instalments, or the buyer neglects or refuses to take delivery of or pay for one or more instalments, it is a question in each case depending on the terms of the contract and the circumstances of the case whether the breach of contract is a repudiation of the whole contract or whether it is a severable breach giving rise to a claim for compensation but not to a right to treat the whole contract as repudiated.

32   Delivery to carrier

(1)   Where, in pursuance of a contract of sale, the seller i
required to send the goods to the buyer, delivery of the goo
(whether named by the buyer or not) for the purpose of tr
the buyer is prima facie deemed to be a delivery of the goods ᴄᴏ ᴄne buyer.

(2)   Unless otherwise authorised by the buyer, the seller must make
such contract with the carrier on behalf of the buyer as may be reasonable
having regard to the nature of the goods and the other circumstances of
the case; and if the seller omits to do so, and the goods are lost or damaged
in course of transit, the buyer may decline to treat the delivery to the
carrier as a delivery to himself or may hold the seller responsible in
damages.

(3)   Unless otherwise agreed, where goods are sent by the seller to the
buyer by a route involving sea transit, under circumstances in which it is
usual to insure, the seller must give such notice to the buyer as may enable
him to insure them during their sea transit; and if the seller fails to do so,
the goods are at his risk during such sea transit.

33   Risk where goods are delivered at distant place

Where the seller of goods agrees to deliver them at his own risk at a place
other than that where they are when sold, the buyer must nevertheless
(unless otherwise agreed) take any risk of deterioration in the goods
necessarily incident to the course of transit.

34   Buyer's right of examining the goods

. . . Unless otherwise agreed, when the seller tenders delivery of goods to
the buyer, he is bound on request to afford the buyer a reasonable opportunity
of examining the goods for the purpose of ascertaining whether they are in
conformity with the contract [and, in the case of a contract for sale by
sample, of comparing the bulk with the sample.]

NOTES

   Words omitted repealed, and words in square brackets added, by the Sale and
Supply of Goods Act 1994, ss 2(2), 7(2), Sch 3.

5 Acceptance

(1)   The buyer is deemed to have accepted the goods [subject to subsection (2) below—
- (a) when he intimates to the seller that he has accepted them, or
- (b) when the goods have been delivered to him and he does any act in relation to them which is inconsistent with the ownership of the seller.

(2)   Where goods are delivered to the buyer, and he has not previously examined them, he is not deemed to have accepted them under subsection (1) above until he has had a reasonable opportunity of examining them for the purpose—
- (a) of ascertaining whether they are in conformity with the contract, and
- (b) in the case of a contract for sale by sample, of comparing the bulk with the sample.

(3)   Where the buyer deals as consumer or (in Scotland) the contract of sale is a consumer contract, the buyer cannot lose his right to rely on subsection (2) above by agreement, waiver or otherwise.

(4)   The buyer is also deemed to have accepted the goods when after the lapse of a reasonable time he retains the goods without intimating to the seller that he has rejected them.

(5)   The questions that are material in determining for the purposes of subsection (4) above whether a reasonable time has elapsed include whether the buyer has had a reasonable opportunity of examining the goods for the purpose mentioned in subsection (2) above.

(6)   The buyer is not by virtue of this section deemed to have accepted the goods merely because—
- (a) he asks for, or agrees to, their repair by or under an arrangement with the seller, or
- (b) the goods are delivered to another under a sub-sale or other disposition.

(7)   Where the contract is for the sale of goods making one or more commercial units, a buyer accepting any goods included in a unit is deemed to have accepted all the goods making the unit; and in this subsection "commercial unit" means a unit division of which would materially impair the value of the goods or the character of the unit.

(8)] Paragraph 10 of Schedule 1 below applies in relation to a contract made before 22nd April 1967 or (in the application of this Act of Northern Ireland) 28th July 1967.

NOTES

Commencement: 3 January 1995 (sub-ss (2)–(7)); 1 January 1980 (remainder).
Words in square brackets substituted by the Sale and Supply of Goods Act 1994, s 2(1).

[35A Right of partial rejection

(1)   If the buyer—
(a) has the right to reject the goods by reason of a breach on the part of the seller that affects some or all of them, but
(b) accepts some of the goods, including, where there are any goods unaffected by the breach, all such goods,

he does not by accepting them lose his right to reject the rest.

(2)   In the case of a buyer having the right to reject an instalment of goods, subsection (1) above applies as if references to the goods were references to the goods comprised in the instalment.

(3)   For the purposes of subsection (1) above, goods are affected by a breach if by reason of the breach they are not in conformity with the contract.

(4)   This section applies unless a contrary intention appears in, or is to be implied from, the contract.]

NOTES

Commencement: 3 January 1995.
Inserted by the Sale and Supply of Goods Act 1994, s 3(1).

36 Buyer not bound to return rejected goods

Unless otherwise agreed, where goods are delivered to the buyer, and he refuses to accept them, having the right to do so, he is not bound to return them to the seller, but it is sufficient if he intimates to the seller that he refuses to accept them.

37 Buyer's liability for not taking delivery of goods

(1)   When the seller is ready and willing to deliver the goods, and requests the buyer to take delivery, and the buyer does not within a reasonable time after such request take delivery of the goods, he is liable to the seller for any loss occasioned by his neglect or refusal to take delivery, and also for a reasonable charge for the care and custody of the goods.

(2)   Nothing in this section affects the rights of the seller where the neglect or refusal of the buyer to take delivery amounts to a repudiation of the contract.

PART V
RIGHTS OF UNPAID SELLER AGAINST THE GOODS

Preliminary

38   Unpaid seller defined

(1)   The seller of goods is an unpaid seller within the meaning of this Act—
  (a) when the whole of the price has not been paid or tendered;
  (b) when a bill of exchange or other negotiable instrument has been received as conditional payment, and the condition on which it was received has not been fulfilled by reason of the dishonour of the instrument or otherwise.

(2)   In this Part of this Act "seller" includes any person who is in the position of a seller, as, for instance, an agent of the seller to whom the bill of lading has been indorsed, or a consignor or agent who has himself paid (or is directly responsible for) the price.

39   Unpaid seller's rights

(1)   Subject to this and any other Act, notwithstanding that the property in the goods may have passed to the buyer, the unpaid seller of goods, as such, has by implication of law—
  (a) a lien on the goods or right to retain them for the price while he is in possession of them;

(b) in case of the insolvency of the buyer, a right of stopping the goods in transit after he has parted with the possession of them;

(c) a right of re-sale as limited by this Act.

(2)   Where the property in goods has not passed to the buyer, the unpaid seller has (in addition to his other remedies) a right of withholding delivery similar to and co-extensive with his rights of lien or retention and stoppage in transit where the property has passed to the buyer.

Unpaid seller's lien

41 Seller's lien

(1)   Subject to this Act, the unpaid seller of goods who is in possession of them is entitled to retain possession of them until payment or tender of the price in the following cases:—

(a) where the goods have been sold without any stipulation as to credit;

(b) where the goods have been sold on credit but the term of credit has expired;

(c) where the buyer becomes insolvent.

(2)   The seller may exercise his lien or right of retention notwithstanding that he is in possession of the goods as agent or bailee or custodier for the buyer.

42 Part delivery

Where an unpaid seller has made part delivery of the goods, he may exercise his lien or right of retention on the remainder, unless such part delivery has been made under such circumstances as to show an agreement to waive the lien or right of retention.

43 Termination of lien

(1)   The unpaid seller of goods loses his lien or right of retention in respect of them—

(a) when he delivers the goods to a carrier or other bailee or custodier for the purpose of transmission to the buyer without reserving the right of disposal of the goods;

(b)   when the buyer or his agent lawfully obtains possession of the goods;

(c)   by waiver of the lien or right of retention.

(2)   An unpaid seller of goods who has a lien or right of retention in respect of them does not lose his lien or right of retention by reason only that he has obtained judgment or decree for the price of the goods.

Stoppage in transit

44   Right of stoppage in transit

Subject to this Act, when the buyer of goods becomes insolvent the unpaid seller who has parted with the possession of the goods has the right of stopping them in transit, that is to say, he may resume possession of the goods as long as they are in course of transit, and may retain them until payment or tender of the price.

45   Duration of transit

(1)   Goods are deemed to be in course of transit from the time when they are delivered to a carrier or other bailee or custodier for the purpose of transmission to the buyer, until the buyer or his agent in that behalf takes delivery of them from the carrier or other bailee or custodier.

(2)   If the buyer or his agent in that behalf obtains delivery of the goods before their arrival at the appointed destination, the transit is at an end.

(3)   If, after the arrival of the goods at the appointed destination, the carrier or other bailee or custodier acknowledges to the buyer or his agent that he holds the goods on his behalf and continues in possession of them as bailee or custodier for the buyer or his agent, the transit is at an end, and it is immaterial that a further destination for the goods may have been indicated by the buyer.

(4)   If the goods are rejected by the buyer, and the carrier or other bailee or custodier continues in possession of them, the transit is not deemed to be at an end, even if the seller has refused to receive them back.

(5)   When goods are delivered to a ship chartered by the buyer it is a question depending on the circumstances of the particular case whether they are in the possession of the master as a carrier or as agent to the buyer.

(6)   Where the carrier or other bailee or custodier wrongfully refuses to deliver the goods to the buyer or his agent in that behalf, the transit is deemed to be at an end.

(7)   Where part delivery of the goods has been made to the buyer or his agent in that behalf, the remainder of the goods may be stopped in transit, unless such part delivery has been made under such circumstances as to show an agreement to give up possession of the whole of the goods.

46   How stoppage in transit is effected

(1)   The unpaid seller may exercise his right of stoppage in transit either by taking actual possession of the goods or by giving notice of his claim to the carrier or other bailee or custodier in whose possession the goods are.

(2)   The notice may be given either to the person in actual possession of the goods or to his principal.

(3)   If given to the principal, the notice is ineffective unless given at such time and under such circumstances that the principal, by the exercise of reasonable diligence, may communicate it to his servant or agent in time to prevent a delivery to the buyer.

(4)   When notice of stoppage in transit is given by the seller to the carrier or other bailee or custodier in possession of the goods, he must re-deliver the goods to, or according to the directions of, the seller; and the expenses of the re-delivery must be borne by the seller.

Re-sale etc by buyer

47   Effect of sub-sale etc by buyer

(1)   Subject to this Act, the unpaid seller's right of lien or retention or stoppage in transit is not affected by any sale or other disposition of the goods which the buyer may have made, unless the seller has assented to it.

(2)   Where a document of title to goods has been lawfully transferred to any person as buyer or owner of the goods, and that person transfers the document to a person who takes it in good faith and for valuable consideration, then—

(a) if the last-mentioned transfer was by way of sale the unpaid seller's right of lien or retention or stoppage in transit is defeated; and

(b) if the last-mentioned transfer was made by way of pledge or other disposition for value, the unpaid seller's right of lien or retention or stoppage in transit can only be exercised subject to the rights of the transferee.

Rescission: and re-sale by seller

48 Rescission: and re-sale by seller

(1)   Subject to this section, a contract of sale is not rescinded by the mere exercise by an unpaid seller of his right of lien or retention or stoppage in transit.

(2)   Where an unpaid seller who has exercised his right of lien or retention or stoppage in transit re-sells the goods, the buyer acquires a good title to them as against the original buyer.

(3)   Where the goods are of a perishable nature, or where the unpaid seller gives notice to the buyer of his intention to re-sell, and the buyer does not within a reasonable time pay or tender the price, the unpaid seller may re-sell the goods and recover from the original buyer damages for any loss occasioned by his breach of contract.

(4)   Where the seller expressly reserves the right of re-sale in case the buyer should make default, and on the buyer making default re-sells the goods, the original contract of sale is rescinded but without prejudice to any claim the seller may have for damages.

PART VI
ACTIONS FOR BREACH OF THE CONTRACT

Seller's remedies

49 Action for price

(1)   Where, under a contract of sale, the property in the goods has passed to the buyer and he wrongfully neglects or refuses to pay for the goods

according to the terms of the contract, the seller may maintain an action against him for the price of the goods.

(2)   Where, under a contract of sale, the price is payable on a day certain irrespective of delivery and the buyer wrongfully neglects or refuses to pay such price, the seller may maintain an action for the price, although the property in the goods has not passed and the goods have not been appropriated to the contract.

(3)   . . .

NOTES

Sub-s (3): applies to Scotland only.

50   Damages for non-acceptance

(1)   Where the buyer wrongfully neglects or refuses to accept and pay for the goods, the seller may maintain an action against him for damages for non-acceptance.

(2)   The measure of damages is the estimated loss directly and naturally resulting, in the ordinary course of events, from the buyer's breach of contract.

(3)   Where there is an available market for the goods in question the measure of damages is prima facie to be ascertained by the difference between the contract price and the market or current price at the time or times when the goods ought to have been accepted or (if no time was fixed for acceptance) at the time of the refusal to accept.

Buyer's remedies

51   Damages for non-delivery

(1)   Where the seller wrongfully neglects or refuses to deliver the goods to the buyer, the buyer may maintain an action against the seller for damages for non-delivery.

(2)   The measure of damages is the estimated loss directly and naturally resulting, in the ordinary course of events, from the seller's breach of contract.

(3)　Where there is an available market for the goods in question the measure of damages is prima facie to be ascertained by the difference between the contract price and the market or current price of the goods at the time or times when they ought to have been delivered or (if no time was fixed) at the time of the refusal to deliver.

52　Specific performance

(1)　In any action for breach of contract to deliver specific or ascertained goods the court may, if it thinks fit, on the plaintiff's application, by its judgment or decree direct that the contract shall be performed specifically, without giving the defendant the option of retaining the goods on payment of damages.

(2)　The plaintiff's application may be made at any time before judgment or decree.

(3)　The judgment or decree may be unconditional, or on such terms and conditions as to damages, payment of the price and otherwise as seem just to the court.

(4)　. . .

NOTES

Sub-s (4): applies to Scotland only.

53　Remedy for breach of warranty

(1)　Where there is a breach of warranty by the seller, or where the buyer elects (or is compelled) to treat any breach of a condition on the part of the seller as a breach of warranty, the buyer is not by reason only of such breach of warranty entitled to reject the goods; but he may—
  (a) set up against the seller the breach of warranty in diminution or extinction of the price, or
  (b) maintain an action against the seller for damages for the breach of warranty.

(2)　The measure of damages for breach of warranty is the estimated loss directly and naturally resulting, in the ordinary course of events, from the breach of warranty.

(3)   In the case of breach of warranty of quality such loss is prima facie the difference between the value of the goods at the time of delivery to the buyer and the value they would have had if they had fulfilled the warranty.

(4)   The fact that the buyer has set up the breach of warranty in diminution or extinction of the price does not prevent him from maintaining an action for the same breach of warranty if he has suffered further damage.

[(5)   This section does not apply to Scotland.]

NOTES

Commencement: 3 January 1995 (sub-s (5)); 1 January 1980 (remainder).
Sub-s (5): substituted by the Sale and Supply of Goods Act 1994, s 7(1), Sch 2, para 5(1), (7).

Interest, etc

54   Interest

Nothing in this Act affects the right of the buyer or the seller to recover interest or special damages in any case where by law interest or special damages may be recoverable, or to recover money paid where the consideration for the payment of it has failed.

PART VII
SUPPLEMENTARY

55   Exclusion of implied terms

(1)   Where a right, duty or liability would arise under a contract of sale of goods by implication of law, it may (subject to the Unfair Contract Terms Act 1977) be negatived or varied by express agreement, or by the course of dealing between the parties, or by such usage as binds both parties to the contract.

(2)   An express [term] does not negative a [term] implied by this Act unless inconsistent with it.

(3)   Paragraph 11 of Schedule 1 below applies in relation to a contract made on or after 18th May 1973 and before 1st February 1978, and paragraph 12 in relation to one made before 18th May 1973.

NOTES

Sub-s (2): words in square brackets substituted by the Sale and Supply of Goods Act 1994, s 7(1), Sch 2, para 5(1), (8).

56   Conflict of laws

Paragraph 13 of Schedule 1 below applies in relation to a contract made on or after 18th May 1973 and before 1st February 1978, so as to make provision about conflict of laws in relation to such a contract.

57   Auction sales

(1)   Where goods are put up for sale by auction in lots, each lot is prima facie deemed to be the subject of a separate contract of sale.

(2)   A sale by auction is complete when the auctioneer announces its completion by the fall of the hammer, or in other customary manner; and until the announcement is made any bidder may retract his bid.

(3)   A sale by auction may be notified to be subject to a reserve or upset price, and a right to bid may also be reserved expressly by or on behalf of the seller.

(4)   Where a sale by auction is not notified to be subject to a right to bid by or on behalf of the seller, it is not lawful for the seller to bid himself or to employ any person to bid at the sale, or for the auctioneer knowingly to take any bid from the seller or any such person.

(5)   A sale contravening subsection (4) above may be treated as fraudulent by the buyer.

(6)   Where, in respect of a sale by auction, a right to bid is expressly reserved (but not otherwise) the seller or any one person on his behalf may bid at the auction.

59 Reasonable time a question of fact

Where a reference is made in this Act to a reasonable time the question what is a reasonable time is a question of fact.

60 Rights etc enforceable by action

Where a right, duty or liability is declared by this Act, it may (unless otherwise provided by this Act) be enforced by action.

61 Interpretation

(1)   In this Act, unless the context or subject matter otherwise requires—
   "action" includes counterclaim and set-off, and in Scotland condescendence and claim and compensation;
   ["bulk" means a mass or collection of goods of the same kind which—
      (a)   is contained in a defined space or area; and
      (b)   is such that any goods in the bulk are interchangeable with any other goods therein of the same number or quantity;]
   "business" includes a profession and the activities of any government department (including a Northern Ireland department) or local or public authority;
   "buyer" means a person who buys or agrees to buy goods;
   ["consumer contract" has the same meaning as in section 25(1) of the Unfair Contract Terms Act 1977; and for the purposes of this Act the onus of proving that a contract is not to be regarded as a consumer contract shall lie on the seller]
   "contract of sale" includes an agreement to sell as well as a sale;
   "credit-broker" means a person acting in the course of a business of credit brokerage carried on by him, that is a business of effecting introductions of individuals desiring to obtain credit—
      (a)   to persons carrying on any business so far as it relates to the provision of credit, or
      (b)   to other persons engaged in credit brokerage;
                    . . . . .
   "delivery" means voluntary transfer of possession from one person to another [except that in relation to sections 20A and 20B above it includes such appropriation of goods to the contract as results in property in the goods being transferred to the buyer;]
   "document of title to goods" has the same meaning as it has in the Factors Acts;

"Factors Acts" means the Factors Act 1889, the Factors (Scotland) Act 1890, and any enactment amending or substituted for the same;

"fault" means wrongful act or default;

"future goods" means goods to be manufactured or acquired by the seller after the making of the contract of sale;

"goods" includes all personal chattels other than things in action and money, and in Scotland all corporeal moveables except money; and in particular "goods" includes emblements, industrial growing crops, and things attached to or forming part of the land which are agreed to be severed before sale or under the contract of sale [and includes an undivided share in goods;]

"plaintiff" includes pursuer, complainer, claimant in a multiplepoinding and defendant or defender counter-claiming;

"property" means the general property in goods, and not merely a special property;

. . . . .

"sale" includes a bargain and sale as well as a sale and delivery;

"seller" means a person who sells or agrees to sell goods;

"specific goods" means goods identified and agreed on at the time a contract of sale is made [and includes an undivided share, specified as a fraction or percentage, of goods identified and agreed on as aforesaid];

"warranty" (as regards England and Wales and Northern Ireland) means an agreement with reference to goods which are the subject of a contract of sale, but collateral to the main purpose of such contract, the breach of which gives rise to a claim for damages, but not to a right to reject the goods and treat the contract as repudiated.

(2)   . . .

(3)   A thing is deemed to be done in good faith within the meaning of this Act when it is in fact done honestly, whether it is done negligently or not.

(4)   A person is deemed to be insolvent within the meaning of this Act if has either ceased to pay his debts in the ordinary course of business or he cannot pay his debts as they become due, . . .

(5)   Goods are in a deliverable state within the meaning of this Act when they are in such a state that the buyer would under the contract be bound to take delivery of them.

[(5A)   References in this Act to dealing as consumer are to be construed in accordance with Part I of the Unfair Contract Terms Act 1977; and, for the purposes of this Act, it is for a seller claiming that the buyer does not deal as consumer to show that he does not.]

(6)   As regards the definition of "business" in subsection (1) above, paragraph 14 of Schedule 1 below applies in relation to a contract made on or after 18th May 1973 and before 1st February 1978, and paragraph 15 in relation to one made before 18th May 1973.

NOTES

Commencement: 3 January 1995 (sub-s (5A)); 1 January 1980 (remainder).
Sub-s (1): definition "bulk" inserted, and words in square brackets in definitions "delivery", "goods" and "specific goods" added, by the Sale of Goods (Amendment) Act 1995, s 2; definition "consumer contract" inserted and definition "quality" repealed by the Sale and Supply of Goods Act 1994, s 7, Sch 2, para 5(1), (9)(a), Sch 3; first definition omitted applies to Scotland only.
Sub-s (2): repealed by the Sale and Supply of Goods Act 1994, s 7, Sch 2, para 5(1), (9)(b), Sch 3.
Sub-s (4): words omitted repealed by the Insolvency Act 1985, s 235, Sch 10, Pt III, the Bankruptcy (Scotland) Act 1985, s 75(2), Sch 8, and the Insolvency Act 1986, s 437, Sch 11, Pt II.
Sub-s (5A): inserted by the Sale and Supply of Goods Act 1994, s 7, Sch 2, para 5(9)(c).

62 Savings: rules of law etc

(1)   The rules in bankruptcy relating to contracts of sale apply to those contracts, notwithstanding anything in this Act.

(2)   The rules of the common law, including the law merchant, except in so far as they are inconsistent with the provisions of this Act, and in particular the rules relating to the law of principal and agent and the effect of fraud, misrepresentation, duress or coercion, mistake, or other invalidating cause, apply to contracts for the sale of goods.

(3)   Nothing in this Act or the Sale of Goods Act 1893 affects the enactments relating to bills of sale, or any enactment relating to the sale of goods which is not expressly repealed or amended by this Act or that.

(4)   The provisions of this Act about contracts of sale do not apply to a transaction in the form of a contract of sale which is intended to operate by way of mortgage, pledge, charge, or other security.

(5)   . . .

Sub-s (5): applies to Scotland only.

64   Short title and commencement

(1)   This Act may be cited as the Sale of Goods Act 1979.

(2)   This Act comes into force on 1 January 1980.

Supply of Goods and Services Act 1982

(C 29)

An Act to amend the law with respect to the terms to be implied in certain contracts for the transfer of the property in goods, in certain contracts for the hire of goods and in certain contracts for the supply of a service; and for connected purposes

[13 July 1982]

PART I
SUPPLY OF GOODS

Contracts for the transfer of property in goods

1   The contracts concerned

(1)   In this Act [in its application to England and Wales and Northern Ireland] a "contract for the transfer of goods" means a contract under which one person transfers or agrees to transfer to another the property in goods, other than an excepted contract.

(2)   For the purposes of this section an excepted contract means any of the following:—
   (a)   a contract of sale of goods;
   (b)   a hire-purchase agreement;
   (c)   a contract under which the property in goods is (or is to be) transferred in exchange for trading stamps on their redemption;
   (d)   a transfer or agreement to transfer which is made by deed and for which there is no consideration other than the presumed consideration imported by the deed;

(e) a contract intended to operate by way of mortgage, pledge, charge or other security.

(3)   For the purposes of this Act [in its application to England and Wales and Northern Ireland] a contract is a contract for the transfer of goods whether or not services are also provided or to be provided under the contract, and (subject to subsection (2) above) whatever is the nature of the consideration for the transfer or agreement to transfer.

NOTES

Sub-ss (1), (3): words in square brackets inserted by the Sale and Supply of Goods Act 1994, s 7(1), Sch 2, para 6(1), (2).

2   Implied terms about title, etc

(1)   In a contract for the transfer of goods, other than one to which subsection (3) below applies, there is an implied condition on the part of the transferor that in the case of a transfer of the property in the goods he has a right to transfer the property and in the case of an agreement to transfer the property in the goods he will have such a right at the time when the property is to be transferred.

(2)   In a contract for the transfer of goods, other than one to which subsection (3) below applies, there is also an implied warranty that—
  (a) the goods are free, and will remain free until the time when the property is to be transferred, from any charge or encumbrance not disclosed or known to the transferee before the contract is made, and
  (b) the transferee will enjoy quiet possession of the goods except so far as it may be disturbed by the owner or other person entitled to the benefit of any charge or encumbrance so disclosed or known.

(3)   This subsection applies to a contract for the transfer of goods in the case of which there appears from the contract or is to be inferred from its circumstances an intention that the transferor should transfer only such title as he or a third person may have.

(4)   In a contract to which subsection (3) above applies there is an implied warranty that all charges or encumbrances known to the transferor and not known to the transferee have been disclosed to the transferee before the contract is made.

(5)   In a contract to which subsection (3) above applies there is also an implied warranty that none of the following will disturb the transferee's quiet possession of the goods, namely—

   (a) the transferor;

   (b) in a case where the parties to the contract intend that the transferor should transfer only such title as a third person may have, that person;

   (c) anyone claiming through or under the transferor or that third person otherwise than under a charge or encumbrance disclosed or known to the transferee before the contract is made.

3 Implied terms where transfer is by description

(1)   This section applies where, under a contract for the transfer of goods, the transferor transfers or agrees to transfer the property in the goods by description.

(2)   In such a case there is an implied condition that the goods will correspond with the description.

(3)   If the transferor transfers or agrees to transfer the property in the goods by sample as well as by description it is not sufficient that the bulk of the goods corresponds with the sample if the goods do not also correspond with the description.

(4)   A contract is not prevented from falling within subsection (1) above by reason only that, being exposed for supply, the goods are selected by the transferee.

4 Implied terms about quality or fitness

(1)   Except as provided by this section and section 5 below and subject to the provisions of any other enactment, there is no implied condition or warranty about the quality or fitness for any particular purpose of goods supplied under a contract for the transfer of goods.

[(2)   Where, under such a contract, the transferor transfers the property in goods in the course of a business, there is an implied condition that the goods supplied under the contract are of satisfactory quality.

(2A)   For the purposes of this section and section 5 below, goods are of satisfactory quality if they meet the standard that a reasonable person would regard as satisfactory, taking account of any description of the goods, the price (if relevant) and all the other relevant circumstances.

(3)   The condition implied by subsection (2) above does not extend to any matter making the quality of goods unsatisfactory—
  (a) which is specifically drawn to the transferee's attention before the contract is made,
  (b) where the transferee examines the goods before the contract is made, which that examination ought to reveal, or
  (c) where the property in the goods is transferred by reference to a sample, which would have been apparent on a reasonable examination of the sample.]

(4)   Subsection (5) below applies where, under a contract for the transfer of goods, the transferor transfers the property in goods in the course of a business and the transferee, expressly or by implication, makes known—
  (a) to the transferor, or
  (b) where the consideration or part of the consideration for the transfer is a sum payable by instalments and the goods were previously sold by a credit-broker to the transferor, to that credit-broker,

any particular purpose for which the goods are being acquired.

(5)   In that case there is (subject to subsection (6) below) an implied condition that the goods supplied under the contract are reasonably fit for that purpose, whether or not that is a purpose for which such goods are commonly supplied.

(6)   Subsection (5) above does not apply where the circumstances show that the transferee does not rely, or that it is unreasonable for him to rely, on the skill or judgment of the transferor or credit-broker.

(7)   An implied condition or warranty about quality or fitness for a particular purpose may be annexed by usage to a contract for the transfer of goods.

(8)   The preceding provisions of this section apply to a transfer by a person who in the course of a business is acting as agent for another as they apply to a transfer by a principal in the course of a business, except where that other is not transferring in the course of a business and either the transferee knows that fact or reasonable steps are taken to bring it to the transferee's notice before the contract concerned is made.

(9)   . . .

Sub-ss (2), (2A), (3): substituted for sub-ss (2), (3), as originally enacted, by the Sale and Supply of Goods Act 1994, s 7(1), Sch 2, para 6(1), (3).

Sub-s (9): repealed by the Sale and Supply of Goods Act 1994, s 7, Sch 2, para 6(1), (3), Sch 3.

5   Implied terms where transfer is by sample

(1)   This section applies where, under a contract for the transfer of goods, the transferor transfers or agrees to transfer the property in the goods by reference to a sample.

(2)   In such a case there is an implied condition—
  (a) that the bulk will correspond with the sample in quality; and
  (b) that the transferee will have a reasonable opportunity of comparing the bulk with the sample; and
  (c) that the goods will be free from any defect, [making their quality unsatisfactory], which would not be apparent on reasonable examination of the sample.

(3)   . . .

(4)   For the purposes of this section a transferor transfers or agrees to transfer the property in goods by reference to a sample where there is an express or implied term to that effect in the contract concerned.

Sub-s (2): words in square brackets in para (c) substituted by the Sale and Supply of Goods Act 1994, s 7(1), Sch 2, para 6(1), (4)(a).

Sub-s (3): repealed by the Sale and Supply of Goods Act 1994, s 7, Sch 2, para 6(1), (4)(b), Sch 3.

[5A   Modification of remedies for breach of statutory condition in non-consumer cases

(1)   Where in the case of a contract for the transfer of goods—
  (a) the transferee would, apart from this subsection, have the right to treat the contract as repudiated by reason of a breach on the part of the transferor of a term implied by section 3, 4 or 5(2)(a) or (c) above, but
  (b) the breach is so slight that it would be unreasonable for him to do so,

then, if the transferee does not deal as consumer, the breach is not to be treated as a breach of condition but may be treated as a breach of warranty.

(2)   This section applies unless a contrary intention appears in, or is to be implied from, the contract.

(3)   It is for the transferor to show that a breach fell within subsection (1)(b) above.]

NOTES

Commencement: 3 January 1995.
Inserted by the Sale and Supply of Goods Act 1994, s 7(1), Sch 2, para 6(1), (5).

Contracts for the hire of goods

6   The contracts concerned

(1)   In this Act [in its application to England and Wales and Northern Ireland] a "contract for the hire of goods" means a contract under which one person bails or agrees to bail goods to another by way of hire, other than an excepted contract.

(2)   For the purposes of this section an excepted contract means any of the following:—
  (a) a hire-purchase agreement;
  (b) a contract under which goods are (or are to be) bailed in exchange for trading stamps on their redemption.

(3)   For the purposes of this Act [in its application to England and Wales and Northern Ireland] a contract is a contract for the hire of goods whether or not services are also provided or to be provided under the contract, and (subject to subsection (2) above) whatever is the nature of the consideration for the bailment or agreement to bail by way of hire.

NOTES

Sub-ss (1), (3): words in square brackets substituted by the Sale and Supply of Goods Act 1994, s 7(1), Sch 2, para 6(1), (6).

7   Implied terms about right to transfer possession, etc

(1)   In a contract for the hire of goods there is an implied condition on the part of the bailor that in the case of a bailment he has a right to transfer possession of the goods by way of hire for the period of the bailment and in the case of an agreement to bail he will have such a right at the time of the bailment.

(2)   In a contract for the hire of goods there is also an implied warranty that the bailee will enjoy quiet possession of the goods for the period of the bailment except so far as the possession may be disturbed by the owner or other person entitled to the benefit of any charge or encumbrance disclosed or known to the bailee before the contract is made.

(3)   The preceding provisions of this section do not affect the right of the bailor to repossess the goods under an express or implied term of the contract.

8   Implied terms where hire is by description

(1)   This section applies where, under a contract for the hire of goods, the bailor bails or agrees to bail the goods by description.

(2)   In such a case there is an implied condition that the goods will correspond with the description.

(3)   If under the contract the bailor bails or agrees to bail the goods by reference to a sample as well as a description it is not sufficient that the bulk of the goods corresponds with the sample if the goods do not also correspond with the description.

(4)   A contract is not prevented from falling within subsection (1) above by reason only that, being exposed for supply, the goods are selected by the bailee.

9   Implied terms about quality or fitness

(1)   Except as provided by this section and section 10 below and subject to the provisions of any other enactment, there is no implied condition or warranty about the quality or fitness for any particular purpose of goods bailed under a contract for the hire of goods.

[(2)   Where, under such a contract, the bailor bails goods in the course of a business, there is an implied condition that the goods supplied under the contract are of satisfactory quality.

(2A)   For the purposes of this section and section 10 below, goods are of satisfactory quality if they meet the standard that a reasonable person would regard as satisfactory, taking account of any description of the goods, the consideration for the bailment (if relevant) and all the other relevant circumstances.

(3)   The condition implied by subsection (2) above does not extend to any matter making the quality of goods unsatisfactory—
  (a) which is specifically drawn to the bailee's attention before the contract is made,
  (b) where the bailee examines the goods before the contract is made, which that examination ought to reveal, or
  (c) where the goods are bailed by reference to a sample, which would have been apparent on a reasonable examination of the sample.]

(4)   Subsection (5) below applies where, under a contract for the hire of goods, the bailor bails goods in the course of a business and the bailee, expressly or by implication, makes known—
  (a) to the bailor in the course of negotiations conducted by him in relation to the making of the contract, or
  (b) to a credit-broker in the course of negotiations conducted by that broker in relation to goods sold by him to the bailor before forming the subject matter of the contract,

any particular purpose for which the goods are being bailed.

(5)   In that case there is (subject to subsection (6) below) an implied condition that the goods supplied under the contract are reasonably fit for that purpose, whether or not that is a purpose for which such goods are commonly supplied.

(6)   Subsection (5) above does not apply where the circumstances show that the bailee does not rely, or that it is unreasonable for him to rely, on the skill or judgment of the bailor or credit-broker.

(7)   An implied condition or warranty about quality or fitness for a particular purpose may be annexed by usage to a contract for the hire of goods.

(8)   The preceding provisions of this section apply to a bailment by a person who in the course of a business is acting as agent for another as they apply to a bailment by a principal in the course of a business, except where that other is not bailing in the course of a business and either the bailee knows that fact or reasonable steps are taken to bring it to the bailee's notice before the contract concerned is made.

(9)   . . .

NOTES

Sub-ss (2), (2A), (3): substituted for sub-ss (2), (3), as originally enacted, by the Sale and Supply of Goods Act 1994, s 7(1), Sch 2, para 6(1), (7).

Sub-s (9): repealed by the Sale and Supply of Goods Act 1994, s 7, Sch 2, para 6(1), (7), Sch 3.

10   Implied terms where hire is by sample

(1)   This section applies where, under a contract for the hire of goods, the bailor bails or agrees to bail the goods by reference to a sample.

(2)   In such a case there is an implied condition—
   (a) that the bulk will correspond with the sample in quality; and
   (b) that the bailee will have a reasonable opportunity of comparing the bulk with the sample; and
   (c) that the goods will be free from any defect, [making their quality unsatisfactory], which would not be apparent on reasonable examination of the sample.

(3)   . . .

(4)   For the purposes of this section a bailor bails or agrees to bail goods by reference to a sample where there is an express or implied term to that effect in the contract concerned.

NOTES

Sub-s (2): words in square brackets in para (c) substituted by the Sale and Supply of Goods Act 1994, s 7(1), Sch 2, para 6(1), (8).

Sub-s (3): repealed by the Sale and Supply of Goods Act 1994, s 7, Sch 2, para 6(1), (8), Sch 3.

[10A   Modification of remedies for breach of statutory condition in non–consumer cases

(1)   Where in the case of a contract for the hire of goods—
  (a) the bailee would, apart from this subsection, have the right to treat the contract as repudiated by reason of a breach on the part of the bailor of a term implied by section 8, 9 or 10(2)(a) or (c) above, but
  (b) the breach is so slight that it would be unreasonable for him to do so,

then, if the bailee does not deal as consumer, the breach is not to be treated as a breach of condition but may be treated as a breach of warranty.

(2)   This section applies unless a contrary intention appears in, or is to be implied from, the contract.

(3)   It is for the bailor to show that a breach fell within subsection (1)(b) above.]

NOTES
Commencement: 3 January 1995.
Inserted by the Sale and Supply of Goods Act 1994, s 7(1), Sch 2, para 6(1), (9).

Exclusion of implied terms, etc

11   Exclusion of implied terms, etc

(1)   Where a right, duty or liability would arise under a contract for the transfer of goods or a contract for the hire of goods by implication of law, it may (subject to subsection (2) below and the 1977 Act) be negatived or varied by express agreement, or by the course of dealing between the parties, or by such usage as binds both parties to the contract.

(2)   An express condition or warranty does not negative a condition or warranty implied by the preceding provisions of this Act unless inconsistent with it.

(3)   Nothing in the preceding provisions of this Act prejudices the operation of any other enactment or any rule of law whereby any condition or warranty (other than one relating to quality or fitness) is to be implied in a contract for the transfer of goods or a contract for the hire of goods.

PART II
SUPPLY OF SERVICES

12 The contracts concerned

(1)   In this Act a "contract for the supply of a service" means, subject to subsection (2) below, a contract under which a person ("the supplier") agrees to carry out a service.

(2)   For the purposes of this Act, a contract of service or apprenticeship is not a contract for the supply of a service.

(3)   Subject to subsection (2) above, a contract is a contract for the supply of a service for the purposes of this Act whether or not goods are also—
   (a) transferred or to be transferred, or
   (b) bailed or to be bailed by way of hire,

under the contract, and whatever is the nature of the consideration for which the service is to be carried out.

(4)   The Secretary of State may by order provide that one or more of sections 13 to 15 below shall not apply to services of a description specified in the order, and such an order may make different provision for different circumstances.

(5)   The power to make an order under subsection (4) above shall be exercisable by statutory instrument subject to annulment in pursuance of a resolution of either House of Parliament.

13 Implied term about care and skill

In a contract for the supply of a service where the supplier is acting in the course of a business, there is an implied term that the supplier will carry out the service with reasonable care and skill.

14 Implied term about time for performance

(1)   Where, under a contract for the supply of a service by a supplier acting in the course of a business, the time for the service to be carried out is not fixed by the contract, left to be fixed in a manner agreed by the contract or determined by the course of dealing between the parties, there is an implied term that the supplier will carry out the service within a reasonable time.

(2)   What is a reasonable time is a question of fact.

15   Implied term about consideration

(1)   Where, under a contract for the supply of a service, the consideration for the service is not determined by the contract, left to be determined in a manner agreed by the contract or determined by the course of dealing between the parties, there is an implied term that the party contracting with the supplier will pay a reasonable charge.

(2)   What is a reasonable charge is a question of fact.

16   Exclusion of implied terms, etc

(1)   Where a right, duty or liability would arise under a contract for the supply of a service by virtue of this Part of this Act, it may (subject to subsection (2) below and the 1977 Act) be negatived or varied by express agreement, or by the course of dealing between the parties, or by such usage as binds both parties to the contract.

(2)   An express term does not negative a term implied by this Part of this Act unless inconsistent with it.

(3)   Nothing in this Part of this Act prejudices—
  (a) any rule of law which imposes on the supplier a duty stricter than that imposed by section 13 or 14 above; or
  (b) subject to paragraph (a) above, any rule of law whereby any term not inconsistent with this Part of this Act is to be implied in a contract for the supply of a service.

(4)   This Part of this Act has effect subject to any other enactment which defines or restricts the rights, duties or liabilities arising in connection with a service of any description.

PART III
SUPPLEMENTARY

18   Interpretation: general

(1)   In the preceding provisions of this Act and this section—

"bailee", in relation to a contract for the hire of goods means (depending on the context) a person to whom the goods are bailed under the contract, or a person to whom they are to be so bailed, or a person to whom the rights under the contract of either of those persons have passed;

"bailor", in relation to a contract for the hire of goods, means (depending on the context) a person who bails the goods under the contract, or a person who agrees to do so, or a person to whom the duties under the contract of either of those persons have passed;

"business" includes a profession and the activities of any government department or local or public authority;

"credit-broker" means a person acting in the course of a business of credit brokerage carried on by him;

"credit brokerage" means the effecting of introductions—

  (a) of individuals desiring to obtain credit to persons carrying on any business so far as it relates to the provision of credit; or

  (b) of individuals desiring to obtain goods on hire to persons carrying on a business which comprises or relates to the bailment [or as regards Scotland the hire] of goods under a contract for the hire of goods; or

  (c) of individuals desiring to obtain credit, or to obtain goods on hire, to other credit-brokers;

"enactment" means any legislation (including subordinate legislation) of the United Kingdom or Northern Ireland;

"goods" [includes all personal chattels, other than things in action and money, and as regards Scotland all corporeal moveables; and in particular "goods" includes] emblements, industrial growing crops, and things attached to or forming part of the land which are agreed to be severed before the transfer [bailment or hire] concerned or under the contract concerned . . . ;

"hire-purchase agreement" has the same meaning as in the 1974 Act;

"property", in relation to goods, means the general property in them and not merely a special property;

. . . . .

"redemption", in relation to trading stamps, has the same meaning as in the Trading Stamps Act 1964 or, as respects Northern Ireland, the Trading Stamps Act (Northern Ireland) 1965;

"trading stamps" has the same meaning as in the said Act of 1964 or, as respects Northern Ireland, the said Act of 1965;

"transferee", in relation to a contract for the transfer of goods, means (depending on the context) a person to whom the property in the goods is transferred under the contract, or a person to whom the property is to be so transferred, or a person to whom the rights under the contract of either of those persons have passed;

"transferor", in relation to a contract for the transfer of goods, means (depending on the context) a person who transfers the property in the goods under the contract, or a person who agrees to do so, or a person to whom the duties under the contract of either of those persons have passed.

(2)   In subsection (1) above, in the definitions of bailee, bailor, transferee and transferor, a reference to rights or duties passing is to their passing by assignment [assignation], operation of law or otherwise.

[(3)   For the purposes of this Act, the quality of goods includes their state and condition and the following (among others) are in appropriate cases aspects of the quality of goods—

    (a) fitness for all the purposes for which goods of the kind in question are commonly supplied,

    (b) appearance and finish,

    (c) freedom from minor defects,

    (d) safety, and

    (e) durability.

(4)   References in this Act to dealing as consumer are to be construed in accordance with Part I of the Unfair Contract Terms Act 1977; and, for the purposes of this Act, it is for the transferor or bailor claiming that the transferee or bailee does not deal as consumer to show that he does not.]

NOTES

Sub-s (1): words in square brackets in definitions "credit brokerage" and "goods" substituted, words omitted from definition "goods" and definition "quality" repealed, by the Sale and Supply of Goods Act 1994, ss 6, 7, Sch 1, para 2, Sch 2, para 6(1), (10).

Sub-s (2): word in square brackets inserted by the Sale and Supply of Goods Act 1994, s 6, Sch 1, para 3.

Sub-ss (3), (4): added by the Sale and Supply of Goods Act 1994, s 7(1), Sch 2, para 6(1), (10).

19   Interpretation: references to Acts

In this Act—

"the 1973 Act" means the Supply of Goods (Implied Terms) Act 1973;

"the 1974 Act" means the Consumer Credit Act 1974;
"the 1977 Act" means the Unfair Contract Terms Act 1977; and
"the 1979 Act" means the Sale of Goods Act 1979.

20 Citation, transitional provisions, commencement and extent

(1) This Act may be cited as the Supply of Goods and Services Act 1982.

(2) The transitional provisions in the Schedule to this Act shall have effect.

(3) Part I of this Act together with section 17 and so much of sections 18 and 19 above as relates to that Part shall not come into operation until 4th January 1983; and Part II of this Act together with so much of sections 18 and 19 above as relates to that Part shall not come into operation until such day as may be appointed by an order made by the Secretary of State.

(4) The power to make an order under subsection (3) above shall be exercisable by statutory instrument.

(5) No provision of this Act applies to a contract made before the provision comes into operation.

(6) This Act [except Part IA, which extends only to Scotland] extends to Northern Ireland [and Parts I and II do not extend] to Scotland.

NOTES

Sub-s (6): words in square brackets inserted or substituted by the Sale and Supply of Goods Act 1994, s 6, Sch 1, para 4.

Companies Act 1985

(C 6)

An Act to consolidate the greater part of the Companies Acts.
[11 March 1985]

PART I
FORMATION AND REGISTRATION OF COMPANIES; JURIDICAL STATUS
AND MEMBERSHIP

CHAPTER III
A COMPANY'S CAPACITY; FORMALITIES OF CARRYING ON BUSINESS

[35 A company's capacity not limited by its memorandum

(1)   The validity of an act done by a company shall not be called into question on the ground of lack of capacity by reason of anything in the company's memorandum.

(2)   A member of a company may bring proceedings to restrain the doing of an act which but for subsection (1) would be beyond the company's capacity; but no such proceedings shall lie in respect of an act to be done in fulfilment of a legal obligation arising from a previous act of the company.

(3)   It remains the duty of the directors to observe any limitations on their powers flowing from the company's memorandum; and action by the directors which but for subsection (1) would be beyond the company's capacity may only be ratified by the company by special resolution.

A resolution ratifying such action shall not affect any liability incurred by the directors or any other person; relief from any such liability must be agreed to separately by special resolution.

(4)   The operation of this section is restricted by [section 65(1) of the Charities Act 1993] and section 112(3) of the Companies Act 1989 in relation to companies which are charities; and section 322A below (invalidity of certain transactions to which directors or their associates are parties) has effect notwithstanding this section.]

NOTES

Substituted, together with ss 35A, 35B, by CA 1989, s 108(1), as from 4 February 1991 subject to transitional provisions.
Sub-s (4): words in square brackets substituted by the Charities Act 1993, s 98(1), Sch 6, para 20(1), (2).

[35 Power of directors to bind the company

(1)   In favour of a person dealing with a company in good faith, the power of the board of directors to bind the company, or authorise others

to do so, shall be deemed to be free of any limitation under the company's constitution.

(2)   For this purpose—
    (a) a person "deals with" a company if he is a party to any transaction or other act to which the company is a party;
    (b) a person shall not be regarded as acting in bad faith by reason only of his knowing that an act is beyond the powers of the directors under the company's constitution; and
    (c) a person shall be presumed to have acted in good faith unless the contrary is proved.

(3)   The references above to limitations on the directors' power under the company's constitution include limitations deriving—
    (a) from a resolution of the company in general meeting or a meeting of any class of shareholders, or
    (b) from any agreement between the members of the company or of any class of shareholders.

(4)   Subsection (1) does not affect any right of a member of the company to bring proceedings to restrain the doing of an act which is beyond the powers of the directors; but no such proceedings shall lie in respect of an act to be done in fulfilment of a legal obligation arising from a previous act of the company.

(5)   Nor does that subsection affect any liability incurred by the directors, or any other person, by reason of the directors' exceeding their powers.

(6)   The operation of this section is restricted by [section 65(1) of the Charities Act 1993] and section 112(3) of the Companies Act 1989 in relation to companies which are charities; and section 322A below (invalidity of certain transactions to which directors or their associates are parties) has effect notwithstanding this section.]

NOTES

Substituted as noted to s 35.
Sub-s (6): words in square brackets substituted by the Charities Act 1993, s 98(1), Sch 6, para 20(1), (2).

[35B   No duty to enquire as to capacity of company or authority of directors

A party to a transaction with a company is not bound to enquire as to whether it is permitted by the company's memorandum or as to any limitation on the powers of the board of directors to bind the company or authorise others to do so.]

[36C Pre-incorporation contracts, deeds and obligations

(1)   A contract which purports to be made by or on behalf of a company at a time when the company has not been formed has effect, subject to any agreement to the contrary, as one made with the person purporting to act for the company or as agent for it, and he is personally liable on the contract accordingly.

(2)   Subsection (1) applies—
  (a) to the making of a deed under the law of England and Wales, and
  (b) to the undertaking of an obligation under the law of Scotland,

as it applies to the making of a contract.]

NOTES

Inserted by CA 1989, s 130(4), as from 31 July 1990.
Modification: this section is modified, in relation to companies incorporated outside Great Britain, by the Foreign Companies (Execution of Documents) Regulations 1994, SI 1994/950, regs 2, 3, as amended by SI 1995/1729, regs 2, 3.

PART V
SHARE CAPITAL, ITS INCREASE, MAINTENANCE AND REDUCTION

CHAPTER VIII
MISCELLANEOUS PROVISIONS ABOUT SHARES AND DEBENTURES

Debentures

[196 Payment of debts out of assets subject to floating charge (England and Wales)

(1)   The following applies in the case of a company registered in England and Wales, where debentures of the company are secured by a charge which, as created, was a floating charge.

(2)   If possession is taken, by or on behalf of the holders of any of the debentures, of any property comprised in or subject to the charge, and the company is not at that time in course of being wound up, the company's preferential debts shall be paid out of assets coming to the hands of the person taking possession in priority to any claims for principal or interest in respect of the debentures.

(3)   "Preferential debts" means the categories of debts listed in Schedule 6 to the Insolvency Act; and for the purposes of that Schedule "the relevant date" is the date of possession being taken as above mentioned.

(4)   Payments made under this section shall be recouped, as far as may be, out of the assets of the company available for payment of general creditors.]

NOTES

Substituted by the Insolvency Act 1986, s 439(1), Sch 13, Pt I, as from 29 December 1986.

PART XI
COMPANY ADMINISTRATION AND PROCEDURE

CHAPTER I
COMPANY IDENTIFICATION

349   Company's name to appear in its correspondence, etc

(1)–(3) . . .

(4)   If an officer of a company or a person on its behalf signs or authorises to be signed on behalf of the company any bill of exchange, promissory note, endorsement, cheque or order for money or goods in which the company's name is not mentioned as required by subsection (1), he is liable to a fine; and he is further personally liable to the holder of the bill of exchange, promissory note, cheque or order for money or goods for the amount of it (unless it is duly paid by the company).

PART XII
REGISTRATION OF CHARGES

CHAPTER I
REGISTRATION OF CHARGES (ENGLAND AND WALES)

395   Certain charges void if not registered

(1)   Subject to the provisions of this Chapter, a charge created by a company registered in England and Wales and being a charge to which this section applies is, so far as any security on the company's property or

undertaking is conferred by the charge, void against the liquidator [or administrator] and any creditor of the company, unless the prescribed particulars of the charge together with the instrument (if any) by which the charge is created or evidenced, are delivered to or received by the registrar of companies for registration in the manner required by this Chapter within 21 days after the date of the charge's creation.

(2)   Subsection (1) is without prejudice to any contract or obligation for repayment of the money secured by the charge; and when a charge becomes void under this section, the money secured by it immediately becomes payable.

NOTES

Substituted by the Companies Act 1989, s 95(1), as from a day to be appointed.
Sub-s (1): words in square brackets inserted by the Insolvency Act 1985, s 109(1), Sch 6, para 10.

396   Charges which have to be registered

(1)   Section 395 applies to the following charges—
   (a) a charge for the purpose of securing any issue of debentures,
   (b) a charge on uncalled share capital of the company,
   (c) a charge created or evidenced by an instrument which, if executed by an individual, would require registration as a bill of sale,
   (d) a charge on land (wherever situated) or any interest in it, but not including a charge for any rent or other periodical sum issuing out of the land,
   (e) a charge on book debts of the company,
   (f) a floating charge on the company's undertaking or property,
   (g) a charge on calls made but not paid,
   (h) a charge on a ship or aircraft, or any share in a ship,
   (j) a charge on goodwill, [or on any intellectual property].

(2)   Where a negotiable instrument has been given to secure the payment of any book debts of a company, the deposit of the instrument for the purpose of securing an advance to the company is not, for purposes of section 395, to be treated as a charge on those book debts.

(3)   The holding of debentures entitling the holder to a charge on land is not for purposes of this section deemed to be an interest in land.

[(3A) The following are "intellectual property" for the purposes of this section—

(a) any patent, trade mark, ... registered design, copyright or design right;

(b) any licence under or in respect of any such right.]

(4) In this Chapter, "charge" includes mortgage.

NOTES

Substituted by the Companies Act 1989, s 95(1), as from a day to be appointed.

Sub-s (1): words in square brackets substituted by the Copyright, Designs and Patents Act 1988, s 303(1), Sch 7, para 31(1), (2).

Sub-s (3A): inserted by the Copyright, Designs and Patents Act 1988, s 303(1), Sch 7, para 31(1), (2), as from 1 August 1989; words omitted repealed by the Trade Marks Act 1994, s 106(2), Sch 5, as from 31 October 1994.

PART XXVII
FINAL PROVISIONS

747   Citation

This Act may be cited as the Companies Act 1985.

Insolvency Act 1986

(C 45)

An Act to consolidate the enactments relating to company insolvency and winding up (including the winding up of companies that are not insolvent, and of unregistered companies); enactments relating to the insolvency and bankruptcy of individuals; and other enactments bearing on those two subject matters, including the functions and qualification of insolvency practitioners, the public administration of insolvency, the penalisation and redress of malpractice and wrongdoing, and the avoidance of certain transactions at an undervalue

[25 July 1986]

NOTES

Modification: The provisions of this Act, except s 413, are applied, with modifications, in relation to a "recognised body" under the Administration of Justice Act 1985, s 9, by the Solicitors' Incorporated Practices Order 1991, SI 1991/2684, arts 2-5, Sch 1.

This Act is also applied, with modifications, in relation to an order or resolution to wind up a society registered under the Industrial and Provident Societies Act 1965; see s 55 of that Act.

THE FIRST GROUP OF PARTS
COMPANY INSOLVENCY; COMPANIES WINDING UP

PART II
ADMINISTRATION ORDERS

Making etc of administration order

8   Power of court to make order

(1)   Subject to this section, if the court—
   (a) is satisfied that a company is or is likely to become unable to pay its debts (within the meaning given to that expression by section 123 of this Act), and
   (b) considers that the making of an order under this section would be likely to achieve one or more of the purposes mentioned below,

the court may make an administration order in relation to the company.

(2)   An administration order is an order directing that, during the period for which the order is in force, the affairs, business and property of the company shall be managed by a person ("the administrator") appointed for the purpose by the court.

(3)   The purposes for whose achievement an administration order may be made are—
   (a) the survival of the company, and the whole or any part of its undertaking, as a going concern;
   (b) the approval of a voluntary arrangement under Part I;
   (c) the sanctioning under section 425 of the Companies Act of a compromise or arrangement between the company and any such persons as are mentioned in that section; and
   (d) a more advantageous realisation of the company's assets than would be effected on a winding up;

and the order shall specify the purpose or purposes for which it is made.

(4)   An administration order shall not be made in relation to a company after it has gone into liquidation, nor where it is—
   (a) an insurance company within the meaning of the Insurance Companies Act 1982, or

[(b) an authorised institution or former authorised institution within the meaning of the Banking Act 1987.]

NOTES

Sub-s (4): para (b) substituted by the Banking Act 1987, s 108(1), Sch 6, para 25(1).

9 Application for order

(1)   An application to the court for an administration order shall be by petition presented either by the company or the directors, or by a creditor or creditors (including any contingent or prospective creditor or creditors), [or by the clerk of a magistrates' court in the exercise of the power conferred by section 87A of the Magistrates' Courts Act 1980 (enforcement of fines imposed on companies)] or by all or any of those parties, together or separately.

(2)   Where a petition is presented to the court—
  (a) notice of the petition shall be given forthwith to any person who has appointed, or is or may be entitled to appoint, an administrative receiver of the company, and to such other persons as may be prescribed, and
  (b) the petition shall not be withdrawn except with the leave of the court.

(3)   Where the court is satisfied that there is an administrative receiver of the company, the court shall dismiss the petition unless it is also satisfied either—
  (a) that the person by whom or on whose behalf the receiver was appointed has consented to the making of the order, or
  (b) that, if an administration order were made, any security by virtue of which the receiver was appointed would—
    [(i) be void against the administrator to any extent by virtue of the provisions of Part XII of the Companies Act 1985 (registration of company charges),]
    (i) be liable to be released or discharged under sections 238 to 240 in Part VI (transactions at an undervalue and preferences),
    (ii) be avoided under section 245 in that Part (avoidance of floating charges), or
    (iii) be challengeable under section 242 (gratuitous alienations) or 243 (unfair preferences) in that Part, or under any rule of law in Scotland.

(4)   Subject to subsection (3), on hearing a petition the court may dismiss it, or adjourn the hearing conditionally or unconditionally, or make an interim order or any other order that it thinks fit.

(5)   Without prejudice to the generality of subsection (4), an interim order under that subsection may restrict the exercise of any powers of the directors or of the company (whether by reference to the consent of the court or of a person qualified to act as an insolvency practitioner in relation to the company, or otherwise).

Sub-s (1): words in square brackets inserted by the Criminal Justice Act 1988, s 62(2)(a).

Sub-s (3): in para (b) sub-para (i) in square brackets inserted, and original sub-paras (i)–(iii) prospectively re-numbered as (ii)–(iv), by the Companies Act 1989, s 107, Sch 16, para 3(2), as from a day to be appointed.

10   Effect of application

(1)   During the period beginning with the presentation of a petition for an administration order and ending with the making of such an order or the dismissal of the petition—
  (a) no resolution may be passed or order made for the winding up of the company;
  (b) no steps may be taken to enforce any security over the company's property, or to repossess goods in the company's possession under any hire-purchase agreement, except with the leave of the court and subject to such terms as the court may impose; and
  (c) no other proceedings and no execution or other legal process may be commenced or continued, and no distress may be levied, against the company or its property except with the leave of the court and subject to such terms as aforesaid.

(2)   Nothing in subsection (1) requires the leave of the court—
  (a) for the presentation of a petition for the winding up of the company,
  (b) for the appointment of an administrative receiver of the company, or
  (c) for the carrying out by such a receiver (whenever appointed) of any of his functions.

(3)   Where—
  (a) a petition for an administration order is presented at a time when there is an administrative receiver of the company, and

(b) the person by or on whose behalf the receiver was appointed has not consented to the making of the order,

the period mentioned in subsection (1) is deemed not to begin unless and until that person so consents.

(4) References in this section and the next to hire-purchase agreements include conditional sale agreements, chattel leasing agreements and retention of title agreements.

(5) ...

11 Effect of order

(1) On the making of an administration order—
  (a) any petition for the winding up of the company shall be dismissed, and
  (b) any administrative receiver of the company shall vacate office.

(2) Where an administration order has been made, any receiver of part of the company's property shall vacate office on being required to do so by the administrator.

(3) During the period for which an administration order is in force—
  (a) no resolution may be passed or order made for the winding up of the company;
  (b) no administrative receiver of the company may be appointed;
  (c) no other steps may be taken to enforce any security over the company's property, or to repossess goods in the company's possession under any hire-purchase agreement, except with the consent of the administrator or the leave of the court and subject (where the court gives leave) to such terms as the court may impose; and
  (d) no other proceedings and no execution or other legal process may be commenced or continued, and no distress may be levied, against the company or its property except with the consent of the administrator or the leave of the court and subject (where the court gives leave) to such terms as aforesaid.

(4)   Where at any time an administrative receiver of the company has vacated office under subsection (1)(b), or a receiver of part of the company's property has vacated office under subsection (2)—

(a) his remuneration and any expenses properly incurred by him, and
(b) any indemnity to which he is entitled out of the assets of the company,

shall be charged on and (subject to subsection (3) above) paid out of any property of the company which was in his custody or under his control at that time in priority to any security held by the person by or on whose behalf he was appointed.

(5)   Neither an administrative receiver who vacates office under subsection (1)(b) nor a receiver who vacates office under subsection (2) is required on or after so vacating office to take any steps for the purpose of complying with any duty imposed on him by section 40 or 59 of this Act (duty to pay preferential creditors).

NOTES

Modified by the Water Industry Act 1991, Sch 3, Parts I, II.

Modified, in relation to railway administration orders, by the Railways Act 1993, s 59, Sch 6, paras 1, 2, 12, 13.

Administrators

15   Power to deal with charged property, etc

(1)   The administrator of a company may dispose of or otherwise exercise his powers in relation to any property of the company which is subject to a security to which this subsection applies as if the property were not subject to the security.

(2)   Where, on an application by the administrator, the court is satisfied that the disposal (with or without other assets) of—

(a) any property of the company subject to a security to which this subsection applies, or
(b) any goods in the possession of the company under a hire-purchase agreement,

would be likely to promote the purpose or one or more of the purposes specified in the administration order, the court may by order authorise the administrator to dispose of the property as if it were not subject to the security or to dispose of the goods as if all rights of the owner under the hire-purchase agreement were vested in the company.

(3)   Subsection (1) applies to any security which, as created, was a floating charge; and subsection (2) applies to any other security.

(4)   Where property is disposed of under subsection (1), the holder of the security has the same priority in respect of any property of the company directly or indirectly representing the property disposed of as he would have had in respect of the property subject to the security.

(5)   It shall be a condition of an order under subsection (2) that—
  (a) the net proceeds of the disposal, and
  (b) where those proceeds are less than such amount as may be determined by the court to be the net amount which would be realised on a sale of the property or goods in the open market by a willing vendor, such sums as may be required to make good the deficiency,

shall be applied towards discharging the sums secured by the security or payable under the hire-purchase agreement.

(6)   Where a condition imposed in pursuance of subsection (5) relates to two or more securities, that condition requires the net proceeds of the disposal and, where paragraph (b) of that subsection applies, the sums mentioned in that paragraph to be applied towards discharging the sums secured by those securities in the order of their priorities.

(7)   An office copy of an order under subsection (2) shall, within 14 days after the making of the order, be sent by the administrator to the registrar of companies.

(8)   If the administrator without reasonable excuse fails to comply with sub-section (7), he is liable to a fine and, for continued contravention, to a daily default fine.

(9)   References in this section to hire-purchase agreements include conditional sale agreements, chattel leasing agreements and retention of title agreements.

NOTES

Modified by the Water Industry Act 1991, Sch 3, Parts I, II.
Modified, in relation to railway administration orders, by the Railways Act 1993, s 59, Sch 6, paras 1, 5, 16.

17 General duties

(1)   The administrator of a company shall, on his appointment, take into his custody or under his control all the property to which the company is or appears to be entitled.

(2)   The administrator shall manage the affairs, business and property of the company—
  (a) at any time before proposals have been approved (with or without modifications) under section 24 below, in accordance with any directions given by the court, and
  (b) at any time after proposals have been so approved, in accordance with those proposals as from time to time revised, whether by him or a predecessor of his.

(3)   The administrator shall summon a meeting of the company's creditors if—
  (a) he is requested, in accordance with the rules, to do so by one-tenth, in value, of the company's creditors, or
  (b) he is directed to do so by the court.

NOTES

Modified by the Water Industry Act 1991, Sch 3, Parts I, II.
Modified, in relation to railway administration orders, by the Railways Act 1993, s 59, Sch 6, paras 1, 6, 17.

Miscellaneous

27 Protection of interests of creditors and members

(1)   At any time when an administration order is in force, a creditor or member of the company may apply to the court by petition for an order under this section on the ground—
  (a) that the company's affairs, business and property are being or have been managed by the administrator in a manner which is unfairly prejudicial to the interests of its creditors or members generally, or of some part of its creditors or members (including at least himself), or
  (b) that any actual or proposed act or omission of the administrator is or would be so prejudicial.

(2)　On an application or an order under this section the court may, subject as follows, make such order as it thinks fit for giving relief in respect of the matters complained of, or adjourn the hearing conditionally or unconditionally, or make an interim order or any other order that it thinks fit.

(3)　An order under this section shall not prejudice or prevent—
  (a) the implementation of a voluntary arrangement approved under section 4 in Part I, or any compromise or arrangement sanctioned under section 425 of the Companies Act; or
  (b) where the application for the order was made more than 28 days after the approval of any proposals or revised proposals under section 24 or 25, the implementation of those proposals or revised proposals.

(4)　Subject as above, an order under this section may in particular—
  (a) regulate the future management by the administrator of the company's affairs, business and property;
  (b) require the administrator to refrain from doing or continuing an act complained of by the petitioner, or to do an act which the petitioner has complained he has omitted to do;
  (c) require the summoning of a meeting of creditors or members for the purpose of considering such matters as the court may direct;
  (d) discharge the administration order and make such consequential provision as the court thinks fit.

(5)　Nothing in section 15 or 16 is to be taken as prejudicing applications to the court under this section.

(6)　Where the administration order is discharged, the administrator shall, within 14 days after the making of the order effecting the discharge, send an office copy of that order to the registrar of companies; and if without reasonable excuse he fails to comply with this subsection, he is liable to a fine and, for continued contravention, to a daily default fine.

NOTES

Modified by the Water Industry Act 1991, Sch 3, Parts I, II.
Modified, in relation to railway administration orders, by the Railways Act 1993, s 59, Sch 6, paras 1, 10.

PART III
RECEIVERSHIP

CHAPTER I
RECEIVERS AND MANAGERS (ENGLAND AND WALES)

Preliminary and general provisions

29 Definitions

(1)   It is hereby declared that, except where the context otherwise requires—
   (a) any reference in the Companies Act or this Act to a receiver or manager of the property of a company, or to a receiver of it, includes a receiver or manager, or (as the case may be) a receiver of part only of that property and a receiver only of the income arising from the property or from part of it; and
   (b) any reference in the Companies Act or this Act to the appointment of a receiver or manager under powers contained in an instrument includes an appointment made under powers which, by virtue of any enactment, are implied in and have effect as if contained in an instrument.

(2)   In this Chapter "administrative receiver" means—
   (a) a receiver or manager of the whole (or substantially the whole) of a company's property appointed by or on behalf of the holders of any debentures of the company secured by a charge which, as created, was a floating charge, or by such a charge and one or more other securities; or
   (b) a person who would be such a receiver or manager but for the appointment of some other person as the receiver of part of the company's property.

Provisions applicable to every receivership

40 Payment of debts out of assets subject to floating charge

(1)   The following applies, in the case of a company, where a receiver is appointed on behalf of the holders of any debentures of the company secured by a charge which, as created, was a floating charge.

(2)   If the company is not at the time in course of being wound up, its preferential debts (within the meaning given to that expression by section 386 in Part XII) shall be paid out of the assets coming to the hands of the receiver in priority to any claims for principal or interest in respect of the debentures.

(3)   Payments made under this section shall be recouped, as far as may be, out of the assets of the company available for payment of general creditors.

Administrative receivers: general

43   Power to dispose of charged property, etc

(1)   Where, on an application by the administrative receiver, the court is satisfied that the disposal (with or without other assets) of any relevant property which is subject to a security would be likely to promote a more advantageous realisation of the company's assets than would otherwise be effected, the court may by order authorise the administrative receiver to dispose of the property as if it were not subject to the security.

(2)   Subsection (1) does not apply in the case of any security held by the person by or on whose behalf the administrative receiver was appointed, or of any security to which a security so held has priority.

(3)   It shall be a condition of an order under this section that—
   (a) the net proceeds of the disposal, and
   (b) where those proceeds are less than such amount as may be determined by the court to be the net amount which would be realised on a sale of the property in the open market by a willing vendor, such sums as may be required to make good the deficiency,

shall be applied towards discharging the sums secured by the security.

(4)   Where a condition imposed in pursuance of subsection (3) relates to two or more securities, that condition shall require the net proceeds of the disposal and, where paragraph (b) of that subsection applies, the sums mentioned in that paragraph to be applied towards discharging the sums secured by those securities in the order of their priorities.

(5)   An office copy of an order under this section shall, within 14 days of the making of the order, be sent by the administrative receiver to the registrar of companies.

(6)   If the administrative receiver without reasonable excuse fails to comply with subsection (5), he is liable to a fine and, for continued contravention, to a daily default fine.

(7)   In this section "relevant property", in relation to the administrative receiver, means the property of which he is or, but for the appointment of some other person as the receiver of part of the company's property, would be the receiver or manager.

44   Agency and liability for contracts

(1)   The administrative receiver of a company—
  (a) is deemed to be the company's agent, unless and until the company goes into liquidation;
  (b) is personally liable on any contract entered into by him in the carrying out of his functions (except in so far as the contract otherwise provides) and[, to the extent of any qualifying liability,] on any contract of employment adopted by him in the carrying out of those functions; and
  (c) is entitled in respect of that liability to an indemnity out of the assets of the company.

(2)   For the purposes of subsection (1)(b) the administrative receiver is not to be taken to have adopted a contract of employment by reason of anything done or omitted to be done within 14 days after his appointment.

[(2A) For the purposes of subsection (1)(b), a liability under a contract of employment is a qualifying liability if—
  (a) it is a liability to pay a sum by way of wages or salary or contribution to an occupational pension scheme,
  (b) it is incurred while the administrative receiver is in office, and
  (c) it is in respect of services rendered wholly or partly after the adoption of the contract.

(2B)   Where a sum payable in respect of a liability which is a qualifying liability for the purposes of subsection (1)(b) is payable in respect of services rendered partly before and partly after the adoption of the contract, liability under subsection (1)(b) shall only extend to so much of the sum as is payable in respect of services rendered after the adoption of the contract.

(2C)   For the purposes of subsections (2A) and (2B)—
  (a) wages or salary payable in respect of a period of holiday or absence from work through sickness or other good cause are deemed to be

wages or (as the case may be) salary in respect of services rendered in that period, and

(b) a sum payable in lieu of holiday is deemed to be wages or (as the case may be) salary in respect of services rendered in the period by reference to which the holiday entitlement arose.

(2D) In subsection (2C)(a), the reference to wages or salary payable in respect of a period of holiday includes any sums which, if they had been paid, would have been treated for the purposes of the enactments relating to social security as earnings in respect of that period.]

(3) This section does not limit any right to indemnity which the administrative receiver would have apart from it, nor limit his liability on contracts entered into or adopted without authority, nor confer any right to indemnity in respect of that liability.

NOTES

Sub-s (1): words in square brackets inserted in relation to contracts of employment adopted on or after 15 March 1994, by the Insolvency Act 1994, s 2(2).

Sub-ss (2A)-(2D): inserted in relation to contracts of employment adopted on or after 15 March 1994, by the Insolvency Act 1994, s 2(3).

PART IV
WINDING UP OF COMPANIES REGISTERED UNDER THE COMPANIES ACTS

CHAPTER II
VOLUNTARY WINDING UP (INTRODUCTORY AND GENERAL)

Resolutions for, and commencement of, voluntary winding up

86 Commencement of winding up

A voluntary winding up is deemed to commence at the time of the passing of the resolution for voluntary winding up.

NOTES

Modified by the Building Societies Act 1986, s 90, Sch 15, and the Friendly Societies Act 1992, s 23, Sch 10.

CHAPTER VI
WINDING UP BY THE COURT

Grounds and effect of winding-up petition

122 Circumstances in which company may be wound up by the court

(1)   A company may be wound up by the court if—
  (a) the company has by special resolution resolved that the company be wound up by the court,
  (b) being a public company which was registered as such on its original incorporation, the company has not been issued with a certificate under section 117 of the Companies Act (public company share capital requirements) and more than a year has expired since it was so registered,
  (c) it is an old public company, within the meaning of the Consequential Provisions Act,
  (d) the company does not commence its business within a year from its incorporation or suspends its business for a whole year,
  (e) [except in the case of a private company limited by shares or by guarantee,] the number of members is reduced below 2,
  (f) the company is unable to pay its debts,
  (g) the court is of the opinion that it is just and equitable that the company should be wound up.

(2)   . . .

NOTES

Sub-s (1): in para (e) words in square brackets inserted by SI 1992/1699, reg 2(b), Schedule, para 8.
Sub-s (2): applies to Scotland only.

123 Definition of inability to pay debts

(1)   A company is deemed unable to pay its debts—
  (a) if a creditor (by assignment or otherwise) to whom the company is indebted in a sum exceeding œ750 then due has served on the company, by leaving it at the company's registered office, a written demand (in the prescribed form) requiring the company to pay the sum so due and the company has for 3 weeks thereafter neglected

to pay the sum or to secure or compound for it to the reasonable satisfaction of the creditor, or

(b) if, in England and Wales, execution or other process issued on a judgment, decree or order of any court in favour of a creditor of the company is returned unsatisfied in whole or in part, or

(c) if, in Scotland, the induciae of a charge for payment on an extract decree, or an extract registered bond, or an extract registered protest, have expired without payment being made, or

(d) if, in Northern Ireland, a certificate of unenforceability has been granted in respect of a judgment against the company, or

(e) if it is proved to the satisfaction of the court that the company is unable to pay its debts as they fall due.

(2)   A company is also deemed unable to pay its debts if it is proved to the satisfaction of the court that the value of the company's assets is less than the amount of its liabilities, taking into account its contingent and prospective liabilities.

(3)   The money sum for the time being specified in subsection (1)(a) is subject to increase or reduction by order under section 416 in Part XV.

NOTES

Modified by the Building Societies Act 1986, s 90, Sch 15, and the Friendly Societies Act 1992, s 23, Sch 10.

127 Avoidance of property dispositions, etc

In a winding up by the court, any disposition of the company's property, and any transfer of shares, or alteration in the status of the company's members, made after the commencement of the winding up is, unless the court otherwise orders, void.

NOTES

Modified by the Building Societies Act 1986, s 90, Sch 15, and the Friendly Societies Act 1992, s 23, Sch 10.

128 Avoidance of attachments, etc

(1)   Where a company registered in England and Wales is being wound up by the court, any attachment, sequestration, distress or execution put

in force against the estate or effects of the company after the commencement of the winding up is void.

(2)   This section, so far as relates to any estate or effects of the company situated in England and Wales, applies in the case of a company registered in Scotland as it applies in the case of a company registered in England and Wales.

NOTES

Modified by the Building Societies Act 1986, s 90, Sch 15, and the Friendly Societies Act 1992, s 23, Sch 10.

Commencement of winding up

129   Commencement of winding up by the court

(1)   If, before the presentation of a petition for the winding up of a company by the court, a resolution has been passed by the company for voluntary winding up, the winding up of the company is deemed to have commenced at the time of the passing of the resolution; and unless the court, on proof of fraud or mistake, directs otherwise, all proceedings taken in the voluntary winding up are deemed to have been validly taken.

(2)   In any other case, the winding up of a company by the court is deemed to commence at the time of the presentation of the petition for winding up.

NOTES

Modified by the Building Societies Act 1986, s 90, Sch 15, and the Friendly Societies Act 1992, s 23, Sch 10.

CHAPTER VIII
PROVISIONS OF GENERAL APPLICATION IN WINDING UP

Preferential debts

175   Preferential debts (general provision)

(1)   In a winding up the company's preferential debts (within the meaning given by section 386 in Part XII) shall be paid in priority to all other debts.

(2)   Preferential debts—
  (a) rank equally among themselves after the expenses of the winding up and shall be paid in full, unless the assets are insufficient to meet them, in which case they abate in equal proportions; and
  (b) so far as the assets of the company available for payment of general creditors are insufficient to meet them, have priority over the claims of holders of debentures secured by, or holders of, any floating charge created by the company, and shall be paid accordingly out of any property comprised in or subject to that charge.

NOTES

Modified by the Building Societies Act 1986, s 90, Sch 15, and the Friendly Societies Act 1992, s 23, Sch 10.

CHAPTER X
MALPRACTICE BEFORE AND DURING LIQUIDATION; PENALISATION OF COMPANIES AND COMPANY OFFICERS; INVESTIGATIONS AND PROSECUTIONS

Penalisation of directors and officers

212 Summary remedy against delinquent directors, liquidators, etc

(1)   This section applies if in the course of the winding up of a company it appears that a person who—
  (a) is or has been an officer of the company,
  (b) has acted as liquidator, administrator or administrative receiver of the company, or
  (c) not being a person falling within paragraph (a) or (b), is or has been concerned, or has taken part, in the promotion, formation or management of the company,

has misapplied or retained, or become accountable for, any money or other property of the company, or been guilty of any misfeasance or breach of any fiduciary or other duty in relation to the company.

(2)   The reference in subsection (1) to any misfeasance or breach of any fiduciary or other duty in relation to the company includes, in the case of a person who has acted as liquidator or administrator of the company, any misfeasance or breach of any fiduciary or other duty in

connection with the carrying out of his functions as liquidator or administrator of the company.

(3)   The court may, on the application of the official receiver or the liquidator, or of any creditor or contributory, examine into the conduct of the person falling within subsection (1) and compel him—
  (a) to repay, restore or account for the money or property or any part of it, with interest at such rate as the court thinks just, or
  (b) to contribute such sum to the company's assets by way of compensation in respect of the misfeasance or breach of fiduciary or other duty as the court thinks just.

(4)   The power to make an application under subsection (3) in relation to a person who has acted as liquidator or administrator of the company is not exercisable, except with the leave of the court, after that person has had his release.

(5)   The power of a contributory to make an application under subsection (3) is not exercisable except with the leave of the court, but is exercisable notwithstanding that he will not benefit from any order the court may make on the application.

NOTES

Modified by the Building Societies Act 1986, s 90, Sch 15, and the Friendly Societies Act 1992, s 23, Sch 10.

213   Fraudulent trading

(1)   If in the course of the winding up of a company it appears that any business of the company has been carried on with intent to defraud creditors of the company or creditors of any other person, or for any fraudulent purpose, the following has effect.

(2)   The court, on the application of the liquidator may declare that any persons who were knowingly parties to the carrying on of the business in the manner above-mentioned are to be liable to make such contributions (if any) to the company's assets as the court thinks proper.

NOTES

Modified by the Building Societies Act 1986, s 90, Sch 15, and the Friendly Societies Act 1992, s 23, Sch 10.

214   Wrongful trading

(1)   Subject to subsection (3) below, if in the course of the winding up of a company it appears that subsection (2) of this section applies in relation to a person who is or has been a director of the company, the court, on the application of the liquidator, may declare that that person is to be liable to make such contribution (if any) to the company's assets as the court thinks proper.

(2)   This subsection applies in relation to a person if—
   (a) the company has gone into insolvent liquidation,
   (b) at some time before the commencement of the winding up of the company, that person knew or ought to have concluded that there was no reasonable prospect that the company would avoid going into insolvent liquidation, and
   (c) that person was a director of the company at that time;

but the court shall not make a declaration under this section in any case where the time mentioned in paragraph (b) above was before 28th April 1986.

(3)   The court shall not make a declaration under this section with respect to any person if it is satisfied that after the condition specified in subsection (2)(b) was first satisfied in relation to him that person took every step with a view to minimising the potential loss to the company's creditors as (assuming him to have known that there was no reasonable prospect that the company would avoid going into insolvent liquidation) he ought to have taken.

(4)   For the purposes of subsections (2) and (3), the facts which a director of a company ought to know or ascertain, the conclusions which he ought to reach and the steps which he ought to take are those which would be known or ascertained, or reached or taken, by a reasonably diligent person having both—
   (a) the general knowledge, skill and experience that may reasonably be expected of a person carrying out the same functions as are carried out by that director in relation to the company, and
   (b) the general knowledge, skill and experience that that director has.

(5)   The reference in subsection (4) to the functions carried out in relation to a company by a director of the company includes any functions which he does not carry out but which have been entrusted to him.

(6)   For the purposes of this section a company goes into insolvent liquidation if it goes into liquidation at a time when its assets are insufficient

for the payment of its debts and other liabilities and the expenses of the winding up.

(7)   In this section "director" includes a shadow director.

(8)   This section is without prejudice to section 213.

NOTES

   Modified by the Building Societies Act 1986, s 90, Sch 15, and the Friendly Societies Act 1992, s 23, Sch 10.

PART VI

MISCELLANEOUS PROVISIONS APPLYING TO COMPANIES WHICH ARE INSOLVENT OR IN LIQUIDATION

Adjustment of prior transactions (administration and liquidation)

238   Transactions at an undervalue (England and Wales)

(1)   This section applies in the case of a company where—
   (a) an administration order is made in relation to the company, or
   (b) the company goes into liquidation;

and "the office-holder" means the administrator or the liquidator, as the case may be.

(2)   Where the company has at a relevant time (defined in section 240) entered into a transaction with any person at an undervalue, the office-holder may apply to the court for an order under this section.

(3)   Subject as follows, the court shall, on such an application, make such order as it thinks fit for restoring the position to what it would have been if the company had not entered into that transaction.

(4)   For the purposes of this section and section 241, a company enters into a transaction with a person at an undervalue if—
   (a) the company makes a gift to that person or otherwise enters into a transaction with that person on terms that provide for the company to receive no consideration, or
   (b) the company enters into a transaction with that person for a consideration the value of which, in money or money's worth, is

significantly less than the value, in money or money's worth, of the consideration provided by the company.

(5)   The court shall not make an order under this section in respect of a transaction at an undervalue if it is satisfied—
- (a) that the company which entered into the transaction did so in good faith and for the purpose of carrying on its business, and
- (b) that at the time it did so there were reasonable grounds for believing that the transaction would benefit the company.

NOTES

Modified by the Building Societies Act 1986, s 90, Sch 15, and the Friendly Societies Act 1992, s 23, Sch 10.

239   Preferences (England and Wales)

(1)   This section applies as does section 238.

(2)   Where the company has at a relevant time (defined in the next section) given a preference to any person, the office-holder may apply to the court for an order under this section.

(3)   Subject as follows, the court shall, on such an application, make such order as it thinks fit for restoring the position to what it would have been if the company had not given that preference.

(4)   For the purposes of this section and section 241, a company gives a preference to a person if—
- (a) that person is one of the company's creditors or a surety or guarantor for any of the company's debts or other liabilities, and
- (b) the company does anything or suffers anything to be done which (in either case) has the effect of putting that person into a position which, in the event of the company going into insolvent liquidation, will be better than the position he would have been in if that thing had not been done.

(5)   The court shall not make an order under this section in respect of a preference given to any person unless the company which gave the preference was influenced in deciding to give it by a desire to produce in relation to that person the effect mentioned in subsection (4)(b).

(6)   A company which has given a preference to a person connected with the company (otherwise than by reason only of being its employee) at the

time the preference was given is presumed, unless the contrary is shown, to have been influenced in deciding to give it by such a desire as is mentioned in subsection (5).

(7)   The fact that something has been done in pursuance of the order of a court does not, without more, prevent the doing or suffering of that thing from constituting the giving of a preference.

NOTES

Modified by the Building Societies Act 1986, s 90, Sch 15, and the Friendly Societies Act 1992, s 23, Sch 10.

240   "Relevant time" under ss 238, 239

(1)   Subject to the next subsection, the time at which a company enters into a transaction at an undervalue or gives a preference is a relevant time if the transaction is entered into, or the preference given—
   (a) in the case of a transaction at an undervalue or of a preference which is given to a person who is connected with the company (otherwise than by reason only of being its employee), at a time in the period of 2 years ending with the onset of insolvency (which expression is defined below),
   (b) in the case of a preference which is not such a transaction and is not so given, at a time in the period of 6 months ending with the onset of insolvency, and
   (c) in either case, at a time between the presentation of a petition for the making of an administration order in relation to the company and the making of such an order on that petition.

(2)   Where a company enters into a transaction at an undervalue or gives a preference at a time mentioned in subsection (1)(a) or (b), that time is not a relevant time for the purposes of section 238 or 239 unless the company—
   (a) is at that time unable to pay its debts within the meaning of section 123 in Chapter VI of Part IV, or
   (b) becomes unable to pay its debts within the meaning of that section in consequence of the transaction or preference;

but the requirements of this subsection are presumed to be satisfied, unless the contrary is shown, in relation to any transaction at an undervalue which is entered into by a company with a person who is connected with the company.

(3)   For the purposes of subsection (1), the onset of insolvency is—
  (a) in a case where section 238 or 239 applies by reason of the making of an administration order or of a company going into liquidation immediately upon the discharge of an administration order, the date of the presentation of the petition on which the administration order was made, and
  (b) in a case where the section applies by reason of a company going into liquidation at any other time, the date of the commencement of the winding up.

NOTES

Modified by the Building Societies Act 1986, s 90, Sch 15, and the Friendly Societies Act 1992, s 23, Sch 10.

241   Orders under ss 238, 239

(1)   Without prejudice to the generality of sections 238(3) and 239(3), an order under either of those sections with respect to a transaction or preference entered into or given by a company may (subject to the next subsection)—
  (a) require any property transferred as part of the transaction, or in connection with the giving of the preference, to be vested in the company,
  (b) require any property to be so vested if it represents in any person's hands the application either of the proceeds of sale of property so transferred or of money so transferred,
  (c) release or discharge (in whole or in part) any security given by the company,
  (d) require any person to pay, in respect of benefits received by him from the company, such sums to the office-holder as the court may direct,
  (e) provide for any surety or guarantor whose obligations to any person were released or discharged (in whole or in part) under the transaction, or by the giving of the preference, to be under such new or revived obligations to that person as the court thinks appropriate,
  (f) provide for security to be provided for the discharge of any obligation imposed by or arising under the order, for such an obligation to be charged on any property and for the security or charge to have the same priority as a security or charge released or discharged (in whole or in part) under the transaction or by the giving of the preference, and

(g) provide for the extent to which any person whose property is vested by the order in the company, or on whom obligations are imposed by the order, is to be able to prove in the winding up of the company for debts or other liabilities which arose from, or were released or discharged (in whole or in part) under or by, the transaction or the giving of the preference.

(2)   An order under section 238 or 239 may affect the property of, or impose any obligation on, any person whether or not he is the person with whom the company in question entered into the transaction or (as the case may be) the person to whom the preference was given; but such an order—

(a) shall not prejudice any interest in property which was acquired from a person other than the company and was acquired [in good faith and for value], or prejudice any interest deriving from such an interest, and

(b) shall not require a person who received a benefit from the transaction or preference [in good faith and for value] to pay a sum to the office-holder, except where that person was a party to the transaction or the payment is to be in respect of a preference given to that person at a time when he was a creditor of the company.

[(2A) Where a person has acquired an interest in property from a person other than the company in question, or has received a benefit from the transaction or preference, and at the time of that acquisition or receipt—

(a) he had notice of the relevant surrounding circumstances and of the relevant proceedings, or

(b) he was connected with, or was an associate of, either the company in question or the person with whom that company entered into the transaction or to whom that company gave the preference,

then, unless the contrary is shown, it shall be presumed for the purposes of paragraph (a) or (as the case may be) paragraph (b) of subsection (2) that the interest was acquired or the benefit was received otherwise than in good faith.]

[(3)   For the purposes of subsection (2A)(a), the relevant surrounding circumstances are (as the case may require)—

(a) the fact that the company in question entered into the transaction at an undervalue; or

(b) the circumstances which amounted to the giving of the preference by the company in question;

and subsections (3A) to (3C) have effect to determine whether, for those purposes, a person has notice of the relevant proceedings.

(3A)   In a case where section 238 or 239 applies by reason of the making of an administration order, a person has notice of the relevant proceedings if he has notice—

   (a) of the fact that the petition on which the administration order is made has been presented; or

   (b) of the fact that the administration order has been made.

(3B)   In a case where section 238 or 239 applies by reason of the company in question going into liquidation immediately upon the discharge of an administration order, a person has notice of the relevant proceedings if he has notice—

   (a) of the fact that the petition on which the administration order is made has been presented;

   (b) of the fact that the administration order has been made; or

   (c) of the fact that the company has gone into liquidation.

(3C)   In a case where section 238 or 239 applies by reason of the company in question going into liquidation at any other time, a person has notice of the relevant proceedings if he has notice—

   (a) where the company goes into liquidation on the making of a winding-up order, of the fact that the petition on which the winding-up order is made has been presented or of the fact that the company has gone into liquidation;

   (b) in any other case, of the fact that the company has gone into liquidation.]

(4)   The provisions of sections 238 to 241 apply without prejudice to the availability of any other remedy, even in relation to a transaction or preference which the company had no power to enter into or give.

NOTES

Sub-s (2): words in square brackets substituted by the Insolvency (No 2) Act 1994, s 1(1).

Sub-s (2A): inserted by the Insolvency (No 2) Act 1994, s 1(2).

Sub-ss (3)-(3C): substituted for sub-s (3) as originally enacted, by the Insolvency (No 2) Act 1994, s 1(3).

Modified by the Building Societies Act 1986, s 90, Sch 15, and the Friendly Societies Act 1992, s 23, Sch 10.

244   Extortionate credit transactions

(1)   This section applies as does section 238, and where the company is, or has been, a party to a transaction for, or involving, the provision of credit to the company.

(2)   The court may, on the application of the office-holder, make an order with respect to the transaction if the transaction is or was extortionate and was entered into in the period of 3 years ending with the day on which the administration order was made or (as the case may be) the company went into liquidation.

(3)   For the purposes of this section a transaction is extortionate if, having regard to the risk accepted by the person providing the credit—
  (a) the terms of it are or were such as to require grossly exorbitant payments to be made (whether unconditionally or in certain contingencies) in respect of the provision of the credit, or
  (b) it otherwise grossly contravened ordinary principles of fair dealing;

and it shall be presumed, unless the contrary is proved, that a transaction with respect to which an application is made under this section is or, as the case may be, was extortionate.

(4)   An order under this section with respect to any transaction may contain such one or more of the following as the court thinks fit, that is to say—
  (a) provision setting aside the whole or part of any obligation created by the transaction,
  (b) provision otherwise varying the terms of the transaction or varying the terms on which any security for the purposes of the transaction is held,
  (c) provision requiring any person who is or was a party to the transaction to pay to the office-holder any sums paid to that person, by virtue of the transaction, by the company,
  (d) provision requiring any person to surrender to the office-holder any property held by him as security for the purposes of the transaction,
  (e) provision directing accounts to be taken between any persons.

(5)   The powers conferred by this section are exercisable in relation to any transaction concurrently with any powers exercisable in relation to that transaction as a transaction at an undervalue or under section 242 (gratuitous alienations in Scotland).

NOTES

Modified by the Building Societies Act 1986, s 90, Sch 15, and the Friendly Societies Act 1992, s 23, Sch 10.

245 Avoidance of certain floating charges

(1)   This section applies as does section 238, but applies to Scotland as well as to England and Wales.

(2)   Subject as follows, a floating charge on the company's undertaking or property created at a relevant time is invalid except to the extent of the aggregate of—
  (a) the value of so much of the consideration for the creation of the charge as consists of money paid, or goods or services supplied, to the company at the same time as, or after, the creation of the charge,
  (b) the value of so much of that consideration as consists of the discharge or reduction, at the same time as, or after the creation of the charge, of any debt of the company, and
  (c) the amount of such interest (if any) as is payable on the amount falling within paragraph (a) or (b) in pursuance of any agreement under which the money was so paid, the goods or services were so supplied or the debt was so discharged or reduced.

(3)   Subject to the next subsection, the time at which a floating charge is created by a company is a relevant time for the purposes of this section if the charge is created—
  (a) in the case of a charge which is created in favour of a person who is connected with the company, at a time in the period of 2 years ending with the onset of insolvency,
  (b) in the case of a charge which is created in favour of any other person, at a time in the period of 12 months ending with the onset of insolvency, or
  (c) in either case, at a time between the presentation of a petition for the making of an administration order in relation to the company and the making of such an order on that petition.

(4)   Where a company creates a floating charge at a time mentioned in subsection (3)(b) and the person in favour of whom the charge is created is not connected with the company, that time is not a relevant time for the purposes of this section unless the company—
  (a) is at that time unable to pay its debts within the meaning of section 123 in Chapter VI of Part IV, or
  (b) becomes unable to pay its debts within the meaning of that section in consequence of the transaction under which the charge is created.

(5)   For the purposes of subsection (3), the onset of insolvency is—

(a) in a case where this section applies by reason of the making of an administration order, the date of the presentation of the petition on which the order was made, and

(b) in a case where this section applies by reason of a company going into liquidation, the date of the commencement of the winding up.

(6)   For the purposes of subsection (2)(a) the value of any goods or services supplied by way of consideration for a floating charge is the amount in money which at the time they were supplied could reasonably have been expected to be obtained for supplying the goods or services in the ordinary course of business and on the same terms (apart from the consideration) as those on which they were supplied to the company.

NOTES

Modified by the Building Societies Act 1986, s 90, Sch 15, and the Friendly Societies Act 1992, s 23, Sch 10.

PART VII
INTERPRETATION FOR FIRST GROUP OF PARTS

247   "Insolvency" and "go into liquidation"

(1)   In this Group of Parts, except in so far as the context otherwise requires, "insolvency", in relation to a company, includes the approval of a voluntary arrangement under Part I, the making of an administration order or the appointment of an administrative receiver.

(2)   For the purposes of any provision in this Group of Parts, a company goes into liquidation if it passes a resolution for voluntary winding up or an order for its winding up is made by the court at a time when it has not already gone into liquidation by passing such a resolution.

NOTES

Modified by the Building Societies Act 1986, s 90, Sch 15, and the Friendly Societies Act 1992, s 23, Sch 10.

249   "Connected" with a company

For the purposes of any provision in this Group of Parts, a person is connected with a company if—

(a) he is a director or shadow director of the company or an associate of such a director or shadow director, or

(b) he is an associate of the company;

and "associate" has the meaning given by section 435 in Part XVIII of this Act.

NOTES

Modified by the Building Societies Act 1986, s 90, Sch 15, and the Friendly Societies Act 1992, s 23, Sch 10.

251 Expressions used generally

In this Group of Parts, except in so far as the context otherwise requires-
"administrative receiver" means—

(a) an administrative receiver as defined by section 29(2) in Chapter I of Part III, or

(b) a receiver appointed under section 51 in Chapter II of that Part in a case where the whole (or substantially the whole) of the company's property is attached by the floating charge;

"business day" means any day other than a Saturday, a Sunday, Christmas Day, Good Friday or a day which is a bank holiday in any part of Great Britain;

"chattel leasing agreement" means an agreement for the bailment or, in Scotland, the hiring of goods which is capable of subsisting for more than 3 months;

"contributory" has the meaning given by section 79;

"director" includes any person occupying the position of director, by whatever name called;

"floating charge" means a charge which, as created, was a floating charge and includes a floating charge within section 462 of the Companies Act (Scottish floating charges);

"office copy", in relation to Scotland, means a copy certified by the clerk of court;

"the official rate", in relation to interest, means the rate payable under section 189(4);

"prescribed" means prescribed by the rules;

"receiver", in the expression "receiver or manager", does not include a receiver appointed under section 51 in Chapter II of Part III;

"retention of title agreement" means an agreement for the sale of goods to a company, being an agreement—

(a) which does not constitute a charge on the goods, but

(b)   under which, if the seller is not paid and the company is wound up, the seller will have priority over all other creditors of the company as respects the goods or any property representing the goods;

"the rules" means rules under section 411 in Part XV; and

"shadow director", in relation to a company, means a person in accordance with whose directions or instructions the directors of the company are accustomed to act (but so that a person is not deemed a shadow director by reason only that the directors act on advice given by him in a professional capacity);

and any expression for whose interpretation provision is made by Part XXVI of the Companies Act, other than an expression defined above in this section, is to be construed in accordance with that provision.

NOTES

Modified by the Building Societies Act 1986, s 90, Sch 15, and the Friendly Societies Act 1992, s 23, Sch 10.

THE SECOND GROUP OF PARTS
INSOLVENCY OF INDIVIDUALS; BANKRUPTCY

PART IX
BANKRUPTCY

CHAPTER V
EFFECT OF BANKRUPTCY ON CERTAIN RIGHTS, TRANSACTIONS, ETC

Adjustment of prior transactions, etc

344   Avoidance of general assignment of book debts

(1)   The following applies where a person engaged in any business makes a general assignment to another person of his existing or future book debts, or any class of them, and is subsequently adjudged bankrupt.

(2)   The assignment is void against the trustee of the bankrupt's estate as regards book debts which were not paid before the presentation of the bankruptcy petition, unless the assignment has been registered under the Bills of Sale Act 1878.

(3)   For the purposes of subsections (1) and (2)—
- (a) "assignment" includes an assignment by way of security or charge on book debts, and
- (b) "general assignment" does not include—
  - (i)   an assignment of book debts due at the date of the assignment from specified debtors or of debts becoming due under specified contracts, or
  - (ii) an assignment of book debts included either in a transfer of a business made in good faith and for value or in an assignment of assets for the benefit of creditors generally.

(4)   For the purposes of registration under the Act of 1878 an assignment of book debts is to be treated as if it were a bill of sale given otherwise than by way of security for the payment of a sum of money; and the provisions of that Act with respect to the registration of bills of sale apply accordingly with such necessary modifications as may be made by rules under that Act.

THE THIRD GROUP OF PARTS
MISCELLANEOUS MATTERS BEARING ON BOTH COMPANY AND INDIVIDUAL INSOLVENCY; GENERAL INTERPRETATION; FINAL PROVISIONS

PART XVI
PROVISIONS AGAINST DEBT AVOIDANCE (ENGLAND AND WALES ONLY)

423   Transactions defrauding creditors

(1)   This section relates to transactions entered into at an undervalue; and a person enters into such a transaction with another person if—
- (a) he makes a gift to the other person or he otherwise enters into a transaction with the other on terms that provide for him to receive no consideration;
- (b) he enters into a transaction with the other in consideration of marriage; or
- (c) he enters into a transaction with the other for a consideration the value of which, in money or money's worth, is significantly less than the value, in money or money's worth, of the consideration provided by himself.

(2)   Where a person has entered into such a transaction, the court may, if satisfied under the next subsection, make such order as it thinks fit for—

(a) restoring the position to what it would have been if the transaction had not been entered into, and

(b) protecting the interests of persons who are victims of the transaction.

(3)   In the case of a person entering into such a transaction, an order shall only be made if the court is satisfied that it was entered into by him for the purpose—

(a) of putting assets beyond the reach of a person who is making, or may at some time make, a claim against him, or

(b) of otherwise prejudicing the interests of such a person in relation to the claim which he is making or may make.

(4)   In this section "the court" means the High Court or—

(a) if the person entering into the transaction is an individual, any other court which would have jurisdiction in relation to a bankruptcy petition relating to him;

(b) if that person is a body capable of being wound up under Part IV or V of this Act, any other court having jurisdiction to wind it up.

(5)   In relation to a transaction at an undervalue, references here and below to a victim of the transaction are to a person who is, or is capable of being, prejudiced by it; and in the following two sections the person entering into the transaction is referred to as "the debtor".

424   Those who may apply for an order under s 423

(1)   An application for an order under section 423 shall not be made in relation to a transaction except—

(a) in a case where the debtor has been adjudged bankrupt or is a body corporate which is being wound up or in relation to which an administration order is in force, by the official receiver, by the trustee of the bankrupt's estate or the liquidator or administrator of the body corporate or (with the leave of the court) by a victim of the transaction;

(b) in a case where a victim of the transaction is bound by a voluntary arrangement approved under Part I or Part VIII of this Act, by the supervisor of the voluntary arrangement or by any person who (whether or not so bound) is such a victim; or

(c) in any other case, by a victim of the transaction.

(2)   An application made under any of the paragraphs of subsection (1) is to be treated as made on behalf of every victim of the transaction.

425 Provision which may be made by order under s 423

(1)   Without prejudice to the generality of section 423, an order made under that section with respect to a transaction may (subject as follows)—

(a) require any property transferred as part of the transaction to be vested in any person, either absolutely or for the benefit of all the persons on whose behalf the application for the order is treated as made;

(b) require any property to be so vested if it represents, in any person's hands, the application either of the proceeds of sale of property so transferred or of money so transferred;

(c) release or discharge (in whole or in part) any security given by the debtor;

(d) require any person to pay to any other person in respect of benefits received from the debtor such sums as the court may direct;

(e) provide for any surety or guarantor whose obligations to any person were released or discharged (in whole or in part) under the transaction to be under such new or revived obligations as the court thinks appropriate;

(f) provide for security to be provided for the discharge of any obligation imposed by or arising under the order, for such an obligation to be charged on any property and for such security or charge to have the same priority as a security or charge released or discharged (in whole or in part) under the transaction.

(2)   An order under section 423 may affect the property of, or impose any obligation on, any person whether or not he is the person with whom the debtor entered into the transaction; but such an order—

(a) shall not prejudice any interest in property which was acquired from a person other than the debtor and was acquired in good faith, for value and without notice of the relevant circumstances, or prejudice any interest deriving from such an interest, and

(b) shall not require a person who received a benefit from the transaction in good faith, for value and without notice of the relevant circumstances to pay any sum unless he was a party to the transaction.

(3)   For the purposes of this section the relevant circumstances in relation to a transaction are the circumstances by virtue of which an order under section 423 may be made in respect of the transaction.

(4)   In this section "security" means any mortgage, charge, lien or other security.

PART XVIII
INTERPRETATION

435   Meaning of "associate"

(1)   For the purposes of this Act any question whether a person is an associate of another person is to be determined in accordance with the following provisions of this section (any provision that a person is an associate of another person being taken to mean that they are associates of each other).

(2)   A person is an associate of an individual if that person is the individual's husband or wife, or is a relative, or the husband or wife of a relative, of the individual or of the individual's husband or wife.

(3)   A person is an associate of any person with whom he is in partnership, and of the husband or wife or a relative of any individual with whom he is in partnership; and a Scottish firm is an associate of any person who is a member of the firm.

(4)   A person is an associate of any person whom he employs or by whom he is employed.

(5)   A person in his capacity as trustee of a trust other than—
   (a)   a trust arising under any of the second Group of Parts or the Bankruptcy (Scotland) Act 1985, or
   (b)   a pension scheme or an employees' share scheme (within the meaning of the Companies Act),

is an associate of another person if the beneficiaries of the trust include, or the terms of the trust confer a power that may be exercised for the benefit of, that other person or an associate of that other person.

(6)   A company is an associate of another company—
   (a)   if the same person has control of both, or a person has control of one and persons who are his associates, or he and persons who are his associates, have control of the other, or
   (b)   if a group of two or more persons has control of each company, and the groups either consist of the same persons or could be regarded as consisting of the same persons by treating (in one or more cases) a member of either group as replaced by a person of whom he is an associate.

(7)   A company is an associate of another person if that person has control of it or if that person and persons who are his associates together have control of it.

(8)   For the purposes of this section a person is a relative of an individual if he is that individual's brother, sister, uncle, aunt, nephew, niece, lineal ancestor or lineal descendant, treating—
  (a) any relationship of the half blood as a relationship of the whole blood and the stepchild or adopted child of any person as his child, and
  (b) an illegitimate child as the legitimate child of his mother and reputed father;

and references in this section to a husband or wife include a former husband or wife and a reputed husband or wife.

(9)   For the purposes of this section any director or other officer of a company is to be treated as employed by that company.

(10)   For the purposes of this section a person is to be taken as having control of a company if—
  (a) the directors of the company or of another company which has control of it (or any of them) are accustomed to act in accordance with his directions or instructions, or
  (b) he is entitled to exercise, or control the exercise of, one third or more of the voting power at any general meeting of the company or of another company which has control of it;

and where two or more persons together satisfy either of the above conditions, they are to be taken as having control of the company.

(11)   In this section "company" includes any body corporate (whether incorporated in Great Britain or elsewhere); and references to directors and other officers of a company and to voting power at any general meeting of a company have effect with any necessary modifications.

PART XIX
FINAL PROVISIONS

444   Citation

This Act may be cited as the Insolvency Act 1986.

Consumer Protection Act 1987

(C 43)

An Act to make provision with respect to the liability of persons for damage caused by defective products; to consolidate with amendments the Consumer Safety Act 1978 and the Consumer Safety (Amendment) Act 1986; to make provision with respect to the giving of price indications; to amend Part I of the Health and Safety at Work etc Act 1974 and sections 31 and 80 of the Explosives Act 1875; to repeal the Trade Descriptions Act 1972 and the Fabrics (Misdescription) Act 1913; and for connected purposes

[15 May 1987]

PART I
PRODUCT LIABILITY

1 Purpose and construction of Part I

(1)   This Part shall have effect for the purpose of making such provision as is necessary in order to comply with the product liability Directive and shall be construed accordingly.

(2)   In this Part, except in so far as the context otherwise requires—
"agricultural produce" means any produce of the soil, of stockfarming or of fisheries;
"dependant" and "relative" have the same meaning as they have in, respectively, the Fatal Accidents Act 1976 and the Damages (Scotland) Act 1976;
"producer", in relation to a product, means—
(a) the person who manufactured it;
(b) in the case of a substance which has not been manufactured but has been won or abstracted, the person who won or abstracted it;
(c) in the case of a product which has not been manufactured, won or abstracted but essential characteristics of which are attributable to an industrial or other process having been carried out (for example, in relation to agricultural produce), the person who carried out that process;
"product" means any goods or electricity and (subject to subsection (3) below) includes a product which is comprised in another product, whether by virtue of being a component part or raw material or otherwise; and

"the product liability Directive" means the Directive of the Council of the European Communities, dated 25th July 1985, (No 85/374/EEC) on the approximation of the laws, regulations and administrative provisions of the member States concerning liability for defective products.

(3)   For the purposes of this Part a person who supplies any product in which products are comprised, whether by virtue of being component parts or raw materials or otherwise, shall not be treated by reason only of his supply of that product as supplying any of the products so comprised.

2   Liability for defective products

(1)   Subject to the following provisions of this Part, where any damage is caused wholly or partly by a defect in a product, every person to whom subsection (2) below applies shall be liable for the damage.

(2)   This subsection applies to—
  (a) the producer of the product;
  (b) any person who, by putting his name on the product or using a trade mark or other distinguishing mark in relation to the product, has held himself out to be the producer of the product;
  (c) any person who has imported the product into a member State from a place outside the member States in order, in the course of any business of his, to supply it to another.

(3)   Subject as aforesaid, where any damage is caused wholly or partly by a defect in a product, any person who supplied the product (whether to the person who suffered the damage, to the producer of any product in which the product in question is comprised or to any other person) shall be liable for the damage if—
  (a) the person who suffered the damage requests the supplier to identify one or more of the persons (whether still in existence or not) to whom subsection (2) above applies in relation to the product;
  (b) that request is made within a reasonable period after the damage occurs and at a time when it is not reasonably practicable for the person making the request to identify all those persons; and
  (c) the supplier fails, within a reasonable period after receiving the request, either to comply with the request or to identify the person who supplied the product to him.

(4)   Neither subsection (2) nor subsection (3) above shall apply to a person in respect of any defect in any game or agricultural produce if the only

supply of the game or produce by that person to another was at a time when it had not undergone an industrial process.

(5)   Where two or more persons are liable by virtue of this Part for the same damage, their liability shall be joint and several.

(6)   This section shall be without prejudice to any liability arising otherwise than by virtue of this Part.

NOTES

Trade marks: references to trade marks or registered trade marks within the meaning of the Trade Marks Act 1938 shall, unless the context otherwise requires, be construed as references to trade marks or registered trade marks within the meaning of the Trade Marks Act 1994; see the Trade Marks Act 1994, Sch 4, para 1.

3   Meaning of "defect"

(1)   Subject to the following provisions of this section, there is a defect in a product for the purposes of this Part if the safety of the product is not such as persons generally are entitled to expect; and for those purposes "safety", in relation to a product, shall include safety with respect to products comprised in that product and safety in the context of risks of damage to property, as well as in the context of risks of death or personal injury.

(2)   In determining for the purposes of subsection (1) above what persons generally are entitled to expect in relation to a product all the circumstances shall be taken into account, including—

(a) the manner in which, and purposes for which, the product has been marketed, its get-up, the use of any mark in relation to the product and any instructions for, or warnings with respect to, doing or refraining from doing anything with or in relation to the product;

(b) what might reasonably be expected to be done with or in relation to the product; and

(c) the time when the product was supplied by its producer to another;

and nothing in this section shall require a defect to be inferred from the fact alone that the safety of a product which is supplied after that time is greater than the safety of the product in question.

4   Defences

(1)   In any civil proceedings by virtue of this Part against any person ("the person proceeded against") in respect of a defect in a product it shall be a defence for him to show—

  (a)   that the defect is attributable to compliance with any requirement imposed by or under any enactment or with any Community obligation; or

  (b)   that the person proceeded against did not at any time supply the product to another; or

  (c)   that the following conditions are satisfied, that is to say—

    (i)   that the only supply of the product to another by the person proceeded against was otherwise than in the course of a business of that person's; and

    (ii)   that section 2(2) above does not apply to that person or applies to him by virtue only of things done otherwise than with a view to profit; or

  (d)   that the defect did not exist in the product at the relevant time; or

  (e)   that the state of scientific and technical knowledge at the relevant time was not such that a producer of products of the same description as the product in question might be expected to have discovered the defect if it had existed in his products while they were under his control; or

  (f)   that the defect—

    (i)   constituted a defect in a product ("the subsequent product") in which the product in question had been comprised; and

    (ii)   was wholly attributable to the design of the subsequent product or to compliance by the producer of the product in question with instructions given by the producer of the subsequent product.

(2)   In this section "the relevant time", in relation to electricity, means the time at which it was generated, being a time before it was transmitted or distributed, and in relation to any other product, means—

  (a)   if the person proceeded against is a person to whom subsection (2) of section 2 above applies in relation to the product, the time when he supplied the product to another;

  (b)   if that subsection does not apply to that person in relation to the product, the time when the product was last supplied by a person to whom that subsection does apply in relation to the product.

5 Damage giving rise to liability

(1)   Subject to the following provisions of this section, in this Part "damage" means death or personal injury or any loss of or damage to any property (including land).

(2)   A person shall not be liable under section 2 above in respect of any defect in a product for the loss of or any damage to the product itself or for the loss of or any damage to the whole or any part of any product which has been supplied with the product in question comprised in it.

(3)   A person shall not be liable under section 2 above for any loss of or damage to any property which, at the time it is lost or damaged, is not—
(a) of a description of property ordinarily intended for private use, occupation or consumption; and
(b) intended by the person suffering the loss or damage mainly for his own private use, occupation or consumption.

(4)   No damages shall be awarded to any person by virtue of this Part in respect of any loss of or damage to any property if the amount which would fall to be so awarded to that person, apart from this subsection and any liability for interest, does not exceed £275.

(5)   In determining for the purposes of this Part who has suffered any loss of or damage to property and when any such loss or damage occurred, the loss or damage shall be regarded as having occurred at the earliest time at which a person with an interest in the property had knowledge of the material facts about the loss or damage.

(6)   For the purposes of subsection (5) above the material facts about any loss of or damage to any property are such facts about the loss or damage as would lead a reasonable person with an interest in the property to consider the loss or damage sufficiently serious to justify his instituting proceedings for damages against a defendant who did not dispute liability and was able to satisfy a judgment.

(7)   For the purposes of subsection (5) above a person's knowledge includes knowledge which he might reasonably have been expected to acquire—
(a) from facts observable or ascertainable by him; or
(b) from facts ascertainable by him with the help of appropriate expert advice which it is reasonable for him to seek;

but a person shall not be taken by virtue of this subsection to have knowledge of a fact ascertainable by him only with the help of expert advice unless he has failed to take all reasonable steps to obtain (and, where appropriate, to act on) that advice.

(8)　　. . .

NOTES

Sub-s (8): applies to Scotland only.

6　Application of certain enactments

(1)　　Any damage for which a person is liable under section 2 above shall be deemed to have been caused—
 (a) for the purposes of the Fatal Accidents Act 1976, by that person's wrongful act, neglect or default;
 (b)–(d)　. . .

(2)　　Where—
 (a) a person's death is caused wholly or partly by a defect in a product, or a person dies after suffering damage which has been so caused;
 (b) a request such as mentioned in paragraph (a) of subsection (3) of section 2 above is made to a supplier of the product by that person's personal representatives or, in the case of a person whose death is caused wholly or partly by the defect, by any dependant or relative of that person; and
 (c) the conditions specified in paragraphs (b) and (c) of that subsection are satisfied in relation to that request,

this Part shall have effect for the purposes of the Law Reform (Miscellaneous Provisions) Act 1934, the Fatal Accidents Act 1976 and the Damages (Scotland) Act 1976 as if liability of the supplier to that person under that subsection did not depend on that person having requested the supplier to identify certain persons or on the said conditions having been satisfied in relation to a request made by that person.

(3)　　Section 1 of the Congenital Disabilities (Civil Liability) Act 1976 shall have effect for the purposes of this Part as if—
 (a) a person were answerable to a child in respect of an occurrence caused wholly or partly by a defect in a product if he is or has been liable under section 2 above in respect of any effect of the occurrence on a parent of the child, or would be so liable if the occurrence caused a parent of the child to suffer damage;

(b)   the provisions of this Part relating to liability under section 2 above applied in relation to liability by virtue of paragraph (a) above under the said section 1; and

(c)   subsection (6) of the said section 1 (exclusion of liability) were omitted.

(4)   Where any damage is caused partly by a defect in a product and partly by the fault of the person suffering the damage, the Law Reform (Contributory Negligence) Act 1945 and section 5 of the Fatal Accidents Act 1976 (contributory negligence) shall have effect as if the defect were the fault of every person liable by virtue of this Part for the damage caused by the defect.

(5)   In subsection (4) above "fault" has the same meaning as in the said Act of 1945.

(6)   Schedule 1 to this Act shall have effect for the purpose of amending the Limitation Act 1980 and the Prescription and Limitation (Scotland) Act 1973 in their application in relation to the bringing of actions by virtue of this Part.

(7)   It is hereby declared that liability by virtue of this Part is to be treated as liability in tort for the purposes of any enactment conferring jurisdiction on any court with respect to any matter.

(8)   Nothing in this Part shall prejudice the operation of section 12 of the Nuclear Installations Act 1965 (rights to compensation for certain breaches of duties confined to rights under that Act).

NOTES

Sub-s (1): paras (b)–(d) apply to Scotland only.

7   Prohibition on exclusions from liability

The liability of a person by virtue of this Part to a person who has suffered damage caused wholly or partly by a defect in a product, or to a dependant or relative of such a person, shall not be limited or excluded by any contract term, by any notice or by any other provision.

8   Power to modify Part I

(1)   Her Majesty may by Order in Council make such modifications of this Part and of any other enactment (including an enactment contained in the following Parts of this Act, or in an Act passed after this Act) as appear to Her Majesty in Council to be necessary or expedient in consequence of

any modification of the product liability Directive which is made at any time after the passing of this Act.

(2)   An Order in Council under subsection (1) above shall not be submitted to Her Majesty in Council unless a draft of the Order has been laid before, and approved by a resolution of, each House of Parliament.

9   Application of Part I to Crown

(1)   Subject to subsection (2) below, this Part shall bind the Crown.

(2)   The Crown shall not, as regards the Crown's liability by virtue of this Part, be bound by this Part further than the Crown is made liable in tort or in reparation under the Crown Proceedings Act 1947, as that Act has effect from time to time.

PART II
CONSUMER SAFETY

10   The general safety requirement

(1)   A person shall be guilty of an offence if he—
(a) supplies any consumer goods which fail to comply with the general safety requirement;
(b) offers or agrees to supply any such goods; or
(c) exposes or possesses any such goods for supply.

(2)   For the purposes of this section consumer goods fail to comply with the general safety requirement if they are not reasonably safe having regard to all the circumstances, including—
(a) the manner in which, and purposes for which, the goods are being or would be marketed, the get-up of the goods, the use of any mark in relation to the goods and any instructions or warnings which are given or would be given with respect to the keeping, use or consumption of the goods;
(b) any standards of safety published by any person either for goods of a description which applies to the goods in question or for matters relating to goods of that description; and
(c) the existence of any means by which it would have been reasonable (taking into account the cost, likelihood and extent of any improvement) for the goods to have been made safer.

(3)   For the purposes of this section consumer goods shall not be regarded as failing to comply with the general safety requirement in respect of—
  (a) anything which is shown to be attributable to compliance with any requirement imposed by or under any enactment or with any Community obligation;
  (b) any failure to do more in relation to any matter than is required by—
     (i)   any safety regulations imposing requirements with respect to that matter;
     (ii) . . .
     (iii) any provision of any enactment or subordinate legislation imposing such requirements with respect to that matter as are designated for the purposes of this subsection by any such regulations.

(4)   In any proceedings against any person for an offence under this section in respect of any goods it shall be a defence for that person to show—
  (a) that he reasonably believed that the goods would not be used or consumed in the United Kingdom; or
  (b) that the following conditions are satisfied, that is to say—
     (i)   that he supplied the goods, offered or agreed to supply them or, as the case may be, exposed or possessed them for supply in the course of carrying on a retail business; and
     (ii) that, at the time he supplied the goods or offered or agreed to supply them or exposed or possessed them for supply, he neither knew nor had reasonable grounds for believing that the goods failed to comply with the general safety requirement; or
  (c) that the terms on which he supplied the goods or agreed or offered to supply them or, in the case of goods which he exposed or possessed for supply, the terms on which he intended to supply them—
     (i)   indicated that the goods were not supplied or to be supplied as new goods; and
     (ii) provided for, or contemplated, the acquisition of an interest in the goods by the persons supplied or to be supplied.

(5)   For the purposes of subsection (4)(b) above goods are supplied in the course of carrying on a retail business if—
  (a) whether or not they are themselves acquired for a person's private use or consumption, they are supplied in the course of carrying on a business of making a supply of consumer goods available to persons who generally acquire them for private use or consumption; and
  (b) the descriptions of goods the supply of which is made available in the course of that business do not, to a significant extent, include manufactured or imported goods which have not previously been supplied in the United Kingdom.

(6)   A person guilty of an offence under this section shall be liable on summary conviction to imprisonment for a term not exceeding six months or to a fine not exceeding level 5 on the standard scale or to both.

(7)   In this section "consumer goods" means any goods which are ordinarily intended for private use or consumption, not being—
   (a) growing crops or things comprised in land by virtue of being attached to it;
   (b) water, food, feeding stuff or fertiliser;
   (c) gas which is, is to be or has been supplied by a person authorised to supply it by or under [section 7A of the Gas Act 1986 (licensing of gas suppliers and gas shippers) or paragraph 5 of Schedule 2A to that Act (supply to very large customers an exception to prohibition on unlicensed activities)] [or under Article 8(1)(c) of the Gas (Northern Ireland) Order 1996];
   (d) aircraft (other than hang-gliders) or motor vehicles;
   (e) controlled drugs or licensed medicinal products;
   (f) tobacco.

NOTES

Sub-s (3): para (b)(ii) repealed by the General Product Safety Regulations 1994, SI 1994/2328.

Sub-s (7): in para (c) words in first pair of square brackets substituted by the Gas Act 1995, s 16(1), Sch 4, para 15(1), words in second pair of square brackets inserted by the Gas (Northern Ireland) Order 1996, SI 1996/275, art 71(1), Sch 6.

11   Safety regulations

(1)   The Secretary of State may by regulations under this section ("safety regulations") make such provision as he considers appropriate for the purposes of section 10(3) above and for the purpose of securing—
   (a) that goods to which this section applies are safe;
   (b) that goods to which this section applies which are unsafe, or would be unsafe in the hands of persons of a particular description, are not made available to persons generally or, as the case may be, to persons of that description; and
   (c) that appropriate information is, and inappropriate information is not, provided in relation to goods to which this section applies.

(2)   Without prejudice to the generality of subsection (1) above, safety regulations may contain provision—

(a) with respect to the composition or contents, design, construction, finish or packing of goods to which this section applies, with respect to standards for such goods and with respect to other matters relating to such goods;

(b) with respect to the giving, refusal, alteration or cancellation of approvals of such goods, of descriptions of such goods or of standards for such goods;

(c) with respect to the conditions that may be attached to any approval given under the regulations;

(d) for requiring such fees as may be determined by or under the regulations to be paid on the giving or alteration of any approval under the regulations and on the making of an application for such an approval or alteration;

(e) with respect to appeals against refusals, alterations and cancellations of approvals given under the regulations and against the conditions contained in such approvals;

(f) for requiring goods to which this section applies to be approved under the regulations or to conform to the requirements of the regulations or to descriptions or standards specified in or approved by or under the regulations;

(g) with respect to the testing or inspection of goods to which this section applies (including provision for determining the standards to be applied in carrying out any test or inspection);

(h) with respect to the ways of dealing with goods of which some or all do not satisfy a test required by or under the regulations or a standard connected with a procedure so required;

(i) for requiring a mark, warning or instruction or any other information relating to goods to be put on or to accompany the goods or to be used or provided in some other manner in relation to the goods, and for securing that inappropriate information is not given in relation to goods either by means of misleading marks or otherwise;

(j) for prohibiting persons from supplying, or from offering to supply, agreeing to supply, exposing for supply or possessing for supply, goods to which this section applies and component parts and raw materials for such goods;

(k) for requiring information to be given to any such person as may be determined by or under the regulations for the purpose of enabling that person to exercise any function conferred on him by the regulations.

(3)   Without prejudice as aforesaid, safety regulations may contain provision—

(a) for requiring persons on whom functions are conferred by or under section 27 below to have regard, in exercising their functions so far as relating to any provision of safety regulations, to matters specified in a direction issued by the Secretary of State with respect to that provision;

(b) for securing that a person shall not be guilty of an offence under section 12 below unless it is shown that the goods in question do not conform to a particular standard;

(c) for securing that proceedings for such an offence are not brought in England and Wales except by or with the consent of the Secretary of State or the Director of Public Prosecutions;

(d) for securing that proceedings for such an offence are not brought in Northern Ireland except by or with the consent of the Secretary of State or the Director of Public Prosecutions for Northern Ireland;

(e) for enabling a magistrates' court in England and Wales or Northern Ireland to try an information or, in Northern Ireland, a complaint in respect of such an offence if the information was laid or the complaint made within twelve months from the time when the offence was committed;

(f) . . .

(g) for determining the persons by whom, and the manner in which, anything required to be done by or under the regulations is to be done.

(4)   Safety regulations shall not provide for any contravention of the regulations to be an offence.

(5)   Where the Secretary of State proposes to make safety regulations it shall be his duty before he makes them—

(a) to consult such organisations as appear to him to be representative of interests substantially affected by the proposal;

(b) to consult such other persons as he considers appropriate; and

(c) in the case of proposed regulations relating to goods suitable for use at work, to consult the Health and Safety Commission in relation to the application of the proposed regulations to Great Britain;

but the preceding provisions of this subsection shall not apply in the case of regulations which provide for the regulations to cease to have effect at the end of a period of not more than twelve months beginning with the day on which they come into force and which contain a statement that it appears to the Secretary of State that the need to protect the public requires that the regulations should be made without delay.

(6)   The power to make safety regulations shall be exercisable by statutory instrument subject to annulment in pursuance of a resolution of either House of Parliament and shall include power—

(a) to make different provision for different cases; and
(b) to make such supplemental, consequential and transitional provision as the Secretary of State considers appropriate.

(7)   This section applies to any goods other than—
(a) growing crops and things comprised in land by virtue of being attached to it;
(b) water, food, feeding stuff and fertiliser;
(c) gas which is, is to be or has been supplied by a person authorised to supply it by or under [section 7A of the Gas Act 1986 (licensing of gas suppliers and gas shippers) or paragraph 5 of Schedule 2A to that Act (supply to very large customers an exception to prohibition on unlicensed activities)] [or under Article 8(1)(c) of the Gas (Northern Ireland) Order 1996];
(d) controlled drugs and licensed medicinal products.

NOTES

Sub-s (3): para (f) applies to Scotland only.
Sub-s (7): in para (c) words in first pair of square brackets substituted by the Gas Act 1995, s 16(1), Sch 4, para 15(2), words in second pair of square brackets inserted by the Gas (Northern Ireland) Order 1996, SI 1996/275, art 71(1), Sch 6.

12   Offences against the safety regulations

(1)   Where safety regulations prohibit a person from supplying or offering or agreeing to supply any goods or from exposing or possessing any goods for supply, that person shall be guilty of an offence if he contravenes the prohibition.

(2)   Where safety regulations require a person who makes or processes any goods in the course of carrying on a business—
(a) to carry out a particular test or use a particular procedure in connection with the making or processing of the goods with a view to ascertaining whether the goods satisfy any requirements of such regulations; or
(b) to deal or not to deal in a particular way with a quantity of the goods of which the whole or part does not satisfy such a test or does not satisfy standards connected with such a procedure,

that person shall be guilty of an offence if he does not comply with the requirement.

(3)   If a person contravenes a provision of safety regulations which prohibits or requires the provision, by means of a mark or otherwise, of information of a particular kind in relation to goods, he shall be guilty of an offence.

(4)   Where safety regulations require any person to give information to another for the purpose of enabling that other to exercise any function, that person shall be guilty of an offence if—
　(a) he fails without reasonable cause to comply with the requirement; or
　(b) in giving the information which is required of him—
　　(i) he makes any statement which he knows is false in a material particular; or
　　(ii) he recklessly makes any statement which is false in a material particular.

(5)   A person guilty of an offence under this section shall be liable on summary conviction to imprisonment for a term not exceeding six months or to a fine not exceeding level 5 on the standard scale or to both.

13   Prohibition notices and notices to warn

(1)   The Secretary of State may—
　(a) serve on any person a notice ("a prohibition notice") prohibiting that person, except with the consent of the Secretary of State, from supplying, or from offering to supply, agreeing to supply, exposing for supply or possessing for supply, any relevant goods which the Secretary of State considers are unsafe and which are described in the notice;
　(b) serve on any person a notice ("a notice to warn") requiring that person at his own expense to publish, in a form and manner and on occasions specified in the notice, a warning about any relevant goods which the Secretary of State considers are unsafe, which that person supplies or has supplied and which are described in the notice.

(2)   Schedule 2 to this Act shall have effect with respect to prohibition notices and notices to warn; and the Secretary of State may by regulations make provision specifying the manner in which information is to be given to any person under that Schedule.

(3)   A consent given by the Secretary of State for the purposes of a prohibition notice may impose such conditions on the doing of anything for

which the consent is required as the Secretary of State considers appropriate.

(4)   A person who contravenes a prohibition notice or a notice to warn shall be guilty of an offence and liable on summary conviction to imprisonment for a term not exceeding six months or to a fine not exceeding level 5 on the standard scale or to both.

(5)   The power to make regulations under subsection (2) above shall be exercisable by statutory instrument subject to annulment in pursuance of a resolution of either House of Parliament and shall include power—
  (a) to make different provision for different cases; and
  (b) to make such supplemental, consequential and transitional provision as the Secretary of State considers appropriate.

(6)   In this section "relevant goods" means—
  (a) in relation to a prohibition notice, any goods to which section 11 above applies; and
  (b) in relation to a notice to warn, any goods to which that section applies or any growing crops or things comprised in land by virtue of being attached to it.

NOTES

    Modification: sub-s (4) is modified by the substitution of "three months" for the words "six months" by the General Product Safety Regulations 1994, SI 1994/2328, reg 11(d), in relation to the enforcement of those Regulations.

14   Suspension notices

(1)   Where an enforcement authority has reasonable grounds for suspecting that any safety provision has been contravened in relation to any goods, the authority may serve a notice ("a suspension notice") prohibiting the person on whom it is served, for such period ending not more than six months after the date of the notice as is specified therein, from doing any of the following things without the consent of the authority, that is to say, supplying the goods, offering to supply them, agreeing to supply them or exposing them for supply.

(2)   A suspension notice served by an enforcement authority in respect of any goods shall—
  (a) describe the goods in a manner sufficient to identify them;

(b) set out the grounds on which the authority suspects that a safety provision has been contravened in relation to the goods; and

(c) state that, and the manner in which, the person on whom the notice is served may appeal against the notice under section 15 below.

(3) A suspension notice served by an enforcement authority for the purpose of prohibiting a person for any period from doing the things mentioned in subsection (1) above in relation to any goods may also require that person to keep the authority informed of the whereabouts throughout that period of any of those goods in which he has an interest.

(4) Where a suspension notice has been served on any person in respect of any goods, no further such notice shall be served on that person in respect of the same goods unless—

(a) proceedings against that person for an offence in respect of a contravention in relation to the goods of a safety provision (not being an offence under this section); or

(b) proceedings for the forfeiture of the goods under section 16 or 17 below,

are pending at the end of the period specified in the first-mentioned notice.

(5) A consent given by an enforcement authority for the purposes of subsection (1) above may impose such conditions on the doing of anything for which the consent is required as the authority considers appropriate.

(6) Any person who contravenes a suspension notice shall be guilty of an offence and liable on summary conviction to imprisonment for a term not exceeding six months or to a fine not exceeding level 5 on the standard scale or to both.

(7) Where an enforcement authority serves a suspension notice in respect of any goods, the authority shall be liable to pay compensation to any person having an interest in the goods in respect of any loss or damage caused by reason of the service of the notice if—

(a) there has been no contravention in relation to the goods of any safety provision; and

(b) the exercise of the power is not attributable to any neglect or default by that person.

(8) Any disputed question as to the right to or the amount of any compensation payable under this section shall be determined by arbitration or, in Scotland, by a single arbiter appointed, failing agreement between the parties, by the sheriff.

NOTES

Modification: sub-s (6) is modified by the substitution of "three months" for the words "six months" by the General Product Safety Regulations 1994, SI 1994/2328, reg 11(d), in relation to the enforcement of those Regulations.

Modified in relation to vessels as consumer goods, by the Simple Pressure Vessels (Safety) Regulations 1991, SI 1991/2749, Sch 5.

15 Appeals against suspension notices

(1)   Any person having an interest in any goods in respect of which a suspension notice is for the time being in force may apply for an order setting aside the notice.

(2)   An application under this section may be made—
  (a) to any magistrates' court in which proceedings have been brought in England and Wales or Northern Ireland—
    (i) for an offence in respect of a contravention in relation to the goods of any safety provision; or
    (ii) for the forfeiture of the goods under section 16 below;
  (b) where no such proceedings have been so brought, by way of complaint to a magistrates' court; or
  (c) . . .

(3)   On an application under this section to a magistrates' court in England and Wales or Northern Ireland the court shall make an order setting aside the suspension notice only if the court is satisfied that there has been no contravention in relation to the goods of any safety provision.

(4)   On an application under this section to the sheriff he shall make an order setting aside the suspension notice only if he is satisfied that at the date of making the order—
  (a) proceedings for an offence in respect of a contravention in relation to the goods of any safety provision; or
  (b) proceedings for the forfeiture of the goods under section 17 below,

have not been brought or, having been brought, have been concluded.

(5)   Any person aggrieved by an order made under this section by a magistrates' court in England and Wales or Northern Ireland, or by a decision of such a court not to make such an order, may appeal against that order or decision—
  (a) in England and Wales, to the Crown Court;
  (b) in Northern Ireland, to the county court;

and an order so made may contain such provision as appears to the court to be appropriate for delaying the coming into force of the order pending the making and determination of any appeal (including any application under section 111 of the Magistrates' Courts Act 1980 or Article 146 of the Magistrates' Courts (Northern Ireland) Order 1981 (statement of case)).

NOTES

Sub-s (2): para (c) applies to Scotland only.

Modified in relation to vessels as consumer goods, by the Simple Pressure Vessels (Safety) Regulations 1991, SI 1991/2749, Sch 5.

16 Forfeiture: England and Wales and Northern Ireland

(1)   An enforcement authority in England and Wales or Northern Ireland may apply under this section for an order for the forfeiture of any goods on the grounds that there has been a contravention in relation to the goods of a safety provision.

(2)   An application under this section may be made—
   (a) where proceedings have been brought in a magistrates' court for an offence in respect of a contravention in relation to some or all of the goods of any safety provision, to that court;
   (b) where an application with respect to some or all of the goods has been made to a magistrates' court under section 15 above or section 33 below, to that court; and
   (c) where no application for the forfeiture of the goods has been made under paragraph (a) or (b) above, by way of complaint to a magistrates' court.

(3)   On an application under this section the court shall make an order for the forfeiture of any goods only if it is satisfied that there has been a contravention in relation to the goods of a safety provision.

(4)   For the avoidance of doubt it is declared that a court may infer for the purposes of this section that there has been a contravention in relation to any goods of a safety provision if it is satisfied that any such provision has been contravened in relation to goods which are representative of those goods (whether by reason of being of the same design or part of the same consignment or batch or otherwise).

(5)   Any person aggrieved by an order made under this section by a magistrates' court, or by a decision of such a court not to make such an order, may appeal against that order or decision—

    (a) in England and Wales, to the Crown Court;
    (b) in Northern Ireland, to the county court;

and an order so made may contain such provision as appears to the court to be appropriate for delaying the coming into force of the order pending the making and determination of any appeal (including any application under section 111 of the Magistrates' Courts Act 1980 or Article 146 of the Magistrates' Courts (Northern Ireland) Order 1981 (statement of case)).

(6)   Subject to subsection (7) below, where any goods are forfeited under this section they shall be destroyed in accordance with such directions as the court may give.

(7)   On making an order under this section a magistrates' court may, if it considers it appropriate to do so, direct that the goods to which the order relates shall (instead of being destroyed) be released, to such person as the court may specify, on condition that that person—
    (a) does not supply those goods to any person otherwise than as mentioned in section 46(7)(a) or (b) below; and
    (b) complies with any order to pay costs or expenses (including any order under section 35 below) which has been made against that person in the proceedings for the order for forfeiture.

18 Power to obtain information

(1)   If the Secretary of State considers that, for the purpose of deciding whether—
    (a) to make, vary or revoke any safety regulations; or
    (b) to serve, vary or revoke a prohibition notice; or
    (c) to serve or revoke a notice to warn,

he requires information which another person is likely to be able to furnish, the Secretary of State may serve on the other person a notice under this section.

(2)   A notice served on any person under this section may require that person—
    (a) to furnish to the Secretary of State, within a period specified in the notice, such information as is so specified;
    (b) to produce such records as are specified in the notice at a time and place so specified and to permit a person appointed by the Secretary of State for the purpose to take copies of the records at that time and place.

(3)   A person shall be guilty of an offence if he—
    (a) fails, without reasonable cause, to comply with a notice served on him under this section; or
    (b) in purporting to comply with a requirement which by virtue of paragraph (a) of subsection (2) above is contained in such a notice—
        (i) furnishes information which he knows is false in a material particular; or
        (ii) recklessly furnishes information which is false in a material particular.

(4)   A person guilty of an offence under subsection (3) above shall—
    (a) in the case of an offence under paragraph (a) of that subsection, be liable on summary conviction to a fine not exceeding level 5 on the standard scale; and
    (b) in the case of an offence under paragraph (b) of that subsection be liable—
        (i) on conviction on indictment, to a fine;
        (ii) on summary conviction, to a fine not exceeding the statutory maximum.

19   Interpretation of Part II

(1)   In this Part—
    "controlled drug" means a controlled drug within the meaning of the Misuse of Drugs Act 1971;
    "feeding stuff" and "fertiliser" have the same meanings as in Part IV of the Agriculture Act 1970;
    "food" does not include anything containing tobacco but, subject to that, has the same meaning as in the [Food Safety Act 1990] or, in relation to Northern Ireland, the same meaning as in the [Food Safety (Northern Ireland) Order 1991];
    "licensed medicinal product" means—
        (a) any medicinal product within the meaning of the Medicines Act 1968 in respect of which a product licence within the meaning of that Act is for the time being in force; or
        (b) any other article or substance in respect of which any such licence is for the time being in force in pursuance of an order under section 104 or 105 of that Act (application of Act to other articles and substances);
    "safe", in relation to any goods, means such that there is no risk, or no risk apart from one reduced to a minimum, that any of the following will (whether immediately or after a definite or indefinite period)

cause the death of, or any personal injury to, any person whatsoever, that is to say—

(a) the goods;

(b) the keeping, use or consumption of the goods;

(c) the assembly of any of the goods which are, or are to be, supplied unassembled;

(d) any emission or leakage from the goods or, as a result of the keeping, use or consumption of the goods, from anything else; or

(e) reliance on the accuracy of any measurement, calculation or other reading made by or by means of the goods,

and "safer" and "unsafe" shall be construed accordingly;

"tobacco" includes any tobacco product within the meaning of the Tobacco Products Duty Act 1979 and any article or substance containing tobacco and intended for oral or nasal use.

(2)   In the definition of "safe" in subsection (1) above, references to the keeping, use or consumption of any goods are references to—

(a) the keeping, use or consumption of the goods by the persons by whom, and in all or any of the ways or circumstances in which, they might reasonably be expected to be kept, used or consumed; and

(b) the keeping, use or consumption of the goods either alone or in conjunction with other goods in conjunction with which they might reasonably be expected to be kept, used or consumed.

NOTES

Sub-s (1): in definition "food" words in first pair of square brackets substituted by the Food Safety Act 1990, s 59(1), Sch 3, para 37, words in second pair of square brackets substituted by the Food Safety (Northern Ireland) Order 1991, SI 1991/762, art 51(1), Sch 2, para 17.

Modification: sub-s (1) modified, in relation to the definition of "licensed medicinal product", by the Medicines for Human Use (Marketing Authorisations Etc) Regulations 1994, SI 1994/3144, reg 9(13).

PART III
MISLEADING PRICE INDICATIONS

20   Offence of giving misleading indication

(1)   Subject to the following provisions of this Part, a person shall be guilty of an offence if, in the course of any business of his, he gives (by any

means whatever) to any consumers an indication which is misleading as to the price at which any goods, services, accommodation or facilities are available (whether generally or from particular persons).

(2)   Subject as aforesaid, a person shall be guilty of an offence if—
    (a) in the course of any business of his, he has given an indication to any consumers which, after it was given, has become misleading as mentioned in subsection (1) above; and
    (b) some or all of those consumers might reasonably be expected to rely on the indication at a time after it has become misleading; and
    (c) he fails to take all such steps as are reasonable to prevent those consumers from relying on the indication.

(3)   For the purposes of this section it shall be immaterial—
    (a) whether the person who gives or gave the indication is or was acting on his own behalf or on behalf of another;
    (b) whether or not that person is the person, or included among the persons, from whom the goods, services, accommodation or facilities are available; and
    (c) whether the indication is or has become misleading in relation to all the consumers to whom it is or was given or only in relation to some of them.

(4)   A person guilty of an offence under subsection (1) or (2) above shall be liable—
    (a) on conviction on indictment, to a fine;
    (b) on summary conviction, to a fine not exceeding the statutory maximum.

(5)   No prosecution for an offence under subsection (1) or (2) above shall be brought after whichever is the earlier of the following, that is to say—
    (a) the end of the period of three years beginning with the day on which the offence was committed; and
    (b) the end of the period of one year beginning with the day on which the person bringing the prosecution discovered that the offence had been committed.

(6)   In this Part—
"consumer"—
    (a) in relation to any goods, means any person who might wish to be supplied with the goods for his own private use or consumption;
    (b) in relation to any services or facilities, means any person who might wish to be provided with the services or facilities otherwise than for the purposes of any business of his; and

(c) in relation to any accommodation, means any person who might wish to occupy the accommodation otherwise than for the purposes of any business of his;

"price", in relation to any goods, services, accommodation or facilities, means—

(a) the aggregate of the sums required to be paid by a consumer for or otherwise in respect of the supply of the goods or the provision of the services, accommodation or facilities; or

(b) except in section 21 below, any method which will be or has been applied for the purpose of determining that aggregate.

21 Meaning of "misleading"

(1) For the purposes of section 20 above an indication given to any consumers is misleading as to a price if what is conveyed by the indication, or what those consumers might reasonably be expected to infer from the indication or any omission from it, includes any of the following, that is to say—

(a) that the price is less than in fact it is;

(b) that the applicability of the price does not depend on facts or circumstances on which its applicability does in fact depend;

(c) that the price covers matters in respect of which an additional charge is in fact made;

(d) that a person who in fact has no such expectation—

(i) expects the price to be increased or reduced (whether or not at a particular time or by a particular amount); or

(ii) expects the price, or the price as increased or reduced, to be maintained (whether or not for a particular period); or

(e) that the facts or circumstances by reference to which the consumers might reasonably be expected to judge the validity of any relevant comparison made or implied by the indication are not what in fact they are.

(2) For the purposes of section 20 above, an indication given to any consumers is misleading as to a method of determining a price if what is conveyed by the indication, or what those consumers might reasonably be expected to infer from the indication or any omission from it, includes any of the following, that is to say—

(a) that the method is not what in fact it is;

(b) that the applicability of the method does not depend on facts or circumstances on which its applicability does in fact depend;

(c) that the method takes into account matters in respect of which an additional charge will in fact be made;

(d) that a person who in fact has no such expectation—
  (i) expects the method to be altered (whether or not at a particular time or in a particular respect); or
  (ii) expects the method, or that method as altered, to remain unaltered (whether or not for a particular period); or
(e) that the facts or circumstances by reference to which the consumers might reasonably be expected to judge the validity of any relevant comparison made or implied by the indication are not what in fact they are.

(3)   For the purposes of subsections (1)(e) and (2)(e) above a comparison is a relevant comparison in relation to a price or method of determining a price if it is made between that price or that method, or any price which has been or may be determined by that method, and—
  (a) any price or value which is stated or implied to be, to have been or to be likely to be attributed or attributable to the goods, services, accommodation or facilities in question or to any other goods, services, accommodation or facilities; or
  (b) any method, or other method, which is stated or implied to be, to have been or to be likely to be applied or applicable for the determination of the price or value of the goods, services, accommodation or facilities in question or of the price or value of any other goods, services, accommodation or facilities.

22   Application to provision of services and facilities

(1)   Subject to the following provisions of this section, references in this Part to services or facilities are references to any services or facilities whatever including, in particular—
  (a) the provision of credit or of banking or insurance services and the provision of facilities incidental to the provision of such services;
  (b) the purchase or sale of foreign currency;
  (c) the supply of electricity;
  (d) the provision of a place, other than on a highway, for the parking of a motor vehicle;
  (e) the making of arrangements for a person to put or keep a caravan on any land other than arrangements by virtue of which that person may occupy the caravan as his only or main residence.

(2)   References in this Part to services shall not include references to services provided to an employer under a contract of employment.

(3)   References in this Part to services or facilities shall not include references to services or facilities which are provided by an authorised

person or appointed representative in the course of the carrying on of an investment business.

(4)   In relation to a service consisting in the purchase or sale of foreign currency, references in this Part to the method by which the price of the service is determined shall include references to the rate of exchange.

(5)   In this section—
   "appointed representative", "authorised person" and "investment business" have the same meanings as in the Financial Services Act 1986;
   "caravan" has the same meaning as in the Caravan Sites and Control of Development Act 1960;
   "contract of employment" and "employer" have the same meanings as in [the Employment Rights Act 1996];
   "credit" has the same meaning as in the Consumer Credit Act 1974.

NOTES

Sub-s (5): words in square brackets substituted by the Employment Rights Act 1996, s 240, Sch 1, para 34.
   Modified by the Banking Coordination (Second Council Directive) Regulations 1992, SI 1992/3218, reg 82(1), Sch 10, para 27.
   Modified by the Investment Services Regulations 1995, SI 1995/3275, reg 57, Sch 10, para 12.

23   Application to provision of accommodation etc

(1)   Subject to subsection (2) below, references in this Part to accommodation or facilities being available shall not include references to accommodation or facilities being available to be provided by means of the creation or disposal of an interest in land except where—
   (a) the person who is to create or dispose of the interest will do so in the course of any business of his; and
   (b) the interest to be created or disposed of is a relevant interest in a new dwelling and is to be created or disposed of for the purpose of enabling that dwelling to be occupied as a residence, or one of the residences, of the person acquiring the interest.

(2)   Subsection (1) above shall not prevent the application of any provision of this Part in relation to—
   (a) the supply of any goods as part of the same transaction as any creation or disposal of an interest in land; or

(b) the provision of any services or facilities for the purposes of, or in connection with, any transaction for the creation or disposal of such an interest.

(3)   In this section—
"new dwelling" means any building or part of a building in Great Britain which—
(a) has been constructed or adapted to be occupied as a residence; and
(b) has not previously been so occupied or has been so occupied only with other premises or as more than one residence,
and includes any yard, garden, out-houses or appurtenances which belong to that building or part or are to be enjoyed with it;
"relevant interest"—
(a) in relation to a new dwelling in England and Wales, means the freehold estate in the dwelling or a leasehold interest in the dwelling for a term of years absolute of more than twenty-one years, not being a term of which twenty-one years or less remains unexpired;
(b) . . .

NOTES

Sub-s (2): in definition "relevant interest" para (b) applies to Scotland only.

24   Defences

(1)   In any proceedings against a person for an offence under subsection (1) or (2) of section 20 above in respect of any indication it shall be a defence for that person to show that his acts or omissions were authorised for the purposes of this subsection by regulations made under section 26 below.

(2)   In proceedings against a person for an offence under subsection (1) or (2) of section 20 above in respect of an indication published in a book, newspaper, magazine [or film or in a programme included in a programme service (within the meaning of the Broadcasting Act 1990),] it shall be a defence for that person to show that the indication was not contained in an advertisement.

(3)   In proceedings against a person for an offence under subsection (1) or (2) of section 20 above in respect of an indication published in an advertisement it shall be a defence for that person to show that—

(a) he is a person who carries on a business of publishing or arranging for the publication of advertisements;

(b) he received the advertisement for publication in the ordinary course of that business; and

(c) at the time of publication he did not know and had no grounds for suspecting that the publication would involve the commission of the offence.

(4)   In any proceedings against a person for an offence under subsection (1) of section 20 above in respect of any indication, it shall be a defence for that person to show that—

(a) the indication did not relate to the availability from him of any goods, services, accommodation or facilities;

(b) a price had been recommended to every person from whom the goods, services, accommodation or facilities were indicated as being available;

(c) the indication related to that price and was misleading as to that price only by reason of a failure by any person to follow the recommendation; and

(d) it was reasonable for the person who gave the indication to assume that the recommendation was for the most part being followed.

(5)   The provisions of this section are without prejudice to the provisions of section 39 below.

(6)   In this section—
        "advertisement" includes a catalogue, a circular and a price list;

. . . . .

NOTES

Sub-s (2): words in square brackets substituted by the Broadcasting Act 1990, s 203(1), Sch 20, para 48.

Sub-s (6): definition "cable programme service" repealed by the Broadcasting Act 1990, s 203(1), (3), Sch 20, para 48, Sch 21.

25 Code of practice

(1)   The Secretary of State may, after consulting the Director General of Fair Trading and such other persons as the Secretary of State considers it appropriate to consult, by order approve any code of practice issued (whether by the Secretary of State or another person) for the purpose of—

(a) giving practical guidance with respect to any of the requirements of section 20 above; and

(b) promoting what appear to the Secretary of State to be desirable practices as to the circumstances and manner in which any person gives an indication as to the price at which any goods, services, accommodation or facilities are available or indicates any other matter in respect of which any such indication may be misleading.

(2) A contravention of a code of practice approved under this section shall not of itself give rise to any criminal or civil liability, but in any proceedings against any person for an offence under section 20(1) or (2) above—

(a) any contravention by that person of such a code may be relied on in relation to any matter for the purpose of establishing that that person committed the offence or of negativing any defence; and

(b) compliance by that person with such a code may be relied on in relation to any matter for the purpose of showing that the commission of the offence by that person has not been established or that that person has a defence.

(3) Where the Secretary of State approves a code of practice under this section he may, after such consultation as is mentioned in subsection (1) above, at any time by order—

(a) approve any modification of the code; or

(b) withdraw his approval;

and references in subsection (2) above to a code of practice approved under this section shall be construed accordingly.

(4) The power to make an order under this section shall be exercisable by statutory instrument subject to annulment in pursuance of a resolution of either House of Parliament.

26 Power to make regulations

(1) The Secretary of State may, after consulting the Director General of Fair Trading and such other persons as the Secretary of State considers it appropriate to consult, by regulations make provision—

(a) for the purpose of regulating the circumstances and manner in which any person—

(i) gives any indication as to the price at which any goods, services, accommodation or facilities will be or are available or have been supplied or provided; or

  (ii) indicates any other matter in respect of which any such indication may be misleading;

 (b) for the purpose of facilitating the enforcement of the provisions of section 20 above or of any regulations made under this section.

(2) The Secretary of State shall not make regulations by virtue of subsection (1)(a) above except in relation to—

 (a) indications given by persons in the course of business; and

 (b) such indications given otherwise than in the course of business as—

  (i) are given by or on behalf of persons by whom accommodation is provided to others by means of leases or licences; and

  (ii) relate to goods, services or facilities supplied or provided to those others in connection with the provision of the accommodation.

(3) Without prejudice to the generality of subsection (1) above, regulations under this section may—

 (a) prohibit an indication as to a price from referring to such matters as may be prescribed by the regulations;

 (b) require an indication as to a price or other matter to be accompanied or supplemented by such explanation or such additional information as may be prescribed by the regulations;

 (c) require information or explanations with respect to a price or other matter to be given to an officer of an enforcement authority and to authorise such an officer to require such information or explanations to be given;

 (d) require any information or explanation provided for the purposes of any regulations made by virtue of paragraph (b) or (c) above to be accurate;

 (e) prohibit the inclusion in indications as to a price or other matter of statements that the indications are not to be relied upon;

 (f) provide that expressions used in any indication as to a price or other matter shall be construed in a particular way for the purposes of this Part;

 (g) provide that a contravention of any provision of the regulations shall constitute a criminal offence punishable—

  (i) on conviction on indictment, by a fine;

  (ii) on summary conviction, by a fine not exceeding the statutory maximum;

 (h) apply any provision of this Act which relates to a criminal offence to an offence created by virtue of paragraph (g) above.

(4) The power to make regulations under this section shall be exercisable by statutory instrument subject to annulment in pursuance of a resolution of either House of Parliament and shall include power—

(a) to make different provision for different cases; and

(b) to make such supplemental, consequential and transitional provision as the Secretary of State considers appropriate.

(5)   In this section "lease" includes a sub-lease and an agreement for a lease and a statutory tenancy (within the meaning of the Landlord and Tenant Act 1985 or the Rent (Scotland) Act 1984).

PART IV
ENFORCEMENT OF PARTS II AND III

27 Enforcement

(1)   Subject to the following provisions of this section—

(a) it shall be the duty of every weights and measures authority in Great Britain to enforce within their area the safety provisions and the provisions made by or under Part III of this Act; and

(b) it shall be the duty of every district council in Northern Ireland to enforce within their area the safety provisions.

(2)   The Secretary of State may by regulations—

(a) wholly or partly transfer any duty imposed by subsection (1) above on a weights and measures authority or a district council in Northern Ireland to such other person who has agreed to the transfer as is specified in the regulations;

(b) relieve such an authority or council of any such duty so far as it is exercisable in relation to such goods as may be described in the regulations.

(3)   The power to make regulations under subsection (2) above shall be exercisable by statutory instrument subject to annulment in pursuance of a resolution of either House of Parliament and shall include power—

(a) to make different provision for different cases; and

(b) to make such supplemental, consequential and transitional provision as the Secretary of State considers appropriate.

(4)   Nothing in this section shall authorise any weights and measures authority, or any person on whom functions are conferred by regulations under subsection (2) above, to bring proceedings in Scotland for an offence.

28 Test purchases

(1)   An enforcement authority shall have power, for the purpose of ascertaining whether any safety provision or any provision made by or under Part III of this Act has been contravened in relation to any goods, services, accommodation or facilities—
  (a) to make, or to authorise an officer of the authority to make, any purchase of any goods; or
  (b) to secure, or to authorise an officer of the authority to secure, the provision of any services, accommodation or facilities.

(2)   Where—
  (a) any goods purchased under this section by or on behalf of an enforcement authority are submitted to a test; and
  (b) the test leads to—
    (i) the bringing of proceedings for an offence in respect of a contravention in relation to the goods of any safety provision or of any provision made by or under Part III of this Act or for the forfeiture of the goods under section 16 or 17 above; or
    (ii) the serving of a suspension notice in respect of any goods; and
  (c) the authority is requested to do so and it is practicable for the authority to comply with the request,

the authority shall allow the person from whom the goods were purchased or any person who is a party to the proceedings or has an interest in any goods to which the notice relates to have the goods tested.

(3)   The Secretary of State may by regulations provide that any test of goods purchased under this section by or on behalf of an enforcement authority shall—
  (a) be carried out at the expense of the authority in a manner and by a person prescribed by or determined under the regulations; or
  (b) be carried out either as mentioned in paragraph (a) above or by the authority in a manner prescribed by the regulations.

(4)   The power to make regulations under subsection (3) above shall be exercisable by statutory instrument subject to annulment in pursuance of a resolution of either House of Parliament and shall include power—
  (a) to make different provision for different cases; and
  (b) to make such supplemental, consequential and transitional provision as the Secretary of State considers appropriate.

(5)   Nothing in this section shall authorise the acquisition by or on behalf of an enforcement authority of any interest in land.

Modified in relation to the regulation of noise emission from household appliances, by the Household Appliances (Noise Emission) Regulations 1990, SI 1990/161, reg 8.

Modified in relation to vessels as consumer goods, by the Simple Pressure Vessels (Safety) Regulations 1991, SI 1991/2749, Sch 5.

29   Powers of search etc

(1)   Subject to the following provisions of this Part, a duly authorised officer of an enforcement authority may at any reasonable hour and on production, if required, of his credentials exercise any of the powers conferred by the following provisions of this section.

(2)   The officer may, for the purposes of ascertaining whether there has been any contravention of any safety provision or of any provision made by or under Part III of this Act, inspect any goods and enter any premises other than premises occupied only as a person's residence.

(3)   The officer may, for the purpose of ascertaining whether there has been any contravention of any safety provision, examine any procedure (including any arrangements for carrying out a test) connected with the production of any goods.

(4)   If the officer has reasonable grounds for suspecting that any goods are manufactured or imported goods which have not been supplied in the United Kingdom since they were manufactured or imported he may—
  (a) for the purpose of ascertaining whether there has been any contravention of any safety provision in relation to the goods, require any person carrying on a business, or employed in connection with a business, to produce any records relating to the business;
  (b) for the purpose of ascertaining (by testing or otherwise) whether there has been any such contravention, seize and detain the goods;
  (c) take copies of, or of any entry in, any records produced by virtue of paragraph (a) above.

(5)   If the officer has reasonable grounds for suspecting that there has been a contravention in relation to any goods of any safety provision or of any provision made by or under Part III of this Act, he may—
  (a) for the purpose of ascertaining whether there has been any such contravention, require any person carrying on a business, or employed in connection with a business, to produce any records relating to the business;

    (b) for the purpose of ascertaining (by testing or otherwise) whether there has been any such contravention, seize and detain the goods;

    (c) take copies of, or of any entry in, any records produced by virtue of paragraph (a) above.

(6)   The officer may seize and detain—

    (a) any goods or records which he has reasonable grounds for believing may be required as evidence in proceedings for an offence in respect of a contravention of any safety provision or of any provision made by or under Part III of this Act;

    (b) any goods which he has reasonable grounds for suspecting may be liable to be forfeited under section 16 or 17 above.

(7)   If and to the extent that it is reasonably necessary to do so to prevent a contravention of any safety provision or of any provision made by or under Part III of this Act, the officer may, for the purpose of exercising his power under subsection (4), (5) or (6) above to seize any goods or records—

    (a) require any person having authority to do so to open any container or to open any vending machine; and

    (b) himself open or break open any such container or machine where a requirement made under paragraph (a) above in relation to the container or machine has not been complied with.

NOTES

Modified in relation to the regulation of noise emission from household appliances, by the Household Appliances (Noise Emission) Regulations 1990, SI 1990/161, reg 8.

Modified in relation to vessels as consumer goods, by the Simple Pressure Vessels (Safety) Regulations 1991, SI 1991/2749, Sch 5.

30 Provisions supplemental to s 29

(1)   An officer seizing any goods or records under section 29 above shall inform the following persons that the goods or records have been so seized, that is to say—

    (a) the person from whom they are seized; and

    (b) in the case of imported goods seized on any premises under the control of the Commissioners of Customs and Excise, the importer of those goods (within the meaning of the Customs and Excise Management Act 1979).

(2)   If a justice of the peace—

(a) is satisfied by any written information on oath that there are reasonable grounds for believing either—

    (i) that any goods or records which any officer has power to inspect under section 29 above are on any premises and that their inspection is likely to disclose evidence that there has been a contravention of any safety provision or of any provision made by or under Part III of this Act; or

    (ii) that such a contravention has taken place, is taking place or is about to take place on any premises; and

(b) is also satisfied by any such information either—

    (i) that admission to the premises has been or is likely to be refused and that notice of intention to apply for a warrant under this subsection has been given to the occupier; or

    (ii) that an application for admission, or the giving of such a notice, would defeat the object of the entry or that the premises are unoccupied or that the occupier is temporarily absent and it might defeat the object of the entry to await his return,

the justice may by warrant under this hand, which shall continue in force for a period of one month, authorise any officer of an enforcement authority to enter the premises, if need be by force.

(3) An officer entering any premises by virtue of section 29 above or a warrant under subsection (2) above may take with him such other persons and such equipment as may appear to him necessary.

(4) On leaving any premises which a person is authorised to enter by a warrant under subsection (2) above, that person shall, if the premises are unoccupied or the occupier is temporarily absent, leave the premises as effectively secured against trespassers as he found them.

(5) If any person who is not an officer of an enforcement authority purports to act as such under section 29 above or this section he shall be guilty of an offence and liable on summary conviction to a fine not exceeding level 5 on the standard scale.

(6) Where any goods seized by an officer under section 29 above are submitted to a test, the officer shall inform the persons mentioned in subsection (1) above of the result of the test and, if—

(a) proceedings are brought for an offence in respect of a contravention in relation to the goods of any safety provision or of any provision made by or under Part III of this Act or for the forfeiture of the goods under section 16 or 17 above, or a suspension notice is served in respect of any goods; and

(b) the officer is requested to do so and it is practicable to comply with the request,

the officer shall allow any person who is a party to the proceedings or, as the case may be, has an interest in the goods to which the notice relates to have the goods tested.

(7)   The Secretary of State may by regulations provide that any test of goods seized under section 29 above by an officer of an enforcement authority shall—
(a) be carried out at the expense of the authority in a manner and by a person prescribed by or determined under the regulations; or
(b) be carried out either as mentioned in paragraph (a) above or by the authority in a manner prescribed by the regulations.

(8)   The power to make regulations under subsection (7) above shall be exercisable by statutory instrument subject to annulment in pursuance of a resolution of either House of Parliament and shall include power—
(a) to make different provision for different cases; and
(b) to make such supplemental, consequential and transitional provision as the Secretary of State considers appropriate.

(9)   . . .

(10)   In the application of this section to Northern Ireland, the references in subsection (2) above to any information on oath shall be construed as references to any complaint on oath.

NOTES

Sub-s (9): applies to Scotland only.
Modified in relation to vessels as consumer goods, by the Simple Pressure Vessels (Safety) Regulations 1991, SI 1991/2749, Sch 5.

31   Power of customs officer to detain goods

(1)   A customs officer may, for the purpose of facilitating the exercise by an enforcement authority or officer of such an authority of any functions conferred on the authority or officer by or under Part II of this Act, or by or under this Part in its application for the purposes of the safety provisions, seize any imported goods and detain them for not more than two working days.

(2)   Anything seized and detained under this section shall be dealt with during the period of its detention in such manner as the Commissioners of Customs and Excise may direct.

(3)   In subsection (1) above the reference to two working days is a reference to a period of forty-eight hours calculated from the time when the goods in question are seized but disregarding so much of any period as falls on a Saturday or Sunday or on Christmas Day, Good Friday or a day which is a bank holiday under the Banking and Financial Dealings Act 1971 in the part of the United Kingdom where the goods are seized.

(4)   In this section and section 32 below "customs officer" means any officer within the meaning of the Customs and Excise Management Act 1979.

NOTES

Modified in relation to vessels as consumer goods, by the Simple Pressure Vessels (Safety) Regulations 1991, SI 1991/2749, Sch 5.

32   Obstruction of authorised officer

(1)   Any person who—
   (a) intentionally obstructs any officer of an enforcement authority who is acting in pursuance of any provision of this Part or any customs officer who is so acting; or
   (b) intentionally fails to comply with any requirement made of him by any officer of an enforcement authority under any provision of this Part; or
   (c) without reasonable cause fails to give any officer of an enforcement authority who is so acting any other assistance or information which the officer may reasonably require of him for the purposes of the exercise of the officer's functions under any provision of this Part,

shall be guilty of an offence and liable on summary conviction to a fine not exceeding level 5 on the standard scale.

(2)   A person shall be guilty of an offence if, in giving any information which is required of him by virtue of subsection (1)(c) above—
   (a) he makes any statement which he knows is false in a material particular; or
   (b) he recklessly makes a statement which is false in a material particular.

(3)   A person guilty of an offence under subsection (2) above shall be liable—
  (a) on conviction on indictment, to a fine;
  (b) on summary conviction, to a fine not exceeding the statutory maximum.

NOTES

Modified in relation to the regulation of noise emission from household appliances, by the Household Appliances (Noise Emission) Regulations 1990, SI 1990/161, reg 8.

Modified in relation to vessels as consumer goods, by the Simple Pressure Vessels (Safety) Regulations 1991, SI 1991/2749, Sch 5.

33   Appeals against detention of goods

(1)   Any person having an interest in any goods which are for the time being detained under any provision of this Part by an enforcement authority or by an officer of such an authority may apply for an order requiring the goods to be released to him or to another person.

(2)   An application under this section may be made—
  (a) to any magistrates' court in which proceedings have been brought in England and Wales or Northern Ireland—
    (i)   for an offence in respect of a contravention in relation to the goods of any safety provision or of any provision made by or under Part III of this Act; or
    (ii) for the forfeiture of the goods under section 16 above;
  (b) where no such proceedings have been so brought, by way of complaint to a magistrates' court; or
  (c) . . .

(3)   On an application under this section to a magistrates' court or to the sheriff, an order requiring goods to be released shall be made only if the court or sheriff is satisfied—
  (a) that proceedings—
    (i)   for an offence in respect of a contravention in relation to the goods of any safety provision or of any provision made by or under Part III of this Act; or
    (ii) for the forfeiture of the goods under section 16 or 17 above,
    have not been brought or, having been brought, have been concluded without the goods being forfeited; and
  (b) where no such proceedings have been brought, that more than six months have elapsed since the goods were seized.

(4)   Any person aggrieved by an order made under this section by a magistrates' court in England and Wales or Northern Ireland, or by a decision of such a court not to make such an order, may appeal against that order or decision—

(a) in England and Wales, to the Crown Court;

(b) in Northern Ireland, to the county court;

and an order so made may contain such provision as appears to the court to be appropriate for delaying the coming into force of the order pending the making and determination of any appeal (including any application under section 111 of the Magistrates' Courts Act 1980 or Article 146 of the Magistrates' Courts (Northern Ireland) Order 1981 (statement of case)).

NOTES

Sub-s (2): para (c) applies to Scotland only.

Modified in relation to vessels as consumer goods, by the Simple Pressure Vessels (Safety) Regulations 1991, SI 1991/2749, Sch 5.

34   Compensation for seizure and detention

(1)   Where an officer of an enforcement authority exercises any power under section 29 above to seize and detain goods, the enforcement authority shall be liable to pay compensation to any person having an interest in the goods in respect of any loss or damage caused by reason of the exercise of the power if—

(a) there has been no contravention in relation to the goods of any safety provision or of any provision made by or under Part III of this Act; and

(b) the exercise of the power is not attributable to any neglect or default by that person.

(2)   Any disputed question as to the right to or the amount of any compensation payable under this section shall be determined by arbitration or, in Scotland, by a single arbiter appointed, failing agreement between the parties, by the sheriff.

NOTES

Modified in relation to vessels as consumer goods, by the Simple Pressure Vessels (Safety) Regulations 1991, SI 1991/2749, Sch 5.

35 Recovery of expenses of enforcement

(1)   This section shall apply where a court—
  (a) convicts a person of an offence in respect of a contravention in relation to any goods of any safety provision or of any provision made by or under Part III of this Act; or
  (b) makes an order under section 16 or 17 above for the forfeiture of any goods.

(2)   The court may (in addition to any other order it may make as to costs or expenses) order the person convicted or, as the case may be, any person having an interest in the goods to reimburse an enforcement authority for any expenditure which has been or may be incurred by that authority—
  (a) in connection with any seizure or detention of the goods by or on behalf of the authority; or
  (b) in connection with any compliance by the authority with directions given by the court for the purposes of any order for the forfeiture of the goods.

NOTES

Modified in relation to vessels as consumer goods, by the Simple Pressure Vessels (Safety) Regulations 1991, SI 1991/2749, Sch 5.

PART V
MISCELLANEOUS AND SUPPLEMENTAL

37 Power of Commissioners of Customs and Excise to disclose information

(1)   If they think it appropriate to do so for the purpose of facilitating the exercise by any person to whom subsection (2) below applies of any functions conferred on that person by or under Part II of this Act, or by or under Part IV of this Act in its application for the purposes of the safety provisions, the Commissioners of Customs and Excise may authorise the disclosure to that person of any information obtained for the purposes of the exercise by the Commissioners of their functions in relation to imported goods.

(2)   This subsection applies to an enforcement authority and to any officer of an enforcement authority.

(3)   A disclosure of information made to any person under subsection (1) above shall be made in such manner as may be directed by the Commissioners of Customs and Excise and may be made through such persons acting on behalf of that person as may be so directed.

(4)   Information may be disclosed to a person under subsection (1) above whether or not the disclosure of the information has been requested by or on behalf of that person.

NOTES

Modified in relation to vessels as consumer goods, by the Simple Pressure Vessels (Safety) Regulations 1991, SI 1991/2749, Sch 5.

38   Restrictions on disclosure of information

(1)   Subject to the following provisions of this section, a person shall be guilty of an offence if he discloses any information—
  (a) which was obtained by him in consequence of its being given to any person in compliance with any requirement imposed by safety regulations or regulations under section 26 above;
  (b) which consists in a secret manufacturing process or a trade secret and was obtained by him in consequence of the inclusion of the information—
    (i) in written or oral representations made for the purposes of Part I or II of Schedule 2 to this Act; or
    (ii) in a statement of a witness in connection with any such oral representations;
  (c) which was obtained by him in consequence of the exercise by the Secretary of State of the power conferred by section 18 above;
  (d) which was obtained by him in consequence of the exercise by any person of any power conferred by Part IV of this Act; or
  (e) which was disclosed to or through him under section 37 above.

(2)   Subsection (1) above shall not apply to a disclosure of information if the information is publicised information or the disclosure is made—
  (a) for the purpose of facilitating the exercise of a relevant person's functions under this Act or any enactment or subordinate legislation mentioned in subsection (3) below;
  (b) for the purposes of compliance with a Community obligation; or
  (c) in connection with the investigation of any criminal offence or for the purposes of any civil or criminal proceedings.

(3)   The enactments and subordinate legislation referred to in subsection (2)(a) above are—
  (a) the Trade Descriptions Act 1968;
  (b) Parts II and III and section 125 of the Fair Trading Act 1973;
  (c) the relevant statutory provisions within the meaning of Part I of the Health and Safety at Work etc Act 1974 or within the meaning of the Health and Safety at Work (Northern Ireland) Order 1978;
  (d) the Consumer Credit Act 1974;
  (e) the Restrictive Trade Practices Act 1976;
  (f) the Resale Prices Act 1976;
  (g) the Estate Agents Act 1979;
  (h) the Competition Act 1980;
  (i) the Telecommunications Act 1984;
  (j) the Airports Act 1986;
  (k) the Gas Act 1986;
  (l) any subordinate legislation made (whether before or after the passing of this Act) for the purpose of securing compliance with the Directive of the Council of the European Communities, dated 10th September 1984 (No 84/450/EEC) on the approximation of the laws, regulations and administrative provisions of the member States concerning misleading advertising;
 [(m) the Electricity Act 1989.]
[(mm) Part IV of the Airports (Northern Ireland) Order 1994;]
  [(n) the Electricity (Northern Ireland) Order 1992]
 [(nn) the Gas (Northern Ireland) Order 1996;]
  [(o) the Railways Act 1993]
  [(p) the Competition Act 1998.]

(4)   In subsection (2)(a) above the reference to a person's functions shall include a reference to any function of making, amending or revoking any regulations or order.

(5)   A person guilty of an offence under this section shall be liable—
  (a) on summary conviction, to a fine not exceeding the statutory maximum;
  (b) on conviction on indictment, to imprisonment for a term not exceeding two years or to a fine or to both.

(6)   In this section—
  "publicised information" means any information which has been disclosed in any civil or criminal proceedings or is or has been required to be contained in a warning published in pursuance of a notice to warn; and

"relevant person" means any of the following, that is to say—
  (a) a Minister of the Crown, Government department or Northern Ireland department;
  (b) the [Competition Commission], the Director General of Fair Trading, the Director General of Telecommunications or the Director General of Gas Supply [or the Director General of Electricity Supply] [or the Director General of Electricity Supply for Northern Ireland] [or the Director General of Gas for Northern Ireland] [or the Rail Regulator];
  (c) the Civil Aviation Authority;
  (d) any weights and measures authority, any district council in Northern Ireland or any person on whom functions are conferred by regulations under section 27(2) above;
  (e) any person who is an enforcing authority for the purposes of Part I of the Health and Safety at Work etc Act 1974 or for the purposes of Part II of the Health and Safety at Work (Northern Ireland) Order 1978.

NOTES

Sub-s (3): paras (e), (f) repealed as from a day to be appointed, and para (p) added by the Competition Act 1998, s 74(1), (3), Sch 12, para 10, Sch 14, Pt I; para (m) inserted by the Electricity Act 1989, s 112(1), Sch 16, para 36; para (mm) inserted by the Airports (Northern Ireland) Order 1994, SI 1994/426, art 71(2), Sch 9, para 11; para (n) inserted by the Electricity (Northern Ireland) Order 1992, SI 1992/231, art 95(1), Sch 12, para 31(a); para (nn) inserted by the Gas (Northern Ireland) Order 1996, SI 1996/275, art 71(1), Sch 6; para (o) inserted by the Railways Act 1993, s 152, Sch 12, para 26(2).

Sub-s (6): in definition "relevant person" in para (b), words in first pair of square brackets substituted by the Competition Act 1998 (Competition Commission) Transitional, Consequential and Supplemental Provisions Order 1999, SI 1999/506, art 22, words in second pair of square brackets added by the Electricity Act 1989, s 112(1), Sch 16, para 36, words in third pair of square brackets added by SI 1992/231, art 95(1), Sch 12, para 31(b), words in fourth pair of square brackets inserted by the Gas (Northern Ireland) Order 1996, SI 1996/275, art 71(1), Sch 6, words in final pair of square brackets added by the Railways Act 1993, s 152, Sch 12, para 26(3).

Modified in relation to the regulation of noise emission from household appliances, by the Household Appliances (Noise Emission) Regulations 1990, SI 1990/161, reg 8.

Modified in relation to vessels as consumer goods, by the Simple Pressure Vessels (Safety) Regulations 1991, SI 1991/2749, Sch 5.

39 Defence of due diligence

(1)   Subject to the following provisions of this section, in proceedings against any person for an offence to which this section applies it shall be a defence for that person to show that he took all reasonable steps and exercised all due diligence to avoid committing the offence.

(2)   Where in any proceedings against any person for such an offence the defence provided by subsection (1) above involves an allegation that the commission of the offence was due—
  (a) to the act or default of another; or
  (b) to reliance on information given by another,

that person shall not, without the leave of the court, be entitled to rely on the defence unless, not less than seven clear days before the hearing of the proceedings, he has served a notice under subsection (3) below on the person bringing the proceedings.

(3)   A notice under this subsection shall give such information identifying or assisting in the identification of the person who committed the act or default or gave the information as is in the possession of the person serving the notice at the time he serves it.

(4)   It is hereby declared that a person shall not be entitled to rely on the defence provided by subsection (1) above by reason of his reliance on information supplied by another, unless he shows that it was reasonable in all the circumstances for him to have relied on the information, having regard in particular—
  (a) to the steps which he took, and those which might reasonably have been taken, for the purpose of verifying the information; and
  (b) to whether he had any reason to disbelieve the information.

(5)   This section shall apply to an offence under section 10, 12(1), (2) or (3), 13(4), 14(6) or 20(1) above.

40 Liability of persons other than principal offender

(1)   Where the commission by any person of an offence to which section 39 above applies is due to an act or default committed by some other person in the course of any business of his, the other person shall be guilty of the offence and may be proceeded against and punished by virtue of this subsection whether or not proceedings are taken against the first-mentioned person.

(2)   Where a body corporate is guilty of an offence under this Act (including where it is so guilty by virtue of subsection (1) above) in respect of any act or default which is shown to have been committed with the consent or connivance of, or to be attributable to any neglect on the part of, any director, manager, secretary or other similar officer of the body corporate or any person who was purporting to act in any such capacity he, as well as the body corporate, shall be guilty of that offence and shall be liable to be proceeded against and punished accordingly.

(3)   Where the affairs of a body corporate are managed by its members, subsection (2) above shall apply in relation to the acts and defaults of a member in connection with his functions of management as if he were a director of the body corporate.

41 Civil proceedings

(1)   An obligation imposed by safety regulations shall be a duty owed to any person who may be affected by a contravention of the obligation and, subject to any provision to the contrary in the regulations and to the defences and other incidents applying to actions for breach of statutory duty, a contravention of any such obligation shall be actionable accordingly.

(2)   This Act shall not be construed as conferring any other right of action in civil proceedings, apart from the right conferred by virtue of Part I of this Act, in respect of any loss or damage suffered in consequence of a contravention of a safety provision or of a provision made by or under Part III of this Act.

(3)   Subject to any provision to the contrary in the agreement itself, an agreement shall not be void or unenforceable by reason only of a contravention of a safety provision or of a provision made by or under Part III of this Act.

(4)   Liability by virtue of subsection (1) above shall not be limited or excluded by any contract term, by any notice or (subject to the power contained in subsection (1) above to limit or exclude it in safety regulations) by any other provision.

(5)   Nothing in subsection (1) above shall prejudice the operation of section 12 of the Nuclear Installations Act 1965 (rights to compensation for certain breaches of duties confined to rights under that Act).

(6)   In this section "damage" includes personal injury and death.

42   Reports etc

(1)   It shall be the duty of the Secretary of State at least once in every five years to lay before each House of Parliament a report on the exercise during the period to which the report relates of the functions which under Part II of this Act, or under Part IV of this Act in its application for the purposes of the safety provisions, are exercisable by the Secretary of State, weights and measures authorities, district councils in Northern Ireland and persons on whom functions are conferred by regulations made under section 27(2) above.

(2)   The Secretary of State may from time to time prepare and lay before each House of Parliament such other reports on the exercise of those functions as he considers appropriate.

(3)   Every weights and measures authority, every district council in Northern Ireland and every person on whom functions are conferred by regulations under subsection (2) of section 27 above shall, whenever the Secretary of State so directs, make a report to the Secretary of State on the exercise of the functions exercisable by that authority or council under that section or by that person by virtue of any such regulations.

(4)   A report under subsection (3) above shall be in such form and shall contain such particulars as are specified in the direction of the Secretary of State.

(5)   The first report under subsection (1) above shall be laid before each House of Parliament not more than five years after the laying of the last report under section 8(2) of the Consumer Safety Act 1978.

43   Financial provisions

(1)   There shall be paid out of money provided by Parliament—
  (a) any expenses incurred or compensation payable by a Minister of the Crown or Government department in consequence of any provision of this Act; and
  (b) any increase attributable to this Act in the sums payable out of money so provided under any other Act.

(2)   Any sums received by a Minister of the Crown or Government department by virtue of this Act shall be paid into the Consolidated Fund.

44   Service of documents etc

(1)   Any document required or authorised by virtue of this Act to be served on a person may be so served—
  (a) by delivering it to him or by leaving it at his proper address or by sending it by post to him at that address; or
  (b) if the person is a body corporate, by serving it in accordance with paragraph (a) above on the secretary or clerk of that body; or
  (c) if the person is a partnership, by serving it in accordance with that paragraph on a partner or on a person having control or management of the partnership business.

(2)   For the purposes of subsection (1) above, and for the purposes of section 7 of the Interpretation Act 1978 (which relates to the service of documents by post) in its application to that subsection, the proper address of any person on whom a document is to be served by virtue of this Act shall be his last known address except that—
  (a) in the case of service on a body corporate or its secretary or clerk, it shall be the address of the registered or principal office of the body corporate;
  (b) in the case of service on a partnership or a partner or a person having the control or management of a partnership business, it shall be the principal office of the partnership;

and for the purposes of this subsection the principal officer of a company registered outside the United Kingdom or of a partnership carrying on business outside the United Kingdom is its principal office within the United Kingdom.

(3)   The Secretary of State may by regulations make provision for the manner in which any information is to be given to any person under any provision of Part IV of this Act.

(4)   Without prejudice to the generality of subsection (3) above regulations made by the Secretary of State may prescribe the person, or manner of determining the person, who is to be treated for the purposes of section 28(2) or 30 above as the person from whom any goods were purchased or seized where the goods were purchased or seized from a vending machine.

(5)   The power to make regulations under subsection (3) or (4) above shall be exercisable by statutory instrument subject to annulment in pursuance of a resolution of either House of Parliament and shall include power—

(a) to make different provision for different cases; and

(b) to make such supplemental, consequential and transitional provision as the Secretary of State considers appropriate.

NOTES

Modified in relation to the regulation of noise emission from household appliances, by the Household Appliances (Noise Emission) Regulations 1990, SI 1990/161, reg 8.

Modified in relation to vessels as consumer goods, by the Simple Pressure Vessels (Safety) Regulations 1991, SI 1991/2749, Sch 5.

45 Interpretation

(1)   In this Act, except in so far as the context otherwise requires—

"aircraft" includes gliders, balloons and hovercraft;

"business" includes a trade or profession and the activities of a professional or trade association or of a local authority or other public authority;

"conditional sale agreement", "credit-sale agreement" and "hire-purchase agreement" have the same meanings as in the Consumer Credit Act 1974 but as if in the definitions in that Act "goods" had the same meaning as in this Act;

"contravention" includes a failure to comply and cognate expressions shall be construed accordingly;

"enforcement authority" means the Secretary of State, any other Minister of the Crown in charge of a Government department, any such department and any authority, council or other person on whom functions under this Act are conferred by or under section 27 above;

"gas" has the same meaning as in Part I of the Gas Act 1986;

"goods" includes substances, growing crops and things comprised in land by virtue of being attached to it and any ship, aircraft or vehicle;

"information" includes accounts, estimates and returns;

"magistrates' court", in relation to Northern Ireland, means a court of summary jurisdiction;

. . . . .

"modifications" includes additions, alterations and omissions, and cognate expressions shall be construed accordingly;

"motor vehicle" has the same meaning as in [the Road Traffic Act 1988];

"notice" means a notice in writing;

"notice to warn" means a notice under section 13(1)(b) above;

"officer", in relation to an enforcement authority, means a person authorised in writing to assist the authority in carrying out its functions under or for the purposes of the enforcement of any of the safety provisions or of any of the provisions made by or under Part III of this Act;

"personal injury" includes any disease and any other impairment of a person's physical or mental condition;

"premises" includes any place and any ship, aircraft or vehicle;

"prohibition notice" means a notice under section 13(1)(a) above;

"records" includes any books or documents and any records in non-documentary form;

"safety provision" means the general safety requirement in section 10 above or any provision of safety regulations, a prohibition notice or a suspension notice;

"safety regulations" means regulations under section 11 above;

"ship" includes any boat and any other description of vessel used in navigation;

"subordinate legislation" has the same meaning as in the Interpretation Act 1978;

"substance" means any natural or artificial substance, whether in solid, liquid or gaseous form or in the form of a vapour, and includes substances that are comprised in or mixed with other goods;

"supply" and cognate expressions shall be construed in accordance with section 46 below;

"suspension notice" means a notice under section 14 above.

(2)   Except in so far as the context otherwise requires, references in this Act to a contravention of a safety provision shall, in relation to any goods, include references to anything which would constitute such a contravention if the goods were supplied to any person.

(3)   References in this Act to any goods in relation to which any safety provision has been or may have been contravened shall include references to any goods which it is not reasonably practicable to separate from any such goods.

(4), (5)   . . .

NOTES

Sub-s (1): definitions "mark" and "trade mark" repealed by the Trade Marks Act 1994, s 106(2), Sch 5; words in square brackets in definition "motor vehicle" substituted by the Road Traffic (Consequential Provisions) Act 1988, s 4, Sch 3, para 35.

Sub-s (4): repealed by the Trade Marks Act 1994, s 106(2), Sch 5.

Sub-s (5): applies to Scotland only.

46   Meaning of "supply"

(1)   Subject to the following provisions of this section, references in this Act to supplying goods shall be construed as references to doing any of the following, whether as principal or agent, that is to say—
- (a)   selling, hiring out or lending the goods;
- (b)   entering into a hire-purchase agreement to furnish the goods;
- (c)   the performance of any contract for work and materials to furnish the goods;
- (d)   providing the goods in exchange for any consideration (including trading stamps) other than money;
- (e)   providing the goods in or in connection with the performance of any statutory function; or
- (f)   giving the goods as a prize or otherwise making a gift of the goods;

and, in relation to gas or water, those references shall be construed as including references to providing the service by which the gas or water is made available for use.

(2)   For the purposes of any reference in this Act to supplying goods, where a person ("the ostensible supplier") supplies goods to another person ("the customer") under a hire-purchase agreement, conditional sale agreement or credit-sale agreement or under an agreement for the hiring of goods (other than a hire-purchase agreement) and the ostensible supplier—
- (a)   carries on the business of financing the provision of goods for others by means of such agreements; and
- (b)   in the course of that business acquired his interest in the goods supplied to the customer as a means of financing the provision of them for the customer by a further person ("the effective supplier"),

the effective supplier and not the ostensible supplier shall be treated as supplying the goods to the customer.

(3)   Subject to subsection (4) below, the performance of any contract by the erection of any building or structure on any land or by the carrying out of any other building works shall be treated for the purposes of this Act as a supply of goods in so far as, but only in so far as, it involves the provision of any goods to any person by means of their incorporation into the building, structure or works.

(4)   Except for the purposes of, and in relation to, notices to warn or any provision made by or under Part III of this Act, references in this Act to supplying goods shall not include references to supplying goods comprised in land where the supply is effected by the creation or disposal of an interest in the land.

(5)   Except in Part I of this Act references in this Act to a person's supplying goods shall be confined to references to that person's supplying goods in the course of a business of his, but for the purposes of this subsection it shall be immaterial whether the business is a business of dealing in the goods.

(6)   For the purposes of subsection (5) above goods shall not be treated as supplied in the course of a business if they are supplied, in pursuance of an obligation arising under or in connection with the insurance of the goods, to the person with whom they were insured.

(7)   Except for the purposes of, and in relation to, prohibition notices or suspension notices, references in Parts II to IV of this Act to supplying goods shall not include—
  (a) references to supplying goods where the person supplied carries on a business of buying goods of the same description as those goods and repairing or reconditioning them;
  (b) references to supplying goods by a sale of articles as scrap (that is to say, for the value of materials included in the articles rather than for the value of the articles themselves).

(8)   Where any goods have at any time been supplied by being hired out or lent to any person, neither a continuation or renewal of the hire or loan (whether on the same or different terms) nor any transaction for the transfer after that time of any interest in the goods to the person to whom they were hired or lent shall be treated for the purposes of this Act as a further supply of the goods to that person.

(9)   A ship, aircraft or motor vehicle shall not be treated for the purposes of this Act as supplied to any person by reason only that services consisting in the carriage of goods or passengers in that ship, aircraft or vehicle, or in its use for any other purpose, are provided to that person in pursuance of an agreement relating to the use of the ship, aircraft or vehicle for a particular period or for particular voyages, flights or journeys.

47   Savings for certain privileges

(1)   Nothing in this Act shall be taken as requiring any person to produce any records if he would be entitled to refuse to produce those records in any proceedings in any court on the grounds that they are the subject of legal professional privilege or, in Scotland, that they contain a confidential communication made by or to an advocate or solicitor in that capacity, or as authorising any person to take possession of any records which are in the possession of a person who would be so entitled.

(2)   Nothing in this Act shall be construed as requiring a person to answer any question or give any information if to do so would incriminate that person or that person's spouse.

Modified in relation to the regulation of noise emission from household appliances, by the Household Appliances (Noise Emission) Regulations 1990, SI 1990/161, reg 8.

Modified in relation to vessels as consumer goods, by the Simple Pressure Vessels (Safety) Regulations 1991, SI 1991/2749, Sch 5.

49   Northern Ireland

(1)   This Act shall extend to Northern Ireland with the exception of—
  (a) the provisions of Parts I and III;
  (b) any provision amending or repealing an enactment which does not so extend; and
  (c) any other provision so far as it has effect for the purposes of, or in relation to, a provision falling within paragraph (a) or (b) above.

(2)   Subject to any Order in Council made by virtue of subsection (1)(a) of section 3 of the Northern Ireland Constitution Act 1973, consumer safety shall not be a transferred matter for the purposes of that Act but shall for the purposes of subsection (2) of that section be treated as specified in Schedule 3 to that Act.

(3)   An Order in Council under paragraph 1(1)(b) of Schedule 1 to the Northern Ireland Act 1974 (exercise of legislative functions for Northern Ireland) which states that it is made only for purposes corresponding to any of the provisions of this Act mentioned in subsection (1)(a) to (c) above—
  (a) shall not be subject to paragraph 1(4) and (5) of that Schedule (affirmative resolution procedure and procedure in cases of urgency); but
  (b) shall be subject to annulment in pursuance of a resolution of either House of Parliament.

Sub-s (2): repealed by the Northern Ireland Act 1998, s 100(2), Sch 15, as from a day to be appointed.

50 Short title, commencement and transitional provision

(1)   This Act may be cited as the Consumer Protection Act 1987.

(2)   This Act shall come into force on such day as the Secretary of State may by order made by statutory instrument appoint, and different days may be so appointed for different provisions or for different purposes.

(3)   The Secretary of State shall not make an order under subsection (2) above bringing into force the repeal of the Trade Descriptions Act 1972, a repeal of any provision of that Act or a repeal of that Act or of any provision of it for any purposes, unless a draft of the order has been laid before, and approved by a resolution of, each House of Parliament.

(4)   An order under subsection (2) above bringing a provision into force may contain such transitional provision in connection with the coming into force of that provision as the Secretary of State considers appropriate.

(5)   Without prejudice to the generality of the power conferred by subsection (4) above, the Secretary of State may by order provide for any regulations made under the Consumer Protection Act 1961 or the Consumer Protection Act (Northern Ireland) 1965 to have effect as if made under section 11 above and for any such regulations to have effect with such modifications as he considers appropriate for that purpose.

(6)   The power of the Secretary of State by order to make such provision as is mentioned in subsection (5) above, shall, in so far as it is not exercised by an order under subsection (2) above, be exercisable by statutory instrument subject to annulment in pursuance of a resolution of either House of Parliament.

(7)   Nothing in this Act or in any order under subsection (2) above shall make any person liable by virtue of Part I of this Act for any damage caused wholly or partly by a defect in a product which was supplied to any person by its producer before the coming into force of Part I of this Act.

(8)   Expressions used in subsection (7) above and in Part I of this Act have the same meanings in that subsection as in that Part.

SCHEDULES

SCHEDULE 2

Section 13

PROHIBITION NOTICES AND NOTICES TO WARN

PART I
PROHIBITION NOTICES

1.   A prohibition notice in respect of any goods shall—
   (a) state that the Secretary of State considers that the goods are unsafe;
   (b) set out the reasons why the Secretary of State considers that the goods are unsafe;
   (c) specify the day on which the notice is to come into force; and
   (d) state that the trader may at any time make representations in writing to the Secretary of State for the purpose of establishing that the goods are safe.

2.—(1)   If representations in writing about a prohibition notice are made by the trader to the Secretary of State, it shall be the duty of the Secretary of State to consider whether to revoke the notice and—
   (a) if he decides to revoke it, to do so;
   (b) in any other case, to appoint a person to consider those representations, any further representations made (whether in writing or orally) by the trader about the notice and the statements of any witnesses examined under this Part of this Schedule.

(2)   Where the Secretary of State has appointed a person to consider representations about a prohibition notice, he shall serve a notification on the trader which—
   (a) states that the trader may make oral representations to the appointed person for the purpose of establishing that the goods to which the notice relates are safe; and
   (b) specifies the place and time at which the oral representations may be made.

(3)   The time specified in a notification served under sub-paragraph (2) above shall not be before the end of the period of twenty-one days beginning with the day on which the notification is served, unless the trader otherwise agrees.

(4)   A person on whom a notification has been served under sub-paragraph (2) above or his representative may, at the place and time specified in the notification—

    (a) make oral representations to the appointed person for the purpose of establishing that the goods in question are safe; and

    (b) call and examine witnesses in connection with the representations.

3.—(1)   Where representations in writing about a prohibition notice are made by the trader to the Secretary of State at any time after a person has been appointed to consider representations about that notice, then, whether or not the appointed person has made a report to the Secretary of State, the following provisions of this paragraph shall apply instead of paragraph 2 above.

(2)   The Secretary of State shall, before the end of the period of one month beginning with the day on which he receives the representations, serve a notification on the trader which states—

    (a) that the Secretary of State has decided to revoke the notice, has decided to vary it or, as the case may be, has decided neither to revoke nor to vary it; or

    (b) that, a person having been appointed to consider representations about the notice, the trader may, at a place and time specified in the notification, make oral representations to the appointed person for the purpose of establishing that the goods to which the notice relates are safe.

(3)   The time specified in a notification served for the purposes of sub-paragraph (2)(b) above shall not be before the end of the period of twenty-one days beginning with the day on which the notification is served, unless the trader otherwise agrees or the time is the time already specified for the purposes of paragraph 2(2)(b) above.

(4)   A person on whom a notification has been served for the purposes of sub-paragraph (2)(b) above or his representative may, at the place and time specified in the notification—

    (a) make oral representations to the appointed person for the purpose of establishing that the goods in question are safe; and

    (b) call and examine witnesses in connection with the representations.

4.—(1)   Where a person is appointed to consider representations about a prohibition notice, it shall be his duty to consider—

    (a) any written representations made by the trader about the notice, other than those in respect of which a notification is served under paragraph 3(2)(a) above;

(b) any oral representations made under paragraph 2(4) or 3(4) above; and

(c) any statements made by witnesses in connection with the oral representations,

and, after considering any matters under this paragraph, to make a report (including recommendations) to the Secretary of State about the matters considered by him and the notice.

(2) It shall be the duty of the Secretary of State to consider any report made to him under sub-paragraph (1) above and, after considering the report, to inform the trader of his decision with respect to the prohibition notice to which the report relates.

5.—(1) The Secretary of State may revoke or vary a prohibition notice by serving on the trader a notification stating that the notice is revoked or, as the case may be, is varied as specified in the notification.

(2) The Secretary of State shall not vary a prohibition notice so as to make the effect of the notice more restrictive for the trader.

(3) Without prejudice to the power conferred by section 13(2) of this Act, the service of a notification under sub-paragraph (1) above shall be sufficient to satisfy the requirement of paragraph 4(2) above that the trader shall be informed of the Secretary of State's decision.

PART II
NOTICES TO WARN

6.—(1) If the Secretary of State proposes to serve a notice to warn on any person in respect of any goods, the Secretary of State, before he serves the notice, shall serve on that person a notification which—
(a) contains a draft of the proposed notice;
(b) states that the Secretary of State proposes to serve a notice in the form of the draft on that person;
(c) states that the Secretary of State considers that the goods described in the draft are unsafe;
(d) sets out the reasons why the Secretary of State considers that those goods are unsafe; and
(e) states that that person may make representations to the Secretary of State for the purpose of establishing that the goods are safe if, before the end of the period of fourteen days beginning with the

day on which the notification is served, he informs the Secretary of State—

   (i) of his intention to make representations; and

   (ii) whether the representations will be made only in writing or both in writing and orally.

(2) Where the Secretary of State has served a notification containing a draft of a proposed notice to warn on any person, he shall not serve a notice to warn on that person in respect of the goods to which the proposed notice relates unless—

  (a) the period of fourteen days beginning with the day on which the notification was served expires without the Secretary of State being informed as mentioned in sub-paragraph (1)(e) above;

  (b) the period of twenty-eight days beginning with that day expires without any written representations being made by that person to the Secretary of State about the proposed notice; or

  (c) the Secretary of State has considered a report about the proposed notice by a person appointed under paragraph 7(1) below.

7.—(1) Where a person on whom a notification containing a draft of a proposed notice to warn has been served—

  (a) informs the Secretary of State as mentioned in paragraph 6(1)(e) above before the end of the period of fourteen days beginning with the day on which the notification was served; and

  (b) makes written representations to the Secretary of State about the proposed notice before the end of the period of twenty-eight days beginning with that day,

the Secretary of State shall appoint a person to consider those representations, any further representations made by that person about the draft notice and the statements of any witnesses examined under this Part of this Schedule.

(2) Where—

  (a) the Secretary of State has appointed a person to consider representations about a proposed notice to warn; and

  (b) the person whose representations are to be considered has informed the Secretary of State for the purposes of paragraph 6(1)(e) above that the representations he intends to make will include oral representations,

the Secretary of State shall inform the person intending to make the representations of the place and time at which oral representations may be made to the appointed person.

(3)   Where a person on whom a notification containing a draft of a proposed notice to warn has been served is informed of a time for the purposes of sub-paragraph (2) above, that time shall not be—
  (a) before the end of the period of twenty-eight days beginning with the day on which the notification was served; or
  (b) before the end of the period of seven days beginning with the day on which that person is informed of the time.

(4)   A person who has been informed of a place and time for the purposes of sub-paragraph (2) above or his representative may, at that place and time—
  (a) make oral representations to the appointed person for the purpose of establishing that the goods to which the proposed notice relates are safe; and
  (b) call and examine witnesses in connection with the representations.

8.—(1)   Where a person is appointed to consider representations about a proposed notice to warn, it shall be his duty to consider—
  (a) any written representations made by the person on whom it is proposed to serve the notice; and
  (b) in a case where a place and time has been appointed under paragraph 7(2) above for oral representations to be made by that person or his representative, any representations so made and any statements made by witnesses in connection with those representations,

and, after considering those matters, to make a report (including recommendations) to the Secretary of State about the matters considered by him and the proposal to serve the notice.

(2)   It shall be the duty of the Secretary of State to consider any report made to him under sub-paragraph (1) above and, after considering the report, to inform the person on whom it was proposed that a notice to warn should be served of his decision with respect to the proposal.

(3)   If at any time after serving a notification on a person under paragraph 6 above the Secretary of State decides not to serve on that person either the proposed notice to warn or that notice with modifications, the Secretary of State shall inform that person of the decision; and nothing done for the purposes of any of the preceding provisions of this Part of this Schedule before that person was so informed shall—
  (a) entitle the Secretary of State subsequently to serve the proposed notice or that notice with modifications; or
  (b) require the Secretary of State, or any person appointed to consider representations about the proposed notice, subsequently

to do anything in respect of, or in consequence of, any such representations.

(4)   Where a notification containing a draft of a proposed notice to warn is served on a person in respect of any goods, a notice to warn served on him in consequence of a decision made under sub-paragraph (2) above shall either be in the form of the draft or shall be less onerous than the draft.

9.   The Secretary of State may revoke a notice to warn by serving on the person on whom the notice was served a notification stating that the notice is revoked.

PART III
GENERAL

10.—(1)   Where in a notification served on any person under this Schedule the Secretary of State has appointed a time for the making of oral representations or the examination of witnesses, he may, by giving that person such notification as the Secretary of State considers appropriate, change that time to a later time or appoint further times at which further representations may be made or the examination of witnesses may be continued; and paragraphs 2(4), 3(4) and 7(4) above shall have effect accordingly.

(2)   For the purposes of this Schedule the Secretary of State may appoint a person (instead of the appointed person) to consider any representations or statements, if the person originally appointed, or last appointed under this sub-paragraph, to consider those representations or statements has died or appears to the Secretary of State to be otherwise unable to act.

11.   In this Schedule—
   "the appointed person" in relation to a prohibition notice or a proposal to serve a notice to warn, means the person for the time being appointed under this Schedule to consider representations about the notice or, as the case may be, about the proposed notice;
   "notification" means a notification in writing;
   "trader", in relation to a prohibition notice, means the person on whom the notice is or was served.

Food Safety Act 1990

(C 16)

An Act to make new provision in place of the Food Act 1984 (except Parts III and V), the Food and Drugs (Scotland) Act 1956 and certain other enactments relating to food; to amend Parts III and V of the said Act of 1984 and Part I of the Food and Environment Protection Act 1985; and for connected purposes

[29 June 1990]

PART I
PRELIMINARY

2 Extended meaning of "sale" etc

(1)   For the purposes of this Act—
   (a) the supply of food, otherwise than on sale, in the course of a business; and
   (b) any other thing which is done with respect to food and is specified in an order made by the Ministers,

shall be deemed to be a sale of the food, and references to purchasers and purchasing shall be construed accordingly.

(2)   This Act shall apply—
   (a) in relation to any food which is offered as a prize or reward or given away in connection with any entertainment to which the public are admitted, whether on payment of money or not, as if the food were, or had been, exposed for sale by each person concerned in the organisation of the entertainment;
   (b) in relation to any food which, for the purpose of advertisement or in furtherance of any trade or business, is offered as a prize or reward or given away, as if the food were, or had been, exposed for sale by the person offering or giving away the food; and
   (c) in relation to any food which is exposed or deposited in any premises for the purpose of being so offered or given away as mentioned in paragraph (a) or (b) above, as if the food were, or had been, exposed for sale by the occupier of the premises;

and in this subsection "entertainment" includes any social gathering, amusement, exhibition, performance, game, sport or trial of skill.

PART II
MAIN PROVISIONS

Food safety

8 Selling food not complying with food safety requirements

(1)   Any person who—
   (a) sells for human consumption, or offers, exposes or advertises for
       sale for such consumption, or has in his possession for the purpose
       of such sale or of preparation for such sale; or
   (b) deposits with, or consigns to, any other person for the purpose of
       such sale or of preparation for such sale,

any food which fails to comply with food safety requirements shall be guilty
of an offence.

(2)   For the purposes of this Part food fails to comply with food safety
requirements if—
   (a) it has been rendered injurious to health by means of any of the
       operations mentioned in section 7(1) above;
   (b) it is unfit for human consumption; or
   (c) it is so contaminated (whether by extraneous matter or otherwise)
       that it would not be reasonable to expect it to be used for human
       consumption in that state;

and references to such requirements or to food complying with such
requirements shall be construed accordingly.

(3)   Where any food which fails to comply with food safety requirements
is part of a batch, lot or consignment of food of the same class or description,
it shall be presumed for the purposes of this section and section 9 below,
until the contrary is proved, that all of the food in that batch, lot or
consignment fails to comply with those requirements.

(4)   For the purposes of this Part, any part of, or product derived wholly
or partly from, an animal—
   (a) which has been slaughtered in a knacker's yard, or of which the
       carcase has been brought into a knacker's yard; or
   (b) in Scotland, which has been slaughtered otherwise than in a
       slaughterhouse,

shall be deemed to be unfit for human consumption.

(5)   In subsection (4) above, in its application to Scotland, "animal" means any description of cattle, sheep, goat, swine, horse, ass or mule; and paragraph (b) of that subsection shall not apply where accident, illness or emergency affecting the animal in question required it to be slaughtered as mentioned in that paragraph.

Consumer protection

14   Selling food not of the nature or substance or quality demanded

(1)   Any person who sells to the purchaser's prejudice any food which is not of the nature or substance or quality demanded by the purchaser shall be guilty of an offence.

(2)   In subsection (1) above the reference to sale shall be construed as a reference to sale for human consumption; and in proceedings under that subsection it shall not be a defence that the purchaser was not prejudiced because he bought for analysis or examination.

15   Falsely describing or presenting food

(1)   Any person who gives with any food sold by him, or displays with any food offered or exposed by him for sale or in his possession for the purpose of sale, a label, whether or not attached to or printed on the wrapper or container, which—
   (a) falsely describes the food; or
   (b) is likely to mislead as to the nature or substance or quality of the food,

shall be guilty of an offence.

(2)   Any person who publishes, or is a party to the publication of, an advertisement (not being such a label given or displayed by him as mentioned in subsection (1) above) which—
   (a) falsely describes any food; or
   (b) is likely to mislead as to the nature or substance or quality of any food,

shall be guilty of an offence.

(3)   Any person who sells, or offers or exposes for sale, or has in his possession for the purpose of sale, any food the presentation of which is likely to mislead as to the nature or substance or quality of the food shall be guilty of an offence.

(4)   In proceedings for an offence under subsection (1) or (2) above, the fact that a label or advertisement in respect of which the offence is alleged to have been committed contained an accurate statement of the composition of the food shall not preclude the court from finding that the offence was committed.

(5)   In this section references to sale shall be construed as references to sale for human consumption.

PART IV
MISCELLANEOUS AND SUPPLEMENTAL

Supplemental

60   Short title, commencement and extent

(1)   This Act may be cited as the Food Safety Act 1990.

(2)   The following provisions shall come into force on the day on which this Act is passed, namely—
    section 13;

    . . .

(3)   Subject to subsection (2) above, this Act shall come into force on such day as the Ministers may by order appoint, and different days may be appointed for different provisions or for different purposes.

(4)   An order under subsection (3) above may make such transitional adaptations of any of the following, namely—
    (a)   the provisions of this Act then in force or brought into force by the order; and
    (b)   the provisions repealed by this Act whose repeal is not then in force or so brought into force,

as appear to the Ministers to be necessary or expedient in consequence of the partial operation of this Act.

(5)   This Act, except—
this section;
. . .

does not extend to Northern Ireland.

Sub-ss (2), (5): words omitted outside the scope of this work.

Property Misdescriptions Act 1991

(C 29)

*An Act to prohibit the making of false or misleading statements about
property matters in the course of estate agency business and property
development business*

[27 June 1991]

1   Offence of property misdescription

(1)   Where a false or misleading statement about a prescribed matter is
made in the course of an estate agency business or a property development
business, otherwise than in providing conveyancing services, the person
by whom the business is carried on shall be guilty of an offence under this
section.

(2)   Where the making of the statement is due to the act or default of an
employee the employee shall be guilty of an offence under this section; and
the employee may be proceeded against and punished whether or not
proceedings are also taken against his employer.

(3)   A person guilty of an offence under this section shall be liable—
  (a) on summary conviction, to a fine not exceeding the statutory
       maximum, and
  (b) on conviction on indictment, to a fine.

(4)   No contract shall be void or unenforceable, and no right of action in
civil proceedings in respect of any loss shall arise, by reason only of the
commission of an offence under this section.

(5)   For the purposes of this section—

(a) "false" means false to a material degree,

(b) a statement is misleading if (though not false) what a reasonable person may be expected to infer from it, or from any omission from it, is false,

(c) a statement may be made by pictures or any other method of signifying meaning as well as by words and, if made by words, may be made orally or in writing,

(d) a prescribed matter is any matter relating to land which is specified in an order made by the Secretary of State,

(e) a statement is made in the course of an estate agency business if (but only if) the making of the statement is a thing done as mentioned in subsection (1) of section 1 of the Estate Agents Act 1979 and that Act either applies to it or would apply to it but for subsection (2)(a) of that section (exception for things done in course of profession by practising solicitor or employee),

(f) a statement is made in the course of a property development business if (but only if) it is made—

   (i) in the course of a business (including a business in which the person making the statement is employed) concerned wholly or substantially with the development of land, and

   (ii) for the purpose of, or with a view to, disposing of an interest in land consisting of or including a building, or a part of a building, constructed or renovated in the course of the business, and

(g) "conveyancing services" means the preparation of any transfer, conveyance, writ, contract or other document in connection with the disposal or acquisition of an interest in land, and services ancillary to that, but does not include anything done as mentioned in section 1(1)(a) of the Estate Agents Act 1979.

(6)   For the purposes of this section any reference in this section or section 1 of the Estate Agents Act 1979 to disposing of or acquiring an interest in land—

(a) in England and Wales and Northern Ireland shall be construed in accordance with section 2 of that Act, and

(b) . . .

(7)   An order under this section may—

(a) make different provision for different cases, and

(b) include such supplemental, consequential and transitional provisions as the Secretary of State considers appropriate;

and the power to make such an order shall be exercisable by statutory instrument which shall be subject to annulment in pursuance of a resolution of either House of Parliament.

Sub-s (6): para (b) applies to Scotland only.

2 Due diligence defence

(1)   In proceedings against a person for an offence under section 1 above it shall be a defence for him to show that he took all reasonable steps and exercised all due diligence to avoid committing the offence.

(2)   A person shall not be entitled to rely on the defence provided by subsection (1) above by reason of his reliance on information given by another unless he shows that it was reasonable in all the circumstances for him to have relied on the information, having regard in particular—
  (a) to the steps which he took, and those which might reasonably have been taken, for the purpose of verifying the information, and
  (b) to whether he had any reason to disbelieve the information.

(3)   Where in any proceedings against a person for an offence under section 1 above the defence provided by subsection (1) above involves an allegation that the commission of the offence was due—
  (a) to the act or default of another, or
  (b) to reliance on information given by another,

the person shall not, without the leave of the court, be entitled to rely on the defence unless he has served a notice under subsection (4) below on the person bringing the proceedings not less than seven clear days before the hearing of the proceedings or, in Scotland, the diet of trial.

(4)   A notice under this subsection shall give such information identifying or assisting in the identification of the person who committed the act or default, or gave the information, as is in the possession of the person serving the notice at the time he serves it.

7 Short title and extent

(1)   This Act may be cited as the Property Misdescriptions Act 1991.

(2)   This Act extends to Northern Ireland.

Timeshare Act 1992

(C 35)

An Act to provide for rights to cancel certain agreements about timeshare accommodation

[16 March 1992]

1 Application of Act

(1)   In this Act—
  (a) "timeshare accommodation" means any living accommodation, in the United Kingdom or elsewhere, used or intended to be used, wholly or partly, for leisure purposes by a class of persons (referred to below in this section as "timeshare users") all of whom have rights to use, or participate in arrangements under which they may use, that accommodation, or accommodation within a pool of accommodation to which that accommodation belongs, for [a specified or ascertainable period of the year], and
  (b) "timeshare rights" means rights by virtue of which a person becomes or will become a timeshare user, being rights exercisable during a period of not less than three years.

(2)   For the purposes of subsection (1)(a) above—
  (a) "accommodation" means accommodation in a building or in a caravan (as defined in section 29(1) of the Caravan Sites and Control of Development Act 1960), ...
  (b) . . .

(3)   Subsection (1)(b) above does not apply to a person's rights—
  (a) . . .
  (b) under a contract of employment ([within the meaning of the Employment Rights Act 1996]) or a policy of insurance,   . . .
  (c) . . .

or to such rights as may be prescribed.

[(3A) For the purposes of sections 1A to 1E, 2(2A) and (2B), 3(3), 5A, 5B and 6A of this Act, subsection (1) above shall be construed as if in paragraph (b), after "become" there were inserted ", on payment of a global price,".]

(4)   In this Act "timeshare agreement" means . . . , an agreement under which timeshare rights are conferred or purport to be conferred on any person and in this Act, in relation to a timeshare agreement—

    (a) references to the offeree are to the person on whom timeshare rights are conferred, or purport to be conferred, and

    (b) references to the offeror are to the other party to the agreement,

and, in relation to any time before the agreement is entered into, references in this Act to the offeree or the offeror are to the persons who become the offeree and offeror when it is entered into.

[(5)   In this Act "timeshare credit agreement" means an agreement, not being a timeshare agreement, under which credit which fully or partly covers the price under a timeshare agreement is granted—

    (a) by the offeror, or

    (b) by another person, under an arrangement between that person and the offeror;

and a person who grants credit under a timeshare credit agreement is in this Act referred to as "the creditor".]

(6)   . . .

[(6A)   No timeshare agreement or timeshare credit agreement to which this Act applies may be cancelled under section 67 of the Consumer Credit Act 1974.]

(7)   This Act applies to any timeshare agreement or timeshare credit agreement if—

    (a) the agreement is to any extent governed by the law of the United Kingdom or of a part of the United Kingdom, or

    (b) when the agreement is entered into, one or both of the parties are in the United Kingdom.

[(7A)   This Act also applies to any timeshare agreement if—

    (a) the relevant accommodation is situated in the United Kingdom, or

    (b) when the agreement is entered into, the offeree is ordinarily resident in the United Kingdom and the relevant accommodation is situated in another EEA State.

(7B)   For the purposes of subsection (7A) above, "the relevant accommodation" means—

    (a) the accommodation which is the subject of the agreement, or

    (b) some or all of the accommodation in the pool of accommodation which is the subject of the agreement,

    as the case may be.]

(8)   In the application of this section to Northern Ireland—
   (a) for the reference in subsection (2)(a) above to section 29(1) of the Caravan Sites and Control of Development Act 1960 there is substituted a reference to section 25(1) of the Caravans Act (Northern Ireland) 1963, and
   (b) for the reference in subsection (3)(b) above to [the Employment Rights Act 1996] there is substituted a reference to article 2(2) of the Industrial Relations (Northern Ireland) Order 1976.

NOTES

Sub-s (1): words in square brackets substituted by the Timeshare Regulations 1997, SI 1997/1081, reg 2(2).

Sub-s (3): paras (a) and (c) and word immediately preceding para (a) repealed by SI 1997/1081, reg 2(3); words in square brackets in para (b) substituted by the Employment Rights Act 1996, s 240, Sch 1, para 53.

Sub-ss (3A), (6A), (7A), (7B): inserted by SI 1997/1081, reg 2(4), (7), (8).

Sub-s (5): substituted by SI 1997/1081, reg 2(5).

Sub-s (6): repealed by SI 1997/1081, reg 2(6).

Sub-s (8): words in square brackets in para (b) substituted by the Employment Rights Act 1996, s 240, Sch 1, para 53.

[1A   Obligations to provide information

(1)   A person who proposes in the course of a business to enter into a timeshare agreement to which this Act applies as offeror (an "operator") must provide any person who requests information on the proposed accommodation with a document complying with subsection (2) below.

(2)   The document shall provide—
   (a) a general description of the proposed accommodation,
   (b) information (which may be brief) on the matters referred to in paragraphs (a) to (g), (i) and (l) of Schedule 1 to this Act, and
   (c) information on how further information may be obtained.

(3)   Where an operator—
   (a) provides a person with a document containing information on the proposed accommodation, and
   (b) subsequently enters as offeror into a timeshare agreement to which this Act applies the subject of which is the proposed accommodation,

subsection (4) below applies.

(4)   If the offeree under the agreement is an individual who—
  (a) is not acting in the course of a business, and
  (b) has received the document mentioned in subsection (3) above,

any information contained in that document which was, or would on request have been, required to be provided under section (2)(b) above shall be deemed to be a term of the agreement.

(5)   If, in a case where subsection (4) above applies, a change in the information contained in the document is communicated to the offeree in writing before the timeshare agreement is entered into, the change shall be deemed for the purposes of this Act always to have been incorporated in the information contained in the document if—
  (a) the change arises from circumstances beyond the offeror's control, or
  (b) the offeror and the offeree expressly agree to the change before entering into the timeshare agreement,

and the change is expressly mentioned in the timeshare agreement.

(6)   A person who contravenes subsection (1) above is guilty of an offence and liable—
  (a) on summary conviction, to a fine not exceeding the statutory maximum, and
  (b) on conviction on indictment, to a fine.

(7)   In this section "the proposed accommodation" means—
  (a) the accommodation which is the subject of the proposed agreement, or
  (b) the accommodation in the pool of accommodation which is the subject of the proposed agreement,

as the case may be.

(8)   This section only applies if—
  (a) the accommodation which is the subject of the proposed agreement or agreement is accommodation in a building, or
  (b) some or all of the accommodation in the pool of accommodation which is the subject of the proposed agreement or agreement is accommodation in a building,

as the case may be.]

NOTES

Commencement: 29 April 1997.
Inserted by the Timeshare Regulations 1997, SI 1997/1081, reg 3(1).

[1B Advertising of timeshare rights

(1)   No person shall advertise timeshare rights in the course of a business unless the advertisement indicates the possibility of obtaining the document referred to in section 1A(1) of this Act and where it may be obtained.

(2)   A person who contravenes this section is guilty of an offence and liable—
- (a) on summary conviction, to a fine not exceeding the statutory maximum, and
- (b) on conviction on indictment, to a fine.

(3)   In proceedings against a person for an offence under this section it shall be a defence for that person to show that at the time when he advertised the timeshare rights—
- (a) he did not know and had no reasonable cause to suspect that he was advertising timeshare rights, or
- (b) he had reasonable cause to believe that the advertisement complied with the requirements of subsection (1) above.

(4)   This section only applies if—
- (a) the timeshare accommodation concerned is, or appears from the advertisement to be, accommodation in a building, or
- (b) some or all of the accommodation in the pool of accommodation concerned is, or appears from the advertisement to be, accommodation in a building,

as the case may be.]

NOTES

Commencement: 29 April 1997.
Inserted by the Timeshare Regulations 1997, SI 1997/1081, reg 4.

[1C Obligatory terms of timeshare agreement

(1)   A person must not in the course of a business enter into a timeshare agreement to which this Act applies as offeror unless the agreement includes, as terms set out in it, the information referred to in Schedule 1 to this Act.

(2)   If and to the extent that any information set out in an agreement in accordance with subsection (1) above is inconsistent with any term (the "deemed term") which is deemed to be included in the agreement under section 1A(4) of this Act, the agreement shall be treated for all purposes of this Act as if the deemed term, and not that information, were set out and included in the agreement.

(3)   A person who contravenes subsection (1) above is guilty of an offence and liable—
(a) on summary conviction, to a fine not exceeding the statutory maximum, and
(b) on conviction on indictment, to a fine.

(4)   This section only applies if the offeree—
(a) is an individual, and
(b) is not acting in the course of a business.

(5)   This section only applies if—
(a) the accommodation which is the subject of the agreement is accommodation in a building, or
(b) some or all of the accommodation in the pool of accommodation which is the subject of the agreement is accommodation in a building,

as the case may be.]

NOTES

Commencement: 29 April 1997.
Inserted by the Timeshare Regulations 1997, SI 1997/1081, reg 5.

[1D   Form of agreement and language of brochure and agreement

(1)   A person must not in the course of a business enter into a timeshare agreement to which this Act applies as offeror unless the agreement is in writing and complies with subsections (3) to (5) below, so far as applicable.

(2)   A person who is required to provide a document under subsection (1) of section 1A of this Act contravenes that subsection if he does not provide a document which complies with subsections (3) and (4) below, so far as applicable.

(3)   If the customer is resident in, or a national of, an EEA State, the agreement or document (as the case may be) must be drawn up in a language which is—

(a) the language, or one of the languages, of the EEA State in which he is resident, or

(b) the language, or one of the languages, of the EEA State of which he is a national,

and is an official language of an EEA State.

(4)   If, in a case falling within subsection (3) above, there are two or more languages in which the agreement or document may be drawn up in compliance with that subsection and the customer nominates one of those languages, the agreement or document must be drawn up in the language he nominates.

(5)   If the offeree is resident in the United Kingdom and the agreement would not, apart from this subsection, be required to be drawn up in English, it must be drawn up in English (in addition to any other language in which it is drawn up).

(6)   A person who contravenes subsection (1) above is guilty of an offence and liable—

(a) on summary conviction, to a fine not exceeding the statutory maximum, and

(b) on conviction on indictment, to a fine.

(7)   In this section "the customer" means—

(a) for the purposes of subsection (1) above, the offeree, and

(b) for the purposes of subsection (2) above, the person to whom the document is required to be provided.

(8)   Subsection (1) above only applies if the offeree—

(a) is an individual, and

(b) is not acting in the course of a business.

(9)   Subsection (1) above only applies if—

(a) the accommodation which is the subject of the agreement is accommodation in a building, or

(b) some or all of the accommodation in the pool of accommodation which is the subject of the agreement is accommodation in a building,

as the case may be.]

Commencement: 29 April 1997.
Inserted by the Timeshare Regulations 1997, SI 1997/1081, reg 6.

[1E   Translation of agreement

(1)   A person must not in the course of a business enter into a timeshare agreement to which this Act applies as offeror unless he complies with subsection (2) below.

(2)   If the timeshare accommodation which is the subject of the agreement, or any of the accommodation in the pool of accommodation which is the subject of the agreement, is situated in an EEA State, the offeror must provide the offeree with a certified translation of the agreement in the language, or one of the languages, of that State.

(3)   The language of the translation must be an official language of an EEA State.

(4)   Subsection (1) above does not apply if the agreement is drawn up in a language in which the translation is required or permitted to be made.

(5)   A person who contravenes subsection (1) above is guilty of an offence and liable—
   (a)   on summary conviction, to a fine not exceeding the statutory maximum, and
   (b)   on conviction on indictment, to a fine.

(6)   In this section "certified translation" means a translation which is certified to be accurate by a person authorised to make or verify translations for the purposes of court proceedings.

(7)   This section only applies if the offeree—
   (a)   is an individual, and
   (b)   is not acting in the course of a business.

(8)   This section only applies if—
   (a)   the accommodation which is the subject of the agreement is accommodation in a building, or
   (b)   some or all of the accommodation in the pool of accommodation which is the subject of the agreement is accommodation in a building,

as the case may be.]

NOTES

Commencement: 29 April 1997.
Inserted by the Timeshare Regulations 1997, SI 1997/1081, reg 7.

2 Obligation to give notice of right to cancel timeshare agreement

(1) A person must not in the course of a business enter into a timeshare agreement to which this Act applies as offeror unless the offeree has received, together with a document setting out the terms of the agreement or the substance of those terms, notice of his right to cancel the agreement.

(2) A notice under this section must state—
  (a) that the offeree is entitled to give notice of cancellation of the agreement to the offeror at any time on or before the date specified in the notice, being a day falling not less than fourteen days after the day on which the agreement is entered into, and
  (b) that if the offeree gives such a notice to the offeror on or before that date he will have no further rights or obligations under the agreement, but will have the right to recover any sums paid under or in contemplation of the agreement.

[(2A) A notice under this section must state—
  (a) that if the offeree is an individual and gives a notice to the offeror as mentioned in subsection (2)(b) above, the notice will have the effect of cancelling any related timeshare credit agreement to which this Act applies, and
  (b) that "related timeshare credit agreement" means a timeshare credit agreement under which credit which fully or partly covers the price under the agreement is granted.

(2B) A notice under this section must state that if the offeree is an individual he may in exceptional circumstances have further rights to cancel the timeshare agreement in addition to those mentioned in subsection (2) above.]

(3) A person who contravenes this section is guilty of an offence and liable—
  (a) on summary conviction, to a fine not exceeding the statutory maximum, and
  (b) on conviction on indictment, to a fine.

[(4)   Subsections (2A) and (2B) above only apply if—
  (a) the accommodation which is the subject of the timeshare agreement is accommodation in a building, or
  (b) some or all of the accommodation in the pool of accommodation which is the subject of the timeshare agreement is accommodation in a building,
as the case may be.]

NOTES

Sub-ss (2A), (2B), (4): inserted by the Timeshare Regulations 1997, SI 1997/1081, reg 8(1), (2).

3   Obligation to give notice of right to cancel timeshare credit agreement

(1)   A person must not in the course of a business enter into a timeshare credit agreement to which this Act applies as creditor unless the offeree has received, together with a document setting out the terms of the agreement or the substance of those terms, notice of his right to cancel the agreement.

(2)   A notice under this section must state—
  (a) that the offeree is entitled to give notice of cancellation of the agreement to the creditor at any time on or before the date specified in the notice, being a day falling not less than fourteen days after the day on which the agreement is entered into, and
  (b) that, if the offeree gives such a notice to the creditor on or before that date, then—
    (i) so far as the agreement relates to repayment of credit and payment of interest, it shall have effect subject to section 7 of this Act, and
    (ii) subject to sub-paragraph (i) above, the offeree will have no further rights or obligations under the agreement.

[(3)   A notice under this section must state that the agreement is a timeshare credit agreement for the purposes of this Act.]

NOTES

Sub-s (3): inserted by the Timeshare Regulations 1997, SI 1997/1081, reg 8(3).

4 Provisions supplementary to sections 2 and 3

(1)   Sections 2 and 3 of this Act do not apply where, in entering into the agreement, the offeree is acting in the course of a business.

(2)   A notice under section 2 or 3 must be accompanied by a blank notice of cancellation and any notice under section 2 or 3 of this Act or blank notice of cancellation must—
  (a) be in such form as may be prescribed, and
  (b) comply with such requirements (whether as to type, size, colour or disposition of lettering, quality or colour of paper, or otherwise) as may be prescribed for securing that the notice is prominent and easily legible.

(3)   An agreement is not invalidated by reason of a contravention of section 2 or 3.

5 Right to cancel timeshare agreement

(1)   Where a person—
  (a) has entered, or proposes to enter, into a timeshare agreement to which this Act applies as offeree, and
  (b) has received the notice required under section 2 of this Act before entering into the agreement,

the agreement may not be enforced against him on or before the date specified in the notice in pursuance of subsection (2)(a) of that section and he may give notice of cancellation of the agreement to the offeror at any time on or before that date.

(2)   Subject to subsection (3) below, where a person who enters into a timeshare agreement to which this Act applies as offeree has not received the notice required under section 2 of this Act before entering into the agreement, the agreement may not be enforced against him and he may give notice of cancellation of the agreement to the offeror at any time.

(3)   If in a case falling within subsection (2) above the offeree affirms the agreement at any time after the expiry of the period of fourteen days beginning with the day on which the agreement is entered into—
  (a) subsection (2) above does not prevent the agreement being enforced against him, and

(b) he may not at any subsequent time give notice of cancellation of the agreement to the offeror [under subsection (2) above].

(4)   The offeree's giving, within the time allowed under this section [or section 5A of this Act], notice of cancellation of the agreement to the offeror at a time when the agreement has been entered into shall have the effect of cancelling the agreement.

(5)   The offeree's giving notice of cancellation of the agreement [under this section] to the offeror before the agreement has been entered into shall have the effect of withdrawing any offer to enter into the agreement.

(6)   Where a timeshare agreement is cancelled under this section [or section 5A of this Act], then, subject to subsection (9) below—
(a) the agreement shall cease to be enforceable, and
(b) subsection (8) below shall apply.

(7)   Subsection (8) below shall also apply where giving a notice of cancellation has the effect of withdrawing an offer to enter into a timeshare agreement.

(8)   Where this subsection applies—
(a) any sum which the offeree has paid under or in contemplation of the agreement to the offeror, or to any person who is the offeror's agent for the purpose of receiving that sum, shall be recoverable from the offeror by the offeree and shall be due and payable at the time the notice of cancellation is given, but
(b) no sum may be recovered by or on behalf of the offeror from the offeree in respect of the agreement.

(9)   Where a timeshare agreement includes provision for providing credit for or in respect of the offeree, then, notwithstanding the giving of notice of cancellation under this section [or section 5A of this Act], so far as the agreement relates to repayment of the credit and payment of interest—
(a) it shall continue to be enforceable, subject to section 7 of this Act, and
(b) the notice required under section 2 of this Act must also state that fact.

NOTES

Sub-ss (3)-(6), (9): words in square brackets inserted by the Timeshare Regulations 1997, SI 1997/1081, reg 9(1)-(5).

[5A Additional right to cancel timeshare agreement

(1)   If a timeshare agreement to which this Act applies does not include, as terms set out in it, the information referred to in paragraph (a), (b), (c), (d)(i), (d)(ii), (h), (i), (k), (l) and (m) of Schedule 1 to this Act, the agreement may not be enforced against the offeree before the end of the period of three months and ten days beginning with the day on which the agreement was entered into, and the offeree may give notice of cancellation of the agreement to the offeror at any time during that period.

(2)   If the information referred to in subsection (1) above is provided to the offeree before the end of the period of three months beginning with the day on which the agreement was entered into—
   (a) the offeree may give notice of cancellation of the agreement to the offeror at any time within the period of ten days beginning with the day on which the information is received by the offeree, but
   (b) the offeree may not at any subsequent time give notice of cancellation of the agreement to the offeror under subsection (1) above.

(3)   If the last day of the period referred to in subsection (1) above or the last day of the period of ten days referred to in subsection (2) above is a public holiday, the period concerned shall not end until the end of the first working day after the public holiday.

(4)   The reference in subsection (1) above to a timeshare agreement to which this Act applies includes a reference to a binding preliminary agreement.

(5)   This section only applies of the offeree—
   (a) is an individual, and
   (b) is not acting in the course of a business.

(6)   This section only applies if—
   (a) the accommodation which is the subject of the agreement is accommodation in a building, or
   (b) some or all of the accommodation in the pool of accommodation which is the subject of the agreement is accommodation in a building,

as the case may be.]

NOTES

Commencement: 29 April 1997.
Inserted by the Timeshare Regulations 1997, SI 1997/1081, reg 9(6).

[5B   Advance payments

(1)   A person who enters, or proposes to enter, in the course of a business into a timeshare agreement to which this Act applies as offeror must not (either in person or through another person) request or accept from the offeree or proposed offeree any advance payment before the end of the period during which notice of cancellation of the agreement may be given under section 5 or 5A of this Act.

(2)   A person who contravenes this section is guilty of an offence and liable—
  (a) on summary conviction, to a fine not exceeding the statutory maximum, and
  (b) on conviction on indictment, to a fine.

(3)   Subsection (1) above only applies if the offeree or proposed offeree—
  (a) is an individual, and
  (b) is not acting in the course of a business.

(4)   Subsection (1) above only applies if—
  (a) the accommodation which is the subject of the agreement or proposed agreement is accommodation in a building, or
  (b) some or all of the accommodation in the pool of accommodation which is the subject of the agreement or proposed agreement is accommodation in a building,

as the case may be.]

NOTES

Commencement: 29 April 1997.
Inserted by the Timeshare Regulations 1997, SI 1997/1081, reg 10.

6   [Right to cancel timeshare credit agreement by giving notice]

(1)   Where a person—
  (a) has entered into a timeshare credit agreement to which this Act applies as offeree, and
  (b) has received the notice required under section 3 of this Act before entering into the agreement,

he may give notice of cancellation of the agreement to the creditor at any time on or before the date specified in the notice in pursuance of subsection (2)(a) of that section.

(2)   Subject to subsection (3) below, where a person who enters into a timeshare credit agreement to which this Act applies as offeree has not received the notice required under section 3 of this Act before entering into the agreement, he may give notice of cancellation of the agreement to the creditor at any time.

(3)   If in a case falling within subsection (2) above the offeree affirms the agreement at any time after the expiry of the period of fourteen days beginning with the day on which the agreement is entered into, he may not at any subsequent time give notice of cancellation of the agreement to the creditor.

(4)   The offeree's giving, within the time allowed under this section, notice of cancellation of the agreement to the creditor at a time when the agreement has been entered into shall have the effect of cancelling the agreement.

(5)   Where a timeshare credit agreement is cancelled under this section [or section 6A of this Act]—
  (a) the agreement shall continue in force, subject to section 7 of this Act, so far as it relates to repayment of the credit and payment of interest, and
  (b) subject to paragraph (a) above, the agreement shall cease to be enforceable.

NOTES

Section heading: substituted by the Timeshare Regulations 1997, SI 1997/1081, reg 11(1).
Sub-s (5): words in square brackets inserted by SI 1997/1081, reg 11(2).

[6A   Automatic cancellation of timeshare credit agreement

(1)   Where—
  (a) a notice of cancellation of a timeshare agreement is given under section 5 or 5A of this Act, and
  (b) the giving of the notice has the effect of cancelling the agreement,

the notice shall also have the effect of cancelling any related timeshare credit agreement to which this Act applies.

(2)   Where a timeshare credit agreement is cancelled as mentioned in subsection (1) above, the offeror shall, if he is not the same person as the

creditor under the related timeshare credit agreement, forthwith on receipt of the notice inform the creditor that the notice has been given.

(3)   A timeshare credit agreement is related to a timeshare agreement for the purposes of this section if credit under the timeshare credit agreement fully or partly covers the price under the timeshare agreement.

(4)   Subsection (1) above only applies if the offeree under the timeshare agreement concerned is an individual.

(5)   Subsection (1) above only applies if—
   (a) the accommodation which is the subject of the timeshare agreement is accommodation in a building, or
   (b) some or all of the accommodation in the pool of accommodation which is the subject of the timeshare agreement is accommodation in a building,

as the case may be.]

NOTES

Commencement: 29 April 1997.
Inserted by the Timeshare Regulations 1997, SI 1997/1081, reg 11(3).

7   Repayment of credit and interest

(1)   This section applies following—
   (a) the giving of notice of cancellation of a timeshare agreement in accordance with section 5 of this Act in a case where subsection (9) of that section applies, . . .
   (b) the giving of notice of cancellation of a timeshare credit agreement in accordance with section 6 of this Act, [or
   (c) the cancellation of a timeshare credit agreement by virtue of section 6A of this Act.]

(2)   If the offeree repays the whole or a portion of the credit—
   (a) before the expiry of one month following the giving of the notice [or the cancellation of the timeshare credit agreement by virtue of section 6A of this Act (as the case may be)], or
   (b) in the case of a credit repayable by instalments, before the date on which the first instalment is due,

no interest shall be payable on the amount repaid.

(3)   If the whole of a credit repayable by instalments is not repaid on or before the date specified in subsection (2)(b) above, the offeree shall not be liable to repay any of the credit except on receipt of a request in writing in such form as may be prescribed, signed by or on behalf of the offeror or (as the case may be) creditor, stating the amounts of the remaining instalments (recalculated by the offeror or creditor as nearly as may be in accordance with the agreement and without extending the repayment period), but excluding any sum other than principal and interest.

8   Defence of due diligence

(1)   In proceedings against a person for an offence under section [1A(6), 1B(2), 1C(3), 1D(6), 1E(5), 2(3) or 5B(2)] of this Act it shall be a defence for that person to show that he took all reasonable steps and exercised all due diligence to avoid committing the offence.

(2)   Where in proceedings against a person for such an offence the defence provided by subsection (1) above involves an allegation that the commission of the offence was due—
  (a) to the act or default of another, or
  (b) to reliance on information given by another,

that person shall not, without the leave of the court, be entitled to rely on the defence unless he has served a notice under subsection (3) below on the person bringing the proceedings not less than seven clear days before the hearing of the proceedings or, in Scotland, the diet of trial.

(3)   A notice under this subsection shall give such information identifying or assisting in the identification of the person who committed the act or default or gave the information as is in the possession of the person serving the notice at the time when he serves it.

9   Liability of person other than principal offender

(1)   Where the commission by a person of an offence under section [1A(6), 1B(2), 1C(3), 1D(6), 1E(5), 2(3) or 5B(2)] of this Act is due to the act or default of some other person, that other person is guilty of the offence and may be proceeded against and punished by virtue of this section whether or not proceedings are taken against the first-mentioned person.

(2)   Where a body corporate is guilty of an offence under section [1A(6), 1B(2), 1C(3), 1D(6), 1E(5), 2(3) or 5B(2)] of this Act (including where it is so guilty by virtue of subsection (1) above) in respect of an act or default which is shown to have been committed with the consent or connivance of, or to be attributable to neglect on the part of, a director, manager, secretary or other similar officer of the body corporate or a person who was purporting to act in such a capacity, he (as well as the body corporate) is guilty of the offence and liable to be proceeded against and punished accordingly.

(3)   Where the affairs of a body corporate are managed by its members, subsection (2) above applies in relation to the acts and defaults of a member in connection with his functions of management as if he were a director of the body corporate.

(4)   . . .

NOTES

Sub-ss (1), (2): words in square brackets substituted by the Timeshare Regulations 1997, SI 1997/1081, reg 13(2).

Sub-s (4): applies to Scotland only.

[10A   Civil proceedings

(1)   The obligation to comply with subsection (1) of section 1A of this Act shall be a duty owed by the person who proposes to enter into a timeshare agreement to any person whom he is required to provide with a document under that subsection and a contravention of the obligation shall be actionable accordingly.

(2)   The obligation to comply with section 1C(1), 1D(1), and 1E(1) of this Act shall in each case be a duty owed by the person who enters into a timeshare agreement as offeror to the offeree and a contravention of the obligation shall be actionable accordingly.

(3)   The obligation to comply with section 6A(2) of this Act shall be a duty owed by the offeror under the timeshare agreement to the creditor under the related timeshare credit agreement and a contravention of the obligation shall be actionable accordingly.]

NOTES

Commencement: 29 April 1997.
Inserted by the Timeshare Regulations 1997, SI 1997/1081, reg 12.

12   General provisions

(1)   For the purposes of this Act, a notice of cancellation of an agreement is a notice (however expressed) showing that the offeree wishes unconditionally to cancel the agreement, whether or not it is in a prescribed form.

(2)   The rights conferred and duties imposed by sections [1A] to 7 of this Act are in addition to any rights conferred or duties imposed by or under any other Act.

(3)   For the purposes of this Act, if the offeree sends a notice by post in a properly addressed and pre-paid letter the notice is to be treated as given at the time of posting.

(4)   This Act shall have effect in relation to any timeshare agreement or timeshare credit agreement notwithstanding any agreement or notice.

(5)   . . .

(6)   In this Act—
   "credit" includes a cash loan and any other form of financial accommodation,
   ["EEA State" means a State which is a Contracting Party to the Agreement on the European Economic Area signed at Oporto on 2nd May 1992 as adjusted by the Protocol signed at Brussels on 17th March 1993,]
   "notice" means notice in writing,
   "order" means an order made by the Secretary of State, and
   "prescribed" means prescribed by an order.

(7)   An order under this Act may make different provision for different cases or circumstances.

(8)   Any power under this Act to make an order shall be exercisable by statutory instrument and a statutory instrument containing an order under this Act (other than an order made for the purposes of section 13(2) of this Act) shall be subject to annulment in pursuance of a resolution of either House of Parliament.

NOTES

Sub-s (2): number in square brackets substituted by the Timeshare Regulations 1997, SI 1997/1081, reg 14(9).
Sub-s (5): repealed by SI 1997/1081, reg 14(10).
Sub-s (6): definition "EEA State" inserted by SI 1997/1081, reg 2(9).

13   Short title, etc

(1)   This Act may be cited as the Timeshare Act 1992.

(2)   This Act shall come into force on such day as may be prescribed.

(3)   This Act extends to Northern Ireland.

[SCHEDULE 1
MINIMUM LIST OF ITEMS TO BE INCLUDED IN A TIMESHARE AGREEMENT TO WHICH SECTION 1C APPLIES

(a) The identities and domiciles of the parties, including specific information on the offeror's legal status at the time of the conclusion of the agreement and the identity and domicile of the owner.
(b) The exact nature of the right which is the subject of the agreement and, if the accommodation concerned, or any of the accommodation in the pool of accommodation concerned, is situated in the territory of an EEA State, a clause setting out the conditions governing the exercise of that right within the territory of that State and if those conditions have been fulfilled or, if they have not, what conditions remain to be fulfilled.
(c) When the timeshare accommodation has been determined, an accurate description of that accommodation and its location.
(d) Where the timeshare accommodation is under construction—
    (i) the state of completion,
    (ii) a reasonable estimate of the deadline for completion of the timeshare accommodation,

    (iii) where it concerns specific timeshare accommodation, the number of the building permit and the name and full address of the competent authority or authorities,

    (iv) the state of completion of the services rendering the timeshare accommodation fully operational (gas, electricity, water and telephone connections),

    (v) a guarantee regarding completion of the timeshare accommodation or a guarantee regarding reimbursement of any payment made if the accommodation is not completed and, where appropriate, the conditions governing the operation of those guarantees.

(e) The services (lighting, water, maintenance, refuse collection) to which the offeree has or will have access and on what conditions.

(f) The common facilities, such as swimming pool, sauna, etc, to which the offeree has or may have access, and where appropriate, on what conditions.

(g) The principles on the basis of which the maintenance of and repairs to the timeshare accommodation and its administration and management will be arranged.

(h) The exact period within which the right which is the subject of the agreement may be exercised and, if necessary, its duration; the date on which the offeree may start to exercise that right.

(i) The price to be paid by the offeree to exercise the right under the agreement; an estimate of the amount to be paid by the offeree for the use of common facilities and services; the basis for the calculation of the amount of charges relating to occupation of the timeshare accommodation, the mandatory statutory charges (for example, taxes and fees) and the administrative overheads (for example, management, maintenance and repairs).

(j) A clause stating that acquisitions will not result in costs, charges or obligations other than those specified in the agreement.

(k) Whether or not it is possible to join a scheme for the exchange or resale of the rights under the agreement, and any costs involved should an exchange or resale scheme be organised by the offeror or by a third party designated by him in the agreement.

(l) Information on the right to cancel or withdraw from the agreement and indication of the person to whom any letter of cancellation or withdrawal should be sent, specifying also the arrangements under which such letters may be sent; where appropriate, information on the arrangements for the cancellation of the credit agreement linked to the agreement in the event of cancellation of the agreement or withdrawal from it.

(m) The date and place of each party's signing of the agreement.]

NOTES

Inserted by the Timeshare Regulations 1997, SI 1997/1081, reg 3(3).

Carriage of Goods by Sea Act 1992

(C 50)

An Act to replace the Bills of Lading Act 1855 with new provision with respect to bills of lading and certain other shipping documents

[16 July 1992]

1 Shipping documents etc to which Act applies

(1)   This Act applies to the following documents, that is to say—
  (a) any bill of lading;
  (b) any sea waybill; and
  (c) any ship's delivery order.

(2)   References in this Act to a bill of lading—
  (a) do not include references to a document which is incapable of transfer either by indorsement or, as a bearer bill, by delivery without indorsement; but
  (b) subject to that, do include references to a received for shipment bill of lading.

(3)   References in this Act to a sea waybill are references to any document which is not a bill of lading but—
  (a) is such a receipt for goods as contains or evidences a contract for the carriage of goods by sea; and
  (b) identifies the person to whom delivery of the goods is to be made by the carrier in accordance with that contract.

(4)   References in this Act to a ship's delivery order are references to any document which is neither a bill of lading nor a sea waybill but contains an undertaking which—
  (a) is given under or for the purposes of a contract for the carriage by sea of the goods to which the document relates, or of goods which include those goods; and
  (b) is an undertaking by the carrier to a person identified in the document to deliver the goods to which the document relates to that person.

(5)   The Secretary of State may by regulations make provision for the application of this Act to cases where a telecommunication system or any other information technology is used for effecting transactions corresponding to—
(a) the issue of a document to which this Act applies;
(b) the indorsement, delivery or other transfer of such a document; or
(c) the doing of anything else in relation to such a document.

(6)   Regulations under subsection (5) above may—
(a) make such modifications of the following provisions of this Act as the Secretary of State considers appropriate in connection with the application of this Act to any case mentioned in that subsection; and
(b) contain supplemental, incidental, consequential and transitional provision;

and the power to make regulations under that subsection shall be exercisable by statutory instrument subject to annulment in pursuance of a resolution of either House of Parliament.

2   Rights under shipping documents

(1)   Subject to the following provisions of this section, a person who becomes—
(a) the lawful holder of a bill of lading;
(b) the person who (without being an original party to the contract of carriage) is the person to whom delivery of the goods to which a sea waybill relates is to be made by the carrier in accordance with that contract; or
(c) the person to whom delivery of the goods to which a ship's delivery order relates is to be made in accordance with the undertaking contained in the order,

shall (by virtue of becoming the holder of the bill or, as the case may be, the person to whom delivery is to be made) have transferred to and vested in him all rights of suit under the contract of carriage as if he had been a party to that contract.

(2)   Where, when a person becomes the lawful holder of a bill of lading, possession of the bill no longer gives a right (as against the carrier) to possession of the goods to which the bill relates, that person shall not have any rights transferred to him by virtue of subsection (1) above unless he becomes the holder of the bill—
(a) by virtue of a transaction effected in pursuance of any contractual or other arrangements made before the time when such a right to possession ceased to attach to possession of the bill; or

(b) as a result of the rejection to that person by another person of goods or documents delivered to the other person in pursuance of any such arrangements.

(3)   The rights vested in any person by virtue of the operation of subsection (1) above in relation to a ship's delivery order—
(a) shall be so vested subject to the terms of the order; and
(b) where the goods to which the order relates form a part only of the goods to which the contract of carriage relates, shall be confined to rights in respect of the goods to which the order relates.

(4)   Where, in the case of any document to which this Act applies—
(a) a person with any interest or right in or in relation to goods to which the document relates sustains loss or damage in consequence of a breach of the contract of carriage; but
(b) subsection (1) above operates in relation to that document so that rights of suit in respect of that breach are vested in another person,

the other person shall be entitled to exercise those rights for the benefit of the person who sustained the loss or damage to the same extent as they could have been exercised if they had been vested in the person for whose benefit they are exercised.

(5)   Where rights are transferred by virtue of the operation of subsection (1) above in relation to any document, the transfer for which that subsection provides shall extinguish any entitlement to those rights which derives—
(a) where that document is a bill of lading, from a person's having been an original party to the contract of carriage; or
(b) in the case of any document to which this Act applies, from the previous operation of that subsection in relation to that document;

but the operation of that subsection shall be without prejudice to any rights which derive from a person's having been an original party to the contract contained in, or evidenced by, a sea waybill and, in relation to a ship's delivery order, shall be without prejudice to any rights deriving otherwise than from the previous operation of that subsection in relation to that order.

3   Liabilities under shipping documents

(1)   Where subsection (1) of section 2 of this Act operates in relation to any document to which this Act applies and the person in whom rights are vested by virtue of that subsection—

(a) takes or demands delivery from the carrier of any of the goods to which the document relates;

(b) makes a claim under the contract of carriage against the carrier in respect of any of those goods; or

(c) is a person who, at a time before those rights were vested in him, took or demanded delivery from the carrier of any of those goods,

that person shall (by virtue of taking or demanding delivery or making the claim or, in a case falling within paragraph (c) above, of having the rights vested in him) become subject to the same liabilities under that contract as if he had been a party to that contract.

(2) Where the goods to which a ship's delivery order relates form a part only of the goods to which the contract of carriage relates, the liabilities to which any person is subject by virtue of the operation of this section in relation to that order shall exclude liabilities in respect of any goods to which the order does not relate.

(3) This section, so far as it imposes liabilities under any contract on any person, shall be without prejudice to the liabilities under the contract of any person as an original party to the contract.

4   Representations in bills of lading

A bill of lading which—

(a) represents goods to have been shipped on board a vessel or to have been received for shipment on board a vessel; and

(b) has been signed by the master of the vessel or by a person who was not the master but had the express, implied or apparent authority of the carrier to sign bills of lading,

shall, in favour of a person who has become the lawful holder of the bill, be conclusive evidence against the carrier of the shipment of the goods or, as the case may be, of their receipt for shipment.

5   Interpretation

(1) In this Act—
   "bill of lading", "sea waybill" and "ship's delivery order" shall be construed in accordance with section 1 above;

"the contract of carriage"—
    (a) in relation to a bill of lading or sea waybill, means the contract contained in or evidenced by that bill or waybill; and
    (b) in relation to a ship's delivery order, means the contract under or for the purposes of which the undertaking contained in the order is given;
"holder", in relation to a bill of lading, shall be construed in accordance with subsection (2) below;
"information technology" includes any computer or other technology by means of which information or other matter may be recorded or communicated without being reduced to documentary form; and
"telecommunication system" has the same meaning as in the Telecommunications Act 1984.

(2)   References in this Act to the holder of a bill of lading are references to any of the following persons, that is to say—
    (a) a person with possession of the bill who, by virtue of being the person identified in the bill, is the consignee of the goods to which the bill relates;
    (b) a person with possession of the bill as a result of the completion, by delivery of the bill, of any indorsement of the bill or, in the case of a bearer bill, of any other transfer of the bill;
    (c) a person with possession of the bill as a result of any transaction by virtue of which he would have become a holder falling within paragraph (a) or (b) above had not the transaction been effected at a time when possession of the bill no longer gave a right (as against the carrier) to possession of the goods to which the bill relates;

and a person shall be regarded for the purposes of this Act as having become the lawful holder of a bill of lading wherever he has become the holder of the bill in good faith.

(3)   References in this Act to a person's being identified in a document include references to his being identified by a description which allows for the identity of the person in question to be varied, in accordance with the terms of the document, after its issue; and the reference in section 1(3)(b) of this Act to a document's identifying a person shall be construed accordingly.

(4)   Without prejudice to sections 2(2) and 4 above, nothing in this Act shall preclude its operation in relation to a case where the goods to which a document relates—
    (a) cease to exist after the issue of the document; or

(b) cannot be identified (whether because they are mixed with other goods or for any other reason);

and references in this Act to the goods to which a document relates shall be construed accordingly.

(5)   The preceding provisions of this Act shall have effect without prejudice to the application, in relation to any case, of the rules (the Hague-Visby Rules) which for the time being have the force of law by virtue of section 1 of the Carriage of Goods by Sea Act 1971.

6   Short title, repeal, commencement and extent

(1)   This Act may be cited as the Carriage of Goods by Sea Act 1992.

(2)   . . .

(3)   This Act shall come into force at the end of the period of two months beginning with the day on which it is passed; but nothing in this Act shall have effect in relation to any document issued before the coming into force of this Act.

(4)   This Act extends to Northern Ireland.

NOTES

Sub-s (2): repeals the Bills of Lading Act 1855.

Arbitration Act 1996

(C 23)

An Act to restate and improve the law relating to arbitration pursuant to an arbitration agreement; to make other provision relating to arbitration and arbitration awards; and for connected purposes

[17 June 1996]

PART I
ARBITRATION PURSUANT TO AN ARBITRATION AGREEMENT

Introductory

1 General principles

The provisions of this Part are founded on the following principles, and shall be construed accordingly—
  (a) the object of arbitration is to obtain the fair resolution of disputes by an impartial tribunal without unnecessary delay or expense;
  (b) the parties should be free to agree how their disputes are resolved, subject only to such safeguards as are necessary in the public interest;
  (c) in matters governed by this Part the court should not intervene except as provided by this Part.

NOTES

Commencement: 31 January 1997.

2 Scope of application of provisions

(1)   The provisions of this Part apply where the seat of the arbitration is in England and Wales or Northern Ireland.

(2)   The following sections apply even if the seat of the arbitration is outside England and Wales or Northern Ireland or no seat has been designated or determined—
  (a) sections 9 to 11 (stay of legal proceedings, &c), and
  (b) section 66 (enforcement of arbitral awards).

(3)   The powers conferred by the following sections apply even if the seat of the arbitration is outside England and Wales or Northern Ireland or no seat has been designated or determined—
  (a) section 43 (securing the attendance of witnesses), and
  (b) section 44 (court powers exercisable in support of arbitral proceedings);

but the court may refuse to exercise any such power if, in the opinion of the court, the fact that the seat of the arbitration is outside England and Wales or Northern Ireland, or that when designated or determined the seat is likely to be outside England and Wales or Northern Ireland, makes it inappropriate to do so.

(4)   The court may exercise a power conferred by any provision of this Part not mentioned in subsection (2) or (3) for the purpose of supporting the arbitral process where—

   (a) no seat of the arbitration has been designated or determined, and
   (b) by reason of a connection with England and Wales or Northern Ireland the court is satisfied that it is appropriate to do so.

(5)   Section 7 (separability of arbitration agreement) and section 8 (death of a party) apply where the law applicable to the arbitration agreement is the law of England and Wales or Northern Ireland even if the seat of the arbitration is outside England and Wales or Northern Ireland or has not been designated or determined.

NOTES

Commencement: 31 January 1997.

3   The seat of the arbitration

In this Part "the seat of the arbitration" means the juridical seat of the arbitration designated—

   (a) by the parties to the arbitration agreement, or
   (b) by any arbitral or other institution or person vested by the parties with powers in that regard, or
   (c) by the arbitral tribunal if so authorised by the parties,

or determined, in the absence of any such designation, having regard to the parties' agreement and all the relevant circumstances.

NOTES

Commencement: 31 January 1997.

4   Mandatory and non-mandatory provisions

(1)   The mandatory provisions of this Part are listed in Schedule 1 and have effect notwithstanding any agreement to the contrary.

(2)   The other provisions of this Part (the "non-mandatory provisions") allow the parties to make their own arrangements by agreement but provide rules which apply in the absence of such agreement.

(3)   The parties may make such arrangements by agreeing to the application of institutional rules or providing any other means by which a matter may be decided.

(4)   It is immaterial whether or not the law applicable to the parties' agreement is the law of England and Wales or, as the case may be, Northern Ireland.

(5)   The choice of a law other than the law of England and Wales or Northern Ireland as the applicable law in respect of a matter provided for by a non-mandatory provision of this Part is equivalent to an agreement making provision about that matter.

For this purpose an applicable law determined in accordance with the parties' agreement, or which is objectively determined in the absence of any express or implied choice, shall be treated as chosen by the parties.

NOTES

Commencement: 31 January 1997.

5   Agreements to be in writing

(1)   The provisions of this Part apply only where the arbitration agreement is in writing, and any other agreement between the parties as to any matter is effective for the purposes of this Part only if in writing.

The expressions "agreement", "agree" and "agreed" shall be construed accordingly.

(2)   There is an agreement in writing—
  (a) if the agreement is made in writing (whether or not it is signed by the parties),
  (b) if the agreement is made by exchange of communications in writing, or
  (c) if the agreement is evidenced in writing.

(3)   Where parties agree otherwise than in writing by reference to terms which are in writing, they make an agreement in writing.

(4)   An agreement is evidenced in writing if an agreement made otherwise than in writing is recorded by one of the parties, or by a third party, with the authority of the parties to the agreement.

(5)   An exchange of written submissions in arbitral or legal proceedings in which the existence of an agreement otherwise than in writing is alleged by one party against another party and not denied by the other party in his response constitutes as between those parties an agreement in writing to the effect alleged.

(6)   References in this Part to anything being written or in writing include its being recorded by any means.

NOTES
Commencement: 31 January 1997.

The arbitration agreement

6   Definition of arbitration agreement

(1)   In this Part an "arbitration agreement" means an agreement to submit to arbitration present or future disputes (whether they are contractual or not).

(2)   The reference in an agreement to a written form of arbitration clause or to a document containing an arbitration clause constitutes an arbitration agreement if the reference is such as to make that clause part of the agreement.

NOTES
Commencement: 31 January 1997.

7   Separability of arbitration agreement

Unless otherwise agreed by the parties, an arbitration agreement which forms or was intended to form part of another agreement (whether or not in writing) shall not be regarded as invalid, non-existent or ineffective because that other agreement is invalid, or did not come into existence or has become ineffective, and it shall for that purpose be treated as a distinct agreement.

NOTES
Commencement: 31 January 1997.

8 Whether agreement discharged by death of a party

(1)   Unless otherwise agreed by the parties, an arbitration agreement is not discharged by the death of a party and may be enforced by or against the personal representatives of that party.

(2)   Subsection (1) does not affect the operation of any enactment or rule of law by virtue of which a substantive right or obligation is extinguished by death.

NOTES

Commencement: 31 January 1997.

Stay of legal proceedings

9 Stay of legal proceedings

(1)   A party to an arbitration agreement against whom legal proceedings are brought (whether by way of claim or counterclaim) in respect of a matter which under the agreement is to be referred to arbitration may (upon notice to the other parties to the proceedings) apply to the court in which the proceedings have been brought to stay the proceedings so far as they concern that matter.

(2)   An application may be made notwithstanding that the matter is to be referred to arbitration only after the exhaustion of other dispute resolution procedures.

(3)   An application may not be made by a person before taking the appropriate procedural step (if any) to acknowledge the legal proceedings against him or after he has taken any step in those proceedings to answer the substantive claim.

(4)   On an application under this section the court shall grant a stay unless satisfied that the arbitration agreement is null and void, inoperative, or incapable of being performed.

(5)   If the court refuses to stay the legal proceedings, any provision that an award is a condition precedent to the bringing of legal proceedings in respect of any matter is of no effect in relation to those proceedings.

NOTES
Commencement: 31 January 1997.

10 Reference of interpleader issue to arbitration

(1)   Where in legal proceedings relief by way of interpleader is granted and any issue between the claimants is one in respect of which there is an arbitration agreement between them, the court granting the relief shall direct that the issue be determined in accordance with the agreement unless the circumstances are such that proceedings brought by a claimant in respect of the matter would not be stayed.

(2)   Where subsection (1) applies but the court does not direct that the issue be determined in accordance with the arbitration agreement, any provision that an award is a condition precedent to the bringing of legal proceedings in respect of any matter shall not affect the determination of that issue by the court.

NOTES
Commencement: 31 January 1997.

11 Retention of security where Admiralty proceedings stayed

(1)   Where Admiralty proceedings are stayed on the ground that the dispute in question should be submitted to arbitration, the court granting the stay may, if in those proceedings property has been arrested or bail or other security has been given to prevent or obtain release from arrest—
  (a) order that the property arrested be retained as security for the satisfaction of any award given in the arbitration in respect of that dispute, or
  (b) order that the stay of those proceedings be conditional on the provision of equivalent security for the satisfaction of any such award.

(2)   Subject to any provision made by rules of court and to any necessary modifications, the same law and practice shall apply in relation to property retained in pursuance of an order as would apply if it were held for the purposes of proceedings in the court making the order.

NOTES
Commencement: 31 January 1997.

Commencement of arbitral proceedings

12   Power of court to extend time for beginning arbitral proceedings, &c

(1)   Where an arbitration agreement to refer future disputes to arbitration provides that a claim shall be barred, or the claimant's right extinguished, unless the claimant takes within a time fixed by the agreement some step—
   (a) to begin arbitral proceedings, or
   (b) to begin other dispute resolution procedures which must be exhausted before arbitral proceedings can be begun,

the court may by order extend the time for taking that step.

(2)   Any party to the arbitration agreement may apply for such an order (upon notice to the other parties), but only after a claim has arisen and after exhausting any available arbitral process for obtaining an extension of time.

(3)   The court shall make an order only if satisfied—
   (a) that the circumstances are such as were outside the reasonable contemplation of the parties when they agreed the provision in question, and that it would be just to extend the time, or
   (b) that the conduct of one party makes it unjust to hold the other party to the strict terms of the provision in question.

(4)   The court may extend the time for such period and on such terms as it thinks fit, and may do so whether or not the time previously fixed (by agreement or by a previous order) has expired.

(5)   An order under this section does not affect the operation of the Limitation Acts (see section 13).

(6)   The leave of the court is required for any appeal from a decision of the court under this section.

NOTES

Commencement: 31 January 1997.

13   Application of Limitation Acts

(1)   The Limitation Acts apply to arbitral proceedings as they apply to legal proceedings.

(2)   The court may order that in computing the time prescribed by the Limitation Acts for the commencement of proceedings (including arbitral proceedings) in respect of a dispute which was the subject matter—

(a) of an award which the court orders to be set aside or declares to be of no effect, or

(b) of the affected part of an award which the court orders to be set aside in part, or declares to be in part of no effect,

the period between the commencement of the arbitration and the date of the order referred to in paragraph (a) or (b) shall be excluded.

(3)   In determining for the purposes of the Limitation Acts when a cause of action accrued, any provision that an award is a condition precedent to the bringing of legal proceedings in respect of a matter to which an arbitration agreement applies shall be disregarded.

(4)   In this Part "the Limitation Acts" means—

(a) in England and Wales, the Limitation Act 1980, the Foreign Limitation Periods Act 1984 and any other enactment (whenever passed) relating to the limitation of actions;

(b) in Northern Ireland, the Limitation (Northern Ireland) Order 1989, the Foreign Limitation Periods (Northern Ireland) Order 1985 and any other enactment (whenever passed) relating to the limitation of actions.

NOTES

Commencement: 31 January 1997.

14   Commencement of arbitral proceedings

(1)   The parties are free to agree when arbitral proceedings are to be regarded as commenced for the purposes of this Part and for the purposes of the Limitation Acts.

(2)   If there is no such agreement the following provisions apply.

(3)   Where the arbitrator is named or designated in the arbitration agreement, arbitral proceedings are commenced in respect of a matter when one party serves on the other party or parties a notice in writing requiring him or them to submit that matter to the person so named or designated.

(4)   Where the arbitrator or arbitrators are to be appointed by the parties, arbitral proceedings are commenced in respect of a matter when one party

serves on the other party or parties notice in writing requiring him or them to appoint an arbitrator or to agree to the appointment of an arbitrator in respect of that matter.

(5)   Where the arbitrator or arbitrators are to be appointed by a person other than a party to the proceedings, arbitral proceedings are commenced in respect of a matter when one party gives notice in writing to that person requesting him to make the appointment in respect of that matter.

NOTES

Commencement: 31 January 1997.

The arbitral tribunal

15   The arbitral tribunal

(1)   The parties are free to agree on the number of arbitrators to form the tribunal and whether there is to be a chairman or umpire.

(2)   Unless otherwise agreed by the parties, an agreement that the number of arbitrators shall be two or any other even number shall be understood as requiring the appointment of an additional arbitrator as chairman of the tribunal.

(3)   If there is no agreement as to the number of arbitrators, the tribunal shall consist of a sole arbitrator.

NOTES

Commencement: 31 January 1997.

16   Procedure for appointment of arbitrators

(1)   The parties are free to agree on the procedure for appointing the arbitrator or arbitrators, including the procedure for appointing any chairman or umpire.

(2)   If or to the extent that there is no such agreement, the following provisions apply.

(3)   If the tribunal is to consist of a sole arbitrator, the parties shall jointly appoint the arbitrator not later than 28 days after service of a request in writing by either party to do so.

(4)   If the tribunal is to consist of two arbitrators, each party shall appoint one arbitrator not later than 14 days after service of a request in writing by either party to do so.

(5)   If the tribunal is to consist of three arbitrators—
   (a) each party shall appoint one arbitrator not later than 14 days after service of a request in writing by either party to do so, and
   (b) the two so appointed shall forthwith appoint a third arbitrator as the chairman of the tribunal.

(6)   If the tribunal is to consist of two arbitrators and an umpire—
   (a) each party shall appoint one arbitrator not later than 14 days after service of a request in writing by either party to do so, and
   (b) the two so appointed may appoint an umpire at any time after they themselves are appointed and shall do so before any substantive hearing or forthwith if they cannot agree on a matter relating to the arbitration.

(7)   In any other case (in particular, if there are more than two parties) section 18 applies as in the case of a failure of the agreed appointment procedure.

NOTES

Commencement: 31 January 1997.

17   Power in case of default to appoint sole arbitrator

(1)   Unless the parties otherwise agree, where each of two parties to an arbitration agreement is to appoint an arbitrator and one party ("the party in default") refuses to do so, or fails to do so within the time specified, the other party, having duly appointed his arbitrator, may give notice in writing to the party in default that he proposes to appoint his arbitrator to act as sole arbitrator.

(2)   If the party in default does not within 7 clear days of that notice being given—
   (a) make the required appointment, and
   (b) notify the other party that he has done so,

the other party may appoint his arbitrator as sole arbitrator whose award shall be binding on both parties as if he had been so appointed by agreement.

(3)   Where a sole arbitrator has been appointed under subsection (2), the party in default may (upon notice to the appointing party) apply to the court which may set aside the appointment.

(4)   The leave of the court is required for any appeal from a decision of the court under this section.

NOTES

Commencement: 31 January 1997.

18   Failure of appointment procedure

(1)   The parties are free to agree what is to happen in the event of a failure of the procedure for the appointment of the arbitral tribunal.

There is no failure if an appointment is duly made under section 17 (power in case of default to appoint sole arbitrator), unless that appointment is set aside.

(2)   If or to the extent that there is no such agreement any party to the arbitration agreement may (upon notice to the other parties) apply to the court to exercise its powers under this section.

(3)   Those powers are—
   (a) to give directions as to the making of any necessary appointments;
   (b) to direct that the tribunal shall be constituted by such appointments (or any one or more of them) as have been made;
   (c) to revoke any appointments already made;
   (d) to make any necessary appointments itself.

(4)   An appointment made by the court under this section has effect as if made with the agreement of the parties.

(5)   The leave of the court is required for any appeal from a decision of the court under this section.

NOTES

Commencement: 31 January 1997.

19 Court to have regard to agreed qualifications

In deciding whether to exercise, and in considering how to exercise, any of its powers under section 16 (procedure for appointment of arbitrators) or section 18 (failure of appointment procedure), the court shall have due regard to any agreement of the parties as to the qualifications required of the arbitrators.

NOTES

Commencement: 31 January 1997.

20 Chairman

(1) Where the parties have agreed that there is to be a chairman, they are free to agree what the functions of the chairman are to be in relation to the making of decisions, orders and awards.

(2) If or to the extent that there is no such agreement, the following provisions apply.

(3) Decisions, orders and awards shall be made by all or a majority of the arbitrators (including the chairman).

(4) The view of the chairman shall prevail in relation to a decision, order or award in respect of which there is neither unanimity nor a majority under subsection (3).

NOTES

Commencement: 31 January 1997.

21 Umpire

(1) Where the parties have agreed that there is to be an umpire, they are free to agree what the functions of the umpire are to be, and in particular—
   (a) whether he is to attend the proceedings, and
   (b) when he is to replace the other arbitrators as the tribunal with power to make decisions, orders and awards.

(2) If or to the extent that there is no such agreement, the following provisions apply.

(3)   The umpire shall attend the proceedings and be supplied with the same documents and other materials as are supplied to the other arbitrators.

(4)   Decisions, orders and awards shall be made by the other arbitrators unless and until they cannot agree on a matter relating to the arbitration.

In that event they shall forthwith give notice in writing to the parties and the umpire, whereupon the umpire shall replace them as the tribunal with power to make decisions, orders and awards as if he were sole arbitrator.

(5)   If the arbitrators cannot agree but fail to give notice of that fact, or if any of them fails to join in the giving of notice, any party to the arbitral proceedings may (upon notice to the other parties and to the tribunal) apply to the court which may order that the umpire shall replace the other arbitrators as the tribunal with power to make decisions, orders and awards as if he were sole arbitrator.

(6)   The leave of the court is required for any appeal from a decision of the court under this section.

NOTES

Commencement: 31 January 1997.

22   Decision-making where no chairman or umpire

(1)   Where the parties agree that there shall be two or more arbitrators with no chairman or umpire, the parties are free to agree how the tribunal is to make decisions, orders and awards.

(2)   If there is no such agreement, decisions, orders and awards shall be made by all or a majority of the arbitrators.

NOTES

Commencement: 31 January 1997.

23   Revocation of arbitrator's authority

(1)   The parties are free to agree in what circumstances the authority of an arbitrator may be revoked.

(2)   If or to the extent that there is no such agreement the following provisions apply.

(3)   The authority of an arbitrator may not be revoked except—
   (a) by the parties acting jointly, or
   (b) by an arbitral or other institution or person vested by the parties with powers in that regard.

(4)   Revocation of the authority of an arbitrator by the parties acting jointly must be agreed in writing unless the parties also agree (whether or not in writing) to terminate the arbitration agreement.

(5)   Nothing in this section affects the power of the court—
   (a) to revoke an appointment under section 18 (powers exercisable in case of failure of appointment procedure), or
   (b) to remove an arbitrator on the grounds specified in section 24.

NOTES

Commencement: 31 January 1997.

24   Power of court to remove arbitrator

(1)   A party to arbitral proceedings may (upon notice to the other parties, to the arbitrator concerned and to any other arbitrator) apply to the court to remove an arbitrator on any of the following grounds—
   (a) that circumstances exist that give rise to justifiable doubts as to his impartiality;
   (b) that he does not possess the qualifications required by the arbitration agreement;
   (c) that he is physically or mentally incapable of conducting the proceedings or there are justifiable doubts as to his capacity to do so;
   (d) that he has refused or failed—
       (i)   properly to conduct the proceedings, or
       (ii) to use all reasonable despatch in conducting the proceedings or making an award,
and that substantial injustice has been or will be caused to the applicant.

(2)   If there is an arbitral or other institution or person vested by the parties with power to remove an arbitrator, the court shall not exercise its power of removal unless satisfied that the applicant has first exhausted any available recourse to that institution or person.

(3)   The arbitral tribunal may continue the arbitral proceedings and make an award while an application to the court under this section is pending.

(4)   Where the court removes an arbitrator, it may make such order as it thinks fit with respect to his entitlement (if any) to fees or expenses, or the repayment of any fees or expenses already paid.

(5)   The arbitrator concerned is entitled to appear and be heard by the court before it makes any order under this section.

(6)   The leave of the court is required for any appeal from a decision of the court under this section.

NOTES

Commencement: 31 January 1997.

25   Resignation of arbitrator

(1)   The parties are free to agree with an arbitrator as to the consequences of his resignation as regards—
   (a)   his entitlement (if any) to fees or expenses, and
   (b)   any liability thereby incurred by him.

(2)   If or to the extent that there is no such agreement the following provisions apply.

(3)   An arbitrator who resigns his appointment may (upon notice to the parties) apply to the court—
   (a)   to grant him relief from any liability thereby incurred by him, and
   (b)   to make such order as it thinks fit with respect to his entitlement (if any) to fees or expenses or the repayment of any fees or expenses already paid.

(4)   If the court is satisfied that in all the circumstances it was reasonable for the arbitrator to resign, it may grant such relief as is mentioned in subsection (3)(a) on such terms as it thinks fit.

(5)   The leave of the court is required for any appeal from a decision of the court under this section.

NOTES

Commencement: 31 January 1997.

26   Death of arbitrator or person appointing him

(1)   The authority of an arbitrator is personal and ceases on his death.

(2)   Unless otherwise agreed by the parties, the death of the person by whom an arbitrator was appointed does not revoke the arbitrator's authority.

NOTES

Commencement: 31 January 1997.

27   Filling of vacancy, &c

(1)   Where an arbitrator ceases to hold office, the parties are free to agree—
   (a) whether and if so how the vacancy is to be filled,
   (b) whether and if so to what extent the previous proceedings should stand, and
   (c) what effect (if any) his ceasing to hold office has on any appointment made by him (alone or jointly).

(2)   If or to the extent that there is no such agreement, the following provisions apply.

(3)   The provisions of sections 16 (procedure for appointment of arbitrators) and 18 (failure of appointment procedure) apply in relation to the filling of the vacancy as in relation to an original appointment.

(4)   The tribunal (when reconstituted) shall determine whether and if so to what extent the previous proceedings should stand.

This does not affect any right of a party to challenge those proceedings on any ground which had arisen before the arbitrator ceased to hold office.

(5)   His ceasing to hold office does not affect any appointment by him (alone or jointly) of another arbitrator, in particular any appointment of a chairman or umpire.

NOTES

Commencement: 31 January 1997.

28 Joint and several liability of parties to arbitrators for fees and expenses

(1)   The parties are jointly and severally liable to pay to the arbitrators such reasonable fees and expenses (if any) as are appropriate in the circumstances.

(2)   Any party may apply to the court (upon notice to the other parties and to the arbitrators) which may order that the amount of the arbitrators' fees and expenses shall be considered and adjusted by such means and upon such terms as it may direct.

(3)   If the application is made after any amount has been paid to the arbitrators by way of fees or expenses, the court may order the repayment of such amount (if any) as is shown to be excessive, but shall not do so unless it is shown that it is reasonable in the circumstances to order repayment.

(4)   The above provisions have effect subject to any order of the court under section 24(4) or 25(3)(b) (order as to entitlement to fees or expenses in case of removal or resignation of arbitrator).

(5)   Nothing in this section affects any liability of a party to any other party to pay all or any of the costs of the arbitration (see sections 59 to 65) or any contractual right of an arbitrator to payment of his fees and expenses.

(6)   In this section references to arbitrators include an arbitrator who has ceased to act and an umpire who has not replaced the other arbitrators.

NOTES

Commencement: 31 January 1997.

29 Immunity of arbitrator

(1)   An arbitrator is not liable for anything done or omitted in the discharge or purported discharge of his functions as arbitrator unless the act or omission is shown to have been in bad faith.

(2)   Subsection (1) applies to an employee or agent of an arbitrator as it applies to the arbitrator himself.

(3)   This section does not affect any liability incurred by an arbitrator by reason of his resigning (but see section 25).

Commencement: 31 January 1997.

Jurisdiction of the arbitral tribunal

30 Competence of tribunal to rule on its own jurisdiction

(1)   Unless otherwise agreed by the parties, the arbitral tribunal may rule on its own substantive jurisdiction, that is, as to—
   (a) whether there is a valid arbitration agreement,
   (b) whether the tribunal is properly constituted, and
   (c) what matters have been submitted to arbitration in accordance with the arbitration agreement.

(2)   Any such ruling may be challenged by any available arbitral process of appeal or review or in accordance with the provisions of this Part.

Commencement: 31 January 1997.

31 Objection to substantive jurisdiction of tribunal

(1)   An objection that the arbitral tribunal lacks substantive jurisdiction at the outset of the proceedings must be raised by a party not later than the time he takes the first step in the proceedings to contest the merits of any matter in relation to which he challenges the tribunal's jurisdiction.

A party is not precluded from raising such an objection by the fact that he has appointed or participated in the appointment of an arbitrator.

(2)   Any objection during the course of the arbitral proceedings that the arbitral tribunal is exceeding its substantive jurisdiction must be made as soon as possible after the matter alleged to be beyond its jurisdiction is raised.

(3)   The arbitral tribunal may admit an objection later than the time specified in subsection (1) or (2) if it considers the delay justified.

(4)   Where an objection is duly taken to the tribunal's substantive jurisdiction and the tribunal has power to rule on its own jurisdiction, it may—

(a) rule on the matter in an award as to jurisdiction, or

(b) deal with the objection in its award on the merits.

If the parties agree which of these courses the tribunal should take, the tribunal shall proceed accordingly.

(5)　The tribunal may in any case, and shall if the parties so agree, stay proceedings whilst an application is made to the court under section 32 (determination of preliminary point of jurisdiction).

NOTES

Commencement: 31 January 1997.

32　Determination of preliminary point of jurisdiction

(1)　The court may, on the application of a party to arbitral proceedings (upon notice to the other parties), determine any question as to the substantive jurisdiction of the tribunal.

A party may lose the right to object (see section 73).

(2)　An application under this section shall not be considered unless—
  (a) it is made with the agreement in writing of all the other parties to the proceedings, or
  (b) it is made with the permission of the tribunal and the court is satisfied—
    (i) that the determination of the question is likely to produce substantial savings in costs,
    (ii) that the application was made without delay, and
    (iii) that there is good reason why the matter should be decided by the court.

(3)　An application under this section, unless made with the agreement of all the other parties to the proceedings, shall state the grounds on which it is said that the matter should be decided by the court.

(4)　Unless otherwise agreed by the parties, the arbitral tribunal may continue the arbitral proceedings and make an award while an application to the court under this section is pending.

(5)　Unless the court gives leave, no appeal lies from a decision of the court whether the conditions specified in subsection (2) are met.

(6)   The decision of the court on the question of jurisdiction shall be treated as a judgment of the court for the purposes of an appeal.

But no appeal lies without the leave of the court which shall not be given unless the court considers that the question involves a point of law which is one of general importance or is one which for some other special reason should be considered by the Court of Appeal.

NOTES

Commencement: 31 January 1997.

The arbitral proceedings

33   General duty of the tribunal

(1)   The tribunal shall—
  (a) act fairly and impartially as between the parties, giving each party a reasonable opportunity of putting his case and dealing with that of his opponent, and
  (b) adopt procedures suitable to the circumstances of the particular case, avoiding unnecessary delay or expense, so as to provide a fair means for the resolution of the matters falling to be determined.

(2)   The tribunal shall comply with that general duty in conducting the arbitral proceedings, in its decisions on matters of procedure and evidence and in the exercise of all other powers conferred on it.

NOTES

Commencement: 31 January 1997.

34   Procedural and evidential matters

(1)   It shall be for the tribunal to decide all procedural and evidential matters, subject to the right of the parties to agree any matter.

(2)   Procedural and evidential matters include—
  (a) when and where any part of the proceedings is to be held;
  (b) the language or languages to be used in the proceedings and whether translations of any relevant documents are to be supplied;

(c) whether any and if so what form of written statements of claim and defence are to be used, when these should be supplied and the extent to which such statements can be later amended;

(d) whether any and if so which documents or classes of documents should be disclosed between and produced by the parties and at what stage;

(e) whether any and if so what questions should be put to and answered by the respective parties and when and in what form this should be done;

(f) whether to apply strict rules of evidence (or any other rules) as to the admissibility, relevance or weight of any material (oral, written or other) sought to be tendered on any matters of fact or opinion, and the time, manner and form in which such material should be exchanged and presented;

(g) whether and to what extent the tribunal should itself take the initiative in ascertaining the facts and the law;

(h) whether and to what extent there should be oral or written evidence or submissions.

(3)   The tribunal may fix the time within which any directions given by it are to be complied with, and may if it thinks fit extend the time so fixed (whether or not it has expired).

NOTES

Commencement: 31 January 1997.

35 Consolidation of proceedings and concurrent hearings

(1)   The parties are free to agree—
   (a) that the arbitral proceedings shall be consolidated with other arbitral proceedings, or
   (b) that concurrent hearings shall be held,

on such terms as may be agreed.

(2)   Unless the parties agree to confer such power on the tribunal, the tribunal has no power to order consolidation of proceedings or concurrent hearings.

NOTES

Commencement: 31 January 1997.

36 Legal or other representation

Unless otherwise agreed by the parties, a party to arbitral proceedings may be represented in the proceedings by a lawyer or other person chosen by him.

NOTES

Commencement: 31 January 1997.

37 Power to appoint experts, legal advisers or assessors

(1)   Unless otherwise agreed by the parties—
  (a) the tribunal may—
     (i)   appoint experts or legal advisers to report to it and the parties, or
     (ii) appoint assessors to assist it on technical matters,
    and may allow any such expert, legal adviser or assessor to attend the proceedings; and
  (b) the parties shall be given a reasonable opportunity to comment on any information, opinion or advice offered by any such person.

(2)   The fees and expenses of an expert, legal adviser or assessor appointed by the tribunal for which the arbitrators are liable are expenses of the arbitrators for the purposes of this Part.

NOTES

Commencement: 31 January 1997.

38 General powers exercisable by the tribunal

(1)   The parties are free to agree on the powers exercisable by the arbitral tribunal for the purposes of and in relation to the proceedings.

(2)   Unless otherwise agreed by the parties the tribunal has the following powers.

(3)   The tribunal may order a claimant to provide security for the costs of the arbitration.

This power shall not be exercised on the ground that the claimant is—
  (a) an individual ordinarily resident outside the United Kingdom, or

(b) a corporation or association incorporated or formed under the law of a country outside the United Kingdom, or whose central management and control is exercised outside the United Kingdom.

(4)   The tribunal may give directions in relation to any property which is the subject of the proceedings or as to which any question arises in the proceedings, and which is owned by or is in the possession of a party to the proceedings—
   (a) for the inspection, photographing, preservation, custody or detention of the property by the tribunal, an expert or a party, or
   (b) ordering that samples be taken from, or any observation be made of or experiment conducted upon, the property.

(5)   The tribunal may direct that a party or witness shall be examined on oath or affirmation, and may for that purpose administer any necessary oath or take any necessary affirmation.

(6)   The tribunal may give directions to a party for the preservation for the purposes of the proceedings of any evidence in his custody or control.

NOTES

Commencement: 31 January 1997.

39   Power to make provisional awards

(1)   The parties are free to agree that the tribunal shall have power to order on a provisional basis any relief which it would have power to grant in a final award.

(2)   This includes, for instance, making—
   (a) a provisional order for the payment of money or the disposition of property as between the parties, or
   (b) an order to make an interim payment on account of the costs of the arbitration.

(3)   Any such order shall be subject to the tribunal's final adjudication; and the tribunal's final award, on the merits or as to costs, shall take account of any such order.

(4)   Unless the parties agree to confer such power on the tribunal, the tribunal has no such power.

This does not affect its powers under section 47 (awards on different issues, &c).

NOTES

Commencement: 31 January 1997.

40 General duty of parties

(1)   The parties shall do all things necessary for the proper and expeditious conduct of the arbitral proceedings.

(2)   This includes—
  (a) complying without delay with any determination of the tribunal as to procedural or evidential matters, or with any order or directions of the tribunal, and
  (b) where appropriate, taking without delay any necessary steps to obtain a decision of the court on a preliminary question of jurisdiction or law (see sections 32 and 45).

NOTES

Commencement: 31 January 1997.

41 Powers of tribunal in case of party's default

(1)   The parties are free to agree on the powers of the tribunal in case of a party's failure to do something necessary for the proper and expeditious conduct of the arbitration.

(2)   Unless otherwise agreed by the parties, the following provisions apply.

(3)   If the tribunal is satisfied that there has been inordinate and inexcusable delay on the part of the claimant in pursuing his claim and that the delay—
  (a) gives rise, or is likely to give rise, to a substantial risk that it is not possible to have a fair resolution of the issues in that claim, or
  (b) has caused, or is likely to cause, serious prejudice to the respondent,

the tribunal may make an award dismissing the claim.

(4)   If without showing sufficient cause a party—
  (a) fails to attend or be represented at an oral hearing of which due notice was given, or
  (b) where matters are to be dealt with in writing, fails after due notice to submit written evidence or make written submissions,

the tribunal may continue the proceedings in the absence of that party or, as the case may be, without any written evidence or submissions on his behalf, and may make an award on the basis of the evidence before it.

(5)   If without showing sufficient cause a party fails to comply with any order or directions of the tribunal, the tribunal may make a peremptory order to the same effect, prescribing such time for compliance with it as the tribunal considers appropriate.

(6)   If a claimant fails to comply with a peremptory order of the tribunal to provide security for costs, the tribunal may make an award dismissing his claim.

(7)   If a party fails to comply with any other kind of peremptory order, then, without prejudice to section 42 (enforcement by court of tribunal's peremptory orders), the tribunal may do any of the following—
   (a) direct that the party in default shall not be entitled to rely upon any allegation or material which was the subject matter of the order;
   (b) draw such adverse inferences from the act of non-compliance as the circumstances justify;
   (c) proceed to an award on the basis of such materials as have been properly provided to it;
   (d) make such order as it thinks fit as to the payment of costs of the arbitration incurred in consequence of the non-compliance.

NOTES

Commencement: 31 January 1997.

Powers of court in relation to arbitral proceedings

42   Enforcement of peremptory orders of tribunal

(1)   Unless otherwise agreed by the parties, the court may make an order requiring a party to comply with a peremptory order made by the tribunal.

(2)   An application for an order under this section may be made—
   (a) by the tribunal (upon notice to the parties),
   (b) by a party to the arbitral proceedings with the permission of the tribunal (and upon notice to the other parties), or
   (c) where the parties have agreed that the powers of the court under this section shall be available.

(3)   The court shall not act unless it is satisfied that the applicant has exhausted any available arbitral process in respect of failure to comply with the tribunal's order.

(4)   No order shall be made under this section unless the court is satisfied that the person to whom the tribunal's order was directed has failed to comply with it within the time prescribed in the order or, if no time was prescribed, within a reasonable time.

(5)   The leave of the court is required for any appeal from a decision of the court under this section.

NOTES

Commencement: 31 January 1997.

43   Securing the attendance of witnesses

(1)   A party to arbitral proceedings may use the same court procedures as are available in relation to legal proceedings to secure the attendance before the tribunal of a witness in order to give oral testimony or to produce documents or other material evidence.

(2)   This may only be done with the permission of the tribunal or the agreement of the other parties.

(3)   The court procedures may only be used if—
    (a) the witness is in the United Kingdom, and
    (b) the arbitral proceedings are being conducted in England and Wales or, as the case may be, Northern Ireland.

(4)   A person shall not be compelled by virtue of this section to produce any document or other material evidence which he could not be compelled to produce in legal proceedings.

NOTES

Commencement: 31 January 1997.

44   Court powers exercisable in support of arbitral proceedings

(1)   Unless otherwise agreed by the parties, the court has for the purposes of and in relation to arbitral proceedings the same power of making orders

about the matters listed below as it has for the purposes of and in relation to legal proceedings.

(2)   Those matters are—
  (a) the taking of the evidence of witnesses;
  (b) the preservation of evidence;
  (c) making orders relating to property which is the subject of the proceedings or as to which any question arises in the proceedings—
    (i) for the inspection, photographing, preservation, custody or detention of the property, or
    (ii) ordering that samples be taken from, or any observation be made of or experiment conducted upon, the property;
    and for that purpose authorising any person to enter any premises in the possession or control of a party to the arbitration;
  (d) the sale of any goods the subject of the proceedings;
  (e) the granting of an interim injunction or the appointment of a receiver.

(3)   If the case is one of urgency, the court may, on the application of a party or proposed party to the arbitral proceedings, make such orders as it thinks necessary for the purpose of preserving evidence or assets.

(4)   If the case is not one of urgency, the court shall act only on the application of a party to the arbitral proceedings (upon notice to the other parties and to the tribunal) made with the permission of the tribunal or the agreement in writing of the other parties.

(5)   In any case the court shall act only if or to the extent that the arbitral tribunal, and any arbitral or other institution or person vested by the parties with power in that regard, has no power or is unable for the time being to act effectively.

(6)   If the court so orders, an order made by it under this section shall cease to have effect in whole or in part on the order of the tribunal or of any such arbitral or other institution or person having power to act in relation to the subject-matter of the order.

(7)   The leave of the court is required for any appeal from a decision of the court under this section.

NOTES

Commencement: 31 January 1997.

45 Determination of preliminary point of law

(1)   Unless otherwise agreed by the parties, the court may on the application of a party to arbitral proceedings (upon notice to the other parties) determine any question of law arising in the course of the proceedings which the court is satisfied substantially affects the rights of one or more of the parties.

An agreement to dispense with reasons for the tribunal's award shall be considered an agreement to exclude the court's jurisdiction under this section.

(2)   An application under this section shall not be considered unless—
  (a) it is made with the agreement of all the other parties to the proceedings, or
  (b) it is made with the permission of the tribunal and the court is satisfied—
    (i)   that the determination of the question is likely to produce substantial savings in costs, and
    (ii) that the application was made without delay.

(3)   The application shall identify the question of law to be determined and, unless made with the agreement of all the other parties to the proceedings, shall state the grounds on which it is said that the question should be decided by the court.

(4)   Unless otherwise agreed by the parties, the arbitral tribunal may continue the arbitral proceedings and make an award while an application to the court under this section is pending.

(5)   Unless the court gives leave, no appeal lies from a decision of the court whether the conditions specified in subsection (2) are met.

(6)   The decision of the court on the question of law shall be treated as a judgment of the court for the purposes of an appeal.

But no appeal lies without the leave of the court which shall not be given unless the court considers that the question is one of general importance, or is one which for some other special reason should be considered by the Court of Appeal.

NOTES

Commencement: 31 January 1997.

The award

46 Rules applicable to substance of dispute

(1)   The arbitral tribunal shall decide the dispute—
  (a) in accordance with the law chosen by the parties as applicable to the substance of the dispute, or
  (b) if the parties so agree, in accordance with such other considerations as are agreed by them or determined by the tribunal.

(2)   For this purpose the choice of the laws of a country shall be understood to refer to the substantive laws of that country and not its conflict of laws rules.

(3)   If or to the extent that there is no such choice or agreement, the tribunal shall apply the law determined by the conflict of laws rules which it considers applicable.

NOTES

Commencement: 31 January 1997.

47 Awards on different issues, &c

(1)   Unless otherwise agreed by the parties, the tribunal may make more than one award at different times on different aspects of the matters to be determined.

(2)   The tribunal may, in particular, make an award relating—
  (a) to an issue affecting the whole claim, or
  (b) to a part only of the claims or cross-claims submitted to it for decision.

(3)   If the tribunal does so, it shall specify in its award the issue, or the claim or part of a claim, which is the subject matter of the award.

NOTES

Commencement: 31 January 1997.

48 Remedies

(1)   The parties are free to agree on the powers exercisable by the arbitral tribunal as regards remedies.

(2)   Unless otherwise agreed by the parties, the tribunal has the following powers.

(3)   The tribunal may make a declaration as to any matter to be determined in the proceedings.

(4)   The tribunal may order the payment of a sum of money, in any currency.

(5)   The tribunal has the same powers as the court—
  (a) to order a party to do or refrain from doing anything;
  (b) to order specific performance of a contract (other than a contract relating to land);
  (c) to order the rectification, setting aside or cancellation of a deed or other document.

NOTES

Commencement: 31 January 1997.

49   Interest

(1)   The parties are free to agree on the powers of the tribunal as regards the award of interest.

(2)   Unless otherwise agreed by the parties the following provisions apply.

(3)   The tribunal may award simple or compound interest from such dates, at such rates and with such rests as it considers meets the justice of the case—
  (a) on the whole or part of any amount awarded by the tribunal, in respect of any period up to the date of the award;
  (b) on the whole or part of any amount claimed in the arbitration and outstanding at the commencement of the arbitral proceedings but paid before the award was made, in respect of any period up to the date of payment.

(4)   The tribunal may award simple or compound interest from the date of the award (or any later date) until payment, at such rates and with such rests as it considers meets the justice of the case, on the outstanding amount of any award (including any award of interest under subsection (3) and any award as to costs).

(5)   References in this section to an amount awarded by the tribunal include an amount payable in consequence of a declaratory award by the tribunal.

(6)   The above provisions do not affect any other power of the tribunal to award interest.

NOTES

Commencement: 31 January 1997.

50   Extension of time for making award

(1)   Where the time for making an award is limited by or in pursuance of the arbitration agreement, then, unless otherwise agreed by the parties, the court may in accordance with the following provisions by order extend that time.

(2)   An application for an order under this section may be made—
   (a) by the tribunal (upon notice to the parties), or
   (b) by any party to the proceedings (upon notice to the tribunal and the other parties),

but only after exhausting any available arbitral process for obtaining an extension of time.

(3)   The court shall only make an order if satisfied that a substantial injustice would otherwise be done.

(4)   The court may extend the time for such period and on such terms as it thinks fit, and may do so whether or not the time previously fixed (by or under the agreement or by a previous order) has expired.

(5)   The leave of the court is required for any appeal from a decision of the court under this section.

NOTES

Commencement: 31 January 1997.

51   Settlement

(1)   If during arbitral proceedings the parties settle the dispute, the following provisions apply unless otherwise agreed by the parties.

(2)   The tribunal shall terminate the substantive proceedings and, if so requested by the parties and not objected to by the tribunal, shall record the settlement in the form of an agreed award.

(3)   An agreed award shall state that it is an award of the tribunal and shall have the same status and effect as any other award on the merits of the case.

(4)   The following provisions of this Part relating to awards (sections 52 to 58) apply to an agreed award.

(5)   Unless the parties have also settled the matter of the payment of the costs of the arbitration, the provisions of this Part relating to costs (sections 59 to 65) continue to apply.

NOTES

Commencement: 31 January 1997.

52   Form of award

(1)   The parties are free to agree on the form of an award.

(2)   If or to the extent that there is no such agreement, the following provisions apply.

(3)   The award shall be in writing signed by all the arbitrators or all those assenting to the award.

(4)   The award shall contain the reasons for the award unless it is an agreed award or the parties have agreed to dispense with reasons.

(5)   The award shall state the seat of the arbitration and the date when the award is made.

NOTES

Commencement: 31 January 1997.

53   Place where award treated as made

Unless otherwise agreed by the parties, where the seat of the arbitration is in England and Wales or Northern Ireland, any award in the proceedings shall be treated as made there, regardless of where it was signed, despatched or delivered to any of the parties.

Commencement: 31 January 1997.

54   Date of award

(1)   Unless otherwise agreed by the parties, the tribunal may decide what is to be taken to be the date on which the award was made.

(2)   In the absence of any such decision, the date of the award shall be taken to be the date on which it is signed by the arbitrator or, where more than one arbitrator signs the award, by the last of them.

Commencement: 31 January 1997.

55   Notification of award

(1)   The parties are free to agree on the requirements as to notification of the award to the parties.

(2)   If there is no such agreement, the award shall be notified to the parties by service on them of copies of the award, which shall be done without delay after the award is made.

(3)   Nothing in this section affects section 56 (power to withhold award in case of non-payment).

Commencement: 31 January 1997.

56   Power to withhold award in case of non-payment

(1)   The tribunal may refuse to deliver an award to the parties except upon full payment of the fees and expenses of the arbitrators.

(2)   If the tribunal refuses on that ground to deliver an award, a party to the arbitral proceedings may (upon notice to the other parties and the tribunal) apply to the court, which may order that—

(a) the tribunal shall deliver the award on the payment into court by the applicant of the fees and expenses demanded, or such lesser amount as the court may specify,

(b) the amount of the fees and expenses properly payable shall be determined by such means and upon such terms as the court may direct, and

(c) out of the money paid into court there shall be paid out such fees and expenses as may be found to be properly payable and the balance of the money (if any) shall be paid out to the applicant.

(3)   For this purpose the amount of fees and expenses properly payable is the amount the applicant is liable to pay under section 28 or any agreement relating to the payment of the arbitrators.

(4)   No application to the court may be made where there is any available arbitral process for appeal or review of the amount of the fees or expenses demanded.

(5)   References in this section to arbitrators include an arbitrator who has ceased to act and an umpire who has not replaced the other arbitrators.

(6)   The above provisions of this section also apply in relation to any arbitral or other institution or person vested by the parties with powers in relation to the delivery of the tribunal's award.

As they so apply, the references to the fees and expenses of the arbitrators shall be construed as including the fees and expenses of that institution or person.

(7)   The leave of the court is required for any appeal from a decision of the court under this section.

(8)   Nothing in this section shall be construed as excluding an application under section 28 where payment has been made to the arbitrators in order to obtain the award.

NOTES

Commencement: 31 January 1997.

57   Correction of award or additional award

(1)   The parties are free to agree on the powers of the tribunal to correct an award or make an additional award.

(2)   If or to the extent there is no such agreement, the following provisions apply.

(3)   The tribunal may on its own initiative or on the application of a party—
    (a) correct an award so as to remove any clerical mistake or error arising from an accidental slip or omission or clarify or remove any ambiguity in the award, or
    (b) make an additional award in respect of any claim (including a claim for interest or costs) which was presented to the tribunal but was not dealt with in the award.

These powers shall not be exercised without first affording the other parties a reasonable opportunity to make representations to the tribunal.

(4)   Any application for the exercise of those powers must be made within 28 days of the date of the award or such longer period as the parties may agree.

(5)   Any correction of an award shall be made within 28 days of the date the application was received by the tribunal or, where the correction is made by the tribunal on its own initiative, within 28 days of the date of the award or, in either case, such longer period as the parties may agree.

(6)   Any additional award shall be made within 56 days of the date of the original award or such longer period as the parties may agree.

(7)   Any correction of an award shall form part of the award.

NOTES
    Commencement: 31 January 1997.

58   Effect of award

(1)   Unless otherwise agreed by the parties, an award made by the tribunal pursuant to an arbitration agreement is final and binding both on the parties and on any persons claiming through or under them.

(2)   This does not affect the right of a person to challenge the award by any available arbitral process of appeal or review or in accordance with the provisions of this Part.

Commencement: 31 January 1997.

Costs of the arbitration

59 Costs of the arbitration

(1)   References in this Part to the costs of the arbitration are to—
   (a) the arbitrators' fees and expenses,
   (b) the fees and expenses of any arbitral institution concerned, and
   (c) the legal or other costs of the parties.

(2)   Any such reference includes the costs of or incidental to any proceedings to determine the amount of the recoverable costs of the arbitration (see section 63).

Commencement: 31 January 1997.

60 Agreement to pay costs in any event

An agreement which has the effect that a party is to pay the whole or part of the costs of the arbitration in any event is only valid if made after the dispute in question has arisen.

Commencement: 31 January 1997.

61 Award of costs

(1)   The tribunal may make an award allocating the costs of the arbitration as between the parties, subject to any agreement of the parties.

(2)   Unless the parties otherwise agree, the tribunal shall award costs on the general principle that costs should follow the event except where it appears to the tribunal that in the circumstances this is not appropriate in relation to the whole or part of the costs.

Commencement: 31 January 1997.

62 Effect of agreement or award about costs

Unless the parties otherwise agree, any obligation under an agreement between them as to how the costs of the arbitration are to be borne, or under an award allocating the costs of the arbitration, extends only to such costs as are recoverable.

Commencement: 31 January 1997.

63 The recoverable costs of the arbitration

(1)   The parties are free to agree what costs of the arbitration are recoverable.

(2)   If or to the extent there is no such agreement, the following provisions apply.

(3)   The tribunal may determine by award the recoverable costs of the arbitration on such basis as it thinks fit.

If it does so, it shall specify—
(a) the basis on which it has acted, and
(b) the items of recoverable costs and the amount referable to each.

(4)   If the tribunal does not determine the recoverable costs of the arbitration, any party to the arbitral proceedings may apply to the court (upon notice to the other parties) which may—
(a) determine the recoverable costs of the arbitration on such basis as it thinks fit, or
(b) order that they shall be determined by such means and upon such terms as it may specify.

(5)   Unless the tribunal or the court determines otherwise—
(a) the recoverable costs of the arbitration shall be determined on the basis that there shall be allowed a reasonable amount in respect of all costs reasonably incurred, and

(b) any doubt as to whether costs were reasonably incurred or were reasonable in amount shall be resolved in favour of the paying party.

(6) The above provisions have effect subject to section 64 (recoverable fees and expenses of arbitrators).

(7) Nothing in this section affects any right of the arbitrators, any expert, legal adviser or assessor appointed by the tribunal, or any arbitral institution, to payment of their fees and expenses.

NOTES

Commencement: 31 January 1997.

64 Recoverable fees and expenses of arbitrators

(1) Unless otherwise agreed by the parties, the recoverable costs of the arbitration shall include in respect of the fees and expenses of the arbitrators only such reasonable fees and expenses as are appropriate in the circumstances.

(2) If there is any question as to what reasonable fees and expenses are appropriate in the circumstances, and the matter is not already before the court on an application under section 63(4), the court may on the application of any party (upon notice to the other parties)—
(a) determine the matter, or
(b) order that it be determined by such means and upon such terms as the court may specify.

(3) Subsection (1) has effect subject to any order of the court under section 24(4) or 25(3)(b) (order as to entitlement to fees or expenses in case of removal or resignation of arbitrator).

(4) Nothing in this section affects any right of the arbitrator to payment of his fees and expenses.

NOTES

Commencement: 31 January 1997.

65   Power to limit recoverable costs

(1)   Unless otherwise agreed by the parties, the tribunal may direct that the recoverable costs of the arbitration, or of any part of the arbitral proceedings, shall be limited to a specified amount.

(2)   Any direction may be made or varied at any stage, but this must be done sufficiently in advance of the incurring of costs to which it relates, or the taking of any steps in the proceedings which may be affected by it, for the limit to be taken into account.

NOTES

Commencement: 31 January 1997.

Powers of the court in relation to award

66   Enforcement of the award

(1)   An award made by the tribunal pursuant to an arbitration agreement may, by leave of the court, be enforced in the same manner as a judgment or order of the court to the same effect.

(2)   Where leave is so given, judgment may be entered in terms of the award.

(3)   Leave to enforce an award shall not be given where, or to the extent that, the person against whom it is sought to be enforced shows that the tribunal lacked substantive jurisdiction to make the award.

The right to raise such an objection may have been lost (see section 73).

(4)   Nothing in this section affects the recognition or enforcement of an award under any other enactment or rule of law, in particular under Part II of the Arbitration Act 1950 (enforcement of awards under Geneva Convention) or the provisions of Part III of this Act relating to the recognition and enforcement of awards under the New York Convention or by an action on the award.

NOTES

Commencement: 31 January 1997.

67   Challenging the award: substantive jurisdiction

(1)   A party to arbitral proceedings may (upon notice to the other parties and to the tribunal) apply to the court—
   (a) challenging any award of the arbitral tribunal as to its substantive jurisdiction; or
   (b) for an order declaring an award made by the tribunal on the merits to be of no effect, in whole or in part, because the tribunal did not have substantive jurisdiction.

A party may lose the right to object (see section 73) and the right to apply is subject to the restrictions in section 70(2) and (3).

(2)   The arbitral tribunal may continue the arbitral proceedings and make a further award while an application to the court under this section is pending in relation to an award as to jurisdiction.

(3)   On an application under this section challenging an award of the arbitral tribunal as to its substantive jurisdiction, the court may by order—
   (a) confirm the award,
   (b) vary the award, or
   (c) set aside the award in whole or in part.

(4)   The leave of the court is required for any appeal from a decision of the court under this section.

NOTES

Commencement: 31 January 1997.

68   Challenging the award: serious irregularity

(1)   A party to arbitral proceedings may (upon notice to the other parties and to the tribunal) apply to the court challenging an award in the proceedings on the ground of serious irregularity affecting the tribunal, the proceedings or the award.

A party may lose the right to object (see section 73) and the right to apply is subject to the restrictions in section 70(2) and (3).

(2)   Serious irregularity means an irregularity of one or more of the following kinds which the court considers has caused or will cause substantial injustice to the applicant—

(a) failure by the tribunal to comply with section 33 (general duty of tribunal);

(b) the tribunal exceeding its powers (otherwise than by exceeding its substantive jurisdiction: see section 67);

(c) failure by the tribunal to conduct the proceedings in accordance with the procedure agreed by the parties;

(d) failure by the tribunal to deal with all the issues that were put to it;

(e) any arbitral or other institution or person vested by the parties with powers in relation to the proceedings or the award exceeding its powers;

(f) uncertainty or ambiguity as to the effect of the award;

(g) the award being obtained by fraud or the award or the way in which it was procured being contrary to public policy;

(h) failure to comply with the requirements as to the form of the award; or

(i) any irregularity in the conduct of the proceedings or in the award which is admitted by the tribunal or by any arbitral or other institution or person vested by the parties with powers in relation to the proceedings or the award.

(3)   If there is shown to be serious irregularity affecting the tribunal, the proceedings or the award, the court may—

(a) remit the award to the tribunal, in whole or in part, for reconsideration,

(b) set the award aside in whole or in part, or

(c) declare the award to be of no effect, in whole or in part.

The court shall not exercise its power to set aside or to declare an award to be of no effect, in whole or in part, unless it is satisfied that it would be inappropriate to remit the matters in question to the tribunal for reconsideration.

(4)   The leave of the court is required for any appeal from a decision of the court under this section.

NOTES

Commencement: 31 January 1997.

69 Appeal on point of law

(1)   Unless otherwise agreed by the parties, a party to arbitral proceedings may (upon notice to the other parties and to the tribunal) appeal to the court on a question of law arising out of an award made in the proceedings.

An agreement to dispense with reasons for the tribunal's award shall be considered an agreement to exclude the court's jurisdiction under this section.

(2)   An appeal shall not be brought under this section except—
    (a) with the agreement of all the other parties to the proceedings, or
    (b) with the leave of the court.

The right to appeal is also subject to the restrictions in section 70(2) and (3).

(3)   Leave to appeal shall be given only if the court is satisfied—
    (a) that the determination of the question will substantially affect the rights of one or more of the parties,
    (b) that the question is one which the tribunal was asked to determine,
    (c) that, on the basis of the findings of fact in the award—
        (i)   the decision of the tribunal on the question is obviously wrong, or
        (ii) the question is one of general public importance and the decision of the tribunal is at least open to serious doubt, and
    (d) that, despite the agreement of the parties to resolve the matter by arbitration, it is just and proper in all the circumstances for the court to determine the question.

(4)   An application for leave to appeal under this section shall identify the question of law to be determined and state the grounds on which it is alleged that leave to appeal should be granted.

(5)   The court shall determine an application for leave to appeal under this section without a hearing unless it appears to the court that a hearing is required.

(6)   The leave of the court is required for any appeal from a decision of the court under this section to grant or refuse leave to appeal.

(7)   On an appeal under this section the court may by order—
    (a) confirm the award,
    (b) vary the award,
    (c) remit the award to the tribunal, in whole or in part, for reconsideration in the light of the court's determination, or
    (d) set aside the award in whole or in part.

The court shall not exercise its power to set aside an award, in whole or in part, unless it is satisfied that it would be inappropriate to remit the matters in question to the tribunal for reconsideration.

(8)   The decision of the court on an appeal under this section shall be treated as a judgment of the court for the purposes of a further appeal.

But no such appeal lies without the leave of the court which shall not be given unless the court considers that the question is one of general importance or is one which for some other special reason should be considered by the Court of Appeal.

NOTES

Commencement: 31 January 1997.

70   Challenge or appeal: supplementary provisions

(1)   The following provisions apply to an application or appeal under section 67, 68 or 69.

(2)   An application or appeal may not be brought if the applicant or appellant has not first exhausted—
   (a) any available arbitral process of appeal or review, and
   (b) any available recourse under section 57 (correction of award or additional award).

(3)   Any application or appeal must be brought within 28 days of the date of the award or, if there has been any arbitral process of appeal or review, of the date when the applicant or appellant was notified of the result of that process.

(4)   If on an application or appeal it appears to the court that the award—
   (a) does not contain the tribunal's reasons, or
   (b) does not set out the tribunal's reasons in sufficient detail to enable the court properly to consider the application or appeal,

the court may order the tribunal to state the reasons for its award in sufficient detail for that purpose.

(5)   Where the court makes an order under subsection (4), it may make such further order as it thinks fit with respect to any additional costs of the arbitration resulting from its order.

(6)   The court may order the applicant or appellant to provide security for the costs of the application or appeal, and may direct that the application or appeal be dismissed if the order is not complied with.

The power to order security for costs shall not be exercised on the ground that the applicant or appellant is—

(a) an individual ordinarily resident outside the United Kingdom, or

(b) a corporation or association incorporated or formed under the law of a country outside the United Kingdom, or whose central management and control is exercised outside the United Kingdom.

(7) The court may order that any money payable under the award shall be brought into court or otherwise secured pending the determination of the application or appeal, and may direct that the application or appeal be dismissed if the order is not complied with.

(8) The court may grant leave to appeal subject to conditions to the same or similar effect as an order under subsection (6) or (7).

This does not affect the general discretion of the court to grant leave subject to conditions.

NOTES

Commencement: 31 January 1997.

71 Challenge or appeal: effect of order of court

(1) The following provisions have effect where the court makes an order under section 67, 68 or 69 with respect to an award.

(2) Where the award is varied, the variation has effect as part of the tribunal's award.

(3) Where the award is remitted to the tribunal, in whole or in part, for reconsideration, the tribunal shall make a fresh award in respect of the matters remitted within three months of the date of the order for remission or such longer or shorter period as the court may direct.

(4) Where the award is set aside or declared to be of no effect, in whole or in part, the court may also order that any provision that an award is a condition precedent to the bringing of legal proceedings in respect of a matter to which the arbitration agreement applies, is of no effect as regards the subject matter of the award or, as the case may be, the relevant part of the award.

Commencement: 31 January 1997.

Miscellaneous

72 Saving for rights of person who takes no part in proceedings

(1) A person alleged to be a party to arbitral proceedings but who takes no part in the proceedings may question—
  (a) whether there is a valid arbitration agreement,
  (b) whether the tribunal is properly constituted, or
  (c) what matters have been submitted to arbitration in accordance with the arbitration agreement,

by proceedings in the court for a declaration or injunction or other appropriate relief.

(2) He also has the same right as a party to the arbitral proceedings to challenge an award—
  (a) by an application under section 67 on the ground of lack of substantive jurisdiction in relation to him, or
  (b) by an application under section 68 on the ground of serious irregularity (within the meaning of that section) affecting him;

and section 70(2) (duty to exhaust arbitral procedures) does not apply in his case.

Commencement: 31 January 1997.

73 Loss of right to object

(1) If a party to arbitral proceedings takes part, or continues to take part, in the proceedings without making, either forthwith or within such time as is allowed by the arbitration agreement or the tribunal or by any provision of this Part, any objection—
  (a) that the tribunal lacks substantive jurisdiction,
  (b) that the proceedings have been improperly conducted,
  (c) that there has been a failure to comply with the arbitration agreement or with any provision of this Part, or

(d) that there has been any other irregularity affecting the tribunal or the proceedings,

he may not raise that objection later, before the tribunal or the court, unless he shows that, at the time he took part or continued to take part in the proceedings, he did not know and could not with reasonable diligence have discovered the grounds for the objection.

(2)   Where the arbitral tribunal rules that it has substantive jurisdiction and a party to arbitral proceedings who could have questioned that ruling—

(a) by any available arbitral process of appeal or review, or

(b) by challenging the award,

does not do so, or does not do so within the time allowed by the arbitration agreement or any provision of this Part, he may not object later to the tribunal's substantive jurisdiction on any ground which was the subject of that ruling.

NOTES

Commencement: 31 January 1997.

74   Immunity of arbitral institutions, &c

(1)   An arbitral or other institution or person designated or requested by the parties to appoint or nominate an arbitrator is not liable for anything done or omitted in the discharge or purported discharge of that function unless the act or omission is shown to have been in bad faith.

(2)   An arbitral or other institution or person by whom an arbitrator is appointed or nominated is not liable, by reason of having appointed or nominated him, for anything done or omitted by the arbitrator (or his employees or agents) in the discharge or purported discharge of his functions as arbitrator.

(3)   The above provisions apply to an employee or agent of an arbitral or other institution or person as they apply to the institution or person himself.

NOTES

Commencement: 31 January 1997.

75 Charge to secure payment of solicitors' costs

The powers of the court to make declarations and orders under section 73 of the Solicitors Act 1974 or Article 71H of the Solicitors (Northern Ireland) Order 1976 (power to charge property recovered in the proceedings with the payment of solicitors' costs) may be exercised in relation to arbitral proceedings as if those proceedings were proceedings in the court.

NOTES

Commencement: 31 January 1997.

Supplementary

76 Service of notices, &c

(1)   The parties are free to agree on the manner of service of any notice or other document required or authorised to be given or served in pursuance of the arbitration agreement or for the purposes of the arbitral proceedings.

(2)   If or to the extent that there is no such agreement the following provisions apply.

(3)   A notice or other document may be served on a person by any effective means.

(4)   If a notice or other document is addressed, pre-paid and delivered by post—
    (a) to the addressee's last known principal residence or, if he is or has been carrying on a trade, profession or business, his last known principal business address, or
    (b) where the addressee is a body corporate, to the body's registered or principal office,

it shall be treated as effectively served.

(5)   This section does not apply to the service of documents for the purposes of legal proceedings, for which provision is made by rules of court.

(6)   References in this Part to a notice or other document include any form of communication in writing and references to giving or serving a notice or other document shall be construed accordingly.

NOTES
Commencement: 31 January 1997.

77 Powers of court in relation to service of documents

(1)   This section applies where service of a document on a person in the manner agreed by the parties, or in accordance with provisions of section 76 having effect in default of agreement, is not reasonably practicable.

(2)   Unless otherwise agreed by the parties, the court may make such order as it thinks fit—
  (a) for service in such manner as the court may direct, or
  (b) dispensing with service of the document.

(3)   Any party to the arbitration agreement may apply for an order, but only after exhausting any available arbitral process for resolving the matter.

(4)   The leave of the court is required for any appeal from a decision of the court under this section.

NOTES
Commencement: 31 January 1997.

78 Reckoning periods of time

(1)   The parties are free to agree on the method of reckoning periods of time for the purposes of any provision agreed by them or any provision of this Part having effect in default of such agreement.

(2)   If or to the extent there is no such agreement, periods of time shall be reckoned in accordance with the following provisions.

(3)   Where the act is required to be done within a specified period after or from a specified date, the period begins immediately after that date.

(4)   Where the act is required to be done a specified number of clear days after a specified date, at least that number of days must intervene between the day on which the act is done and that date.

(5)   Where the period is a period of seven days or less which would include a Saturday, Sunday or a public holiday in the place where

anything which has to be done within the period falls to be done, that day shall be excluded.

In relation to England and Wales or Northern Ireland, a "public holiday" means Christmas Day, Good Friday or a day which under the Banking and Financial Dealings Act 1971 is a bank holiday.

NOTES

Commencement: 31 January 1997.

79 Power of court to extend time limits relating to arbitral proceedings

(1)   Unless the parties otherwise agree, the court may by order extend any time limit agreed by them in relation to any matter relating to the arbitral proceedings or specified in any provision of this Part having effect in default of such agreement.

This section does not apply to a time limit to which section 12 applies (power of court to extend time for beginning arbitral proceedings, &c).

(2)   An application for an order may be made—
    (a) by any party to the arbitral proceedings (upon notice to the other parties and to the tribunal), or
    (b) by the arbitral tribunal (upon notice to the parties).

(3)   The court shall not exercise its power to extend a time limit unless it is satisfied—
    (a) that any available recourse to the tribunal, or to any arbitral or other institution or person vested by the parties with power in that regard, has first been exhausted, and
    (b) that a substantial injustice would otherwise be done.

(4)   The court's power under this section may be exercised whether or not the time has already expired.

(5)   An order under this section may be made on such terms as the court thinks fit.

(6)   The leave of the court is required for any appeal from a decision of the court under this section.

NOTES
Commencement: 31 January 1997.

80 Notice and other requirements in connection with legal proceedings

(1)   References in this Part to an application, appeal or other step in relation to legal proceedings being taken "upon notice" to the other parties to the arbitral proceedings, or to the tribunal, are to such notice of the originating process as is required by rules of court and do not impose any separate requirement.

(2)   Rules of court shall be made—
  (a) requiring such notice to be given as indicated by any provision of this Part, and
  (b) as to the manner, form and content of any such notice.

(3)   Subject to any provision made by rules of court, a requirement to give notice to the tribunal of legal proceedings shall be construed—
  (a) if there is more than one arbitrator, as a requirement to give notice to each of them; and
  (b) if the tribunal is not fully constituted, as a requirement to give notice to any arbitrator who has been appointed.

(4)   References in this Part to making an application or appeal to the court within a specified period are to the issue within that period of the appropriate originating process in accordance with rules of court.

(5)   Where any provision of this Part requires an application or appeal to be made to the court within a specified time, the rules of court relating to the reckoning of periods, the extending or abridging of periods, and the consequences of not taking a step within the period prescribed by the rules, apply in relation to that requirement.

(6)   Provision may be made by rules of court amending the provisions of this Part—
  (a) with respect to the time within which any application or appeal to the court must be made,
  (b) so as to keep any provision made by this Part in relation to arbitral proceedings in step with the corresponding provision of rules of court applying in relation to proceedings in the court, or
  (c) so as to keep any provision made by this Part in relation to legal proceedings in step with the corresponding provision of rules of court applying generally in relation to proceedings in the court.

(7)   Nothing in this section affects the generality of the power to make rules of court.

NOTES

Commencement: 31 January 1997.

81   Saving for certain matters governed by common law

(1)   Nothing in this Part shall be construed as excluding the operation of any rule of law consistent with the provisions of this Part, in particular, any rule of law as to—
  (a) matters which are not capable of settlement by arbitration;
  (b) the effect of an oral arbitration agreement; or
  (c) the refusal of recognition or enforcement of an arbitral award on grounds of public policy.

(2)   Nothing in this Act shall be construed as reviving any jurisdiction of the court to set aside or remit an award on the ground of errors of fact or law on the face of the award.

NOTES

Commencement: 31 January 1997.

82   Minor definitions

(1)   In this Part—
  "arbitrator", unless the context otherwise requires, includes an umpire;
  "available arbitral process", in relation to any matter, includes any process of appeal to or review by an arbitral or other institution or person vested by the parties with powers in relation to that matter;
  "claimant", unless the context otherwise requires, includes a counterclaimant, and related expressions shall be construed accordingly;
  "dispute" includes any difference;
  "enactment" includes an enactment contained in Northern Ireland legislation;
  "legal proceedings" means civil proceedings in the High Court or a county court;
  "peremptory order" means an order made under section 41(5) or made in exercise of any corresponding power conferred by the parties;

"premises" includes land, buildings, moveable structures, vehicles, vessels, aircraft and hovercraft;

"question of law" means—

    (a) for a court in England and Wales, a question of the law of England and Wales, and

    (b) for a court in Northern Ireland, a question of the law of Northern Ireland;

"substantive jurisdiction", in relation to an arbitral tribunal, refers to the matters specified in section 30(1)(a) to (c), and references to the tribunal exceeding its substantive jurisdiction shall be construed accordingly.

(2)   References in this Part to a party to an arbitration agreement include any person claiming under or through a party to the agreement.

NOTES

Commencement: 31 January 1997.

83   Index of defined expressions: Part I

In this Part the expressions listed below are defined or otherwise explained by the provisions indicated—

agreement, agree and agreed	section 5(1)
agreement in writing	section 5(2) to (5)
arbitration agreement	sections 6 and 5(1)
arbitrator	section 82(1)
available arbitral process	section 82(1)
claimant	section 82(1)
commencement (in relation to arbitral proceedings)	section 14
costs of the arbitration	section 59
the court	section 105
dispute	section 82(1)
enactment	section 82(1)
legal proceedings	section 82(1)
Limitation Acts	section 13(4)
notice (or other document)	section 76(6)
party—	
— in relation to an arbitration agreement	section 82(2)
— where section 106(2) or (3) applies	section 106(4)
peremptory order	section 82(1) (and see section 41(5))
premises	section 82(1)
question of law	section 82(1)
recoverable costs	sections 63 and 64
seat of the arbitration	section 3

serve and service (of notice or other document)	section 76(6)
substantive jurisdiction (in relation to an arbitral tribunal)	section 82(1) (and see section 30(1)(a) to(c))
upon notice (to the parties or the tribunal)	section 80
written and in writing	section 5(6)

NOTES

Commencement: 31 January 1997.

84 Transitional provisions

(1)   The provisions of this Part do not apply to arbitral proceedings commenced before the date on which this Part comes into force.

(2)   They apply to arbitral proceedings commenced on or after that date under an arbitration agreement whenever made.

(3)   The above provisions have effect subject to any transitional provision made by an order under section 109(2) (power to include transitional provisions in commencement order).

NOTES

Commencement: 31 January 1997.

PART II
OTHER PROVISIONS RELATING TO ARBITRATION

Domestic arbitration agreements

85 Modification of Part I in relation to domestic arbitration agreement

(1)   In the case of a domestic arbitration agreement the provisions of Part I are modified in accordance with the following sections.

(2)   For this purpose a "domestic arbitration agreement" means an arbitration agreement to which none of the parties is—
   (a) an individual who is a national of, or habitually resident in, a state other than the United Kingdom, or

(b) a body corporate which is incorporated in, or whose central control and management is exercised in, a state other than the United Kingdom,

and under which the seat of the arbitration (if the seat has been designated or determined) is in the United Kingdom.

(3)   In subsection (2) "arbitration agreement" and "seat of the arbitration" have the same meaning as in Part I (see sections 3, 5(1) and 6).

NOTES
Commencement: to be appointed.

86   Staying of legal proceedings

(1)   In section 9 (stay of legal proceedings), subsection (4) (stay unless the arbitration agreement is null and void, inoperative, or incapable of being performed) does not apply to a domestic arbitration agreement.

(2)   On an application under that section in relation to a domestic arbitration agreement the court shall grant a stay unless satisfied—
   (a) that the arbitration agreement is null and void, inoperative, or incapable of being performed, or
   (b) that there are other sufficient grounds for not requiring the parties to abide by the arbitration agreement.

(3)   The court may treat as a sufficient ground under subsection (2)(b) the fact that the applicant is or was at any material time not ready and willing to do all things necessary for the proper conduct of the arbitration or of any other dispute resolution procedures required to be exhausted before resorting to arbitration.

(4)   For the purposes of this section the question whether an arbitration agreement is a domestic arbitration agreement shall be determined by reference to the facts at the time the legal proceedings are commenced.

NOTES
Commencement: to be appointed.

87   Effectiveness of agreement to exclude court's jurisdiction

(1)   In the case of a domestic arbitration agreement any agreement to exclude the jurisdiction of the court under—

(a) section 45 (determination of preliminary point of law), or

(b) section 69 (challenging the award: appeal on point of law),

is not effective unless entered into after the commencement of the arbitral proceedings in which the question arises or the award is made.

(2)   For this purpose the commencement of the arbitral proceedings has the same meaning as in Part I (see section 14).

(3)   For the purposes of this section the question whether an arbitration agreement is a domestic arbitration agreement shall be determined by reference to the facts at the time the agreement is entered into.

NOTES

Commencement: to be appointed.

88   Power to repeal or amend sections 85 to 87

(1)   The Secretary of State may by order repeal or amend the provisions of sections 85 to 87.

(2)   An order under this section may contain such supplementary, incidental and transitional provisions as appear to the Secretary of State to be appropriate.

(3)   An order under this section shall be made by statutory instrument and no such order shall be made unless a draft of it has been laid before and approved by a resolution of each House of Parliament.

NOTES

Commencement: 31 January 1997.

Consumer arbitration agreements

89   Application of unfair terms regulations to consumer arbitration agreements

(1)   The following sections extend the application of the Unfair Terms in Consumer Contracts Regulations 1994 in relation to a term which constitutes an arbitration agreement.

For this purpose "arbitration agreement" means an agreement to submit to arbitration present or future disputes or differences (whether or not contractual).

(2)   In those sections "the Regulations" means those regulations and includes any regulations amending or replacing those regulations.

(3)   Those sections apply whatever the law applicable to the arbitration agreement.

NOTES

Commencement: 31 January 1997.

90   Regulations apply where consumer is a legal person

The Regulations apply where the consumer is a legal person as they apply where the consumer is a natural person.

NOTES

Commencement: 31 January 1997.

91   Arbitration agreement unfair where modest amount sought

(1)   A term which constitutes an arbitration agreement is unfair for the purposes of the Regulations so far as it relates to a claim for a pecuniary remedy which does not exceed the amount specified by order for the purposes of this section.

(2)   Orders under this section may make different provision for different cases and for different purposes.

(3)   The power to make orders under this section is exercisable—
   (a) for England and Wales, by the Secretary of State with the concurrence of the Lord Chancellor,
   (b) for Scotland, by the Secretary of State . . . , and
   (c) for Northern Ireland, by the Department of Economic Development for Northern Ireland with the concurrence of the Lord Chancellor.

(4)   Any such order for England and Wales or Scotland shall be made by statutory instrument which shall be subject to annulment in pursuance of a resolution of either House of Parliament.

(5)   Any such order for Northern Ireland shall be a statutory rule for the purposes of the Statutory Rules (Northern Ireland) Order 1979 and shall be subject to negative resolution, within the meaning of section 41(6) of the Interpretation Act (Northern Ireland) 1954.

NOTES

Commencement: 19 December 1996 (certain purposes); 31 January 1997 (remaining purposes).

Sub-s (3): words omitted from para (b) repealed by the Transfer of Functions (Lord Advocate and Secretary of State) Order 1999, SI 1999/678, art 6.

Transfer of functions: functions of the Lord Advocate are transferred to the Secretary of State by virtue of SI 1999/678, art 2, Schedule.

Small claims arbitration in the county court

92   Exclusion of Part I in relation to small claims arbitration in the county court

Nothing in Part I of this Act applies to arbitration under section 64 of the County Courts Act 1984.

NOTES

Commencement: 31 January 1997.

Appointment of judges as arbitrators

93   Appointment of judges as arbitrators

(1)   A judge of the Commercial Court or an official referee may, if in all the circumstances he thinks fit, accept appointment as a sole arbitrator or as umpire by or by virtue of an arbitration agreement.

(2)   A judge of the Commercial Court shall not do so unless the Lord Chief Justice has informed him that, having regard to the state of business in the High Court and the Crown Court, he can be made available.

(3)   An official referee shall not do so unless the Lord Chief Justice has informed him that, having regard to the state of official referees' business, he can be made available.

(4)   The fees payable for the services of a judge of the Commercial Court or official referee as arbitrator or umpire shall be taken in the High Court.

(5)   In this section—
   "arbitration agreement" has the same meaning as in Part I; and
   "official referee" means a person nominated under section 68(1)(a) of the Supreme Court Act 1981 to deal with official referees' business.

(6)   The provisions of Part I of this Act apply to arbitration before a person appointed under this section with the modifications specified in Schedule 2.

NOTES
   Commencement: 31 January 1997.

Statutory arbitrations

94   Application of Part I to statutory arbitrations

(1)   The provisions of Part I apply to every arbitration under an enactment (a "statutory arbitration"), whether the enactment was passed or made before or after the commencement of this Act, subject to the adaptations and exclusions specified in sections 95 to 98.

(2)   The provisions of Part I do not apply to a statutory arbitration if or to the extent that their application—
   (a) is inconsistent with the provisions of the enactment concerned, with any rules or procedure authorised or recognised by it, or
   (b) is excluded by any other enactment.

(3)   In this section and the following provisions of this Part "enactment"—
   (a) in England and Wales, includes an enactment contained in subordinate legislation within the meaning of the Interpretation Act 1978;
   (b) in Northern Ireland, means a statutory provision within the meaning of section 1(f) of the Interpretation Act (Northern Ireland) 1954.

NOTES
   Commencement: 31 January 1997.

95 General adaptation of provisions in relation to statutory arbitrations

(1)   The provisions of Part I apply to a statutory arbitration—
  (a) as if the arbitration were pursuant to an arbitration agreement and as if the enactment were that agreement, and
  (b) as if the persons by and against whom a claim subject to arbitration in pursuance of the enactment may be or has been made were parties to that agreement.

(2)   Every statutory arbitration shall be taken to have its seat in England and Wales or, as the case may be, in Northern Ireland.

NOTES

Commencement: 31 January 1997.

96 Specific adaptations of provisions in relation to statutory arbitrations

(1)   The following provisions of Part I apply to a statutory arbitration with the following adaptations.

(2)   In section 30(1) (competence of tribunal to rule on its own jurisdiction), the reference in paragraph (a) to whether there is a valid arbitration agreement shall be construed as a reference to whether the enactment applies to the dispute or difference in question.

(3)   Section 35 (consolidation of proceedings and concurrent hearings) applies only so as to authorise the consolidation of proceedings, or concurrent hearings in proceedings, under the same enactment.

(4)   Section 46 (rules applicable to substance of dispute) applies with the omission of subsection (1)(b) (determination in accordance with considerations agreed by parties).

NOTES

Commencement: 31 January 1997.

97 Provisions excluded from applying to statutory arbitrations

The following provisions of Part I do not apply in relation to a statutory arbitration—

(a) section 8 (whether agreement discharged by death of a party);

(b) section 12 (power of court to extend agreed time limits);

(c) sections 9(5), 10(2) and 71(4) (restrictions on effect of provision that award condition precedent to right to bring legal proceedings).

NOTES

Commencement: 31 January 1997.

98 Power to make further provision by regulations

(1) The Secretary of State may make provision by regulations for adapting or excluding any provision of Part I in relation to statutory arbitrations in general or statutory arbitrations of any particular description.

(2) The power is exercisable whether the enactment concerned is passed or made before or after the commencement of this Act.

(3) Regulations under this section shall be made by statutory instrument which shall be subject to annulment in pursuance of a resolution of either House of Parliament.

NOTES

Commencement: 31 January 1997.

PART III
RECOGNITION AND ENFORCEMENT OF CERTAIN FOREIGN AWARDS

Enforcement of Geneva Convention awards

99 Continuation of Part II of the Arbitration Act 1950

Part II of the Arbitration Act 1950 (enforcement of certain foreign awards) continues to apply in relation to foreign awards within the meaning of that Part which are not also New York Convention awards.

NOTES

Commencement: 31 January 1997.

Recognition and enforcement of New York Convention awards

100 New York Convention awards

(1)   In this Part a "New York Convention award" means an award made, in pursuance of an arbitration agreement, in the territory of a state (other than the United Kingdom) which is a party to the New York Convention.

(2)   For the purposes of subsection (1) and of the provisions of this Part relating to such awards—
  (a) "arbitration agreement" means an arbitration agreement in writing, and
  (b) an award shall be treated as made at the seat of the arbitration, regardless of where it was signed, despatched or delivered to any of the parties.

  In this subsection "agreement in writing" and "seat of the arbitration" have the same meaning as in Part I.

(3)   If Her Majesty by Order in Council declares that a state specified in the Order is a party to the New York Convention, or is a party in respect of any territory so specified, the Order shall, while in force, be conclusive evidence of that fact.

(4)   In this section "the New York Convention" means the Convention on the Recognition and Enforcement of Foreign Arbitral Awards adopted by the United Nations Conference on International Commercial Arbitration on 10th June 1958.

NOTES

Commencement: 31 January 1997.

101 Recognition and enforcement of awards

(1)   A New York Convention award shall be recognised as binding on the persons as between whom it was made, and may accordingly be relied on by those persons by way of defence, set-off or otherwise in any legal proceedings in England and Wales or Northern Ireland.

(2)   A New York Convention award may, by leave of the court, be enforced in the same manner as a judgment or order of the court to the same effect.

  As to the meaning of "the court" see section 105.

(3)   Where leave is so given, judgment may be entered in terms of the award.

Commencement: 31 January 1997.

102   Evidence to be produced by party seeking recognition or enforcement

(1)   A party seeking the recognition or enforcement of a New York Convention award must produce—
  (a) the duly authenticated original award or a duly certified copy of it, and
  (b) the original arbitration agreement or a duly certified copy of it.

(2)   If the award or agreement is in a foreign language, the party must also produce a translation of it certified by an official or sworn translator or by a diplomatic or consular agent.

Commencement: 31 January 1997.

103   Refusal of recognition or enforcement

(1)   Recognition or enforcement of a New York Convention award shall not be refused except in the following cases.

(2)   Recognition or enforcement of the award may be refused if the person against whom it is invoked proves—
  (a) that a party to the arbitration agreement was (under the law applicable to him) under some incapacity;
  (b) that the arbitration agreement was not valid under the law to which the parties subjected it or, failing any indication thereon, under the law of the country where the award was made;
  (c) that he was not given proper notice of the appointment of the arbitrator or of the arbitration proceedings or was otherwise unable to present his case;
  (d) that the award deals with a difference not contemplated by or not falling within the terms of the submission to arbitration or contains

decisions on matters beyond the scope of the submission to arbitration (but see subsection (4));

(e) that the composition of the arbitral tribunal or the arbitral procedure was not in accordance with the agreement of the parties or, failing such agreement, with the law of the country in which the arbitration took place;

(f) that the award has not yet become binding on the parties, or has been set aside or suspended by a competent authority of the country in which, or under the law of which, it was made.

(3)   Recognition or enforcement of the award may also be refused if the award is in respect of a matter which is not capable of settlement by arbitration, or if it would be contrary to public policy to recognise or enforce the award.

(4)   An award which contains decisions on matters not submitted to arbitration may be recognised or enforced to the extent that it contains decisions on matters submitted to arbitration which can be separated from those on matters not so submitted.

(5)   Where an application for the setting aside or suspension of the award has been made to such a competent authority as is mentioned in subsection (2)(f), the court before which the award is sought to be relied upon may, if it considers it proper, adjourn the decision on the recognition or enforcement of the award.

It may also on the application of the party claiming recognition or enforcement of the award order the other party to give suitable security.

NOTES

Commencement: 31 January 1997.

104   Saving for other bases of recognition or enforcement

Nothing in the preceding provisions of this Part affects any right to rely upon or enforce a New York Convention award at common law or under section 66.

NOTES

Commencement: 31 January 1997.

PART IV
GENERAL PROVISIONS

105 Meaning of "the court": jurisdiction of High Court and county court

(1)   In this Act "the court" means the High Court or a county court, subject to the following provisions.

(2)   The Lord Chancellor may by order make provision—
  (a) allocating proceedings under this Act to the High Court or to county courts; or
  (b) specifying proceedings under this Act which may be commenced or taken only in the High Court or in a county court.

(3)   The Lord Chancellor may by order make provision requiring proceedings of any specified description under this Act in relation to which a county court has jurisdiction to be commenced or taken in one or more specified county courts.

Any jurisdiction so exercisable by a specified county court is exercisable throughout England and Wales or, as the case may be, Northern Ireland.

(4)   An order under this section—
  (a) may differentiate between categories of proceedings by reference to such criteria as the Lord Chancellor sees fit to specify, and
  (b) may make such incidental or transitional provision as the Lord Chancellor considers necessary or expedient.

(5)   An order under this section for England and Wales shall be made by statutory instrument which shall be subject to annulment in pursuance of a resolution of either House of Parliament.

(6)   An order under this section for Northern Ireland shall be a statutory rule for the purposes of the Statutory Rules (Northern Ireland) Order 1979 which shall be subject to annulment in pursuance of a resolution of either House of Parliament in like manner as a statutory instrument and section 5 of the Statutory Instruments Act 1946 shall apply accordingly.

NOTES

Commencement: 17 December 1996.

106   Crown application

(1)   Part I of this Act applies to any arbitration agreement to which Her Majesty, either in right of the Crown or of the Duchy of Lancaster or otherwise, or the Duke of Cornwall, is a party.

(2)   Where Her Majesty is party to an arbitration agreement otherwise than in right of the Crown, Her Majesty shall be represented for the purposes of any arbitral proceedings—
  (a) where the agreement was entered into by Her Majesty in right of the Duchy of Lancaster, by the Chancellor of the Duchy or such person as he may appoint, and
  (b) in any other case, by such person as Her Majesty may appoint in writing under the Royal Sign Manual.

(3)   Where the Duke of Cornwall is party to an arbitration agreement, he shall be represented for the purposes of any arbitral proceedings by such person as he may appoint.

(4)   References in Part I to a party or the parties to the arbitration agreement or to arbitral proceedings shall be construed, where subsection (2) or (3) applies, as references to the person representing Her Majesty or the Duke of Cornwall.

NOTES

Commencement: 31 January 1997.

108   Extent

(1)   The provisions of this Act extend to England and Wales and, except as mentioned below, to Northern Ireland.

(2)   The following provisions of Part II do not extend to Northern Ireland—
  section 92 (exclusion of Part I in relation to small claims arbitration in the county court), and
  section 93 and Schedule 2 (appointment of judges as arbitrators).

(3)   Sections 89, 90 and 91 (consumer arbitration agreements) extend to Scotland and the provisions of Schedules 3 and 4 (consequential amendments and repeals) extend to Scotland so far as they relate to enactments which so extend, subject as follows.

(4)   The repeal of the Arbitration Act 1975 extends only to England and Wales and Northern Ireland.

Commencement: 17 December 1996.

109   Commencement

(1)   The provisions of this Act come into force on such day as the Secretary of State may appoint by order made by statutory instrument, and different days may be appointed for different purposes.

(2)   An order under subsection (1) may contain such transitional provisions as appear to the Secretary of State to be appropriate.

Commencement: 17 December 1996.

110   Short title

This Act may be cited as the Arbitration Act 1996.

Commencement: 17 December 1996.

SCHEDULES

SCHEDULE I

Section 4(1)

MANDATORY PROVISIONS OF PART I

sections 9 to 11 (stay of legal proceedings);

section 12 (power of court to extend agreed time limits);

section 13 (application of Limitation Acts);

section 24 (power of court to remove arbitrator);

section 26(1) (effect of death of arbitrator);

section 28 (liability of parties for fees and expenses of arbitrators);

section 29 (immunity of arbitrator);

section 31 (objection to substantive jurisdiction of tribunal);

section 32 (determination of preliminary point of jurisdiction);

section 33 (general duty of tribunal);

section 37(2) (items to be treated as expenses of arbitrators);

section 40 (general duty of parties);

section 43 (securing the attendance of witnesses);

section 56 (power to withhold award in case of non-payment);

section 60 (effectiveness of agreement for payment of costs in any event);

section 66 (enforcement of award);

sections 67 and 68 (challenging the award: substantive jurisdiction and serious irregularity), and sections 70 and 71 (supplementary provisions; effect of order of court) so far as relating to those sections;

section 72 (saving for rights of person who takes no part in proceedings);

section 73 (loss of right to object);

section 74 (immunity of arbitral institutions, &c);

section 75 (charge to secure payment of solicitors' costs).

NOTES
Commencement: 31 January 1997.

SCHEDULE 2

Section 93(6)

MODIFICATIONS OF PART I IN RELATION TO JUDGE-ARBITRATORS

Introductory

1.   In this Schedule "judge-arbitrator" means a judge of the Commercial Court or official referee appointed as arbitrator or umpire under section 93.

General

2.—(1) Subject to the following provisions of this Schedule, references in Part I to the court shall be construed in relation to a judge-arbitrator, or in relation to the appointment of a judge-arbitrator, as references to the Court of Appeal.

(2)   The references in sections 32(6), 45(6) and 69(8) to the Court of Appeal shall in such a case be construed as references to the House of Lords.

Arbitrator's fees

3.—(1) The power of the court in section 28(2) to order consideration and adjustment of the liability of a party for the fees of an arbitrator may be exercised by a judge-arbitrator.

(2)   Any such exercise of the power is subject to the powers of the Court of Appeal under sections 24(4) and 25(3)(b) (directions as to entitlement to fees or expenses in case of removal or resignation).

Exercise of court powers in support of arbitration

4.—(1) Where the arbitral tribunal consists of or includes a judge-arbitrator the powers of the court under sections 42 to 44 (enforcement of peremptory orders, summoning witnesses, and other court powers) are exercisable by the High Court and also by the judge-arbitrator himself.

(2)   Anything done by a judge-arbitrator in the exercise of those powers shall be regarded as done by him in his capacity as judge of the High Court and have effect as if done by that court.

Nothing in this sub-paragraph prejudices any power vested in him as arbitrator or umpire.

Extension of time for making award

5.—(1) The power conferred by section 50 (extension of time for making award) is exercisable by the judge-arbitrator himself.

(2)   Any appeal from a decision of a judge-arbitrator under that section lies to the Court of Appeal with the leave of that court.

Withholding award in case of non-payment

6.—(1) The provisions of paragraph 7 apply in place of the provisions of section 56 (power to withhold award in the case of non-payment) in relation to the withholding of an award for non-payment of the fees and expenses of a judge-arbitrator.

(2)   This does not affect the application of section 56 in relation to the delivery of such an award by an arbitral or other institution or person vested by the parties with powers in relation to the delivery of the award.

7.—(1) A judge-arbitrator may refuse to deliver an award except upon payment of the fees and expenses mentioned in section 56(1).

(2)   The judge-arbitrator may, on an application by a party to the arbitral proceedings, order that if he pays into the High Court the fees and expenses demanded, or such lesser amount as the judge-arbitrator may specify—
   (a) the award shall be delivered,
   (b) the amount of the fees and expenses properly payable shall be determined by such means and upon such terms as he may direct, and
   (c) out of the money paid into court there shall be paid out such fees and expenses as may be found to be properly payable and the balance of the money (if any) shall be paid out to the applicant.

(3)   For this purpose the amount of fees and expenses properly payable is the amount the applicant is liable to pay under section 28 or any agreement relating to the payment of the arbitrator.

(4)   No application to the judge-arbitrator under this paragraph may be made where there is any available arbitral process for appeal or review of the amount of the fees or expenses demanded.

(5)   Any appeal from a decision of a judge-arbitrator under this paragraph lies to the Court of Appeal with the leave of that court.

(6)   Where a party to arbitral proceedings appeals under sub-paragraph (5), an arbitrator is entitled to appear and be heard.

Correction of award or additional award

8.   Subsections (4) to (6) of section 57 (correction of award or additional award: time limit for application or exercise of power) do not apply to a judge-arbitrator.

Costs

9.   Where the arbitral tribunal consists of or includes a judge-arbitrator the powers of the court under section 63(4) (determination of recoverable costs) shall be exercised by the High Court.

10.—(1) The power of the court under section 64 to determine an arbitrator's reasonable fees and expenses may be exercised by a judge-arbitrator.

(2)   Any such exercise of the power is subject to the powers of the Court of Appeal under sections 24(4) and 25(3)(b) (directions as to entitlement to fees or expenses in case of removal or resignation).

Enforcement of award

11.   The leave of the court required by section 66 (enforcement of award) may in the case of an award of a judge-arbitrator be given by the judge-arbitrator himself.

Solicitors' costs

12.   The powers of the court to make declarations and orders under the provisions applied by section 75 (power to charge property recovered in arbitral proceedings with the payment of solicitors' costs) may be exercised by the judge-arbitrator.

Powers of court in relation to service of documents

13.—(1) The power of the court under section 77(2) (powers of court in relation to service of documents) is exercisable by the judge-arbitrator.

(2)   Any appeal from a decision of a judge-arbitrator under that section lies to the Court of Appeal with the leave of that court.

Powers of court to extend time limits relating to arbitral proceedings

14.—(1) The power conferred by section 79 (power of court to extend time limits relating to arbitral proceedings) is exercisable by the judge-arbitrator himself.

(2)   Any appeal from a decision of a judge-arbitrator under that section lies to the Court of Appeal with the leave of that court.

NOTES

Commencement: 31 January 1997.

Late Payment of Commercial Debts (Interest) Act 1998

(C 20)

An Act to make provision with respect to interest on the late payment of certain debts arising under commercial contracts for the supply of goods or services; and for connected purposes

[11 June 1998]

PART I
STATUTORY INTEREST ON QUALIFYING DEBTS

1   Statutory interest

(1)   It is an implied term in a contract to which this Act applies that any qualifying debt created by the contract carries simple interest subject to and in accordance with this Part.

(2)   Interest carried under that implied term (in this Act referred to as "statutory interest") shall be treated, for the purposes of any rule of law or enactment (other than this Act) relating to interest on debts, in the same way as interest carried under an express contract term.

(3)   This Part has effect subject to Part II (which in certain circumstances permits contract terms to oust or vary the right to statutory interest that would otherwise be conferred by virtue of the term implied by subsection (1)).

NOTES

Commencement: 1 November 1998 (certain purposes); to be appointed (remaining purposes).

2   Contracts to which Act applies

(1)   This Act applies to a contract for the supply of goods or services where the purchaser and the supplier are each acting in the course of a business, other than an excepted contract.

(2)   In this Act "contract for the supply of goods or services" means—
   (a) a contract of sale of goods; or
   (b) a contract (other than a contract of sale of goods) by which a person does any, or any combination, of the things mentioned in subsection (3) for a consideration that is (or includes) a money consideration.

(3)   Those things are—
   (a) transferring or agreeing to transfer to another the property in goods;
   (b) bailing or agreeing to bail goods to another by way of hire or, in Scotland, hiring or agreeing to hire goods to another; and
   (c) agreeing to carry out a service.

(4)   For the avoidance of doubt a contract of service or apprenticeship is not a contract for the supply of goods or services.

(5)   The following are excepted contracts—
   (a) a consumer credit agreement;
   (b) a contract intended to operate by way of mortgage, pledge, charge or other security; and
   (c) a contract of a description specified in an order made by the Secretary of State.

(6)   An order under subsection (5)(c) may specify a description of contract by reference to any feature of the contract (including the parties).

(7)   In this section—

"business" includes a profession and the activities of any government department or local or public authority;

"consumer credit agreement" has the same meaning as in the Consumer Credit Act 1974;

"contract of sale of goods" and "goods" have the same meaning as in the Sale of Goods Act 1979;

"property in goods" means the general property in them and not merely a special property.

NOTES

Commencement: 1 November 1998 (certain purposes); to be appointed (remaining purposes).

3   Qualifying debts

(1)   A debt created by virtue of an obligation under a contract to which this Act applies to pay the whole or any part of the contract price is a "qualifying debt" for the purposes of this Act, unless (when created) the whole of the debt is prevented from carrying statutory interest by this section.

(2)   A debt does not carry statutory interest if or to the extent that it consists of a sum to which a right to interest or to charge interest applies by virtue of any enactment (other than section 1 of this Act).

This subsection does not prevent a sum from carrying statutory interest by reason of the fact that a court, arbitrator or arbiter would, apart from this Act, have power to award interest on it.

(3)   A debt does not carry (and shall be treated as never having carried) statutory interest if or to the extent that a right to demand interest on it, which exists by virtue of any rule of law, is exercised.

(4)   A debt does not carry statutory interest if or to the extent that it is of a description specified in an order made by the Secretary of State.

(5)   Such an order may specify a description of debt by reference to any feature of the debt (including the parties or any other feature of the contract by which it is created).

NOTES

Commencement: 1 November 1998 (certain purposes); to be appointed (remaining purposes).

4 Period for which statutory interest runs

(1)   Statutory interest runs in relation to a qualifying debt in accordance with this section (unless section 5 applies).

(2)   Statutory interest starts to run on the day after the relevant day for the debt, at the rate prevailing under section 6 at the end of the relevant day.

(3)   Where the supplier and the purchaser agree a date for payment of the debt (that is, the day on which the debt is to be created by the contract), that is the relevant day unless the debt relates to an obligation to make an advance payment.

A date so agreed may be a fixed one or may depend on the happening of an event or the failure of an event to happen.

(4)   Where the debt relates to an obligation to make an advance payment, the relevant day is the day on which the debt is treated by section 11 as having been created.

(5)   In any other case, the relevant day is the last day of the period of 30 days beginning with—
  (a) the day on which the obligation of the supplier to which the debt relates is performed; or
  (b) the day on which the purchaser has notice of the amount of the debt or (where that amount is unascertained) the sum which the supplier claims is the amount of the debt,

whichever is the later.

(6)   Where the debt is created by virtue of an obligation to pay a sum due in respect of a period of hire of goods, subsection (5)(a) has effect as if it referred to the last day of that period.

(7)   Statutory interest ceases to run when the interest would cease to run if it were carried under an express contract term.

(8)   In this section "advance payment" has the same meaning as in section 11.

NOTES

Commencement: 1 November 1998 (certain purposes); to be appointed (remaining purposes).

5 Remission of statutory interest

(1) This section applies where, by reason of any conduct of the supplier, the interests of justice require that statutory interest should be remitted in whole or part in respect of a period for which it would otherwise run in relation to a qualifying debt.

(2) If the interests of justice require that the supplier should receive no statutory interest for a period, statutory interest shall not run for that period.

(3) If the interests of justice require that the supplier should receive statutory interest at a reduced rate for a period, statutory interest shall run at such rate as meets the justice of the case for that period.

(4) Remission of statutory interest under this section may be required—
  (a) by reason of conduct at any time (whether before or after the time at which the debt is created); and
  (b) for the whole period for which statutory interest would otherwise run or for one or more parts of that period.

(5) In this section "conduct" includes any act or omission.

NOTES

Commencement: 1 November 1998 (certain purposes); to be appointed (remaining purposes).

6 Rate of statutory interest

(1) The Secretary of State shall by order made with the consent of the Treasury set the rate of statutory interest by prescribing—
  (a) a formula for calculating the rate of statutory interest; or
  (b) the rate of statutory interest.

(2) Before making such an order the Secretary of State shall, among other things, consider the extent to which it may be desirable to set the rate so as to—

(a) protect suppliers whose financial position makes them particularly vulnerable if their qualifying debts are paid late; and

(b) deter generally the late payment of qualifying debts.

NOTES

Commencement: 1 November 1998 (certain purposes); to be appointed (remaining purposes).

PART II
CONTRACT TERMS RELATING TO LATE PAYMENT OF QUALIFYING DEBTS

7   Purpose of Part II

(1)   This Part deals with the extent to which the parties to a contract to which this Act applies may by reference to contract terms oust or vary the right to statutory interest that would otherwise apply when a qualifying debt created by the contract (in this Part referred to as "the debt") is not paid.

(2)   This Part applies to contract terms agreed before the debt is created; after that time the parties are free to agree terms dealing with the debt.

(3)   This Part has effect without prejudice to any other ground which may affect the validity of a contract term.

NOTES

Commencement: 1 November 1998 (certain purposes); to be appointed (remaining purposes).

8   Circumstances where statutory interest may be ousted or varied

(1)   Any contract terms are void to the extent that they purport to exclude the right to statutory interest in relation to the debt, unless there is a substantial contractual remedy for late payment of the debt.

(2)   Where the parties agree a contractual remedy for late payment of the debt that is a substantial remedy, statutory interest is not carried by the debt (unless they agree otherwise).

(3)   The parties may not agree to vary the right to statutory interest in relation to the debt unless either the right to statutory interest as varied or the overall remedy for late payment of the debt is a substantial remedy.

(4)   Any contract terms are void to the extent that they purport to—
  (a) confer a contractual right to interest that is not a substantial remedy for late payment of the debt, or
  (b) vary the right to statutory interest so as to provide for a right to statutory interest that is not a substantial remedy for late payment of the debt,

unless the overall remedy for late payment of the debt is a substantial remedy.

(5)   Subject to this section, the parties are free to agree contract terms which deal with the consequences of late payment of the debt.

NOTES

Commencement: 1 November 1998 (certain purposes); to be appointed (remaining purposes).

9 Meaning of "substantial remedy"

(1)   A remedy for the late payment of the debt shall be regarded as a substantial remedy unless—
  (a) the remedy is insufficient either for the purpose of compensating the supplier for late payment or for deterring late payment; and
  (b) it would not be fair or reasonable to allow the remedy to be relied on to oust or (as the case may be) to vary the right to statutory interest that would otherwise apply in relation to the debt.

(2)   In determining whether a remedy is not a substantial remedy, regard shall be had to all the relevant circumstances at the time the terms in question are agreed.

(3)   In determining whether subsection (1)(b) applies, regard shall be had (without prejudice to the generality of subsection (2)) to the following matters—
  (a) the benefits of commercial certainty;
  (b) the strength of the bargaining positions of the parties relative to each other;
  (c) whether the term was imposed by one party to the detriment of the other (whether by the use of standard terms or otherwise); and
  (d) whether the supplier received an inducement to agree to the term.

NOTES

Commencement: 1 November 1998 (certain purposes); to be appointed (remaining purposes).

10   Interpretation of Part II

(1)   In this Part—

"contract term" means a term of the contract creating the debt or any other contract term binding the parties (or either of them);

"contractual remedy" means a contractual right to interest or any contractual remedy other than interest;

"contractual right to interest" includes a reference to a contractual right to charge interest;

"overall remedy", in relation to the late payment of the debt, means any combination of a contractual right to interest, a varied right to statutory interest or a contractual remedy other than interest;

"substantial remedy" shall be construed in accordance with section 9.

(2)   In this Part a reference (however worded) to contract terms which vary the right to statutory interest is a reference to terms altering in any way the effect of Part I in relation to the debt (for example by postponing the time at which interest starts to run or by imposing conditions on the right to interest).

(3)   In this Part a reference to late payment of the debt is a reference to late payment of the sum due when the debt is created (excluding any part of that sum which is prevented from carrying statutory interest by section 3).

NOTES

Commencement: 1 November 1998 (certain purposes); to be appointed (remaining purposes).

PART III
GENERAL AND SUPPLEMENTARY

11   Treatment of advance payments of the contract price

(1)   A qualifying debt created by virtue of an obligation to make an advance payment shall be treated for the purposes of this Act as if it was created on the day mentioned in subsection (3), (4) or (5) (as the case may be).

(2)   In this section "advance payment" means a payment falling due before the obligation of the supplier to which the whole contract price relates ("the supplier's obligation") is performed, other than a payment of a part of the contract price that is due in respect of any part performance of that obligation and payable on or after the day on which that part performance is completed.

(3)   Where the advance payment is the whole contract price, the debt shall be treated as created on the day on which the supplier's obligation is performed.

(4)   Where the advance payment is a part of the contract price, but the sum is not due in respect of any part performance of the supplier's obligation, the debt shall be treated as created on the day on which the supplier's obligation is performed.

(5)   Where the advance payment is a part of the contract price due in respect of any part performance of the supplier's obligation, but is payable before that part performance is completed, the debt shall be treated as created on the day on which the relevant part performance is completed.

(6)   Where the debt is created by virtue of an obligation to pay a sum due in respect of a period of hire of goods, this section has effect as if—
   (a) references to the day on which the supplier's obligation is performed were references to the last day of that period; and
   (b) references to part performance of that obligation were references to part of that period.

(7)   For the purposes of this section an obligation to pay the whole outstanding balance of the contract price shall be regarded as an obligation to pay the whole contract price and not as an obligation to pay a part of the contract price.

NOTES

Commencement: 1 November 1998 (certain purposes); to be appointed (remaining purposes).

12 Conflict of laws

(1)   This Act does not have effect in relation to a contract governed by the law of a part of the United Kingdom by choice of the parties if—

(a) there is no significant connection between the contract and that part of the United Kingdom; and

(b) but for that choice, the applicable law would be a foreign law.

(2)　This Act has effect in relation to a contract governed by a foreign law by choice of the parties if—

(a) but for that choice, the applicable law would be the law of a part of the United Kingdom; and

(b) there is no significant connection between the contract and any country other than that part of the United Kingdom.

(3)　In this section—

"contract" means a contract falling within section 2(1); and

"foreign law" means the law of a country outside the United Kingdom.

Commencement: 1 November 1998 (certain purposes); to be appointed (remaining purposes).

13　Assignments, etc

(1)　The operation of this Act in relation to a qualifying debt is not affected by—

(a) any change in the identity of the parties to the contract creating the debt; or

(b) the passing of the right to be paid the debt, or the duty to pay it (in whole or in part) to a person other than the person who is the original creditor or the original debtor when the debt is created.

(2)　Any reference in this Act to the supplier or the purchaser is a reference to the person who is for the time being the supplier or the purchaser or, in relation to a time after the debt in question has been created, the person who is for the time being the creditor or the debtor, as the case may be.

(3)　Where the right to be paid part of a debt passes to a person other than the person who is the original creditor when the debt is created, any reference in this Act to a debt shall be construed as (or, if the context so requires, as including) a reference to part of a debt.

(4)　A reference in this section to the identity of the parties to a contract changing, or to a right or duty passing, is a reference to it changing or passing by assignment or assignation, by operation of law or otherwise.

NOTES

Commencement: 1 November 1998 (certain purposes); to be appointed (remaining purposes).

14 Contract terms relating to the date for payment of the contract price

(1)   This section applies to any contract term which purports to have the effect of postponing the time at which a qualifying debt would otherwise be created by a contract to which this Act applies.

(2)   Sections 3(2)(b) and 17(1)(b) of the Unfair Contract Terms Act 1977 (no reliance to be placed on certain contract terms) shall apply in cases where such a contract term is not contained in written standard terms of the purchaser as well as in cases where the term is contained in such standard terms.

(3)   In this section "contract term" has the same meaning as in section 10(1).

NOTES

Commencement: 1 November 1998 (certain purposes); to be appointed (remaining purposes).

15 Orders and regulations

(1)   Any power to make an order or regulations under this Act is exercisable by statutory instrument.

(2)   Any statutory instrument containing an order or regulations under this Act, other than an order under section 17(2), shall be subject to annulment in pursuance of a resolution of either House of Parliament.

NOTES

Commencement: 1 November 1998 (certain purposes); to be appointed (remaining purposes).

16 Interpretation

(1)   In this Act—

"contract for the supply of goods or services" has the meaning given in section 2(2);

"contract price" means the price in a contract of sale of goods or the money consideration referred to in section 2(2)(b) in any other contract for the supply of goods or services;

"purchaser" means (subject to section 13(2)) the buyer in a contract of sale or the person who contracts with the supplier in any other contract for the supply of goods or services;

"qualifying debt" means a debt falling within section 3(1);

"statutory interest" means interest carried by virtue of the term implied by section 1(1); and

"supplier" means (subject to section 13(2)) the seller in a contract of sale of goods or the person who does one or more of the things mentioned in section 2(3) in any other contract for the supply of goods or services.

(2)   In this Act any reference (however worded) to an agreement or to contract terms includes a reference to both express and implied terms (including terms established by a course of dealing or by such usage as binds the parties).

NOTES

Commencement: 1 November 1998 (certain purposes); to be appointed (remaining purposes).

17 Short title, commencement and extent

(1)   This Act may be cited as the Late Payment of Commercial Debts (Interest) Act 1998.

(2)   This Act (apart from this section) shall come into force on such day as the Secretary of State may by order appoint; and different days may be appointed for different descriptions of contract or for other different purposes.

An order under this subsection may specify a description of contract by reference to any feature of the contract (including the parties).

(3)   The Secretary of State may by regulations make such transitional, supplemental or incidental provision (including provision modifying any provision of this Act) as the Secretary of State may

consider necessary or expedient in connection with the operation of this Act while it is not fully in force.

(4)   This Act does not affect contracts of any description made before this Act comes into force for contracts of that description.

(5)   This Act extends to Northern Ireland.

NOTES

Commencement: 1 November 1998 (certain purposes); to be appointed (remaining purposes).

Contracts (Rights of Third Parties) Bill

NOTES

This is the text of the Contracts (Rights of Third Parties) Bill [HL Bill 65], as amended on report and printed on 27 May 1999.

A bill to make provision for the enforcement of contractual terms by third parties.

1 Right of third party to enforce contractual term

(1)   Subject to the provisions of this Act, a person who is not a party to a contract (a "third party") may in his own right enforce a term of the contract if—
  (a)   the contract expressly provides that he may, or
  (b)   subject to subsection (2), the term purports to confer a benefit on him.

(2)   Subsection (1)(b) does not apply if on a proper construction of the contract it appears that the parties did not intend the term to be enforceable by the third party.

(3)   The third party must be expressly identified in the contract by name, as a member of a class or as answering a particular description but need not be in existence when the contract is entered into.

(4)   This section does not confer a right on a third party to enforce a term of a contract otherwise than subject to and in accordance with any other relevant terms of the contract.

(5)   For the purpose of exercising his right to enforce a term of the contract, there shall be available to the third party any remedy that would have been available to him in an action for breach of contract if he had been a party to the contract (and the rules relating to damages, injunctions, specific performance and other relief shall apply accordingly).

(6)   Where a term of a contract excludes or limits liability in relation to any matter references in this Act to the third party enforcing the term shall be construed as references to his availing himself of the exclusion or limitation.

(7)   In this Act, in relation to a term of a contract which is enforceable by a third party—
"the promisor" means the party to the contract against whom the term is enforceable by the third party, and
"the promisee" means the party to the contract by whom the term is enforceable against the promisor.

2   Variation and cancellation of contract

(1)   Subject to the provisions of this section, where a third party has a right under section 1 to enforce a term of the contract, the parties to the contract may not, by agreement, rescind the contract, or vary it in such a way as to extinguish or alter his entitlement under that right, without his consent, if—
(a) the third party has communicated his assent to the term to the promisor,
(b) the promisor is aware that the third party has relied on the term, or
(c) the promisor can reasonably be expected to have foreseen that the third party would rely on the term and the third party has in fact relied on it.

(2)   The assent referred to in subsection (1)(a)—
(a) may be by words or conduct, and
(b) if sent to the promisor by post or other means, shall not be regarded as communicated to the promisor until received by him.

(3)   Subsection (1) is subject to any express term of the contract under which—
(a) the parties to the contract may by agreement rescind or vary the contract without the consent of the third party, or
(b) the consent of the third party is required in circumstances specified in the contract instead of those set out in subsection (1)(a) to (c).

(4)   Where the consent of a third party is required under subsection (1) or (3), the court or arbitral tribunal may, on the application of the parties to the contract, dispense with his consent if satisfied—
    (a) that his consent cannot be obtained because his whereabouts cannot reasonably be ascertained, or
    (b) that he is mentally incapable of giving his consent.

(5)   The court or arbitral tribunal may, on the application of the parties to a contract, dispense with any consent that may be required under subsection (1)(c) if satisfied that it cannot reasonably be ascertained whether or not the third party has in fact relied on the term.

(6)   If the court or arbitral tribunal dispenses with a third party's consent, it may impose such conditions as it thinks fit, including a condition requiring the payment of compensation to the third party.

(7)   The jurisdiction conferred on the court by subsections (4) to (6) is exercisable by both the High Court and a county court.

3   Defences etc available to promisor

(1)   Subsections (2) to (5) apply where, in reliance on section 1, proceedings for the enforcement of a term of a contract are brought by a third party.

(2)   The promisor shall have available to him by way of defence or set-off any matter that—
    (a) arises from or in connection with the contract and is relevant to the term, and
    (b) would have been available to him by way of defence or set-off if the proceedings had been brought by the promisee.

(3)   The promisor shall also have available to him by way of defence or set-off any matter if—
    (a) an express term of the contract provides for it to be available to him in proceedings brought by the third party, and
    (b) it would have been available to him by way of defence or set-off if the proceedings had been brought by the promisee.

(4)   The promisor shall also have available to him—
    (a) by way of defence or set-off any matter, and
    (b) by way of counterclaim any matter not arising from the contract,

that would have been available to him by way of defence or set-off or, as the case may be, by way of counterclaim against the third party if the third party had been a party to the contract.

(5)   Subsections (2) and (4) are subject to any express term of the contract as to the matters that are not to be available to the promisor by way of defence, set-off or counterclaim.

(6)   Where in any proceedings brought against him a third party seeks in reliance on section 1 to enforce a term of a contract (including, in particular, a term purporting to exclude or limit liability), he may not do so if he could not have done so (whether by reason of any particular circumstances relating to him or otherwise) had he been a party to the contract.

4   Enforcement of contract by promisee

Section 1 does not affect any right of the promisee to enforce any term of the contract.

5   Protection of promisor from double liability

Where under section 1 a term of a contract is enforceable by a third party, and the promisee has recovered from the promisor a sum in respect of—
  (a) the third party's loss in respect of the term, or
  (b) the expense to the promisee of making good to the third party the default of the promisor,

then, in any proceedings brought in reliance on that section by the third party, the court or arbitral tribunal shall reduce any award to the third party to such extent as it thinks appropriate to take account of the sum recovered by the promisee.

6   Exceptions

(1)   Section 1 confers no rights on a third party in the case of a contract on a bill of exchange, promissory note or other negotiable instrument.

(2)   Section 1 confers no rights on a third party in the case of any contract binding on a company and its members under section 14 of the Companies Act 1985.

(3)   Section 1 confers no right on a third party to enforce—
   (a) any term of a contract of employment against an employee,
   (b) any term of a worker's contract against a worker (including a home worker), or
   (c) any term of a relevant contract against an agency worker.

(4)   In subsection (3)—
   (a) "contract of employment", "employee", "worker's contract", and "worker" have the meaning given by section 54 of the National Minimum Wage Act 1998,
   (b) "home worker" has the meaning given by section 35(2) of that Act,
   (c) "agency worker" has the same meaning as in section 34(1) of that Act, and
   (d) "relevant contract" means a contract entered into, in a case where section 34 of that Act applies, by the agency worker as respects work falling within subsection (1)(a) of that section.

(5)   Section 1 confers no rights on a third party in the case of—
   (a) a contract for the carriage of goods by sea, or
   (b) a contract for the carriage of goods by rail or road, or for the carriage of cargo by air, which is subject to the rules of the appropriate international transport convention,

except that a third party may in reliance on that section avail himself of an exclusion or limitation of liability in such a contract.

(6)   In subsection (5) "contract for the carriage of goods by sea" means a contract of carriage—
   (a) contained in or evidenced by a bill of lading, sea waybill or a corresponding electronic transaction, or
   (b) under or for the purposes of which there is given an undertaking which is contained in a ship's delivery order or a corresponding electronic transaction.

(7)   For the purposes of subsection (6)—
   (a) "bill of lading", "sea waybill" and "ship's delivery order" have the same meaning as in the Carriage of Goods by Sea Act 1992, and
   (b) a corresponding electronic transaction is a transaction within section 1(5) of that Act which corresponds to the issue, indorsement, delivery or transfer of a bill of lading, sea waybill or ship's delivery order.

(8)   In subsection (5) "the appropriate international transport convention" means—

   (a) in relation to a contract for the carriage of goods by rail, the Convention which has the force of law in the United Kingdom under section 1 of the International Transport Conventions Act 1983,

   (b) in relation to a contract for the carriage of goods by road, the Convention which has the force of law in the United Kingdom under section 1 of the Carriage of Goods by Road Act 1965, and

   (c) in relation to a contract for the carriage of cargo by air—

      (i) the Convention which has the force of law in the United Kingdom under section 1 of the Carriage by Air Act 1961, or

      (ii) the Convention which has the force of law under section 1 of the Carriage by Air (Supplementary Provisions) Act 1962, or

     (iii) either of the amended Conventions set out in Part B of Schedule 2 to the Carriage by Air Acts (Application of Provisions) Order 1967.

7 Supplementary provisions relating to third party

(1)   Section 1 does not affect any right or remedy of a third party that exists or is available apart from this Act.

(2)   Section 2(2) of the Unfair Contract Terms Act 1977 (restriction on exclusion etc of liability for negligence) shall not apply where the negligence consists of the breach of an obligation arising from a term of a contract and the person seeking to enforce it is a third party acting in reliance on section 1.

(3)   In sections 5 and 8 of the Limitation Act 1980 the references to an action founded on a simple contract and an action upon a specialty shall respectively include references to an action brought in reliance on section 1 relating to a simple contract and an action brought in reliance on that section relating to a specialty.

(4)   Where—

   (a) a third party has a right under section 1 to enforce a term that disputes between himself and the promisor are to be submitted to arbitration, and

   (b) the term is an agreement in writing for the purposes of Part I of the Arbitration Act 1996,

then, as regards any matter which the third party requires to be referred to arbitration in exercise of the right, Part I of the Arbitration Act 1996 has effect as if the right were under an arbitration agreement in writing (within the meaning of that Part of that Act) between the third party and the promisor.

(5)   A third party shall not, by virtue of section 1(5) or 3(4) or (6), be treated as a party to the contract for the purposes of any other Act (or any instrument made under any other Act).

8 Short title, commencement and extent

(1)   This Act may be cited as the Contracts (Rights of Third Parties) Act 1999.

(2)   This Act comes into force on the day on which it is passed but, subject to subsection (3), does not apply in relation to a contract entered into before the end of the period of six months beginning with that day.

(3)   The restriction in subsection (2) does not apply in relation to a contract which—
   (a) is entered into on or after the day on which this Act is passed, and
   (b) expressly provides for the application of this Act.

(4)   This Act extends to England and Wales only.

PART II

STATUTORY INSTRUMENTS

Consumer Transactions (Restrictions on Statements) Order 1976

(SI 1976/1813)

NOTES

Made: 1 November 1976.

Authority: Fair Trading Act 1973, s 22.

Commencement: partly on 1 December 1976, partly on 1 November 1977 and fully on 1 November 1978.

1 This Order may be cited as the Consumer Transactions (Restrictions on Statements) Order 1976, and shall come into operation as respects—

(a) this Article, Article 2 and Article 3(a), at the expiry of the period of 1 month beginning with the date on which this Order is made;

(b) the remainder of Article 3, at the expiry of the period of 12 months beginning with that date; and

(c) the remainder of this Order, at the expiry of the period of 2 years beginning with that date.

2.—(1)   In this Order—

"advertisement" includes a catalogue and a circular;

"consumer" means a person acquiring goods otherwise than in the course of a business but does not include a person who holds himself out as acquiring them in the course of a business;

"consumer transaction" means—

[(a) a consumer sale, that is a sale of goods (other than an excepted sale) by a seller where the goods—

(i) are of a type ordinarily bought for private use or consumption, and

(ii) are sold to a person who does not buy or hold himself out as buying them in the course of a business.

For the purposes of this paragraph an excepted sale is a sale by auction, a sale by competitive tender and a sale arising by virtue of a contract for the international sale of goods as originally defined in section 62(1) of the Sale of Goods Act 1893 as amended by the Supply of Goods (Implied Terms) Act 1973;

(b) a hire-purchase agreement (within the meaning of section 189(1) of the Consumer Credit Act 1974) where the owner makes the agreement in the course of a business and the goods to which the agreement relates—

> > (i) are of a type ordinarily supplied for private use or consumption, and
> >
> > (ii) are hired to a person who does not hire or hold himself out as hiring them in the course of a business;]
>
> (c) an agreement for the redemption of trading stamps under a trading stamp scheme within section 10(1) of the Trading Stamps Act 1964 or, as the case may be, within section 9 of the Trading Stamps Act (Northern Ireland) 1965;
>
> "container" includes any form of packaging of goods whether by way of wholly or partly enclosing the goods or by way of attaching the goods to, or winding the goods round, some other article, and in particular includes a wrapper or confining band;
>
> "statutory rights" means the rights arising by virtue of sections 13 to 15 of the Sale of Goods Act 1893 as amended by the Act of 1973, sections 9 to 11 of the Act of 1973, or section 4(1)(c) of the Trading Stamps Act 1964 or section 4(1)(c) of the Trading Stamps Act (Northern Ireland) 1965 both as amended by the Act of 1973.

(2) The Interpretation Act 1889 shall apply for the interpretation of this Order as it applies for the interpretation of an Act of Parliament.

NOTES

Para (1): in definition "consumer transaction" words in square brackets substituted by the Consumer Transactions (Restrictions on Statements) (Amendment) Order 1978, SI 1978/127, art 4.

3 A person shall not, in the course of a business—

(a) display, at any place where consumer transactions are effected (whether wholly or partly), a notice containing a statement which purports to apply, in relation to consumer transactions effected there, a term which would—

> [(i) be void by virtue of section 6 or 20 of the Unfair Contract Terms Act 1977,] or
>
> (ii) be inconsistent with a warranty (in Scotland a stipulation) implied by section 4(1)(c) of the Trading Stamps Act 1964 or section 4(1)(c) of the Trading Stamps Act (Northern Ireland) 1965 both as amended by the Act of 1973,

if applied to some or all such consumer transactions;

(b) publish or cause to be published any advertisement which is intended to induce persons to enter into consumer transactions and which contains a statement purporting to apply in relation to such consumer transactions such a term as is mentioned in paragraph (a)(i) or (ii), being a term which would be void by virtue of, or as the case may be, inconsistent with, the provisions so mentioned if applied to some or all of those transactions;

(c) supply to a consumer pursuant to a consumer transaction goods bearing, or goods in a container bearing, a statement which is a term of that consumer transaction and which is void by virtue of, or inconsistent with, the said provisions, or if it were a term of that transaction, would be so void or inconsistent;

(d) furnish to a consumer in connection with the carrying out of a consumer transaction or to a person likely, as a consumer, to enter into such a transaction, a document which includes a statement which is a term of that transaction and is void or inconsistent as aforesaid, or, if it were a term of that transaction or were to become a term of a prospective transaction, would be so void or inconsistent.

NOTES

Para (a): sub-para (i) substituted by the Consumer Transactions (Restrictions on Statements) (Amendment) Order 1978, SI 1978/127, art 5.

4   A person shall not in the course of a business—
  (i) supply to a consumer pursuant to a consumer transaction goods bearing, or goods in a container bearing, a statement about the rights that the consumer has against that person or about the obligations to the consumer accepted by that person in relation to the goods (whether legally enforceable or not), being rights or obligations that arise if the goods are defective or are not fit for a purpose or do not correspond with a description;
  (ii) furnish to a consumer in connection with the carrying out of a consumer transaction or to a person likely, as a consumer, to enter into such a transaction with him or through his agency a document containing a statement about such rights and obligations,

unless there is in close proximity to any such statement another statement which is clear and conspicuous and to the effect that the first mentioned statement does not or will not affect the statutory rights of a consumer.

5.—(1)   This Article applies to goods which are supplied in the course of a business by one person ("the supplier") to another where, at the time of the supply, the goods were intended by the supplier to be, or might reasonably be expected by him to be, the subject of a subsequent consumer transaction.

(2)   A supplier shall not—
  (a) supply goods to which this Article applies if the goods bear, or are in a container bearing, a statement which sets out or describes or limits

obligations (whether legally enforceable or not) accepted or to be accepted by him in relation to the goods; or

(b) furnish a document in relation to the goods which contains such a statement,

unless there is in close proximity to any such statement another statement which is clear and conspicuous and to the effect that the first mentioned statement does not or will not affect the statutory rights of a consumer.

(3)　A person does not contravene paragraph (2) above—
  (i) in a case to which sub-paragraph (a) of that paragraph applies, unless the goods have become the subject of a consumer transaction;
  (ii) in a case to which sub-paragraph (b) applies, unless the document has been furnished to a consumer in relation to goods which were the subject of a consumer transaction, or to a person likely to become a consumer pursuant to such a transaction; or
  (iii) by virtue of any statement if before the date on which this Article comes into operation the document containing, or the goods or container bearing, the statement has ceased to be in his possession.

Business Advertisements (Disclosure) Order 1977

(SI 1977/1918)

NOTES

Made: 21 November 1977.
Authority: Fair Trading Act 1973, s 22.
Commencement: 1 January 1978.

1.—(1) This Order may be cited as the Business Advertisements (Disclosure) Order 1977 and shall come into operation on 1st January 1978.

(2)　The Interpretation Act 1889 shall apply for the interpretation of this Order as it applies for the interpretation of an Act of Parliament.

2.—(1) Subject to paragraphs (2) and (3) below, a person who is seeking to sell goods that are being sold in the course of a business shall not publish or cause to be published an advertisement—
  (a) which indicates that the goods are for sale, and
  (b) which is likely to induce consumers to buy the goods,

unless it is reasonably clear whether from the contents of the advertisement, its format or size, the place or manner of its publication or otherwise that the goods are to be sold in the course of a business.

(2)   Paragraph (1) applies whether the person who is seeking to sell the goods is acting on his own behalf or that of another, and where he is acting as agent, whether he is acting in the course of a business carried on by him or not; but the reference in that paragraph to a business does not include any business carried on by the agent.

(3)   Paragraph (1) above shall not apply in relation to advertisements—
   (a) which are concerned only with sales by auction or competitive tender; or
   (b) which are concerned only with the sale of flowers, fruit or vegetables, eggs or dead animals, fish or birds, gathered, produced or taken by the person seeking to sell the goods.

Consumer Protection (Cancellation of Contracts Concluded away from Business Premises) Regulations 1987

(SI 1987/2117)

NOTES
Made: 7 December 1987.
Authority: European Communities Act 1972, s 2(2).
Commencement: 31 December 1998 (regs 3(3), 4A-4H); 1 July 1988 (remainder).

1   Citation and commencement

These Regulations may be cited as the Consumer Protection (Cancellation of Contracts Concluded away from Business Premises) Regulations 1987 and shall come into force on 1st July 1988.

2   Interpretation

(1)   In these Regulations—
   "business" includes a trade or profession;

"consumer" means a person, other than a body corporate, who, in making a contract to which these Regulations apply, is acting for purposes which can be regarded as outside his business;

["enforcement authority" means, in Great Britain, a weights and measures authority and, in Northern Ireland, the Department of Economic Development;]

"goods" has the meaning given by section 61(1) of the Sale of Goods Act 1979;

"land mortgage" includes any security charged on land and in relation to Scotland includes any heritable security;

"notice of cancellation" has the meaning given by regulation 4(5) below;

"security" in relation to a contract means a mortgage, charge, pledge, bond, debenture, indemnity, guarantee, bill, note or other right provided by the consumer, or at his request (express or implied), to secure the carrying out of his obligations under the contract;

"signed" has the same meaning as in the Consumer Credit Act 1974; and

"trader" means a person who, in making a contract to which these Regulations apply, is acting for the purposes of his business, and anyone acting in the name or on behalf of such a person.

(2) . . .

NOTES

Para (1): definition "enforcement authority" inserted by the Consumer Protection (Cancellation of Contracts Concluded away from Business Premises) (Amendment) Regulations 1998, SI 1998/3050, reg 2(a).

Para (2): applies to Scotland only.

3 Contracts to which the Regulations apply

(1) These Regulations apply to a contract, other than an excepted contract, for the supply by a trader of goods or services to a consumer which is made—

(a) during an unsolicited visit by a trader—
 (i) to the consumer's home or to the home of another person; or
 (ii) to the consumer's place of work;

(b) during a visit by a trader as mentioned in paragraph (a)(i) or (ii) above at the express request of the consumer where the goods or services to which the contract relates are other than those concerning which the consumer requested the visit of the trader, provided that when the visit was requested the consumer did not know, or could not reasonably have known, that the supply of those other goods or services formed part of the trader's business activities;

(c) after an offer was made by the consumer in respect of the supply by a trader of the goods or services in the circumstances mentioned in paragraph (a) or (b) above or (d) below; or

(d) during an excursion organised by the trader away from premises on which he is carrying on any business (whether on a permanent or temporary basis).

(2)   For the purposes of this regulation an excepted contract means—

(a) any contract—

   (i) for the sale or other disposition of land, or for a lease or land mortgage;

   (ii) to finance the purchase of land;

   (iii) for a bridging loan in connection with the purchase of land; or

   (iv) for the construction or extension of a building or other erection on land:

Provided that these Regulations shall apply to a contract for the supply of goods and their incorporation in any land or a contract for the repair or improvement of a building or other erection on land, where the contract is not financed by a loan secured by a land mortgage;

(b) any contract for the supply of food, drink or other goods intended for current consumption by use in the household and supplied by regular roundsmen;

(c) any contract for the supply of goods or services which satisfies all the following conditions, namely—

   (i) terms of the contract are contained in a trader's catalogue which is readily available to the consumer to read in the absence of the trader or his representative before the conclusion of the contract;

   (ii) the parties to the contract intend that there shall be maintained continuity of contact between the trader or his representative and the consumer in relation to the transaction in question or any subsequent transaction; and

   (iii) both the catalogue and the contract contain or are accompanied by a prominent notice indicating that the consumer has a right to return to the trader or his representative goods supplied to him within the period of not less than 7 days from the day on which the goods are received by the consumer and otherwise to cancel the contract within that period without the consumer incurring any liability, other than any liability which may arise from the failure of the consumer to take reasonable care of the goods while they are in his possession;

(d) contracts of insurance to which the Insurance Companies Act 1982 applies;

(e) investment agreements within the meaning of the Financial Services Act 1986, and agreements for the making of deposits within the meaning of the Banking Act 1987 in respect of which Regulations have been made for regulating the making of unsolicited calls under section 34 of that Act;

(f) any contract not falling within sub-paragraph (g) below under which the total payments to be made by the consumer do not exceed £35; and

(g) any contract under which credit within the meaning of the Consumer Credit Act 1974 is provided not exceeding £35 other than a hire-purchase or conditional sale agreement.

[(3) In this regulation "unsolicited visit" means a visit by a trader, whether or not he is the trader who supplies the goods or services, which does not take place at the express request of the consumer and includes—

(a) a visit by a trader which takes place after he, or a person acting in his name or on his behalf, telephones the consumer (otherwise than at the consumer's express request) and indicates during the course of the telephone call (either expressly or by implication) that he, or the trader in whose name or on whose behalf he is acting, is willing to visit the consumer; and

(b) a visit by a trader which takes place after he, or a person acting in his name or on his behalf, visits the consumer (otherwise than at the consumer's express request) and indicates during the course of that visit (either expressly or by implication) that he, or the trader in whose name or on whose behalf he is acting, is willing to make a subsequent visit to the consumer.]

NOTES

Commencement: 31 December 1998 (sub-para (3)); 1 July 1988 (remainder).

Para (3): substituted by the Consumer Protection (Cancellation of Contracts Concluded away from Business Premises) (Amendment) Regulations 1998, SI 1998/3050, reg 2(b).

4 Cancellation of Contract

(1) No contract to which these Regulations apply shall be enforceable against the consumer unless the trader has delivered to the consumer notice in writing in accordance with paragraphs (3) and (4) below indicating the right of the consumer to cancel the contract within the period of 7 days mentioned in paragraph (5) below containing both the information set out in Part I of the Schedule to these Regulations and a Cancellation Form in the form set out in Part II of the Schedule and completed in accordance with the footnotes.

(2)   Paragraph (1) above does not apply to a cancellable agreement within the meaning of the Consumer Credit Act 1974 or to an agreement which may be cancelled by the consumer in accordance with terms of the agreement conferring upon him similar rights as if the agreement were such a cancellation agreement.

(3)   The information to be contained in the notice under paragraph (1) above shall be easily legible and if incorporated in the contract or other document shall be afforded no less prominence than that given to any other information in the document apart from the heading to the document and the names of the parties to the contract and any information inserted in handwriting.

(4)   The notice shall be dated and delivered to the consumer—
    (a) in the cases mentioned in regulation 3(1)(a), (b) and (d) above, at the time of the making of the contract; and
    (b) in the case mentioned in regulation 3(1)(c) above, at the time of the making of the offer by the consumer.

(5)   If within the period of 7 days following the making of the contract the consumer serves a notice in writing (a "notice of cancellation") on the trader or any other person specified in a notice referred to in paragraph (1) above as a person to whom notice of cancellation may be given which, however expressed and whether or not conforming to the cancellation form set out in Part II of the Schedule to these Regulations, indicates the intention of the consumer to cancel the contract, the notice of cancellation shall operate to cancel the contract.

(6)   Except as otherwise provided under these Regulations, a contract cancelled under paragraph (5) above shall be treated as if it had never been entered into by the consumer.

(7)   Notwithstanding anything in section 7 of the Interpretation Act 1978, a notice of cancellation sent by post by a consumer shall be deemed to have been served at the time of posting, whether or not it is actually received.

[4A   Offence relating to the failure to provide notice of cancellation rights

(1)   A trader is guilty of an offence if he enters into a contract to which these Regulations apply (other than an agreement referred to in regulation 4(2) above) with a consumer but fails (or, in the case mentioned in regulation 3(1)(c) above, has failed) to deliver to the consumer the

notice in writing referred to in regulation 4(1) above in accordance with paragraph (2) below.

(2)   A notice is delivered in accordance with this paragraph if it—
   (a)   contains what is required by regulation 4(1) above;
   (b)   complies with the requirements of regulation 4(3) above; and
   (c)   is dated and delivered to the consumer at the time specified in regulation 4(4) above.

(3)   A person who is guilty of an offence under paragraph (1) above shall be liable on summary conviction to a fine not exceeding level 4 on the standard scale.]

NOTES

Commencement: 31 December 1998.
Inserted, together with regs 4B-4H, by the Consumer Protection (Cancellation of Contracts Concluded away from Business Premises) (Amendment) Regulations 1998, SI 1998/3050, reg 2(c).

[4B   Defence of due diligence

(1)   In proceedings against any person for an offence under regulation 4A above it shall be a defence for that person to show that he took all reasonable steps and exercised all due diligence to avoid committing the offence.

(2)   Where in proceedings against a person for such an offence the defence provided for by paragraph (1) above involves an allegation that the commission of the offence was due—
   (a)   to the act or default of another, or
   (b)   to reliance on information given by another,

that person shall not, without the leave of the court, be entitled to rely on the defence unless he has served a notice under paragraph (3) below on the person bringing the proceedings not less than seven clear days before the hearing of the proceedings or, in Scotland, the diet of the trial.

(3)   A notice under this paragraph shall give such information identifying or assisting in the identification of the person who committed the act or default or gave the information as is in the possession of the person serving the notice at the time when he serves it.]

NOTES
Commencement: 31 December 1998.
Inserted as noted to reg 4A.

[4C   Liability of persons other than the principal offender

(1)   Where the commission by a person of an offence under regulation 4A above is due to the act or default of some other person, that other person is guilty of the offence and may be proceeded against and punished by virtue of this regulation whether or not proceedings are taken against the first-mentioned person.

(2)   Where a body corporate is guilty of an offence under regulation 4A above in respect of any act or default which is shown to have been committed with the consent or connivance of, or to be attributable to any neglect on the part of, any director, manager, secretary or other similar officer of the body corporate or any person who was purporting to act in any such capacity he, as well as the body corporate, shall be guilty of that offence and shall be liable to be proceeded against and punished accordingly.

(3)   Where the affairs of a body corporate are managed by its members, paragraph (2) above shall apply in relation to the acts and defaults of a member in connection with his functions of management as if he were a director of the body corporate.

(4)   Where an offence under regulation 4A above committed in Scotland by a Scottish partnership is proved to have been committed with the consent or connivance of, or to be attributable to neglect on the part of, a partner, he (as well as the partnership) is guilty of the offence and liable to be proceeded against and punished accordingly.]

NOTES
Commencement: 31 December 1998.
Inserted as noted to reg 4A.

[4D   Duty to enforce regulation 4A

(1)   Subject to paragraph (2) below—
   (a) it shall be the duty of every weights and measures authority in Great Britain to enforce regulation 4A above within its area; and
   (b) it shall be the duty of the Department of Economic Development to enforce regulation 4A above in Northern Ireland.

(2)　Nothing in paragraph (1) above shall authorise any weights and measures authority to bring proceedings in Scotland for an offence.]

NOTES

Commencement: 31 December 1998.
Inserted as noted to reg 4A.

[ENFORCEMENT POWERS, OBSTRUCTION OF AUTHORISED OFFICERS AND RESTRICTIONS ON DISCLOSURE OF INFORMATION

4E.—(1)　If a duly authorised officer of an enforcement authority has reasonable grounds for suspecting that an offence has been committed under regulation 4A above, he may—

(a) require a person carrying on or employed in a business to produce any book, document or record in non-documentary form relating to the business, and take copies of it or any entry in it, or

(b) require such a person to produce in a visible and legible documentary form any information so relating which is contained in a computer, and take copies of it,

for the purposes of ascertaining whether such an offence has been committed.

(2)　If such an officer has reasonable grounds for believing that any books, documents or records may be required as evidence in proceedings for such an offence, he may seize and detain them and shall, if he does so, inform the person from whom they are seized.

(3)　The powers of an officer under this regulation may be exercised by him only at a reasonable hour and on production (if required) of his credentials.

(4)　Nothing in this regulation requires a person to produce, or authorises the taking from a person of, a book, document or record which he could not be compelled to produce in civil proceedings before the High Court or (in Scotland) the Court of Session.]

NOTES

Commencement: 31 December 1998.
Inserted as noted to reg 4A.

[**4F.**—(1)　A person who—

(a) intentionally obstructs an officer of an enforcement authority acting in pursuance of his functions under these Regulations,

(b) without reasonable cause fails to comply with the requirement made of him by regulation 4E(1) above, or

(c) without reasonable excuse fails to give an officer of an enforcement authority acting in pursuance of his functions under these Regulations any other assistance or information which the officer has reasonably required of him for the purpose of the performance of the officer's functions under these Regulations,

is guilty of an offence.

(2)   If a person, in giving information to an officer of an enforcement authority who is acting in pursuance of his functions under these Regulations—

(a) makes a statement which he knows is false in a material particular, or

(b) recklessly makes a statement which is false in a material particular,

he is guilty of an offence.

(3)   A person guilty of an offence under paragraph (1) or (2) above shall be liable on summary conviction to a fine not exceeding level 3 on the standard scale.]

NOTES

Commencement: 31 December 1998.
Inserted as noted to reg 4A.

[**4G.**—(1)   If a person discloses to another any information obtained in the exercise of his functions under regulations 4D and 4E above, he is guilty of an offence unless the information has already been disclosed in any civil or criminal proceedings or the disclosure is made—

(a) in or for the purpose of the performance by him or any other person of any such function, or

(b) for a purpose specified in section 38(2)(a) or (b) of the Consumer Protection Act 1987 (enforcement of various enactments and compliance with Community obligations) or in the circumstances or for a purpose described in section 38(2)(c) of that Act (criminal investigations and civil or criminal proceedings).

(2)   A person guilty of an offence under paragraph (1) above shall be liable on summary conviction to a fine not exceeding level 3 on the standard scale.]

NOTES
Commencement: 31 December 1998.
Inserted as noted to reg 4A.

[**4H** Nothing in regulations 4E or 4F above requires a person to answer any question or give any information if to do so might incriminate him.]

NOTES
Commencement: 31 December 1998.
Inserted as noted to reg 4A.

5   Recovery of money paid by consumer

(1)   Subject to regulation 7(2) below, on the cancellation of a contract under regulation 4 above, any sum paid by or on behalf of the consumer under or in contemplation of the contract shall become repayable.

(2)   If under the terms of the cancelled contract the consumer or any person on his behalf is in possession of any goods, he shall have a lien on them for any sum repayable to him under paragraph (1) above.

(3)   Where any security has been provided in relation to the cancelled contract, the security, so far as it is so provided, shall be treated as never having had effect and any property lodged with the trader solely for the purposes of the security as so provided shall be returned by him forthwith.

6   Repayment of credit

(1)   Notwithstanding the cancellation of a contract under regulation 4 above under which credit is provided, the contract shall continue in force so far as it relates to repayment of credit and payment of interest.

(2)   If, following the cancellation of the contract, the consumer repays the whole or a portion of the credit—
  (a) before the expiry of one month following service of the notice of cancellation, or
  (b) in the case of a credit repayable by instalments, before the date on which the first instalment is due,

no interest shall be payable on the amount repaid.

(3)   If the whole of a credit repayable by instalments is not repaid on or before the date specified in paragraph (2)(b) above, the consumer shall not be liable to repay any of the credit except on receipt of a request in writing signed by the trader stating the amounts of the remaining instalments (recalculated by the trader as nearly as may be in accordance with the contract and without extending the repayment period), but excluding any sum other than principal and interest.

(4)   Repayment of a credit, or payment of interest, under a cancelled contract shall be treated as duly made if it is made to any person on whom, under regulation 4(5) above, a notice of cancellation could have been served.

(5)   Where any security has been provided in relation to the contract, the duty imposed on the consumer by this regulation shall not be enforceable before the trader has discharged any duty imposed on him by regulation 5(3) above.

[(6)   In this regulation, the following expressions have the meanings hereby assigned to them:—
    "cash" includes money in any form;
    "credit" means a cash loan and any facility enabling the consumer to overdraw on a current account;
    "current account" means an account under which the customer may, by means of cheques or similar orders payable to himself or to any other person, obtain or have the use of money held or made available by the person with whom the account is kept and which records alterations in the financial relationship between the said person and the customer; and
    "repayment", in relation to credit, means the repayment of money—
        (a) paid to a consumer before the cancellation of the contract; or
        (b) to the extent that he has overdrawn on his current account before the cancellation.

NOTES

Para (6): substituted by the Consumer Protection (Cancellation of Contracts Concluded away from Business Premises) (Amendment) Regulations 1988, SI 1988/958, reg 2.

7 Return of goods by consumer after cancellation

(1)   Subject to paragraph (2) below, a consumer who has before cancelling a contract under regulation 4 above acquired possession of any goods by virtue

of the contract shall be under a duty, subject to any lien, on the cancellation to restore the goods to the trader in accordance with this regulation, and meanwhile to retain possession of the goods and take reasonable care of them.

(2)   The consumer shall not be under a duty to restore—
 (i)   perishable goods;
 (ii) goods which by their nature are consumed by use and which, before the cancellation, were so consumed;
 (iii) goods supplied to meet an emergency; or
 (iv) goods which, before the cancellation, had become incorporated in any land or thing not comprised in the cancelled contract,

but he shall be under a duty to pay in accordance with the cancelled contract for the supply of the goods and for the provision of any services in connection with the supply of the goods before the cancellation.

(3)   The consumer shall not be under any duty to deliver the goods except at his own premises and in pursuance of a request in writing signed by the trader and served on the consumer either before, or at the time when, the goods are collected from those premises.

(4)   If the consumer—
 (i)   delivers the goods (whether at his own premises or elsewhere) to any person on whom, under regulation 4(5) above, a notice of cancellation could have been served; or
 (ii) sends the goods at his own expense to such a person,

he shall be discharged from any duty to retain possession of the goods or restore them to the trader.

(5)   Where the consumer delivers the goods as mentioned in paragraph (4)(i) above, his obligation to take care of the goods shall cease; and if he sends the goods as mentioned in paragraph (4)(ii) above, he shall be under a duty to take reasonable care to see that they are received by the trader and not damaged in transit, but in other respects his duty to take care of the goods shall cease.

(6)   Where, at any time during the period of 21 days following the cancellation, the consumer receives such a request as is mentioned in paragraph (3) above and unreasonably refuses or unreasonably fails to comply with it, his duty to retain possession and take reasonable care of the goods shall continue until he delivers or sends the goods as mentioned in paragraph (4) above, but if within that period he does not receive such a request his duty to take reasonable care of the goods shall cease at the end of that period.

(7)   Where any security has been provided in relation to the cancelled contract, the duty imposed on the consumer to restore goods by this regulation shall not be enforceable before the trader has discharged any duty imposed on him by regulation 5(3) above.

(8)   Breach of a duty imposed by this regulation on a consumer is actionable as a breach of statutory duty.

8   Goods given in part-exchange

(1)   This regulation applies on the cancellation of a contract under regulation 4 above where the trader agreed to take goods in part-exchange (the "part-exchange goods") and those goods have been delivered to him.

(2)   Unless, before the end of the period of ten days beginning with the date of cancellation, the part-exchange goods are returned to the consumer in a condition substantially as good as when they were delivered to the trader, the consumer shall be entitled to recover from the trader a sum equal to the part-exchange allowance.

(3)   During the period of ten days beginning with the date of cancellation, the consumer, if he is in possession of goods to which the cancelled contract relates, shall have a lien on them for—
  (a) delivery of the part-exchange goods in a condition substantially as good as when they were delivered to the trader; or
  (b) a sum equal to the part-exchange allowance;

and if the lien continues to the end of that period it shall thereafter subsist only as a lien for a sum equal to the part-exchange allowance.

(4)   In this regulation the part-exchange allowance means the sum agreed as such in the cancelled contract, or if no such sum was agreed, such sum as it would have been reasonable to allow in respect of the part-exchange goods if no notice of cancellation had been served.

10   No contracting-out

(1)   A term contained in a contract to which these Regulations apply is void if, and to the extent that, it is inconsistent with a provision for the protection of the consumer contained in these Regulations.

(2)   Where a provision of these Regulations specifies the duty or liability of the consumer in certain circumstances a term contained in a contract to

which these Regulations apply is inconsistent with that provision if it purports to impose, directly or indirectly, an additional duty or liability on him in those circumstances.

11 Service of documents

(1)   A document to be served under these Regulations on a person may be so served—
  (a) by delivering it to him, or by sending it by post to him, or by leaving it with him, at his proper address addressed to him by name;
  (b) if the person is a body corporate, by serving it in accordance with paragraph (a) above on the secretary or clerk of that body; or
  (c) if the person is a partnership, by serving it in accordance with paragraph (a) above on a partner or on a person having the control or management of the partnership business.

(2)   For the purposes of these Regulations, a document sent by post to, or left at, the address last known to the server of the document as the address of a person shall be treated as sent by post to, or left at, his proper address.

SCHEDULE

Regulation 4(i)

PART I
INFORMATION TO BE CONTAINED IN NOTICE OF CANCELLATION RIGHTS

1.   The name of the trader.

2.   The trader's reference number, code or other details to enable the contract or offer to be identified.

3.   A statement that the consumer has a right to cancel the contract if he wishes and that this right can be exercised by sending or taking a written notice of cancellation to the person mentioned in paragraph 4 within the period of 7 days following the making of the contract.

4.   The name and address of a person to whom notice of cancellation may be given.

5.   A statement that the consumer can use the cancellation form provided if he wishes.

PART II
CANCELLATION FORM TO BE INCLUDED IN NOTICE OF CANCELLATION RIGHTS

(Complete, detach and return this form ONLY IF YOU WISH TO CANCEL THE CONTRACT.)

To: 1

I/We* hereby give notice that I/we* wish to cancel my/our* contract

      2

Signed

Date

*Delete as appropriate

Notes:

1. Trader to insert name and address of person to whom notice may be given.

2. Trader to insert reference number, code or other details to enable the contract or offer to be identified. He may also insert the name and address of the consumer.

Control of Misleading Advertisements Regulations 1988

(SI 1988/915)

NOTES

Made: 23 May 1988.
Authority: European Communities Act 1972, s 2(2).
Commencement: 20 June 1988.

1 Citation and commencement

These Regulations may be cited as the Control of Misleading Advertisements Regulations 1988 and shall come into force on 20th June 1988.

2 Interpretation

(1)   In these Regulations—
"advertisement" means any form of representation which is made in connection with a trade, business, craft or profession in order to promote the supply or transfer of goods or services, immovable property, rights or obligations;

. . . . .

. . . . .

["the Commission" means the Independent Television Commission;]
"court", in relation to England and Wales and Northern Ireland, means the High Court, and, in relation to Scotland, the Court of Session;
"Director" means the Director General of Fair Trading;

. . . . .

. . . . .

["licensed service" means—
   (a) in relation to a complaint made to the Commission, a service in respect of which the Commission have granted a licence under Part I or II of the Broadcasting Act 1990; and
   (b) in relation to a complaint made to the Radio Authority, a service in respect of which the Radio Authority have granted a licence under Part III of that Act;
   and "licensed local delivery service" means a service in respect of which the Commission have granted a licence under Part II of that Act;]
"publication" in relation to an advertisement means the dissemination of that advertisement whether to an individual person or a number of persons and whether orally or in writing or in any other way whatsoever, and "publish" shall be construed accordingly.
["relevant body" means the Commission or the Radio Authority;
"on S4C" has the same meaning as in Part I of the Broadcasting Act 1990;
"the Welsh Authority" has the same meaning as in that Act.]

(2)   For the purposes of these Regulations an advertisement is misleading if in any way, including its presentation, it deceives or is likely to deceive the persons to whom it is addressed or whom it reaches and if, by reason of its deceptive nature, it is likely to affect their economic behaviour or, for those reasons, injures or is likely to injure a competitor of the person whose interests the advertisement seeks to promote.

(3)   . . .

NOTES

Para (1): definitions omitted repealed, definitions "the Commission", "relevant body", "on S4C" and "the Welsh Authority" inserted, definition "licensed service" substituted, by the Broadcasting Act 1990, s 203(1), (3), Sch 20, para 51(1), Sch 21.

Para (3): applies to Scotland only.

3 Application

(1)   These Regulations do not apply to—

  (a) the following advertisements issued or caused to be issued by or on behalf of an authorised person or appointed representative, that is to say—

    (i)   investment advertisements; and

    (ii) any other advertisements in respect of investment business,

except where any such advertisements relate exclusively to any matter in relation to which the authorised person in question is an exempted person; and

  (b) advertisements of a description referred to in section 58(1)(d) of the Financial Services Act 1986, . . .

(2)   In this regulation "appointed representative", . . . , "authorised person", "exempted person", "investment advertisement" and "investment business" have the same meanings as in the Financial Services Act 1986.

NOTES

Words omitted revoked by the Public Offers of Securities Regulations 1995, SI 1995/1537, reg 17, Sch 2, para 11.

Modification: by virtue of the Banking Coordination (Second Council Directive) Regulations 1992, SI 1992/3218, reg 82(1), Sch 10, para 51, this regulation has effect as if reference to an authorised institution within the meaning of the Banking Act included a reference to a European deposit-taker.

4 Complaints to the Director

(1)   Subject to paragraphs (2) and (3) below, it shall be the duty of the Director to consider any complaint made to him that an advertisement is misleading, unless the complaint appears to the Director to be frivolous or vexatious.

(2)   The Director shall not consider any complaint which these Regulations require or would require, leaving aside any question as to the

frivolous or vexatious nature of the complaint, [the Commission, the Radio Authority or the Welsh Authority] to consider.

(3)   Before considering any complaint under paragraph (1) above the Director may require the person making the complaint to satisfy him that—

    (a)   there have been invoked in relation to the same or substantially the same complaint about the advertisement in question such established means of dealing with such complaints as the Director may consider appropriate, having regard to all the circumstances of the particular case;

    (b)   a reasonable opportunity has been allowed for those means to deal with the complaint in question; and

    (c)   those means have not dealt with the complaint adequately.

(4)   In exercising the powers conferred on him by these Regulations the Director shall have regard to—

    (a)   all the interests involved and in particular the public interest; and

    (b)   the desirability of encouraging the control, by self-regulatory bodies, of advertisements.

NOTES

Para (2): words in square brackets substituted by the Broadcasting Act 1990, s 203(1), Sch 20, para 51(2).

5   Applications to the Court by the Director

(1)   If, having considered a complaint about an advertisement pursuant to regulation 4(1) above, he considers that the advertisement is misleading, the Director may, if he thinks it appropriate to do so, bring proceedings for an injunction (in which proceedings he may also apply for an interlocutory injunction) against any person appearing to him to be concerned or likely to be concerned with the publication of the advertisement.

(2)   The Director shall give reasons for his decision to apply or not to apply, as the case may be, for an injunction in relation to any complaint which these Regulations require him to consider.

6   Functions of the Court

(1)   The court on an application by the Director may grant an injunction on such terms as it may think fit but (except where it grants an interlocutory

injunction) only if the court is satisfied that the advertisement to which the application relates is misleading. Before granting an injunction the court shall have regard to all the interests involved and in particular the public interest.

(2)   An injunction may relate not only to a particular advertisement but to any advertisement in similar terms or likely to convey a similar impression.

(3)   In considering an application for an injunction the court may, whether or not on the application of any party to the proceedings, require any person appearing to the court to be responsible for the publication of the advertisement to which the application relates to furnish the court with evidence of the accuracy of any factual claim made in the advertisement. The court shall not make such a requirement unless it appears to the court to be appropriate in the circumstances of the particular case, having regard to the legitimate interests of the person who would be the subject of or affected by the requirement and of any other person concerned with the advertisement.

(4)   If such evidence is not furnished to it following a requirement made by it under paragraph (3) above or if it considers such evidence inadequate, the court may decline to consider the factual claim mentioned in that paragraph accurate.

(5)   The court shall not refuse to grant an injunction for lack of evidence that—
   (a) the publication of the advertisement in question has given rise to loss or damage to any person; or
   (b) the person responsible for the advertisement intended it to be misleading or failed to exercise proper care to prevent its being misleading.

(6)   An injunction may prohibit the publication or the continued or further publication of an advertisement.

7   Powers of the Director to obtain and disclose information and disclosure of information generally

(1)   For the purpose of facilitating the exercise by him of any functions conferred on him by these Regulations, the Director may, by notice in writing signed by him or on his behalf, require any person to furnish to him such information as may be specified or described in the notice or to produce to him any documents so specified or described.

(2)   A notice under paragraph (1) above may—
    (a) specify the way in which and the time within which it is to be complied with; and
    (b) be varied or revoked by a subsequent notice.

(3)   Nothing in this regulation compels the production or furnishing by any person of a document or of information which he would in an action in a court be entitled to refuse to produce or furnish on grounds of legal professional privilege or, in Scotland, on the grounds of confidentiality as between client and professional legal adviser.

(4)   If a person makes default in complying with a notice under paragraph (1) above the court may, on the application of the Director, make such order as the court thinks fit for requiring the default to be made good, and any such order may provide that all the costs or expenses of and incidental to the application shall be borne by the person in default or by any officers of a company or other association who are responsible for its default.

(5)   Subject to any provision to the contrary made by or under any enactment, where the Director considers it appropriate to do so for the purpose of controlling misleading advertisements, he may refer to any person any complaint (including any related documentation) about an advertisement or disclose to any person any information (whether or not obtained by means of the exercise of the power conferred by paragraph (1) above).

(6)   . . .

(7)   Subject to paragraph (5) above, any person who knowingly discloses, otherwise than for the purposes of any legal proceedings or of a report of such proceedings or the investigation of any criminal offence, any information obtained by means of the exercise of the power conferred by paragraph (1) above without the consent either of the person to whom the information relates, or, if the information relates to a business, the consent of the person for the time being carrying on that business, shall be guilty of an offence and liable on summary conviction to imprisonment for a term not exceeding 3 months or to a fine not exceeding [level 5 on the standard scale] or to both.

(8)   The Director may arrange for the dissemination in such form and manner as he considers appropriate of such information and advise concerning the operation of these Regulations as may appear to him to be expedient to give to the public and to all persons likely to be affected by these Regulations.

Para (7): words in square brackets substituted by virtue of the Criminal Justice Act 1988, s 52.

[8 Complaints to the Commission and the Radio Authority

(1)   Subject to paragraph (2) below, it shall be the duty of a relevant body to consider any complaint made to it that any advertisement included or proposed to be included in a licensed service is misleading, unless the complaint appears to the body to be frivolous or vexatious.

(2)   The Commission shall not consider any complaint about an advertisement included or proposed to be included in a licensed local delivery service by the reception and immediate re-transmission of broadcasts made by the British Broadcasting Corporation.

(3)   A relevant body shall give reasons for its decisions.

(4)   In exercising the powers conferred on it by these Regulations a relevant body shall have regard to all the interests involved and in particular the public interest.]

Substituted, together with regs 9-11, by the Broadcasting Act 1990, s 203(1), Sch 20, para 51(3).

[9 Control by the Commission and the Radio Authority of misleading advertisements

(1)   If, having considered a complaint about an advertisement pursuant to regulation 8(1) above, it considers that the advertisement is misleading, a relevant body may, if it thinks it appropriate to do so, exercise in relation to the advertisement the power conferred on it—
   (a) where the relevant body is the Commission, by section 9(6) of the Broadcasting Act 1990 (power of Commission to give directions about advertisements), or
   (b) where the relevant body is the Radio Authority, by section 93(6) of that Act (power of Radio Authority to give directions about advertisements).

(2)   A relevant body may require any person appearing to it to be responsible for an advertisement which the body believes may be misleading

to furnish it with evidence as to the accuracy of any factual claim made in the advertisement. In deciding whether or not to make such a requirement the body shall have regard to the legitimate interests of any person who would be the subject of or affected by the requirement.

(3)   If such evidence is not furnished to it following a requirement made by it under paragraph (2) above or if it considers such evidence inadequate, a relevant body may consider the factual claim inaccurate.]

NOTES

Substituted as noted to reg 8 above.

[10   Complaints to the Welsh Authority

(1)   Subject to paragraph (2) below, it shall be the duty of the Welsh Authority to consider any complaint made to them that any advertisement broadcast or proposed to be broadcast on S4C is misleading, unless the complaint appears to the Authority to be frivolous or vexatious.

(2)   The Welsh Authority shall not consider any complaint about an advertisement broadcast or proposed to be broadcast on S4C by the reception and immediate re-transmission of broadcasts made by the British Broadcasting Corporation.

(3)   The Welsh Authority shall give reasons for their decisions.

(4)   In exercising the powers conferred on them by these Regulations the Welsh Authority shall have regard to all the interests involved and in particular the public interest.]

NOTES

Substituted as noted to reg 8 above.

[11   Control by the Welsh Authority of misleading advertisements

(1)   If, having considered a complaint about an advertisement pursuant to regulation 10(1) above, they consider that the advertisement is misleading, the Welsh Authority may, if they think it appropriate to do so, refuse to broadcast the advertisement.

(2)   The Welsh Authority may require any person appearing to them to be responsible for an advertisement which the Authority believe may be misleading to furnish them with evidence as to the accuracy of any factual claim made in the advertisement. In deciding whether or not to make such a requirement the Authority shall have regard to the legitimate interests of any person who would be the subject of or affected by the requirement.

(3)   If such evidence is not furnished to them following a requirement made by them under paragraph (2) above or if they consider such evidence inadequate, the Welsh Authority may consider the factual claim inaccurate.]

NOTES

Substituted as noted to reg 8 above.

Consumer Protection (Code of Practice for Traders on Price Indications) Approval Order 1988

(SI 1988/2078)

NOTES

Made: 29 November 1988.
Authority: Consumer Protection Act 1987, s 25(1).
Commencement: 1 March 1989.

1   This Order may be cited as the Consumer Protection (Code of Practice for Traders on Price Indications) Approval Order 1988 and shall come into force on 1 March 1989.

2   The code of practice, as set out in the Schedule to this Order, issued by the Secretary of State for the purpose of—
  (a) giving practical guidance with respect to the requirements of section 20 of the Consumer Protection Act 1987; and
  (b) promoting what appear to the Secretary of State to be desirable practices as to the circumstances and manner in which a person gives an indication as to the price at which goods, services, accommodation or facilities are available or indicates any other matter in respect of which any such indication may be misleading—

is hereby approved.

SCHEDULE

Article 2

CODE OF PRACTICE FOR TRADERS ON PRICE INDICATIONS

INTRODUCTION

The Consumer Protection Act

1.   The Consumer Protection Act 1987 makes it a criminal offence to give consumers a misleading price indication about goods, services, accommodation (including the sale of new homes) or facilities. It applies however you give the price indication—whether in a TV or press advertisement, in a catalogue or leaflet, on notices, price tickets or shelf-edge marking in stores, or if you give it orally, for example on the telephone. The term "price indication" includes price comparisons as well as indications of a single price.

2.   This code of practice is approved under section 25 of the Act which gives the Secretary of State power to approve codes of practice to give practical guidance to traders. It is addressed to traders and sets out what is good practice to follow in giving price indications in a wide range of different circumstances, so as to avoid giving misleading price indications. But the Act does not require you to do as this code tells you. You may still give price indications which do not accord with this code, provided they are not misleading. "Misleading" is defined in section 21 of the Act. The definition covers indications about any conditions attached to a price, about what you expect to happen to a price in future and what you say in price comparisons, as well as indications about the actual price the consumer will have to pay. It also applies in the same way to any indications you give about the way in which a price will be calculated.

Price comparisons

3.   If you want to make price comparisons, you should do so only if you can show that they are accurate and valid. Indications which give only the price of the product are unlikely to be misleading if they are accurate and cover the total charge you will make. Comparisons with prices which you can show have been or are being charged for the same or similar goods, services, accommodation or facilities and have applied for a reasonable period are also unlikely to be misleading. Guidance on these matters is contained in this code.

Enforcement

4.   Enforcement of the Consumer Protection Act 1987 is the responsibility of officers of the local weights and measures authority (in Northern Ireland, the Department of Economic Development)—usually called Trading Standards Officers. If a Trading Standards Officer has reasonable grounds to suspect that you have given a misleading price indication, the Act gives the Officer power to require you to produce any records relating to your business and to seize and detain goods or records which the Officer has reasonable grounds for believing may be required as evidence in court proceedings.

5.   It may only be practicable for Trading Standards Officers to obtain from you the information necessary to carry out their duties under the Act. In these circumstances the Officer may seek information and assistance about both the claim and the supporting evidence from you. Be prepared to co-operate with Trading Standards Officers and respond to reasonable requests for information and assistance. The Act makes it an offence to obstruct a Trading Standards Officer intentionally or to fail (without good cause) to give any assistance or information the Officer may reasonably require to carry out duties under the Act.

Court proceedings

6.   If you are taken to court for giving a misleading price indication, the court can take into account whether or not you have followed the code. If you have done as the code advises, that will not be an absolute defence but it will tend to show that you have not committed an offence. Similarly if you have done something the code advises against doing it may tend to show that the price indication was misleading. If you do something which is not covered by the code, your price indication will need to be judged only against the terms of the general offence. The Act provides for a defence of due diligence, that is, that you have taken all reasonable steps to avoid committing the offence of giving a misleading price indication, but failure to follow the code of practice may make it difficult to show this.

Regulations

7.   The Act also provides power to make regulations about price indications and you should ensure that your price indications comply with any such regulations. There are none at present.

Other legislation

8.   This code deals only with the requirements of Part III of the Consumer Protection Act 1987. In some sectors there will be other relevant legislation. For example, price indications about credit terms must comply with the Consumer Credit Act 1974 and the regulations made under it, as well as with the Consumer Protection Act 1987.

Definitions

In this code—

Accommodation includes hotel and other holiday accommodation and new homes for sale freehold or on a lease of over 21 years but does not include rented homes.

Consumer means anyone who might want the goods, services, accommodation or facilities, other than for business use.

Price means both the total amount the consumer will have to pay to get the goods, services, accommodation or facilities and any method which has been or will be used to calculate that amount.

Price comparison means any indication given to consumers that the price at which something is offered to consumers is less than or equal to some other price.

Product means goods, services, accommodation and facilities (but not credit facilities, except where otherwise specified).

Services and Facilities means any services or facilities whatever (including credit, banking and insurance services, purchase or sale of foreign currency, supply of electricity, off-street car parking and caravan sites) except those provided by a person who is an authorised person or appointed representative under the Financial Services Act 1986 in the course of an investment business, services provided by an employee to his employer and facilities for a caravan which is the occupier's main or only home.

Shop means any shop, store, stall or other place (including a vehicle or the consumer's home) at which goods, services, accommodation or facilities are offered to consumers.

Trader means anyone (retailers, manufacturers, agents, service providers and others) who is acting in the course of a business.

PART I: PRICE COMPARISONS

1.1 Price comparisons generally

1.1.1. Always make the meaning of price indications clear. Do not leave consumers to guess whether or not a price comparison is being made. If no price comparison is intended, do not use words or phrases which, in their normal, everyday use and in the context in which they are used, are likely to give your customers the impression that price comparison is being made.

1.1.2. Price comparisons should always state the higher price as well as the price you intend to charge for the product (goods, services, accommodation or facilities). Do not make statements like "sale price £5" or "reduced to £39" without quoting the higher price to which they refer.

1.1.3. It should be clear what sort of price the higher price is. For example, comparisons with something described by words like "regular price", "usual price" or "normal price" should say whose regular, usual or normal price it is (eg "our normal price"). Descriptions like "reduced from" and crossed out higher prices should be used only if they refer to your own previous price. Words should not be used in price indications other than with their normal everyday meanings.

1.1.4. Do not use initials or abbreviations to describe the higher price in a comparison, except for the initials "RRP" to describe a recommended retail price or the abbreviation "man rec price" to describe a manufacturer's recommended price (see paragraph 1.6.2 below).

1.1.5. Follow the part of the code (sections 1.2 to 1.6 as appropriate) which applies to the type of comparison you intend to make.

1.2 Comparisons with the trader's own previous price

General

1.2.1. In any comparison between your present selling price and another price at which you have in the past offered the product, you should state the previous price as well as the new lower price.

1.2.2. In any comparison with your own previous price—
  (a) the previous price should be the last price at which the product was available to consumers in the previous 6 months;

(b)   the product should have been available to consumers at that price for at least 28 consecutive days in the previous 6 months; and

(c)   the previous price should have applied (as above) for that period at the same shop where the reduced price is now being offered.

The 28 days at (b) above may include bank holidays, Sundays or other days of religious observance when the shop was closed; and up to 4 days when, for reasons beyond your control, the product was not available for supply. The product must not have been offered at a different price between that 28 day period and the day when the reduced price is first offered.

1.2.3. If the previous price in a comparison does not meet one or more of the conditions set out in paragraph 1.2.2 above—

(i)   the comparison should be fair and meaningful; and

(ii)   give a clear and positive explanation of the period for which and the circumstances in which that higher price applied.

For example "these goods were on sale here at the higher price from 1 February to 26 February" or "these goods were on sale at the higher price in 10 of our 95 stores only". Display the explanation clearly, and as prominently as the price indication. You should not use general disclaimers saying for example that the higher prices used in comparisons have not necessarily applied for 28 consecutive days.

Food, drink and perishable goods

1.2.4. For any food and drink, you need not give a positive explanation if the previous price in a comparison has not applied for 28 consecutive days, provided it was the last price at which the goods were on sale in the previous 6 months and applied in the same shop where the reduced price is now being offered. This also applies to non-food perishables, if they have a shelf-life of less than 6 weeks.

Catalogue and Mail order traders

1.2.5. Where products are sold only through a catalogue, advertisement or leaflet, any comparison with a previous price should be with the price in your own last catalogue, advertisement or leaflet. If you sell the same products both in shops and through catalogues etc, the previous price should be the last price at which you offered the product. You should also follow the guidance in paragraphs 1.2.2(a) and (b). If your price comparison does not meet these conditions, you should follow the guidance in paragraph 1.2.3.

Making a series of reductions

1.2.6. If you advertise a price reduction and then want to reduce the price further during the same sale or special offer period, the intervening price (or prices) need not have applied for 28 days. In these circumstances unless you use a positive explanation (paragraph 1.2.3)—

the highest price in the series must have applied for 28 consecutive days in the last 6 months at the same shop: and

you must show the highest price, the intervening price(s) and the current selling price (eg "£40, £20, £10, £5").

1.3 Introductory offers, after-sale or after-promotion prices

Introductory Offers

1.3.1. Do not call a promotion an introductory offer unless you intend to continue to offer the product for sale after the offer period is over and to do so at a higher price.

1.3.2. Do not allow an offer to run on so long that it becomes misleading to describe it as an introductory or other special offer. What is a reasonable period will depend on the circumstances (but, depending on the shelf-life of the product, it is likely to be a matter of weeks, not months). An offer is unlikely to be misleading if you state the date the offer will end and keep to it. If you then extend the offer period, make it clear that you have done so.

Quoting a future price

1.3.3. If you indicate an after-sale or after-promotion price, do so only if you are certain that, subject only to circumstances beyond your control, you will continue to offer identical products at that price for at least 28 days in the 3 months after the end of the offer period or after the offer stocks run out.

1.3.4. If you decide to quote a future price, write what you mean in full. Do not use initials to describe it (eg "ASP", "APP"). The description should be clearly and prominently displayed, with the price indication.

1.4 Comparisons with prices related to different circumstances

1.4.1. This section covers comparisons with prices—
  (a) for different quantities (eg "15p each, 4 for 50p");
  (b) for goods in a different condition (eg "seconds £20, when perfect £30");
  (c) for a different availability (eg "price £50, price when ordered specially £60");
  (d) for goods in a totally different state (eg "price in kit form £50, price ready-assembled £70"); or
  (e) for special groups of people (eg "senior citizens' price £2.50, others £5").

General

1.4.2. Do not make such comparisons unless the product is available in the different quantity, conditions etc at the price you quote. Make clear to consumers the different circumstances which apply and show them prominently with the price indication. Do not use initials (eg "RAP" for "ready- assembled price") to describe the different circumstances, but write what you mean in full.

"When perfect" comparisons

1.4.3. If you do not have the perfect goods on sale in the same shop—
  (a) follow section 1.2 if the "when perfect" price is your own previous price for the goods;
  (b) follow section 1.5 if the "when perfect" price is another trader's price; or
  (c) follow section 1.6 if the "when perfect" price is one recommended by the manufacturer or supplier.

Goods in a different state

1.4.4. Only make comparisons with goods in a totally different state if—
  (a) a reasonable proportion (say a third (by quantity)) of your stock of those goods is readily available for sale to consumers in that different state (for example, ready assembled) at the quoted price and from the shop where the price comparison is made; or
  (b) another trader is offering those goods in that state at the quoted price and you follow section 1.5 below.

Prices for special groups of people

1.4.5. If you want to compare different prices which you charge to different groups of people (eg one price for existing customers and another for new customers, or one price for people who are members of a named organisation (other than the trader) and another for those who are not), do not use words like "our normal" or "our regular" to describe the higher price, unless it applies to at least half your customers.

1.5 Comparisons with another trader's prices

1.5.1. Only compare your prices with another trader's price if—
  (a) you know that his price which you quote is accurate and up-to-date;
  (b) you give the name of the other trader clearly and prominently, with the price comparison;
  (c) you identify the shop where the other trader's price applies, if that other trader is a retailer; and
  (d) the other trader's price which you quote applies to the same products – or to substantially similar products and you state any differences clearly.

1.5.2. Do not make statements like "if you can buy this product elsewhere for less, we will refund the difference" about your "own brand" products which other traders do not stock, unless your offer will also apply to other traders' equivalent goods. If there are any conditions attached to the offer (eg it only applies to goods on sale in the same town) you should show them clearly and prominently, with the statement.

1.6 Comparisons with "Recommended Retail Price" or similar

General

1.6.1. This Section covers comparisons with recommended retail prices, manufacturers' recommended prices, suggested retail prices, suppliers' suggested retail prices and similar descriptions. It also covers prices given to co-operative and voluntary group organisations by their wholesalers or headquarters organisations.

1.6.2. Do not use initials or abbreviations to describe the higher price in a comparison unless—
  (a) you use the initials "RRP" to describe a recommended retail price; or
  (b) you use the abbreviation "man. rec. price" to describe a manufacturer's recommended price.

Write all other descriptions out in full and show them clearly and prominently with the price indication.

1.6.3. Do not use a recommended price in a comparison unless—

(a) it has been recommended to you by the manufacturer or supplier as a price at which the product might be sold to consumers;

(b) you deal with that manufacturer or supplier on normal commercial terms. (This will generally be the case for members of co-operative or voluntary group organisations in relation to their wholesalers or headquarters organisations); and

(c) the price is not significantly higher than prices at which the product is generally sold at the time you first make that comparison.

1.7 Pre-printed prices

1.7.1. Make sure you pass on to consumers any reduction stated on the manufacturer's packaging (eg "flash packs" such as "10p off RRP").

1.7.2. You are making a price comparison if goods have a clearly visible price already printed on the packaging which is higher than the price you will charge for them. Such pre-printed prices are, in effect, recommended prices (except for retailers' own label goods) and you should follow paragraphs 1.6.1 to 1.6.4. You need not state that the price is a recommended price.

1.8 References to value or worth

1.8.1. Do not compare your prices with an amount described only as "worth" or "value".

1.8.2. Do not present general advertising slogans which refer to "value" or "worth" in a way which is likely to be seen by consumers as a price comparison.

1.9 Sales or special events

1.9.1. If you have bought in items specially for a sale, and you make this clear, you should not quote a higher price when indicating that they are special purchases. Otherwise, your price indications for individual items in the sale which are reduced should comply with section 1.1 of the code and whichever of sections 1.2 to 1.6 applies to the type of comparison you are making.

1.9.2. If you just have a general notice saying, for example, that all products are at "half marked price", the marked price on the individual items should be your own previous price and you should follow section 1.2 of the code.

1.9.3. Do not use general notices saying, eg "up to 50% off" unless the maximum reduction quoted applies to at least 10% (by quantity) of the range of products on offer.

1.10 Free offers

1.10.1. Make clear to consumers, at the time of the offer for sale, exactly what they will have to buy to get the "free offer".

1.10.2. If you give any indication of the monetary value of the "free offer", and that sum is not your own present price for the product, follow whichever of sections 1.2 to 1.6 covers the type of price it is.

1.10.3. If there are any conditions attached to the "free offer", give at least the main points of those conditions with the price indication and make clear to consumers where, before they are committed to buy, they can get full details of the conditions.

1.10.4. Do not claim that an offer is free if—
  (a) you have imposed additional charges that you would not normally make;
  (b) you have inflated the price of any product the consumer must buy or the incidental charges (for example, postage) the consumer must pay to get the "free offer"; or
  (c) you will reduce the price to consumers who do not take it up.

PART 2: ACTUAL PRICE TO THE CONSUMER

2.1 Indicating two different prices

2.1.1. The Consumer Protection Act makes it an offence to indicate a price for goods or services which is lower than the one that actually applies, for example, showing one price in an advertisement, window display, shelf marking or on the item itself, and then charging a higher price at the point of sale or checkout.

2.2.1. Make clear in your price indications the full price consumers will have to pay for the product. Some examples of how to do so in particular circumstances are set out below.

Limited availability of product

2.2.2. Where the price you are quoting for products only applies to a limited number of, say, orders, sizes or colours, you should make this clear in your price indication (eg "available in other colours or sizes at additional cost").

Prices relating to differing forms of products

2.2.3. If the price you are quoting for particular products does not apply to the products in the form they are displayed or advertised, say so clearly in your price indication. For example, advertisements for self-assembly furniture and the like should make it clear that the price refers to a kit of parts.

Postage, packing and delivery charges

2.2.4. If you sell by mail order, make clear any additional charges for postage, packing or delivery on the order form or similar document, so that consumers are fully aware of them before being committed to buying. Where you cannot determine these charges in advance, show clearly on the order form how they will be calculated (eg "Post Office rates apply"), or the place in the catalogue etc where the information is given.

2.2.5. If you sell goods from a shop and offer a delivery service for certain items, make it clear whether there are any separate delivery charges (eg for delivery outside a particular area) and what those charges are, before the consumer is committed to buying.

Value Added Tax

(i)   Price indications to consumers

2.2.6. All price indications you give to private consumers, by whatever means, should include VAT.

(ii)   Price indications to business customers

2.2.7. Prices may be indicated exclusive of VAT in shops where or advertisements from which most of your business is with business customers. If you also carry out business with private consumers at those shops or from those advertisements you should make clear that the prices exclude VAT and—
   (i)   display VAT-inclusive prices with equal prominence, or
   (ii) display prominent statements that on top of the quoted price customers will also have to pay VAT at 15% (or the current rate).

(iii) Professional fees

2.2.8. Where you indicate a price (including estimates) for a professional fee, make clear what it covers. The price should generally include VAT. In cases where the fee is based on an as-yet-unknown sum of money (for example, the sale price of a house), either—
   (i)   quote a fee which includes VAT; or
   (ii) make it clear that in addition to your fee the consumer would have to pay VAT at the current rate (eg "fee of 1.5 % of purchase price, plus VAT at 15%).

   Make sure that whichever method you choose is used for both estimates and final bills.

(iv)   Building work

2.2.9. In estimates for building work, either include VAT in the price indication or indicate with equal prominence the amount or rate of VAT payable in addition to your basic figure. If you give a separate amount for VAT, make it clear that if any provisional sums in estimates vary then the amount of VAT payable would also vary.

Service, cover and minimum charges in hotels, restaurants and similar establishments

2.2.10. If your customers in hotels, restaurants or similar places must pay a non-optional extra charge, eg a "service charge"—
   (i)   incorporate the charge within fully inclusive prices wherever practicable; and
   (ii) display the fact clearly on any price list or priced menu, whether displayed inside or outside (eg by using statements like "all prices include service").

Do not include suggested optional sums, whether for service or any other item, in the bill presented to the customer.

2.2.11. It will not be practical to include some non-optional extra charges in a quoted price; for instance, if you make a flat charge per person or per table in a restaurant (often referred to as a "cover charge") or a minimum charge. In such cases the charge should be shown as prominently as other prices on any list or menu, whether displayed inside or outside.

Holiday and travel prices

2.2.12. If you offer a variety of prices to give consumers a choice, (for example, paying more or less for a holiday depending on the time of year or the standard of accommodation), make clear in your brochure—or any other price indication—what the basic price is and what it covers. Give details of any optional additional charges and what those charges cover, or of the place where this information can be found, clearly and close to the basic price.

2.2.13. Any non-optional extra charges which are for fixed amounts should be included in the basic price and not shown as additions, unless they are only payable by some consumers. In that case you should specify, near to the details of the basic price, either what the amounts are and the circumstances in which they are payable, or where in the brochure etc the information is given.

2.2.14. Details of non-optional extra charges which may vary, (such as holiday insurance) or of where in the brochure etc the information is given should be made clear to consumers near to the basic price.

2.2.15. If you reserve the right to increase prices after consumers have made their booking, state this clearly with all indications of prices, and include prominently in your brochure full information on the circumstances in which a surcharge is payable.

Ticket prices

2.2.16. If you sell tickets, whether for sporting events, cinema, theatre etc and your prices are higher than the regular price that would be charged to the public at the box office, ie higher than the "face value", you should make clear in any price indication what the "face value" of the ticket is.

Call-out charges

2.2.17. If you make a minimum call-out charge or other flat-rate charge (for example, for plumbing, gas or electrical appliance repairs etc carried out in consumers' homes), ensure that the consumer is made aware of the charge and whether the actual price may be higher (eg if work takes longer than a specific time) before being committed to using your services.

Credit facilities

2.2.18. Price indications about consumer credit should comply with the relevant requirements of regulations under the Consumer Credit Act 1974 governing the form and content of advertisements.

Insurance

2.2.19. Where actual premium rates for a particular consumer or the availability of insurance cover depend on an individual assessment, this should be made clear when any indication of the premium or the method of determining it is given to consumers.

PART 3: PRICE INDICATIONS WHICH BECOME MISLEADING AFTER THEY HAVE BEEN GIVEN

3.1 General

3.1.1. The Consumer Protection Act makes it an offence to give a price indication which, although correct at the time, becomes misleading after you have given it, if—

(i) consumers could reasonably be expected still to be relying on it; and
(ii) you do not take reasonable steps to prevent them doing so.

Clearly it will not be necessary or even possible in many instances to inform all those who may have been given the misleading price indication. However, you should always make sure consumers are given the correct information before they are committed to buying a product and be prepared to cancel any transaction which a consumer has entered into on the basis of a price indication which has become misleading.

3.1.2. Do not give price indications which you know or intend will only apply for a limited period, without making this fact clear in the advertisement or price indication.

3.1.3. The following paragraphs set out what you should do in some particular circumstances.

3.2 Newspaper and magazine advertisements

3.2.1. If the advertisement does not say otherwise, the price indication should apply for a reasonable period (as a general guide, at least 7 days or until the next issue of the newspaper or magazine in which the advertisement was published, whichever is longer). If the price indication becomes misleading within this period make sure consumers are given the correct information before they are committed to buying the product.

3.3 Mail order advertisements, catalogues and leaflets

3.3.1. Paragraph 3.2.1 above also applies to the time for which price indications in mail order advertisements and in regularly published catalogues or brochures should apply. If a price indication becomes misleading within this period, make the correct price indication clear to anyone who orders the product to which it relates. Do so before the consumer is committed to buying the product and, wherever practicable, before the goods are sent to the consumer.

3.4 Selling through agents

Holiday brochures and travel agents

3.4.1. Surcharges are covered in paragraph 2.2.15. If a price indication becomes misleading for any other reason, tour operators who sell direct to consumers should follow paragraph 3.3.1 above; and tour operators who sell through travel agents should follow paragraphs 3.4.2 and 3.4.3 below.

3.4.2. If a price indication becomes misleading while your brochure is still current, make this clear to the travel agents to whom you distributed the brochure. Be prepared to cancel any holiday bookings consumers have made on the basis of a misleading price indication.

3.4.3. In the circumstances set out in paragraph 3.4.2, travel agents should ensure that the correct price indication is made clear to consumers before they make a booking.

Insurance and independent intermediaries

3.4.4. Insurers who sell their products through agents or independent intermediaries should take all reasonable steps to ensure that all such agents who are known to hold information on the insurer's premium rates and terms of the cover provided are told clearly of any changes in those rates or terms.

3.4.5. Agents, independent intermediaries and providers of quotation systems should ensure that they act on changes notified to them by an insurer.

3.5 Changes in the rate of value added tax

3.5.1. If your price indications become misleading because of a change in the general rate of VAT, or other taxes paid at point of sale, make the correct price indication clear to any consumers who order products. Do so before the consumer is committed to buying the product and, wherever practicable, before the goods are sent to the consumer.

PART 4: SALE OF NEW HOMES

4.1.   A "new home" is any building, or part of building to be used only as a private dwelling which is either—
(i)   a newly-built house or flat, or
(ii) a newly-converted existing building which has not previously been used in that form as a private home.

4.2.   The Consumer Protection Act and this code apply to new homes which are either for sale freehold or covered by a long lease, ie with more than 21 years to run. In this context the term "trader" covers not only a business vendor, such as a developer, but also an estate agent acting on behalf of such a vendor.

4.3.   You should follow the relevant provision of Part 1 of the code if—
(i)   you want to make a comparison between the price at which you offer new homes for sale and any other price;

(ii)   you offer an inclusive price for new homes which also covers such items as furnishings, domestic appliances and insurance and you compare their value with, for example, High Street prices for similar items.

4.4.   Part 2 of the code gives details of the provisions you should follow if—

  (i)   the new houses you are selling, or any goods or services which apply to them, are only available in limited numbers or range;

 (ii)   the sale price you give does not apply to the houses as displayed; or

(iii)   there are additional non-optional charges payable.

Consumer Credit (Exempt Agreements) Order 1989

(SI 1989/869)

NOTES

Made: 19 May 1989.
Authority: Consumer Credit Act 1974, ss 16(1), (4)-(6), 182(2), (4).
Commencement: 19 June 1989.

1   Citation, commencement, interpretation and revocation

(1)   This Order may be cited as the Consumer Credit (Exempt Agreements) Order 1989 and shall come into force on 19th June 1989.

(2)   In this Order—

"the Act" means the Consumer Credit Act 1974;

"business premises" means premises for occupation for the purposes of a business (including any activity carried on by a body of persons, whether corporate or unincorporate) or for those and other purposes;

and references to the total charge for credit and the rate thereof are respectively references to the total charge for credit and the rate thereof calculated in accordance with the Consumer Credit (Total Charge for Credit) Regulations 1980.

(3)   The Orders specified in Schedule 2 to this Order are hereby revoked.

2 Exemption of certain consumer credit agreements secured on land

(1)   The Act shall not regulate a consumer credit agreement which falls within section 16(2) of the Act, being an agreement to which this paragraph applies.

(2)   Where the creditor is a body specified in Part I of Schedule 1 to this Order, or a building society authorised under the Building Societies Act 1986, or an authorised institution under the Banking Act 1987 or a wholly-owned subsidiary of such an institution, paragraph (1) above applies only to—

  (a) a debtor-creditor-supplier agreement falling within section 16(2)(a) or (c) of the Act;
  (b) a debtor-creditor agreement secured by any land mortgage to finance—
      (i)   the purchase of land; or
      (ii) the provision of dwellings or business premises on any land; or
      (iii) subject to paragraph (3) below, the alteration, enlarging, repair or improvement of a dwelling or business premises on any land;
  (c) a debtor-creditor agreement secured by any land mortgage to refinance any existing indebtedness of the debtor, whether to the creditor or another person, under any agreement by which the debtor was provided with credit for any of the purposes specified in heads (i) to (iii) of sub-paragraph (b) above.

(3)   Head (iii) of sub-paragraph (b) of paragraph (2) above applies only—
  (i) where the creditor is the creditor under—
      (a) an agreement (whenever made) by which the debtor is provided with credit for any of the purposes specified in head (i) and head (ii) of that sub-paragraph; or
      (b) an agreement (whenever made) refinancing an agreement under which the debtor is provided with credit for any of the said purposes,

being, in either case, an agreement relating to the land referred to in the said head (iii) and secured by a land mortgage on that land; or
  (ii) where a debtor-creditor agreement to finance the alteration, enlarging, repair or improvement of a dwelling, secured by a land mortgage on that dwelling, is made as a result of any such services as are described in [section 4(3)(e) of the Housing Associations Act 1985] which are certified as having been provided by—
      (a) a local authority;
      (b) a housing association within the meaning of section 1 of the Housing Associations Act 1985 or [Article 3 of the Housing (Northern Ireland) Order 1992];

(c)   a body established by such a housing association for the purpose of providing such services as are described in the said [section 4(3)(e) of the Housing Associations Act 1985];

(d)   a charity;

(e)   the National Home Improvement Council; . . .

(f)   the Northern Ireland Housing Executive[; or

(g)   a body, or a body of any description, that has been approved by the Secretary of State under section 169(4)(c) of the Local Government and Housing Act 1989 [or the Department of the Environment for Northern Ireland under article 103(4)(c) of the Housing (Northern Ireland) Order 1992.]]

(4)   Where the creditor is a body specified in Part II of Schedule 1 to this Order, paragraph (1) above applies only to an agreement of a description specified in that Part in relation to that body and made pursuant to an enactment or for a purpose so specified.

(5)   Where the creditor is a body specified in Part III of Schedule 1 to this Order, paragraph (1) above applies only to an agreement of a description falling within Article 2(2)(a) to (c) above, being an agreement advancing money on the security of a dwelling-house.

NOTES

Para (3): words in first, second and third pairs of square brackets substituted by the Consumer Credit (Exempt Agreements) (Amendment) (No 2) Order 1993, SI 1993/2922, art 2(a); word omitted revoked and words in fourth pair of square brackets added, by the Consumer Credit (Exempt Agreements) (Amendment) (No 3) Order 1991, SI 1991/2844, art 2(a); words in square brackets therein added by SI 1993/2922, art 2(a).

Modification: by virtue of the Banking Coordination (Second Council Directive) Regulations 1992, SI 1992/3218, reg 82(1), Sch 10, para 54, para (2) has effect as if reference to an authorised institution within the meaning of the Banking Act included a reference to a European deposit-taker.

3 Exemption of certain consumer credit agreements by reference to the number of payments to be made by the debtor

(1)   The Act shall not regulate a consumer credit agreement which is an agreement of one of the following descriptions, that is to say—

(a) a debtor-creditor-supplier agreement being either—

(i)   an agreement for fixed-sum credit under which the total number of payments to be made by the debtor does not exceed four, and those payments are required to be made within a period not exceeding 12 months beginning with the date of the agreement; or

(ii) an agreement for running-account credit which provides for the making of payments by the debtor in relation to specified periods and requires that the number of payments to be made by the debtor in repayment of the whole amount of the credit provided in each such period shall not exceed one;

not being, in either case, an agreement of a description specified in paragraph (2) below; and in this sub-paragraph, "payment" means a payment comprising an amount in respect of credit with or without any other amount;

(b) a debtor-creditor-supplier agreement financing the purchase of land being an agreement under which the number of payments to be made by the debtor does not exceed four; and in this sub-paragraph, "payment" means a payment comprising or including an amount in respect of credit or the total charge for credit (if any);

(c) a debtor-creditor-supplier agreement for fixed-sum credit to finance a premium under a contract of insurance relating to any land or to anything thereon where—

(i) the creditor is the creditor under an agreement secured by a land mortgage on that land which either is an exempt agreement by virtue of section 16(1) of the Act or of article 2 above, or is a personal credit agreement which would be an exempt agreement by virtue of either of those provisions if the credit provided were not to exceed £15,000;

(ii) the amount of the credit is to be repaid within the period to which the premium relates, not being a period exceeding 12 months; and

(iii) there is no charge forming part of the total charge for credit under the agreement other than interest at a rate not exceeding the rate of interest from time to time payable under the agreement mentioned in head (i) above,

and the number of payments to be made by the debtor does not exceed twelve; and in this sub-paragraph "payment" has the same meaning as it has in paragraph (1)(b) above; and

(d) a debtor-creditor-supplier agreement for fixed-sum credit where—

(i) the creditor is the creditor under an agreement secured by a land mortgage on any land which either is an exempt agreement by virtue of section 16(1) of the Act or of article 2 above, or is a personal credit agreement which would be an exempt agreement by virtue of either of those provisions if the credit provided were not to exceed £15,000;

(ii) the agreement is to finance a premium under a contract of life insurance which provides, in the event of the death before the credit under the agreement referred to in head (i) above has been repaid of the person on whose life the contract is effected, for payment of a sum not exceeding the amount sufficient to defray the sums which, immediately after that credit has been advanced, would be payable to the creditor in respect of that credit and of the total charge for that credit; and

(iii) there is no charge forming part of the total charge for credit under the agreement other than interest at a rate not exceeding the rate of interest from time to time payable under the agreement referred to in head (i) above,

and the number of payments to be made by the debtor does not exceed twelve; and in this sub-paragraph, "payment" has the same meaning as it has in sub-paragraph (1)(b) above.

(2)   The descriptions of agreement referred to in sub-paragraph (a) of paragraph (1) above and to which accordingly that sub-paragraph does not apply are—

(a) agreements financing the purchase of land;
(b) agreements which are conditional sale agreements or hire-purchase agreements; and
(c) agreements secured by a pledge (other than a pledge of documents of title or of bearer bonds).

[4   Exemption of certain consumer credit agreements by reference to the rate of the total charge for credit

(1)   The Act shall not regulate a debtor-creditor agreement—

(a) which is an agreement of a type offered to a certain class or classes of persons and not offered to the public generally; and
(b) under the terms of which the only charge included in the total charge for credit is interest which cannot at any time exceed the sum of one per cent and the highest of the base rates named in paragraph (2) below, being the latest rates in operation on the date 28 days before any such time.

(2)   The banks referred to in paragraph (1) above are—
Bank of England
Bank of Scotland
Barclays Bank PLC
Clydesdale Bank PLC
Co-operative Bank Public Limited Company
Coutts & Co
Lloyds Bank PLC

Midland Bank Public Limited Company
National Westminster Bank Public Limited Company
the Royal Bank of Scotland plc
TSB Bank Plc.

(3)   For the purposes of sub-paragraph (1)(b) above, "interest" means interest at a rate determined in accordance with the formula set out in paragraph (1) of regulation 7 of the Consumer Credit (Total Charge for Credit) Regulations 1980, and in that formula as applied by that sub-paragraph "period rate of charge" has the meaning assigned to it in paragraph (2) of that regulation.]

NOTES
    Commencement: 1 September 1998.
    Substituted by the Consumer Credit (Exempt Agreements) (Amendment) Order 1998, SI 1998/1944, art 2.

5 Exemption of certain consumer credit agreements having a connection with a country outside the United Kingdom

The Act shall not regulate a consumer credit agreement made—
  (a) in connection with trade in goods or services between the United Kingdom and a country outside the United Kingdom or within a country or between countries outside the United Kingdom, being an agreement under which credit is provided to the debtor in the course of a business carried on by him; or
  (b) between a creditor listed in Part IV of Schedule 1 to this Order and a debtor who is—
    (i)   a member of any of the armed forces of the United States of America;
    (ii) an employee not habitually resident in the United Kingdom of any of those forces; or
    (iii) any such member's or employee's wife or husband or any other person (whether or not a child of his) whom he wholly or partly maintains and treats as a child of the family.

6 Exemption of certain consumer hire agreements

The Act shall not regulate a consumer hire agreement where the owner is a body corporate authorised by or under any enactment to supply [gas,] electricity or water and the subject of the agreement is a meter or metering equipment used or to be used in connection with the supply of [gas,] electricity or water, as the case may be.

NOTES

Words in square brackets inserted by the Consumer Credit (Exempt Agreements) (Amendment) Order 1991, SI 1991/1393, art 2(b).

Property Misdescriptions (Specified Matters) Order 1992

(SI 1992/2834)

NOTES

Made: 11 November 1992.
Authority: Property Misdescriptions Act 1991, s 1.
Commencement: 4 April 1993.

1 This Order may be cited as the Property Misdescriptions (Specified Matters) Order 1992 and shall come into force on 4th April 1993.

2 The matters contained in the Schedule to this Order are hereby specified to the extent described in that Schedule for the purposes of section 1(1) of the Property Misdescriptions Act 1991.

SCHEDULE

Article 2

SPECIFIED MATTERS

1. Location or address.

2. Aspect, view, outlook or environment.

3. Availability and nature of services, facilities or amenities.

4. Proximity to any services, places, facilities or amenities.

5. Accommodation, measurements or sizes.

6.   Fixtures and fittings.

7.   Physical or structural characteristics, form of construction or condition.

8.   Fitness for any purpose or strength of any buildings or other structures on land or of land itself.

9.   Treatments, processes, repairs or improvements or the effects thereof.

10.   Conformity or compliance with any scheme, standard, test or regulations or the existence of any guarantee.

11.   Survey, inspection, investigation, valuation or appraisal by any person or the results thereof.

12.   The grant or giving of any award or prize for design or construction.

13.   History, including the age, ownership or use of land or any building or fixture and the date of any alterations thereto.

14.   Person by whom any building, (or part of any building), fixture or component was designed, constructed, built, produced, treated, processed, repaired, reconditioned or tested.

15.   The length of time during which land has been available for sale either generally or by or through a particular person.

16.   Price (other than the price at which accommodation or facilities are available and are to be provided by means of the creation or disposal of an interest in land in the circumstances specified in section 23(1)(a) and (b) of the Consumer Protection Act 1987 or Article 16(1)(a) and (b) of the Consumer Protection (NI) Order 1987 (which relate to the creation or disposal of certain interests in new dwellings)) and previous price.

17.   Tenure or estate.

18.   Length of any lease or of the unexpired term of any lease and the terms and conditions of a lease (and, in relation to land in Northern Ireland, any fee farm grant creating the relation of landlord and tenant shall be treated as a lease).

19.   Amount of any ground-rent, rent or premium and frequency of any review.

20.   Amount of any rent-charge.

21.   Where all or any part of any land is let to a tenant or is subject to a licence, particulars of the tenancy or licence, including any rent, premium or other payment due and frequency of any review.

22.   Amount of any service or maintenance charge or liability for common repairs.

23.   Council tax payable in respect of a dwelling within the meaning of section 3, or in Scotland section 72, of the Local Government Finance Act 1992 or the basis or any part of the basis on which that tax is calculated.

24.   Rates payable in respect of a non-domestic hereditament within the meaning of section 64 of the Local Government Finance Act 1988 or, in Scotland, in respect of lands and heritages shown on a valuation roll or the basis or any part of the basis on which those rates are calculated.

25.   Rates payable in respect of a hereditament within the meaning of the Rates (Northern Ireland) Order 1977 or the basis or any part of the basis on which those rates are calculated.

26.   Existence or nature of any planning permission or proposals for development, construction or change of use.

27.   In relation to land in England and Wales, the passing or rejection of any plans of proposed building work in accordance with section 16 of the Building Act 1984 and the giving of any completion certificate in accordance with regulation 15 of the Building Regulations 1991.

28.   In relation to land in Scotland, the granting of a warrant under section 6 of the Building (Scotland) Act 1959 or the granting of a certificate of completion under section 9 of that Act.

29.   In relation to land in Northern Ireland, the passing or rejection of any plans of proposed building work in accordance with Article 13 of the Building Regulations (Northern Ireland) Order 1979 and the giving of any completion certificate in accordance with building regulations made under that Order.

30.   Application of any statutory provision which restricts the use of land or which requires it to be preserved or maintained in a specified manner.

31.   Existence or nature of any restrictive covenants, or of any restrictions on resale, restrictions on use, or pre-emption rights and, in relation to land in Scotland, (in addition to the matters mentioned previously in this paragraph) the existence or nature of any reservations or real conditions.

32.   Easements, servitudes or wayleaves.

33.   Existence and extent of any public or private right of way.

Banking Coordination (Second Council Directive) Regulations 1992

(SI 1992/3218)

NOTES

Made: 16 December 1992.
Authority: European Communities Act 1972, s 2(2).
Commencement: 1 January 1993.

PART I
GENERAL

1 Citation and commencement

(1)   These Regulations may be cited as the Banking Coordination (Second Council Directive) Regulations 1992.

(2)   These Regulations shall come into force on 1st January 1993.

2 Interpretation: general

(1)   In these Regulations—
    "the Banking Act" means the Banking Act 1987;
    "the Building Societies Act" means the Building Societies Act 1986;
    "the Consumer Credit Act" means the Consumer Credit Act 1974;
    "the Financial Services Act" means the Financial Services Act 1986;
    "the Insurance Companies Act" means the Insurance Companies Act 1982;

"another member State" means a member State other than the United Kingdom;

"appointed representative" has the same meaning as in the Financial Services Act;

"authorised or permitted", in relation to the carrying on of a listed activity, shall be construed in accordance with regulation 4 or, as the case may be, regulation 21 below;

["the Authority" means the Financial Services Authority (formerly known as the Securities and Investments Board);]

"the Bank" means the Bank of England;

"the Board" means The Securities and Investments Board [(now known as the Financial Services Authority)];

"branch" means one or more places of business established or proposed to be established in the same member State for the purpose of carrying on home-regulated activities;

"the commencement date" means 1st January 1993 [except in relation to the application of these Regulations as they have effect by virtue of section 2(1) of the European, Economic Area Act 1993 to the carrying on by credit institutions and financial institutions incorporated in or formed under the law of a member State of the Communities of listed activities in a relevant EFTA State and to the carrying on by credit institutions and financial institutions incorporated in or formed under the law of a relevant EFTA State of listed activities in the European Economic Area, where it means 1st January 1994] [in relation to a relevant EFTA State other than Liechtenstein and 1st June 1995 in relation to Liechtenstein;]

"the Commission" means the Building Societies Commission;

"connected UK authority", in relation to a credit or financial institution carrying on or proposing to carry on a listed activity in the United Kingdom, means an authority in the United Kingdom which has regulatory functions in relation to that activity;

"constituent instrument", in relation to an institution, includes any memorandum or articles of the institution;

"Consumer Credit Act business" means consumer credit business, consumer hire business or ancillary credit business;

"consumer credit business", "consumer hire business" and "ancillary credit business" have the same meanings as in the Consumer Credit Act;

"credit institution" means a credit institution as defined in article 1 of the First Council Directive, that is to say, an undertaking whose business is to receive deposits or other repayable funds from the public and to grant credits for its own account;

"delegation order" and "designated agency" have the same meanings as in the Financial Services Act;

"deposit" has the same meaning as in the Banking Act;

"the Director" means the Director General of Fair Trading;

"ecu" means the European currency unit as defined in Article 1 of Council Regulation No 3180/78/EEC;

"establish", in relation to a branch, means establish the place of business or, as the case may be, the first place of business which constitutes the branch;

"the European Commission" means the Commission of the Communities;

"European institution", "European authorised institution" and "European subsidiary" have the meanings given by regulation 3 below;

"financial institution" means a financial institution as defined in article 1 of the Second Council Directive, that is to say, an undertaking other than a credit institution the principal activity of which is to acquire holdings or to carry on one or more of the activities listed in points 2 to 12 in the Annex (the text of which is set out in Schedule 1 to these Regulations);

"the First Council Directive" means the First Council Directive on the coordination of laws, regulations and administrative provisions relating to the taking up and pursuit of the business of credit institutions (No 77/780/EEC);

"home-regulated activity" shall be construed in accordance with regulation 3(7) or, as the case may be, regulation 20(6) below;

"home-regulated investment business", in relation to a European institution or quasi-European authorised institution, means investment business which consists in carrying on one or more listed activities—

(a) in relation to which a supervisory authority in its home State has regulatory functions; and

(b) which, in the case of a European subsidiary, it is carrying on its home State;

"home State", in relation to an institution incorporated in or formed under the law of another member State, means that State;

"initial capital" means capital as defined in points 1 and 2 of article 2(1) of the Council Directive on the own funds of credit institutions (No 89/299/EEC);

"investment business" has the same meaning as in the Financial Services Act;

"listed activity" means an activity listed in the Annex to the Second Council Directive (list of activities subject to mutual recognition), the text of which is set out in Schedule 1 to these Regulations;

"member" and "rules", in relation to a recognised self-regulating organisation, have the same meanings as in the Financial Services Act;

["member State" means a member State of the Communities or a relevant EFTA State;]

"own funds" means own funds as defined in the Council Directive on the own funds of credit institutions (No 89/299/EEC);

"principal", in relation to an appointed representative, has the same meaning as in the Financial Services Act;

"quasi-European institution", "quasi-European authorised institution" and "quasi-European subsidiary" have the meanings given by regulation 3(4) below;

"recognised self-regulating organisation" has the same meaning as in the Financial Services Act;

["relevant EFTA State" means any of Austria, Finland, Iceland [Liechtenstein], Norway and Sweden;]

"the relevant supervisory authority", in relation to another member State, means the authority in that State which has regulatory functions in relation to the acceptance of deposits from the public, whether or not it also has such functions in relation to one or more other listed activities;

"requisite details", in relation to a branch in the United Kingdom or another member State (whether established or proposed to be established), means—

(a) particulars of the programme of operations of the business to be carried on from the branch, including a description of the particular home-regulated activities to be carried on and of the structural organisation of the branch;

(b) the name under which the business is to be carried on and the address in the member State from which information about the business may be obtained; and

(c) the names of the managers of the business;

"the Second Council Directive" means the Second Council Directive on the coordination of laws, regulations and administrative provisions relating to the taking up and pursuit of the business of credit institutions and amending the First Council Directive (No 89/646/EEC);

"the Solvency Ratio Directive" means the Council Directive on a solvency ratio for credit institutions (No 89/647/EEC);

"supervisory authority", in relation to another member State, means an authority in that State which has regulatory functions in relation to one or more listed activities;

"the UK authority", "UK institution", "UK authorised institution" and "UK subsidiary" have the meanings given by regulation 20 below;

"voting rights", in relation to an undertaking, shall be construed in accordance with paragraph 2 of Schedule 10A to the Companies Act 1985 or paragraph 2 of Schedule 10A to the Companies (Northern Ireland) Order 1986.

[(2B) Any reference in these Regulations to the First Council Directive or the Second Council Directive is a reference to that Directive as amended by the Prudential Supervision Directive (within the meaning of the Financial Institutions (Prudential Supervision) Regulations 1996).]

(2) In these Regulations "parent undertaking", "share", "subsidiary undertaking" and "undertaking" have the same meanings as in Part VII of the Companies Act 1985 or Part VIII of the Companies (Northern Ireland) Order 1986 except that—
  (a) "subsidiary undertaking" also includes, in relation to an institution incorporated in or formed under the law of another member State, any undertaking which is a subsidiary undertaking within the meaning of any rule of law in force in that State for purposes connected with the implementation of the Seventh Company Law Directive based on article 54(3)(g) of the Treaty on consolidated accounts (No 83/349/EEC); and
  (b) "parent undertaking" shall be construed accordingly.

(3) For the purposes of these Regulations a subsidiary undertaking of an institution is a 90 per cent subsidiary undertaking of the institution if the institution holds 90 per cent. or more of the voting rights in the subsidiary undertaking.

(4) Any reference in these Regulations to the carrying on of home-regulated investment business in the United Kingdom—
  (a) is a reference to the carrying on of such business in reliance on regulation 5(1)(b) below; and
  (b) shall be construed in accordance with section 1(3) of the Financial Services Act.

NOTES

Para (1): definition "the Authority" inserted and in definition "the Board" words in square brackets added by the Bank of England Act 1998, s 23, Sch 5, Pt I, Ch II, paras 21, 22; words in first pair of square brackets in definition "the commencement date", and definitions "member State" and "relevant EFTA State" inserted by the Banking Coordination (Second Council Directive) (Amendment) Regulations 1993, SI 1993/3225, reg 2(a)–(c); words in second pair of square brackets in definition "the commencement date" and word in square brackets in definition "relevant EFTA State" inserted by the Banking Coordination (Second Council Directive) (Amendment) Regulations 1995, SI 1995/1217, reg 2.

Para (2B): inserted by the Financial Institutions (Prudential Supervision) Regulations 1996, SI 1996/1669, reg 2, Sch 5, para 10.

Note: it is believed that the numbering of para (2B) is a mistake in the Queen's Printer's Copy and it should read "para (1A)".

PART II
RECOGNITION OF EUROPEAN INSTITUTIONS

Functions of Director

18 Power to prohibit the carrying on of Consumer Credit Act business

(1)   If it appears to the Director that paragraph (2) below has been or is likely to be contravened as respects a European institution, he may impose on the institution a prohibition under this regulation, that is to say, a prohibition on carrying on, or purporting to carry on, in the United Kingdom any Consumer Credit Act business which consists of or includes carrying on one or more home-regulated activities.

(2)   This paragraph is contravened as respects a European institution if—
   (a) the institution or any of the institution's employees, agents or associates (whether past or present); or
   (b) where the institution is a body corporate, any controller of the institution or an associate of any such controller,

does any of the things specified in paragraphs (a) to (d) of section 25(2) of the Consumer Credit Act.

(3)   A prohibition under this regulation may be absolute or may be imposed for a specified period or until the occurrence of a specified event or until specified conditions are complied with; and any period, event or conditions specified in the case of a prohibition may be varied by the Director on the application of the institution concerned.

(4)   Any prohibition imposed under this regulation may be withdrawn by written notice served by the Director on the institution concerned; and any such notice shall take effect on such date as is specified in the notice.

(5)   In this regulation "associate" has the same meaning as in section 25(2) of the Consumer Credit Act and "controller" has the meaning given by section 189(1) of that Act.

(6)   Schedule 5 to these Regulations (which makes supplemental provision with respect to prohibitions imposed under this regulation and restrictions imposed under regulation 19 below) shall have effect.

19 Power to restrict the carrying on of Consumer Credit Act business

(1)   In this regulation "restriction" means a direction that a European institution may not carry on in the United Kingdom, otherwise than in accordance with such condition or conditions as may be specified in the direction, any Consumer Credit Act business which—
  (a) consists of or includes carrying on one or more home-regulated activities; and
  (b) is specified in the direction.

(2)   Where it appears to the Director that the situation as respects a European institution is such that the powers conferred by paragraph (1) of regulation 18 above are exercisable, the Director may, instead of imposing a prohibition, impose such restriction as appears to him desirable.

(3)   Any restriction imposed under this regulation—
  (a) may be withdrawn; or
  (b) may be varied with the agreement of the institution concerned,

by written notice served by the Director on the institution; and any such notice shall take effect on such date as is specified in the notice.

(4)   An institution which contravenes or fails to comply with a restriction shall be guilty of an offence and liable—
  (a) on conviction on indictment, to a fine;
  (b) on summary conviction, to a fine not exceeding the statutory maximum.

(5)   The fact that a restriction has not been complied with (whether or not constituting an offence under paragraph (4) above) shall be a ground for the imposition of a prohibition under regulation 18 above.

PART VI
AMENDMENTS OF CONSUMER CREDIT ACT

57 Effect of standard licence

(1)   Section 22 of the Consumer Credit Act (standard and group licences) shall have effect as if it included provision that a standard licence held by a European institution or quasi-European authorised institution does not cover the carrying on by that institution of any home-regulated activities.

(2)   In this regulation and regulation 58 below "standard licence" has the meaning given by section 22(1)(a) of the Consumer Credit Act.

58   Grant of standard licence

(1)   Section 25 of the Consumer Credit Act (licensee to be a fit person) shall have effect as if—
   (a) it included provision that a standard licence shall not be issued to a European institution or quasi-European authorised institution in respect of any home-regulated activities; and
   (b) the reference in subsection (2)(b) to any provision made by or under that Act, or by or under any enactment regulating the provision of credit to individuals or other transactions with individuals, included a reference to any corresponding provision in force in another member State.

(2)   That section shall also have effect as if it included provision that where—
   (a) a UK authorised institution applies for a standard licence; and
   (b) the institution states in its application that it proposes to carry on a Consumer Credit Act business which consists of or includes one or more listed activities, the Director shall not grant the licence unless the UK authority has notified the Director that, were the licence granted, the UK authority would not by reason of that proposal exercise any of its relevant powers.

(3)   In paragraph (2) above "relevant powers" means—
   (a) in relation to the [Authority], the powers conferred on it by section 11 or 12 of the Banking Act (power to revoke or restrict authorisations);
   (b) in relation to the Commission, the powers conferred on it by section 42 or 43 of the Building Societies Act (power to impose conditions on or revoke authorisations).

NOTES

Para (3): in sub-para (a) word "Authority" in square brackets substituted by the Bank of England Act 1998, s 23(1), Sch 5, para 27.

59   Conduct of business

(1)   Section 26 of the Consumer Credit Act (conduct of business), and any existing regulations made otherwise than by virtue of section 54 of

that Act, shall have effect as if any reference to a licensee included a reference to a European institution carrying on a Consumer Credit Act business.

(2)   Section 54 of that Act (conduct of business regulations), and any existing regulations made by virtue of that section, shall have effect as if any reference to a licensee who carries on a consumer credit business, a consumer hire business or a business of credit brokerage, debt-adjusting or debt-counselling included a reference to a European institution who carries on such a business.

(3)   In this regulation "existing regulations" means regulations made under section 26 of that Act before the commencement date.

60   The register

Section 35 of the Consumer Credit Act (the register) shall have effect as if the particulars to be included in the register included—
   (a) particulars of information received by the Director under regulation 13 above;
   (b) particulars of prohibitions and restrictions imposed by him under regulation 18 or 19 above;
   (c) such particulars of documents received by him under paragraph 3(3), 4(3) or 5(4) of Schedule 2 to these Regulations as he thinks fit; and
   (d) particulars of such other matters (if any) arising under these Regulations as he thinks fit.

61   Enforcement of agreements

(1)   Section 40 of the Consumer Credit Act (enforcement of agreements by unlicensed trader) shall have effect as if the reference in subsection (1) to a regulated agreement, other than a non-commercial agreement, made when the creditor or owner was unlicensed did not include a reference to such an agreement made when the creditor or owner was a relevant institution.

(2)   Section 148 of that Act (enforcement of agreement for services of unlicensed trader) shall have effect as if the reference in subsection (1) to an agreement for the services of a person carrying on an ancillary credit business made when that person was unlicensed did not include a reference to such an agreement made when that person was a relevant institution.

(3)   Section 149 of that Act (enforcement of regulated agreements made on the introduction of an unlicensed credit-broker) shall have effect as if references in subsections (1) and (2) to introductions by an unlicensed credit-broker did not include references to introductions by a credit-broker who was a relevant institution.

(4)   In this regulation "relevant institution" means a European institution—
- (a) to which regulation 5(1)(c) above applies; and
- (b) which is not precluded from making the agreement or introductions in question by a restriction imposed under regulation 19 above.

62   Restrictions on disclosure of information

Section 174 of the Consumer Credit Act (restrictions on disclosure of information) shall have effect as if in subsection (3A)—
- (a) the reference to the [Authority's] functions under the Banking Act included a reference to its functions under these Regulations; and
- (b) the reference to the Director's functions under the Consumer Credit Act included a reference to his functions under these Regulations.

NOTES

In para (a) word "Authority's" in square brackets substituted by the Bank of England Act 1998, s 23(1), Sch 5, para 28.

63   Power to modify subordinate legislation in relation to European institutions

(1)   If the Secretary of State is satisfied that it is necessary to do so for the purpose of implementing the Second Council Directive so far as relating to any particular European institution, he may, on the application or with the consent of the institution, by order direct that all or any of the provisions of—
- (a) any regulations made under section 26 of the Consumer Credit Act; or
- (b) any regulations or orders made under Parts IV to VIII of that Act,

shall not apply to the institution or shall apply to it with such modifications as may be specified in the order.

(2)   An order under this regulation may be subject to conditions.

(3)   An order under this regulation may be revoked at any time by the Secretary of State; and the Secretary of State may at any time vary any such order on the application or with the consent of the European institution to which it applies.

SCHEDULES

SCHEDULE 5

Regulation 18(6)

PROHIBITIONS AND RESTRICTIONS BY THE DIRECTOR

Preliminary

1.   In this Schedule—
"appeal period" has the same meaning as in the Consumer Credit Act;
"prohibition" means a prohibition under regulation 18 of these Regulations;
"restriction" means a restriction under regulation 19 of these Regulations.

Notice of prohibition or restriction

2.—(1)   This paragraph applies where the Director proposes, in relation to a European institution—
  (a) to impose a prohibition;
  (b) to impose a restriction; or
  (c) to vary a restriction otherwise than with the agreement of the institution.

(2)   The Director shall, by notice—
  (a) inform the institution that, as the case may be, the Director proposes to impose the prohibition or restriction or vary the restriction, stating his reasons; and
  (b) invite the institution to submit representations to the proposal in accordance with paragraph 4 below.

(3)   If he imposes the prohibition or restriction or varies the restriction, the Director may give directions authorising the institution to carry into effect agreements made before the coming into force of the prohibition, restriction or variation.

(4)   A prohibition, restriction or variation shall not come into force before the end of the appeal period.

(5)   Where the Director imposes a prohibition or restriction or varies a restriction, he shall serve a copy of the prohibition, restriction or variation—
    (a) on the [Authority]; and
    (b) on the relevant supervisory authority in the institution's home State.

Application to revoke prohibition or restriction

3.—(1)   This paragraph applies where the Director proposes to refuse an application made by a European institution for the revocation of a prohibition or restriction.

(2)   The Director shall, by notice—
    (a) inform the institution that the Director proposes to refuse the application, stating his reasons; and
    (b) invite the institution to submit representations in support of the application in accordance with paragraph 4 below.

Representations to Director

4.—(1)   Where this paragraph applies to an invitation by the Director to an institution to submit representations, the Director shall invite the institution, within 21 days after the notice containing the invitation is given to it, or such longer period as the Director may allow—
    (a) to submit its representations in writing to the Director; and
    (b) to give notice to the Director, if it thinks fit, that it wishes to make representations orally;

and where notice is given under paragraph (b) above the Director shall arrange for the oral representations to be heard.

(2)   In reaching his determination the Director shall take into account any representations submitted or made under this paragraph.

(3)   The Director shall give notice of his determination to the institution.

Appeals

5.   Section 41 of the Consumer Credit Act (appeals to the Secretary of State) shall have effect as if—

(a) the following determinations were mentioned in column 1 of the table set out at the end of that section, namely—
  (i) imposition of a prohibition or restriction or the variation of a restriction; and
  (ii) refusal of an application for the revocation of a prohibition or restriction; and
(b) the European institution concerned were mentioned in column 2 of that table in relation to those determinations.

NOTES

Para 2: in sub-para (5)(a) word "Authority" in square brackets substituted by the Bank of England Act 1998, s 23(1), Sch 5, para 31.

SCHEDULE 10

Regulation 82(1)

MINOR AND CONSEQUENTIAL AMENDMENTS

PART I
PRIMARY LEGISLATION

. . .

Consumer Credit Act 1974 (c 39)

7.   Section 16(1)(h) of the Consumer Credit Act 1974 (exclusion of authorised institution's agreements from Act) shall have effect as if the reference to an institution authorised under the Banking Act included a reference to a European deposit-taker.

. . .

Consumer Protection Act 1987 (c 43)

27.   Section 22 of the Consumer Protection Act 1987 (application to provision of services and facilities) shall have effect as if it included provision that references in Part III of that Act to services or facilities shall not include references to services or facilities which are provided by a European institution in the course of carrying on home-regulated investment business in the United Kingdom.

. . .

Consumer Credit (Exempt Agreements) Order 1989 (SI 1989/869)

54.   Article 2(2) of the Consumer Credit (Exempt Agreements) Order 1989 (exemption of agreements secured on land) shall have effect as if the reference to an authorised institution under the Banking Act included a reference to a European deposit-taker.

. . .

NOTES

Commencement: 1 January 1993.

Package Travel, Package Holidays and Package Tours Regulations 1992

(SI 1992/3288)

NOTES

Made: 22 December 1992.
Authority: European Communities Act 1972, s 2(2).
Commencement: 23 December 1992.

1   Citation and commencement

These Regulations may be cited as the Package Travel, Package Holidays and Package Tours Regulations 1992 and shall come into force on the day after the day on which they are made.

2   Interpretation

(1)   In these Regulations—
    "brochure" means any brochure in which packages are offered for sale;
    "contract" means the agreement linking the consumer to the organiser or to the retailer, or to both, as the case may be;

"the Directive" means Council Directive 90/314/EEC on package travel, package holidays and package tours;

["member State" means a member State of the European Community or another State in the European Economic Area;]

"offer" includes an invitation to treat whether by means of advertising or otherwise, and cognate expressions shall be construed accordingly;

"organiser" means the person who, otherwise than occasionally, organises packages and sells or offers them for sale, whether directly or through a retailer;

"the other party to the contract" means the party, other than the consumer, to the contract, that is, the organiser or the retailer, or both, as the case may be;

"package" means the pre-arranged combination of at least two of the following components when sold or offered for sale at an inclusive price and when the service covers a period of more than twenty-four hours or includes overnight accommodation:—

(a) transport;

(b) accommodation;

(c) other tourist services not ancillary to transport or accommodation and accounting for a significant proportion of the package, and

    (i) the submission of separate accounts for different components shall not cause the arrangements to be other than a package;

    (ii) the fact that a combination is arranged at the request of the consumer and in accordance with his specific instructions (whether modified or not) shall not of itself cause it to be treated as other than pre-arranged;

and

"retailer" means the person who sells or offers for sale the package put together by the organiser.

(2)   In the definition of "contract" in paragraph (1) above, "consumer" means the person who takes or agrees to take the package ("the principal contractor") and elsewhere in these Regulations "consumer" means, as the context requires, the principal contractor, any person on whose behalf the principal contractor agrees to purchase the package ("the other beneficiaries") or any person to whom the principal contractor or any of the other beneficiaries transfers the package ("the transferee").

NOTES

Para (1): definition "member State" inserted by the Package Travel, Package Holidays and Package Tours (Amendment) Regulations 1995, SI 1995/1648, reg 2(a).

3 Application of Regulations

(1)   These Regulations apply to packages sold or offered for sale in the territory of the United Kingdom.

(2)   Regulations 4 to 15 apply to packages so sold or offered for sale on or after 31st December 1992.

(3)   Regulations 16 to 22 apply to contracts which, in whole or part, remain to be performed on 31st December 1992.

4 Descriptive matter relating to packages must not be misleading

(1)   No organiser or retailer shall supply to a consumer any descriptive matter concerning a package, the price of a package or any other conditions applying to the contract which contains any misleading information.

(2)   If an organiser or retailer is in breach of paragraph (1) he shall be liable to compensate the consumer for any loss which the consumer suffers in consequence.

5 Requirements as to brochures

(1)   Subject to paragraph (4) below, no organiser shall make available a brochure to a possible consumer unless it indicates in a legible, comprehensible and accurate manner the price and adequate information about the matters specified in Schedule 1 to these Regulations in respect of the packages offered for sale in the brochure to the extent that those matters are relevant to the packages so offered.

(2)   Subject to paragraph (4) below, no retailer shall make available to a possible consumer a brochure which he knows or has reasonable cause to believe does not comply with the requirements of paragraph (1).

(3)   An organiser who contravenes paragraph (1) of this regulation and a retailer who contravenes paragraph (2) thereof shall be guilty of an offence and liable:—
  (a) on summary conviction, to a fine not exceeding level 5 on the standard scale; and
  (b) on conviction on indictment, to a fine.

(4)   Where a brochure was first made available to consumers generally before 31st December 1992 no liability shall arise under this

regulation in respect of an identical brochure being made available to a consumer at any time.

6 Circumstances in which particulars in brochure are to be binding

(1)   Subject to paragraphs (2) and (3) of this regulation, the particulars in the brochure (whether or not they are required by regulation 5(1) above to be included in the brochure) shall constitute implied warranties (or, as regards Scotland, implied terms) for the purposes of any contract to which the particulars relate.

(2)   Paragraph (1) of this regulation does not apply—
  (a) in relation to information required to be included by virtue of paragraph 9 of Schedule 1 to these Regulations; or
  (b) where the brochure contains an express statement that changes may be made in the particulars contained in it before a contract is concluded and changes in the particulars so contained are clearly communicated to the consumer before a contract is concluded.

(3)   Paragraph (1) of this regulation does not apply when the consumer and the other party to the contract agree after the contract has been made that the particulars in the brochure, or some of those particulars, should not form part of the contract.

7 Information to be provided before contract is concluded

(1)   Before a contract is concluded, the other party to the contract shall provide the intending consumer with the information specified in paragraph (2) below in writing or in some other appropriate form.

(2)   The information referred to in paragraph (1) is:—
  (a) general information about passport and visa requirements which apply to [nationals of the member State or States concerned] who purchase the package in question, including information about the length of time it is likely to take to obtain the appropriate passports and visas;
  (b) information about health formalities required for the journey and the stay; and
  (c) the arrangements for security for the money paid over and (where applicable) for the repatriation of the consumer in the event of insolvency.

(3)   If the intending consumer is not provided with the information required by paragraph (1) in accordance with that paragraph the other party to the contract shall be guilty of an offence and liable:—
- (a) on summary conviction, to a fine not exceeding level 5 on the standard scale; and
- (b) on conviction on indictment, to a fine.

NOTES

Para (2): words in square brackets substituted by the Package Travel, Package Holidays and Package Tours (Amendment) Regulations 1998, SI 1998/1208, reg 5.

8   Information to be provided in good time

(1)   The other party to the contract shall in good time before the start of the journey provide the consumer with the information specified in paragraph (2) below in writing or in some other appropriate form.

(2)   The information referred to in paragraph (1) is the following:—
- (a) the times and places of intermediate stops and transport connections and particulars of the place to be occupied by the traveller (for example, cabin or berth on ship, sleeper compartment on train);
- (b) the name, address and telephone number—
  - (i)   of the representative of the other party to the contract in the locality where the consumer is to stay,

    or, if there is no such representative,
  - (ii) of an agency in that locality on whose assistance a consumer in difficulty would be able to call,

    or, if there is no such representative or agency, a telephone number or other information which will enable the consumer to contact the other party to the contract during the stay; and
- (c) in the case of a journey or stay abroad by a child under the age of 16 on the day when the journey or stay is due to start, information enabling direct contact to be made with the child or the person responsible at the place where he is to stay; and
- (d) except where the consumer is required as a term of the contract to take out an insurance policy in order to cover the cost of cancellation by the consumer or the cost of assistance, including repatriation, in the event of accident or illness, information about an insurance policy which the consumer may, if he wishes, take out in respect of the risk of those costs being incurred.

(3)   If the consumer is not provided with the information required by paragraph (1) in accordance with that paragraph the other party to the contract shall be guilty of an offence and liable:—

(a) on summary conviction, to a fine not exceeding level 5 on the standard scale; and

(b) on conviction on indictment, to a fine.

9   Contents and form of contract

(1)   The other party to the contract shall ensure that—

(a) depending on the nature of the package being purchased, the contract contains at least the elements specified in Schedule 2 to these Regulations;

(b) subject to paragraph (2) below, all the terms of the contract are set out in writing or such other form as is comprehensible and accessible to the consumer and are communicated to the consumer before the contract is made; and

(c) a written copy of these terms is supplied to the consumer.

(2)   Paragraph (1)(b) above does not apply when the interval between the time when the consumer approaches the other party to the contract with a view to entering into a contract and the time of departure under the proposed contract is so short that it is impracticable to comply with the sub-paragraph.

(3)   It is an implied condition (or, as regards Scotland, an implied term) of the contract that the other party to the contract complies with the provisions of paragraph (1).

(4)   . . .

NOTES

Para (4): applies to Scotland only.

10   Transfer of bookings

(1)   In every contract there is an implied term that where the consumer is prevented from proceeding with the package the consumer may transfer his booking to a person who satisfies all the conditions applicable to the package, provided that the consumer gives reasonable notice to the other

party to the contract of his intention to transfer before the date when departure is due to take place.

(2)   Where a transfer is made in accordance with the implied term set out in paragraph (1) above, the transferor and the transferee shall be jointly and severally liable to the other party to the contract for payment of the price of the package (or, if part of the price has been paid, for payment of the balance) and for any additional costs arising from such transfer.

11   Price revision

(1)   Any term in a contract to the effect that the prices laid down in the contract may be revised shall be void and of no effect unless the contract provides for the possibility of upward or downward revision and satisfies the conditions laid down in paragraph (2) below.

(2)   The conditions mentioned in paragraph (1) are that—
    (a) the contract states precisely how the revised price is to be calculated;
    (b) the contract provides that price revisions are to be made solely to allow for variations in:—
       (i)   transportation costs, including the cost of fuel,
      (ii)   dues, taxes or fees chargeable for services such as landing taxes or embarkation or disembarkation fees at ports and airports, or
      (iii) the exchange rates applied to the particular package; and

(3)   Notwithstanding any terms of a contract,
    (i)   no price increase may be made in a specified period which may not be less than 30 days before the departure date stipulated; and
    (ii) as against an individual consumer liable under the contract, no price increase may be made in respect of variations which would produce an increase of less than 2 per cent, or such greater percentage as the contract may specify, ("non-eligible variations") and that the non-eligible variations shall be left out of account in the calculation.

12   Significant alterations to essential terms

In every contract there are implied terms to the effect that—
    (a) where the organiser is constrained before the departure to alter significantly an essential term of the contract, such as the price

(so far as regulation 11 permits him to do so), he will notify the consumer as quickly as possible in order to enable him to take appropriate decisions and in particular to withdraw from the contract without penalty or to accept a rider to the contract specifying the alterations made and their impact on the price; and

(b) the consumer will inform the organiser or the retailer of his decision as soon as possible.

13 Withdrawal by consumer pursuant to regulation 12 and cancellation by organiser

(1)   The terms set out in paragraphs (2) and (3) below are implied in every contract and apply where the consumer withdraws from the contract pursuant to the term in it implied by virtue of regulation 12(a), or where the organiser, for any reason other than the fault of the consumer, cancels the package before the agreed date of departure.

(2)   The consumer is entitled—
   (a) to take a substitute package of equivalent or superior quality if the other party to the contract is able to offer him such a substitute; or
   (b) to take a substitute package of lower quality if the other party to the contract is able to offer him one and to recover from the organiser the difference in price between the price of the package purchased and that of the substitute package; or
   (c) to have repaid to him as soon as possible all the monies paid by him under the contract.

(3)   The consumer is entitled, if appropriate, to be compensated by the organiser for non-performance of the contract except where—
   (a) the package is cancelled because the number of persons who agree to take it is less than the minimum number required and the consumer is informed of the cancellation, in writing, within the period indicated in the description of the package; or
   (b) the package is cancelled by reason of unusual and unforeseeable circumstances beyond the control of the party by whom this exception is pleaded, the consequences of which could not have been avoided even if all due care had been exercised.

(4)   Overbooking shall not be regarded as a circumstance falling within the provisions of sub-paragraph (b) of paragraph (3) above.

14 Significant proportion of services not provided

(1)   The terms set out in paragraphs (2) and (3) below are implied in every contract and apply where, after departure, a significant proportion of the services contracted for is not provided or the organiser becomes aware that he will be unable to procure a significant proportion of the services to be provided.

(2)   The organiser will make suitable alternative arrangements, at no extra cost to the consumer, for the continuation of the package and will, where appropriate, compensate the consumer for the difference between the services to be supplied under the contract and those supplied.

(3)   If it is impossible to make arrangements as described in paragraph (2), or these are not accepted by the consumer for good reasons, the organiser will, where appropriate, provide the consumer with equivalent transport back to the place of departure or to another place to which the consumer has agreed and will, where appropriate, compensate the consumer.

15 Liability of other party to the contract for proper performance of obligations under contract

(1)   The other party to the contract is liable to the consumer for the proper performance of the obligations under the contract, irrespective of whether such obligations are to be performed by that other party or by other suppliers of services but this shall not affect any remedy or right of action which that other party may have against those other suppliers of services.

(2)   The other party to the contract is liable to the consumer for any damage caused to him by the failure to perform the contract or the improper performance of the contract unless the failure or the improper performance is due neither to any fault of that other party nor to that of another supplier of services, because—
-   (a) the failures which occur in the performance of the contract are attributable to the consumer;
-   (b) such failures are attributable to a third party unconnected with the provision of the services contracted for, and are unforeseeable or unavoidable; or
-   (c) such failures are due to—
    -   (i) unusual and unforeseeable circumstances beyond the control of the party by whom this exception is pleaded, the consequences of which could not have been avoided even if all due care had been exercised; or

(ii)   an event which the other party to the contract or the supplier of services, even with all due care, could not foresee or forestall.

(3)   In the case of damage arising from the non-performance or improper performance of the services involved in the package, the contract may provide for compensation to be limited in accordance with the international conventions which govern such services.

(4)   In the case of damage other than personal injury resulting from the non-performance or improper performance of the services involved in the package, the contract may include a term limiting the amount of compensation which will be paid to the consumer, provided that the limitation is not unreasonable.

(5)   Without prejudice to paragraph (3) and paragraph (4) above, liability under paragraphs (1) and (2) above cannot be excluded by any contractual term.

(6)   The terms set out in paragraphs (7) and (8) below are implied in every contract.

(7)   In the circumstances described in paragraph (2)(b) and (c) of this regulation, the other party to the contract will give prompt assistance to a consumer in difficulty.

(8)   If the consumer complains about a defect in the performance of the contract, the other party to the contract, or his local representative, if there is one, will make prompt efforts to find appropriate solutions.

(9)   The contract must clearly and explicitly oblige the consumer to communicate at the earliest opportunity, in writing or any other appropriate form, to the supplier of the services concerned and to the other party to the contract any failure which he perceives at the place where the services concerned are supplied.

16   Security in event of insolvency-requirements and offences

(1)   The other party to the contract shall at all times be able to provide sufficient evidence of security for the refund of money paid over and for the repatriation of the consumer in the event of insolvency.

(2)   Without prejudice to paragraph (1) above, and subject to paragraph (4) below, save to the extent that—

(a) the package is covered by measures adopted or retained by the member State where he is established for the purpose of implementing Article 7 of the Directive; or

(b) the package is one in respect of which he is required to hold a licence under the Civil Aviation (Air Travel Organisers' Licensing) Regulations 1972 or the package is one that is covered by the arrangements he has entered into for the purposes of those Regulations,

the other party to the contract shall at least ensure that there are in force arrangements as described in regulations 17,18, 19 or 20 or, if that party is acting otherwise than in the course of business, as described in any of those regulations or in regulation 21.

(3)   Any person who contravenes paragraph (1) or (2) of this regulation shall be guilty of an offence and liable:—

(a) on summary conviction to a fine not exceeding level 5 on the standard scale; and

(b) on conviction on indictment, to a fine.

(4)   A person shall not be guilty of an offence under paragraph (3) above by reason only of the fact that arrangements such as are mentioned in paragraph (2) above are not in force in respect of any period before 1 April 1993 unless money paid over is not refunded when it is due or the consumer is not repatriated in the event of insolvency.

(5)   For the purposes of regulations 17 to 21 below a contract shall be treated as having been fully performed if the package or, as the case may be, the part of the package has been completed irrespective of whether the obligations under the contract have been properly performed for the purposes of regulation 15.

17   Bonding

(1)   The other party to the contract shall ensure that a bond is entered into by an authorised institution under which the institution binds itself to pay to an approved body of which that other party is a member a sum calculated in accordance with paragraph (3) below in the event of the insolvency of that other party.

(2)   Any bond entered into pursuant to paragraph (1) above shall not be expressed to be in force for a period exceeding eighteen months.

(3)   The sum referred to in paragraph (1) above shall be such sum as may reasonably be expected to enable all monies paid over by consumers

under or in contemplation of contracts for relevant packages which have not been fully performed to be repaid and shall not in any event be a sum which is less than the minimum sum calculated in accordance with paragraph (4) below.

(4)   The minimum sum for the purposes of paragraph (3) above shall be a sum which represents:—

    (a) not less than 25 per cent of all the payments which the other party to the contract estimates that he will receive under or in contemplation of contracts for relevant packages in the twelve month period from the date of entry into force of the bond referred to in paragraph (1) above; or

    (b) the maximum amount of all the payments which the other party to the contract expects to hold at any one time, in respect of contracts which have not been fully performed,

whichever sum is the smaller.

(5)   Before a bond is entered into pursuant to paragraph (1) above, the other party to the contract shall inform the approved body of which he is a member of the minimum sum which he proposes for the purposes of paragraphs (3) and (4) above and it shall be the duty of the approved body to consider whether such sum is sufficient for the purpose mentioned in paragraph (3) and, if it does not consider that this is the case, it shall be the duty of the approved body so to inform the other party to the contract and to inform him of the sum which, in the opinion of the approved body, is sufficient for that purpose.

(6)   Where an approved body has informed the other party to the contract of a sum pursuant to paragraph (5) above, the minimum sum for the purposes of paragraphs (3) and (4) above shall be that sum.

(7)   In this regulation—

    "approved body" means a body which is for the time being approved by the Secretary of State for the purposes of this regulation;

    "authorised institution" means a person authorised under the law of a member State[, of the Channel Islands or of the Isle of Man] to carry on the business of entering into bonds of the kind required by this regulation.

NOTES

Para (7): words in square brackets in definition "authorised institution" inserted by the Package Travel, Package Holidays and Package Tours (Amendment) Regulations 1995, SI 1995/1648, reg 2(b).

18   Bonding where approved body has reserve fund or insurance

(1)   The other party to the contract shall ensure that a bond is entered into by an authorised institution, under which the institution agrees to pay to an approved body of which that other party is a member a sum calculated in accordance with paragraph (3) below in the event of the insolvency of that other party.

(2)   Any bond entered into pursuant to paragraph (1) above shall not be expressed to be in force for a period exceeding eighteen months.

(3)   The sum referred to in paragraph (1) above shall be such sum as may be specified by the approved body as representing the lesser of—
   (a) the maximum amount of all the payments which the other party to the contract expects to hold at any one time in respect of contracts which have not been fully performed; or
   (b) the minimum sum calculated in accordance with paragraph (4) below.

(4)   The minimum sum for the purposes of paragraph (3) above shall be a sum which represents not less than 10 per cent of all the payments which the other party to the contract estimates that he will receive under or in contemplation of contracts for relevant packages in the twelve month period from the date of entry referred to in paragraph (1) above.

(5)   In this regulation "approved body" means a body which is for the time being approved by the Secretary of State for the purposes of this regulation and no such approval shall be given unless the conditions mentioned in paragraph (6) below are satisfied in relation to it.

(6)   A body may not be approved for the purposes of this regulation unless—
   (a) it has a reserve fund or insurance cover with an insurer authorised in respect of such business in a member State[, the Channel Islands or the Isle of Man] of an amount in each case which is designed to enable all monies paid over to a member of the body of consumers under or in contemplation of contracts for relevant packages which have not been fully performed to be repaid to those consumers in the event of the insolvency of the member; and
   (b) where it has a reserve fund, it agrees that the fund will be held by persons and in a manner approved by the Secretary of State.

(7)   In this regulation, authorised institution has the meaning given to that expression by paragraph (7) of regulation 17.

NOTES

Para (6): words in square brackets inserted by the Package Travel, Package Holidays and Package Tours (Amendment) Regulations 1995, SI 1995/1648, reg 2(c).

19 Insurance

(1)   The other party to the contract shall have insurance under one or more appropriate policies with an insurer authorised in respect of such business in a member State under which the insurer agrees to indemnify consumers, who shall be insured persons under the policy, against the loss of money paid over by them under or in contemplation of contracts for packages in the event of the insolvency of the contractor.

(2)   The other party to the contract shall ensure that it is a term of every contract with a consumer that the consumer acquires the benefit of a policy of a kind mentioned in paragraph (1) above in the event of the insolvency of the other party to the contract.

(3)   In this regulation—
"appropriate policy" means one which does not contain a condition which provides (in whatever terms) that no liability shall arise under the policy, or that any liability so arising shall cease:—
  (i)   in the event of some specified thing being done or omitted to be done after the happening of the event giving rise to a claim under the policy;
  (ii) in the event of the policy holder not making payments under or in connection with other policies; or
  (iii) unless the policy holder keeps specified records or provides the insurer with or makes available to him information therefrom.

20 Monies in trust

(1)   The other party to the contract shall ensure that all monies paid over by a consumer under or in contemplation of a contract for a relevant package are held in the United Kingdom by a person as trustee for the consumer until the contract has been fully performed or any sum of money paid by the consumer in respect of the contract has been repaid to him or has been forfeited on cancellation by the consumer.

(2)   The costs of administering the trust mentioned in paragraph (1) above shall be paid for by the other party to the contract.

(3)   Any interest which is earned on the monies held by the trustee pursuant to paragraph (1) shall be held for the other party to the contract and shall be payable to him on demand.

(4)   Where there is produced to the trustee a statement signed by the other party to the contract to the effect that—

   (a)   a contract for a package the price of which is specified in that statement has been fully performed;

   (b)   the other party to the contract has repaid to the consumer a sum of money specified in that statement which the consumer had paid in respect of a contract for a package; or

   (c)   the consumer has on cancellation forfeited a sum of money specified in that statement which he had paid in respect of a contract for a relevant package,

the trustee shall (subject to paragraph (5) below) release to the other party to the contract the sum specified in the statement.

(5)   Where the trustee considers it appropriate to do so, he may require the other party to the contract to provide further information or evidence of the matters mentioned in sub-paragraph (a), (b) or (c) of paragraph (4) above before he releases any sum to that other party pursuant to that paragraph.

(6)   Subject to paragraph (7) below, in the event of the insolvency of the other party to the contract the monies held in trust by the trustee pursuant to paragraph (1) of this regulation shall be applied to meet the claims of consumers who are creditors of that other party in respect of contracts for packages in respect of which the arrangements were established and which have not been fully performed and, if there is a surplus after those claims have been met, it shall form part of the estate of that insolvent other party for the purposes of insolvency law.

(7)   If the monies held in trust by the trustee pursuant to paragraph (1) of this regulation are insufficient to meet the claims of consumers as described in paragraph (6), payments to those consumers shall be made by the trustee on a pari passu basis.

21   Monies in trust where other party to contract is acting otherwise than in the course of business

(1)   The other party to the contract shall ensure that all monies paid over by a consumer under or in contemplation of a contract for a relevant

package are held in the United Kingdom by a person as trustee for the consumer for the purpose of paying for the consumer's package.

(2)   The costs of administering the trust mentioned in paragraph (1) shall be paid for out of the monies held in trust and the interest earned on those monies.

(3)   Where there is produced to the trustee a statement signed by the other party to the contract to the effect that—

    (a)   the consumer has previously paid over a sum of money specified in that statement in respect of a contract for a package and that sum is required for the purpose of paying for a component (or part of a component) of the package;

    (b)   the consumer has previously paid over a sum of money specified in that statement in respect of a contract for a package and the other party to the contract has paid that sum in respect of a component (or part of a component) of the package;

    (c)   the consumer requires the repayment to him of a sum of money specified in that statement which was previously paid over by the consumer in respect of a contract for a package; or

    (d)   the consumer has on cancellation forfeited a sum of money specified in that statement which he had paid in respect of a contract for a package,

the trustee shall (subject to paragraph (4) below) release to the other party to the contract the sum specified in the statement.

(4)   Where the trustee considers it appropriate to do so, he may require the other party to the contract to provide further information or evidence of the matters mentioned in sub-paragraph (a), (b), (c) or (d) of paragraph (3) above before he releases to that other party any sum from the monies held in trust for the consumer.

(5)   Subject to paragraph (6) below, in the event of the insolvency of the other party to the contract and of contracts for packages not being fully performed (whether before or after the insolvency) the monies held in trust by the trustee pursuant to paragraph (1) of this regulation shall be applied to meet the claims of consumers who are creditors of that other party in respect of amounts paid over by them and remaining in the trust fund after deductions have been made in respect of amounts released to that other party pursuant to paragraph (3) and, if there is a surplus after those claims have been met, it shall be divided amongst those consumers pro rata.

(6)   If the monies held in trust by the trustee pursuant to paragraph (1) of this regulation are insufficient to meet the claims of consumers as

described in paragraph (5) above, payments to those consumers shall be made by the trustee on a pari passu basis.

(7)   Any sums remaining after all the packages in respect of which the arrangements were established have been fully performed shall be dealt with as provided in the arrangements or, in default of such provision, may be paid to the other party to the contract.

22   Offences arising from breach of regulations 20 and 21

(1)   If the other party to the contract makes a false statement under paragraph (4) of regulation 20 or paragraph (3) of regulation 21 he shall be guilty of an offence.

(2)   If the other party to the contract applies monies released to him on the basis of a statement made by him under regulation 21(3)(a) or (c) for a purpose other than that mentioned in the statement he shall be guilty of an offence.

(3)   If the other party to the contract is guilty of an offence under paragraph (1) or (2) of this regulation shall be liable—
  (a) on summary conviction to a fine not exceeding level 5 on the standard scale; and
  (b) on conviction on indictment, to a fine.

23   Enforcement

Schedule 3 to these Regulations (which makes provision about the enforcement of regulations 5, 7, 8, 16 and 22 of these Regulations) shall have effect.

24   Due diligence defence

(1)   Subject to the following provisions of this regulation, in proceedings against any person for an offence under regulation 5, 7, 8, 16 or 22 of these Regulations, it shall be a defence for that person to show that he took all reasonable steps and exercised all due diligence to avoid committing the offence.

(2)   Where in any proceedings against any person for such an offence the defence provided by paragraph (1) above involves an allegation that the commission of the offence was due—
  (a) to the act or default of another; or
  (b) to reliance on information given by another,

that person shall not, without the leave of the court, be entitled to rely on the defence unless, not less than seven clear days before the hearing of the proceedings, or, in Scotland, the trial diet, he has served a notice under paragraph (3) below on the person bringing the proceedings.

(3)   A notice under this paragraph shall give such information identifying or assisting in the identification of the person who committed the act or default or gave the information as is in the possession of the person serving the notice at the time he serves it.

(4)   It is hereby declared that a person shall not be entitled to rely on the defence provided by paragraph (1) above by reason of his reliance on information supplied by another, unless he shows that it was reasonable in all the circumstances for him to have relied on the information, having regard in particular—
   (a) to the steps which he took, and those which might reasonably have been taken, for the purpose of verifying the information; and
   (b) to whether he had any reason to disbelieve the information.

25   Liability of persons other than principal offender

(1)   Where the commission by any person of an offence under regulation 5, 7, 8, 16 or 22 of these Regulations is due to an act or default committed by some other person in the course of any business of his, the other person shall be guilty of the offence and may be proceeded against and punished by virtue of this paragraph whether or not proceedings are taken against the first-mentioned person.

(2)   Where a body corporate is guilty of an offence under any of the provisions mentioned in paragraph (1) above (including where it is so guilty by virtue of the said paragraph (1)) in respect of any act or default which is shown to have been committed with the consent or connivance of, or to be attributable to any neglect on the part of, any director, manager, secretary or other similar officer of the body corporate or any person who was purporting to act in any such capacity he, as well as the body corporate, shall be guilty of that offence and shall be liable to be proceeded against and punished accordingly.

(3)   Where the affairs of a body corporate are managed by its members, paragraph (2) above shall apply in relation to the acts and defaults of a member in connection with his functions of management as if he were a director of the body corporate.

(4)   . . .

(5)   On proceedings for an offence under regulation 5 by virtue of paragraph (1) above committed by the making available of a brochure it shall be a defence for the person charged to prove that he is a person whose business it is to publish or arrange for the publication of brochures and that he received the brochure for publication in the ordinary course of business and did not know and had no reason to suspect that its publication would amount to an offence under these Regulations.

Para (4): applies to Scotland only.

26   Prosecution time limit

(1)   No proceedings for an offence under regulation 5, 7, 8, 16 or 22 of these Regulations or under paragraphs 5(3), 6 or 7 of Schedule 3 thereto shall be commenced after—
  (a) the end of the period of three years beginning with the date of the commission of the offence; or
  (b) the end of the period of one year beginning with the date of the discovery of the offence by the prosecutor,

whichever is the earlier.

(2)   For the purposes of this regulation a certificate signed by or on behalf of the prosecutor and stating the date on which the offence was discovered by him shall be conclusive evidence of that fact; and a certificate stating that matter and purporting to be so signed shall be treated as so signed unless the contrary is proved.

(3)   . . .

Para (3): applies to Scotland only.

27   Saving for civil consequences

No contract shall be void or unenforceable, and no right of action in civil proceedings in respect of any loss shall arise, by reason only of the commission of an offence under regulations 5, 7, 8, 16 or 22 of these Regulations.

28 Terms implied in contract

Where it is provided in these Regulations that a term (whether so described or whether described as a condition or warranty) is implied in the contract it is so implied irrespective of the law which governs the contract.

SCHEDULES

SCHEDULE 1

Regulation 5

INFORMATION TO BE INCLUDED (IN ADDITION TO THE PRICE) IN BROCHURES WHERE RELEVANT TO PACKAGES OFFERED

1.　The destination and the means, characteristics and categories of transport used.

2.　The type of accommodation, its location, category or degree of comfort and its main features and, where the accommodation is to be provided in a member State, its approval or tourist classification under the rules of that member State.

3.　The meals which are included in the package.

4.　The itinerary.

5.　General information about passport and visa requirements which apply for [nationals of the member State or States in which the brochure is made available] and health formalities required for the journey and the stay.

6.　Either the monetary amount or the percentage of the price which is to be paid on account and the timetable for payment of the balance.

7.　Whether a minimum number of persons is required for the package to take place and, if so, the deadline for informing the consumer in the event of cancellation.

8.   The arrangements (if any) which apply if consumers are delayed at the outward or homeward points of departure.

9.   The arrangements for security for money paid over and for the repatriation of the consumer in the event of insolvency.

NOTES

Para 5: words in square brackets substituted with savings by the Package Travel, Package Holidays and Package Tours (Amendment) Regulations 1998, SI 1998/1208, reg 4; for savings see reg 3 thereof.

SCHEDULE 2

Regulation 9

ELEMENTS TO BE INCLUDED IN THE CONTRACT IF RELEVANT TO THE PARTICULAR PACKAGE

1.   The travel destination(s) and, where periods of stay are involved, the relevant periods, with dates.

2.   The means, characteristics and categories of transport to be used and the dates, times and points of departure and return.

3.   Where the package includes accommodation, its location, its tourist category or degree of comfort, its main features and, where the accommodation is to be provided in a member State, its compliance with the rules of that member State.

4.   The meals which are included in the package.

5.   Whether a minimum number of persons is required for the package to take place and, if so, the deadline for informing the consumer in the event of cancellation.

6.   The itinerary.

7.   Visits, excursions or other services which are included in the total price agreed for the package.

8.   The name and address of the organiser, the retailer and, where appropriate, the insurer.

9.   The price of the package, if the price may be revised in accordance with the term which may be included in the contract under regulation 11, an indication of the possibility of such price revisions, and an indication of any dues, taxes or fees chargeable for certain services (landing, embarkation or disembarkation fees at ports and airports and tourist taxes) where such costs are not included in the package.

10.   The payment schedule and method of payment.

11.   Special requirements which the consumer has communicated to the organiser or retailer when making the booking and which both have accepted.

12.   The periods within which the consumer must make any complaint about the failure to perform or the inadequate performance of the contract.

SCHEDULE 3

Regulation 23

ENFORCEMENT

Enforcement authority

1.—(1)   Every local weights and measures authority in Great Britain shall be an enforcement authority for the purposes of regulations 5, 7, 8, 16 and 22 of these Regulations ("the relevant regulations"), and it shall be the duty of each such authority to enforce those provisions within their area.

(2)   The Department of Economic Development in Northern Ireland shall be an enforcement authority for the purposes of the relevant regulations, and it shall be the duty of the Department to enforce those provisions within Northern Ireland.

Prosecutions

2.—(1)   Where an enforcement authority in England or Wales proposes to institute proceedings for an offence under any of the relevant regulations, it shall as between the enforcement authority and the Director General of Fair Trading be the duty of the enforcement authority to give to the Director

General of Fair Trading notice of the intended proceedings, together with a summary of the facts on which the charges are to be founded, and to postpone institution of the proceedings until either—

(a) twenty-eight days have elapsed since the giving of that notice; or

(b) the Director General of Fair Trading has notified the enforcement authority that he has received the notice and the summary of the facts.

(2) Nothing in paragraph 1 above shall authorise a local weights and measures authority to bring proceedings in Scotland for an offence.

. . .

Commercial Agents (Council Directive) Regulations 1993

(SI 1993/3053)

NOTES

Made: 7 December 1993.
Authority: European Communities Act 1972, s 2(2).
Commencement: 1 January 1994.

PART I
GENERAL

1 Citation, commencement and applicable law

(1) These Regulations may be cited as the Commercial Agents (Council Directive) Regulations 1993 and shall come into force on 1st January 1994.

(2) These Regulations govern the relations between commercial agents and their principals and, subject to paragraph (3), apply in relation to the activities of commercial agents in Great Britain.

[(3) A court or tribunal shall—

(a) apply the law of the other member State concerned in place of regulations 3 to 22 where the parties have agreed that the agency contract is to be governed by the law of that member State;

(b) (whether or not it would otherwise be required to do so) apply these regulations where the law of another member State corresponding to these regulations enables the parties to agree that the agency contract is to be governed by the law of a different member State and the parties have agreed that it is to be governed by the law of England and Wales or Scotland.]

NOTES

Commencement: 16 December 1998 (para (3)); 1 January 1994 (remainder).

Para (3): substituted by the Commercial Agents (Council Directive) (Amendment) Regulations 1998, SI 1998/2868, reg 2(a).

2 Interpretation, application and extent

(1)   In these Regulations—

"commercial agent" means a self-employed intermediary who has continuing authority to negotiate the sale or purchase of goods on behalf of another person (the "principal"), or to negotiate and conclude the sale or purchase of goods on behalf of and in the name of that principal; but shall be understood as not including in particular—

    (i)   a person who, in his capacity as an officer of a company or association, is empowered to enter into commitments binding on that company or association;

    (ii)   a partner who is lawfully authorised to enter into commitments binding on his partners;

    (iii)   a person who acts as an insolvency practitioner (as that expression is defined in section 388 of the Insolvency Act 1986) or the equivalent in any other jurisdiction;

"commission" means any part of the remuneration of a commercial agent which varies with the number or value of business transactions;

["EEA Agreement" means the Agreement on the European Economic Area signed at Oporto on 2nd May 1992 as adjusted by the Protocol signed at Brussels on 17th March 1993;

"member State" includes a State which is a contracting party to the EEA Agreement;]

"restraint of trade clause" means an agreement restricting the business activities of a commercial agent following termination of the agency contract.

(2)   These Regulations do not apply to—

(a) commercial agents whose activities are unpaid;

    (b) commercial agents when they operate on commodity exchanges or in the commodity market;

    (c) the Crown Agents for Overseas Governments and Administrations, as set up under the Crown Agents Act 1979, or its subsidiaries.

(3)   The provisions of the Schedule to these Regulations have effect for the purpose of determining the persons whose activities as commercial agents are to be considered secondary.

(4)   These Regulations shall not apply to the persons referred to in paragraph (3) above.

(5)   These Regulations do not extend to Northern Ireland.

NOTES

    Para (1): definitions "EEA Agreement" and "member State" inserted by the Commercial Agents (Council Directive) (Amendment) Regulations 1998, SI 1998/2868, reg 2(b).

PART II
RIGHTS AND OBLIGATIONS

3 Duties of a commercial agent to his principal

(1)   In performing his activities a commercial agent must look after the interests of his principal and act dutifully and in good faith.

(2)   In particular, a commercial agent must—

    (a) make proper efforts to negotiate and where appropriate conclude the transactions he is instructed to take care of;

    (b) communicate to his principal all the necessary information available to him;

    (c) comply with reasonable instructions given by his principal.

4 Duties of a principal to his commercial agent

(1)   In his relations with his commercial agent a principal must act dutifully and in good faith.

(2)   In particular, a principal must—

(a) provide his commercial agent with the necessary documentation relating to the goods concerned;

(b) obtain for his commercial agent the information necessary for the performance of the agency contract, and in particular notify his commercial agent within a reasonable period once he anticipates that the volume of commercial transactions will be significantly lower than that which the commercial agent could normally have expected.

(3)   A principal shall, in addition, inform his commercial agent within a reasonable period of his acceptance or refusal of, and of any non-execution by him of, a commercial transaction which the commercial agent has procured for him.

5   Prohibition on derogation from regulations 3 and 4 and consequence of breach

(1)   The parties may not derogate from regulations 3 and 4 above.

(2)   The law applicable to the contract shall govern the consequence of breach of the rights and obligations under regulations 3 and 4 above.

PART III
REMUNERATION

6   Form and amount of remuneration in absence of agreement

(1)   In the absence of any agreement as to remuneration between the parties, a commercial agent shall be entitled to the remuneration that commercial agents appointed for the goods forming the subject of his agency contract are customarily allowed in the place where he carries on his activities and, if there is no such customary practice, a commercial agent shall be entitled to reasonable remuneration taking into account all the aspects of the transaction.

(2)   This regulation is without prejudice to the application of any enactment or rule of law concerning the level of remuneration.

(3)   Where a commercial agent is not remunerated (wholly or in part) by commission, regulations 7 to 12 below shall not apply.

7   Entitlement to commission on transactions concluded during agency contract

(1)   A commercial agent shall be entitled to commission on commercial transactions concluded during the period covered by the agency contract—
- (a) where the transaction has been concluded as a result of his action; or
- (b) where the transaction is concluded with a third party whom he has previously acquired as a customer for transactions of the same kind.

(2)   A commercial agent shall also be entitled to commission on transactions concluded during the period covered by the agency contract where he has an exclusive right to a specific geographical area or to a specific group of customers and where the transaction has been entered into with a customer belonging to that area or group.

8   Entitlement to commission on transactions concluded after agency contract has terminated

Subject to regulation 9 below, a commercial agent shall be entitled to commission on commercial transactions concluded after the agency contract has terminated if—
- (a) the transaction is mainly attributable to his efforts during the period covered by the agency contract and if the transaction was entered into within a reasonable period after that contract terminated; or
- (b) in accordance with the conditions mentioned in regulation 7 above, the order of the third party reached the principal or the commercial agent before the agency contract terminated.

9   Apportionment of commission between new and previous commercial agents

(1)   A commercial agent shall not be entitled to the commission referred to in regulation 7 above if that commission is payable, by virtue of regulation 8 above, to the previous commercial agent, unless it is equitable because of the circumstances for the commission to be shared between the commercial agents.

(2)   The principal shall be liable for any sum due under paragraph (1) above to the person entitled to it in accordance with that paragraph, and any sum which the other commercial agent receives to which he is not entitled shall be refunded to the principal.

10   When commission due and date for payment

(1)   Commission shall become due as soon as, and to the extent that, one of the following circumstances occurs—
   (a) the principal has executed the transaction; or
   (b) the principal should, according to his agreement with the third party, have executed the transaction; or
   (c) the third party has executed the transaction.

(2)   Commission shall become due at the latest when the third party has executed his part of the transaction or should have done so if the principal had executed his part of the transaction, as he should have.

(3)   The commission shall be paid not later than on the last day of the month following the quarter in which it became due, and, for the purposes of these Regulations, unless otherwise agreed between the parties, the first quarter period shall run from the date the agency contract takes effect, and subsequent periods shall run from that date in the third month thereafter or the beginning of the fourth month, whichever is the sooner.

(4)   Any agreement to derogate from paragraphs (2) and (3) above to the detriment of the commercial agent shall be void.

11   Extinction of right to commission

(1)   The right to commission can be extinguished only if and to the extent that—
   (a) it is established that the contract between the third party and the principal will not be executed; and
   (b) that fact is due to a reason for which the principal is not to blame.

(2)   Any commission which the commercial agent has already received shall be refunded if the right to it is extinguished.

(3)   Any agreement to derogate from paragraph (1) above to the detriment of the commercial agent shall be void.

12   Periodic supply of information as to commission due and right of inspection of principal's books

(1)   The principal shall supply his commercial agent with a statement of the commission due, not later than the last day of the month following the

quarter in which the commission has become due, and such statement shall set out the main components used in calculating the amount of the commission.

(2)   A commercial agent shall be entitled to demand that he be provided with all the information (and in particular an extract from the books) which is available to his principal and which he needs in order to check the amount of the commission due to him.

(3)   Any agreement to derogate from paragraphs (1) and (2) above shall be void.

(4)   Nothing in this regulation shall remove or restrict the effect of, or prevent reliance upon, any enactment or rule of law which recognises the right of an agent to inspect the books of a principal.

PART IV
CONCLUSION AND TERMINATION OF THE AGENCY CONTRACT

13   Right to signed written statement of terms of agency contract

(1)   The commercial agent and principal shall each be entitled to receive from the other on request, a signed written document setting out the terms of the agency contract including any terms subsequently agreed.

(2)   Any purported waiver of the right referred to in paragraph (1) above shall be void.

14   Conversion of agency contract after expiry of fixed period

An agency contract for a fixed period which continues to be performed by both parties after that period has expired shall be deemed to be converted into an agency contract for an indefinite period.

15   Minimum periods of notice for termination of agency contract

(1)   Where an agency contract is concluded for an indefinite period either party may terminate it by notice.

(2)   The period of notice shall be—
  (a) 1 month for the first year of the contract;

    (b) 2 months for the second year commenced;

    (c) 3 months for the third year commenced and for the subsequent years;

and the parties may not agree on any shorter periods of notice.

(3)   If the parties agree on longer periods than those laid down in paragraph (2) above, the period of notice to be observed by the principal must not be shorter than that to be observed by the commercial agent.

(4)   Unless otherwise agreed by the parties, the end of the period of notice must coincide with the end of a calendar month.

(5)   The provisions of this regulation shall also apply to an agency contract for a fixed period where it is converted under regulation 14 above into an agency contract for an indefinite period subject to the proviso that the earlier fixed period must be taken into account in the calculation of the period of notice.

16 Savings with regard to immediate termination

These Regulations shall not affect the application of any enactment or rule of law which provides for the immediate termination of the agency contract—

    (a) because of the failure of one party to carry out all or part of his obligations under that contract; or

    (b) where exceptional circumstances arise.

17 Entitlement of commercial agent to indemnity or compensation on termination of agency contract

(1)   This regulation has effect for the purpose of ensuring that the commercial agent is, after termination of the agency contract, indemnified in accordance with paragraphs (3) to (5) below or compensated for damage in accordance with paragraphs (6) and (7) below.

(2)   Except where the agency [contract] otherwise provides, the commercial agent shall be entitled to be compensated rather than indemnified.

(3)   Subject to paragraph (9) and to regulation 18 below, the commercial agent shall be entitled to an indemnity if and to the extent that—

    (a) he has brought the principal new customers or has significantly increased the volume of business with existing customers and the principal continues to derive substantial benefits from the business with such customers; and

  (b) the payment of this indemnity is equitable having regard to all the circumstances and, in particular, the commission lost by the commercial agent on the business transacted with such customers.

(4)   The amount of the indemnity shall not exceed a figure equivalent to an indemnity for one year calculated from the commercial agent's average annual remuneration over the preceding five years and if the contract goes back less than five years the indemnity shall be calculated on the average for the period in question.

(5)   The grant of an indemnity as mentioned above shall not prevent the commercial agent from seeking damages.

(6)   Subject to paragraph (9) and to regulation 18 below, the commercial agent shall be entitled to compensation for the damage he suffers as a result of the termination of his relations with his principal.

(7)   For the purpose of these Regulations such damage shall be deemed to occur particularly when the termination takes place in either or both of the following circumstances, namely circumstances which—

  (a) deprive the commercial agent of the commission which proper performance of the agency contract would have procured for him whilst providing his principal with substantial benefits linked to the activities of the commercial agent; or

  (b) have not enabled the commercial agent to amortize the costs and expenses that he had incurred in the performance of the agency contract on the advice of his principal.

(8)   Entitlement to the indemnity or compensation for damage as provided for under paragraphs (2) to (7) above shall also arise where the agency contract is terminated as a result of the death of the commercial agent.

(9)   The commercial agent shall lose his entitlement to the indemnity or compensation for damage in the instances provided for in paragraphs (2) to (8) above if within one year following termination of his agency contract he has not notified his principal that he intends pursuing his entitlement.

NOTES

Para (2): word in square brackets substituted by the Commercial Agents (Council Directive) (Amendment) Regulations 1998, SI 1998/2868, reg 2(c).

18 Grounds for excluding payment of indemnity or compensation under regulation 17

The [indemnity or] compensation referred to in regulation 17 above shall not be payable to the commercial agent where—
  (a) the principal has terminated the agency contract because of default attributable to the commercial agent which would justify immediate termination of the agency contract pursuant to regulation 16 above; or
  (b) the commercial agent has himself terminated the agency contract, unless such termination is justified—
    (i) by circumstances attributable to the principal, or
    (ii) on grounds of the age, infirmity or illness of the commercial agent in consequence of which he cannot reasonably be required to continue his activities; or
  (c) the commercial agent, with the agreement of his principal, assigns his rights and duties under the agency contract to another person.

NOTES

Words in square brackets inserted by the Commercial Agents (Council Directive) (Amendment) Regulations 1993, SI 1993/3173, reg 2.

19 Prohibition on derogation from regulations 17 and 18

The parties may not derogate from regulations 17 and 18 to the detriment of the commercial agent before the agency contract expires.

20 Restraint of trade clauses

(1)   A restraint of trade clause shall be valid only if and to the extent that—
  (a) it is concluded in writing; and
  (b) it relates to the geographical area or the group of customers and the geographical area entrusted to the commercial agent and to the kind of goods covered by his agency under the contract.

(2)   A restraint of trade clause shall be valid for not more than two years after termination of the agency contract.

(3)   Nothing in this regulation shall affect any enactment or rule of law which imposes other restrictions on the validity or enforceability of restraint

of trade clauses or which enables a court to reduce the obligations on the parties resulting from such clauses.

PART V
MISCELLANEOUS AND SUPPLEMENTAL

21 Disclosure of information

Nothing in these Regulations shall require information to be given where such disclosure would be contrary to public policy.

22 Service of notice etc

(1)   Any notice, statement or other document to be given or supplied to a commercial agent or to be given or supplied to the principal under these Regulations may be so given or supplied—
  (a) by delivering it to him;
  (b) by leaving it at his proper address addressed to him by name;
  (c) by sending it by post to him addressed either to his registered address or to the address of his registered or principal office;

or by any other means provided for in the agency contract.

(2)   Any such notice, statement or document may—
  (a) in the case of a body corporate, be given or served on the secretary or clerk of that body;
  (b) in the case of a partnership, be given to or served on any partner or on any person having the control or management of the partnership business.

23 Transitional provisions

(1)   Notwithstanding any provision in an agency contract made before 1st January 1994, these Regulations shall apply to that contract after that date and, accordingly any provision which is inconsistent with these Regulations shall have effect subject to them.

(2)   Nothing in these Regulations shall affect the rights and liabilities of a commercial agent or a principal which have accrued before 1st January 1994.

THE SCHEDULE

Regulation 2(3)

1. The activities of a person as a commercial agent are to be considered secondary where it may reasonably be taken that the primary purpose of the arrangement with his principal is other than as set out in paragraph 2 below.

2. An arrangement falls within this paragraph if—
  (a) the business of the principal is the sale, or as the case may be purchase, of goods of a particular kind; and
  (b) the goods concerned are such that—
      (i) transactions are normally individually negotiated and concluded on a commercial basis, and
      (ii) procuring a transaction on one occasion is likely to lead to further transactions in those goods with that customer on future occasions, or to transactions in those goods with other customers in the same geographical area or among the same group of customers, and

    that accordingly it is in the commercial interests of the principal in developing the market in those goods to appoint a representative to such customers with a view to the representative devoting effort, skill and expenditure from his own resources to that end.

3. The following are indications that an arrangement falls within paragraph 2 above, and the absence of any of them is an indication to the contrary—
  (a) the principal is the manufacturer, importer or distributor of the goods;
  (b) the goods are specifically identified with the principal in the market in question rather than, or to a greater extent than, with any other person;
  (c) the agent devotes substantially the whole of his time to representative activities (whether for one principal or for a number of principals whose interests are not conflicting);
  (d) the goods are not normally available in the market in question other than by means of the agent;
  (e) the arrangement is described as one of commercial agency.

4. The following are indications that an arrangement does not fall within paragraph 2 above—
  (a) promotional material is supplied direct to potential customers;
  (b) persons are granted agencies without reference to existing agents in a particular area or in relation to a particular group;
  (c) customers normally select the goods for themselves and merely place their orders through the agent.

5.   The activities of the following categories of persons are presumed, unless the contrary is established, not to fall within paragraph 2 above—
Mail order catalogue agents for consumer goods.
Consumer credit agents.

General Product Safety Regulations 1994

(SI 1994/2328)

NOTES
Made: 5 September 1994.
Authority: European Communities Act 1972, s 2(2).

1   Citation and commencement

[(1)]   These Regulations may be cited as the General Product Safety Regulations 1994 and shall come into force on 3rd October 1994.

[(2)   Nothing in these Regulations applies to a medicinal product for human use to which the Medicines for Human Use (Marketing Authorizations Etc) Regulations 1994 apply.]

NOTES
Commencement: 1 January 1995 (para (2)); 3 October 1994 (remainder).
Para (1): numbered as such by the Medicines for Human Use (Marketing Authorisations Etc) Regulations 1994, SI 1994/3144, reg 11, Sch 7, para 21(b).
Para (2): inserted by SI 1994/3144, reg 11, Sch 7, para 21(c).

2   Interpretation

(1)   In these Regulations—
"the 1968 Act" means the Medicines Act 1968;
"the 1987 Act" means the Consumer Protection Act 1987;
"the 1990 Act" means the Food Safety Act 1990;
"commercial activity" includes a business and a trade;
"consumer" means a consumer acting otherwise than in the course of a commercial activity;

"dangerous product" means any product other than a safe product;

"distributor" means any professional in the supply chain whose activity does not affect the safety properties of a product;

"enforcement authority" means the Secretary of State, any other Minister of the Crown in charge of a Government Department, any such department and any authority, council and other person on whom functions under these Regulations are imposed by or under regulation 11;

"general safety requirement" means the requirement in regulation 7;

"the GPS Directive" means Council Directive 92/59/EEC on general product safety;

"the 1991 Order" means the Food Safety (Northern Ireland) Order 1991;

"producer" means

(a) the manufacturer of the product, when he is established in the Community, and includes any person presenting himself as the manufacturer by affixing to the product his name, trade mark or other distinctive mark, or the person who reconditions the product;

(b) when the manufacturer is not established in the Community—

(i) if the manufacturer does not have a representative established in the Community, the importer of the product;

(ii) in all other cases, the manufacturer's representative; and

(c) other professionals in the supply chain, insofar as their activities may affect the safety properties of a product placed on the market;

"product" means any product intended for consumers or likely to be used by consumers, supplied whether for consideration or not in the course of a commercial activity and whether new, used or reconditioned; provided, however, a product which is used exclusively in the context of a commercial activity even if it is used for or by a consumer shall not be regarded as a product for the purposes of these Regulations provided always and for the avoidance of doubt this exception shall not extend to the supply of such a product to a consumer;

"safe product" means any product which, under normal or reasonably foreseeable conditions of use, including duration, does not present any risk or only the minimum risks compatible with the product's use, considered as acceptable and consistent with a high level of protection for the safety and health of persons, taking into account in particular—

(a) the characteristics of the product, including its composition, packaging, instructions for assembly and maintenance;

(b) the effect on other products, where it is reasonably foreseeable that it will be used with other products;

(c) the presentation of the product, the labelling, any instructions for its use and disposal and any other indication or information provided by the producer; and

(d) the categories of consumers at serious risk when using the product, in particular children,

and the fact that higher levels of safety may be obtained or other products presenting a lesser degree of risk may be available shall not of itself cause the product to be considered other than a safe product.

(2) References in these Regulations to the "Community" are references to the European Economic Area established under the Agreement signed at Oporto on 2nd May 1992 as adjusted by the Protocol signed at Brussels on 17th March 1993.

NOTES

Commencement: 3 October 1994.

Application and revocation

3 These Regulations do not apply to—
(a) second-hand products which are antiques;
(b) products supplied for repair or reconditioning before use, provided the supplier clearly informs the person to whom he supplies the product to that effect; or
(c) any product where there are specific provisions in rules of Community law governing all aspects of the safety of the product.

NOTES

Commencement: 3 October 1994.

4 The requirements of these Regulations apply to a product where the product is the subject of provisions of Community law other than the GPS Directive insofar as those provisions do not make specific provision governing an aspect of the safety of the product.

NOTES

Commencement: 3 October 1994.

5 For the purposes of these Regulations the provisions of section 10 of the 1987 Act to the extent that they impose general safety requirements which must be complied with if products are to be—

(i)   placed on the market, offered or agreed to be placed on the market or exposed or possessed to be placed on the market by producers; or

(ii) supplied, offered or agreed to be supplied or exposed or possessed to be supplied by distributors,

are hereby disapplied.

NOTES

Commencement: 3 October 1994.

7   General safety requirement

No producer shall place a product on the market unless the product is a safe product.

NOTES

Commencement: 3 October 1994.

8   Requirement as to information

(1)   Within the limits of his activity, a producer shall—

  (a) provide consumers with the relevant information to enable them to assess the risks inherent in a product throughout the normal or reasonably foreseeable period of its use, where such risks are not immediately obvious without adequate warnings, and to take precautions against those risks; and

  (b) adopt measures commensurate with the characteristics of the products which he supplies, to enable him to be informed of the risks which these products might present and to take appropriate action, including, if necessary, withdrawing the product in question from the market to avoid those risks.

(2)   The measures referred to in sub-paragraph (b) of paragraph (1) above may include, whenever appropriate—

  (i)   marking of the products or product batches in such a way that they can be identified;

  (ii) sample testing of marketed products;

  (iii) investigating complaints; and

  (iv) keeping distributors informed of such monitoring.

Commencement: 3 October 1994.

9 Requirements of distributors

A distributor shall act with due care in order to help ensure compliance
with the requirements of regulation 7 above and, in particular, without
limiting the generality of the foregoing—
- (a) a distributor shall not supply products to any person which he knows,
or should have presumed, on the basis of the information in his
possession and as a professional, are dangerous products; and
- (b) within the limits of his activities, a distributor shall participate in
monitoring the safety of products placed on the market, in particular
by passing on information on the product risks and cooperating in
the action taken to avoid those risks.

Commencement: 3 October 1994.

10 Presumption of conformity and product assessment

(1)   Where in relation to any product such product conforms to the specific
rules of the law of the United Kingdom laying down the health and safety
requirements which the product must satisfy in order to be marketed there
shall be a presumption that, until the contrary is proved, the product is a
safe product.

(2)   Where no specific rules as are mentioned or referred to in
paragraph (1) exist, the conformity of a product to the general safety
requirement shall be assessed taking into account—
- (i) voluntary national standards of the United Kingdom giving effect to
a European standard; or
- (ii) Community technical specifications; or
- (iii) if there are no such voluntary national standards of the United
Kingdom or Community technical specifications—
  - (aa) standards drawn up in the United Kingdom; or
  - (bb) the codes of good practice in respect of health and safety in the
product sector concerned; or
  - (cc) the state of the art and technology

and the safety which consumers may reasonably expect.

NOTES
Commencement: 3 October 1994.

11   Enforcement

For the purposes of providing for the enforcement of these Regulations—
- (a) section 13 of the 1987 Act (prohibition notices and notices to warn) shall (to the extent that it does not already do so) apply to products as it applies to relevant goods under that section;
- (b) the requirements of these Regulations shall constitute safety provisions for the purposes of sections 14 (suspension notices), 15 (appeals against suspension notices), 16 (forfeiture: England, Wales and Northern Ireland), 17 (forfeiture: Scotland) and 18 (power to obtain information) of the 1987 Act;
- (c) (i)   subject to paragraph (ii) below a weights and measures authority in Great Britain and a district council in Northern Ireland shall have the same duty to enforce these Regulations as they have in relation to Part II of the 1987 Act, and Part IV, sections 37 and 38 and subsections (3) and (4) of section 42 of that Act shall apply accordingly;
-      (ii)   without prejudice to the provisions of paragraphs (a) and (b) above and sub-paragraph (i) above, insofar as these Regulations apply:—
-          (aa)   to products licensed in accordance with the provisions of the 1968 Act [or authorised in accordance with the provisions of the Marketing Authorisations for Veterinary Medicinal Products Regulations 1994] [or which are the subject of a marketing authorization within the meaning of the Medicines for Human Use (Marketing Authorizations Etc) Regulations 1994], it shall be the duty of the enforcement authority as defined in section 132(1) of the 1968 Act to enforce or to secure the enforcement of these Regulations and sections 108 to 115 and section 119 of and Schedule 3 to that Act shall apply accordingly as if these Regulations were regulations made under the said Act;
-          (bb)   in relation to food within the meaning of section 1 of the 1990 Act, it shall be the duty of each food authority as defined in section 5 of the 1990 Act to enforce or to secure the enforcement of these Regulations, within its area, in Great Britain and sections 9, 29, 30 and 32 of that Act shall apply accordingly as if these Regulations were food safety requirements made under the said Act and section 10 of that Act shall apply as if these Regulations were regulations made under Part II of that Act; and

(cc) in relation to food within the meaning of article 2 of the 1991 Order, it shall be the duty of the relevant enforcement authority as provided for in article 26 of that Order to enforce or to secure enforcement of these Regulations in Northern Ireland and articles 8, 29, 30, 31 and 33 of that Order shall apply accordingly as if these Regulations were food safety requirements made under that Order and article 9 of that Order shall apply as if these Regulations were regulations made under Part II of that Order;

(d) in sections 13(4) and 14(6) of the 1987 Act for the words "six months" there shall be substituted "three months"; and

(e) nothing in this regulation shall authorise any enforcement authority to bring proceedings in Scotland for an offence.

NOTES

Commencement: 3 October 1994.

Words in first pair of square brackets inserted by the Marketing Authorisations for Veterinary Medicinal Products Regulations 1994, SI 1994/3142, reg 21, Sch 5, para 29; words in second pair of square brackets inserted by the Medicines for Human Use (Marketing Authorisations Etc) Regulations 1994, SI 1994/3144, reg 11, Sch 7, para 21.

Offences and preparatory acts

12 Any person who contravenes regulation 7 or 9(a) shall be guilty of an offence.

NOTES

Commencement: 3 October 1994.

13 No producer or distributor shall—

(a) offer or agree to place on the market any dangerous product or expose or possess any such product for placing on the market; or

(b) offer or agree to supply any dangerous product or expose or possess any such product for supply,

and any person who contravenes the requirements of this regulation shall be guilty of an offence.

NOTES

Commencement: 3 October 1994.

14 Defence of due diligence

(1)   Subject to the following paragraphs of this regulation, in proceedings against any person for an offence under these Regulations it shall be a defence for that person to show that he took all reasonable steps and exercised all due diligence to avoid committing the offence.

(2)   Where in any proceedings against any person for such an offence the defence provided by paragraph (1) above involves an allegation that the commission of the offence was due—
  (a) to the act or default of another, or
  (b) to reliance on information given by another,

that person shall not, without leave of the court, be entitled to rely on the defence unless, not less than seven days before, in England, Wales and Northern Ireland, the hearing of the proceedings or, in Scotland, the trial diet, he has served a notice under paragraph (3) below on the person bringing the proceedings.

(3)   A notice under this paragraph shall give such information identifying or assisting in the identification of the person who committed the act or default or gave the information as is in the possession of the person serving the notice at the time he serves it.

(4)   It is hereby declared that a person shall not be entitled to rely on the defence provided in paragraph (1) above by reason of his reliance on information supplied by another, unless he shows that it was reasonable in all the circumstances for him to have relied on the information, having regard in particular—
  (a) to the steps which he took, and those which might reasonably have been taken, for the purpose of verifying the information; and
  (b) to whether he had any reason to disbelieve the information.

(5)   It is hereby declared that a person shall not be entitled to rely on the defence provided by paragraph (1) above or by section 39(1) of the 1987 Act (defence of due diligence) if he has contravened regulation 9(b).

NOTES
Commencement: 3 October 1994.

15 Liability of persons other than principal offender

(1)   Where the commission by any person of an offence to which regulation 14 above applies is due to the act or default committed by some

other person in the course of a commercial activity of his, the other person shall be guilty of an offence and may be proceeded against and punished by virtue of this paragraph whether or not proceedings are taken against the first-mentioned person.

(2)   Where a body corporate is guilty of an offence under these Regulations (including where it is so guilty by virtue of paragraph (1) above) in respect of any act or default which is shown to have been committed with the consent or connivance of, or to be attributable to any neglect on the part of any director, manager, secretary or other similar officer of the body corporate or any person who was purporting to act in any such capacity he, as well as the body corporate, shall be guilty of that offence and shall be liable to be proceeded against and punished accordingly.

(3)   Where the affairs of a body corporate are managed by its members, paragraph (2) above shall apply in relation to the acts and defaults of a member in connection with his functions of management as if he were a director of the body corporate.

(4)   Where a Scottish partnership is guilty of an offence under regulation 14 above (including where it is so guilty by virtue of paragraph (1) above) in respect of any act or default which is shown to have been committed with the consent or connivance or, or be attributable to any neglect on the part of, a partner in the partnership, he, as well as the partnership, shall be guilty of that offence and shall be liable to be proceeded against and punished accordingly.

NOTES

Commencement: 3 October 1994.

16   Extension of the time for bringing summary proceedings

(1)   Notwithstanding section 127 of the Magistrates' Courts Act 1980 and article 19 of the Magistrates' Courts (Northern Ireland) Order 1981, in England, Wales and Northern Ireland a magistrates' court may try an information (in the case of England and Wales) or a complaint (in the case of Northern Ireland) in respect of proceedings for an offence under regulation 12 or 13 above if (in the case of England and Wales) the information is laid or (in the case of Northern Ireland) the complaint is made within twelve months from the date of the offence.

(2)   Notwithstanding section 331 of the Criminal Procedure (Scotland) Act 1975, in Scotland summary proceedings for an offence under regulation 12 or 13 above may be commenced at any time within twelve months from the date of the offence.

(3)   For the purposes of paragraph (2) above, section 331(3) of the Criminal Procedure (Scotland) Act 1975 shall apply as it applies for the purposes of that section.

NOTES

Commencement: 3 October 1994.

17   Penalties

A person guilty of an offence under regulation 12 or 13 above shall be liable on summary conviction to—
    (a) imprisonment for a term not exceeding three months; or
    (b) a fine not exceeding level 5 on the standard scale;

or to both.

NOTES

Commencement: 3 October 1994.

18   Duties of enforcement authorities

(1)   Every enforcement authority shall give immediate notice to the Secretary of State of any action taken by it to prohibit or restrict the supply of any product or forfeit or do any other thing in respect of any product for the purposes of these Regulations.

(2)   The requirements of paragraph (1) above shall not apply in the case of any action taken in respect of any second-hand product.

NOTES

Commencement: 3 October 1994.

Unfair Terms in Consumer Contracts Regulations 1994

(SI 1994/3159)

NOTES

Made: 8 December 1994.
Authority: European Communities Act 1972, s 2(2).

1   Citation and commencement

These Regulations may be cited as the Unfair Terms in Consumer Contracts Regulations 1994 and shall come into force on 1st July 1995.

NOTES

Commencement: 1 July 1995.

2   Interpretation

(1)   In these Regulations—
  "business" includes a trade or profession and the activities of any government department or local or public authority;
  "the Community" means the European Economic Community and the other States in the European Economic Area;
  "consumer" means a natural person who, in making a contract to which these Regulations apply, is acting for purposes which are outside his business;
  "court" in relation to England and Wales and Northern Ireland means the High Court, and in relation to Scotland, the Court of Session;
  "Director" means the Director General of Fair Trading;
  "EEA Agreement" means the Agreement on the European Economic Area signed at Oporto on 2 May 1992 as adjusted by the protocol signed at Brussels on 17 March 1993;
  "member State" shall mean a State which is a contracting party to the EEA Agreement but until the EEA Agreement comes into force in relation to Liechtenstein does not include the State of Liechtenstein;

"seller" means a person who sells goods and who, in making a contract to which these Regulations apply, is acting for purposes relating to his business; and

"supplier" means a person who supplies goods or services and who, in making a contract to which these Regulations apply, is acting for purposes relating to his business.

(2)   . . .

NOTES

Commencement: 1 July 1995.
Para (2): applies to Scotland only.

3   Terms to which these Regulations apply

(1)   Subject to the provisions of Schedule 1, these Regulations apply to any term in a contract concluded between a seller or supplier and a consumer where the said term has not been individually negotiated.

(2)   In so far as it is in plain, intelligible language, no assessment shall be made of the fairness of any term which—
  (a) defines the main subject matter of the contract, or
  (b) concerns the adequacy of the price or remuneration, as against the goods or services sold or supplied.

(3)   For the purposes of these Regulations, a term shall always be regarded as not having been individually negotiated where it has been drafted in advance and the consumer has not been able to influence the substance of the term.

(4)   Notwithstanding that a specific term or certain aspects of it in a contract has been individually negotiated, these Regulations shall apply to the rest of a contract if an overall assessment of the contract indicates that it is a pre-formulated standard contract.

(5)   It shall be for any seller or supplier who claims that a term was individually negotiated to show that it was.

NOTES

Commencement: 1 July 1995.

4 Unfair terms

(1)   In these Regulations, subject to paragraphs (2) and (3) below, "unfair term" means any term which contrary to the requirement of good faith causes a significant imbalance in the parties' rights and obligations under the contract to the detriment of the consumer.

(2)   An assessment of the unfair nature of a term shall be made taking into account the nature of the goods or services for which the contract was concluded and referring, as at the time of the conclusion of the contract, to all circumstances attending the conclusion of the contract and to all the other terms of the contract or of another contract on which it is dependent.

(3)   In determining whether a term satisfies the requirement of good faith, regard shall be had in particular to the matters specified in Schedule 2 to these Regulations.

(4)   Schedule 3 to these Regulations contains an indicative and non-exhaustive list of the terms which may be regarded as unfair.

NOTES

Commencement: 1 July 1995.

5 Consequence of inclusion of unfair terms in contracts

(1)   An unfair term in a contract concluded with a consumer by a seller of supplier shall not be binding on the consumer.

(2)   The contract shall continue to bind the parties if it is capable of continuing in existence without the unfair term.

NOTES

Commencement: 1 July 1995.

6 Construction of written contracts

A seller or supplier shall ensure that any written term of a contract is expressed in plain, intelligible language, and if there is doubt about the meaning of a written term, the interpretation most favourable to the consumer shall prevail.

NOTES

Commencement: 1 July 1995.

7 Choice of law clauses

These Regulations shall apply notwithstanding any contract term which applies or purports to apply the law of a non member State, if the contract has a close connection with the territory of the member States.

NOTES

Commencement: 1 July 1995.

8 Prevention of continued use of unfair terms

(1)   shall be the duty of the Director to consider any complaint made to him that any contract term drawn up for general use is unfair, unless the complaint appears to the Director to be frivolous or vexatious.

(2)   If having considered a complaint about any contract term pursuant to paragraph (1) above the Director considers that the contract term is unfair he may, if he considers it appropriate to do so, bring proceedings for an injunction (in which proceedings he may also apply for an interlocutory injunction) against any person appearing to him to be using or recommending use of such a term in contracts concluded with consumers.

(3)   The Director may, if he considers it appropriate to do so, have regard to any undertakings given to him by or on behalf of any person as to the continued use of such a term in contracts concluded with consumers.

(4)   The Director shall give reasons for his decision to apply or not to apply, as the case may be, for an injunction in relation to any complaint which these Regulations require him to consider.

(5)   The court on an application by the Director may grant an injunction on such terms as it thinks fit.

(6)   An injunction may relate not only to use of a particular contract term drawn up for general use but to any similar term, or a term having like effect, used or recommended for use by any party to the proceedings.

(7)   The Director may arrange for the dissemination in such form and manner as he considers appropriate of such information and advice concerning the operation of these Regulations as may appear to him to be expedient to give to the public and to all persons likely to be affected by these Regulations.

NOTES

Commencement: 1 July 1995.

SCHEDULES

SCHEDULE I

Regulation 3(1)

CONTRACTS AND PARTICULAR TERMS EXCLUDED FROM THE SCOPE OF THESE REGULATIONS

These Regulations do not apply to—
- (a) any contract relating to employment;
- (b) any contract relating to succession rights;
- (c) any contract relating to rights under family law;
- (d) any contract relating to the incorporation and organisation of companies or partnerships; and
- (e) any term incorporated in order to comply with or which reflects—
  - (i)   statutory or regulatory provisions of the United Kingdom; or
  - (ii) the provisions or principles of international conventions to which the member States or the Community are party.

NOTES

Commencement: 1 July 1995.

SCHEDULE 2

Regulation 4(3)

ASSESSMENT OF GOOD FAITH

In making an assessment of good faith, regard shall be had in particular to—

(a) the strength of the bargaining positions of the parties;
(b) whether the consumer had an inducement to agree to the term;
(c) whether the goods or services were sold or supplied to the special order of the consumer, and
(d) the extent to which the seller or supplier has dealt fairly and equitably with the consumer.

NOTES
Commencement: 1 July 1995.

SCHEDULE 3

Regulation 4(4)

INDICATIVE AND ILLUSTRATIVE LIST OF TERMS WHICH MAY BE REGARDED AS UNFAIR

1. Terms which have the object or effect of—
   (a) excluding or limiting the legal liability of a seller or supplier in the event of the death of a consumer or personal injury to the latter resulting from an act or omission of that seller or supplier;
   (b) inappropriately excluding or limiting the legal rights of the consumer vis-à-vis the seller or supplier or another party in the event of total or partial non-performance or inadequate performance by the seller or supplier of any of the contractual obligations, including the option of offsetting a debt owed to the seller or supplier against any claim which the consumer may have against him;
   (c) making an agreement binding on the consumer whereas provision of services by the seller or supplier is subject to a condition whose realisation depends on his own will alone;
   (d) permitting the seller or supplier to retain sums paid by the consumer where the latter decides not to conclude or perform the contract, without providing for the consumer to receive compensation of an equivalent amount from the seller or supplier where the latter is the party cancelling the contract;
   (e) requiring any consumer who fails to fulfil his obligation to pay a disproportionately high sum in compensation;
   (f) authorising the seller or supplier to dissolve the contract on a discretionary basis where the same facility is not granted to the consumer, or permitting the seller or supplier to retain the sums paid for services not yet supplied by him where it is the seller or supplier himself who dissolves the contract;

(g) enabling the seller or supplier to terminate a contract of indeterminate duration without reasonable notice except where there are serious grounds for doing so;

(h) automatically extending a contract of fixed duration where the consumer does not indicate otherwise, when the deadline fixed for the consumer to express this desire not to extend the contract is unreasonably early;

(i) irrevocably binding the consumer to terms with which he had no real opportunity of becoming acquainted before the conclusion of the contract;

(j) enabling the seller or supplier to alter the terms of the contract unilaterally without a valid reason which is specified in the contract;

(k) enabling the seller or supplier to alter unilaterally without a valid reason any characteristics of the product or service to be provided;

(l) providing for the price of goods to be determined at the time of delivery or allowing a seller of goods or supplier of services to increase their price without in both cases giving the consumer the corresponding right to cancel the contract if the final price is too high in relation to the price agreed when the contract was concluded;

(m) giving the seller or supplier the right to determine whether the goods or services supplied are in conformity with the contract, or giving him the exclusive right to interpret any term of the contract;

(n) limiting the seller's or supplier's obligation to respect commitments undertaken by his agents or making his commitments subject to compliance with a particular formality;

(o) obliging the consumer to fulfil all his obligations where the seller or supplier does not perform his;

(p) giving the seller or supplier the possibility of transferring his rights and obligations under the contract, where this may serve to reduce the guarantees for the consumer, without the latter's agreement;

(q) excluding or hindering the consumer's right to take legal action or exercise any other legal remedy, particularly by requiring the consumer to take disputes exclusively to arbitration not covered by legal provisions, unduly restricting the evidence available to him or imposing on him a burden of proof which, according to the applicable law, should lie with another party to the contract.

2. Scope of subparagraphs 1(g), (j) and (l)—

(a) Subparagraph 1(g) is without hindrance to terms by which a supplier of financial services reserves the right to terminate unilaterally a contract of indeterminate duration without notice where there is a valid reason, provided that the supplier is required to inform the other contracting party or parties thereof immediately.

(b) Subparagraph 1(j) is without hindrance to terms under which a supplier of financial services reserves the right to alter the rate of interest payable by the consumer or due to the latter, or the amount of other charges for financial services without notice where there is a valid reason, provided that the supplier is required to inform the other contracting party or parties thereof at the earliest opportunity and that the latter are free to dissolve the contract immediately.

Subparagraph 1(j) is also without hindrance to terms under which a seller or supplier reserves the right to alter unilaterally the conditions of a contract of indeterminate duration, provided that he is required to inform the consumer with reasonable notice and that the consumer is free to dissolve the contract.

(c) Subparagraphs 1(g), (j) and (1)—
  — transactions in transferable securities, financial instruments and other products or services where the price is linked to fluctuations in a stock exchange quotation or index or a financial market rate that the seller or supplier does not control;
  — contracts for the purchase or sale of foreign currency, traveller's cheques or international money orders denominated in foreign currency;
(d) Subparagraph 1(1) is without hindrance to price indexation clauses, where lawful, provided that the method by which prices vary is explicitly described.

NOTES

Commencement: 1 July 1995.

Price Indications (Resale of Tickets) Regulations 1994

(SI 1994/3248)

NOTES

Made: 14 December 1994.
Authority: Consumer Protection Act 1987, s 26.

1   Citation and commencement

These Regulations may be cited as the Price Indications (Resale of Tickets) Regulations 1994 and shall come into force on 20th February 1995.

NOTES
Commencement: 20 February 1995.

2 Interpretation

In these Regulations—
"ticket" means a card, badge or document giving to its holder—
(a) the right of admission to a place of entertainment; or
(b) the said right of admission to a place of entertainment and the right to use a seat or space in such a place
and the fact that those rights are subject to the condition that the holder may be refused admission to or may be removed from the place of entertainment shall not cause it to be treated as other than a ticket; and
"entertainment" includes any gathering, amusement, exhibition, performance, game, sport or trial of skill or other similar event.

NOTES
Commencement: 20 February 1995.

3 Scope of application

(1) Subject to paragraph (3) below, the provisions of these Regulations have effect when a person to whom paragraph (2) below applies gives to consumers, in the course of business, an indication of the price at which a ticket, or a ticket in combination with another element, is or will be available ("a price indication").

(2) This paragraph applies to any person save for the holder or promoter of the entertainment to which the ticket relates or a person acting on behalf of such holder or promoter who is prepared or may be prepared to supply a ticket by way of resale.

(3) These Regulations do not apply where a person gives a price indication in relation to a package to which the Package Travel, Package Holidays and Package Tours Regulations 1992 apply.

NOTES
Commencement: 20 February 1995.

4 Price indication information

Where a person gives a price indication, the following information shall be given to consumers—

(a) the price (if any) and any other detail which appears on the ticket which relates to or affects the rights conferred or to be conferred on the holder of the ticket (including the location of the seat or space) and which has been caused to be placed thereon by the holder or promoter of the entertainment to which the ticket relates; and

(b) the location of the seat or space (if any) which the holder of the ticket will have the right to use and any features of such seat or space which would adversely affect the holder's use or enjoyment of it and which are known or could reasonably be expected to be known to the person giving the price indication.

NOTES

Commencement: 20 February 1995.

5 Requirements relating to price indication information

(1) The information required to be given by regulation 4 above shall be given before the person who gives a price indication enters into any contract with a consumer under which the ticket is to be supplied.

(2) Except in cases where the contract to supply the ticket by way of resale is concluded by telephone, the information required to be given by regulation 4(a) above shall be given in writing.

(3) The requirement of paragraph (2) of this regulation shall be deemed to be satisfied if the consumer is shown the ticket in accordance with paragraph (1) above and in such a manner that the details appearing on the ticket are visible by and legible to the consumer.

NOTES

Commencement: 20 February 1995.

6 Manner of giving price indication information

The information required to be given by regulation 4 need not be given in the same manner as the price indication but—

(a) if the information is given orally, it shall be given audibly and in a manner that is comprehensible to the consumer, and

(b) if it is given in writing, it shall be given clearly, prominently and legibly

and in any case the information shall be given in such a way that it comes to the attention of the consumer before he enters into any contract under which the ticket is to be supplied to him.

NOTES

Commencement: 20 February 1995.

7 Price indication information to be accurate

Any information which is given pursuant to the requirements of regulation 4 above shall be accurate.

NOTES

Commencement: 20 February 1995.

8 Offences and defences

(1)   Any contravention of a requirement of these Regulations shall constitute a criminal offence punishable—
   (a) on conviction on indictment, by a fine; or
   (b) on summary conviction, by a fine not exceeding the statutory maximum.

(2)   In relation to an offence under this regulation—
   (a) section 24(2) of the Act (defence that indication was not contained in an advertisement) shall apply as it applies to an offence under subsection (1) or (2) of section 20 of the Act;
   (b) section 39 of the Act (defence of due diligence) shall apply as it applies to an offence mentioned in subsection (5) of that section; and
   (c) subsection (1) of section 40 of the Act (liability of persons other than principal offender) shall apply as it applies to an offence mentioned in section 39(5) of the Act and subsections (2) and (3) of the said section 40 shall apply as they apply to an offence under the Act.

(3)   In this regulation, "the Act" means the Consumer Protection Act 1987.

NOTES

Commencement: 20 February 1995.

Trading Schemes Regulations 1997

(SI 1997/30)

NOTES

Made: 13 January 1997.
Authority: Fair Trading Act 1973, s 119.

1 Citation, commencement and application

(1)   These Regulations may be cited as the Trading Schemes Regulations 1997 and shall come into force on 6th February 1997.

(2)   Subject to paragraph (3) below, these Regulations shall apply—
   (a) from the date of their coming into force to any trading scheme to which Part XI of the Fair Trading Act 1973 applies and which came into existence on or after the date of coming into force of these Regulations, and to any agreement made under such a trading scheme;
   (b) after a period of six months from the date of their coming into force to any trading scheme in existence prior to the coming into force of the Act and to which Part XI of the Fair Trading Act 1973 did not apply prior to that date.

(3)   Where an agreement is made after the date of coming into force of these Regulations but prior to the expiry of a six months period after that date under a trading scheme to which Part XI of the Fair Trading Act 1973 applied prior to the coming into force of the Act such agreement shall comply either with the 1989 Regulations or these Regulations.

(4)   Subject to paragraph (3) above the 1989 Regulations shall not apply to any trading scheme coming into operation after the date of the coming into force of these Regulations or to any agreement made after that date under any trading scheme to which Part XI of the Fair Trading Act 1973 applies.

NOTES

Commencement: 6 February 1997.

2 Interpretation

In these Regulations—

"the Act" means the Trading Schemes Act 1996;

"advertisement" means any advertisement, document, prospectus, circular or notice, whether transmitted in electronic or any other form, which promotes a trading scheme;

"the 1989 Regulations" means the Pyramid Selling Schemes Regulations 1989;

"the 1990 Regulations" means the Pyramid Selling Schemes (Amendment) Regulations 1990;

"participant" has the same meaning as in Part XI of the Fair Trading Act 1973;

"security" means a mortgage, charge, pledge, bond, debenture, indemnity, guarantee, bill, note or other right provided by the participant, or at his request (expressed or implied), to secure the carrying out of the obligations of the participant under an agreement referred to in regulation 4.

"trading scheme" has the same meaning as in Part XI of the Fair Trading Act 1973.

NOTES

Commencement: 6 February 1997.

3 Contents of advertisements

(1)   Subject to paragraph (2) of this regulation, a promoter of, or a participant in, a trading scheme shall not issue, circulate or distribute any advertisement which contains information likely to lead directly or indirectly to persons becoming participants in a trading scheme by any means unless such advertisement

    (a) states the name and address of the promoter, or in the case of a scheme promoted by more than one person, the names and addresses of all of the promoters;

    (b) describes the goods or services acquired or supplied under the trading scheme; and

    (c) contains the words set out in Schedule 1 to these Regulations which must

        (i) not appear at the beginning or the end of the advertisement;

        (ii) insofar as the advertisement contains any information as to the sources of income for participants from participation in the trading scheme, appear together with such information and be given no less prominence than such information;

(iii) be easily legible or audible; and

(iv) be afforded no less prominence than that given to any other information in the advertisement apart from the heading of the advertisement.

(2) This regulation does not apply to any advertisement which—

(a) forms part of a newspaper or magazine; or

(b) is transmitted by way of a radio or television broadcast.

NOTES

Commencement: 6 February 1997.

4 Pre-performance requirements

(1) Save where the requirements set out in paragraph (2) below are satisfied, no promoter of, nor participant in, a trading scheme shall—

(a) supply goods or services to a participant in the trading scheme;

(b) provide any goods or services under a transaction effected by such a participant;

(c) be a party to any arrangement under which goods or services are supplied or provided as aforesaid; or

(d) accept from any such participant any payment or undertaking to make a payment in respect of any goods or services supplied or provided as mentioned in any of the preceding paragraphs (a) to (c) above or in respect of any goods or services to be so supplied or provided.

(2) The requirements referred to in paragraph (1) above are that—

(a) the arrangements with a participant do not include a statement or promise that the participant will receive a payment or benefit in respect of the continued participation of another person in the trading scheme to which such arrangements relate or in any other trading scheme;

(b) the promoter or a participant and the participant joining the trading scheme shall have signed a written agreement which contains all the terms under which the participant joining the trading scheme is participating in the trading scheme and which complies with regulation 5;

(c) a copy of that agreement shall have been furnished to the participant joining the trading scheme.

NOTES

Commencement: 6 February 1997.

5   Contents of contracts

The agreement referred to in regulation 4 shall include:—

(a) the name and address of the promoter or, in the case of a scheme promoted by more than one person, the names and addresses of all the promoters;

(b) a description of the goods or services to be acquired by or supplied to the participant by the promoter or promoters, other participants or suppliers nominated by the promoter or promoters or any other person under the trading scheme;

(c) a statement describing the capacity in which the participant shall act for the purposes of any transaction which he may effect under the trading scheme;

(d) a statement describing the financial obligation of the participant during the period of twelve months from the commencement date of the agreement. The promoter shall give to the participant at least 60 days advance written notice of any subsequent changes in such financial obligation.

(e) a statement describing the right of the participant to cancel the agreement:—

(i) within 14 days of entering into the agreement without penalty and with the right to recover any monies which he had paid to or for the benefit of the promoter or any of the promoters or any other participant in connection with his participation in the trading scheme or paid to any other participant in accordance with the provisions of the trading scheme and the manner in which that cancellation and recovery shall be effected;

(ii) within 14 days of entering into the agreement the right to return to an address specified in the agreement which must be an address in the United Kingdom, any goods the participant has purchased within that period under the trading scheme and which remain unsold provided that such unsold goods remain in the condition in which they were in at the time of purchase, whether or not their external wrappings have been broken and to recover any monies paid in respect of such goods;

(iii) within 14 days of entering into the agreement the right to cancel any services ordered within that period under the trading scheme and to recover any monies paid in respect of such services not yet supplied to the participant;

and that the promoter or any other person who has supplied goods to the participant under the trading scheme shall not be entitled to make a handling charge in respect of goods returned under sub-paragraph (ii) above or services cancelled under sub-paragraph (iii) above;

(f) a statement describing the rights of the participant to terminate the agreement at any time without penalty by giving 14 days written notice to the promoter or any of the promoters at an address which is specified in the agreement;

(g) a statement describing the rights of the participant following termination of the agreement by the promoter or the participant as set out in these Regulations;

(h) the written warnings in the form set out in Part I and Part II of Schedule 2 hereto which comply with the following:—

   (i) the words are easily legible; and

   (ii) the words in Part II are printed immediately above the space for the participant's signature.

(i) a statement setting out the conditions under which the participant shall be entitled to return goods to the promoter or any promoters or any other participant which shall include at least the rights conferred on the participant by regulation 6 below and which must include an address in the United Kingdom to which such goods can be returned.

(j) a statement setting out the conditions when commission already paid by the promoter or another participant will be recoverable from the participant which shall include at least the rights conferred on the participant by regulation 9.

(k) where the agreement comprises more than one document, a statement setting out all documents which form part of the contract between the parties and that those documents form the entire agreement between the parties.

NOTES

Commencement: 6 February 1997.

6 Right to return goods to promoter on termination

(1) The rights referred to in regulation 5(i) are, that if a participant or the promoter or any of the promoters terminates an agreement referred to in regulation 4 or any agreement entered into in consequence of such an agreement with a participant, the participant shall, subject to subsection (2) below, have the right to be released from all future contractual obligations and to return to the promoter or any of the promoters or any other participant any goods the participant has purchased within a period of 90 days prior to such termination under the scheme and which remain unsold and to recover from the promoter or such other participant who supplied the goods—

(a) where the participant has terminated the agreement, the price (inclusive of Value Added Tax) which the participant paid for them less—

    (i) in the case of any goods the condition of which has deteriorated due to an act or default on the part of the participant, an amount equal to the diminution in their value resulting from such deterioration; and

    (ii) a reasonable handling charge;

(b) where the promoter or any of the promoters or any other participant has terminated the agreement the price (inclusive of Value Added Tax) which the participant paid for them together with any costs incurred by the participant for returning the goods to the promoter or any other participant;

(c) on terms whereby the purchase price is payable upon delivery of the goods or, if the goods are already held by the promoter or any of the promoters, forthwith, and

(d) on terms whereby the goods not already held by the promoter or any of the promoters will be delivered within 21 days of such termination at the promoter's expense to the address stated in the agreement.

(2) Where an agreement referred to in regulation 4 contains an obligation on the participant not to compete with the business of the promoter after termination of such agreement, such non-competition provision shall continue in force after the date of termination.

NOTES

Commencement: 6 February 1997.

7 Securities and guarantees

A promoter of, or a participant in, a trading scheme shall not accept from a participant any guarantee or security in whatever form in respect of goods or services supplied or to be supplied or in respect of the payment of the price for goods or services supplied or to be supplied or an undertaking to provide such a guarantee or such security unless the creditor or a promoter or other supplier who is not a creditor has agreed in writing to refund the amount of that payment to the debtor upon his returning the relevant goods in an undamaged condition to the creditor or to any promoter or supplier.

NOTES
Commencement: 6 February 1997.

8 Supply of goods and services

A promoter of, or a participant in, a trading scheme shall not make a supply of goods or services to the participant unless, in respect of every supply of goods or services under a trading scheme, such promoter or participant has provided the participant to whom the goods are supplied or to be supplied with an adequate record of the transaction in respect of which payment is due from that participant. For the purposes of this regulation an itemised order form, invoice or receipt shall constitute an adequate record.

NOTES
Commencement: 6 February 1997.

9 Recovery of commission

The rights referred to in regulation 5(j) are the right to retain, after termination of an agreement referred to in regulation 4 or any agreement made thereunder, any commission paid to the participant under a trading scheme unless—

(a) the commission was paid in respect of goods returned to the promoter or another participant who paid the commission;

(b) the promoter has refunded all monies due to the participant under the agreement referred to in regulation 4 in respect of goods returned to him by the participant;

(c) the commission payment is claimed within 120 days of the date of having been made; and

(d) the promoter has entered into an agreement with the participant that complies with the requirements in regulation 5 and that agreement and any subsequent agreement contains a statement describing when commission becomes repayable to the promoter and the terms upon which recovery of that payment may be made; and

(e) the promoter recovers the commission payment in accordance with the terms referred to in paragraph (d) above.

NOTES
Commencement: 6 February 1997.

10 £200 liability limit

A promoter of, or a participant in, a trading scheme shall not accept from a participant joining the trading scheme any payment or an undertaking to make a payment of any sum exceeding £200 unless 7 days have expired from the making of the agreement relating to goods or services supplied or to be supplied under that agreement to the participant by the promoter or any other participant under the trading scheme.

NOTES

Commencement: 6 February 1997.

11 Civil consequences of contraventions

(1)　Where a participant makes a payment to or for the benefit of a promoter of, or to a participant in, a trading scheme and the acceptance of that payment involves a contravention of these Regulations, that contravention shall be actionable at the suit of the participant who suffers loss as a result of the contravention subject to the defences and other incidents applying to actions for breach of statutory duty.

(2)　No undertaking to make any payment given by a participant in a trading scheme involving a contravention of sub-paragraph (d) of paragraph (1) of regulation 4 or regulation 10 shall be enforceable against him in any civil proceedings or recoverable in any other way.

(3)　A participant in a trading scheme shall be under no liability to pay for any goods or services as the case may be—
  (a) supplied to him in circumstances involving a contravention of regulations 4 to 10; or
  (b) unless it was clearly explained to him by a promoter or a participant supplying or seeking to supply goods or services under the trading scheme, before he purchased the goods or services, that he had a free choice whether or not to purchase those goods or services and the purchase price for those goods or services and his annual financial obligation under the agreement was clearly stated.

NOTES

Commencement: 6 February 1997.

SCHEDULES

SCHEDULE 1

Regulation 3(1)(c)

Warning for use in advertisements—

1.   It is illegal for a promoter or a participant in a trading scheme to persuade anyone to make a payment by promising benefits from getting others to join a scheme.

2.   Do not be misled by claims that high earnings are easily achieved.

NOTES

Commencement: 6 February 1997.

SCHEDULE 2

Regulation 5

Warning for use in contracts—

Part I

1.   It is illegal for a promoter or a participant in a trading scheme to persuade anyone to make a payment by promising benefits from getting others to join a scheme.

2.   Do not be misled by claims that high earnings are easily achieved.

Part II

3.   If you sign this contract, you have 14 days in which to cancel and get your money back.

NOTES

Commencement: 6 February 1997.

Trading Schemes (Exclusion) Regulations 1997

(SI 1997/31)

NOTES

Made: 13 January 1997.
Authority: Fair Trading Act 1973, s 118(6)(b).

1 Citation and Commencement

These Regulations may be cited as the Trading Schemes (Exclusion) Regulations 1997 and shall come into force on 6th February 1997.

NOTES

Commencement: 6 February 1997.

2 Interpretation

In these Regulations—
"the Act" means the Fair Trading Act 1973.
"annual profit of the trading scheme" means for each financial year the net profit of the promoter or promoters of the trading scheme as shown in the accounts of the trading scheme.
"chain letter" means any trading scheme under which a letter is sent to participants or prospective participants directly or indirectly instructing or requesting them to—
    (a) send monies or other benefits to at least one of the individuals on a list of individuals, shown with their mailing addresses, which is contained in or accompanying that letter; and
    (b) carry on the chain by sending copies of the letter to other individuals not on the list and removing from the list any one name and address and adding their own to it.
"participant" has the same meaning as in section 118(8) of the Act.
"single tier trading scheme" means a trading scheme the only members of which are the promoter or promoters and one or more participants and under which, in the United Kingdom, either a single promoter or a single participant operates at one level and any other participant or participants of the trading scheme

operate at the same level below such promoter or participant aforesaid.

"trading scheme" has the same meaning as in section 118(8) of the Act.

Commencement: 6 February 1997.

3 Disapplication of Part XI of the Act

For the purpose of section 118(6)(b) of the Act the Secretary of State hereby prescribes trading schemes of the following description, that is to say—

(a) [any trading scheme which is a single tier trading scheme under which a participant operating at a level immediately below that of the promoter or single participant in the UK, who introduces another participant to the scheme at that level,] does not receive any payment or benefit, or can only receive a single benefit or payment, in respect of the introduction of that participant, such payment or benefit not exceeding £50 and can receive no other benefit or payment in respect of or flowing directly or indirectly from the membership or activities of that participant in that or any other trading scheme, unless such other benefit or payment results from—

  (i) a sharing of expenses of the operation of the trading scheme;

  (ii) a share in the annual profit of the trading scheme; or

  (iii) the sale of the participant's business, being a business in respect of which a registration under the Value Added Taxes Act 1994 was in force at the date of sale.

[(b) any trading scheme all of the participants in which are making or have the intention of making taxable supplies in the UK and are registered for Value Added Tax; or]

(c) any trading scheme which is a chain letter provided there is no requirement on the participant to send monies or other benefits

  (i) to a central address or the promoter of the trading scheme for onward distribution; or

  (ii) to any person or organisation other than or additional to the person whose name and address is to be deleted from the list when the participant sends the letter to others; or

  (iii) to an organisation or person for onward transmission to a participant (whether or not that participant is identified on the list); and

where the promoter does not benefit from the provision of any other service or facilities offered or provided either by him or any other person or organisation to participants.

Commencement: 6 February 1997.

Words in square brackets substituted by the Trading Schemes (Exclusion) (Amendment) Regulations 1997, SI 1997/1887, reg 2.

Foreign Package Holidays (Tour Operators and Travel Agents) Order 1998

(SI 1998/1945)

NOTES

Made: 7 August 1998.
Authority: Fair Trading Act 1973, s 91(2).

1 Citation, commencement and interpretation

(1) This Order may be cited as the Foreign Package Holidays (Tour Operators and Travel Agents) Order 1998 and shall come into force on 16th November 1998.

(2) In this Order—

"accommodation" means the provision of a place to sleep, including the provision of a site for the erection of a tent or a parking place for a caravan, mobile home or other similar vehicle, but does not include the provision of sleeping accommodation in a means of transport unless that accommodation represents a substantial proportion of the accommodation for the holiday;

"foreign package holiday" means services, accommodation and facilities provided under a contract, made within the United Kingdom, by a tour operator for a holiday outside the United Kingdom provided transport to or from the United Kingdom and accommodation outside the United Kingdom (whether or not for the duration of the holiday) are included;

"inducement" means a benefit, whether pecuniary or not, offered by a travel agent as an incentive to acquire a foreign package holiday through him;

"tour operator" means a person who, otherwise than occasionally, organises foreign package holidays and supplies or offers them for supply, whether directly or through a travel agent;

"travel agent" means a person who supplies or offers for supply a foreign package holiday put together by a tour operator; and

"travel insurance" means any policy of insurance against the risks to any person arising during or in connection with a foreign package holiday.

NOTES

Commencement: 16 November 1998.

2   Price reductions and travel insurance

It shall be unlawful for a travel agent or a tour operator (where the tour operator and a travel agent are interconnected bodies corporate and when the tour operator is supplying or offering to supply foreign package holidays directly to the public or any class of persons) to discriminate either in respect of the price charged for a foreign package holiday or by requiring payment of an additional charge against a person who does not acquire travel insurance in respect of that holiday from that travel agent or, as the case may be, tour operator.

NOTES

Commencement: 16 November 1998.

"Most favoured customer" agreements and related conduct

3   It shall be unlawful for a tour operator to make or carry out an agreement (whenever made) with a travel agent which—

(a) imposes any restriction, whether as to charges or other terms or conditions or otherwise, in respect of the supply or offer of supply by the travel agent of foreign package holidays of another tour operator; or

(b) requires a travel agent, when supplying or offering to supply foreign package holidays of that operator, to offer inducements at least equal in value to or marginally less in value than the inducements which the travel agent applies when supplying or offering to supply the foreign package holidays of another tour operator.

NOTES

Commencement: 16 November 1998.

4   It shall be unlawful for a tour operator to withhold or threaten to withhold supplies of foreign package holidays from, or to discriminate in respect of the supply of foreign package holidays to, a travel agent who does not, or does not propose to, offer inducements at least equal in value to or marginally less in value than the inducements which the travel agent applies, or proposes to apply, when supplying or offering to supply the foreign package holidays of another tour operator.

NOTES

Commencement: 16 November 1998.

5   This Order shall not apply in respect of an agreement in so far as it is, or if made would be, an agreement to which the Restrictive Trade Practices Act 1976 applies or, as the case may be, would apply.

NOTES

Commencement: 16 November 1998.

Telecommunications (Data Protection and Privacy) (Direct Marketing) Regulations 1998

(SI 1998/3170)

NOTES

Made: 16 December 1998.
European Communities Act 1972, s 2(2)..

PART I
GENERAL

I   Citation and commencement

These Regulations may be cited as the Telecommunications (Data Protection and Privacy) (Direct Marketing) Regulations 1998 and shall come into force on 1st May 1999.

NOTES
Commencement: 1 May 1999.

2 Interpretation

(1)   In these Regulations—

"corporate subscriber" means a subscriber who is not an individual, that is to say, a subscriber who is—

(a) a company within the meaning of section 735(1) of the Companies Act 1985;

(b) a company incorporated in pursuance of a royal charter or letters patent;

(c) a partnership in Scotland;

(d) a corporation sole, or

(e) any other body corporate or other entity which is a legal person distinct from the persons (if any) of which it is composed;

"the Data Protection Registrar" and "the Registrar" both mean the Registrar appointed under section 3 of the Data Protection Act 1984;

"the Directive" means Directive 97/66/EC of the European Parliament and of the Council of the European Union;

"the Director" means the Director General of Telecommunications appointed under section 1 of the Telecommunications Act 1984;

"individual" means a living individual and includes an unincorporate body of such individuals;

"public telecommunications network" means any transmission system and, where applicable, switching equipment and other resources which—

(a) permit the conveyance of signals between defined termination points by wire, by radio, by optical or by other electromagnetic means, and

(b) are used, in whole or in part, for the provision of publicly available telecommunications services;

"subscriber" means a person who is a party to a contract with a telecommunications service provider for the supply of publicly available telecommunications services;

"telecommunications network provider" means a person who provides a public telecommunications network (whether or not he is also a telecommunications service provider);

"telecommunications service provider" means a person who provides publicly available telecommunications services (whether or not he is also a telecommunications network provider);

"telecommunications services" means services the provision of which consists, in whole or in part, of the transmission and routing of signals on telecommunications networks, not being services by way of radio or television broadcasting.

(2)   Subject to paragraph (1) and except where the context otherwise requires, expressions used in these Regulations which are also used in the Directive have the same meanings in these Regulations as they have in the Directive.

(3)   In a case in which signals are conveyed to telecommunications equipment used by a subscriber, wholly or partly otherwise than by line, any reference in these Regulations to a line shall be construed as including a reference to what, in that case, functionally corresponds to a line.

NOTES

Commencement: 1 May 1999.

3   Incidental and consequential amendments and modification of contracts

(1)   The amendments set out in Schedule 1 shall have effect.

(2)   To the extent that any term in a contract between a subscriber to, and the provider of, publicly available telecommunications services would be inconsistent with a requirement of these Regulations, that term shall be void.

NOTES

Commencement: 1 May 1999.

4   Consents and notifications for purposes of Regulations

(1)   Except where the context otherwise requires, a consent or notification for the purposes of these Regulations may be in general or more limited terms and may be subject to conditions and, so long as it remains in force, shall have effect according to its tenor.

(2)   A notification for the purposes of these Regulations may (without prejudice to any other method of transmission) be sent by post.

NOTES
Commencement: 1 May 1999.

PART II
USE OF TELECOMMUNICATIONS SERVICES FOR DIRECT MARKETING
PURPOSES

5 Application and interpretation of Part II

(1)   This Part shall apply in relation to the use of publicly available telecommunications services for direct marketing purposes.

(2)   Any reference in this Part to direct marketing is a reference to the communication of any advertising or marketing material on a particular line.

(3)   In this Part, "caller" means a person using publicly available telecommunications services for direct marketing purposes, except that where such services are so used at the instigation of some other person "caller" means that other person.

NOTES
Commencement: 1 May 1999.

6 Use of automated calling systems for direct marketing purposes-communications on lines of individual or corporate subscribers

(1)   This regulation applies in relation to the use of publicly available telecommunications services for the communication of material, for direct marketing purposes, by means of an automated calling system, that is to say, a system which, when activated, operates to make calls without human intervention, whether the called line is that of a subscriber who is an individual or that of a corporate subscriber.

(2)   A person shall not use, or instigate the use of, publicly available telecommunications services, and a subscriber to such services shall not permit his line to be used, as mentioned in paragraph (1), except where the called line is that of a subscriber who has previously notified the caller that

he consents to such communications as are there mentioned being made by, or at the instigation of, the caller in question on that line.

NOTES

Commencement: 1 May 1999.

7 Use of fax for direct marketing purposes-unsolicited communications on lines of individual or corporate subscribers

(1)   This regulation applies in relation to the use of publicly available telecommunications services for the unsolicited communication of material, for direct marketing purposes, by means of facsimile transmission, whether the called line is that of a subscriber who is an individual or that of a corporate subscriber.

(2)   A person shall not use, or instigate the use of, publicly available telecommunications services, and a subscriber to such services shall not permit his line to be used, as mentioned in paragraph (1) where—
  (a) the called line is that of a subscriber who has previously notified the caller (notwithstanding, in the case of a subscriber who is an individual, that he enjoys the benefit of regulation 8) that such unsolicited communications as are there mentioned should not be sent on that line, or
  (b) the number allocated to a subscriber in respect of the called line is one listed in the record kept under paragraph (4).

(3)   For the purposes of paragraphs (1) and (2), the communication of material as mentioned in paragraph (1) shall not be treated as unsolicited where the called line is that of a subscriber who has notified the caller that he does not object to receiving on that line such communications as are so mentioned from the caller in question.

(4)   For the purposes of this regulation—
  (a) the Director shall maintain and keep up-to-date, in printed form or in electronic form, a record of the numbers allocated to subscribers, in respect of particular lines, who have notified him (notwithstanding, in the case of individuals, that they enjoy the benefit of regulation 8) that they do not for the time being wish to receive such communications as are mentioned in paragraph (1) on the lines in question, and he shall remove a number from the record where he has reason to believe that it has ceased to be allocated to the subscriber by whom he was so notified, and

(b) on the request of—
   (i) a person wishing to send, or instigate the sending of, such communications, or
   (ii) a subscriber wishing to permit the use of his line for the sending of such communications,
   for information derived from that record, the Director shall, unless it is not reasonably practicable so to do, on the payment to him of such fee as is applicable and is, subject to paragraph (5), required by him, make the information requested available to that person or that subscriber.

(5) For the purposes of paragraph (4)(b) the Director may require different fees—
   (a) for making available information derived from the record in different forms or manners, or
   (b) for making available information derived from the whole or from different parts of the record,

but the fees required by him shall be ones in relation to which the Secretary of State has notified the Director that he is satisfied that they are designed to secure, as nearly as may be and taking one year with another, that the aggregate fees received, or reasonably expected to be received, equal the costs incurred, or reasonably expected to be incurred, by the Director, in discharging his duties under paragraph (4).

(6) The functions of the Director under paragraph (4), other than the function of determining the fees to be required for the purposes of sub-paragraph (b) thereof, may be discharged on his behalf by some other person in pursuance of arrangements in that behalf made by the Director with that other person.

NOTES

Commencement: 1 May 1999.

8 Use of fax for direct marketing purposes-communications on lines of subscribers who are individuals

(1) This regulation applies in relation to the use of publicly available telecommunications services for the communication of material, for direct marketing purposes, by means of facsimile transmission where the called line is that of a subscriber who is an individual; and the provisions of this regulation and those of regulation 7 are without prejudice to each other.

(2)   A person shall not use, or instigate the use of, publicly available telecommunications services, and a subscriber to such services shall not permit his line to be used, as mentioned in paragraph (1), except where the called line is that of a subscriber who has previously notified the caller that he consents to such communications as are there mentioned being sent by the caller in question on that line.

NOTES

Commencement: 1 May 1999.

9   Unsolicited calls for direct marketing purposes on lines of subscribers who are individuals

(1)   This regulation applies in relation to the use of publicly available telecommunications services for the purposes of making unsolicited calls, for direct marketing purposes, otherwise than by means of an automated calling system within the meaning of regulation 6(1) or by means of facsimile transmission, where the called line is that of a subscriber who is an individual.

(2)   A person shall not use, or instigate the use of, publicly available telecommunications services, and a subscriber to such services shall not permit his line to be used, as mentioned in paragraph (1) where—
  (a)   the called line is that of a subscriber who has previously notified the caller that such unsolicited calls as are there mentioned should not be made on that line, or
  (b)   the number allocated to a subscriber in respect of the called line is one listed in the record kept under paragraph (4).

(3)   For the purposes of paragraphs (1) and (2), a call on a subscriber's line shall not be treated as an unsolicited call if that subscriber has notified the caller that he does not object to receiving on that line calls made by, or at the instigation of, the caller in question for direct marketing purposes.

(4)   For the purposes of this regulation—
  (a)   the Director shall maintain and keep up-to-date, in printed form or in electronic form, a record of the numbers allocated to subscribers who are individuals, in respect of particular lines, who have notified him that they do not for the time being wish to receive unsolicited calls made for direct marketing purposes on the lines in question, and he shall remove a number from the record where he has reason to believe that it has ceased to be allocated to the subscriber by whom he was so notified, and

(b) on the request of—
- (i) a person wishing to make, or instigate the making of, such calls, or
- (ii) a subscriber wishing to permit the use of his line for the making of such calls,

for information derived from that record, the Director shall, unless it is not reasonably practicable so to do, on the payment to him of such fee as is applicable and is, subject to paragraph (5), required by him, make the information requested available to that person or that subscriber.

(5)   For the purpose of paragraph (4)(b) the Director may require different fees—
- (a) for making available information derived from the record in different forms or manners, or
- (b) for making available information derived from the whole or from different parts of the record,

but the fees required by him shall be ones in relation to which the Secretary of State has notified the Director that he is satisfied that they are designed to secure, as nearly as may be and taking one year with another, that the aggregate fees received, or reasonably expected to be received, equal the costs incurred, or reasonably expected to be incurred, by the Director in discharging his duties under paragraph (4).

(6)   The functions of the Director under paragraph (4), other than the function of determining the fees to be required for the purposes of sub-paragraph (b) thereof, may be discharged on his behalf by some other person in pursuance of arrangements in that behalf made by the Director with that other person.

NOTES

Commencement: 1 May 1999.

10   Notifications for the purposes of regulation 7(4)(a) or 9(4)(a)

(1)   Where any such person as is mentioned in paragraph (3) has in his possession such a notification as is mentioned in regulation 7(4)(a) or regulation 9(4)(a) (to whomsoever it is addressed) or a copy or record of such a notification—
- (a) he shall, without undue delay, transmit a copy of that notification or a copy of that record to the Director, and

(b) subject to receipt by the Director of a copy of a notification or of a record thereof so transmitted, the notification in question shall be treated for the purposes of regulation 7(4)(a) or, as the case may be, regulation 9(4)(a) as if it had been given to the Director.

(2)   Where the Director has made arrangements in pursuance of paragraph (6) of regulation 7 or, as the case may be, paragraph (6) of regulation 9 for the discharge of functions under paragraph (4) of the regulation in question by some other person on his behalf, paragraph (1) of this regulation shall have effect, in relation to such a notification as is mentioned in paragraph (4)(a) of the regulation in question, as if for the reference to the Director in sub-paragraph (a) and the first reference to him in sub-paragraph (b) there were substituted references to that other person.

(3)   The persons referred to in paragraph (1) are—
   (a) a telecommunications service provider;
   (b) the producer of a directory of subscribers, and
   (c) where, in connection with the production of such a directory, information relating to a particular subscriber is supplied to the producer thereof by some other person, that other person.

(4)   In paragraph (3), "directory of subscribers" means a directory of subscribers to publicly available telecommunications services, whether in printed form or in electronic form, which is made available to the public or a section of the public and, in relation to such a directory, "producer" means the person by whom the directory is published or prepared.

NOTES

Commencement: 1 May 1999.

11   Supplementary provisions

(1)   Where publicly available telecommunications services are used for the communication of material for direct marketing purposes—
   (a) by means of an automated calling system within the meaning of regulation 6(1) or by means of facsimile transmission, the caller shall ensure that the material communicated includes the particulars mentioned in paragraph (2)(a) and (b) below;
   (b) otherwise than as mentioned in sub-paragraph (a), the caller shall ensure that the material communicated includes the particulars mentioned in paragraph (2)(a) below and, if the recipient of the call so requests, those mentioned in paragraph (2)(b) below.

(2)   The particulars referred to in paragraph (1) are—
   (a) the name of the caller;
   (b) either the address of the caller or a freephone telephone number on which he can be reached.

(3)   Where a person by whom numbers are allocated to subscribers is requested by or on behalf of the Director, for the purposes of his functions under regulation 7(4) or 9(4), to furnish information as to when a particular number ceases to be allocated to a particular subscriber, that person shall comply with the request.

(4)   A caller shall not be held to have contravened regulation 7 or regulation 9 by reason of the making, or instigating the making, of a call and a subscriber shall not be held to have contravened regulation 7 or regulation 9 by permitting his line to be used for the making of a call, notwithstanding that the number of the called line is one listed in the record kept under paragraph (4) of the regulation in question, if that number was not so listed at any time within the 28 days preceding that on which the call is made.

NOTES

Commencement: 1 May 1999.

PART III
COMPENSATION AND ENFORCEMENT

12 Compensation for failure to comply with requirements of Regulations

(1)   A person who suffers damage by reason of any contravention of any of the requirements of these Regulations by any other person shall be entitled to compensation from the other person for that damage.

(2)   In proceedings brought against a person by virtue of this regulation it shall be a defence to prove that he had taken such care as in all the circumstances was reasonably required to comply with the requirement concerned.

NOTES

Commencement: 1 May 1999.

13 Enforcement-application of sections 10, 13, 14 and 16 of the Data Protection Act 1984

(1)   Subject to the omissions and other modifications set out in Schedule 2, the provisions of sections 10, 13, 14 and 16 of the Data Protection Act 1984 and of Schedules 3 and 4 thereto shall apply for the purposes of the enforcement of these Regulations and connected purposes.

(2)   In regulations 14 and 15, "enforcement functions" means the functions of the Data Protection Registrar under the said provisions as so applied.

(3)   The provisions of this regulation and those of regulation 12 are without prejudice to each other.

NOTES

Commencement: 1 May 1999.

14 Request that Registrar should exercise his enforcement functions

Where it is alleged that there has been a contravention of any of the requirements of these Regulations either the Director or a person aggrieved by the alleged contravention may request the Registrar to exercise his enforcement functions in respect of that contravention; but those functions shall be exerciseable by him whether or not he has been so requested.

NOTES

Commencement: 1 May 1999.

15 Technical advice to Registrar

The Director shall comply with any reasonable request made by the Registrar, in connection with his enforcement functions, for advice on technical and similar matters relating to telecommunications.

NOTES

Commencement: 1 May 1999.

PART III

EUROPEAN MATERIAL

Directive of the European Parliament and of the Council of 27 January 1997 on cross-border credit transfers

(97/5/EC)

NOTES

Date of publication in OJ: OJ L43, 14.2.97, p 25.

THE EUROPEAN PARLIAMENT AND THE COUNCIL OF THE EUROPEAN UNION,

Having regard to the Treaty establishing the European Community, and in particular Article 100a thereof,

Having regard to the proposal from the Commission,[1]

Having regard to the opinion of the Economic and Social Committee,[2]

Having regard to the opinion of the European Monetary Institute,

Acting in accordance with the procedure laid down in Article 189b of the Treaty[3] in the light of the joint text approved on 22 November 1996 by the Conciliation Committee,

(1)   Whereas the volume of cross-border payments is growing steadily as completion of the internal market and progress towards full economic and monetary union lead to greater trade and movement of people within the Community; whereas cross-border credit transfers account for a substantial part of the volume and value of cross-border payments;

(2)   Whereas it is essential for individuals and businesses, especially small and medium-sized enterprises, to be able to make credit transfers rapidly, reliably and cheaply from one part of the Community to another; whereas, in conformity with the Commission Notice on the application of the EC competition rules to cross-border credit transfers,[4] greater competition in the market for cross-border credit transfers should lead to improved services and reduced prices;

(3)   Whereas this Directive seeks to follow up the progress made towards completion of the internal market, in particular towards liberalisation of capital movements, with a view to the implementation of economic and monetary union; whereas its provisions must apply to credit transfers in the currencies of the Member States and in ecus;

(4)   Whereas the European Parliament, in its resolution of 12 February 1993,[5] called for a Council Directive to lay down rules in the area of transparency and performance of cross-border payments;

(5)   Whereas the issues covered by this Directive must be dealt with separately from the systemic issues which remain under consideration within the Commission; whereas it may become necessary to make a further proposal to cover these systemic issues, particularly the problem of settlement finality;

(6)   Whereas the purpose of this Directive is to improve cross-border credit transfer services and thus assist the European Monetary Institute (EMI) in its task of promoting the efficiency of cross-border payments with a view to the preparation of the third stage of economic and monetary union;

(7)   Whereas, in line with the objectives set out in the second recital, this Directive should apply to any credit transfer of an amount of less than ECU 50 000;

(8)   Whereas, having regard to the third paragraph of Article 3b of the Treaty, and with a view to ensuring transparency, this Directive lays down the minimum requirements needed to ensure an adequate level of customer information both before and after the execution of a cross-border credit transfer; whereas these requirements include indication of the complaints and redress procedures offered to customers, together with the arrangements for access thereto; whereas this Directive lays down minimum execution requirements, in particular in terms of performance, which institutions offering cross-border credit transfer services should adhere to, including the obligation to execute a cross-border credit transfer in accordance with the customer's instructions; whereas this Directive fulfils the conditions deriving from the principles set out in Commission Recommendation 90/109/EEC of 14 February 1990 on the transparency of banking conditions relating to cross-border financial transactions;[6] whereas this Directive is without prejudice to Council Directive 91/308/EEC of 10 June 1991 on prevention of the use of the financial system for the purpose of money laundering;[7]

(9)   Whereas this Directive should contribute to reducing the maximum time taken to execute a cross-border credit transfer and encourage those institutions which already take a very short time to do so to maintain that practice;

(10)   Whereas the Commission, in the report it will submit to the European Parliament and the Council within two years of implementation of this Directive, should particularly examine the time-limit to be applied in the absence of a time-limit agreed between the originator and his institution, taking into account both technical developments and the situation existing in each Member State;

(11)   Whereas there should be an obligation upon institutions to refund in the event of a failure to successfully complete a credit transfer; whereas the obligation to refund imposes a contingent liability on institutions which might, in the absence of any limit, have a prejudicial effect on solvency requirements; whereas that obligation to refund should therefore be applicable up to ECU 12 500;

(12)   Whereas Article 8 does not affect the general provisions of national law whereby an institution has responsibility towards the originator when a cross-border credit transfer has not been completed because of an error committed by that institution;

(13)   Whereas it is necessary to distinguish, among the circumstances with which institutions involved in the execution of a cross-border credit transfer may be confronted, including circumstances relating to insolvency, those caused by force majeure; whereas for that purpose the definition of force majeure given in Article 4(6) of Directive 90/314/EEC of 13 June 1990 on package travel, package holidays and package tours[8] should be taken as a basis;

(14)   Whereas there need to be adequate and effective complaints and redress procedures in the Member States for the settlement of possible disputes between customers and institutions, using existing procedures where appropriate,

HAVE ADOPTED THIS DIRECTIVE—

NOTES

[1]   OJ C360, 17.12.1994, p 13, and OJ C199, 3.8.1995, p 16.
[2]   OJ C236, 11.9.1995, p 1.

³ Opinion of the European Parliament of 19 May 1995 (OJ C151, 19.6.1995, p 370), Council common position of 4 December 1995 (OJ C353, 30.12.1995, p 52) and Decision of the European Parliament of 13 March 1996 (OJ C96, 1.4.1996, p 74). Decision of the Council of 19 December 1996 and Decision of the European Parliament of 16 January 1997.

⁴ OJ C251, 27.9.1995, p 3.

⁵ OJ C72, 15.3.1993, p 158.

⁶ OJ L67, 15.3.1990, p 39.

⁷ OJ L166, 28.6.1991, p 77.

⁸ OJ L158, 23.6.1990. p 59.

SECTION I
SCOPE AND DEFINITIONS

Article 1   Scope

The provisions of this Directive shall apply to cross-border credit transfers in the currencies of the Member States and the ECU up to the equivalent of ECU 50 000 ordered by persons other than those referred to in Article 2(a), (b) and (c) and executed by credit institutions or other institutions.

Article 2   Definitions

For the purposes of this Directive-
   (a) 'credit institution' means an institution as defined in Article 1 of Council Directive 77/780/EEC,[1] and includes branches, within the meaning of the third indent of that Article and located in the Community, of credit institutions which have their head offices outside the Community and which by way of business execute cross-border credit transfers;
   (b) 'other institution' means any natural or legal person, other than a credit institution, that by way of business executes cross-border credit transfers;
   (c) 'financial institution' means an institution as defined in Article 4(1) of Council Regulation (EC) No 3604/93 of 13 December 1993 specifying definitions for the application of the prohibition of privileged access referred to in Article 104a of the Treaty;[2]
   (d) 'institution' means a credit institution or other institution; for the purposes of Articles 6, 7 and 8, branches of one credit institution situated in different Member States which participate in the execution of a cross-border credit transfer shall be regarded as separate institutions;

(e) 'intermediary institution' means an institution which is neither that of the originator nor that of the beneficiary and which participates in the execution of a cross-border credit transfer;

(f) 'cross-border credit transfer' means a transaction carried out on the initiative of an originator via an institution or its branch in one Member State, with a view to making available an amount of money to a beneficiary at an institution or its branch in another Member State; the originator and the beneficiary may be one and the same person;

(g) 'cross-border credit transfer order' means an unconditional instruction in any form, given directly by an originator to an institution to execute a cross-border credit transfer;

(h) 'originator' means a natural or legal person that orders the making of a cross-border credit transfer to a beneficiary;

(i) 'beneficiary' means the final recipient of a cross-border credit transfer for whom the corresponding funds are made available in an account to which he has access;

(j) 'customer' means the originator or the beneficiary, as the context may require;

(k) 'reference interest rate' means an interest rate representing compensation and established in accordance with the rules laid down by the Member State in which the establishment which must pay the compensation to the customer is situated;

(l) 'date of acceptance' means the date of fulfilment of all the conditions required by the institution as to the execution of the cross-border credit transfer order and relating to the availability of adequate financial cover and the information required to execute that order.

NOTES

[1]   OJ L322, 17.12.1977, p 30. Directive as last amended by Directive 95/26/EC (OJ L168, 18.7.1995, p 7).

[2]   OJ L332, 31.12.1993, p 4.

SECTION II
TRANSPARENCY OF CONDITIONS FOR CROSS-BORDER CREDIT
TRANSFERS

Article 3   Prior information on conditions for cross-border credit transfers

The institutions shall make available to their actual and prospective customers in writing, including where appropriate by electronic means, and in a readily comprehensible form, information on conditions for cross-border credit transfers. This information shall include at least—

— indication of the time needed, when a cross-border credit transfer order given to the institution is executed, for the funds to be credited to the account of the beneficiary's institution; the start of that period must be clearly indicated,

— indication of the time needed, upon receipt of a cross-border credit transfer, for the funds credited to the account of the institution to be credited to the beneficiary's account,

— the manner of calculation of any commission fees and charges payable by the customer to the institution, including where appropriate the rates,

— the value date, if any, applied by the institution,

— details of the complaint and redress procedures available to the customer and arrangements for access to them,

— indication of the reference exchange rates used.

Article 4 Information subsequent to a cross-border credit transfer

The institutions shall supply their customers, unless the latter expressly forgo this, subsequent to the execution or receipt of a cross-border credit transfer, with clear information in writing, including where appropriate by electronic means, and in a readily comprehensible form. This information shall include at least—

— a reference enabling the customer to identify the cross-border credit transfer,

— the original amount of the cross-border credit transfer,

— the amount of all charges and commission fees payable by the customer,

— the value date, if any, applied by the institution.

Where the originator has specified that the charges for the cross-border credit transfer are to be wholly or partly borne by the beneficiary, the latter shall be informed thereof by his own institution.

Where any amount has been converted, the institution which converted it shall inform its customer of the exchange rate used.

SECTION III
MINIMUM OBLIGATIONS OF INSTITUTIONS IN RESPECT OF CROSS-BORDER
CREDIT TRANSFERS

Article 5 Specific undertakings by the institution

Unless it does not wish to do business with that customer, an institution must at a customer's request, for a cross-border credit transfer with stated specifications, give an undertaking concerning the time needed for execution of the transfer and the commission fees and charges payable, apart from those relating to the exchange rate used.

Article 6 Obligations regarding time taken

1.   The originator's institution shall execute the cross-border credit transfer in question within the time limit agreed with the originator.

Where the agreed time limit is not complied with or, in the absence of any such time limit, where, at the end of the fifth banking business day following the date of acceptance of the cross-border credit transfer order, the funds have not been credited to the account of the beneficiary's institution, the originator's institution shall compensate the originator.

Compensation shall comprise the payment of interest calculated by applying the reference rate of interest to the amount of the cross-border credit transfer for the period from—
— the end of the agreed time limit or, in the absence of any such time limit, the end of the fifth banking business day following the date of acceptance of the cross-border credit transfer order, to
— the date on which the funds are credited to the account of the beneficiary's institution.

Similarly, where non-execution of the cross-border credit transfer within the time limit agreed or, in the absence of any such time limit, before the end of the fifth banking business day following the date of acceptance of the cross-border credit transfer is attributable to an intermediary institution, that institution shall be required to compensate the originator's institution.

2.   The beneficiary's institution shall make the funds resulting from the cross-border credit transfer available to the beneficiary within the time limit agreed with the beneficiary.

Where the agreed time limit is not complied with or, in the absence of any such time limit, where, at the end of the banking business day following the day on which the funds were credited to the account of the beneficiary's institution, the funds have not been credited to the beneficiary's account, the beneficiary's institution shall compensate the beneficiary.

Compensation shall comprise the payment of interest calculated by applying the reference rate of interest to the amount of the cross-border credit transfer for the period from—
— the end of the agreed time limit or, in the absence of any such time limit, the end of the banking business day following the day on which the funds were credited to the account of the beneficiary's institution, to
— the date on which the funds are credited to the beneficiary's account.

3.   No compensation shall be payable pursuant to paragraphs 1 and 2 where the originator's institution or, as the case may be, the beneficiary's institution can establish that the delay is attributable to the originator or, as the case may be, the beneficiary.

4.   Paragraphs 1, 2 and 3 shall be entirely without prejudice to the other rights of customers and institutions that have participated in the execution of a cross-border credit transfer order.

Article 7   Obligation to execute the cross-border transfer in accordance with instructions

1.   The originator's institution, any intermediary institution and the beneficiary's institution, after the date of acceptance of the cross-border credit transfer order, shall each be obliged to execute that credit transfer for the full amount thereof unless the originator has specified that the costs of the cross-border credit transfer are to be borne wholly or partly by the beneficiary.

The first subparagraph shall be without prejudice to the possibility of the beneficiary's institution levying a charge on the beneficiary relating to the administration of his account, in accordance with the relevant rules and customs. However, such a charge may not be used by the institution to avoid the obligations imposed by the said subparagraph.

2.   Without prejudice to any other claim which may be made, where the originator's institution or an intermediary institution has made a deduction from the amount of the cross-border credit transfer in breach of paragraph 1,

the originator's institution shall, at the originator's request, credit, free of all deductions and at its own cost, the amount deducted to the beneficiary unless the originator requests that the amount be credited to him.

Any intermediary institution which has made a deduction in breach of paragraph 1 shall credit the amount deducted, free of all deductions and at its own cost, to the originator's institution or, if the originator's institution so requests, to the beneficiary of the cross-border credit transfer.

3.   Where a breach of the duty to execute the cross-border credit transfer order in accordance with the originator's instructions has been caused by the beneficiary's institution, and without prejudice to any other claim which may be made, the beneficiary's institution shall be liable to credit to the beneficiary, at its own cost, any sum wrongly deducted.

Article 8   Obligation upon institutions to refund in the event of non-execution of transfers

1.   If, after a cross-border credit transfer order has been accepted by the originator's institution, the relevant amounts are not credited to the account of the beneficiary's institution, and without prejudice to any other claim which may be made, the originator's institution shall credit the originator, up to ECU 12 500, with the amount of the cross-border credit transfer plus—

— interest calculated by applying the reference interest rate to the amount of the cross-border credit transfer for the period between the date of the cross-border credit transfer order and the date of the credit, and

— the charges relating to the cross-border credit transfer paid by the originator.

These amounts shall be made available to the originator within fourteen banking business days following the date of his request, unless the funds corresponding to the cross-border credit transfer have in the meantime been credited to the account of the beneficiary's institution.

Such a request may not be made before expiry of the time limit agreed between the originator's institution and the originator for the execution of the cross-border credit transfer order or, in the absence of any such time limit, before expiry of the time limit laid down in the second subparagraph of Article 6(1).

Similarly, each intermediary institution which has accepted the cross-border credit transfer order owes an obligation to refund at its own cost the amount of the credit transfer, including the related costs and interest, to the institution which instructed it to carry out the order. If the cross-border credit transfer was not completed because of errors or omissions in the instructions given by that institution, the intermediary institution shall endeavour as far as possible to refund the amount of the transfer.

2.   By way of derogation from paragraph 1, if the cross-border credit transfer was not completed because of its non-execution by an intermediary institution chosen by the beneficiary's institution, the latter institution shall be obliged to make the funds available to the beneficiary up to ECU 12 500.

3.   By way of derogation from paragraph 1, if the cross-border credit transfer was not completed because of an error or omission in the instructions given by the originator to his institution or because of non-execution of the cross-border credit transfer by an intermediary institution expressly chosen by the originator, the originator's institution and the other institutions involved shall endeavour as far as possible to refund the amount of the transfer.

Where the amount has been recovered by the originator's institution, it shall be obliged to credit it to the originator. The institutions, including the originator's institution, are not obliged in this case to refund the charges and interest accruing, and can deduct the costs arising from the recovery if specified.

Article 9   Situation of force majeure

Without prejudice to the provisions of Directive 91/308/EEC, institutions participating in the execution of a cross-border credit transfer order shall be released from the obligations laid down in this Directive where they can adduce reasons of force majeure, namely abnormal and unforeseeable circumstances beyond the control of the person pleading force majeure, the consequences of which would have been unavoidable despite all efforts to the contrary, which are relevant to its provisions.

Article 10   Settlement of disputes

Member States shall ensure that there are adequate and effective complaints and redress procedures for the settlement of disputes between an originator and his institution or between a beneficiary and his institution, using existing procedures where appropriate.

SECTION IV
FINAL PROVISIONS

Article 11   Implementation

1.   Member States shall bring into force the laws, regulations and administrative provisions necessary to comply with this Directive by 14 August 1999 at the latest. They shall forthwith inform the Commission thereof.

When Member States adopt these provisions, they shall contain a reference to this Directive or shall be accompanied by such reference on the occasion of their official publication. The methods of making such reference shall be laid down by Member States.

2.   Member States shall communicate to the Commission the text of the main laws, regulations or administrative provisions which they adopt in the field governed by this Directive.

Article 12   Report to the European Parliament and the Council

No later than two years after the date of implementation of this Directive, the Commission shall submit a report to the European Parliament and the Council on the application of this Directive, accompanied where appropriate by proposals for its revision.

This report shall, in the light of the situation existing in each Member State and of the technical developments that have taken place, deal particularly with the question of the time limit set in Article 6(1).

Article 13   Entry into force

This Directive shall enter into force on the date of its publication in the *Official Journal of the European Communities*.

Article 14   Addressees

This Directive is addressed to the Member States.

Done at Brussels, 27 January 1997.

Directive of the European Parliament of 20 May 1997 on the protection of consumers in respect of distance contracts

(97/7/EC)

NOTES

Date of publication in OJ: OJ L144, 4.6.97, p 19.

THE EUROPEAN PARLIAMENT AND THE COUNCIL OF THE EUROPEAN UNION,

Having regard to the Treaty establishing the European Community, and in particular Article 100a thereof,

Having regard to the proposal from the Commission,[1]

Having regard to the opinion of the Economic and Social Committee,[2]

Acting in accordance with the procedure laid down in Article 189b of the Treaty,[3] in the light of the joint text approved by the Conciliation Committee on 27 November 1996,

(1)   Whereas, in connection with the attainment of the aims of the internal market, measures must be taken for the gradual consolidation of that market;

(2)   Whereas the free movement of goods and services affects not only the business sector but also private individuals; whereas it means that consumers should be able to have access to the goods and services of another Member State on the same terms as the population of that State;

(3)   Whereas, for consumers, cross-border distance selling could be one of the main tangible results of the completion of the internal market, as noted, inter alia, in the communication from the Commission to the Council entitled 'Towards a single market in distribution'; whereas it is essential to the smooth operation of the internal market for consumers to be able to

have dealings with a business outside their country, even if it has a subsidiary in the consumer's country of residence;

(4)   Whereas the introduction of new technologies is increasing the number of ways for consumers to obtain information about offers anywhere in the Community and to place orders; whereas some Member States have already taken different or diverging measures to protect consumers in respect of distance selling, which has had a detrimental effect on competition between businesses in the internal market; whereas it is therefore necessary to introduce at Community level a minimum set of common rules in this area;

(5)   Whereas paragraphs 18 and 19 of the Annex to the Council resolution of 14 April 1975 on a preliminary programme of the European Economic Community for a consumer protection and information policy[4] point to the need to protect the purchasers of goods or services from demands for payment for unsolicited goods and from high-pressure selling methods;

(6)   Whereas paragraph 33 of the communication from the Commission to the Council entitled 'A new impetus for consumer protection policy', which was approved by the Council resolution of 23 June 1986,[5] states that the Commission will submit proposals regarding the use of new information technologies enabling consumers to place orders with suppliers from their homes;

(7)   Whereas the Council resolution of 9 November 1989 on future priorities for relaunching consumer protection policy[6] calls upon the Commission to give priority to the areas referred to in the Annex to that resolution; whereas that Annex refers to new technologies involving teleshopping; whereas the Commission has responded to that resolution by adopting a three-year action plan for consumer protection policy in the European Economic Community (1990–1992); whereas that plan provides for the adoption of a Directive;

(8)   Whereas the languages used for distance contracts are a matter for the Member States;

(9)   Whereas contracts negotiated at a distance involve the use of one or more means of distance communication; whereas the various means of communication are used as part of an organised distance sales or service-provision scheme not involving the simultaneous presence of the supplier and the consumer; whereas the constant development of those means of communication does not allow an exhaustive list to be compiled but does require principles to be defined which are valid even for those which are not as yet in widespread use;

(10)   Whereas the same transaction comprising successive operations or a series of separate operations over a period of time may give rise to different legal descriptions depending on the law of the Member States; whereas the provisions of this Directive cannot be applied differently according to the law of the Member States, subject to their recourse to Article 14; whereas, to that end, there is therefore reason to consider that there must at least be compliance with the provisions of this Directive at the time of the first of a series of successive operations or the first of a series of separate operations over a period of time which may be considered as forming a whole, whether that operation or series of operations are the subject of a single contract or successive, separate contracts;

(11)   Whereas the use of means of distance communication must not lead to a reduction in the information provided to the consumer; whereas the information that is required to be sent to the consumer should therefore be determined, whatever the means of communication used; whereas the information supplied must also comply with the other relevant Community rules, in particular those in Council Directive 84/450/EEC of 10 September 1984 relating to the approximation of the laws, regulations and administrative provisions of the Member States concerning misleading advertising;[7] whereas, if exceptions are made to the obligation to provide information, it is up to the consumer, on a discretionary basis, to request certain basic information such as the identity of the supplier, the main characteristics of the goods or services and their price;

(12)   Whereas in the case of communication by telephone it is appropriate that the consumer receive enough information at the beginning of the conversation to decide whether or not to continue;

(13)   Whereas information disseminated by certain electronic technologies is often ephemeral in nature insofar as it is not received on a permanent medium; whereas the consumer must therefore receive written notice in good time of the information necessary for proper performance of the contract;

(14)   Whereas the consumer is not able actually to see the product or ascertain the nature of the service provided before concluding the contract; whereas provision should be made, unless otherwise specified in this Directive, for a right of withdrawal from the contract; whereas, if this right is to be more than formal, the costs, if any, borne by the consumer when exercising the right of withdrawal must be limited to the direct costs for returning the goods; whereas this right of withdrawal shall be without prejudice to the consumer's rights under national laws, with particular regard to the receipt of damaged products and services or of products and

services not corresponding to the description given in the offer of such products or services; whereas it is for the Member States to determine the other conditions and arrangements following exercise of the right of withdrawal;

(15)   Whereas it is also necessary to prescribe a time limit for performance of the contract if this is not specified at the time of ordering;

(16)   Whereas the promotional technique involving the dispatch of a product or the provision of a service to the consumer in return for payment without a prior request from, or the explicit agreement of, the consumer cannot be permitted, unless a substitute product or service is involved;

(17)   Whereas the principles set out in Articles 8 and 10 of the European Convention for the Protection of Human Rights and Fundamental Freedoms of 4 November 1950 apply; whereas the consumer's right to privacy, particularly as regards freedom from certain particularly intrusive means of communication, should be recognised; whereas specific limits on the use of such means should therefore be stipulated; whereas Member States should take appropriate measures to protect effectively those consumers, who do not wish to be contacted through certain means of communication, against such contacts, without prejudice to the particular safeguards available to the consumer under Community legislation concerning the protection of personal data and privacy;

(18)   Whereas it is important for the minimum binding rules contained in this Directive to be supplemented where appropriate by voluntary arrangements among the traders concerned, in line with Commission recommendation 92/295/EEC of 7 April 1992 on codes of practice for the protection of consumers in respect of contracts negotiated at a distance;[8]

(19)   Whereas in the interest of optimum consumer protection it is important for consumers to be satisfactorily informed of the provisions of this Directive and of codes of practice that may exist in this field;

(20)   Whereas non-compliance with this Directive may harm not only consumers but also competitors; whereas provisions may therefore be laid down enabling public bodies or their representatives, or consumer organisations which, under national legislation, have a legitimate interest in consumer protection, or professional organisations which have a legitimate interest in taking action, to monitor the application thereof;

(21)   Whereas it is important, with a view to consumer protection, to address the question of cross-border complaints as soon as this is feasible; whereas

the Commission published on 14 February 1996 a plan of action on consumer access to justice and the settlement of consumer disputes in the internal market; whereas that plan of action includes specific initiatives to promote out-of-court procedures; whereas objective criteria (Annex II) are suggested to ensure the reliability of those procedures and provision is made for the use of standardised claims forms (Annex III);

(22)   Whereas in the use of new technologies the consumer is not in control of the means of communication used; whereas it is therefore necessary to provide that the burden of proof may be on the supplier;

(23)   Whereas there is a risk that, in certain cases, the consumer may be deprived of protection under this Directive through the designation of the law of a non-member country as the law applicable to the contract; whereas provisions should therefore be included in this Directive to avert that risk;

(24)   Whereas a Member State may ban, in the general interest, the marketing on its territory of certain goods and services through distance contracts; whereas that ban must comply with Community rules; whereas there is already provision for such bans, notably with regard to medicinal products, under Council Directive 89/552/EEC of 3 October 1989 on the coordination of certain provisions laid down by law, regulation or administrative action in Member States concerning the pursuit of television broadcasting activities[9] and Council Directive 92/28/EEC of 31 March 1992 on the advertising of medicinal products for human use,[10]

NOTES

[1]   OJ C156, 23.6.92, p 14 and OJ C308, 15.11.93, p 18.
[2]   OJ C19, 25.1.93, p 111.
[3]   Opinion of the European Parliament of 26 May 1993 (OJ C176, 28.6.93, p 95), Council common position of 29 June 1995 (OJ C288, 30.10.95, p.1) and Decision of the European Parliament of 13 December 1995 (OJ C17, 22.1.96, p 51). Decision of the European Parliament of 16 January 1997 and Council Decision of 20 January 1997.
[4]   OJ C92, 25.4.75, p 1.
[5]   OJ C167, 5.7.86, p 1.
[6]   OJ C294, 22.11.89, p 1.
[7]   OJ L250, 19.9.84, p 17.
[8]   OJ L156, 10.6.92, p 21.
[9]   OJ L298, 17.10.89, p 23.
[10]   OJ L113, 30.4.92, p 13.

HAVE ADOPTED THIS DIRECTIVE—

Article 1 Object

The object of this Directive is to approximate the laws, regulations and administrative provisions of the Member States concerning distance contracts between consumers and suppliers.

Article 2 Definitions

For the purposes of this Directive-
  (1) 'distance contract' means any contract concerning goods or services concluded between a supplier and a consumer under an organised distance sales or service-provision scheme run by the supplier, who, for the purpose of the contract, makes exclusive use of one or more means of distance communication up to and including the moment at which the contract is concluded;
  (2) 'consumer' means any natural person who, in contracts covered by this Directive, is acting for purposes which are outside his trade, business or profession;
  (3) 'supplier' means any natural or legal person who, in contracts covered by this Directive, is acting in his commercial or professional capacity;
  (4) 'means of distance communication' means any means which, without the simultaneous physical presence of the supplier and the consumer, may be used for the conclusion of a contract between those parties. An indicative list of the means covered by this Directive is contained in Annex I;
  (5) 'operator of a means of communication' means any public or private natural or legal person whose trade, business or profession involves making one or more means of distance communication available to suppliers.

Article 3 Exemptions

1.   This Directive shall not apply to contracts—
  — relating to financial services, a non-exhaustive list of which is given in Annex II,
  — concluded by means of automatic vending machines or automated commercial premises
  — concluded with telecommunications operators through the use of public payphones,
  — concluded for the construction and sale of immovable property or relating to other immovable property rights, except for rental,
  — concluded at an auction.

2.   Articles 4, 5, 6 and 7(1) shall not apply—
   — to contracts for the supply of foodstuffs, beverages or other goods intended for everyday consumption supplied to the home of the consumer, to his residence or to his workplace by regular roundsmen,
   — to contracts for the provision of accommodation, transport, catering or leisure services, where the supplier undertakes, when the contract is concluded, to provide these services on a specific date or within a specific period; exceptionally, in the case of outdoor leisure events, the supplier can reserve the right not to apply Article 7(2) in specific circumstances.

Article 4   Prior information

1.   In good time prior to the conclusion of any distance contract, the consumer shall be provided with the following information—
   (a) the identity of the supplier and, in the case of contracts requiring payment in advance, his address;
   (b) the main characteristics of the goods or services;
   (c) the price of the goods or services including all taxes;
   (d) delivery costs, where appropriate;
   (e) the arrangements for payment, delivery or performance;
   (f) the existence of a right of withdrawal, except in the cases referred to in Article 6(3);
   (g) the cost of using the means of distance communication, where it is calculated other than at the basic rate;
   (h) the period for which the offer or the price remains valid;
   (i) where appropriate, the minimum duration of the contract in the case of contracts for the supply of products or services to be performed permanently or recurrently.

2.   The information referred to in paragraph 1, the commercial purpose of which must be made clear, shall be provided in a clear and comprehensible manner in any way appropriate to the means of distance communication used, with due regard, in particular, to the principles of good faith in commercial transactions, and the principles governing the protection of those who are unable, pursuant to the legislation of the Member States, to give their consent, such as minors.

3.   Moreover, in the case of telephone communications, the identity of the supplier and the commercial purpose of the call shall be made explicitly clear at the beginning of any conversation with the consumer.

Article 5   Written confirmation of information

1.   The consumer must receive written confirmation or confirmation in another durable medium available and accessible to him of the information referred to in Article 4(1)(a) to (f), in good time during the performance of the contract, and at the latest at the time of delivery where goods not for delivery to third parties are concerned, unless the information has already been given to the consumer prior to conclusion of the contract in writing or on another durable medium available and accessible to him.

In any event the following must be provided—
— written information on the conditions and procedures for exercising the right of withdrawal, within the meaning of Article 6, including the cases referred to in the first indent of Article 6(3),
— the geographical address of the place of business of the supplier to which the consumer may address any complaints,
— information on after-sales services and guarantees which exist,
— the conclusion for cancelling the contract, where it is of unspecified duration or a duration exceeding one year.

2.   Paragraph 1 shall not apply to services which are performed through the use of a means of distance communication, where they are supplied on only one occasion and are invoiced by the operator of the means of distance communication. Nevertheless, the consumer must in all cases be able to obtain the geographical address of the place of business of the supplier to which he may address any complaints.

Article 6   Right of withdrawal

1.   For any distance contract the consumer shall have a period of at least seven working days in which to withdraw from the contract without penalty and without giving any reason. The only charge that may be made to the consumer because of the exercise of his right of withdrawal is the direct cost of returning the goods.

The period for exercise of this right shall begin—
— in the case of goods, from the day of receipt by the consumer where the obligations laid down in Article 5 have been fulfilled,
— in the case of services, from the day of conclusion of the contract or from the day on which the obligations laid down in Article 5 were fulfilled if they are fulfilled after conclusion of the contract, provided that this period does not exceed the three-month period referred to in the following subparagraph.

If the supplier has failed to fulfil the obligations laid down in Article 5, the period shall be three months. The period shall begin—
— in the case of goods, from the day of receipt by the consumer,
— in the case of services, from the day of conclusion of the contract.

If the information referred to in Article 5 is supplied within this three-month period, the seven working day period referred to in the first subparagraph shall begin as from that moment.

2.   Where the right of withdrawal has been exercised by the consumer pursuant to this Article, the supplier shall be obliged to reimburse the sums paid by the consumer free of charge. The only charge that may be made to the consumer because of the exercise of his right of withdrawal is the direct cost of returning the goods. Such reimbursement must be carried out as soon as possible and in any case within 30 days.

3.   Unless the parties have agreed otherwise, the consumer may not exercise the right of withdrawal provided for in paragraph 1 in respect of contracts—
— for the provision of services if performance has begun, with the consumer's agreement, before the end of the seven working day period referred to in paragraph 1,
— for the supply of goods or services the price of which is dependent on fluctuations in the financial market which cannot be controlled by the supplier,
— for the supply of goods made to the consumer's specifications or clearly personalised or which, by reason of their nature, cannot be returned or are liable to deteriorate or expire rapidly,
— for the supply of audio or video recordings or computer software which were unsealed by the consumer,
— for the supply of newspapers, periodicals and magazines,
— for gaming and lottery services.

4.   The Member States shall make provision in their legislation to ensure that—
— if the price of goods or services is fully or partly covered by credit granted by the supplier, or
— if that price is fully or partly covered by credit granted to the consumer by a third party on the basis of an agreement between the third party and the supplier,

the credit agreement shall be cancelled, without any penalty, if the consumer exercises his right to withdraw from the contract in accordance with paragraph 1.

Member States shall determine the detailed rules for cancellation of the credit agreement.

Article 7   Performance

1.   Unless the parties have agreed otherwise, the supplier must execute the order within a maximum of 30 days from the day following that on which the consumer forwarded his order to the supplier.

2.   Where a supplier fails to perform his side of the contract on the grounds that the goods or services ordered are unavailable, the consumer must be informed of this situation and must be able to obtain a refund of any sums he has paid as soon as possible and in any case within 30 days.

3.   Nevertheless, Member States may lay down that the supplier may provide the consumer with goods or services of equivalent quality and price provided that this possibility was provided for prior to the conclusion of the contract or in the contract. The consumer shall be informed of this possibility in a clear and comprehensible manner. The cost of returning the goods following exercise of the right of withdrawal shall, in this case, be borne by the supplier, and the consumer must be informed of this. In such cases the supply of goods or services may not be deemed to constitute inertia selling within the meaning of Article 9.

Article 8   Payment by card

Member States shall ensure that appropriate measures exist to allow a consumer—
— to request cancellation of a payment where fraudulent use has been made of his payment card in connection with distance contracts covered by this Directive,
— in the event of fraudulent use, to be recredited with the sums paid or have them returned.

Article 9   Inertia selling

Member States shall take the measures necessary to—
— prohibit the supply of goods or services to a consumer without their being ordered by the consumer beforehand, where such supply involves a demand for payment,

— exempt the consumer from the provision of any consideration in cases of unsolicited supply, the absence of a response not constituting consent.

Article 10   Restrictions on the use of certain means of distance communication

1.   Use by a supplier of the following means requires the prior consent of the consumer—
   — automated calling system without human intervention (automatic calling machine),
   — facsimile machine (fax).

2.   Member States shall ensure that means of distance communication, other than those referred to in paragraph 1, which allow individual communications may be used only where there is no clear objection from the consumer.

Article 11   Judicial or administrative redress

1.   Member States shall ensure that adequate and effective means exist to ensure compliance with this Directive in the interests of consumers.

2.   The means referred to in paragraph 1 shall include provisions whereby one or more of the following bodies, as determined by national law, may take action under national law before the courts or before the competent administrative bodies to ensure that the national provisions for the implementation of this Directive are applied—
   (a) public bodies or their representatives;
   (b) consumer organisations having a legitimate interest in protecting consumers;
   (c) professional organisations having a legitimate interest in acting.

3. (a) Member States may stipulate that the burden of proof concerning the existence of prior information, written confirmation, compliance with time-limits or consumer consent can be placed on the supplier.
   (b) Member States shall take the measures needed to ensure that suppliers and operators of means of communication, where they are able to do so, cease practices which do not comply with measures adopted pursuant to this Directive.

4.   Member States may provide for voluntary supervision by self-regulatory bodies of compliance with the provisions of this Directive and recourse to such bodies for the settlement of disputes to be added to the means which Member States must provided to ensure compliance with the provisions of this Directive.

Article 12   Binding nature

1.   The consumer may not waive the rights conferred on him by the transposition of this Directive into national law.

2.   Member States shall take the measures needed to ensure that the consumer does not lose the protection granted by this Directive by virtue of the choice of the law of a non-member country as the law applicable to the contract if the latter has close connection with the territory of one or more Member States.

Article 13   Community rules

1.   The provisions of this Directive shall apply insofar as there are no particular provisions in rules of Community law governing certain types of distance contracts in their entirety.

2.   Where specific Community rules contain provisions governing only certain aspects of the supply of goods or provision of services, those provisions, rather than the provisions of this Directive, shall apply to these specific aspects of the distance contracts.

Article 14   Minimal clause

Member States may introduce or maintain, in the area covered by this Directive, more stringent provisions compatible with the Treaty, to ensure a higher level of consumer protection. Such provisions shall, where appropriate, include a ban, in the general interest, on the marketing of certain goods or services, particularly medicinal products, within their territory by means of distance contracts, with due regard for the Treaty.

Article 15   Implementation

1.   Member States shall bring into force the laws, regulations and administrative provisions necessary to comply with this Directive no later

than three years after it enters into force. They shall forthwith inform the Commission thereof.

2.   When Member States adopt the measures referred to in paragraph 1, these shall contain a reference to this Directive or shall be accompanied by such reference on the occasion of their official publication. The procedure for such reference shall be laid down by Member States.

3.   Member States shall communicate to the Commission the text of the provisions of national law which they adopt in the field governed by this Directive.

4.   No later than four years after the entry into force of this Directive the Commission shall submit a report to the European Parliament and the Council on the implementation of this Directive, accompanied if appropriate by a proposal for the revision thereof.

Article 16   Consumer information

Member States shall take appropriate measures to inform the consumer of the national law transposing this Directive and shall encourage, where appropriate, professional organisations to inform consumers of their codes of practice.

Article 17   Complaints systems

The Commission shall study the feasibility of establishing effective means to deal with consumers' complaints in respect of distance selling. Within two years after the entry into force of this Directive the Commission shall submit a report to the European Parliament and the Council on the results of the studies, accompanied if appropriate by proposals.

Article 18

This Directive shall enter into force on the day of its publication in the *Official Journal of the European Communities*.

Article 19

This Directive is addressed to the Member States.

Done at Brussels, 20 May 1997.

ANNEX I
MEANS OF COMMUNICATION COVERED BY ARTICLE 2(4)

— Unaddressed printed matter
— Addressed printed matter
— Standard letter
— Press advertising with order form
— Catalogue
— Telephone with human intervention
— Telephone without human intervention (automatic calling machine, audiotext)
— Radio
— Videophone (telephone with screen)
— Videotex (microcomputer and television screen) with keyboard or touch screen
— Electronic mail
— Facsimile machine (fax)
— Television (teleshopping).

ANNEX II
FINANCIAL SERVICES WITHIN THE MEANING OF ARTICLE 3(1)

— Investment services
— Insurance and reinsurance operations
— Banking services
— Operations relating to dealings in futures or options.

Such services include in particular—
— investment services referred to in the Annex to Directive 93/22/EEC;[1] services of collective investment undertakings,
— services covered by the activities subject to mutual recognition referred to in the Annex to Directive 89/646/EEC;[2]
— operations covered by the insurance and reinsurance activities referred to in—
— Article 1 of Directive 73/239/EEC,[3]

— the Annex to Directive 79/267/EEC,[4]
— Directive 64/225/EEC,[5]
— Directives 92/49/EEC[6] and 92/96/EEC.[7]

NOTES

[1] OJ L141, 11.6.93, p 27.

[2] OJ L386, 30.12.89, p 1. Directive as amended by Directive 92/30/EEC (OJ L110, 28.4.92, p 52).

[3] OJ L228, 16.8.73, p 3. Directive as last amended by Directive 92/49/EEC (OJ L228, 11.8. 92, p 1).

[4] OJ L63, 13.3.79, p 1. Directive as last amended by Directive 90/619/EEC (OJ L330, 29.11.90, p 50).

[5] OJ L56, 4.4.64, p 878. Directive as amended by the 1973 Act of Accession.

[6] OJ L228, 11 8.92, p 1.

[7] OJ L360, 9.12.92, p 1.

Statement by the Council and the Parliament re Article 6(1)

The Council and the Parliament note that the Commission will examine the possibility and desirability of harmonising the method of calculating the cooling-off period under existing consumer-protection legislation, notably Directive 85/577/EEC of 20 December 1985 on the protection of consumers in respect of contracts negotiated away from commercial establishments ('door-to-door sales').[1]

NOTES

[1] OJ L372, 31.12.85, p 31.

Statement by the Commission re Article 3(1), first indent

The Commission recognises the importance of protecting consumers in respect of distance contracts concerning financial services and has published a Green Paper entitled 'Financial services: meeting consumers' expectations'. In the light of reactions to the Green Paper the Commission will examine ways of incorporating consumer protection into the policy on financial services and the possible legislative implications and, if need be, will submit appropriate proposals.

Commission Recommendation of 30 July 1997 concerning transactions by electronic payment instruments and in particular the relationship between issuer and holder

(Text with EEA relevance)

(97/489/EC)

NOTES

Date of publication in OJ: OJ L208, 2.8.97, p 52.

THE COMMISSION OF THE EUROPEAN COMMUNITIES,

Having regard to the Treaty establishing the European Community and in particular Article 155, second indent, thereof,

(1)   Whereas one of the main objectives of the Community is to ensure the full functioning of the internal market of which payment systems are essential parts; whereas transactions made by electronic payment instruments account for an increasing proportion of the volume and the value of domestic and cross-border payments; whereas, given the current context of rapid innovation and technological progress, this trend is expected to accelerate notably as a consequence of the wide array of innovative businesses, markets and trading communities engendered by electronic commerce;

(2)   Whereas it is important for individuals and businesses to be able to use electronic payment instruments throughout the Community; whereas this recommendation seeks to follow up progress made towards the completion of the internal market, notably in the light of the liberalization of capital movements, and will also contribute to the implementation of economic and monetary union;

(3)   Whereas this recommendation covers transactions effected by electronic payment instruments; whereas, for the purposes of this recommendation,

these include instruments allowing for (remote) access to a customer's account, notably payment cards and phone- and home-banking applications; whereas transactions by means of a payment card shall cover electronic and non-electronic payment by means of a payment card, including processes for which a signature is required and a voucher is produced; whereas, for the purposes of this recommendation, means of payment instruments also include reloadable electronic money instruments in the form of stored-value cards and electronic tokens stored on network computer memory; whereas reloadable electronic money instruments, because of their features, in particular the possible link to the holder's account, are those for which the need for customer protection is strongest; whereas, as far as electronic money instruments are concerned, coverage under this recommendation is therefore limited to instruments of the reloadable type;

(4) Whereas this recommendation is intended to contribute to the advent of the information society and, in particular, electronic commerce by promoting customer confidence in and retailer acceptance of these instruments; whereas, to this end, the Commission will also consider the possibility of modernizing and updating its recommendation 87/598/EEC,[1] with a view to establishing a clear framework for the relationship between acquirers and acceptors in respect of electronic payment instruments; whereas, in line with those objectives, this recommendation sets out minimum information requirements which should be contained in the terms and conditions applied to transactions made by electronic payment instruments, as well as the minimum obligations and liabilities of the parties concerned; whereas such terms and conditions should be set out in writing, including where appropriate by electronic means, and maintain a fair balance between the interests of the parties concerned; whereas, in compliance with Council Directive 93/13/EEC of 5 April 1993 on unfair terms in consumer contracts,[2] such terms and conditions should in particular be in an understandable and comprehensible form;

(5) Whereas, with a view to ensuring transparency, this recommendation sets out the minimum requirements needed to ensure an adequate level of customer information upon conclusion of a contract as well as subsequent to transactions effected by means of a payment instrument, including information on charges, exchange rates and interest rates; whereas, for the purpose of informing the holder of the manner of calculation of the interest rate, reference is to be made to Council Directive 87/102/EEC of 22 December 1986 for the approximation of the laws, regulations and administrative provisions of the Member States concerning consumer credit,[3] as amended by Directive 90/88/EEC;[4]

(6)   Whereas this recommendation sets out minimum requirements concerning the obligations and liabilities of the parties concerned; whereas information to a holder should include a clear statement of the extent of the customer's obligation as holder of an electronic payment instrument enabling him/her to make payments in favour in third persons, as well as to perform certain financial transactions for himself/herself;

(7)   Whereas, to improve customer's access to redress, this recommendation calls on Member States to ensure that there are adequate and effective means for the settlement of disputes between a holder and an issuer; whereas the Commission published on 14 February 1996 a plan of action on consumer access to justice and the settlement of consumer disputes in the internal market; whereas that plan of action includes specific initiatives to promote out-of-court procedures; whereas objective criteria (Annex II) are suggested to ensure the reliability of those procedures and provision is made for the use of standardized claims forms (Annex III);

(8)   Whereas this recommendation seeks to ensure a high level of consumer protection in the field of electronic payment instruments;

(9)   Whereas it is essential that transactions effected by means of electronic payment instruments should be the subject of records in order that transactions can be traced and errors can be rectified; whereas the burden of proof to show that a transaction was accurately recorded and entered into the accounts and was not affected by technical breakdown or other deficiency should lie upon the issuer;

(10)   Whereas, without prejudice to any rights of a holder under national law, payment instructions given by a holder in respect of transactions effected by means of an electronic payment instrument should be irrevocable, except if the amount was not determined when the order was given;

(11)   Whereas rules need to be specified concerning the issuer's liability for non-execution or for defective execution of a holder's payment instructions and for transactions which have not been authorized by him/her, subject always to the holder's own obligations in the case of lost or stolen electronic payment instruments;

(12)   Whereas the Commission will monitor the implementation of this Recommendation and, if it finds the implementation unsatisfactory, it intends to propose the appropriate binding legislation covering the issues dealt with in this recommendation,

NOTES

[1] OJ L365, 24.12.87, p 72.
[2] OJ L95, 21.4.93, p 29.
[3] OJ L42, 12.2.87, p 48.
[4] OJ L61, 10.3.90, p 14.

HEREBY RECOMMENDS—

SECTION I
SCOPE AND DEFINITIONS

Article 1   Scope

1.   This Recommendation applies to the following transactions—
   (a) transfers of funds, other than those ordered and executed by financial institutions, effected by means of an electronic payment instrument;
   (b) cash withdrawals by means of an electronic payment instrument and the loading (and unloading) of an electronic money instrument, at devices such as cash dispensing machines and automated teller machines and at the premises of the issuer or an institution who is under contract to accept the payment instrument.

2.   By way of derogation from paragraph 1, Article 4(1), the second and third indents of Article 5(b), Article 6, Article 7(2)(c), (d) and the first indent of (e), Article 8(1), (2) and (3) and Article 9(2) do not apply to transactions effected by means of an electronic money instrument. However, where the electronic money instrument is used to load (and unload) value through remote access to the holder's account, this Recommendation is applicable in its entirety.

3.   This recommendation does not apply to—
   (a) payments by cheques;
   (b) the guarantee function of certain cards in relation to payments by cheques.

Article 2   Definitions

For the purpose of this recommendation, the following definitions apply—
   (a) 'electronic payment instrument' means an instrument enabling its holder to effect transactions of the kind specified in Article 1(1).

This covers both remote access payment instruments and electronic money instruments;

(b) 'remote access payment instrument' means an instrument enabling a holder to access funds held on his/her account at an institution, whereby payment is allowed to be made to a payee and usually requiring a personal identification code and/or any other similar proof of identity. This includes in particular payment cards (whether credit, debit, deferred debit or charge cards) and phone- and home-banking applications;

(c) 'electronic money instrument' means a reloadable payment instrument other than a remote access payment instrument, whether a stored-value card or a computer memory, on which value units are stored electronically, enabling its holder to effect transactions of the kind specified in Article 1(1);

(d) 'financial institution' means an institution as defined in Article 4(1) of Council Regulation (EC) No 3604/93;[1]

(e) 'issuer' means a person who, in the course of his business, makes available to another person a payment instrument pursuant to a contract concluded with him/her;

(f) 'holder' means a person who, pursuant to a contract concluded between him/her and an issuer, holds a payment instrument.

NOTES

[1] OJ L332, 31.12.93, p 4.

SECTION II
TRANSPARENCY OF CONDITIONS FOR TRANSACTIONS

Article 3 Minimum information contained in the terms and conditions governing the issuing and use of an electronic payment instrument

1.   Upon signature of the contract or in any event in good time prior to delivering an electronic payment instrument, the issuer communicates to the holder the contractual terms and conditions (hereinafter referred to as 'the terms') governing the issue and use of that electronic payment instrument. The terms indicate the law applicable to the contract.

2.   The terms are set out in writing, including where appropriate by electronic means, in easily understandable words and in a readily comprehensive form, and are available at least in the official language or

languages of the Member State in which the electronic payment instrument is offered.

3. The terms include at least—
   (a) a description of the electronic payment instrument, including where appropriate the technical requirements with respect to the holder's communication equipment authorized for use, and the way in which it can be used, including the financial limits applied, if any;
   (b) a description of the holder's and issuer's respective obligations and liabilities; they include a description of the reasonable steps that the holder must take to keep safe the electronic payment instrument and the means (such as a personal identification number or other code) which enable it to be used;
   (c) where applicable, the normal period within which the holder's account will be debited or credited, including the value date, or, where the holder has no account with the issuer, the normal period within which he/she will be invoiced;
   (d) the types of any charges payable by the holder. In particular, this includes where applicable details of the following charges—
       — the amount of any initial and annual fees,
       — any commission fees and charges payable by the holder to the issuer for particular types of transactions,
       — any interest rate, including the manner of its calculation, which may be applied;
   (e) the period of time during which a given transaction can be contested by the holder and an indication of the redress and complaints procedures available to the holder and the method of gaining access to them.

4. If the electronic payment instrument is usable for transactions abroad (outside the country of issuing/affiliation), the following information is also communicated to the holder—
   (a) an indication of the amount of any fees and charges levied for foreign currency transactions, including where appropriate the rates;
   (b) the reference exchange rate used for converting foreign currency transactions, including the relevant date for determining such a rate.

Article 4   Information subsequent to a transaction

1. The issuer supplies the holder with information relating to the transactions effected by means of an electronic payment instrument. This

information, set out in writing, including where appropriate by electronic means, and in a readily comprehensible form, includes at least—
- (a) a reference enabling the holder to identify the transaction, including, where appropriate, the information relating to the acceptor at/with which the transaction took place;
- (b) the amount of the transaction debited to the holder in billing currency and, where applicable, the amount in foreign currency;
- (c) the amount of any fees and charges applied for particular types of transactions.

The issuer also provides the holder with the exchange rate used for converting foreign currency transactions.

2. The issuer of an electronic money instrument provides the holder with the possibility of verifying the last five transactions executed with the instrument and the outstanding value stored thereon.

SECTION III
OBLIGATIONS AND LIABILITIES OF THE PARTIES TO A CONTRACT

Article 5   Obligations of the holder

The holder—
- (a) uses the electronic payment instrument in accordance with the terms governing the issuing and use of a payment instrument; in particular, the holder takes all reasonable steps to keep safe the electronic payment instrument and the means (such as a personal identification number or other code) which enable it to be used;
- (b) notifies the issuer (or the entity specified by the latter) without delay after becoming aware of—
  - — the loss or theft of the electronic payment instrument or of the means which enable it to be used,
  - — the recording on his/her account of any unauthorized transaction,
  - — any error or other irregularity in the maintaining of that account by the issuer;
- (c) does not record his personal identification number or other code in any easily recognizable form, in particular on the electronic payment instrument or on any item which he/she keeps or carries with the electronic payment instrument;
- (d) does not countermand an order which he/she has given by means of his/her electronic payment instrument, except if the amount was not determined when the order was given.

Article 6 Liabilities of the holder

1.   Up to the time of notification, the holder bears the loss sustained in consequence of the loss or theft of the electronic payment instrument up to a limit, which may not exceed ECU 150, except where he/she acted with extreme negligence, in contravention of relevant provisions under Article 5(a), (b) or (c), or fraudulently, in which case such a limit does not apply.

2.   As soon as the holder has notified the issuer (or the entity specified by the latter) as required by Article 5(b), except where he/she acted fraudulently, he/she is not thereafter liable for the loss arising in consequence of the loss or theft of his/her electronic payment instrument.

3.   By derogation from paragraphs 1 and 2, the holder is not liable if the payment instrument has been used, without physical presentation or electronic identification (of the instrument itself). The use of a confidential code or any other similar proof of identity is not, by itself, sufficient to entail the holder's liability.

Article 7 Obligations of the issuer

1.   The issuer may alter the terms, provided that sufficient notice of the change is given individually to the holder to enable him/her to withdraw if he/she so chooses. A period of not less than one month is specified after which time the holder is deemed to have accepted the terms if he/she has not withdrawn.

However, any significant change to the actual interest rate is not subject to the provisions of the first subparagraph and comes into effect upon the date specified in the publication of such a change. In this event, and without prejudice to the right of the holder to withdraw from the contract, the issuer informs the holder individually thereof as soon as possible.

2.   The issuer—
   (a) does not disclose the holder's personal identification number or other code, except to the holder;
   (b) does not dispatch an unsolicited electronic payment instrument, except where it is a replacement for an electronic payment instrument already held by the holder;
   (c) keeps for a sufficient period of time, internal records to enable the transactions referred to in Article 1(1) to be traced and errors to be rectified;

(d) ensures that appropriate means are available to enable the holder to make the notification required under Article 5(b). Where notification is made by telephone, the issuer (or the entity specified by the latter) provides the holder with the means of proof that he/she has made such a notification;

(e) proves, in any dispute with the holder concerning a transaction referred to in Article 1(1), and without prejudice to any proof to the contrary that may be produced by the holder, that the transaction—

— was accurately recorded and entered into accounts,

— was not affected by technical breakdown or other deficiency.

Article 8   Liabilities of the issuer

1.   The issuer is liable, subject to Article 5, Article 6 and Article 7(2)(a) and (e)—

(a) for the non-execution or defective execution of the holder's transactions referred to in Article 1(1), even if a transaction is initiated at devices/terminals or through equipment which are not under the issuer's direct or exclusive control, provided that the transaction is not initiated at devices/terminals or through equipment unauthorized for use by the issuer;

(b) for transactions not authorized by the holder, as well as for any error or irregularity attributable to the issuer in the maintaining of the holder's account.

2.   Without prejudice to paragraph 3, the amount of the liability indicated in paragraph 1 consists of—

(a) the amount of the unexecuted or defectively executed transaction and, if any, interest thereon;

(b) the sum required to restore the holder to the position he/she was in before the unauthorized transaction took place.

3.   Any further financial consequences, and, in particular, those concerning the extent of the damage for which compensation is to be paid, are borne by the issuer in accordance with the law applicable to the contract concluded between the issuer and the holder.

4.   The issuer is liable to the holder of an electronic money instrument for the lost amount of value stored on the instrument and for the defective execution of the holder's transactions, where the loss or

defective execution is attributable to a malfunction of the instrument, of the device/terminal or any other equipment authorized for use, provided that the malfunction was not caused by the holder knowingly or in breach of Article 3(3)(a).

SECTION IV
NOTIFICATION, SETTLEMENT OF DISPUTES AND FINAL PROVISION

Article 9   Notification

1.   The issuer (or the entity specified by him) provides means whereby a holder may at any time of day or night notify the loss or theft of his/her electronic payment instrument.

2.   The issuer (or the entity specified by him), upon receipt of notification, is under the obligation, even if the holder acted with extreme negligence or fraudulently, to take all reasonable action open to him to stop any further use of the electronic payment instrument.

Article 10   Settlement of disputes

Member States are invited to ensure that there are adequate and effective means for the settlement of disputes between a holder and an issuer.

Article 11   Final provision

Member States are invited to take the measures necessary to ensure that the issuers of electronic payment instruments conduct their activities in accordance with Articles 1 to 9 by not later than 31 December 1998.

Done at Brussels, 30 July 1997.

Directive of the European Parliament and of the Council of 16 February 1998 on consumer protection in the indication of the prices of products offered to consumers

(98/6/EC)

NOTES

Date of publication in OJ: OJ L80, 18.3.98, p 27.

THE EUROPEAN PARLIAMENT AND THE COUNCIL OF THE EUROPEAN UNION,

Having regard to the Treaty establishing the European Community, and in particular Article 129a(2) thereof,

Having regard to the proposal from the Commission,[1]

Having regard to the opinion of the Economic and Social Committee,[2]

Acting in accordance with the procedure laid down in Article 189b of the Treaty,[3] in the light of the joint text approved by the Conciliation Committee on 9 December 1997,

(1) Whereas transparent operation of the market and correct information is of benefit to consumer protection and healthy competition between enterprises and products;

(2) Whereas consumers must be guaranteed a high level of protection; whereas the Community should contribute thereto by specific action which supports and supplements the policy pursued by the Member States regarding precise, transparent and unambiguous information for consumers on the prices of products offered to them;

(3) Whereas the Council Resolution of 14 April 1975 on a preliminary programme of the European Economic Community for a consumer protection and information policy[4] and the Council Resolution of 19 May 1981 on a

second programme of the European Economic Community for a consumer protection and information policy[5] provide for the establishment of common principles for indicating prices;

(4)   Whereas these principles have been established by Directive 79/581/EEC concerning the indication of prices of certain foodstuffs[6] and Directive 88/314/EEC concerning the indication of prices of non-food products;[7]

(5)   Whereas the link between indication of the unit price of products and their pre-packaging in pre-established quantities or capacities corresponding to the values of the ranges adopted at Community level has proved overly complex to apply; whereas it is thus necessary to abandon this link in favour of a new simplified mechanism and in the interest of the consumer, without prejudice to the rules governing packaging standardisation;

(6)   Whereas the obligation to indicate the selling price and the unit price contributes substantially to improving consumer information, as this is the easiest way to enable consumers to evaluate and compare the price of products in an optimum manner and hence to make informed choices on the basis of simple comparisons;

(7)   Whereas, therefore, there should be a general obligation to indicate both the selling price and the unit price for all products except for products sold in bulk, where the selling price cannot be determined until the consumer indicates how much of the product is required;

(8)   Whereas it is necessary to take into account the fact that certain products are customarily sold in quantities different from one kilogramme, one litre, one metre, one square metre or one cubic metre; whereas it is thus appropriate to allow Member States to authorise that the unit price refer to a different single unit of quantity, taking into account the nature of the product and the quantities in which it is customarily sold in the Member State concerned;

(9)   Whereas the obligation to indicate the unit price may entail an excessive burden for certain small retail businesses under certain circumstances; whereas Member States should therefore be allowed to refrain from applying this obligation during an appropriate transitional period;

(10)   Whereas Member States should also remain free to waive the obligation to indicate the unit price in the case of products for which such price indication would not be useful or would be liable to cause confusion for instance when indication of the quantity is not relevant

for price comparison purposes, or when different products are marketed in the same packaging;

(11)   Whereas in the case of non-food products, Member States, with a view to facilitating application of the mechanism implemented, are free to draw up a list of products or categories of products for which the obligation to indicate the unit price remains applicable;

(12)   Whereas Community-level rules can ensure homogenous and transparent information that will benefit all consumers in the context of the internal market; whereas the new, simplified approach is both necessary and sufficient to achieve this objective;

(13)   Whereas Member States must make sure that the system is effective; whereas the transparency of the system should also be maintained when the euro is introduced; whereas, to that end, the maximum number of prices to be indicated should be limited;

(14)   Whereas particular attention should be paid to small retail businesses; whereas, to this end, the Commission should, in its report on the application of this Directive to be presented no later than three years after the date referred to in Article 11(1), take particular account of the experience gleaned in the application of the Directive by small retail businesses, inter alia, regarding technological developments and the introduction of the single currency; whereas this report, having regard to the transitional period referred to in Article 6, should be accompanied by a proposal,

NOTES

[1]   OJ C260, 5.10.95, p 5 and OJ C249, 27.8.96, p 2.
[2]   OJ C82, 19.3.96, p 32.
[3]   Opinion of the European Parliament of 18 April 1996 (OJ C141, 13.5.96, p 191). Council Common Position of 27 September 1996 (OJ C333, 7.11.96, p 7) and Decision of the European Parliament of 18 February 1997 (OJ C85, 17.3.97, p 26). Decision of the European Parliament of 16 December 1997 and Decision of the Council of 18 December 1997.
[4]   OJ C92, 25.4.75, p 1.
[5]   OJ C133, 3.6.81, p 1.
[6]   OJ L158, 26.6.79, p 19. Directive as last amended by Directive 95/58/EC (OJ L299, 12.12.95, p 11).
[7]   OJ L142, 9.6.88, p 19. Directive as last amended by Directive 95/58/EC (OJ L299, 12.12.95, p 11).

HAVE ADOPTED THIS DIRECTIVE—

Article 1

The purpose of this Directive is to stipulate indication of the selling price and the price per unit of measurement of products offered by traders to consumers in order to improve consumer information and to facilitate comparison of prices.

Article 2

For the purposes of this Directive—
  (a) selling price shall mean the final price for a unit of the product, or a given quantity of the product, including VAT and all other taxes;
  (b) unit price shall mean the final price, including VAT and all other taxes, for one kilogramme, one litre, one metre, one square metre or one cubic metre of the product or a different single unit of quantity which is widely and customarily used in the Member State concerned in the marketing of specific products;
  (c) products sold in bulk shall mean products which are not pre-packaged and are measured in the presence of the consumer;
  (d) trader shall mean any natural or legal person who sells or offers for sale products which fall within his commercial or professional activity;
  (e) consumer shall mean any natural person who buys a product for purposes that do not fall within the sphere of his commercial or professional activity.

Article 3

1.   The selling price and the unit price shall be indicated for all products referred to in Article 1, the indication of the unit price being subject to the provisions of Article 5. The unit price need not be indicated if it is identical to the sales price.

2.   Member States may decide not to apply paragraph 1 to—
  — products supplied in the course of the provision of a service,
  — sales by auction and sales of works of art and antiques.

3.   For products sold in bulk, only the unit price must be indicated.

4.   Any advertisement which mentions the selling price of products referred to in Article 1 shall also indicate the unit price subject to Article 5.

Article 4

1.   The selling price and the unit price must be unambiguous, easily identifiable and clearly legible. Member States may provide that the maximum number of prices to be indicated be limited.

2.   The unit price shall refer to a quantity declared in accordance with national and Community provisions.

Where national or Community provisions require the indication of the net weight and the net drained weight for certain pre-packed products, it shall be sufficient to indicate the unit price of the net drained weight.

Article 5

1.   Member States may waive the obligation to indicate the unit price of products for which such indication would not be useful because of the products' nature or purpose or would be liable to create confusion.

2.   With a view to implementing paragraph 1, Member States may, in the case of non-food products, establish a list of the products or product categories to which the obligation to indicate the unit price shall remain applicable.

Article 6

If the obligation to indicate the unit price were to constitute an excessive burden for certain small retail businesses because of the number of products on sale, the sales area, the nature of the place of sale, specific conditions of sale where the product is not directly accessible for the consumer or certain forms of business, such as certain types of itinerant trade, Member States may, for a transitional period following the date referred to in Article 11(1), provide that the obligation to indicate the unit price of products other than those sold in bulk, which are sold in the said businesses, shall not apply, subject to Article 12.

Article 7

Member States shall provide appropriate measures to inform all persons concerned of the national law transposing this Directive.

Article 8

Member States shall lay down penalties for infringements of national provisions adopted in application of this Directive, and shall take all necessary measures to ensure that these are enforced. These penalties must be effective, proportionate and dissuasive.

Article 9

1.   The transition period of nine years referred to in Article 1 of Directive 95/58/EC of the European Parliament and of the Council of 29 November 1995 amending Directive 79/581/EEC on consumer protection in the indication of the prices of foodstuffs and Directive 88/314/EEC on consumer protection in the indication of the prices of non-food products[1] shall be extended until the date referred to in Article 11(1) of this Directive.

2.   Directives 79/581/EEC and 88/314/EEC shall be repealed with effect from the date referred to in Article 11(1) of this Directive.

NOTES

   [1]   OJ L299, 12.12.95, p 11.

Article 10

This Directive shall not prevent Member States from adopting or maintaining provisions which are more favourable as regards consumer information and comparison of prices, without prejudice to their obligations under the Treaty.

Article 11

1.   Member States shall bring into force the laws, regulations and administrative provisions necessary to comply with this Directive not later than 18 March 2000. They shall forthwith inform the Commission thereof. The provisions adopted shall be applicable as of that date.

When Member States adopt these measures, they shall contain a reference to this Directive or shall be accompanied by such reference at the time of their official publication. The methods of making such reference shall be laid down by Member States.

2.   Member States shall communicate to the Commission the text of the provisions of national law which they adopt in the field governed by this Directive.

3.   Member States shall communicate the provisions governing the penalties provided for in Article 8, and any later amendments thereto.

Article 12

The Commission shall, not later than three years after the date referred to in Article 11(1), submit to the European Parliament and the Council a comprehensive report on the application of this Directive, in particular on the application of Article 6, accompanied by a proposal.

The European Parliament and the Council shall, on this basis, re-examine the provisions of Article 6 and shall act, in accordance with the Treaty, within three years of the presentation by the Commission of the proposal referred to in the first paragraph.

Article 13

This Directive shall enter into force on the day of its publication in the Official Journal of the European Communities.

Article 14

This Directive is addressed to the Member States.

Done at Brussels, 16 February 1998.

Commission Declaration

Article 2(b)—

The Commission takes the view that the expression 'for one kilogramme, one litre, one metre, one square metre or cubic metre of the product or a different single unit of quantity' in Article 2(b) also applies to products sold by individual item or singly.

Commission Declaration

Article 12, first paragraph—

The Commission considers that Article 12, first paragraph, of the Directive cannot be construed as calling into question its right of initiative.

Directive of the European Parliament and of the Council of 19 May 1998 on settlement finality in payment and securities settlement systems

(98/26/EC)

NOTES

Date of publication in OJ: OJ L166, 11.6.98, p 45.

THE EUROPEAN PARLIAMENT AND THE COUNCIL OF THE EUROPEAN UNION,

Having regard to the Treaty establishing the European Community, and in particular Article 100a thereof,

Having regard to the proposal from the Commission,[1]

Having regard to the opinion of the European Monetary Institute,[2]

Having regard to the opinion of the Economic and Social Committee,[3]

Acting in accordance with the procedure laid down in Article 189b of the Treaty,[4]

(1) Whereas the Lamfalussy report of 1990 to the Governors of the central banks of the Group of Ten Countries demonstrated the important systemic risk inherent in payment systems which operate on the basis of several legal types of payment netting, in particular multilateral netting; whereas the reduction of legal risks associated with participation in real time gross

settlement systems is of paramount importance, given the increasing development of these systems;

(2) Whereas it is also of the utmost importance to reduce the risk associated with participation in securities settlement systems, in particular where there is a close connection between such systems and payment systems;

(3) Whereas this Directive aims at contributing to the efficient and cost effective operation of cross-border payment and securities settlement arrangements in the Community, which reinforces the freedom of movement of capital in the internal market; whereas this Directive thereby follows up the progress made towards completion of the internal market, in particular towards the freedom to provide services and liberalisation of capital movements, with a view to the realisation of Economic and Monetary Union;

(4) Whereas it is desirable that the laws of the Member States should aim to minimise the disruption to a system caused by insolvency proceedings against a participant in that system;

(5) Whereas a proposal for a Directive on the reorganisation and winding-up of credit institutions submitted in 1985 and amended on 8 February 1988 is still pending before the Council; whereas the Convention on Insolvency Proceedings drawn up on 23 November 1995 by the Member States meeting within the Council explicitly excludes insurance undertakings, credit institutions and investment firms;

(6) Whereas this Directive is intended to cover payment and securities settlement systems of a domestic as well as of a cross-border nature; whereas the Directive is applicable to Community systems and to collateral security constituted by their participants, be they Community or third country participants, in connection with participation in these systems;

(7) Whereas Member States may apply the provisions of this Directive to their domestic institutions which participate directly in third country systems and to collateral security provided in connection with participation in such systems;

(8) Whereas Member States should be allowed to designate as a system covered by this Directive a system whose main activity is the settlement of securities even if the system to a limited extent also deals with commodity derivatives;

(9)   Whereas the reduction of systemic risk requires in particular the finality of settlement and the enforceability of collateral security; whereas collateral security is meant to comprise all means provided by a participant to the other participants in the payment and/or securities settlement systems to secure rights and obligations in connection with that system, including repurchase agreements, statutory liens and fiduciary transfers; whereas regulation in national law of the kind of collateral security which can be used should not be affected by the definition of collateral security in this Directive;

(10)   Whereas this Directive, by covering collateral security provided in connection with operations of the central banks of the Member States functioning as central banks, including monetary policy operations, assists the European Monetary Institute in its task of promoting the efficiency of cross-border payments with a view to the preparation of the third stage of Economic and Monetary Union and thereby contributes to developing the necessary legal framework in which the future European central bank may develop its policy;

(11)   Whereas transfer orders and their netting should be legally enforceable under all Member States' jurisdictions and binding on third parties;

(12)   Whereas rules on finality of netting should not prevent systems testing, before the netting takes place, whether orders that have entered the system comply with the rules of that system and allow the settlement of that system to take place;

(13)   Whereas nothing in this Directive should prevent a participant or a third party from exercising any right or claim resulting from the underlying transaction which they may have in law to recovery or restitution in respect of a transfer order which has entered a system, eg in case of fraud or technical error, as long as this leads neither to the unwinding of netting nor to the revocation of the transfer order in the system;

(14)   Whereas it is necessary to ensure that transfer orders cannot be revoked after a moment defined by the rules of the system;

(15)   Whereas it is necessary that a Member State should immediately notify other Member States of the opening of insolvency proceedings against a participant in the system;

(16)   Whereas insolvency proceedings should not have a retroactive effect on the rights and obligations of participants in a system;

(17)   Whereas, in the event of insolvency proceedings against a participant in a system, this Directive furthermore aims at determining which insolvency law is applicable to the rights and obligations of that participant in connection with its participation in a system;

(18)   Whereas collateral security should be insulated from the effects of the insolvency law applicable to the insolvent participant;

(19)   Whereas the provisions of Article 9(2) should only apply to a register, account or centralized deposit system which evidences the existence of proprietary rights in or for the delivery or transfer of the securities concerned;

(20)   Whereas the provisions of Article 9(2) are intended to ensure that if the participant, the central bank of a Member State or the future European central bank has a valid and effective collateral security as determined under the law of the Member State where the relevant register, account or centralized deposit system is located, then the validity and enforceability of that collateral security as against that system (and the operator thereof) and against any other person claiming directly or indirectly through it, should be determined solely under the law of that Member State;

(21)   Whereas the provisions of Article 9(2) are not intended to prejudice the operation and effect of the law of the Member State under which the securities are constituted or of the law of the Member State where the securities may otherwise be located (including, without limitation, the law concerning the creation, ownership or transfer of such securities or of rights in such securities) and should not be interpreted to mean that any such collateral security will be directly enforceable or be capable of being recognised in any such Member State otherwise than in accordance with the law of that Member State;

(22)   Whereas it is desirable that Member States endeavour to establish sufficient links between all the securities settlement systems covered by this Directive with a view towards promoting maximum transparency and legal certainty of transactions relating to securities;

(23)   Whereas the adoption of this Directive constitutes the most appropriate way of realising the abovementioned objectives and does not go beyond what is necessary to achieve them,

NOTES

   [1]   OJ C207, 18.7.96, p 13, and OJ C259, 26.8.97, p 6.
   [2]   Opinion delivered on 21 November 1996.

[3] OJ C56, 24.2.97, p 1.

[4] Opinion of the European Parliament of 9 April 1997 (OJ C132, 28.4.97, p 74), Council Common Position of 13 October 1997 (OJ C375, 10.12.97, p 34) and Decision of the European Parliament of 29 January 1998 (OJ C56, 23.2.98). Council Decision of 27 April 1998.

HAVE ADOPTED THIS DIRECTIVE—

SECTION I
SCOPE AND DEFINITIONS

Article 1

The provisions of this Directive shall apply to—
  (a) any system as defined in Article 2(a), governed by the law of a Member State and operating in any currency, the ecu or in various currencies which the system converts one against another;
  (b) any participant in such a system;
  (c) collateral security provided in connection with—
    — participation in a system, or
    — operations of the central banks of the Member States in their functions as central banks.

Article 2

For the purpose of this Directive—
  (a) 'system' shall mean a formal arrangement—
    — between three or more participants, without counting a possible settlement agent, a possible central counterparty, a possible clearing house or a possible indirect participant, with common rules and standardised arrangements for the execution of transfer orders between the participants,
    — governed by the law of a Member State chosen by the participants; the participants may, however, only choose the law of a Member State in which at least one of them has its head office, and
    — designated, without prejudice to other more stringent conditions of general application laid down by national law, as a system and notified to the Commission by the Member State whose law is applicable, after that Member State is satisfied as to the adequacy of the rules of the system.

Subject to the conditions in the first subparagraph, a Member State may designate as a system such a formal arrangement whose business consists of the execution of transfer orders as defined in the second indent of (i) and which to a limited extent executes orders relating to other financial instruments, when that Member State considers that such a designation is warranted on grounds of systemic risk.

A Member State may also on a case-by-case basis designate as a system such a formal arrangement between two participants, without counting a possible settlement agent, a possible central counterparty, a possible clearing house or a possible indirect participant, when that Member State considers that such a designation is warranted on grounds of systemic risk;

    (b) 'institution' shall mean—

— a credit institution as defined in the first indent of Article 1 of Directive 77/780/EEC[1] including the institutions set out in the list in Article 2(2) thereof, or

— an investment firm as defined in point 2 of Article 1 of Directive 93/22/EEC[2] excluding the institutions set out in the list in Article 2(2)(a) to (k) thereof, or

— public authorities and publicly guaranteed undertakings, or

— any undertaking whose head office is outside the Community and whose functions correspond to those of the Community credit institutions or investment firms as defined in the first and second indent,

which participates in a system and which is responsible for discharging the financial obligations arising from transfer orders within that system.

If a system is supervised in accordance with national legislation and only executes transfer orders as defined in the second indent of (i), as well as payments resulting from such orders, a Member State may decide that undertakings which participate in such a system and which have responsibility for discharging the financial obligations arising from transfer orders within this system, can be considered institutions, provided that at least three participants of this system are covered by the categories referred to in the first subparagraph and that such a decision is warranted on grounds of systemic risk;

    (c) 'central counterparty' shall mean an entity which is interposed between the institutions in a system and which acts as the exclusive counterparty of these institutions with regard to their transfer orders;

    (d) 'settlement agent' shall mean an entity providing to institutions and/or a central counterparty participating in systems, settlement accounts through which transfer orders within such systems are settled and, as the case may be, extending credit to those institutions and/or central counterparties for settlement purposes;

(e) 'clearing house' shall mean an entity responsible for the calculation of the net positions of institutions, a possible central counterparty and/or a possible settlement agent;

(f) 'participant' shall mean an institution, a central counterparty, a settlement agent or a clearing house.

According to the rules of the system, the same participant may act as a central counterparty, a settlement agent or a clearing house or carry out part or all of these tasks.

A Member State may decide that for the purposes of this Directive an indirect participant may be considered a participant if it is warranted on the grounds of systemic risk and on condition that the indirect participant is known to the system;

(g) 'indirect participant' shall mean a credit institution as defined in the first indent of (b) with a contractual relationship with an institution participating in a system executing transfer orders as defined in the first indent of (i) which enables the abovementioned credit institution to pass transfer orders through the system;

(h) 'securities' shall mean all instruments referred to in section B of the Annex to Directive 93/22/EEC;

(i) 'transfer' order shall mean—

— any instruction by a participant to place at the disposal of a recipient an amount of money by means of a book entry on the accounts of a credit institution, a central bank or a settlement agent, or any instruction which results in the assumption or discharge of a payment obligation as defined by the rules of the system, or

— an instruction by a participant to transfer the title to, or interest in, a security or securities by means of a book entry on a register, or otherwise;

(j) 'insolvency proceedings' shall mean any collective measure provided for in the law of a Member State, or a third country, either to wind up the participant or to reorganise it, where such measure involves the suspending of, or imposing limitations on, transfers or payments;

(k) 'netting' shall mean the conversion into one net claim or one net obligation of claims and obligations resulting from transfer orders which a participant or participants either issue to, or receive from, one or more other participants with the result that only a net claim can be demanded or a net obligation be owed;

(l) 'settlement account' shall mean an account at a central bank, a settlement agent or a central counterparty used to hold funds and securities and to settle transactions between participants in a system;

(m) 'collateral security' shall mean all realisable assets provided under a pledge (including money provided under a pledge), a repurchase or similar agreement, or otherwise, for the purpose of securing rights and obligations potentially arising in connection with a system, or provided to central banks of the Member States or to the future European central bank.

NOTES

[1] First Council Directive 77/780/EEC of 12 December 1977 on the coordination of the laws, regulations and administrative provisions relating to the taking up and pursuit of the business of credit institutions (OJ L322, 17.12.77, p 30). Directive as last amended by Directive 96/13/EC (OJ L66, 16.3.96, p 15).

[2] Council Directive 93/22/EEC of 10 May 1993 on investment services in the securities field (OJ L141, 11.6.93, p 27). Directive as last amended by Directive 97/9/EC (OJ L84, 26.3.1997, p 22).

SECTION II
NETTING AND TRANSFER ORDERS

Article 3

1.	Transfer orders and netting shall be legally enforceable and, even in the event of insolvency proceedings against a participant, shall be binding on third parties, provided that transfer orders were entered into a system before the moment of opening of such insolvency proceedings as defined in Article 6(1).

Where, exceptionally, transfer orders are entered into a system after the moment of opening of insolvency proceedings and are carried out on the day of opening of such proceedings, they shall be legally enforceable and binding on third parties only if, after the time of settlement, the settlement agent, the central counterparty or the clearing house can prove that they were not aware, nor should have been aware, of the opening of such proceedings.

2.	No law, regulation, rule or practice on the setting aside of contracts and transactions concluded before the moment of opening of insolvency proceedings, as defined in Article 6(1) shall lead to the unwinding of a netting.

3.	The moment of entry of a transfer order into a system shall be defined by the rules of that system. If there are conditions laid down in the national law governing the system as to the moment of entry, the rules of that system must be in accordance with such conditions.

Article 4

Member States may provide that the opening of insolvency proceedings against a participant shall not prevent funds or securities available on the settlement account of that participant from being used to fulfil that participant's obligations in the system on the day of the opening of the insolvency proceedings. Furthermore, Member States may also provide that such a participant's credit facility connected to the system be used against available, existing collateral security to fulfil that participant's obligations in the system.

Article 5

A transfer order may not be revoked by a participant in a system, nor by a third party, from the moment defined by the rules of that system.

SECTION III
PROVISIONS CONCERNING INSOLVENCY PROCEEDINGS

Article 6

1.   For the purpose of this Directive, the moment of opening of insolvency proceedings shall be the moment when the relevant judicial or administrative authority handed down its decision.

2.   When a decision has been taken in accordance with paragraph 1, the relevant judicial or administrative authority shall immediately notify that decision to the appropriate authority chosen by its Member State.

3.   The Member State referred to in paragraph 2 shall immediately notify other Member States.

Article 7

Insolvency proceedings shall not have retroactive effects on the rights and obligations of a participant arising from, or in connection with, its participation in a system earlier than the moment of opening of such proceedings as defined in Article 6(1).

Article 8

In the event of insolvency proceedings being opened against a participant in a system, the rights and obligations arising from, or in connection with, the participation of that participant shall be determined by the law governing that system.

SECTION IV
INSULATION OF THE RIGHTS OF HOLDERS OF COLLATERAL
SECURITY FROM THE EFFECTS OF THE INSOLVENCY OF THE
PROVIDER

Article 9

1.   The rights of—
    — a participant to collateral security provided to it in connection with a system, and
    — central banks of the Member States or the future European central bank to collateral security provided to them,

shall not be affected by insolvency proceedings against the participant or counterparty to central banks of the Member States or the future European central bank which provided the collateral security. Such collateral security may be realised for the satisfaction of these rights.

2.   Where securities (including rights in securities) are provided as collateral security to participants and/or central banks of the Member States or the future European central bank as described in paragraph 1, and their right (or that of any nominee, agent or third party acting on their behalf) with respect to the securities is legally recorded on a register, account or centralised deposit system located in a Member State, the determination of the rights of such entities as holders of collateral security in relation to those securities shall be governed by the law of that Member State.

SECTION V
FINAL PROVISIONS

Article 10

Member States shall specify the systems which are to be included in the scope of this Directive and shall notify them to the Commission and inform the Commission of the authorities they have chosen in accordance with Article 6(2).

The system shall indicate to the Member State whose law is applicable the participants in the system, including any possible indirect participants, as well as any change in them.

In addition to the indication provided for in the second subparagraph, Member States may impose supervision or authorisation requirements on systems which fall under their jurisdiction.

Anyone with a legitimate interest may require an institution to inform him of the systems in which it participates and to provide information about the main rules governing the functioning of those systems.

Article 11

1.   Member States shall bring into force the laws, regulations and administrative provisions necessary to comply with this Directive before 11 December 1999. They shall forthwith inform the Commission thereof.

When Member States adopt these measures, they shall contain a reference to this Directive or shall be accompanied by such reference on the occasion of their official publication. The methods of making such a reference shall be laid down by the Member States.

2.   Member States shall communicate to the Commission the text of the provisions of domestic law which they adopt in the field governed by this Directive. In this Communication, Member States shall provide a table of correspondence showing the national provisions which exist or are introduced in respect of each Article of this Directive.

Article 12

No later than three years after the date mentioned in Article 11(1), the Commission shall present a report to the European Parliament and the

Council on the application of this Directive, accompanied where appropriate by proposals for its revision.

Article 13

This Directive shall enter into force on the day of its publication in the Official Journal of the European Communities.

Article 14

This Directive is addressed to the Member States.

Done at Brussels, 19 May 1998.

Amended proposal for a European Parliament and Council Directive combating late payment in commercial transactions[1]

(98/C 374/04)

(Text with EEA relevance)

COM(1998) 615 final — 98/0099(COD)

(Submitted by the Commission pursuant to Article 189a(2) of EC Treaty on 30 October 1998)

NOTES

Date of Publication in OJ: OJ C374, 3.12.98, p 4.
[1]   OJ C168, 3.6.98, p 13.

THE EUROPEAN PARLIAMENT AND THE COUNCIL OF THE EUROPEAN UNION,

Having regard to the Treaty establishing the European Community, and in particular Article 100a thereof,

Having regard to the proposal from the Commission,

Having regard to the opinion of the Economic and Social Committee,[1]

Acting in accordance with the procedure laid down in Article 189b of the Treaty,[2]

(1) Whereas the European Parliament in its Resolution[3] on the Integrated Programme in favour of SMEs and the craft sector[4] emphasised that the Commission should forward proposals to deal with the problem of late payment;

(2) Whereas on 12 May 1995 the Commission adopted a Recommendation on payment periods in commercial transactions;[5]

(3) Whereas the European Parliament in its Resolution on the Commission Recommendation on payment periods in commercial transactions[6] called on the Commission to consider transforming its recommendation into a proposal for a Council Directive to be submitted as soon as possible;

(4) Whereas on 29 May 1997 the Economic and Social Committee adopted an opinion on the Commission's Green Paper on public procurement in the European Union: Exploring the Way Forward,[7] recommending maximum payment periods and interest on late payments by public authorities;

(5) Whereas on 4 June 1997 the Commission published an Action Plan for the Single Market,[8] which underlined that late payment represents an increasingly serious obstacle for the success of the Single Market;

(6) Whereas on 17 July 1997 the Commission published a Report on late payments in commercial transactions,[9] summarising the results of an evaluation of the effects of the Commission's Recommendation of 12 May 1995;

(7) Whereas heavy administrative and financial burdens are placed on businesses, particularly small and medium-sized ones, as a result of the excessive payment periods and late payment; whereas moreover, these problems are a major cause of insolvencies threatening the survival of businesses and result in numerous job losses;

(8) Whereas the differences between the payment rules and practices in the Member States constitute an obstacle to the proper functioning of the internal market; whereas a creditor who needs to collect receivables from debtors situated in several Member States is confronted with widely differing rules of national legislation making it difficult, time consuming and costly for him to do so;

(9) Whereas this has the effect of considerably limiting commercial transactions between Member States; whereas this is in contradiction with Article 7a of the Treaty as entrepreneurs should be able to trade throughout the Internal Market under conditions which ensure that transborder operations do not entail greater risks than domestic sales; whereas it would lead to distortions of competition if different rules applied to domestic and transborder operations;

(10) Whereas the most recent statistics indicate that there has been, at best, no improvement in late payments in many Member States since the adoption of the Recommendation of 12 May 1995;

(11) Whereas, in accordance with the principle of subsidiarity and the principle of proportionality as set out in Article 3b of the Treaty, the objective of combating late payments in the internal market cannot be sufficiently achieved by the Member States acting individually and can, therefore, be better achieved by the Community; whereas this Directive confines itself to the minimum required in order to achieve those objectives and does not go beyond what is necessary for that purpose;

(12) Whereas late payment constitutes a breach of contract which has been made financially attractive to debtors in most Member States by low interest rates on late payments and/or slow redress procedures; whereas a decisive shift is necessary to reverse this trend and the consequences of late payments must be such as both to discourage late payment and to fully compensate creditors for the costs incurred;

(13) Whereas the use of retention of title clauses as a means of speeding up payment is at present constrained by a number of differences in national law; whereas it is necessary to ensure that creditors are in a position to exercise the retention of title throughout the Community, using a single clause recognised by all Member States, and that excessive length of payment periods and late payments do not distort commercial transactions in the functioning of the internal market;

(14) Whereas the consequences of late payment can be dissuasive only if they are accompanied by redress procedures which are rapid, effective and

inexpensive for the creditor; whereas in conformity with the principle of non-discrimination contained in Article 6 of the Treaty, these procedures should be available to creditors from all Member States irrespective of their residence;

(15) Whereas public authorities handle a considerable volume of payments to businesses; whereas strict payment discipline on the part of these authorities would have a beneficial trickle-down effect on the economy as a whole; whereas, with regard to public contracts, contracting enterprises in turn likewise delay payments to their suppliers and subcontractors, habitually imposing disproportionate payment periods—practices which seriously damage the interests of many businesses, especially SMEs; whereas for payments executed by the Commission it has already been decided to give certain creditors the right to receive default interest on late payments;

(16) Whereas for the purposes of the implementation of this Directive, the Commission should be assisted by a committee of an advisory nature,

(17) Whereas it could be necessary, when this Directive is reviewed, to take into consideration the possibility of addressing the consequences of long contractual payment periods;

(18) Whereas the term 'contracting authorities' should correspond to the definition laid down in Directive 92/50/EEC[10] and Directive 93/37/EEC[11] and should include, for the purposes of the present Directive, the 'contracting entities' as defined in Directive 93/38/EEC.[12]

NOTES

[1]   Adopted on 10.9.1998, not yet published in the Official Journal.
[2]   Opinion of the European Parliament adopted on 17.9.1998, not yet published in the Official Journal.
[3]   OJ C323, 21.11.94, p 19.
[4]   COM(94) 207 final, 3.6.94.
[5]   OJ L127, 10.6.95, p 19.
[6]   OJ C211, 22.7.96, p 43.
[7]   OJ C287, 22.9.97, p 92.
[8]   CSE(97) 1 final, 4.6.97, pp 8 and 38.
[9]   OJ C216, 17.7.97, p 10.
[10]   OJ L209, 24.7.92, p 1.
[11]   OJ L199, 9.8.93, p 54.
[12]   OJ L199, 9.8.93, p 84.

HAVE ADOPTED THIS DIRECTIVE—

CHAPTER I

Article 1   Scope

The provisions of this Directive shall apply to all payments made in commercial transactions.

Article 2   Definitions

For the purposes of this Directive—
1. 'commercial transactions' means transactions between undertakings which lead to delivery of goods or provision of services for remuneration; an undertaking is any organisation set up on a permanent basis with an independent economic activity, even where it is carried on by a single person and even where it is not intended to make a profit; contracting authorities shall in every case be deemed to be undertakings for the purposes of this Directive;
2. 'late payment' means failure to observe the contractual or statutory terms of payment;
3. 'retention of title' means the agreement, irrespective of any formal requirements, that the seller remains the owner of the goods in question until the price has been paid in full;
4. 'contracting authorities' corresponds to the definition laid down in Directive 92/50/EEC[1] and Directive 93/37/EEC[2] and includes the 'contracting entities' as defined in Directive 93/38/EEC.[3]
5. 'public procurement contracts' means contracts for pecuniary interest concluded in writing between a contracting authority within the meaning of paragraph 4 and an undertaking which is not a contracting authority.

NOTES
[1]   OJ L209, 24.7.92, p 1.
[2]   OJ L199, 9.8.93, p 54.
[3]   OJ L199, 9.8.93, p 84.

CHAPTER II

Article 3   Default date, interest and compensation for the damage incurred

1.   Member States shall enact the necessary legislation and amend their procedural rules in such a way that, subject to the goods or services having

been duly provided and the underlying legal conditions being correctly fulfilled, the following is ensured—

(a) the default date for the payment of debts shall not be more than 21 calendar days from the date of receipt of the invoice, unless otherwise specified in the contract or in the sellers's general conditions of sale;

(b) the invoice shall be deemed to have been received no later than the fifth calendar day following the date of the invoice, unless the buyer or seller is able to furnish proof of receipt at another time;

(c) in the absence of an invoice or if the date of its receipt cannot be determined with certainty or if the date of receipt is earlier than the date of supply of the goods or services concerned, the default date shall be calculated from the latter date;

(d) where the default date specified in the contract or in the seller's general conditions of sale is more than 45 calendar days from the date of receipt of the invoice, the buyer shall provide the seller, at the buyer's cost, with a bill of exchange, specifying explicitly the date for its payment and guaranteed by an accepted credit institution;

(e) where the buyer fails to provide the seller with a bill of exchange in accordance with point (d) above, the normal default date and level of interest as foreseen in this article shall be applicable and any contractual derogations therefrom to the detriment of the seller shall be automatically null and void; the remainder of the contract shall remain in force;

(f) the creditor shall be entitled to claim interest from the debtor on any outstanding amount when the default date as determined under points (a) to (e) above has been exceeded without the creditor having received the amount due;

(g) interest shall accrue automatically from the day after the default date without the necessity of a reminder;

(h) the level of interest for late payment (the 'statutory rate'), which the creditor is entitled to claim, shall be the sum of the tender (repo) interest rate of the European Central Bank (the 'reference rate') plus at least 8 percentage points (the 'margin'), unless otherwise specified in the contract or in the seller's general conditions of sale; for Member States which do not participate in the third phase of Economic and Monetary Union, the reference rates referred to above shall be the equivalent rates set by their central banks;

(i) the statutory rate for interest on late payment shall change automatically in accordance with changes to the reference rate mentioned in point (e);

(j) in addition to the right to interest, the creditor shall be entitled to claim full compensation from the debtor for the damage incurred.

2. The margin referred to in paragraph 1(h) may be modified by the Commission in accordance with the procedure referred to in Article 9 if it

becomes apparent that the statutory rate is no longer sufficiently high to discourage the buyer from paying late and to compensate the seller for any loss incurred as a result of late payment, in particular for any interest he would have to pay on overdraft credit.

3.   Three years after the end of the period defined in Article 10(1) of the Directive, the Commission, having been advised by the Committee referred to in Article 9, shall undertake a review of, *inter alia*, the statutory rate to assess the impact on commercial transactions and the operation of the legislation in practice. The results of this reviews and of other reviews will be made known to the European Parliament.

Article 4   Retention of title

1.   Member States shall ensure that the seller retains title if a retention of title clause has been agreed. Apart from an individual contract, such an agreement shall be considered valid if the retention of title clause is contained in the seller's standard contract, on the invoice, or on delivery documents accompanying the goods, which the buyer has received no later than at the time of delivery, and to which he has not objected. No other formality shall be required.

2.   Member States shall recognise the validity of the clauses contained in the Annex or of clauses having equivalent effect.

3.   Once the default date has passed without the buyer having paid, the seller may claim that the goods in question be returned to him. Member States shall provide for the retention of title to be enforceable against third parties, even in the case of bankruptcy of the debtor or in the case of any other procedure recognised as being similar under the legislation of the Member States. No later than when the buyer takes possession of the goods, he becomes responsible for any damage to or loss of the goods.

4.   Member States may adopt provisions for the protection of third parties acting in good faith, and as regards down payments already made by the debtor. They may also adopt provisions concerning goods which are incorporated in other movable or immovable property.

Article 5   Accelerated recovery procedures for undisputed debts

1.   Member States shall ensure that there is an accelerated debt recovery procedure for undisputed debts.

2.   This procedure shall apply irrespective of the amount of the debt.

3.   This procedure shall be available to creditors from all Member States, irrespective of their place of residence.

4.   The creditor shall be able to choose whether or not he wishes to be represented by a third person.

5.   The procedure before the court shall be formulated in such a way that a period of 60 calendar days is not exceeded from the receipt of the creditor's request to the time when the writ of execution or equivalent document becomes enforceable.

This period is without prejudice to—
(a) the application of the rules governing notification or service and
(b) the rights of the defendant to dispute the debt.

Article 6   Simplified legal procedures for small debts

Member States shall ensure that simplified procedures are available for debts up to a ceiling, which shall not be less than ECU 20 000. These procedures shall provide for simple, low-cost methods for taking legal action for the settlement of debts.

This sum can if necessary be modified by the Commission to reflect changing economic conditions in accordance with the procedure referred to in Article 9.

These procedures shall be available to creditors from all Member States irrespective of their place of residence.

CHAPTER III

Article 7   Transparency in public procurement contracts

Member States shall ensure that public procurement contracts contain precise details of the default dates and deadlines applied by the contracting authorities, even if these default dates and deadlines are determined in general contract conditions laid down by law. In particular, time limits shall be fixed for the completion of pre-payment administrative formalities, such as public works reception procedures. A similar obligation of transparency shall apply in the relationship between a main contractor and a subcontractor carrying out public works.

Article 8   Prompt payment, default date and automatic interest

Member States shall ensure that—
1. the default date for the payment of contractual debts by the contracting authorities as determined under Article 3(1)(a) to (c) shall not be more than 45 calendar days except where the value of the contract exceeds ECU 100 000 where the maximum default date will be 60 calendar days; the contract shall in no circumstances override these maximum default dates; in a public contract, the main contractor has to grant conditions to the suppliers and subcontractors which are at least as favourable as those granted to the main contractor by the contracting authority;

In order to guarantee these conditions to suppliers and subcontractors, the main contractor shall be required to provide a guarantee made out to the supplier or subcontractor covering payment of all the amounts owed. This guarantee shall be executable upon expiry of 60 calendar days from the date of submission of the invoice to the main contractor by the supplier or subcontractor;
2. a creditor shall be entitled to interest from the contracting authority on any outstanding amount when the default date has been exceeded; the interest shall be calculated as set out in Article 3(1)(g) and (h), and shall be paid automatically by the contracting authority without the necessity of a claim;
3. the contracting authority is not permitted to request or require that the creditor waives any of the rights referred to in this Article, nor may the main contractor request or require that his suppliers or subcontractors waive those rights.

CHAPTER IV

Article 9   Committee

For the purposes of reviewing the functioning of this Directive and in particular for the cases mentioned in Article 3(2) and Article 6, the Commission shall be assisted by a committee of an advisory nature composed of the representatives of the Member States and chaired by the representative of the Commission.

The representative of the Commission shall submit to the committee a draft of the measures to be taken. The committee shall deliver its opinion on the draft, within a time limit which the chairman may lay down according to the urgency of the matter, if necessary by taking a vote.

The opinion shall be recorded in the minutes; in addition, each Member State shall have the right to ask to have its position recorded in the minutes.

The Commission shall take the utmost account of the opinion delivered by the committee. It shall inform the committee of the manner in which its opinion has been taken into account.

Article 10   Transposition

1.   Member States shall bring into force the laws, regulations and administrative provisions necessary to comply with this Directive by 31 December 2000 at the latest. They shall forthwith inform the Commission thereof.

When Member States adopt these provisions, these shall contain a reference to this Directive or shall be accompanied by such reference at the time of their official publication. The procedure for such reference shall be adopted by Member States.

2.   Member States may maintain or bring into force provisions which are more favourable to the creditor than the provisions necessary to comply with this Directive.

3.   Member States shall communicate to the Commission the text of the mains laws, regulations or administrative provisions which they adopt in the field covered by this Directive.

Article 11   Entry into force

This Directive shall enter into force on the twentieth day following that of its publication in the *Official Journal of the European Communities*.

Article 12   Addressees

This Directive is addressed to the Member States.

ANNEX
LIST OF CLAUSES TO BE RECOGNISED BY MEMBER STATES FOR THE PURPOSES OF ARTICLE 4

ES: «El vendedor conservará la propiedad de los bienes hasta el pago final.»

DA: »Varen forbliver sælgerens ejendom, indtil den er fuldstændig betalt.«

DE: „Die Ware bleibt bis zur vollständigen Bezahlung im Eigentum des Verkäufers.“

EL: «Ο πωλητης παρακρατει την κυριοτητα των αγαθων μεχρι την πληρη εξοψληοη του τιμηυατος.»

EN: 'The goods remain the property of the seller until fully paid.'

FR: «Les marchandises restent la propriété du vendeur jusqu'au paiement complet.»

IT: «Le merci restano di proprietà del venditore fino al pieno pagamento.»

NL: ,,De waren blijven tot de volledige betaling eigendom van de verkoper."

PT: «O vendedor conservará a propriedade dos bens até ao momento do pagamento final. »

FI: "Tavara on myyjän omaisuutta, kunnes kauppahinta on kokonaisuudessaan maksettu."

SV: "Varorna förblir säljarens egendom tills de betalats helt och hället."

Common Position (EC) No 51/98 adopted by the Council on 24 September 1998 with a view to adopting European Parliament and Council Directive 98/ /EC, of ... on certain aspects of the sale of consumer goods and associated guarantees

(98/C 333/04)

NOTES

Date of Publication in OJ: OJ C333, 30.10.98, p 46.
See further Commission Opinion COM(1999) 16.

THE EUROPEAN PARLIAMENT AND THE COUNCIL OF THE EUROPEAN UNION,

Having regard to the Treaty establishing the European Community, and in particular Article 100a thereof,

Having regard to the proposal from the Commission,[1]

Having regard to the opinion of the Economic and Social Committee,[2]

Acting in accordance with the procedure laid down in Article 189b of the Treaty,[3]

(1)   Whereas the internal market comprises an area without internal frontiers in which the free movement of goods, persons, services and capital is guaranteed; whereas free movement of goods concerns not only transactions by persons acting in the course of a business but also transactions by private individuals; whereas it implies that consumers resident in one Member State should be free to purchase goods in the territory of another Member State on the basis of a uniform minimum set of fair rules governing the sale of consumer goods;

(2)   Whereas the laws of the Member States concerning the sale of consumer goods are somewhat disparate, with the result that national consumer goods markets differ from one another and that competition between sellers may be distorted;

(3)   Whereas consumers who are keen to benefit from the large market by purchasing goods in Member States other than their State of residence play a fundamental role in the completion of the internal market; whereas the artificial reconstruction of frontiers and the compartmentalisation of markets should be prevented; whereas the opportunities available to consumers have been greatly broadened by new communication technologies which allow ready access to distribution systems in other Member States or in non-member countries; whereas, in the absence of minimum harmonisation of the rules governing the sale of consumer goods, the development of the sale of goods through the medium of new distance communication technologies risks being impeded;

(4)   Whereas the creation of a common set of minimum rules of consumer law, valid no matter where goods are purchased within the Community, will strengthen consumer confidence and enable consumers to make the most of the internal market;

(5)   Whereas the main difficulties encountered by consumers and the main source of disputes with sellers concern the non-conformity of goods with the contract; whereas it is therefore appropriate to approximate national legislation governing the sale of consumer goods in this respect, without however impinging on provisions and principles of national law relating to contractual and non-contractual liability;

(6)   Whereas the goods must, above all, conform with the contractual specifications; whereas the principle of conformity with the contract may be considered as common to the different national legal traditions; whereas in certain national legal traditions it may not be possible to rely solely on this principle to ensure a minimum level of protection for the consumer; whereas under such legal traditions, in particular, additional national provisions may be useful to ensure that the consumer is protected in cases where the parties have agreed no specific contractual terms or where the parties have concluded contractual terms or agreements which directly or indirectly waive or restrict the rights of the consumer and which, to the extent that these rights result from this Directive, are not binding on the consumer;

(7)   Whereas, in order to facilitate the application of the principle of conformity with the contract, it is useful to introduce a rebuttable presumption of conformity with the contract covering the most common

situations; whereas that presumption does not restrict the principle of freedom of contract; whereas, furthermore, in the absence of specific contractual terms, as well as where the minimum protection clause is applied, the elements mentioned in this presumption may be used to determine the lack of conformity of the goods with the contract; whereas the quality and performance which consumers can reasonably expect will depend on the nature of the goods, including whether they are new or second-hand; whereas the elements mentioned in the presumption are cumulative; whereas, if the circumstances of the case render any particular element manifestly inappropriate, the remaining elements of the presumption nevertheless still apply;

(8)   Whereas the seller should be directly liable to the consumer for the conformity of the goods with the contract; whereas this is the traditional solution enshrined in the legal orders of the Member States; whereas nevertheless the seller should be free, as provided for by national law, to pursue remedies against the producer, a previous seller in the same chain of contracts or any other intermediary, unless he has renounced that entitlement; whereas the rules governing against whom and how the seller may pursue such remedies are to be determined by national law;

(9)   Whereas, in the case of non-conformity of the goods with the contract, consumers should be entitled to have the goods restored to conformity with the contract free of charge, choosing either repair or replacement, or, failing this, to have the price reduced or the contract rescinded;

(10)   Whereas the consumer in the first place may require the seller to repair the goods or to replace them unless those remedies are impossible or disproportionate; whereas whether a remedy is disproportionate should be determined objectively; whereas a remedy would be disproportionate if it imposed, in comparison with the other remedy, unreasonable costs; whereas, in order to determine whether the costs are unreasonable, the costs of one remedy should be significantly higher than the costs of the other remedy;

(11)   Whereas in cases of a lack of conformity, the seller may always offer the consumer, by way of settlement, any available remedy; whereas it is for the consumer to decide whether to accept or reject this proposal;

(12)   Whereas the references to the time of delivery do not imply that Member States have to change their rules on the passing of the risk;

(13)   Whereas Member States may provide that any reimbursement to the consumer may be reduced to take account of the use the consumer has had of the goods since they were delivered to him; whereas the detailed

arrangements whereby rescission of the contract is effected may be laid down in national law;

(14)   Whereas the specific nature of second-hand goods makes it generally impossible to replace them; whereas therefore the consumer's right of replacement is generally not available for these goods; whereas for such goods, Member States may enable the parties to agree a shortened period of liability;

(15)   Whereas it is appropriate to limit in time the period during which the seller is liable for any lack of conformity which exists at the time of delivery of the goods; whereas Member States may also provide for a limitation on the period during which consumers can exercise their rights, provided such a period does not expire within two years from the time of delivery; whereas where, under national legislation, the time when a limitation period starts is not the time of delivery of the goods, the total duration of the limitation period provided for by national law may not be shorter than two years from the time of delivery;

(16)   Whereas Member States may provide for suspension or interruption of the period during which any lack of conformity must become apparent and of the limitation period, where applicable and in accordance with their national law, in the event of repair, replacement or negotiations between seller and consumer with a view to an amicable settlement;

(17)   Whereas Member States should be allowed to set a period within which the consumer must inform the seller of any lack of conformity; whereas Member States may ensure a higher level of protection for the consumer by not introducing such an obligation; whereas in any case consumers throughout the Community should have at least two months in which to inform the seller that a lack of conformity exists;

(18)   Whereas Member States should guard against such a period placing at a disadvantage consumers shopping across borders; whereas all Member States should inform the Commission of their use of this provision; whereas the Commission should monitor the effect of the varied application of this provision on consumers and on the internal market; whereas information on the use made of this provision by a Member State should be available to the other Member States and to consumers and consumer organisations throughout the Community; whereas a summary of the situation in all Member States should therefore be published in the *Official Journal of the European Communities*;

(19)   Whereas, for certain categories of goods, it is current practice for sellers and producers to offer guarantees on goods against any defect which

becomes apparent within a certain period; whereas this practice can stimulate competition; whereas, while such guarantees are legitimate marketing tools, they should not mislead the consumer; whereas, to ensure that consumers are not misled, guarantees should contain certain information, including a statement that the guarantee does not affect the consumer's legal rights;

(20) Whereas the parties may not, by common consent, restrict or waive the rights granted to consumers, since otherwise the legal protection afforded would be thwarted; whereas this principle should apply also to clauses which imply that the consumer was aware of any lack of conformity of the consumer goods existing at the time the contract was concluded; whereas the protection granted to consumers under this Directive should not be reduced on the grounds that the law of a non-member State has been chosen as being applicable to the contract;

(21) Whereas legislation and case-law in this area in the various Member States show that there is growing concern to ensure a high level of consumer protection; whereas, in the light of this trend and the experience acquired in implementing this Directive, it may be necessary to envisage more far-reaching harmonisation, notably by providing for the producer's direct liability for defects for which he is responsible;

(22) Whereas Member States should be allowed to adopt or maintain in force more stringent provisions in the field covered by this Directive to ensure an even higher level of consumer protection,

NOTES

[1] OJ C307, 16.10.96, p 8, and OJ C148, 14.5.1998, p 12.
[2] OJ C66, 3.3.97, p 5.
[3] Opinion of the European Parliament of 10 March 1998 (OJ C104, 6.4.98, p 30), Council Common Position of 24 September 1998 and European Parliament Decision of . . . (not yet published in the Official Journal).

HAVE ADOPTED THIS DIRECTIVE—

Article I Scope and definitions

1. The purpose of this Directive is the approximation of the laws, regulations and administrative provisions of the Member States on certain aspects of the sale of consumer goods and associated guarantees in order to ensure a uniform minimum level of consumer protection in the context of the internal market.

2.   For the purposes of this Directive—

(a) 'consumer' shall mean any natural person who, in the contracts covered by this Directive, is acting for purposes which are not related to his trade, business or profession;

(b) 'consumer goods' shall mean any tangible movable item, with the exception of—
   — goods sold by way of execution or otherwise by authority of law,
   — water and gas where they are not put up for sale in a limited volume or set quantity,
   — electricity;

(c) 'seller' shall mean any natural or legal person who, under a contract, sells consumer goods in the course of his trade, business or profession;

(d) 'producer' shall mean the manufacturer of consumer goods, the importer of consumer goods into the territory of the Community or any person purporting to be a producer by placing his name, trade mark or other distinctive sign on the consumer goods;

(e) 'guarantee' shall mean any undertaking by a seller or producer to the consumer, given without extra charge, to reimburse the price paid or to replace, repair or handle consumer goods in any way if they do not meet the specifications set out in the guarantee statement or in the relevant advertising;

(f) 'repair' shall mean, in the event of lack of conformity, bringing consumer goods into conformity with the contract of sale.

3.   Member States may provide that the expression 'consumer goods' does not cover second-hand goods sold at public auction where consumers have the opportunity of attending the sale in person.

4.   Contracts for the supply of consumer goods to be manufactured or produced shall also be deemed contracts of sale for the purpose of this Directive unless the consumer has to supply a substantial part of the materials necessary for manufacture or production.

Article 2   Conformity with the contract

1.   The seller must deliver goods to the consumer which are in conformity with the contract of sale.

2.   Consumer goods are presumed to be in conformity with the contract if they—

(a) comply with the description given by the seller and possess the qualities of the goods which the seller has held out to the consumer as a sample or model;

(b) are fit for any particular purpose for which the consumer requires them and which he made known to the seller at the time of conclusion of the contract except where the circumstances show that the consumer did not rely on the seller's explanations;

(c) are fit for the purposes for which goods of the same type are normally used;

(d) show the quality and performance which are normal in goods of the same type and which the consumer can reasonably expect, given the nature of the goods and taking into account any public statements on the specific characteristics of the goods made about them by the seller, the producer or his representative, particularly in advertising or on labelling.

3.   There shall be deemed not to be a lack of conformity for the purposes of this Article if, at the time the contract was concluded, the consumer was aware, or could not reasonably be unaware of, the lack of conformity.

4.   The seller shall not be bound by public statements, as referred to in paragraph 2(d) if he—

— shows that he was not, and could not reasonably have been, aware of the statement in question,

— shows that by the time of conclusion of the contract the statement had been corrected, or

— shows that the decision to buy the consumer goods could not have been influenced by the statement.

5.   Any lack of conformity resulting from incorrect installation of the consumer goods shall be deemed to be equivalent to lack of conformity of the goods if installation forms part of the contract of sale of the goods and the goods were installed by the seller or under his responsibility.

Article 3   Rights of the consumer

1.   The seller shall be liable to the consumer for any lack of conformity which exists at the time the goods were delivered.

2.   In the case of a lack of conformity, the consumer shall be entitled to have the goods brought into conformity free of charge by repair or replacement, in accordance with paragraph 3, or to have an appropriate reduction made in the price or the contract rescinded with regard to those goods, in accordance with paragraphs 4 and 5.

3.   In the first place, the consumer may require the seller to repair the goods or he may require the seller to replace them, in either case free of charge, unless this is impossible or disproportionate.

A remedy shall be deemed to be disproportionate if it imposes costs on the seller which, in comparison with the alternative remedy, are unreasonable, taking into account—
— the value the goods would have if there were no lack of conformity,
— the significance of the lack of conformity, and
— whether the alternative remedy could be completed without significant inconvenience to the consumer.

Any repair or replacement shall be completed within a reasonable time and without any significant inconvenience to the consumer, taking account of the nature of the goods and the purpose for which the consumer required the goods.

4.   If the consumer is entitled to require neither repair nor replacement, or if the seller has not completed the remedy within a reasonable time and without any significant inconvenience to the consumer, the consumer may require an appropriate reduction of the price or have the contract rescinded.

5.   The consumer is not entitled to have the contract rescinded if the lack of conformity is minor.

Article 4   Right of redress

Where the final seller is liable to the consumer because of a lack of conformity resulting from an act or omission by the producer, a previous seller in the same chain of contracts or any other intermediary, the final seller shall be entitled to pursue remedies against the person or persons liable in the contractual chain unless he has renounced that entitlement. The person or persons liable against whom the final seller may pursue remedies, together with the relevant actions and conditions of exercise, shall be determined by national law.

Article 5   Time limits

1.   The seller shall be held liable under Article 3 where the lack of conformity becomes apparent within two years as from delivery of the goods. If, under national legislation, the rights laid down in Article 3(2) are

subject to a limitation period, that period shall not expire within a period of two years from the time of delivery.

2.   Member States may provide that, in order to benefit from his rights, the consumer must inform the seller of the lack of conformity within a period of two months from the date on which he detected such lack of conformity.

Member States shall inform the Commission of their use of this paragraph. The Commission shall monitor the effect of the existence of this option for the Member States on consumers and on the internal market.

Not later than . . . ,[1] the Commission shall prepare a report on the use made by Member States of this paragraph. This report shall be published in the Official Journal of the European Communities.

3.   Unless proved otherwise, any lack of conformity which becomes apparent within six months of delivery of the goods shall be presumed to have existed at the time of delivery unless this presumption is incompatible with the nature of the goods or the nature of the lack of conformity.

NOTES

[1]   42 months after entry into force of this Directive.

Article 6   Guarantees

1.   A guarantee shall be legally binding on the offerer under the conditions laid down in the guarantee statement and the associated advertising.

2.   The guarantee shall—
— state that the consumer has legal rights under applicable national legislation governing the sale of consumer goods and make clear that those rights are not affected by the guarantee,
— set our in plain intelligible language the contents of the guarantee and the essential particulars necessary for making claims under the guarantee, notably the duration and territorial scope of the guarantee as well as the name and address of the guarantor.

3.   On request from the consumer, the guarantee shall be made available in writing or feature in another durable medium available and accessible to him.

4.   Within its own territory, the Member State in which the consumer goods are marketed may, in accordance with the rules of the Treaty, provide that the guarantee be drafted in one or more languages which it shall determine from among the official languages of the Community.

5.   Should a guarantee infringe the requirements of paragraphs 2, 3 or 4, the validity of this guarantee shall in no way be affected, and the consumer can still rely on the guarantee and require that it be honoured.

Article 7   Binding nature

1.   Any contractual terms or agreements concluded with the seller before the lack of conformity is brought to the seller's attention which directly or indirectly waive or restrict the rights resulting from this Directive shall, as provided for by national law, not be binding on the consumer.

Member States may provide that, in the case of second-hand goods, the seller and consumer may agree contractual terms or agreements which have a shorter time period for the liability of the seller than that set down in Article 5(1). Such period may not be less than one year.

2.   Member States shall take the necessary measures to ensure that consumers are not deprived of the protection afforded by this Directive as a result of opting for the law of a non-member State as the law applicable to the contract where the contract has a close connection with the territory of the Member States.

Article 8   National law and minimum protection

1.   The rights resulting from this Directive shall be exercised without prejudice to other rights which the consumer may invoke under the national rules governing contractual or non-contractual liability.

2.   Member States may adopt or maintain in force more stringent provisions, compatible with the Treaty in the field covered by this Directive, to ensure a higher level of consumer protection.

Article 9   Transposition

1.   Member States shall bring into force the laws, regulations and administrative provisions necessary to comply with this Directive not later than ...[1] They shall forthwith inform the Commission thereof.

When Member States adopt these measures, they shall contain a reference to this Directive, or shall be accompanied by such reference at the time of their official publication. The procedure for such reference shall be adopted by Member States.

2.   Member States shall communicate to the Commission the provisions of national law which they adopt in the field covered by this Directive.

NOTES

  [1]   36 months after entry into force of this Directive.

Article 10   Review

The Commission shall, not later than . . . ,[1] review the application of this Directive and submit to the European Parliament and the Council a report. The report shall examine, inter alia, the case for introducing the producer's direct liability and, if appropriate, shall be accompanied by proposals.

NOTES

  [1]   Seven years after entry into force of this Directive.

Article 11   Entry into force

This Directive shall enter into force on the day of its publication in the Official Journal of the European Communities.

Article 12

This Directive is addressed to the Member States.

Proposal for a Directive of the European Parliament and of the Council concerning the distance marketing of consumer financial services and amending Council Directive 90/619/EEC and Directives 97/7/EC and 98/27/EC

(98/C 385/10)

(Text with EEA relevance)

COM(1998) 468 final—98/0245(COD)

(Submitted by the Commission on 19 November 1998)

NOTES

Date of Publication in OJ: OJ C385, 11.12.98, p 10.

THE EUROPEAN PARLIAMENT AND THE COUNCIL OF THE EUROPEAN UNION,

Having regard to the Treaty establishing the European Community, and in particular Article 57(2) and Articles 66 and 100a thereof,

Having regard to the proposal from the Commission,

Having regard to the opinion of the Economic and Social Committee,

Acting in accordance with the procedure laid down in Article 189b of the Treaty,

(1)   Whereas it is important, in the context of achieving the aims of the single market, to adopt measures designed to progressively consolidate this market and those measures must contribute to attaining a high level of consumer protection, in accordance with Article 129a of the Treaty;

(2)   Whereas, both for consumers and suppliers of financial services, the distance marketing of financial services will constitute one of the main tangible results of the completion of the internal market;

(3)   Whereas, within the framework of the internal market, it is in the interest of consumers to have access without discrimination to the widest possible range of financial services available in the Community so that they can choose those that are best suited to their needs; whereas in order to safeguard freedom of choice, which is an essential consumer right, a certain degree of protection is required in order to enhance their confidence in distance selling;

(4)   Whereas it is essential to the smooth operation of the internal market for consumers to be able to negotiate and conclude contracts with a supplier established outside their country, regardless of whether the supplier is also established in the consumer's country of residence;

(5)   Whereas the establishment of a legal framework governing the distance marketing of financial services should contribute to promoting the advent of the information society and the development of electronic commerce;

(6)   Whereas Directive 97/7/EC of the European Parliament and of the Council of 20 May 1997 on the protection of consumers in respect of distance contracts,[1] lays down the main rules applicable to distance contracts for goods or services concluded between a supplier and a consumer; whereas, however, that Directive does not cover financial services;

(7)   Whereas, in the context of the analysis conducted by the Commission with a view to ascertaining the need for specific measures in this field, the Commission invited all the interested parties to transmit their comments, notably in connection with the preparation of its Green Paper entitled 'Financial Services-Meeting Consumers' Expectations';[2] whereas the consultations in this context showed that there is a need to strengthen consumer protection in this area; whereas the Commission therefore decided to present a specific proposal concerning the distance marketing of financial services;

(8)   Whereas the adoption by the Member States of conflicting or different consumer protection rules governing the distance marketing of consumer financial services would impede the functioning of the internal market and competition between firms in the market; whereas it is therefore necessary to enact common rules at Community level in this area;

(9)   Whereas, given the high level of consumer protection guaranteed by this Directive, with a view to ensuring the free movement of financial

services, Member States may not adopt provisions other than those laid down in this Directive in the fields harmonised by this Directive;

(10) Whereas this Directive covers all financial services liable to be provided at a distance; whereas, however, certain financial services are governed by specific provisions of Community law; whereas those specific provisions continue to apply to those financial services; whereas that applies in particular to provisions governing prior information of the consumer; whereas, however, it is advisable to lay down principles governing the distance marketing of such services;

(11) Whereas, in accordance with the principles of subsidiarity and proportionality as set out in Article 3b of the Treaty, the objectives of this Directive cannot be sufficiently achieved by the Member States and can therefore be better achieved by the Community; whereas it is necessary by also sufficient to enact measures which allow consumers to inform themselves and to consider the proposed contractual terms and conditions, as well as measures to ensure that those rights are respected; whereas it is also appropriate to enact measures to protect consumers against the high-pressure selling of financial services and against certain unsolicited uses of means of distance communication; whereas consumers cannot fully enjoy the rights vested in them by this Directive unless appropriate arrangements are made for settling disputes;

(12) Whereas contracts negotiated at a distance involve the use of means of distance communication; whereas the various means of communication are used as part of a distance sales or service-provision scheme not involving the simultaneous presence of the supplier and the consumer; whereas the constant development of those means of communication requires principles to be defined that are valid even for those means that are not yet in widespread use; whereas, therefore, distance contracts are to be those the offer, negotiation and conclusion of which are carried out at a distance;

(13) Whereas a single contract involving successive operations may be subject to different legal treatment in the different Member States, whereas, however, it is important that this Directive be applied in the same way in all the Member States; whereas, to this end, it is appropriate that this Directive should be considered to apply to the first of a series of successive operations, or to the first of a series of separate operations over a period of time which may be considered as forming a whole, irrespective of whether that operation or series of operations are the subject of a single contract or several successive contracts;

(14)   Whereas by covering a service-provision scheme organised by the financial services provider, this Directive aims to exclude from its scope services provided on a strictly occasional basis and outside a commercial structure dedicated to the conclusion of distance contracts;

(15)   Whereas the supplier is the person providing services at a distance; whereas this Directive should however also apply when one of the marketing stages involves an intermediary; whereas, having regard to the nature and degree of that involvement, the pertinent provisions of this Directive should apply to such an intermediary, irrespective of his legal status;

(16)   Whereas the use of means of distance communications must not lead to an unwarranted restriction on the information provided to the client; whereas in the interest of transparency this Directive lays down the requirements needed to ensure that an appropriate level of information is provided the consumer both before and after conclusion of the contract; whereas the consumer must receive, before conclusion of the contract, the contractual terms and conditions so that he can properly appraise the offer and hence make a well informed choice; whereas the contractual terms and conditions may not be unilaterally modified for a period of 14 days in order to give the consumer time for reflection;

(17)   Whereas provision should be made for a right of withdrawal on the part of the consumer, without penalty and without having to furnish grounds, whenever the contract has been concluded by the consumer without his having received, at the time of conclusion of the contract, the contractual terms and conditions applicable to it, or whenever he has been unfairly induced to conclude the contract during the reflection period set out in this Directive;

(18)   Whereas provision should be made for a reinforcement of the right of consumers to withdraw from contracts relating to mortgages, life insurance and personal pension operations;

(19)   Whereas consumers should be protected against unsolicited sales; whereas consumers should be exempt from any obligation in the case of unsolicited supplies, the absence of a reply not being construed as signifying consent on their part; whereas, however, this rule should be without prejudice to the tacit renewal of contracts validly concluded between the parties;

(20)   Whereas Member States should take appropriate measures to effectively protect consumers who do not wish to be contacted through

certain means of communication; whereas this Directive is without prejudice to the particular safeguards available to consumers under Community legislation concerning the protection of personal data and privacy;

(21) Whereas, with a view to protecting consumers, it is important to make arrangements for resolving disputes; whereas there is a need for suitable and effective complaint and redress procedures in the Member States with a view to settling potential disputes between suppliers and consumers, by using, where appropriate, existing procedures;

(22) Whereas, as regards consumer access to justice and in particular to courts and tribunals in the case of cross-border disputes, account should be taken of the Communication from the Commission to the Council and European Parliament entitled 'Towards greater effectiveness in the adoption and enforcement of decisions within the European Union';[3]

(23) Whereas Member States should encourage public or private bodies established with a view to settling disputes out of court to cooperate to in resolving cross-border disputes; whereas such cooperation could in particular entail allowing consumers to submit to extra-judicial bodies in the Member State of their residence complaints concerning suppliers established in other Member States;

(24) Whereas the Community and the Member States have entered into commitments in the context of the WTO-General Agreement on Trade in Services (GATS) concerning the possibility for European consumers to purchase banking and investment services abroad; whereas the GATS entitles Member States to adopt measures for prudential reasons, including measures to protect investors, depositors, policy-holders and persons to whom a financial service is owed by the supplier of the financial service; whereas such measures should not impose restrictions going beyond what is required to ensure consumer protection;

(25) Whereas it is consequently necessary to amend Council Directive 90/619/EEC of 8 November 1990 on the coordination of laws, regulations and administrative provisions relating to direct life insurance, laying down provisions to facilitate the effective exercise of freedom to provide services and amending Directive 79/267/EEC,[4] as amended by Directive 92/96/EC;[5]

(26) Whereas, in view of the adoption of this Directive, it is necessary to adapt the scope of Directive 97/7/EC and Directive 98/27/EC of the European Parliament and of the Council of 19 May 1998 on injunctions for the protection of consumers' interests,[6]

NOTES
1   OJ L144, 4.6.97, p 19.
2   COM(96) 209 final, 22.5.96.
3   OJ C33, 31.1.98, p 3.
4   OJ L330, 29.11.90, p 50.
5   OJ L360, 9.12.92, p 1.
6   OJ L166, 11.6.98, p 51.

HAVE ADOPTED THIS DIRECTIVE—

CHAPTER I
SCOPE AND DEFINITIONS

Article 1   Scope

1.   The object of this Directive is to approximate the laws, regulations and administrative provisions of the Member States concerning the distance marketing of consumer financial services.

2.   In the case of contracts for financial services comprising successive operations or a series of separate operations performed over time, the provisions of this Directive shall apply only to the first operation, irrespective of whether those operations are deemed by national law to form part of a single contract or individual separate contracts.

Article 2   Definitions

For the purposes of this Directive—
- (a) 'distance contract' means any contract concerning financial services concluded between a supplier and a consumer under an organised distance sales or service-provision scheme run by the supplier, who, for the purpose of that contract, makes use of means of distance communication up to and including the time at which the contract is concluded;
- (b) 'financial service' means any service relating to the activities of credit institutions, insurance companies or investment firms, as referred to in Council Directives 89/646/EEC,[1] 93/22/EEC,[2] 73/239/EEC[3] and 79/267/EEC;[4] an indicative list of those services is provided in the Annex;
- (c) 'supplier' means any natural or legal person who, acting in his commercial or professional capacity, is the actual provider of services

subject to contracts covered by this Directive or acts as intermediary in the supply of those services or in the conclusion of a distance contract between those parties;

(d) 'consumer' means any natural person resident in the territory of the Community who, in contracts covered by this Directive, is acting for purposes which are outside his trade, business or profession;

(e) 'means of distance communication' refers to any means which, without the simultaneous physical presence of the supplier and the consumer, may be used for the distance marketing of a service between those parties;

(f) 'durable medium' means any instrument enabling the consumer to store information, without himself having to record this information, and in particular floppy disks, CD-ROMs, and the hard drive of the consumer's computer on which electronic mail is stored;

(g) 'operator or supplier of a means of distance communication' means any public or private, natural or legal person whose trade, business or profession involves making one or more means of distance communication available to suppliers.

NOTES

1. OJ L386, 30.12.89, p 1.
2. OJ L141, 11.6. 93, p 27.
3. OJ L228, 16.8.73, p 3.
4. OJ L63, 13.3.79, p 1.

CHAPTER II
RIGHTS AND OBLIGATIONS OF THE PARTIES

Article 3   Right of reflection before conclusion of the contract

1.   Before conclusion of a distance contract, the supplier shall communicate all the contractual terms and conditions to the consumer in writing or in a durable medium available and accessible to him. The supplier may not unilaterally modify these terms for a period of fourteen days.

The parties may agree on a longer period.

However, consumers may conclude the contract before expiry of the period referred to in the first subparagraph or the agreed period.

The consumer's silence at the end of the reflection period shall not be construed as signifying his consent.

2.   The periods referred to in paragraph 1 shall be calculated from the day on which the consumer receives the contractual terms and conditions in writing or in a durable medium available and accessible to him.

3.   Without prejudice to paragraph 1, in the case of contracts concerning the financial services referred to in points 5 and 7 of the Annex, when the supplier communicates the contractual terms and conditions to the consumer before conclusion of the contract, any price which depends on fluctuations in the financial market outside the supplier's control shall be established with the consumer's express consent when the contract is concluded.

4.   The provisions of paragraphs 1 and 2 shall be without prejudice to the rules of the Member States concerning the conclusion of contracts, and in particular the rules governing the manner in which parties express their consent to the contract.

Article 4   Right of withdrawal after conclusion of the contract

1.   Where the contract has been concluded at the consumer's request before the contractual terms and conditions have been communicated to him by the supplier, the supplier shall communicate the contract to the consumer in writing or in a durable medium available and accessible to him once the contract has been concluded.

The consumer has a right of withdrawal for 14 days, without incurring any penalty and without having to indicate his grounds. This period shall be extended to 30 days in the case of contracts relating to mortgages, life assurance or personal pension operations.

The withdrawal period shall be calculated from the day on which the consumer receives the contractual terms and conditions.

The right of withdrawal shall not apply to contracts concerning—
(a) the financial services referred to in points 5 and 7 of the Annex, whose price depends on market fluctuations outside the supplier's control;
(b) non-life insurance policies of less than one month's duration.

2.   Where the contract is concluded by the consumer during the reflection period provided for in Article 3 and he has been unfairly induced to do so by the supplier, the consumer shall have a right of withdrawal for fourteen days without incurring any charge or penalty, and without prejudice to his right to seek compensation for the damage he has suffered.

When suppliers communicate objective information to the consumer on prices of financial services that depend on market fluctuations, this shall not be considered as an unfair inducement.

The withdrawal period shall run from the conclusion of the contract.

3.   The consumer shall exercise his right of withdrawal by notifying the supplier to this effect in writing or in a durable medium available and accessible to the supplier.

4.   Member States shall provide in their legislation that if the price of financial services is fully or partly covered by credit granted to the consumer by the supplier, or by a third party on the basis of an agreement between the third party and the supplier, the credit agreement is cancelled, where the right referred to in paragraph 1 is exercised, without any penalty being imposed on the consumer.

5.   The other legal effects and conditions of withdrawal shall be governed by the law applicable to the contract.

Article 5   Payment for the service provided before withdrawal

1.   Where the consumer exercises his right of withdrawal under Article 4(1), he may be required to pay, without any undue delay, only—
   (a) the price of the financial service actually provided by the supplier, where that price can be determined by the supplier before conclusion of the contract;
   (b) the part of the total price of the financial service covered by the contract on a pro rata basis for the period between the day on which the contract was concluded and the day on which he exercises his right of withdrawal, where the price cannot be determined by the supplier before conclusion of the contract.

2.   The supplier shall inform the consumer, before conclusion of the contract, in any way appropriate to the means of distance communication used, of the price or the amount used as a basis for calculating the price which he will be required to pay pursuant to paragraph 1 if he exercises his right of withdrawal.

Unless he can prove that the consumer was duly informed about the price, the supplier may not require the consumer to pay any amount where he exercises his right of withdrawal.

3. The supplier shall, without any undue delay, return to the consumer any sums he has received from him on conclusion of the distance contract, except for the sums referred to in paragraph 1.

Article 6 Consumer information

The supplier shall inform the consumer, in a clear and comprehensible manner, of his rights pursuant to Articles 3 and 4, prior to conclusion of the contract, in any manner appropriate to the means of distance communication used.

Article 7 Communication using a durable medium

Communication of the contractual terms and conditions provided for in Articles 3 and 4 may be effected in writing or in a durable medium available and accessible to the consumer, notwithstanding any other provision which provides that such communication may only be in writing.

Article 8 Unavailability of the service

1. If the financial service which is the subject of the contract is partly or totally unavailable, the supplier shall duly inform the consumer without any undue delay.

2. If the financial service is totally unavailable, the supplier shall, without any undue delay, reimburse any sum paid by the consumer.

3. If the financial service is only partly available, the contract may only be performed with the express consent of the consumer and the supplier.

Failing this, the supplier shall return to the consumer any sums paid by him.

Where the service is only partly performed, the supplier shall return to the consumer all sums relating to the part of the service that has not been performed.

Article 9 Unsolicited services

1. Without prejudice to the legal rules of the Member States concerning the tacit extension of contracts, the distance supply of unsolicited financial services to consumers shall be prohibited.

2.   Consumers shall be exempt from any obligations in cases of unsolicited supplies and the absence of a response shall not constitute consent by them.

Article 10   Unsolicited communications

1.   The use of automated calling systems without human intervention (automatic calling machines) or fax machines in marketing financial services at a distance may be authorised only in respect of consumers who have already given their consent.

2.   Member States shall take appropriate measures to ensure that communications not solicited by consumers and made with a view to selling distance financial services by means other than those referred to in paragraph 1,
    (a) shall not be authorised if the consent of the consumers in question has not been given, or
    (b) may only be used in the absence of express prior objection from the consumers.

The measures referred to in the first paragraph shall not entail costs for the consumer.

Article 11   Imperative nature of the provisions of this directive

1.   Consumers may not waive the rights conferred on them by this Directive.

2.   Member States shall provide for appropriate penalties in the event of the supplier's failure to comply with Articles 6 and 10.

In such cases they shall give consumers the right to cancel the contract at any time, without charge or penalties, and ensure that they are compensated for any damage they have suffered without any undue delay. Such compensation may in particular include reimbursement of sums paid by the consumer to the supplier in performance of the contract.

3.   Consumers may not be deprived of the protection granted by this Directive where the law governing the contract is that of a third country if the consumer is resident on the territory of a Member State and the contract has a close link with the Community.

CHAPTER III
DISPUTES

Article 12   Settlement of disputes

1.   Member States shall ensure that adequate and effective complaints and redress procedures for the settlement of disputes between suppliers and consumers are put in place, using existing procedures where appropriate.

2.   The procedures referred to in paragraph 1 shall include provisions whereby one or more of the following bodies, as determined by national law, may take action under national law before the courts or competent administrative bodies to ensure that the national provisions for the implementation of this directive are applied—
   (a) public bodies or their representatives;
   (b) consumer organisations having a legitimate interest in protecting consumers;
   (c) professional organisations having a legitimate interest in acting.

3.   Member States shall encourage the public or private bodies established for the out-of-court settlement of disputes to co-operate in the resolution of cross-border disputes.

4.   Member States shall take the measures necessary to ensure that operators and suppliers of means of distance communication put an end to practices that have been declared to be contrary to this Directive, on the basis of a judicial decision, an administrative decision or a decision issued by a supervisory authority notified to them, where those operators or suppliers are in a position to do so.

Article 13   Burden of proof

The burden of proof in respect of the supplier's obligations to inform the consumer and the consumer's consent to conclusion of the contract and, where appropriate, its performance, shall lie with the supplier.

Any contractual term or condition providing that the burden of proof of the respect by the supplier of all or part of the obligations incumbent on him pursuant to this Directive should lie with the consumer shall be an unfair term within the meaning of Council Directive 93/13/EEC.[1]

NOTES
¹   OJ L95, 21.4.93, p 29.

(Arts 14-16 (Ch IV) amend Directive 90/619/EEC, 97/7/EC and 98/27/EC.)

CHAPTER V
FINAL PROVISIONS

Article 17   Transposition

1.   Member States shall bring into force the laws, regulations and administrative provisions necessary to comply with this Directive by 30 June 2002 at the latest. They shall forthwith inform the Commission thereof.

At the time of their official publication, these provisions shall refer to this Directive or shall be accompanied by such a reference. Member States shall determine how such reference is to be made.

2.   Member States shall communicate to the Commission the text of the main laws, regulations or administrative provisions which they adopt in the field governed by this Directive. In that communication, Member States shall provide a table showing the national provisions corresponding to each article of this Directive.

Article 18   Entry into force

This Directive shall enter into force on the twentieth day following that of its publication in the Official Journal of the European Communities.

Article 19   Addressees

This Directive is addressed to the Member States.

ANNEX
INDICATIVE LIST OF FINANCIAL SERVICES

1.   Acceptance of deposits and other repayable funds

2.   Lending, in particular consumer credit and mortgage loans

3.   Financial leasing

4.   Money transfers, issuing and administering means of payment

5.   Foreign exchange services

6.   Guarantees and commitments

7.   Reception, transmission and/or execution of orders related to, and services in respect of or related to the following financial products—
    (a) money market instruments
    (b) transferable securities
    (c) UCITS and other collective investment schemes
    (d) financial futures and options
    (e) exchange and interest rate instruments

8.   Portfolio management and investment advice concerning any of the instruments listed under 7

9.   Safe-keeping and administration of securities

10.   Safe custody services

11   Non-life insurance

12.   Life assurance

13.   Life assurance linked to investment funds

14.   Permanent health insurance

15.   Capital redemption operations

16.   Individual pension schemes

Proposal for a European Parliament and Council Directive on certain legal aspects of electronic commerce in the internal market

(1999/C 30/04)

(Text with EEA relevance)

COM(1998) 586 final—98/0325(COD)

(Submitted by the Commission on 23 December 1998)

NOTES

Date of Publication in OJ: OJ C30, 5.2.99, p 4.

THE EUROPEAN PARLIAMENT AND THE COUNCIL OF THE EUROPEAN UNION,

Having regard to the Treaty establishing the European Community, and in particular Articles 57(2), 66 and 100a thereof,

Having regard to the proposal from the Commission,

Having regard to the opinion of the Economic and Social Committee,

Acting in accordance with the procedure referred to in Article 189b of the Treaty,

(1) Whereas the European Union is seeking to forge ever closer links between the States and peoples of Europe, to ensure economic and social progress; whereas, in accordance with Article 7a of the Treaty, the internal market comprises an area without internal frontiers in which the free movement of goods, services and the freedom of establishment are ensured; whereas the development of Information Society services within the area without internal frontiers is vital to eliminating the barriers which divide the European peoples;

(2) Whereas the development of electronic commerce within the Information Society offers significant employment opportunities in the Community, particularly in small and medium-sized enterprises, and will stimulate economic growth and investment in innovation by European companies;

(3) Whereas Information Society services span a wide range of economic activities which can, in particular, consist of selling goods online; whereas they are not solely restricted to services giving rise to online contracting but also, in so far as they represent an economic activity, extend to services which are not remunerated by those who receive them, such as those offering online information; whereas Information Society services also include online activities via telephony and telefax;

(4) Whereas the development of Information Society services within the Community is restricted by a number of legal obstacles to the proper functioning of the internal market which hamper or make less attractive the exercise of the freedom of establishment and the freedom to provide services; whereas these obstacles arise from divergences in legislation and from the legal uncertainty as to which national rules apply to such services; whereas, in the absence of coordination and adjustment of legislation in the relevant areas, obstacles might be justified in the light of the case-law of the Court of Justice of the European Communities; whereas legal uncertainty exists with regard to the extent to which Member States may control services originating from another Member State;

(5) Whereas, in the light of Community objectives, of Articles 52 and 59 of the Treaty and of secondary Community law, these obstacles should be eliminated by coordinating certain national laws and by clarifying certain legal concepts at Community level to the extent necessary for the proper functioning of the internal market; whereas, by dealing only with certain specific matters which give rise to problems for the internal market, this Directive is fully consistent with the need to respect the principle of subsidiarity as set out in Article 3b of the Treaty;

(6) Whereas, in accordance with the principle of proportionality, the measures provided for in this Directive are strictly limited to the minimum needed to achieve the objective of the proper functioning of the internal market; whereas, where action at Community level is necessary, and in order to guarantee an area which is truly without internal frontiers as far as electronic commerce is concerned, the Directive must ensure a high level of protection of objectives of general interest, in particular consumer protection and the protection of public health; whereas according to Article 129 of the Treaty, the protection of public health is an essential component of other

Community policies; whereas this Directive does not impact on the legal requirements applicable to the delivery of goods as such, nor those applicable to services which are not Information Society services;

(7)   Whereas this Directive does not aim to establish specific rules on international private law relating to conflicts of law or jurisdiction and is therefore without prejudice to the relevant international conventions;

(8)   Whereas Information Society services should be supervised at the source of the activity, in order to ensure an effective protection of public interest objectives; whereas, to that end, it is necessary to ensure that the competent authority provides such protection not only for the citizens of its own country but for all Community citizens; whereas, moreover, in order to effectively guarantee freedom to provide services and legal certainty for suppliers and recipients of services, such Information Society services should only be subject to the law of the Member State in which the service provider is established; whereas, in order to improve mutual trust between Member States, it is essential to state clearly this responsibility on the part of the Member State whence the services originate;

(9)   Whereas the place at which a service provider is established should be determined in accordance with the case-law of the Court of Justice; whereas the place of establishment of a company providing services via an internet website is not the place at which the technology supporting its website is located or the place at which its website is accessible; whereas, where the same supplier has a number of establishments the competent Member State will be the one in which the supplier has the centre of his activities; whereas in cases where it is particularly difficult to assess in which Member States the supplier is established, cooperative procedures should be established between the Member States and the consultative committee should be capable of being convened in urgent cases to examine such difficulties;

(10)   Whereas commercial communications are essential for the financing of Information Society services and for developing a wide variety of new, charge-free services; whereas in the interests of consumer protection and fair trading, commercial communications, including discounts, promotional offers and promotional competitions, must meet a number of transparency requirements and that these requirements are without prejudice to Directive 97/7/EC of the European Parliament and of the Council on the protection of consumers in respect of distance contracts;[1] whereas this Directive should not affect existing directives on commercial communications, in particular Directive 98/43/EC of the European Parliament and of the Council[2] on tobacco advertising;

(11)   Whereas Article 10(2) of Directive 97/7/EC and Article 12(2) of European Parliament and Council Directive 97/66/EC of 15 December 1997 concerning the processing of personal data and the protection of privacy in the telecommunications sector[3] address the issue of consent by receivers to certain forms of unsolicited commercial communication and are fully applicable to Information Society services;

(12)   Whereas, in order to remove barriers to the development of cross-border services within the Community which professional practitioners might offer on the internet, it is necessary that compliance be guaranteed at Community level with professional rules aiming, in particular, to protect consumers or public health; whereas codes of conduct at Community level would be the best means of determining the rules on professional ethics applicable to commercial communication; whereas the drawing-up or, where appropriate, the adaptation of such rules should in the first place be encouraged by, rather than laid down in, this Directive; whereas the regulated professional activities governed by this Directive should be understood in the light of the definition set out in Article 1(d) of Council Directive 89/48/EEC of 21 December 1988 on a general system for the recognition of higher-education diplomas awarded on completion of professional education and training of at least three years' duration;[4]

(13)   Whereas each Member State should amend its legislation containing requirements, and in particular requirements as to form, which are likely to curb the use of contracts by electronic means, subject to any Community measure in the field of taxation that could be adopted on electronic invoicing; whereas the examination of the legislation requiring such adjustment should be systematic and should cover all the necessary stages and acts of the contractual process, including the filing of the contract; whereas the result of this amendment should be to make contracts concluded electronically genuinely and effectively workable in law and in practice; whereas the legal effect of electronic signatures is dealt with by European Parliament and Council Directive 98/. . ./EC [on a common framework for electronic signatures];[5] whereas it is necessary to clarify at what point in time a contract entered into electronically is considered to be actually concluded; whereas the service recipient's agreement to enter into a contract may take the form of an online payment; whereas the acknowledgement of receipt by a service provider may take the form of the online provision of the service paid for;

(14)   Whereas, amongst others, Council Directive 93/13/EEC[6] regarding unfair contract terms and Directive 97/7/EC, from a vital element for protecting consumers in contractual matters; whereas those directives also

apply in their entirety to Information Society services; whereas that same Community acquis also embraces Council Directive 84/450/EEC[7] on misleading advertising, as amended by European Parliament and Council Directive 97/55/EC,[8] Council Directive 87/102/EEC[9] on consumer credit; as last amended by European Parliament and Council Directive 98/7/EC,[10] Council Directive 90/314/EEC[11] on package travel, package holidays and package tours, and European Parliament and Council Directive 98/6/EC[12] on the indication of prices of products offered to consumers; whereas this Directive should be without prejudice to Directive 98/43/EC, adopted within the framework of the internal market, or to other directives on the protection of public health;

(15)   Whereas the confidentiality of electronic messages is guaranteed by Article 5 of Directive 97/66/EC; whereas, in accordance with that Directive, Member States must prohibit any kind of interception or surveillance of such electronic messages by others than the senders and receivers;

(16)   Whereas, both existing and emerging disparities in Member States' legislation and case-law concerning civil and criminal liability of service providers acting as intermediaries prevent the smooth functioning of the internal market, in particular by impairing the development of cross-border services and producing distortions of competition; whereas service providers have a duty to act, under certain circumstances, with a view to preventing or ceasing illegal activities; whereas the provisions of this Directive should constitute the appropriate basis for the development of rapid and reliable procedures for removing and disabling access to illegal information; whereas such mechanisms could be developed on the basis of voluntary agreements between all parties concerned; whereas it is in the interest of all parties involved in the provision of Information Society services to adopt and implement such procedures; whereas the provisions of this Directive relating to liability should not preclude the development and effective operation, by the different interested parties, of technical systems of protection and identification;

(17)   Whereas each Member State should be required, where necessary, to amend any legislation which is liable to hamper the use of schemes for the out-of-court settlement of disputes through electronic channels; whereas the result of this amendment must be to make the functioning of such schemes genuinely and effectively possible in law and in practice, even across borders; whereas the bodies responsible for such out-of-court settlement of consumer disputes must comply with certain essential principles, as set out in Commission Recommendation 98/257/EC of 30 March 1998 on the principles applicable to the bodies responsible for such settlement of consumer disputes.[13]

(18)   Whereas it is necessary to exclude certain activities from the scope of this Directive, on the grounds that the freedom to provide services in these fields cannot, at this stage, be guaranteed under the Treaty or existing secondary legislation; whereas excluding these activities does not preclude any instruments which might prove necessary for the proper functioning of the internal market; whereas taxation, particularly value-added tax imposed on a large number of the services covered by this Directive, must be excluded from the scope of this Directive; whereas, in this respect, the Commission also intends to extend the application of the principle of taxation at source to the provision of services within the internal market, thus giving its approach a general coherence;

(19)   Whereas as regards the derogation contained in this Directive regarding contractual obligations concerning contracts concluded by consumers, those obligations should be interpreted as including information on the essential elements of the content of the contract, including consumer rights, which have a determining influence on the decision to contract;

(20)   Whereas this Directive should not apply to services supplied by service providers established in a third country; whereas, in view of the global dimension of electronic commerce, it is, however, appropriate to ensure that the Community rules are consistent with international rules; whereas this Directive is without prejudice to the results of discussions within international organisations (WTO, OECD, UNCITRAL) on legal issues; whereas this Directive should also be without prejudice to the discussions within the Global Business Dialogue which were launched on the basis of the Commission Communication of 4 February 1998 on 'Globalisation and the Information Society-The need for strengthened international coordination';[14]

(21)   Whereas the Member States need to ensure, that, when Community acts are transposed into national legislation, Community law is duly applied with the same effectiveness and thoroughness as national law;

(22)   Whereas the adoption of this Directive will not prevent the Member States from taking into account the various social, societal and cultural implications which are inherent in the advent of the Information Society nor hinder cultural, and notably audiovisual, policy measures, which the Member States might adopt, in conformity with Community law, taking into account their linguistic diversity, national and regional specificities and their cultural heritage; whereas, in any case, the development of the Information Society must ensure that Community citizens can have access to the cultural European heritage provided in the digital environment;

(23) Whereas the Council, in its Resolution of 3 November 1998 on the consumer aspects of the Information Society, stressed that the protection of consumers deserved special attention in this field; whereas the Commission will examine the degree to which existing consumer protection rules provide insufficient protection in the context of the Information Society and will identify, where necessary, the deficiencies of this legislation and those issues which could require additional measures; whereas, if need be, the Commission should make specific additional proposals to resolve such deficiencies that will thereby have been identified;

(24) Whereas this Directive should be without prejudice to Council Regulation (EEC) No 2299/89 of 24 July 1989 on a code of conduct for computerised reservation systems,[15] as amended by Regulation (EEC) No 3089/93;[16]

(25) Whereas Commission Regulation (EC) No 2027/97[17] and the Warsaw Convention of 12 October 1929 place various obligations upon air carriers regarding the provision of information to their passengers, including information about the liability of the carrier; whereas this Directive is without prejudice to the requirements of those instruments.

NOTES

[1]   OJ L144, 4.6.97, p 19.
[2]   OJ L213, 30.7.98, p 9.
[3]   OJ L24, 30.1.98, p 1.
[4]   OJ L19, 24.1.89, p 16.
[5]   COM(1998) 297 final, 13.5.98.
[6]   OJ L95, 21.4.93, p 29.
[7]   OJ L250, 19.9.84, p 17.
[8]   OJ L290, 23.10.97, p 18.
[9]   OJ L42, 12. 2.87, p 48.
[10]   OJ L101, 1.4.98, p 17.
[11]   OJ L158, 23.6.98, p 59.
[12]   OJ L80, 18.3.98, p 27.
[13]   OJ L115, 17.4.98, p 31.
[14]   COM(1998) 50 final.
[15]   OJ L220, 29.7.89, p 1.
[16]   OJ L278, 11.11.93, p 1.
[17]   OJ L285, 17.10.97, p 1.

HAVE ADOPTED THIS DIRECTIVE—

CHAPTER I
GENERAL PROVISIONS

Article 1 Objective and scope

1.   This Directive seeks to ensure the proper functioning of the internal market, particularly the free movement of Information Society services between the Member States.

2.   This Directive approximates, to the extent necessary for the achievement of the objective set out in paragraph 1, national provisions on Information Society services relating to the internal market arrangements, the establishment of service providers, commercial communications, electronic contracts, the liability of intermediaries, codes of conduct, out-of-court dispute settlements, court actions and cooperation between Member States.

3.   This Directive complements Community law applicable to Information Society services without prejudice to the existing level of protection for public health and consumer interests, as established by Community acts, including those adopted for the functioning of the internal market.

Article 2 Definitions

For the purpose of this Directive, the following terms shall bear the following meanings—
  (a) 'Information Society services': any service normally provided for remuneration, at a distance, by electronic means and at the individual request of a recipient of services;

For the purpose of this definition—
  — 'at a distance' means that the service is provided without the parties being simultaneously present;
  — 'by electronic means' means that a service is sent initially and received at its destination by means of electronic equipment for the processing (including digital compression) and storage of data, and entirely transmitted, conveyed and received by wire, by radio, by optical means or by other electromagnetic means;
  — 'at the individual request of a recipient of services' means a service provided through the transmission of data on individual request.

(b) 'service provider': any natural or legal person providing an Information Society service;

(c) 'established service provider': a service provider who effectively pursues an economic activity using a fixed establishment for an indeterminate duration. The presence and use of the technical means and technologies required to provide the service do not constitute an establishment of the provider;

(d) 'recipient of the service': any natural or legal person who, for professional ends or otherwise, uses an Information Society service, in particular for the purpose of seeking information or making it accessible;

(e) 'commercial communications': any form of communication designed to promote, directly or indirectly, the goods, services or image of a company, organisation or person pursuing a commercial, industrial or craft activity or exercising a liberal profession. The following do not as such constitute commercial communications—

— information allowing direct access to the activity of the company, organisation or person, in particular a domain name or an e-mail address,

— communications relating to the goods, services or image of the company, organisation or person compiled in an independent manner, in particular without financial consideration.

(f) 'coordinated field': the requirements applicable to Information Society service providers and Information Society services.

Article 3   Internal market

1.   Each Member State shall ensure that the Information Society services provided by a service provider established on its territory comply with the national provisions applicable in the Member State in question which fall within this Directive's coordinated field.

2.   Member States may not, for reasons falling within this Directive's coordinated field, restrict the freedom to provide Information Society services from another Member State.

3.   Paragraph 1 shall cover the provisions set out in Articles 9, 10 and 11 only in so far as the law of the Member State applies by virtue of its rules of international private law.

CHAPTER II
PRINCIPLES

SECTION 2
COMMERCIAL COMMUNICATIONS

Article 6   Information to be provided

Member States shall lay down in their legislation that commercial communication shall comply with the following conditions—
  (a) the commercial communication shall be clearly identifiable as such;
  (b) the natural or legal person on whose behalf the commercial communication is made shall be clearly identifiable;
  (c) promotional offers, such as discounts, premium and gifts, where authorised, shall be clearly identifiable as such, and the conditions which are to be met to qualify for them shall be easily accessible and be presented accurately and unequivocally;
  (d) promotional competitions or games, where authorised, shall be clearly identifiable as such, and the conditions for participation shall be easily accessible and be presented accurately and unequivocally.

Article 7   Unsolicited commercial communication

Member States shall lay down in their legislation that unsolicited commercial communication by e-mail must be clearly and unequivocally identifiable as such as soon as it is received by the recipient.

Article 8   Regulated professions

1.   Member States shall lay down in their legislation relating to commercial communication by regulated professions that the provision of Information Society services is authorised provided that the professional rules regarding the independence, dignity and honour of the profession, professional secrecy and fairness towards clients and other members of the profession are met.

2.   Member States and the Commission shall encourage professional associations and bodies to establish codes of conduct at Community level in order to determine the types of information that can be given for the purpose of providing the Information Society service in conformity with the rules referred to in paragraph 1.

3.   Where necessary, in order to ensure the proper functioning of the internal market, and in the light of the internal market, and in the light of the codes of conduct applicable at Community level, the Commission may stipulate, in accordance with the procedure laid down in Article 23, the information referred to in paragraph 2.

SECTION 3
ELECTRONIC CONTRACTS

Article 9   Treatment of electronic contracts

1.   Member States shall ensure that their legislation allows contracts to be concluded electronically. Member States shall in particular ensure that the legal requirements applicable to the contractual process neither prevent the effective use of electronic contracts nor result in such contracts being deprived of legal effect and validity on account of their having been made electronically.

2.   Member States may lay down that paragraph 1 shall not apply to the following contracts—
  (a) contracts requiring the involvement of a notary;
  (b) contracts which, in order to be valid, are required to be registered with a public authority;
  (c) contracts governed by family law;
  (d) contracts governed by the law of succession.

3.   The list of categories of contract provided for in paragraph 2 may be amended by the Commission in accordance with the procedure laid down in Article 23.

4.   Member States shall submit to the Commission a complete list of the categories of contracts covered by the derogations provided for in paragraph 2.

Article 10   Information to be provided

1.   Member States shall lay down in their legislation that, except when otherwise agreed by professional persons, the manner of the formation of a contract by electronic means shall be explained by the service provider clearly and unequivocally, and prior to the conclusion of the contract. The information to be provided shall include, in particular—

(a) the different stages to follow to conclude the contract;
(b) whether or not the concluded contract will be filed and whether it will be accessible;
(c) the expedients for correcting handling errors.

2.   Member States shall provide in their legislation that the different steps to be followed for concluding a contract electronically shall be set out in such a way as to ensure that parties can give their full and informed consent.

3.   Member States shall lay down in their legislation that, except when otherwise agreed by professional parties, the service providers shall indicate any codes of conduct to which they subscribe and information on how those codes can be consulted electronically.

Article 11   Moment at which the contract is concluded

1.   Member States shall lay down in their legislation that, save where otherwise agreed by professional persons, in cases where a recipient, in accepting a service provider's offer, is required to give his consent through technological means, such as clicking on an icon, the following principles apply—
(a) the contract is concluded when the recipient of the service—
    — has received from the service provider, electronically, an acknowledgement of receipt of the recipient's acceptance, and
    — has confirmed receipt of the acknowledgement of receipt;
(b) acknowledgement of receipt is deemed to be received and conformation is deemed to have given when the parties to whom they are addressed are able to access them;
(c) acknowledgement of receipt by the service provider and confirmation of the service recipient shall be sent as quickly as possible.

2.   Member States shall lay down in their legislation that, save where otherwise agreed by professional persons, the service provider shall make available to the recipient of the service appropriate means allowing him to identify and correct handling errors.

CHAPTER IV
EXCLUSIONS FROM SCOPE AND DEROGATIONS

Article 22   Exclusions and derogations

1.   This Directive shall not apply to—

    (a) taxation;

    (b) the field covered by Directive 95/46/EC of the European Parliament and of the Council;[1]

    (c) the activities of Information Society services referred to in Annex I. This list of activities may be amended by the Commission in accordance with the procedure laid down by Article 23.

2.   Article 3 shall not apply to the fields referred to in Annex II.

3.   By way of derogation from Article 3(2), and without prejudice to court action, the competent authorities of Member States may take such measures restricting the freedom to provide an Information Society service as are consistent with Community law and with the following provisions—

    (a) the measures shall be—

        (i) necessary for one of the following reasons—
- public policy, in particular the protection of minors, or the fight against any incitement to hatred on grounds of race, sex, religion or nationality,
- the protection of public health,
- public security,
- consumer protection;

        (ii) taken against an Information Society service which prejudices the objectives referred to in point (i) or which presents a serious and grave risk of prejudice to those objectives,

        (iii) proportionate to those objectives;

    (b) prior to taking the measures in question, the Member State has—
- asked the Member State referred to in Article 3(1) to take measures and the latter did not take such measures, or the latter were inadequate;
- notified the Commission and the Member State in which the service provider is established of its intention to take such measures;

    (c) Member States may lay down in their legislation that, in the case of urgency, the conditions stipulated in point (b) do not apply. Where this is the case, the measures shall be notified in the shortest possible time to the Commission and to the Member State in which the service provider is established, indicating the reasons for which the Member State considers that there is urgency;

    (d) the Commission may decide on the compatibility of the measures with Community law. Where it adopts a negative decision, the Member States shall refrain from taking any proposed measures or shall be required to urgently put an end to the measures in question.

NOTES

[1]   OJ L281, 23.11.95, p 31.

CHAPTER V
ADVISORY COMMITTEE AND FINAL PROVISIONS

Article 25 Implementation

Member States shall bring into force the laws, regulations and administrative provisions necessary to comply with this Directive within one year of its entry into force. They shall forthwith inform the Commission thereof.

When Member States adopt these provisions, these shall contain a reference to this Directive or shall be accompanied by such reference at the time of their official publication. The methods of making such reference shall be laid down by Member States.

Article 26 Entry into force

This Directive shall enter into force on the twentieth day following that of its publication in the Official Journal of the European Communities.

Article 27 Addressees

This Directive is addressed to the Member States.

ANNEX I
ACTIVITIES EXCLUDED FROM THE SCOPE OF APPLICATION OF THE
DIRECTIVE

Information Society services' activities, as referred to in Article 22(1), which are not covered by this Directive—
— the activities of notaries;
— the representation of a client and defence of his interests before the courts;
— gambling activities, excluding those carried out for commercial communication purposes.

ANNEX II
DEROGATIONS FROM ARTICLE 3

As referred to in Article 22(2) in which Article 3 does not apply—
— copyright, neighbouring rights, rights referred to in Directive 87/54/EEC[1] and Directive 96/6/EC[2] as well as industrial property rights;

— the emission of electronic money by institutions in respect of which Member States have applied one of the derogations provided for in Article 7(1) of Directive . . ./. . ./EC;[3]
— Article 44 paragraph 2 of Directive 85/611/EEC;[4]
— Article 30 and Title IV of Directive 92/49/EEC;[5] Title IV of Directive 92/96/EEC,[6] Articles 7 and 8 of Directive 88/357/EEC[7] and Article 4 of Directive 90/619/EEC;[8]
— contractual obligations concerning consumer contracts;
— unsolicited commercial communications by e-mail, or by an equivalent individual communication.

NOTES

[1]   Council Directive 87/54/EEC of 16 December 1986 on the legal protection of topographies of semiconductor products; OJ L24, 27.1.87, p 36.

[2]   Directive 96/9/EC of the European Parliament and of the Council of 11 March 1996 on the legal protection of databases; OJ L77, 27.3.96, p 20.

[3]   European Parliament and Council Directive [on the taking up and the prudential supervision of the business of electronic money institutions].

[4]   Council Directive 85/611/EEC of 20 December 1985 on the coordination of laws, regulations and administrative provisions relating to undertaking for collective investment in transferable securities (UCITS), OJ L375, 31.12.85, p 3, as last amended by Directive 95/26/EC of the European Parliament and of the Council (OJ L168, 18.7.95, p 7).

[5]   Council Directive 92/49/EEC of 18 June 1992 on the coordination of laws, regulations and administrative provisions relating to direct insurance other than life assurance and amending Directives 73/239/EEC and 88/357/EEC (third non-life insurance Directive) OJ L228, 11.8.92, p 1, as amended by Directive 95/26/EC.

[6]   Council Directive 92/56/EEC of 10 November 1992 on the coordination of laws, regulations and administrative provisions relating to direct life insurance and amending Directives 79/267/EEC and 90/619/EEC (third life assurance Directive), OJ L360, 9.12.92, p 1, as amended by Directive 95/26/EC.

[7]   Second Council Directive 88/357/EEC of 22 June 1988 on the coordination of laws, regulations and administrative provisions relating to direct insurance other than life assurance and laying down provisions to facilitate the effective exercise of freedom to provide services and amending Directive 73/239/EEC, OJ L172, 4.7.88, p 1, as last amended by Directive 92/49/EEC.

[8]   Council Directive 90/619/EEC of 8 November 1990 on the coordination of laws, regulations and administrative provisions relating to direct life assurance laying down provisions to facilitate the effective exercise of freedom to provide services and amending Directive 79/267/EEC, OJ L330, 29.11.90, p 50, as amended by Directive 92/96/EEC.

PART IV

INTERNATIONAL MATERIALS

United Nations Convention on the Carriage of Goods by Sea, 1978

(Hamburg, 31 March 1978)

NOTES

This Convention was adopted by the United Nations Conference on the Carriage of Goods by Sea convened in Hamburg, at the invitation of the Government of the Federal Republic of Germany, from 6 to 31 March 1978.

PREAMBLE

THE STATES PARTIES TO THIS CONVENTION,

HAVING RECOGNISED the desirability of determining by agreement certain rules relating to the carriage of goods by sea,

HAVE DECIDED to conclude a Convention for this purpose and have thereto agreed as follows—

PART I
GENERAL PROVISIONS

Article 1   Definitions

In this Convention—
1. "Carrier" means any person by whom or in whose name a contract of carriage of goods by sea has been concluded with a shipper.
2. "Actual carrier" means any person to whom the performance of the carriage of the goods, or of part of the carriage, has been entrusted by the carrier, and includes any other person to whom such performance has been entrusted.
3. "Shipper" means any person by whom or in whose name or on whose behalf a contract of carriage of goods by sea has been concluded with a carrier, or any person by whom or in whose name or on whose behalf the goods are actually delivered to the carrier in relation to the contract of carriage by sea.
4. "Consignee" means the person entitled to take delivery of the goods.

5. "Goods" includes live animals; where the goods are consolidated in a container, pallet or similar article of transport or where they are packed, goods includes such article of transport or packaging if supplied by the shipper.

6. "Contract of carriage by sea" means any contract whereby the carrier undertakes against payment of freight to carry goods by sea from one port to another; however, a contract which involves carriage by sea and also carriage by some other means is deemed to be a contract of carriage by sea for the purposes of this Convention only in so far as it relates to the carriage by sea.

7. "Bill of lading" means a document which evidences a contract of carriage by sea and the taking over or loading of the goods by the carrier, and by which the carrier undertakes to deliver the goods against surrender of the document. A provision in the document that the goods are to be delivered to the order of a named person, or to order, or to bearer, constitutes such an undertaking.

8. "Writing" includes, *inter alia*, telegram and telex.

Article 2   Scope of application

1.   The provisions of this Convention are applicable to all contracts of carriage by sea between two different States, if—
   (a) the port of loading as provided for in the contract of carriage by sea is located in a Contracting State, or
   (b) the port of discharge as provided for in the contract of carriage by sea is located in a Contracting State, or
   (c) one of the optional ports of discharge provided for in the contract of carriage by sea is the actual port of discharge and such port is located in a Contracting State, or
   (d) the bill of lading or other document evidencing the contract of carriage by sea is issued in a Contracting State, or
   (e) the bill of lading or other document evidencing the contract of carriage by sea provides that the provisions of this Convention or the legislation of any State giving effect to them are to govern the contract.

2.   The provisions of this Convention are applicable without regard to the nationality of the ship, the carrier, the actual carrier, the shipper, the consignee or any other interested person.

3.   The provisions of this Convention are not applicable to charter-parties. However, where a bill of lading is issued pursuant to a charter-party, the

provisions of the Convention apply to such a bill of lading if it governs the relation between the carrier and the holder of the bill of lading, not being the charterer.

4.   If a contract provides for future carriage of goods in a series of shipments during an agreed period, the provisions of this Convention apply to each shipment. However, where a shipment is made under a charter-party, the provisions of paragraph 3 of this article apply.

Article 3   Interpretation of the Convention

In the interpretation and application of the provisions of this Convention regard shall be had to its international character and to the need to promote uniformity.

PART II
LIABILITY OF THE CARRIER

Article 4   Period of responsibility

1.   The responsibility of the carrier for the goods under this Convention covers the period during which the carrier is in charge of the goods at the port of loading, during the carriage and at the port of discharge.

2.   For the purpose of paragraph 1 of this article, the carrier is deemed to be in charge of the goods—
    (a) from the time he has taken over the goods from—
        (i)   the shipper, or a person acting on his behalf; or
        (ii) an authority or other third party to whom, pursuant to law or regulations applicable at the port of loading, the goods must be handed over for shipment;
    (b) until the time he has delivered the goods—
        (i)   by handing over the goods to the consignee; or
        (ii) in cases where the consignee does not receive the goods from the carrier, by placing them at the disposal of the consignee in accordance with the contract or with the law or with the usage of the particular trade, applicable at the port of discharge; or
        (iii) by handing over the goods to an authority or other third party to whom, pursuant to law or regulations applicable at the port of discharge, the goods must be handed over.

3.   In paragraphs 1 and 2 of this article, reference to the carrier or to the consignee means, in addition to the carrier or the consignee, the servants or agents, respectively of the carrier or the consignee.

Article 5   Basis of liability

1.   The carrier is liable for loss resulting from loss of or damage to the goods, as well as from delay in delivery, if the occurrence which caused the loss, damage or delay took place while the goods were in his charge as defined in article 4, unless the carrier proves that he, his servants or agents took all measures that could reasonably be required to avoid the occurrence and its consequences.

2.   Delay in delivery occurs when the goods have not been delivered at the port of discharge provided for in the contract of carriage by sea within the time expressly agreed upon or, in the absence of such agreement, within the time which it would be reasonable to require of a diligent carrier, having regard to the circumstances of the case.

3.   The person entitled to make a claim for the loss of goods may treat the goods as lost if they have not been delivered as required by article 4 within 60 consecutive days following the expiry of the time for delivery according to paragraph 2 of this article.

4.—(a)   The carrier is liable
    (i)   for loss of or damage to the goods or delay in delivery caused by fire, if the claimant proves that the fire arose from fault or neglect on the part of the carrier, his servants or agents;
    (ii)   for such loss, damage or delay in delivery which is proved by the claimant to have resulted from the fault or neglect of the carrier, his servants or agents in taking all measures that could reasonably be required to put out the fire and avoid or mitigate its consequences.
  (b)   In case of fire on board the ship affecting the goods, if the claimant or the carrier so desires, a survey in accordance with shipping practices must be held into the cause and circumstances of the fire, and a copy of the surveyor's report shall be made available on demand to the carrier and the claimant.

5.   With respect to live animals, the carrier is not liable for loss, damage or delay in delivery resulting from any special risks inherent in that kind of carriage. If the carrier proves that he has complied with any special

instructions given to him by the shipper respecting the animals and that, in the circumstances of the case, the loss, damage or delay in delivery could be attributed to such risks, it is presumed that the loss, damage or delay in delivery was so caused, unless there is proof that all or a part of the loss, damage or delay in delivery resulted from fault or neglect on the part of the carrier, his servants or agents.

6. The carrier is not liable, except in general average, where loss, damage or delay in delivery resulted from measures to save life or from reasonable measures to save property at sea.

7. Where fault or neglect on the part of the carrier, his servants or agents combines with another cause to produce loss, damage or delay in delivery, the carrier is liable only to the extent that the loss, damage or delay in delivery is attributable to such fault or neglect, provided that the carrier proves the amount of the loss, damage or delay in delivery not attributable thereto.

Article 6   Limits of liability

1.—(a) The liability of the carrier for loss resulting from loss of or damage to goods according to the provisions of article 5 is limited to an amount equivalent to 835 units of account per package or other shipping unit or 2.5 units of account per kilogram of gross weight of the goods lost or damaged, whichever is the higher.

   (b) The liability of the carrier for delay in delivery according to the provisions of article 5 is limited to an amount equivalent to two and a half times the freight payable for the goods delayed, but not exceeding the total freight payable under the contract of carriage of goods by sea.

   (c) In no case shall the aggregate liability of the carrier, under both subparagraphs (a) and (b) of this paragraph, exceed the limitation which would be established under subparagraph (a) of this paragraph for total loss of the goods with respect to which such liability was incurred.

2. For the purpose of calculating which amount is the higher in accordance with paragraph 1(a) of this article, the following rules apply—

   (a) Where a container, pallet or similar article of transport is used to consolidate goods, the package or other shipping units enumerated in the bill of lading, if issued, or otherwise in any other document evidencing the contract of carriage by sea, as packed in such article of transport are deemed packages or shipping units. Except as aforesaid the goods in such article of transport are deemed one shipping unit.

(b) In cases where the article of transport itself has been lost or damaged, that article of transport, if not owned or otherwise supplied by the carrier, is considered one separate shipping unit.

3.   Unit of account means the unit of account mentioned in article 26.

4.   By agreement between the carrier and the shipper, limits of liability exceeding those provided for in paragraph 1 may be fixed.

Article 7   Application to non-contractual claims

1.   The defences and limits of liability provided for in this Convention apply in any action against the carrier in respect of loss or damage to the goods covered by the contract of carriage by sea, as well as of delay in delivery whether the action is founded in contract, in tort or otherwise.

2.   If such an action is brought against a servant or agent of the carrier, such servant or agent, if he proves that he acted within the scope of his employment, is entitled to avail himself of the defences and limits of liability which the carrier is entitled to invoke under this Convention.

3.   Except as provided in article 8, the aggregate of the amounts recoverable from the carrier and from any persons referred to in paragraph 2 of this article shall not exceed the limits of liability provided for in this Convention.

Article 8   Loss of right to limit responsibility

1.   The carrier is not entitled to the benefit of the limitation of liability provided for in article 6 if it is proved that the loss, damage or delay in delivery resulted from an act or omission of the carrier done with the intent to cause such loss, damage or delay, or recklessly and with knowledge that such loss, damage or delay would probably result.

2.   Notwithstanding the provisions of paragraph 2 of article 7, a servant or agent of the carrier is not entitled to the benefit of the limitation of liability provided for in article 6 if it is proved that the loss, damage or delay in delivery resulted from an act or omission of such servant or agent, done with the intent to cause such loss, damage or delay, or recklessly and with knowledge that such loss, damage or delay would probably result.

Article 9   Deck cargo

1.   The carrier is entitled to carry the goods on deck only if such carriage is in accordance with an agreement with the shipper or with the usage of the particular trade or is required by statutory rules or regulations.

2.   If the carrier and the shipper have agreed that the goods shall or may be carried on deck, the carrier must insert in the bill of lading or other document evidencing the contract of carriage by sea a statement to that effect. In the absence of such a statement the carrier has the burden of proving that an agreement for carriage on deck has been entered into; however, the carrier is not entitled to invoke such an agreement against a third party, including a consignee, who has acquired the bill of lading in good faith.

3.   Where the goods have been carried on deck contrary to the provisions of paragraph 1 of this article or where the carrier may not under paragraph 2 of this article invoke an agreement for carriage on deck, the carrier, notwithstanding the provisions of paragraph 1 of article 5, is liable for loss of or damage to the goods, as well as for delay in delivery, resulting solely from the carriage on deck, and the extent of his liability is to be determined in accordance with the provisions of article 6 or article 8 of this Convention, as the case may be.

4.   Carriage of goods on deck contrary to express agreement for carriage under deck is deemed to be an act or omission of the carrier within the meaning of article 8.

Article 10   Liability of the carrier and actual carrier

1.   Where the performance of the carriage or part thereof has been entrusted to an actual carrier, whether or not in pursuance of a liberty under the contract of carriage by sea to do so, the carrier nevertheless remains responsible for the entire carriage according to the provisions of this Convention. The carrier is responsible, in relation to the carriage performed by the actual carrier, for the acts and omissions of the actual carrier and of his servants and agents acting within the scope of their employment.

2.   All the provisions of this Convention governing the responsibility of the carrier also apply to the responsibility of the actual carrier for the carriage performed by him. The provisions of paragraphs 2 and 3 of article 7 and of paragraph 2 of article 8 apply if an action is brought against a servant or agent of the actual carrier.

3.   Any special agreement under which the carrier assumes obligations not imposed by this Convention or waives rights conferred by this Convention affects the actual carrier only if agreed to by him expressly and in writing. Whether or not the actual carrier has so agreed, the carrier nevertheless remains bound by the obligations or waivers resulting from such special agreement.

4.   Where and to the extent that both the carrier and the actual carrier are liable, their liability is joint and several.

5.   The aggregate of the amounts recoverable from the carrier, the actual carrier and their servants and agents shall not exceed the limits of liability provided for in this Convention.

6.   Nothing in this article shall prejudice any right of recourse as between the carrier and the actual carrier.

Article 11   Through carriage

1.   Notwithstanding the provisions of paragraph 1 of article 10, where a contract of carriage by sea provides explicitly that a specified part of the carriage covered by the said contract is to be performed by a named person other than the carrier, the contract may also provide that the carrier is not liable for loss, damage or delay in delivery caused by an occurrence which takes place while the goods are in the charge of the actual carrier during such part of the carriage. Nevertheless, any stipulation limiting or excluding such liability is without effect if no judicial proceedings can be instituted against the actual carrier in a court competent under paragraph 1 or 2 of article 21. The burden of proving that any loss, damage or delay in delivery has been caused by such an occurrence rests upon the carrier.

2.   The actual carrier is responsible in accordance with the provisions of paragraph 2 of article 10 for loss, damage or delay in delivery caused by an occurrence which takes place while the goods are in his charge.

PART III
LIABILITY OF THE SHIPPER

Article 12   General rule

The shipper is not liable for loss sustained by the carrier or the actual carrier, or for damage sustained by the ship, unless such loss or damage was caused by the fault or neglect of the shipper, his servants or agents. Nor is any servant or agent of the shipper liable for such loss or damage unless the loss or damage was caused by fault or neglect on his part.

Article 13   Special rules on dangerous goods

1.   The shipper must mark or label in a suitable manner dangerous goods as dangerous.

2.   Where the shipper hands over dangerous goods to the carrier or an actual carrier, as the case may be, the shipper must inform him of the dangerous character of the goods and, if necessary, of the precautions to be taken. If the shipper fails to do so and such carrier or actual carrier does not otherwise have knowledge of their dangerous character—
   (a) the shipper is liable to the carrier and any actual carrier for the loss resulting from the shipment of such goods, and
   (b) the goods may at any time be unloaded, destroyed or rendered innocuous, as the circumstances may require, without payment of compensation.

3.   The provisions of paragraph 2 of this article may not be invoked by any person if during the carriage he has taken the goods in his charge with knowledge of their dangerous character.

4.   If, in cases where the provisions of paragraph 2, subparagraph (b), of this article do not apply or may not be invoked, dangerous goods become an actual danger to life or property, they may be unloaded, destroyed or rendered innocuous, as the circumstances may require, without payment of compensation except where there is an obligation to contribute in general average or where the carrier is liable in accordance with the provisions of article 5.

PART IV
TRANSPORT DOCUMENTS

Article 14   Issue of bill of lading

1.   When the carrier or the actual carrier takes the goods in his charge, the carrier must, on demand of the shipper, issue to the shipper a bill of lading.

2.   The bill of lading may be signed by a person having authority from the carrier. A bill of lading signed by the master of the ship carrying the goods is deemed to have been signed on behalf of the carrier.

3.   The signature on the bill of lading may be in handwriting, printed in facsimile, perforated, stamped, in symbols, or made by any other mechanical or electronic means, if not inconsistent with the law of the country where the bill of lading is issued.

Article 15   Contents of bill of lading

1.   The bill of lading must include, inter alia, the following particulars—
  (a) the general nature of the goods, the leading marks necessary for identification of the goods, an express statement, if applicable, as to the dangerous character of the goods, the number of packages or pieces, and the weight of the goods or their quantity otherwise expressed, all such particulars as furnished by the shipper;
  (b) the apparent condition of the goods;
  (c) the name and principal place of business of the carrier;
  (d) the name of the shipper;
  (e) the consignee if named by the shipper;
  (f) the port of loading under the contract of carriage by sea and the date on which the goods were taken over by the carrier at the port of loading;
  (g) the port of discharge under the contract of carriage by sea;
  (h) the number of originals of the bill of lading, if more than one;
  (i) the place of issuance of the bill of lading;
  (j) the signature of the carrier or a person acting on his behalf;
  (k) the freight to the extent payable by the consignee or other indication that freight is payable by him;
  (l) the statement referred to in paragraph 3 of article 23;
  (m) the statement, if applicable, that the goods shall or may be carried on deck;
  (n) the date or the period of delivery of the goods at the port of discharge if expressly agreed upon between the parties; and
  (o) any increased limit or limits of liability where agreed in accordance with paragraph 4 of article 6.

2.   After the goods have been loaded on board, if the shipper so demands, the carrier must issue to the shipper a "shipped" bill of lading which, in addition to the particulars required under paragraph 1 of this article, must state that the goods are on board a named ship or ships, and the date or

dates of loading. If the carrier has previously issued to the shipper a bill of lading or other document of title with respect to any of such goods, on request of the carrier the shipper must surrender such document in exchange for a "shipped" bill of lading. The carrier may amend any previously issued document in order to meet the shipper's demand for a "shipped" bill of lading if, as amended, such document includes all the information required to be contained in a "shipped" bill of lading.

3.   The absence in the bill of lading of one or more particulars referred to in this article does not affect the legal character of the document as a bill of lading provided that it nevertheless meets the requirements set out in paragraph 7 of article 1.

Article 16   Bills of lading: reservations and evidentiary effect

1.   If the bill of lading contains particulars concerning the general nature, leading marks, number of packages or pieces, weight or quantity of the goods which the carrier or other person issuing the bill of lading on his behalf knows or has reasonable grounds to suspect do not accurately represent the goods actually taken over or, where a "shipped" bill of lading is issued, loaded, or if he had no reasonable means of checking such particulars, the carrier or such other person must insert in the bill of lading a reservation specifying these inaccuracies, grounds of suspicion or the absence of reasonable means of checking.

2.   If the carrier or other person issuing the bill of lading on his behalf fails to note on the bill of lading the apparent condition of the goods, he is deemed to have noted on the bill of lading that the goods were in apparent good condition.

3.   Except for particulars in respect of which and to the extent to which a reservation permitted under paragraph 1 of this article has been entered—
   (a) the bill of lading is prima facie evidence of the taking over or, where a "shipped" bill of lading is issued, loading, by the carrier of the goods as described in the bill of lading; and
   (b) proof to the contrary by the carrier is not admissible if the bill of lading has been transferred to a third party, including a consignee, who in good faith has acted in reliance on the description of the goods therein.

4.   A bill of lading which does not, as provided in paragraph 1, subparagraph (k), of article 15, set forth the freight or otherwise indicate

that freight is payable by the consignee or does not set forth demurrage incurred at the port of loading payable by the consignee, is prima facie evidence that no freight or such demurrage is payable by him. However, proof to the contrary by the carrier is not admissible when the bill of lading has been transferred to a third party, including a consignee, who in good faith has acted in reliance on the absence in the bill of lading of any such indication.

Article 17　Guarantees by the shipper

1.　　The shipper is deemed to have guaranteed to the carrier the accuracy of particulars relating to the general nature of the goods, their marks, number, weight and quantity as furnished by him for insertion in the bill of lading. The shipper must indemnify the carrier against the loss resulting from inaccuracies in such particulars. The shipper remains liable even if the bill of lading has been transferred by him. The right of the carrier to such indemnity in no way limits his liability under the contract of carriage by sea to any person other than the shipper.

2.　　Any letter of guarantee or agreement by which the shipper undertakes to indemnify the carrier against loss resulting from the issuance of the bill of lading by the carrier, or by a person acting on his behalf, without entering a reservation relating to particulars furnished by the shipper for insertion in the bill of lading, or to the apparent condition of the goods, is void and of no effect as against any third party, including a consignee, to whom the bill of lading has been transferred.

3.　　Such a letter of guarantee or agreement is valid as against the shipper unless the carrier or the person acting on his behalf, by omitting the reservation referred to in paragraph 2 of this article, intends to defraud a third party, including a consignee, who acts in reliance on the description of the goods in the bill of lading. In the latter case, if the reservation omitted relates to particulars furnished by the shipper for insertion in the bill of lading, the carrier has no right of indemnity from the shipper pursuant to paragraph 1 of this article.

4.　　In the case of intended fraud referred to in paragraph 3 of this article, the carrier is liable, without the benefit of the limitation of liability provided for in this Convention, for the loss incurred by a third party, including a consignee, because he has acted in reliance on the description of the goods in the bill of lading.

Article 18 Documents other than bills of lading

Where a carrier issues a document other than a bill of lading to evidence the receipt of the goods to be carried, such a document is prima facie evidence of the conclusion of the contract of carriage by sea and the taking over by the carrier of the goods as therein described.

PART V
CLAIMS AND ACTIONS

Article 19 Notice of loss, damage or delay

1.   Unless notice of loss or damage, specifying the general nature of such loss or damage, is given in writing by the consignee to the carrier not later than the working day after the day when the goods were handed over to the consignee, such handing over is prima facie evidence of the delivery by the carrier of the goods as described in the document of transport or, if no such document has been issued, in good condition.

2.   Where the loss or damage is not apparent, the provisions of paragraph 1 of this article apply correspondingly if notice in writing is not given within 15 consecutive days after the day when the goods were handed over to the consignee.

3.   If the state of the goods at the time they were handed over to the consignee has been the subject of a joint survey or inspection by the parties, notice in writing need not be given of loss or damage ascertained during such survey or inspection.

4.   In the case of any actual or apprehended loss or damage, the carrier and the consignee must give all reasonable facilities to each other for inspecting and tallying the goods.

5.   No compensation shall be payable for loss resulting from delay in delivery unless a notice has been given in writing to the carrier within 60 consecutive days after the day when the goods were handed over to the consignee.

6.   If the goods have been delivered by an actual carrier, any notice given under this article to him shall have the same effect as if it had been given to the carrier; and any notice given to the carrier shall have effect as if given to such actual carrier.

7.   Unless notice of loss or damage, specifying the general nature of the loss or damage, is given in writing by the carrier or actual carrier to the shipper not later than 90 consecutive days after the occurrence of such loss or damage or after the delivery of the goods in accordance with paragraph 2 of article 4, whichever is later, the failure to give such notice is prima facie evidence that the carrier or the actual carrier has sustained no loss or damage due to the fault or neglect of the shipper, his servants or agents.

8.   For the purpose of this article, notice given to a person acting on the carrier's or the actual carrier's behalf, including the master or the officer in charge of the ship, or to a person acting on the shipper's behalf is deemed to have been given to the carrier, to the actual carrier or to the shipper, respectively.

Article 20   Limitation of actions

1.   Any action relating to carriage of goods under this Convention is time-barred if judicial or arbitral proceedings have not been instituted within a period of two years.

2.   The limitation period commences on the day on which the carrier has delivered the goods or part thereof or, in cases where no goods have been delivered, on the last day on which the goods should have been delivered.

3.   The day on which the limitation period commences is not included in the period.

4.   The person against whom a claim is made may at any time during the running of the limitation period extend that period by a declaration in writing to the claimant. This period may be further extended by another declaration or declarations.

5.   An action for indemnity by a person held liable may be instituted even after the expiration of the limitation period provided for in the preceding paragraphs if instituted within the time allowed by the law of the State where proceedings are instituted. However, the time allowed shall not be less than 90 days commencing from the day when the person instituting such action for indemnity has settled the claim or has been served with process in the action against himself.

Article 21   Jurisdiction

1.   In judicial proceedings relating to carriage of goods under this Convention the plaintiff, at his option, may institute an action in a court which according to the law of the State where the court is situated, is competent and within the jurisdiction of which is situated one of the following places—
  (a) the principal place of business or, in the absence thereof, the habitual residence of the defendant; or
  (b) the place where the contract was made, provided that the defendant has there a place of business, branch or agency through which the contract was made; or
  (c) the port of loading or the port of discharge; or
  (d) any additional place designated for that purpose in the contract of carriage by sea.

2.—(a) Notwithstanding the preceding provisions of this article, an action may be instituted in the courts of any port or place in a Contracting State at which the carrying vessel or any other vessel of the same ownership may have been arrested in accordance with applicable rules of the law of that State and of international law. However, in such a case, at the petition of the defendant, the claimant must remove the action, at his choice, to one of the jurisdictions referred to in paragraph 1 of this article for the determination of the claim, but before such removal the defendant must furnish security sufficient to ensure payment of any judgment that may subsequently be awarded to the claimant in the action.
  (b) All questions relating to the sufficiency or otherwise of the security shall be determined by the court of the port or place of the arrest.

3.   No judicial proceedings relating to carriage of goods under this Convention may be instituted in a place not specified in paragraph 1 or 2 of this article. The provisions of this paragraph do not constitute an obstacle to the jurisdiction of the Contracting States for provisional or protective measures.

4.—(a) Where an action has been instituted in a court competent under paragraphs 1 and 2 of this article or where judgment has been delivered by such a court, no new action may be started between the same parties on the same grounds unless the judgment of the court before which the first action was instituted is not enforceable in the country in which the new proceedings are instituted;

(b) for the purpose of this article, the institution of measures with a view to obtaining enforcement of a judgment is not to be considered as the starting of a new action;

(c) for the purpose of this article, the removal of an action to a different court within the same country, or to a court in another country, in accordance with paragraph 2(a) of this article, is not to be considered as the starting of a new action.

5.   Notwithstanding the provisions of the preceding paragraphs, an agreement made by the parties, after a claim under the contract of carriage by sea has arisen, which designates the place where the claimant may institute an action, is effective.

Article 22   Arbitration

1.   Subject to the provisions of this article, parties may provide by agreement evidenced in writing that any dispute that may arise relating to carriage of goods under this Convention shall be referred to arbitration.

2.   Where a charter-party contains a provision that disputes arising thereunder shall be referred to arbitration and a bill of lading issued pursuant to the charter-party does not contain a special annotation providing that such provision shall be binding upon the holder of the bill of lading, the carrier may not invoke such provision as against a holder having acquired the bill of lading in good faith.

3.   The arbitration proceedings shall, at the option of the claimant, be instituted at one of the following places—

(a) a place in a State within whose territory is situated—

(i)   the principal place of business of the defendant or, in the absence thereof, the habitual residence of the defendant; or

(ii) the place where the contract was made, provided that the defendant has there a place of business, branch or agency through which the contract was made; or

(iii) the port of loading or the port of discharge; or

(b) any place designated for that purpose in the arbitration clause or agreement.

4.   The arbitrator or arbitration tribunal shall apply the rules of this Convention.

5.   The provisions of paragraphs 3 and 4 of this article are deemed to be part of every arbitration clause or agreement, and any term of such clause or agreement which is inconsistent therewith is null and void.

6.   Nothing in this article affects the validity of an agreement relating to arbitration made by the parties after the claim under the contract of carriage by sea has arisen.

PART VI
SUPPLEMENTARY PROVISIONS

Article 23   Contractual stipulations

1.   Any stipulation in a contract of carriage by sea, in a bill of lading, or in any other document evidencing the contract of carriage by sea is null and void to the extent that it derogates, directly or indirectly, from the provisions of this Convention. The nullity of such a stipulation does not affect the validity of the other provisions of the contract or document of which it forms a part. A clause assigning benefit of insurance of goods in favour of the carrier, or any similar clause, is null and void.

2.   Notwithstanding the provisions of paragraph 1 of this article, a carrier may increase his responsibilities and obligations under this Convention.

3.   Where a bill of lading or any other document evidencing the contract of carriage by sea is issued, it must contain a statement that the carriage is subject to the provisions of this Convention which nullify any stipulation derogating therefrom to the detriment of the shipper or the consignee.

4.   Where the claimant in respect of the goods has incurred loss as a result of a stipulation which is null and void by virtue of the present article, or as a result of the omission of the statement referred to in paragraph 3 of this article, the carrier must pay compensation to the extent required in order to give the claimant compensation in accordance with the provisions of this Convention for any loss of or damage to the goods as well as for delay in delivery. The carrier must, in addition, pay compensation for costs incurred by the claimant for the purpose of exercising his right, provided that costs incurred in the action where the foregoing provision is invoked are to be determined in accordance with the law of the State where proceedings are instituted.

Article 24   General average

1.   Nothing in this Convention shall prevent the application of provisions in the contract of carriage by sea or national law regarding the adjustment of general average.

2.   With the exception of article 20, the provisions of this Convention relating to the liability of the carrier for loss of or damage to the goods also determine whether the consignee may refuse contribution in general average and the liability of the carrier to indemnify the consignee in respect of any such contribution made or any salvage paid.

Article 25   Other conventions

1.   This Convention does not modify the rights or duties of the carrier, the actual carrier and their servants and agents, provided for in international conventions or national law relating to the limitation of liability of owners of seagoing ships.

2.   The provisions of articles 21 and 22 of this Convention do not prevent the application of the mandatory provisions of any other multilateral convention already in force at the date of this Convention relating to matters dealt with in the said articles, provided that the dispute arises exclusively between parties having their principal place of business in States members of such other convention. However, this paragraph does not affect the application of paragraph 4 of article 22 of this Convention.

3.   No liability shall arise under the provisions of this Convention for damage caused by a nuclear incident if the operator of a nuclear installation is liable for such damage—
   (a) under either the Paris Convention of 29 July 1960 on Third Party Liability in the Field of Nuclear Energy as amended by the Additional Protocol of 28 January 1964, or the Vienna Convention of 21 May 1963 on Civil Liability for Nuclear Damage, or
   (b) by virtue of national law governing the liability for such damage, provided that such law is in all respects as favourable to persons who may suffer damage as either the Paris Convention or the Vienna Convention.

4.   No liability shall arise under the provisions of this Convention for any loss of or damage to or delay in delivery of luggage for which the carrier is responsible under any international convention or national law relating to the carriage of passengers and their luggage by sea.

5.   Nothing contained in this Convention prevents a Contracting State from applying any other international convention which is already in force at the date of this Convention and which applies mandatorily to contracts of carriage of goods primarily by a mode of transport other than transport

by sea. This provision also applies to any subsequent revision or amendment of such international convention.

Article 26   Unit of account

1.   The unit of account referred to in article 6 of this Convention is the special drawing right as defined by the International Monetary Fund. The amounts mentioned in article 6 are to be converted into the national currency of a State according to the value of such currency at the date of judgment or the date agreed upon by the parties. The value of a national currency, in terms of the special drawing right, of a Contracting State which is a member of the International Monetary Fund is to be calculated in accordance with the method of valuation applied by the International Monetary Fund in effect at the date in question for its operations and transactions. The value of a national currency, in terms of the special drawing right, of a Contracting State which is not a member of the International Monetary Fund is to be calculated in a manner determined by that State.

2.   Nevertheless, those States which are not members of the International Monetary Fund and whose law does not permit the application of the provisions of paragraph 1 of this article may, at the time of signature, or at the time of ratification, acceptance, approval or accession or at any time thereafter, declare that the limits of liability provided for in this Convention to be applied in their territories shall be fixed as 12,500 monetary units per package or other shipping unit or 37.5 monetary units per kilogram of gross weight of the goods.

3.   The monetary unit referred to in paragraph 2 of this article corresponds to sixty-five and a half milligrams of gold of millesimal fineness nine hundred. The conversion of the amounts referred to in paragraph 2 into the national currency is to be made according to the law of the State concerned.

4.   The calculation mentioned in the last sentence of paragraph 1 and the conversion mentioned in paragraph 3 of this article is to be made in such a manner as to express in the national currency of the Contracting State as far as possible the same real value for the amounts in article 6 as is expressed there in units of account. Contracting States must communicate to the depositary the manner of calculation pursuant to paragraph 1 of this article, or the result of the conversion mentioned in paragraph 3 of this article, as the case may be, at the time of signature or when depositing their instruments of ratification, acceptance, approval or accession, or when

availing themselves of the option provided for in paragraph 2 of this article and whenever there is a change in the manner of such calculation or in the result of such conversion.

PART VII
FINAL CLAUSES

Article 29   Reservations

No reservations may be made to this Convention.

Article 30   Entry into force

1.   This Convention enters into force on the first day of the month following the expiration of one year from the date of deposit of the twentieth instrument of ratification, acceptance, approval or accession.

2.   For each State which becomes a Contracting State to this Convention after the date of the deposit of the twentieth instrument of ratification, acceptance, approval or accession, this Convention enters into force on the first day of the month following the expiration of one year after the deposit of the appropriate instrument on behalf of that State.

3.   Each Contracting State shall apply the provisions of this Convention to contracts of carriage by sea concluded on or after the date of the entry into force of this Convention in respect of that State.

Article 31   Denunciation of other conventions

1.   Upon becoming a Contracting State to this Convention, any State party to the International Convention for the Unification of certain Rules relating to Bills of Lading signed at Brussels on 25 August 1924 (1924 Convention) must notify the Government of Belgium as the depositary of the 1924 Convention of its denunciation of the said Convention with a declaration that the denunciation is to take effect as from the date when this Convention enters into force in respect of that State.

2.   Upon the entry into force of this Convention under paragraph 1 of article 30, the depositary of this Convention must notify the Government

of Belgium as the depositary of the 1924 Convention of the date of such entry into force, and of the names of the Contracting States in respect of which the Convention has entered into force.

3.   The provisions of paragraphs 1 and 2 of this article apply correspondingly in respect of States parties to the Protocol signed on 23 February 1968 to amend the International Convention for the Unification of certain Rules relating to Bills of Lading signed at Brussels on 25 August 1924.

4.   Notwithstanding article 2 of this Convention, for the purposes of paragraph 1 of this article, a Contracting State may, if it deems it desirable, defer the denunciation of the 1924 Convention and of the 1924 Convention as modified by the 1968 Protocol for a maximum period of five years from the entry into force of this Convention. It will then notify the Government of Belgium of its intention. During this transitory period, it must apply to the Contracting States this Convention to the exclusion of any other one.

Done at Hamburg, this thirty-first day of March, one thousand nine hundred and seventy-eight, in a single original, of which the Arabic, Chinese, English, French, Russian and Spanish texts are equally authentic.

In witness whereof the undersigned plenipotentiaries, being duly authorised by their respective Governments, have signed the present Convention.

NOTES

The text of this Convention was prepared by the United Nations Commission on International Trade Law.

United Nations Convention on Contracts for the International Sale of Goods

The States Parties to this Convention,

Bearing in mind the broad objectives in the resolutions adopted by the sixth special session of the General Assembly of the United Nations on the establishment of a New International Economic Order,

Considering that the development of international trade on the basis of equality and mutual benefit is an important element in promoting friendly relations among States,

Being of the opinion that the adoption of uniform rules which govern contracts for the international sale of goods and take into account the different social, economic and legal systems would contribute to the removal of legal barriers in international trade and promote the development of international trade,

Have agreed as follows—

PART I
SPHERE OF APPLICATION AND GENERAL PROVISIONS

CHAPTER I
SPHERE OF APPLICATION

Article 1

(1)   This Convention applies to contracts of sale of goods between parties whose places of business are in different States—
    (a) when the States are Contracting States; or
    (b) when the rules of private international law lead to the application of the law of a Contracting State.

(2)   The fact that the parties have their places of business in different States is to be disregarded whenever this fact does not appear either from the contract or from any dealings between, or from information disclosed by, the parties at any time before or at the conclusion of the contract.

(3)   Neither the nationality of the parties nor the civil or commercial character of the parties or of the contract is to be taken into consideration in determining the application of this Convention.

Article 2

This Convention does not apply to sales—
    (a) of goods bought for personal, family or household use, unless the seller, at any time before or at the conclusion of the contract,

neither knew nor ought to have known that the goods were bought for any such use;

(b) by auction;

(c) on execution or otherwise by authority of law;

(d) of stocks, shares, investment securities, negotiable instruments or money;

(e) of ships, vessels, hovercraft or aircraft;

(f) of electricity.

Article 3

(1)   Contracts for the supply of goods to be manufactured or produced are to be considered sales unless the party who orders the goods undertakes to supply a substantial part of the materials necessary for such manufacture or production.

(2)   This Convention does not apply to contracts in which the preponderant part of the obligations of the party who furnishes the goods consists in the supply of labour or other services.

Article 4

This Convention governs only the formation of the contract of sale and the rights and obligations of the seller and the buyer arising from such a contract. In particular, except as otherwise expressly provided in this Convention, it is not concerned with—

(a) the validity of the contract or of any of its provisions or of any usage;

(b) the effect which the contract may have on the property in the goods sold.

Article 5

This Convention does not apply to the liability of the seller for death or personal injury caused by the goods to any person.

Article 6

The parties may exclude the application of this Convention or, subject to article 12, derogate from or vary the effect of any of its provisions.

CHAPTER II
GENERAL PROVISIONS

Article 7

(1)   In the interpretation of this Convention, regard is to be had to its international character and to the need to promote uniformity in its application and the observance of good faith in international trade.

(2)   Questions concerning matters governed by this Convention which are not expressly settled in it are to be settled in conformity with the general principles on which it is based or, in the absence of such principles, in conformity with the law applicable by virtue of the rules of private international law.

Article 8

(1)   For the purposes of this Convention statements made by and other conduct of a party are to be interpreted according to his intent where the other party knew or could not have been unaware what that intent was.

(2)   If the preceding paragraph is not applicable, statements made by and other conduct of a party are to be interpreted according to the understanding that a reasonable person of the same kind as the other party would have had in the same circumstances.

(3)   In determining the intent of a party or the understanding a reasonable person would have had, due consideration is to be given to all relevant circumstances of the case including the negotiations, any practices which the parties have established between themselves, usages and any subsequent conduct of the parties.

Article 9

(1)   The parties are bound by any usage to which they have agreed and by any practices which they have established between themselves.

(2)   The parties are considered, unless otherwise agreed, to have impliedly made applicable to their contract or its formation a usage of which the parties knew or ought to have known and which in international trade is widely known to, and regularly observed by, parties to contracts of the type involved in the particular trade concerned.

Article 10

For the purposes of this Convention—
   (a) if a party has more than one place of business, the place of business is that which has the closest relationship to the contract and its performance, having regard to the circumstances known to or contemplated by the parties at any time before or at the conclusion of the contract;
   (b) if a party does not have a place of business, reference is to be made to his habitual residence.

Article 11

A contract of sale need not be concluded in or evidenced by writing and is not subject to any other requirement as to form. It may be proved by any means, including witnesses.

Article 12

Any provision of article 11, article 29 or Part II of this Convention that allows a contract of sale or its modification or termination by agreement or any offer, acceptance or other indication of intention to be made in any form other than in writing does not apply where any party has his place of business in a Contracting State which has made a declaration under article 96 of this Convention. The parties may not derogate from or vary the effect of this article.

Article 13

For the purposes of this Convention "writing" includes telegram and telex.

PART II
FORMATION OF THE CONTRACT

Article 14

(1)   A proposal for concluding a contract addressed to one or more specific persons constitutes an offer if it is sufficiently definite and indicates the intention of the offeror to be bound in case of acceptance. A proposal is sufficiently definite if it indicates the goods and expressly

or implicitly fixes or makes provision for determining the quantity and the price.

(2)   A proposal other than one addressed to one or more specific persons is to be considered merely as an invitation to make offers, unless the contrary is clearly indicated by the person making the proposal.

Article 15

(1)   An offer becomes effective when it reaches the offeree.

(2)   An offer, even if it is irrevocable, may be withdrawn if the withdrawal reaches the offeree before or at the same time as the offer.

Article 16

(1)   Until a contract is concluded an offer may be revoked if the revocation reaches the offeree before he has dispatched an acceptance.

(2)   However, an offer cannot be revoked—
   (a) if it indicates, whether by stating a fixed time for acceptance or otherwise, that it is irrevocable; or
   (b) if it was reasonable for the offeree to rely on the offer as being irrevocable and the offeree has acted in reliance on the offer.

Article 17

An offer, even if it is irrevocable, is terminated when a rejection reaches the offeror.

Article 18

(1)   A statement made by or other conduct of the offeree indicating assent to an offer is an acceptance. Silence or inactivity does not in itself amount to acceptance.

(2)   An acceptance of an offer becomes effective at the moment the indication of assent reaches the offeror. An acceptance is not effective if the indication of assent does not reach the offeror within the time he has

fixed or, if no time is fixed, within a reasonable time, due account being taken of the circumstances of the transaction, including the rapidity of the means of communication employed by the offeror. An oral offer must be accepted immediately unless the circumstances indicate otherwise.

(3)   However, if, by virtue of the offer or as a result of practices which the parties have established between themselves or of usage, the offeree may indicate assent by performing an act, such as one relating to the dispatch of the goods or payment of the price, without notice to the offeror, the acceptance is effective at the moment the act is performed, provided that the act is performed within the period of time laid down in the preceding paragraph.

Article 19

(1)   A reply to an offer which purports to be an acceptance but contains additions, limitations or other modifications is a rejection of the offer and constitutes a counter-offer.

(2)   However, a reply to an offer which purports to be an acceptance but contains additional or different terms which do not materially alter the terms of the offer constitutes an acceptance, unless the offeror, without undue delay, objects orally to the discrepancy or dispatches a notice to that effect. If he does not so object, the terms of the contract are the terms of the offer with the modifications contained in the acceptance.

(3)   Additional or different terms relating, among other things, to the price, payment, quality and quantity of the goods, place and time of delivery, extent of one party's liability to the other or the settlement of disputes are considered to alter the terms of the offer materially.

Article 20

(1)   A period of time for acceptance fixed by the offeror in a telegram or a letter begins to run from the moment the telegram is handed in for dispatch or from the date shown on the letter or, if no such date is shown, from the date shown on the envelope. A period of time for acceptance fixed by the offeror by telephone, telex or other means of instantaneous communication, begins to run from the moment that the offer reaches the offeree.

(2)   Official holidays or non-business days occurring during the period for acceptance are included in calculating the period. However, if a notice of acceptance cannot be delivered at the address of the offeror on the last day of the period because that day falls on an official holiday or a non-business day at the place of business of the offeror, the period is extended until the first business day which follows.

Article 21

(1)   A late acceptance is nevertheless effective as an acceptance if without delay the offeror orally so informs the offeree or dispatches a notice to that effect.

(2)   If a letter or other writing containing a late acceptance shows that it has been sent in such circumstances that if its transmission had been normal it would have reached the offeror in due time, the late acceptance is effective as an acceptance unless, without delay, the offeror orally informs the offeree that he considers his offer as having lapsed or dispatches a notice to that effect.

Article 22

An acceptance may be withdrawn if the withdrawal reaches the offeror before or at the same time as the acceptance would have become effective.

Article 23

A contract is concluded at the moment when an acceptance of an offer becomes effective in accordance with the provisions of this Convention.

Article 24

For the purposes of this Part of the Convention, an offer, declaration of acceptance or any other indication of intention "reaches" the addressee when it is made orally to him or delivered by any other means to him personally, to his place of business or mailing address or, if he does not have a place of business or mailing address, to his habitual residence.

PART III
SALE OF GOODS

CHAPTER I
GENERAL PROVISIONS

Article 25

A breach of contract committed by one of the parties is fundamental if it results in such detriment to the other party as substantially to deprive him of what he is entitled to expect under the contract, unless the party in breach did not foresee and a reasonable person of the same kind in the same circumstances would not have forseen such a result.

Article 26

A declaration of avoidance of the contract is effective only if made by notice to the other party.

Article 27

Unless otherwise expressly provided in this Part of the Convention, if any notice, request or other communication is given or made by a party in accordance with this Part and by means appropriate in the circumstances, a delay or error in the transmission of the communication or its failure to arrive does not deprive that party of the right to rely on the communication.

Article 28

If, in accordance with the provisions of this Convention, one party is entitled to require performance of any obligation by the other party, a court is not bound to enter a judgment for specific performance unless the court would do so under its own law in respect of similar contracts of sale not governed by this Convention.

Article 29

(1)   A contract may be modified or terminated by the mere agreement of the parties.

(2)   A contract in writing which contains a provision requiring any modification or termination by agreement to be in writing may not be otherwise modified or terminated by agreement. However, a party may be precluded by his conduct from asserting such a provision to the extent that the other party has relied on that conduct.

CHAPTER II
OBLIGATIONS OF THE SELLER

Article 30

The seller must deliver the goods, hand over any documents relating to them and transfer the property in the goods, as required by the contract and this Convention.

SECTION I
DELIVERY OF THE GOODS AND HANDING OVER OF DOCUMENTS

Article 31

If the seller is not bound to deliver the goods at any other particular place, his obligation to deliver consists—
- (a) if the contract of sale involves carriage of the goods-in handing the goods over to the first carrier for transmission to the buyer;
- (b) if, in cases not within the preceding subparagraph, the contract relates to specific goods, or unidentified goods to be drawn from a specific stock or to be manufactured or produced, and at the time of the conclusion of the contract the parties knew that the goods were at, or were to be manufactured or produced at, a particular place-in placing the goods at the buyer's disposal at that place;
- (c) in other cases-in placing the goods at the buyer's disposal at the place where the seller had his place of business at the time of the conclusion of the contract.

Article 32

(1)   If the seller, in accordance with the contract or this Convention, hands the goods over to a carrier and if the goods are not clearly identified to the contract by markings on the goods, by shipping documents or

otherwise, the seller must give the buyer notice of the consignment specifying the goods.

(2)   If the seller is bound to arrange for carriage of the goods, he must make such contracts as are necessary for carriage to the place fixed by means of transportation appropriate in the circumstances and according to the usual terms for such transportation.

(3)   If the seller is not bound to effect insurance in respect of the carriage of the goods, he must, at the buyer's request, provide him with all available information necessary to enable him to effect such insurance.

Article 33

The seller must deliver the goods—
  (a) if a date is fixed by or determinable from the contract, on that date;
  (b) if a period of time is fixed by or determinable from the contract, at any time within that period unless circumstances indicate that the buyer is to choose a date; or
  (c) in any other case, within a reasonable time after the conclusion of the contract.

Article 34

If the seller is bound to hand over documents relating to the goods, he must hand them over at the time and place and in the form required by the contract. If the seller has handed over documents before that time, he may, up to that time, cure any lack of conformity in the documents, if the exercise of this right does not cause the buyer unreasonable inconvenience or unreasonable expense. However, the buyer retains any right to claim damages as provided for in this Convention.

SECTION II
CONFORMITY OF THE GOODS AND THIRD PARTY CLAIMS

Article 35

(1)   The seller must deliver goods which are of the quantity, quality and description required by the contract and which are contained or packaged in the manner required by the contract.

(2)   Except where the parties have agreed otherwise, the goods do not conform with the contract unless they—

(a) are fit for the purposes for which goods of the same description would ordinarily be used;

(b) are fit for any particular purpose expressly or impliedly made known to the seller at the time of the conclusion of the contract, except where the circumstances show that the buyer did not rely, or that it was unreasonable for him to rely, on the seller's skill and judgment;

(c) possess the qualities of goods which the seller has held out to the buyer as a sample or model;

(d) are contained or packaged in the manner usual for such goods or, where there is no such manner, in a manner adequate to preserve and protect the goods.

(3)   The seller is not liable under subparagraphs (a) to (d) of the preceding paragraph for any lack of conformity of the goods if at the time of the conclusion of the contract the buyer knew or could not have been unaware of such lack of conformity.

Article 36

(1)   The seller is liable in accordance with the contract and this Convention for any lack of conformity which exists at the time when the risk passes to the buyer, even though the lack of conformity becomes apparent only after that time.

(2)   The seller is also liable for any lack of conformity which occurs after the time indicated in the preceding paragraph and which is due to a breach of any of his obligations, including a breach of any guarantee that for a period of time the goods will remain fit for their ordinary purpose or for some particular purpose or will retain specified qualities or characteristics.

Article 37

If the seller has delivered goods before the date for delivery, he may, up to that date, deliver any missing part or make up any deficiency in the quantity of the goods delivered, or deliver goods in replacement of any non-conforming goods delivered or remedy any lack of conformity in the goods delivered, provided that the exercise of this right does not cause the buyer unreasonable inconvenience or unreasonable expense. However, the buyer retains any right to claim damages as provided for in this Convention.

Article 38

(1)   The buyer must examine the goods, or cause them to be examined, within as short a period as is practicable in the circumstances.

(2)   If the contract involves carriage of the goods, examination may be deferred until after the goods have arrived at their destination.

(3)   If the goods are redirected in transit or redispatched by the buyer without a reasonable opportunity for examination by him and at the time of the conclusion of the contract the seller knew or ought to have known of the possibility of such redirection or redispatch, examination may be deferred until after the goods have arrived at the new destination.

Article 39

(1)   The buyer loses the right to rely on a lack of conformity of the goods if he does not give notice to the seller specifying the nature of the lack of conformity within a reasonable time after he has discovered it or ought to have discovered it.

(2)   In any event, the buyer loses the right to rely on a lack of conformity of the goods if he does not give the seller notice thereof at the latest within a period of two years from the date on which the goods were actually handed over to the buyer, unless this time-limit is inconsistent with a contractual period of guarantee.

Article 40

The seller is not entitled to rely on the provisions of articles 38 and 39 if the lack of conformity relates to facts of which he knew or could not have been unaware and which he did not disclose to the buyer.

Article 41

The seller must deliver goods which are free from any right or claim of a third party, unless the buyer agreed to take the goods subject to that right or claim. However, if such right or claim is based on industrial property or other intellectual property, the seller's obligation is governed by article 42.

Article 42

(1)   The seller must deliver goods which are free from any right or claim of a third party based on industrial property or other intellectual property, of which at the time of the conclusion of the contract the seller knew or could not have been unaware, provided that the right or claim is based on industrial property or other intellectual property—
   (a) under the law of the State where the goods will be resold or otherwise used, if it was contemplated by the parties at the time of the conclusion of the contract that the goods would be resold or otherwise used in that State; or
   (b) in any other case, under the law of the State where the buyer has his place of business.

(2)   The obligation of the seller under the preceding paragraph does not extend to cases where—
   (a) at the time of the conclusion of the contract the buyer knew or could not have been unaware of the right or claim; or
   (b) the right or claim results from the seller's compliance with technical drawings, designs, formulae or other such specifications furnished by the buyer.

Article 43

(1)   The buyer loses the right to rely on the provisions of article 41 or article 42 if he does not give notice to the seller specifying the nature of the right or claim of the third party within a reasonable time after he has become aware or ought to have become aware of the right or claim.

(2)   The seller is not entitled to rely on the provisions of the preceding paragraph if he knew of the right or claim of the third party and the nature of it.

Article 44

Notwithstanding the provisions of paragraph (1) of article 39 and paragraph (1) of article 43, the buyer may reduce the price in accordance with article 50 or claim damages, except for loss of profit, if he has a reasonable excuse for his failure to give the required notice.

SECTION III
REMEDIES FOR BREACH OF CONTRACT BY THE SELLER

Article 45

(1)   If the seller fails to perform any of his obligations under the contract or this Convention, the buyer may—
   (a) exercise the rights provided in articles 46 to 52;
   (b) claim damages as provided in articles 74 to 77.

(2)   The buyer is not deprived of any right he may have to claim damages by exercising his right to other remedies.

(3)   No period of grace may be granted to the seller by a court or arbitral tribunal when the buyer resorts to a remedy for breach of contract.

Article 46

(1)   The buyer may require performance by the seller of his obligations unless the buyer has resorted to a remedy which is inconsistent with this requirement.

(2)   If the goods do not conform with the contract, the buyer may require delivery of substitute goods only if the lack of conformity constitutes a fundamental breach of contract and a request for substitute goods is made either in conjunction with notice given under article 39 or within a reasonable time thereafter.

(3)   If the goods do not conform with the contract, the buyer may require the seller to remedy the lack of conformity by repair, unless this is unreasonable having regard to all the circumstances. A request for repair must be made either in conjunction with notice given under article 39 or within a reasonable time thereafter.

Article 47

(1)   The buyer may fix an additional period of time of reasonable length for performance by the seller of his obligations.

(2)   Unless the buyer has received notice from the seller that he will not perform within the period so fixed, the buyer may not, during that period, resort to any remedy for breach of contract. However, the buyer is not deprived thereby of any right he may have to claim damages for delay in performance.

Article 48

(1)   Subject to article 49, the seller may, even after the date for delivery, remedy at his own expense any failure to perform his obligations, if he can do so without unreasonable delay and without causing the buyer unreasonable inconvenience or uncertainty of reimbursement by the seller of expenses advanced by the buyer. However, the buyer retains any right to claim damages as provided for in this Convention.

(2)   If the seller requests the buyer to make known whether he will accept performance and the buyer does not comply with the request within a reasonable time, the seller may perform within the time indicated in his request. The buyer may not, during that period of time, resort to any remedy which is inconsistent with performance by the seller.

(3)   A notice by the seller that he will perform within a specified period of time is assumed to include a request, under the preceding paragraph, that the buyer make known his decision.

(4)   A request or notice by the seller under paragraph (2) or (3) of this article is not effective unless received by the buyer.

Article 49

(1)   The buyer may declare the contract avoided—
   (a) if the failure by the seller to perform any of his obligations under the contract or this Convention amounts to a fundamental breach of contract; or
   (b) in case of non-delivery, if the seller does not deliver the goods within the additional period of time fixed by the buyer in accordance with paragraph (1) of article 47 or declares that he will not deliver within the period so fixed.

(2)   However, in cases where the seller has delivered the goods, the buyer loses the right to declare the contract avoided unless he does so—

(a) in respect of late delivery, within a reasonable time after he has become aware that delivery has been made;

(b) in respect of any breach other than late delivery, within a reasonable time—

    (i) after he knew or ought to have known of the breach;

    (ii) after the expiration of any additional period of time fixed by the buyer in accordance with paragraph (1) of article 47, or after the seller has declared that he will not perform his obligations within such an additional period; or

    (iii) after the expiration of any additional period of time indicated by the seller in accordance with paragraph (2) of article 48, or after the buyer has declared that he will not accept performance.

Article 50

If the goods do not conform with the contract and whether or not the price has already been paid, the buyer may reduce the price in the same proportion as the value that the goods actually delivered had at the time of the delivery bears to the value that conforming goods would have had at that time. However, if the seller remedies any failure to perform his obligations in accordance with article 37 or article 48 or if the buyer refuses to accept performance by the seller in accordance with those articles, the buyer may not reduce the price.

Article 51

(1) If the seller delivers only a part of the goods or if only a part of the goods delivered is in conformity with the contract, articles 46 to 50 apply in respect of the part which is missing or which does not conform.

(2) The buyer may declare the contract avoided in its entirety only if the failure to make delivery completely or in conformity with the contract amounts to a fundamental breach of the contract.

Article 52

(1) If the seller delivers the goods before the date fixed, the buyer may take delivery or refuse to take delivery.

(2) If the seller delivers a quantity of goods greater than that provided for in the contract, the buyer may take delivery or refuse to take delivery

of the excess quantity. If the buyer takes delivery of all or part of the excess quantity, he must pay for it at the contract rate.

CHAPTER III
OBLIGATIONS OF THE BUYER

Article 53

The buyer must pay the price for the goods and take delivery of them as required by the contract and this Convention.

SECTION I
PAYMENT OF THE PRICE

Article 54

The buyer's obligation to pay the price includes taking such steps and complying with such formalities as may be required under the contract or any laws and regulations to enable payment to be made.

Article 55

Where a contract has been validly concluded but does not expressly or implicitly fix or make provision for determining the price, the parties are considered, in the absence of any indication to the contrary, to have impliedly made reference to the price generally charged at the time of the conclusion of the contract for such goods sold under comparable circumstances in the trade concerned.

Article 56

If the price is fixed according to the weight of goods, in case of doubt it is to be determined by the net weight.

Article 57

(1)   If the buyer is not bound to pay the price at any other particular place, he must pay it to the seller—

    (a) at the seller's place of business; or

    (b) if the payment is to be made against the handing over of the goods or of documents, at the place where the handing over takes place.

(2)   The seller must bear any increase in the expenses incidental to payment which is caused by a change in his place of business subsequent to the conclusion of the contract.

Article 58

(1)   If the buyer is not bound to pay the price at any other specific time he must pay it when the seller places either the goods or documents controlling their disposition at the buyer's disposal in accordance with the contract and this Convention. The seller may make such payment a condition for handing over the goods or documents.

(2)   If the contract involves carriage of the goods, the seller may dispatch the goods on terms whereby the goods, or documents controlling their disposition, will not be handed over to the buyer except against payment of the price.

(3)   The buyer is not bound to pay the price until he has had an opportunity to examine the goods, unless the procedures for delivery or payment agreed upon by the parties are inconsistent with his having such an opportunity.

Article 59

The buyer must pay the price on the date fixed by or determinable from the contract and this Convention without the need for any request or compliance with any formality on the part of the seller.

SECTION II
TAKING DELIVERY

Article 60

The buyer's obligation to take delivery consists—

    (a) in doing all the acts which could reasonably be expected of him in order to enable the seller to make delivery; and

    (b) in taking over the goods.

SECTION III
REMEDIES FOR BREACH OF CONTRACT BY THE BUYER

Article 61

(1)   If the buyer fails to perform any of his obligations under the contract or this Convention, the seller may—
   (a) exercise the rights provided in articles 62 to 65;
   (b) claim damages as provided in articles 74 to 77.

(2)   The seller is not deprived of any right he may have to claim damages by exercising his right to other remedies.

(3)   No period of grace may be granted to the buyer by a court or arbitral tribunal when the seller resorts to a remedy for breach of contract.

Article 62

The seller may require the buyer to pay the price, take delivery or perform his other obligations, unless the seller has resorted to a remedy which is inconsistent with this requirement.

Article 63

(1)   The seller may fix an additional period of time of reasonable length for performance by the buyer of his obligations.

(2)   Unless the seller has received notice from the buyer that he will not perform within the period so fixed, the seller may not, during that period, resort to any remedy for breach of contract. However, the seller is not deprived thereby of any right he may have to claim damages for delay in performance.

Article 64

(1)   The seller may declare the contract avoided—
   (a) if the failure by the buyer to perform any of his obligations under the contract or this Convention amounts to a fundamental breach of contract; or

(b) if the buyer does not, within the additional period of time fixed by the seller in accordance with paragraph (1) of article 63, perform his obligation to pay the price or take delivery of the goods, or if he declares that he will not do so within the period so fixed.

(2)   However, in cases where the buyer has paid the price, the seller loses the right to declare the contract avoided unless he does so—

(a) in respect of late performance by the buyer, before the seller has become aware that performance has been rendered; or

(b) in respect of any breach other than late performance by the buyer, within a reasonable time—

(i) after the seller knew or ought to have known of the breach; or

(ii) after the expiration of any additional period of time fixed by the seller in accordance with paragraph (1) of article 63, or after the buyer has declared that he will not perform his obligations within such an additional period.

Article 65

(1)   If under the contract the buyer is to specify the form, measurement or other features of the goods and he fails to make such specification either on the date agreed upon or within a reasonable time after receipt of a request from the seller, the seller may, without prejudice to any other rights he may have, make the specification himself in accordance with the requirements of the buyer that may be known to him.

(2)   If the seller makes the specification himself, he must inform the buyer of the details thereof and must fix a reasonable time within which the buyer may make a different specification. If, after receipt of such a communication, the buyer fails to do so within the time so fixed, the specification made by the seller is binding.

CHAPTER IV
PASSING OF RISK

Article 66

Loss of or damage to the goods after the risk has passed to the buyer does not discharge him from his obligation to pay the price, unless the loss or damage is due to an act or omission of the seller.

Article 67

(1)   If the contract of sale involves carriage of the goods and the seller is not bound to hand them over at a particular place, the risk passes to the buyer when the goods are handed over to the first carrier for transmission to the buyer in accordance with the contract of sale. If the seller is bound to hand the goods over to a carrier at a particular place, the risk does not pass to the buyer until the goods are handed over to the carrier at that place. The fact that the seller is authorised to retain documents controlling the disposition of the goods does not affect the passage of the risk.

(2)   Nevertheless, the risk does not pass to the buyer until the goods are clearly identified to the contract, whether by markings on the goods, by shipping documents, by notice given to the buyer or otherwise.

Article 68

The risk in respect of goods sold in transit passes to the buyer from the time of the conclusion of the contract. However, if the circumstances so indicate, the risk is assumed by the buyer from the time the goods were handed over to the carrier who issued the documents embodying the contract of carriage. Nevertheless, if at the time of the conclusion of the contract of sale the seller knew or ought to have known that the goods had been lost or damaged and did not disclose this to the buyer, the loss or damage is at the risk of the seller.

Article 69

(1)   In cases not within articles 67 and 68, the risk passes to the buyer when he takes over the goods or, if he does not do so in due time, from the time when the goods are placed at his disposal and he commits a breach of contract by failing to take delivery.

(2)   However, if the buyer is bound to take over the goods at a place other than a place of business of the seller, the risk passes when delivery is due and the buyer is aware of the fact that the goods are placed at his disposal at that place.

(3)   If the contract relates to goods not then identified, the goods are considered not to be placed at the disposal of the buyer until they are clearly identified to the contract.

Article 70

If the seller has committed a fundamental breach of contract, articles 67, 68 and 69 do not impair the remedies available to the buyer on account of the breach.

CHAPTER V
PROVISIONS COMMON TO THE OBLIGATIONS OF THE SELLER AND OF
THE BUYER

SECTION I
ANTICIPATORY BREACH AND INSTALMENT CONTRACTS

Article 71

(1)   A party may suspend the performance of his obligations if, after the conclusion of the contract, it becomes apparent that the other party will not perform a substantial part of his obligations as a result of—
  (a) a serious deficiency in his ability to perform or in his creditworthiness; or
  (b) his conduct in preparing to perform or in performing the contract.

(2)   If the seller has already dispatched the goods before the grounds described in the preceding paragraph become evident, he may prevent the handing over of the goods to the buyer even though the buyer holds a document which entitles him to obtain them. The present paragraph relates only to the rights in the goods as between the buyer and the seller.

(3)   A party suspending performance, whether before or after dispatch of the goods, must immediately give notice of the suspension to the other party and must continue with performance if the other party provides adequate assurance of his performance.

Article 72

(1)   If prior to the date for performance of the contract it is clear that one of the parties will commit a fundamental breach of contract, the other party may declare the contract avoided.

(2)   If time allows, the party intending to declare the contract avoided must give reasonable notice to the other party in order to permit him to provide adequate assurance of his performance.

(3)   The requirements of the preceding paragraph do not apply if the other party has declared that he will not perform his obligations.

Article 73

(1)   In the case of a contract for delivery of goods by instalments, if the failure of one party to perform any of his obligations in respect of any instalment constitutes a fundamental breach of contract with respect to that instalment, the other party may declare the contract avoided with respect to that instalment.

(2)   If one party's failure to perform any of his obligations in respect of any instalment gives the other party good grounds to conclude that a fundamental breach of contract will occur with respect to future instalments, he may declare the contract avoided for the future, provided that he does so within a reasonable time.

(3)   A buyer who declares the contract avoided in respect of any delivery may, at the same time, declare it avoided in respect of deliveries already made or of future deliveries if, by reason of their interdependence, those deliveries could not be used for the purpose contemplated by the parties at the time of the conclusion of the contract.

SECTION II
DAMAGES

Article 74

Damages for breach of contract by one party consist of a sum equal to the loss, including loss of profit, suffered by the other party as a consequence of the breach. Such damages may not exceed the loss which the party in breach foresaw or ought to have foreseen at the time of the conclusion of the contract, in the light of the facts and matters of which he then knew or ought to have known, as a possible consequence of the breach of contract.

Article 75

If the contract is avoided and if, in a reasonable manner and within a reasonable time after avoidance, the buyer has bought goods in replacement or the seller has resold the goods, the party claiming damages may recover the difference between the contract price and the price in the substitute transaction as well as any further damages recoverable under article 74.

Article 76

(1)   If the contract is avoided and there is a current price for the goods, the party claiming damages may, if he has not made a purchase or resale under article 75, recover the difference between the price fixed by the contract and the current price at the time of avoidance as well as any further damages recoverable under article 74. If, however, the party claiming damages has avoided the contract after taking over the goods, the current price at the time of such taking over shall be applied instead of the current price at the time of avoidance.

(2)   For the purposes of the preceding paragraph, the current price is the price prevailing at the place where delivery of the goods should have been made or, if there is no current price at that place, the price at such other place as serves as a reasonable substitute, making due allowance for differences in the cost of transporting the goods.

Article 77

A party who relies on a breach of contract must take such measures as are reasonable in the circumstances to mitigate the loss, including loss of profit, resulting from the breach. If he fails to take such measures, the party in breach may claim a reduction in the damages in the amount by which the loss should have been mitigated.

SECTION III
INTEREST

Article 78

If a party fails to pay the price or any other sum that is in arrears, the other party is entitled to interest on it, without prejudice to any claim for damages recoverable under article 74.

SECTION IV
EXEMPTION

Article 79

(1)   A party is not liable for a failure to perform any of his obligations if he proves that the failure was due to an impediment beyond his control and that he could not reasonably be expected to have taken the impediment into account at the time of the conclusion of the contract or to have avoided or overcome it or its consequences.

(2)   If the party's failure is due to the failure by a third person whom he has engaged to perform the whole or a part of the contract, that party is exempt from liability only if—
   (a) he is exempt under the preceding paragraph; and
   (b) the person whom he has so engaged would be so exempt if the provisions of that paragraph were applied to him.

(3)   The exemption provided by this article has effect for the period during which the impediment exists.

(4)   The party who fails to perform must give notice to the other party of the impediment and its effect on his ability to perform. If the notice is not received by the other party within a reasonable time after the party who fails to perform knew or ought to have known of the impediment, he is liable for damages resulting from such non receipt.

(5)   Nothing in this article prevents either party from exercising any right other than to claim damages under this Convention.

Article 80

A party may not rely on a failure of the other party to perform, to the extent that such failure was caused by the first party's act or omission.

SECTION V
EFFECTS OF AVOIDANCE

Article 81

(1)   Avoidance of the contract releases both parties from their obligations under it, subject to any damages which may be due. Avoidance does not

affect any provision of the contract for the settlement of disputes or any other provision of the contract governing the rights and obligations of the parties consequent upon the avoidance of the contract.

(2)   A party who has performed the contract either wholly or in part may claim restitution from the other party of whatever the first party has supplied or paid under the contract. If both parties are bound to make restitution, they must do so concurrently.

Article 82

(1)   The buyer loses the right to declare the contract avoided or to require the seller to deliver substitute goods if it is impossible for him to make restitution of the goods substantially in the condition in which he received them.

(2)   The preceding paragraph does not apply—
  (a) if the impossibility of making restitution of the goods or of making restitution of the goods substantially in the condition in which the buyer received them is not due to his act or omission;
  (b) if the goods or part of the goods have perished or deteriorated as a result of the examination provided for in article 38; or
  (c) if the goods or part of the goods have been sold in the normal course of business or have been consumed or transformed by the buyer in the course of normal use before he discovered or ought to have discovered the lack of conformity.

Article 83

A buyer who has lost the right to declare the contract avoided or to require the seller to deliver substitute goods in accordance with article 82 retains all other remedies under the contract and this Convention.

Article 84

(1)   If the seller is bound to refund the price, he must also pay interest on it, from the date on which the price was paid.

(2)   The buyer must account to the seller for all benefits which he has derived from the goods or part of them—

(a) if he must make restitution of the goods or part of them; or

(b) if it is impossible for him to make restitution of all or part of the goods or to make restitution of all or part of the goods substantially in the condition in which he received them, but he has nevertheless declared the contract avoided or required the seller to deliver substitute goods.

SECTION VI
PRESERVATION OF THE GOODS

Article 85

If the buyer is in delay in taking delivery of the goods or, where payment of the price and delivery of the goods are to be made concurrently, if he fails to pay the price, and the seller is either in possession of the goods or otherwise able to control their disposition, the seller must take such steps as are reasonable in the circumstances to preserve them. He is entitled to retain them until he has been reimbursed his reasonable expenses by the buyer.

Article 86

(1)   If the buyer has received the goods and intends to exercise any right under the contract or this Convention to reject them, he must take such steps to preserve them as are reasonable in the circumstances. He is entitled to retain them until he has been reimbursed his reasonable expenses by the seller.

(2)   If goods dispatched to the buyer have been placed at his disposal at their destination and he exercises the right to reject them, he must take possession of them on behalf of the seller, provided that this can be done without payment of the price and without unreasonable inconvenience or unreasonable expense. This provision does not apply if the seller or a person authorised to take charge of the goods on his behalf is present at the destination. If the buyer takes possession of the goods under this paragraph, his rights and obligations are governed by the preceding paragraph.

Article 87

A party who is bound to take steps to preserve the goods may deposit them in a warehouse of a third person at the expense of the other party provided that the expense incurred is not unreasonable.

Article 88

(1)   A party who is bound to preserve the goods in accordance with article 85 or 86 may sell them by any appropriate means if there has been an unreasonable delay by the other party in taking possession of the goods or in taking them back or in paying the price or the cost of preservation, provided that reasonable notice of the intention to sell has been given to the other party.

(2)   If the goods are subject to rapid deterioration or their preservation would involve unreasonable expense, a party who is bound to preserve the goods in accordance with article 85 or 86 must take reasonable measures to sell them. To the extent possible he must give notice to the other party of his intention to sell.

(3)   A party selling the goods has the right to retain out of the proceeds of sale an amount equal to the reasonable expenses of preserving the goods and of selling them. He must account to the other party for the balance.

PART IV
FINAL PROVISIONS

Article 90

This Convention does not prevail over any international agreement which has already been or may be entered into and which contains provisions concerning the matters governed by this Convention, provided that the parties have their places of business in States parties to such agreement.

Article 92

(1)   A Contracting State may declare at the time of signature, ratification, acceptance, approval or accession that it will not be bound by Part II of this Convention or that it will not be bound by Part III of this Convention.

(2)   A Contracting State which makes a declaration in accordance with the preceding paragraph in respect of Part II or Part III of this Convention is not to be considered a Contracting State within paragraph (1) of article 1 of this Convention in respect of matters governed by the Part to which the declaration applies.

Article 93

(1)   If a Contracting State has two or more territorial units in which, according to its constitution, different systems of law are applicable in relation to the matters dealt with in this Convention, it may, at the time of signature, ratification, acceptance, approval or accession, declare that this Convention is to extend to all its territorial units or only to one or more of them, and may amend its declaration by submitting another declaration at any time.

(2)   These declarations are to be notified to the depositary and are to state expressly the territorial units to which the Convention extends.

(3)   If, by virtue of a declaration under this article, this Convention extends to one or more but not all of the territorial units of a Contracting State, and if the place of business of a party is located in that State, this place of business, for the purposes of this Convention, is considered not to be in a Contracting State, unless it is in a territorial unit to which the Convention extends.

(4)   If a Contracting State makes no declaration under paragraph (1) of this article, the Convention is to extend to all territorial units of that State.

Article 94

(1)   Two or more Contracting States which have the same or closely related legal rules on matters governed by this Convention may at any time declare that the Convention is not to apply to contracts of sale or to their formation where the parties have their places of business in those States. Such declarations may be made jointly or by reciprocal unilateral declarations.

(2)   A Contracting State which has the same or closely related legal rules on matters governed by this Convention as one or more non-Contracting States may at any time declare that the Convention is not to apply to contracts of sale or to their formation where the parties have their places of business in those States.

(3)   If a State which is the object of a declaration under the preceding paragraph subsequently becomes a Contracting State, the declaration made will, as from the date on which the Convention enters into force in respect of the new Contracting State, have the effect of a declaration made under

paragraph (1), provided that the new Contracting State joins in such declaration or makes a reciprocal unilateral declaration.

Article 95

Any State may declare at the time of the deposit of its instrument of ratification, acceptance, approval or accession that it will not be bound by subparagraph (1)(b) of article 1 of this Convention.

Article 96

A Contracting State whose legislation requires contracts of sale to be concluded in or evidenced by writing may at any time make a declaration in accordance with article 12 that any provision of article 11, article 29, or Part II of this Convention, that allows a contract of sale or its modification or termination by agreement or any offer, acceptance, or other indication of intention to be made in any form other than in writing, does not apply where any party has his place of business in that State.

Article 98

No reservations are permitted except those expressly authorised in this Convention.

Article 99

(1)   This Convention enters into force, subject to the provisions of paragraph (6) of this article, on the first day of the month following the expiration of twelve months after the date of deposit of the tenth instrument of ratification, acceptance, approval or accession, including an instrument which contains a declaration made under article 92.

(2)   When a State ratifies, accepts, approves or accedes to this Convention after the deposit of the tenth instrument of ratification, acceptance, approval or accession, this Convention, with the exception of the Part excluded, enters into force in respect of that State, subject to the provisions of paragraph (6) of this article, on the first day of the month following the expiration of twelve months after the date of the deposit of its instrument of ratification, acceptance, approval or accession.

(3)   A State which ratifies, accepts, approves or accedes to this Convention and is a party to either or both the Convention relating to a Uniform Law on the Formation of Contracts for the International Sale of Goods done at The Hague on 1 July 1964 (1964 Hague Formation Convention) and the Convention relating to a Uniform Law on the International Sale of Goods done at The Hague on 1 July 1964 (1964 Hague Sales Convention) shall at the same time denounce, as the case may be, either or both the 1964 Hague Sales Convention and the 1964 Hague Formation Convention by notifying the Government of the Netherlands to that effect.

(4)   A State party to the 1964 Hague Sales Convention which ratifies, accepts, approves or accedes to the present Convention and declares or has declared under article 92 that it will not be bound by Part II of this Convention shall at the time of ratification, acceptance, approval or accession denounce the 1964 Hague Sales Convention by notifying the Government of the Netherlands to that effect.

(5)   A State party to the 1964 Hague Formation Convention which ratifies, accepts, approves or accedes to the present Convention and declares or has declared under article 92 that it will not be bound by Part III of this Convention shall at the time of ratification, acceptance, approval or accession denounce the 1964 Hague Formation Convention by notifying the Government of the Netherlands to that effect.

(6)   For the purpose of this article, ratifications, acceptances, approvals and accessions in respect of this Convention by States parties to the 1964 Hague Formation Convention or to the 1964 Hague Sales Convention shall not be effective until such denunciations as may be required on the part of those States in respect of the latter two Conventions have themselves become effective. The depositary of this Convention shall consult with the Government of the Netherlands, as the depositary of the 1964 Conventions, so as to ensure necessary co-ordination in this respect.

DONE at Vienna, this day of eleventh day of April, one thousand nine hundred and eighty, in a single original, of which the Arabic, Chinese, English, French, Russian and Spanish texts are equally authentic.

IN WITNESS WHEREOF the undersigned plenipotentiaries, being duly authorised by their respective Governments, have signed this Convention.

NOTES

The text of this Convention was prepared by the United Nations Commission on International Trade Law.

ICC Uniform Customs and Practice for Documentary Credits

(1993 Revision)

(ICC Publication No 500—ISBN 92.842.1155.7 (E))

NOTES

Published in its official English version by the International Chamber of Commerce

Copyright c.1993—International Chamber of Commerce (ICC), Paris.

Available from: ICC Publishing SA, 38 Cours Albert 1er, 75008 Paris, France or ICC United Kingdom, 14/15 Belgrave Square, London SW1X 8PS, United Kingdom.

A. GENERAL PROVISIONS AND DEFINITIONS

Article 1   Application of UCP

The Uniform Customs and Practice for Documentary Credits, 1993 Revision, ICC Publication N0500, shall apply to all Documentary Credits (including to the extent to which they may be applicable, Standby Letter(s) of Credit) where they are incorporated into the text of the Credit. They are binding on all parties thereto, unless otherwise expressly stipulated in the Credit.

Article 2   Meaning of Credit

For the purposes of these Articles, the expressions "Documentary Credit(s)" and "Standby Letter(s) of Credit" (hereinafter referred to as "Credit(s)"), mean any arrangement, however named or described, whereby a bank (the "Issuing Bank") acting at the request and on the instructions of a customer (the "Applicant") or on its own behalf,

   (i)   is to make a payment to or to the order of a third party ("the Beneficiary"), or is to accept and pay bills of exchange (Draft(s)) drawn by the Beneficiary, or

   (ii)   authorises another bank to effect such payment, or to accept and pay such bills of exchange (Draft(s)), or

   (iii)   authorises another bank to negotiate,

against stipulated document(s), provided that the terms and conditions of the Credit are complied with.

For the purposes of these Articles, branches of a bank in different countries are considered another bank.

Article 3   Credits v Contracts

(a)   Credits, by their nature, are separate transactions from the sales or other contract(s) on which they may be based and banks are in no way concerned with or bound by such contract(s), even if any reference whatsoever to such contract(s) is included in the Credit. Consequently, the undertaking of a bank to pay, accept and pay Draft(s) or negotiate and/or to fulfil any other obligation under the Credit, is not subject to claims or defences by the Applicant resulting from his relationships with the Issuing Bank or the Beneficiary.

(b)   A Beneficiary can in no case avail himself of the contractual relationships existing between the banks or between the Applicant and the Issuing Bank.

Article 4   Documents v Goods/Services/Performances

In Credit operations all parties concerned deal with documents, and not with goods, services and/or other performances to which the documents may relate.

Article 5   Instructions to Issue/Amend Credits

(a)   Instructions for the issuance of a Credit, the Credit itself, instructions for an amendment thereto, and the amendment itself, must be complete and precise.

In order to guard against confusion and misunderstanding, banks should discourage any attempt—
   (i) to include excessive detail in the Credit or in any amendment thereto;
   (ii) to give instructions to issue, advise or confirm a Credit by reference to a Credit previously issued (similar Credit) where such previous Credit has been subject to accepted amendment(s), and/or unaccepted amendment(s).

(b)   All instructions for the issuance of a Credit and the Credit itself and, where applicable, all instructions for an amendment thereto and the amendment itself, must state precisely the document(s) against which payment, acceptance or negotiation is to be made.

B. FORM AND NOTIFICATION OF CREDITS

Article 6   Revocable v Irrevocable Credits

(a)   A Credit may be either
 (i)   revocable, or
 (ii) irrevocable.

(b)   The Credit, therefore, should clearly indicate whether it is revocable or irrevocable.

(c)   In the absence of such indication the Credit shall be deemed to be irrevocable.

Article 7   Advising Bank's Liability

(a)   A Credit may be advised to a Beneficiary through another bank (the "Advising Bank") without engagement on the part of the Advising Bank, but that bank, if it elects to advise the Credit, shall take reasonable care to check the apparent authenticity of the Credit which it advises. If the bank elects not to advise the Credit, it must so inform the Issuing Bank without delay.

(b)   If the Advising Bank cannot establish such apparent authenticity it must inform, without delay, the bank from which the instructions appear to have been received that it has been unable to establish the authenticity of the Credit and if it elects nonetheless to advise the Credit it must inform the Beneficiary that it has not been able to establish the authenticity of the Credit.

Article 8   Revocation of a Credit

(a)   A revocable Credit may be amended or cancelled by the Issuing Bank at any moment and without prior notice to the Beneficiary.

(b)   However, the Issuing Bank must—
  (i)   reimburse another bank with which a revocable Credit has been made available for sight payment, acceptance or negotiation for any payment, acceptance or negotiation made by such bank prior to receipt by it of notice of amendment or cancellation, against documents which appear on their face to be in compliance with the terms and conditions of the Credit;
  (ii) reimburse another bank with which a revocable Credit has been made available for deferred payment, if such a bank has, prior to receipt by it of notice of amendment or cancellation, taken up documents which appear on their face to be in compliance with the terms and conditions of the Credit.

Article 9   Liability of Issuing and Confirming Banks

(a)   An irrevocable Credit constitutes a definite undertaking of the Issuing Bank, provided that the stipulated documents are presented to the Nominated Bank or to the Issuing Bank and that the terms and conditions of the Credit are complied with—
  (i)   if the Credit provides for sight payment to pay at sight;
  (ii) if the Credit provides for deferred payment to pay on the maturity date(s) determinable in accordance with the stipulations of the Credit;
  (iii) if the Credit provides for acceptance—
        (a) by the Issuing Bank to accept Draft(s) drawn by the Beneficiary on the Issuing Bank and pay them at maturity, or
        (b) by another drawee bank to accept and pay at maturity Draft(s) drawn by the Beneficiary on the Issuing Bank in the event the drawee bank stipulated in the Credit does not accept Draft(s) drawn on it, or to pay Draft(s) accepted but not paid by such drawee bank at maturity;
  (iv) if the Credit provides for negotiation to pay without recourse to drawers and/or bona fide holders, Draft(s) drawn by the Beneficiary and/or document(s) presented under the Credit. A Credit should not be issued available by Draft(s) on the Applicant. If the Credit nevertheless calls for Draft(s) on the Applicant, banks will consider such Draft(s) as an additional document(s).

(b)   A confirmation of an irrevocable Credit by another bank (the "Confirming Bank") upon the authorisation or request of the Issuing Bank, constitutes a definite undertaking of the Confirming Bank, in addition to that of the Issuing Bank, provided that the stipulated documents are presented to the Confirming Bank or to any other Nominated Bank and that the terms and conditions of the Credit are complied with—

(i) if the Credit provides for sight payment to pay at sight;

(ii) if the Credit provides for deferred payment to pay on the maturity date(s) determinable in accordance with the stipulations of the Credit;

(iii) if the Credit provides for acceptance—

(a) by the Confirming Bank to accept Draft(s) drawn by the Beneficiary on the Confirming Bank and pay them at maturity, or

(b) by another drawee bank to accept and pay at maturity Draft(s) drawn by the Beneficiary on the Confirming Bank, in the event the drawee bank stipulated in the Credit does not accept Draft(s) drawn on it, or to pay Draft(s) accepted but not paid by such drawee bank at maturity;

(iv) if the Credit provides for negotiation to negotiate without recourse to drawers and/or bona fide holders, Draft(s) drawn by the Beneficiary and/or document(s) presented under the Credit. A Credit should not be issued available by Draft(s) on the Applicant. If the Credit nevertheless calls for Draft(s) on the Applicant, banks will consider such Draft(s) as an additional document(s).

(c) (i) If another bank is authorised or requested by the Issuing Bank to add its confirmation to a Credit but is not prepared to do so, it must so inform the Issuing Bank without delay.

(ii) Unless the Issuing Bank specifies otherwise in its authorisation or request to add confirmation, the Advising Bank may advise the Credit to the Beneficiary without adding its confirmation.

(d) (i) Except as otherwise provided by Article 48, an irrevocable Credit can neither be amended nor cancelled without the agreement of the Issuing Bank, the Confirming Bank, if any, and the Beneficiary.

(ii) The Issuing Bank shall be irrevocably bound by an amendment(s) issued by it from the time of the issuance of such amendment(s). A Confirming Bank may extend its confirmation to an amendment and shall be irrevocably bound as of the time of its advice of the amendment. A Confirming Bank may, however, choose to advise an amendment to the Beneficiary without extending its confirmation and if so, must inform the Issuing Bank and the Beneficiary without delay.

(iii) The terms of the original Credit (or a Credit incorporating previously accepted amendment(s)) will remain in force for the Beneficiary until the Beneficiary communicates his acceptance of the amendment to the bank that advised such amendment. The Beneficiary should give notification of acceptance or rejection of amendment(s). If the Beneficiary fails to give such notification, the tender of documents to the Nominated Bank or Issuing Bank, that conform to the Credit and to not yet accepted amendment(s), will be deemed to be notification of acceptance by the Beneficiary of such amendment(s) and as of that moment the Credit will be amended.

(iv) Partial acceptance of amendments contained in one and the same advice of amendment is not allowed and consequently will not be given any effect.

Article 10   Types of Credit

(a)   All Credits must clearly indicate whether they are available by sight payment, by deferred payment, by acceptance or by negotiation.

(b)(i) Unless the Credit stipulates that it is available only with the Issuing Bank, all Credits must nominate the bank (the "Nominated Bank") which is authorised to pay, to incur a deferred payment undertaking, to accept Draft(s) or to negotiate. In a freely negotiable Credit, any bank is a Nominated Bank.

Presentation of documents must be made to the Issuing Bank or the Confirming Bank, if any, or any other Nominated Bank.

(ii) Negotiation means the giving of value for Draft(s) and/or document(s) by the bank authorised to negotiate. Mere examination of the documents without giving of value does not constitute a negotiation.

(c)   Unless the Nominated Bank is the Confirming Bank, nomination by the Issuing Bank does not constitute any undertaking by the Nominated Bank to pay, to incur a deferred payment undertaking, to accept Draft(s), or to negotiate. Except where expressly agreed to by the Nominated Bank and so communicated to the Beneficiary, the Nominated Bank's receipt of and/or examination and/or forwarding of the documents does not make that bank liable to pay, to incur a deferred payment undertaking, to accept Draft(s), or to negotiate.

(d)   By nominating another bank, or by allowing for negotiation by any bank, or by authorising or requesting another bank to add its confirmation, the Issuing Bank authorises such bank to pay, accept Draft(s) or negotiate as the case may be, against documents which appear on their face to be in compliance with the terms and conditions of the Credit and undertakes to reimburse such bank in accordance with the provisions of these Articles.

Article 11 Teletransmitted and Pre-Advised Credits

(a)(i) When an Issuing Bank instructs an Advising Bank by an authenticated teletransmission to advise a Credit or an amendment to a Credit, the teletransmission will be deemed to be the operative Credit instrument or the operative amendment, and no mail confirmation should be sent. Should a mail confirmation nevertheless be sent, it will have no effect and the Advising Bank will have no obligation to check such mail confirmation against the operative Credit instrument or the operative amendment received by teletransmission.

(ii) If the teletransmission states "full details to follow" (or words of similar effect) or states that the mail confirmation is to be the operative Credit instrument or the operative amendment, then the teletransmission will not be deemed to be the operative Credit instrument or the operative amendment. The Issuing Bank must forward the operative Credit instrument or the operative amendment to such Advising Bank without delay.

(b) If a bank uses the services of an Advising Bank to have the Credit advised to the Beneficiary, it must also use the services of the same bank for advising an amendment(s).

(c) A preliminary advice of the issuance or amendment of an irrevocable Credit (pre-advice), shall only be given by an Issuing Bank if such bank is prepared to issue the operative Credit instrument or the operative amendment thereto. Unless otherwise stated in such preliminary advice by the Issuing Bank, an Issuing Bank having given such pre-advice shall be irrevocably committed to issue or amend the Credit, in terms not inconsistent with the pre-advice, without delay.

Article 12 Incomplete or Unclear Instructions

If incomplete or unclear instructions are received to advise, confirm or amend a Credit, the bank requested to act on such instructions may give preliminary notification to the Beneficiary for information only and without responsibility. This preliminary notification should state clearly that the notification is provided for information only and without the responsibility of the Advising Bank. In any event, the Advising Bank must inform the Issuing Bank of the action taken and request it to provide the necessary information.

The Issuing Bank must provide the necessary information without delay. The Credit will be advised, confirmed or amended, only when complete and clear instructions have been received and if the Advising Bank is then prepared to act on the instructions.

C. LIABILITIES AND RESPONSIBILITIES

Article 13   Standard for Examination of Documents

(a)   Banks must examine all documents stipulated in the Credit with reasonable care, to ascertain whether or not they appear, on their face, to be in compliance with the terms and conditions of the Credit. Compliance of the stipulated documents on their face with the terms and conditions of the Credit, shall be determined by international standard banking practice as reflected in these Articles. Documents which appear on their face to be inconsistent with one another will be considered as not appearing on their face to be in compliance with the terms and conditions of the Credit.

Documents not stipulated in the Credit will not be examined by banks. If they receive such documents, they shall return them to the presenter or pass them on without responsibility.

(b)   The Issuing Bank, the Confirming Bank, if any, or a Nominated Bank acting on their behalf, shall each have a reasonable time, not to exceed seven banking days following the day of receipt of the documents, to examine the documents and determine whether to take up or refuse the documents and to inform the party from which it received the documents accordingly.

(c)   If a Credit contains conditions without stating the document(s) to be presented in compliance therewith, banks will deem such conditions as not stated and will disregard them.

Article 14   Discrepant Documents and Notice

(a)   When the Issuing Bank authorises another bank to pay, incur a deferred payment undertaking, accept Draft(s), or negotiate against documents which appear on their face to be in compliance with the terms and conditions of the Credit, the Issuing Bank and the Confirming Bank, if any, are bound—

   (i) to reimburse the Nominated Bank which has paid, incurred a deferred payment undertaking, accepted Draft(s), or negotiated,

   (ii) to take up the documents.

(b)   Upon receipt of the documents the Issuing Bank and/or Confirming Bank, if any, or a Nominated Bank acting on their behalf, must determine on the basis of the documents alone whether or not they appear on their face to be in compliance with the terms and conditions of the Credit. If the documents appear on their face not to be in compliance with the terms and conditions of the Credit, such banks may refuse to take up the documents.

(c)   If the Issuing Bank determines that the documents appear on their face not to be in compliance with the terms and conditions of the Credit, it may in its sole judgment approach the Applicant for a waiver of the discrepancy(ies). This does not, however, extend the period mentioned in sub-Article 13(b).

(d)(i) If the Issuing Bank and/or Confirming Bank, if any, or a Nominated Bank acting on their behalf, decides to refuse the documents, it must give notice to that effect by telecommunication or, if that is not possible, by other expeditious means, without delay but no later than the close of the seventh banking day following the day of receipt of the documents. Such notice shall be given to the bank from which it received the documents, or to the Beneficiary, if it received the documents directly from him.

   (ii) Such notice must state all discrepancies in respect of which the bank refuses the documents and must also state whether it is holding the documents at the disposal of, or is returning them to, the presenter.

   (iii) The Issuing Bank and/or Confirming Bank, if any, shall then be entitled to claim from the remitting bank refund, with interest, of any reimbursement which has been made to that bank.

(e)   If the Issuing Bank and/or Confirming Bank, if any, fails to act in accordance with the provisions of this Article and/or fails to hold the documents at the disposal of, or return them to the presenter, the Issuing Bank and/or Confirming Bank, if any, shall be precluded from claiming that the documents are not in compliance with the terms and conditions of the Credit.

(f)   If the remitting bank draws the attention of the Issuing Bank and/or Confirming Bank, if any, to any discrepancy(ies) in the document(s) or advises such banks that it has paid, incurred a deferred payment undertaking, accepted Draft(s) or negotiated under reserve or against an indemnity in respect of such discrepancy(ies), the Issuing Bank and/or Confirming Bank, if any, shall not be thereby relieved from any of their obligations under any provision of this Article. Such reserve or indemnity concerns only the relations between the remitting bank and the party towards whom the reserve was made, or from whom, or on whose behalf, the indemnity was obtained.

Article 15   Disclaimer on Effectiveness of Documents

Banks assume no liability or responsibility for the form, sufficiency, accuracy, genuineness, falsification or legal effect of any document(s), or for the general and/or particular conditions stipulated in the document(s) or superimposed thereon; nor do they assume any liability or responsibility for the description, quantity, weight, quality, condition, packing, delivery, value or existence of the goods represented by any document(s), or for the good faith or acts and/or omissions, solvency, performance or standing of the consignors, the carriers, the forwarders, the consignees or the insurers of the goods, or any other person whomsoever.

Article 16   Disclaimer on the Transmission of Messages

Banks assume no liability or responsibility for the consequences arising out of delay and/or loss in transit of any message(s), letter(s) or document(s), or for delay, mutilation or other error(s) arising in the transmission of any telecommunication. Banks assume no liability or responsibility for errors in translation and/or interpretation of technical terms, and reserve the right to transmit Credit terms without translating them.

Article 17   Force Majeure

Banks assume no liability or responsibility for the consequences arising out of the interruption of their business by Acts of God, riots, civil commotions, insurrections, wars or any other causes beyond their control, or by any strikes or lockouts. Unless specifically authorised, banks will not, upon resumption of their business, pay, incur a deferred payment undertaking, accept Draft(s) or negotiate under Credits which expired during such interruption of their business.

Article 18 Disclaimer for Acts of an Instructed Party

(a)   Banks utilizing the services of another bank or other banks for the purpose of giving effect to the instructions of the Applicant do so for the account and at the risk of such Applicant.

(b)   Banks assume no liability or responsibility should the instructions they transmit not be carried out, even if they have themselves taken the initiative in the choice of such other bank(s).

(c) (i) A party instructing another party to perform services is liable for any charges, including commissions, fees, costs or expenses incurred by the instructed party in connection with its instructions.
   (ii) Where a Credit stipulates that such charges are for the account of a party other than the instructing party, and charges cannot be collected, the instructing party remains ultimately liable for the payment thereof.

(d)   The Applicant shall be bound by and liable to indemnify the banks against all obligations and responsibilities imposed by foreign laws and usages.

Article 19 Bank-to-Bank Reimbursement Arrangements

(a)   If an Issuing Bank intends that the reimbursement to which a paying, accepting or negotiating bank is entitled, shall be obtained by such bank (the "Claiming Bank"), claiming on another party (the "Reimbursing Bank"), it shall provide such Reimbursing Bank in good time with the proper instructions or authorisation to honour such reimbursement claims.

(b)   Issuing Banks shall not require a Claiming Bank to supply a certificate of compliance with the terms and conditions of the Credit to the Reimbursing Bank.

(c)   An Issuing Bank shall not be relieved from any of its obligations to provide reimbursement if and when reimbursement is not received by the Claiming Bank from the Reimbursing Bank.

(d)   The Issuing Bank shall be responsible to the Claiming Bank for any loss of interest if reimbursement is not provided by the Reimbursing Bank on first demand, or as otherwise specified in the Credit, or mutually agreed, as the case may be.

(e) The Reimbursing Bank's charges should be for the account of the Issuing Bank. However, in cases where the charges are for the account of another party, it is the responsibility of the Issuing Bank to so indicate in the original Credit and in the reimbursement authorisation. In cases where the Reimbursing Bank's charges are for the account of another party they shall be collected from the Claiming Bank when the Credit is drawn under. In cases where the Credit is not drawn under, the Reimbursing Bank's charges remain the obligation of the Issuing Bank.

D. DOCUMENTS

Article 20   Ambiguity as to the Issuers of Documents

(a) Terms such as "first class", "well known", "qualified", "independent", "official", "competent", "local" and the like, shall not be used to describe the issuers of any document(s) to be presented under a Credit. If such terms are incorporated in the Credit, banks will accept the relative document(s) as presented, provided that it appears on its face to be in compliance with the other terms and conditions of the Credit and not to have been issued by the Beneficiary.

(b) Unless otherwise stipulated in the Credit, banks will also accept as an original document(s), a document(s) produced or appearing to have been produced—

  (i) by reprographic, automated or computerized systems;

  (ii) as carbon copies;

provided that it is marked as original and, where necessary, appears to be signed.

A document may be signed by handwriting, by facsimile signature, by perforated signature, by stamp, by symbol, or by any other mechanical or electronic method of authentication.

(c)(i) Unless otherwise stipulated in the Credit, banks will accept as a copy(ies), a document(s) either labelled copy or not marked as an original a copy(ies) need not be signed.

  (ii) Credits that require multiple document(s) such as "duplicate", "two fold", "two copies" and the like, will be satisfied by the presentation of one original and the remaining number in copies except where the document itself indicates otherwise.

(d)   Unless otherwise stipulated in the Credit, a condition under a Credit calling for a document to be authenticated, validated, legalised, visaed, certified or indicating a similar requirement, will be satisfied by any signature, mark, stamp or label on such document that on its face appears to satisfy the above condition.

Article 21   Unspecified Issuers or Contents of Documents

When documents other than transport documents, insurance documents and commercial invoices are called for, the Credit should stipulate by whom such documents are to be issued and their wording or data content. If the Credit does not so stipulate, banks will accept such documents as presented, provided that their data content is not inconsistent with any other stipulated document presented.

Article 22   Issuance Date of Documents v Credit Date

Unless otherwise stipulated in the Credit, banks will accept a document bearing a date of issuance prior to that of the Credit, subject to such document being presented within the time limits set out in the Credit and in these Articles.

Article 23   Marine/Ocean Bill of Lading

(a)   If a Credit calls for a bill of lading covering a port-to-port shipment, banks will, unless otherwise stipulated in the Credit, accept a document, however named, which—
  (i) appears on its face to indicate the name of the carrier and to have been signed or otherwise authenticated by—
    — the carrier or a named agent for or on behalf of the carrier, or
    — the master or a named agent for or on behalf of the master.
    Any signature or authentication of the carrier or master must be identified as carrier or master, as the case may be. An agent signing or authenticating for the carrier or master must also indicate the name and the capacity of the party, ie carrier or master, on whose behalf that agent is acting, and
  (ii) indicates that the goods have been loaded on board, or shipped on a named vessel.
    Loading on board or shipment on a named vessel may be indicated by pre-printed wording on the bill of lading that the goods have been

loaded on board a named vessel or shipped on a named vessel, in which case the date of issuance of the bill of lading will be deemed to be the date of loading on board and the date of shipment.

In all other cases loading on board a named vessel must be evidenced by a notation on the bill of lading which gives the date on which the goods have been loaded on board, in which case the date of the on board notation will be deemed to be the date of shipment.

If the bill of lading contains the indication "intended vessel", or similar qualification in relation to the vessel, loading on board a named vessel must be evidenced by an on board notation on the bill of lading which, in addition to the date on which the goods have been loaded on board, also includes the name of the vessel on which the goods have been loaded, even if they have been loaded on the vessel named as the "intended vessel".

If the bill of lading indicates a place of receipt or taking in charge different from the port of loading, the on board notation must also include the port of loading stipulated in the Credit and the name of the vessel on which the goods have been loaded, even if they have been loaded on the vessel named in the bill of lading. This provision also applies whenever loading on board the vessel is indicated by pre-printed wording on the bill of lading, and

(iii) indicates the port of loading and the port of discharge stipulated in the Credit, notwithstanding that it—

  (a) indicates a place of taking in charge different from the port of loading, and/or a place of final destination different from the port of discharge, and/or

  (b) contains the indication "intended" or similar qualification in relation to the port of loading and/or port of discharge, as long as the document also states the ports of loading and/or discharge stipulated in the Credit, and

(iv) consists of a sole original bill of lading or, if issued in more than one original, the full set as so issued, and

(v) appears to contain all of the terms and conditions of carriage, or some of such terms and conditions by reference to a source or document other than the bill of lading (short form/ blank back bill of lading); banks will not examine the contents of such terms and conditions, and

(vi) contains no indication that it is subject to a charter party and/or no indication that the carrying vessel is propelled by sail only, and

(vii) in all other respects meets the stipulations of the Credit.

(b) For the purpose of this Article, transhipment means unloading and reloading from one vessel to another vessel during the course of ocean carriage from the port of loading to the port of discharge stipulated in the Credit.

(c)   Unless transhipment is prohibited by the terms of the Credit, banks will accept a bill of lading which indicates that the goods will be transhipped, provided that the entire ocean carriage is covered by one and the same bill of lading.

(d)   Even if the Credit prohibits transhipment, banks will accept a bill of lading which—
  (i)   indicates that transhipment will take place as long as the relevant cargo is shipped in Container(s), Trailer(s) and/or "LASH" barge(s) as evidenced by the bill of lading, provided that the entire ocean carriage is covered by one and the same bill of lading, and/or
  (ii) incorporates clauses stating that the carrier reserves the right to tranship.

Article 24   Non-Negotiable Sea Waybill

(a)   If a Credit calls for a non-negotiable sea waybill covering a port-to-port shipment, banks will, unless otherwise stipulated in the Credit, accept a document, however named, which—
  (i)   appears on its face to indicate the name of the carrier and to have been signed or otherwise authenticated by—
        — the carrier or a named agent for or on behalf of the carrier, or
        — the master or a named agent for or on behalf of the master,
        Any signature or authentication of the carrier or master must be identified as carrier or master, as the case may be. An agent signing or authenticating for the carrier or master must also indicate the name and the capacity of the party, ie carrier or master, on whose behalf that agent is acting, and
  (ii) indicates that the goods have been loaded on board, or shipped on a named vessel.
        Loading on board or shipment on a named vessel may be indicated by pre-printed wording on the non-negotiable sea waybill that the goods have been loaded on board a named vessel or shipped on a named vessel, in which case the date of issuance of the non-negotiable sea waybill will be deemed to be the date of loading on board and the date of shipment.
        In all other cases loading on board a named vessel must be evidenced by a notation on the non-negotiable sea waybill which gives the date on which the goods have been loaded on board, in which case the date of the on board notation will be deemed to be the date of shipment.
        If the non-negotiable sea waybill contains the indication "intended vessel", or similar qualification in relation to the vessel, loading on

board a named vessel must be evidenced by an on board notation on the non-negotiable sea waybill which, in addition to the date on which the goods have been loaded on board, includes the name of the vessel on which the goods have been loaded, even if they have been loaded on the vessel named as the "intended vessel".

If the non-negotiable sea waybill indicates a place of receipt or taking in charge different from the port of loading, the on board notation must also include the port of loading stipulated in the Credit and the name of the vessel on which the goods have been loaded, even if they have been loaded on a vessel named in the non-negotiable sea waybill. This provision also applies whenever loading on board the vessel is indicated by pre-printed wording on the non-negotiable sea waybill, and

(iii) indicates the port of loading and the port of discharge stipulated in the Credit, notwithstanding that it—

    (a) indicates a place of taking in charge different from the port of loading, and/or a place of final destination different from the port of discharge, and/or

    (b) contains the indication "intended" or similar qualification in relation to the port of loading and/or port of discharge, as long as the document also states the ports of loading and/or discharge stipulated in the Credit, and

(iv) consists of a sole original non-negotiable sea waybill, or if issued in more than one original, the full set as so issued, and

(v) appears to contain all of the terms and conditions of carriage, or some of such terms and conditions by reference to a source or document other than the non-negotiable sea waybill (short form/ blank back non-negotiable sea waybill); banks will not examine the contents of such terms and conditions, and

(vi) contains no indication that it is subject to a charter party and/or no indication that the carrying vessel is propelled by sail only, and

(vii) in all other respects meets the stipulations of the Credit.

(b) For the purpose of this Article, transhipment means unloading and reloading from one vessel to another vessel during the course of ocean carriage from the port of loading to the port of discharge stipulated in the Credit.

(c) Unless transhipment is prohibited by the terms of the Credit, banks will accept a non-negotiable sea waybill which indicates that the goods will be transhipped, provided that the entire ocean carriage is covered by one and the same non-negotiable sea waybill.

(d) Even if the Credit prohibits transhipment, banks will accept a non-negotiable sea waybill which—

(i)   indicates that transhipment will take place as long as the relevant cargo is shipped in Container(s), Trailer(s) and/or "LASH" barge(s) as evidenced by the non-negotiable sea waybill, provided that the entire ocean carriage is covered by one and the same non-negotiable sea waybill, and/or

(ii)   incorporates clauses stating that the carrier reserves the right to tranship.

Article 25   Charter Party Bill of Lading

(a)   If a Credit calls for or permits a charter party bill of lading, banks will, unless otherwise stipulated in the Credit, accept a document, however named, which—

   (i)   contains any indication that it is subject to a charter party, and

  (ii)   appears on its face to have been signed or otherwise authenticated by—

     — the master or a named agent for or on behalf of the master, or

     — the owner or a named agent for or on behalf of the owner.

Any signature or authentication of the master or owner must be identified as master or owner as the case may be. An agent signing or authenticating for the master or owner must also indicate the name and the capacity of the party, ie master or owner, on whose behalf that agent is acting, and

 (iii)   does or does not indicate the name of the carrier, and

 (iv)   indicates that the goods have been loaded on board or shipped on a named vessel.

Loading on board or shipment on a named vessel may be indicated by pre-printed wording on the bill of lading that the goods have been loaded on board a named vessel or shipped on a named vessel, in which case the date of issuance of the bill of lading will be deemed to be the date of loading on board and the date of shipment.

In all other cases loading on board a named vessel must be evidenced by a notation on the bill of lading which gives the date on which the goods have been loaded on board, in which case the date of the on board notation will be deemed to be the date of shipment, and

  (v)   indicates the port of loading and the port of discharge stipulated in the Credit, and

 (vi)   consists of a sole original bill of lading or, if issued in more than one original, the full set as so issued, and

 (vii)   contains no indication that the carrying vessel is propelled by sail only, and

(viii)   in all other respects meets the stipulations of the Credit.

(b)   Even if the Credit requires the presentation of a charter party contract in connection with a charter party bill of lading, banks will not examine such charter party contract, but will pass it on without responsibility on their part.

Article 26   Multimodal Transport Document

(a)   If a Credit calls for a transport document covering at least two different modes of transport (multimodal transport), banks will, unless otherwise stipulated in the Credit, accept a document, however named, which—
  (i) appears on its face to indicate the name of the carrier or multimodal transport operator and to have been signed or otherwise authenticated by—
      — the carrier or multimodal transport operator or a named agent for or on behalf of the carrier or multimodal transport operator, or
      — the master or a named agent for or on behalf of the master.
      Any signature or authentication of the carrier, multimodal transport operator or master must be identified as carrier, multimodal transport operator or master, as the case may be. An agent signing or authenticating for the carrier, multimodal transport operator or master must also indicate the name and the capacity of the party, ie carrier, multimodal transport operator or master, on whose behalf that agent is acting, and
  (ii) indicates that the goods have been dispatched, taken in charge or loaded on board.
      Dispatch, taking in charge or loading on board may be indicated by wording to that effect on the multimodal transport document and the date of issuance will be deemed to be the date of dispatch, taking in charge or loading on board and the date of shipment. However, if the document indicates, by stamp or otherwise, a date of dispatch, taking in charge or loading on board, such date will be deemed to be the date of shipment, and
  (iii) (a) indicates the place of taking in charge stipulated in the Credit which may be different from the port, airport or place of loading, and the place of final destination stipulated in the Credit which may be different from the port, airport or place of discharge, and/or
       (b) contains the indication "intended" or similar qualification in relation to the vessel and/or port of loading and/or port of discharge, and
  (iv) consists of a sole original multimodal transport document or, if issued in more than one original, the full set as so issued, and

(v)   appears to contain all of the terms and conditions of carriage, or some of such terms and conditions by reference to a source or document other than the multimodal transport document (short form/ blank back multimodal transport document); banks will not examine the contents of such terms and conditions, and

(vi)   contains no indication that it is subject to a charter party and/or no indication that the carrying vessel is propelled by sail only, and

(vii)   in all other respects meets the stipulations of the Credit.

(b)   Even if the Credit prohibits transhipment, banks will accept a multimodal transport document which indicates that transhipment will or may take place, provided that the entire carriage is covered by one and the same multimodal transport document.

Article 27   Air Transport Document

(a)   If a Credit calls for an air transport document, banks will, unless otherwise stipulated in the Credit, accept a document, however named, which—

(i)   appears on its face to indicate the name of the carrier and to have been signed or otherwise authenticated by—

— the carrier, or

— a named agent for or on behalf of the carrier.

Any signature or authentication of the carrier must be identified as carrier. An agent signing or authenticating for the carrier must also indicate the name and the capacity of the party, ie carrier, on whose behalf that agent is acting, and

(ii)   indicates that the goods have been accepted for carriage, and

(iii)   where the Credit calls for an actual date of dispatch, indicates a specific notation of such date, the date of dispatch so indicated on the air transport document will be deemed to be the date of shipment. For the purpose of this Article, the information appearing in the box on the air transport document (marked 'For Carrier Use Only" or similar expression) relative to the flight number and date will not be considered as a specific notation of such date of dispatch.

In all other cases, the date of issuance of the air transport document will be deemed to be the date of shipment, and

(iv)   indicates the airport of departure and the airport of destination stipulated in the Credit, and

(v)   appears to be the original for consignor/shipper even if the Credit stipulates a full set of originals, or similar expressions, and

(vi)   appears to contain all of the terms and conditions of carriage, or some of such terms and conditions, by reference to a source or document other than the air transport document; banks will not examine the contents of such terms and conditions, and

(vii) in all other respects meets the stipulations of the Credit.

(b) For the purpose of this Article, transhipment means unloading and reloading from one aircraft to another aircraft during the course of carriage from the airport of departure to the airport of destination stipulated in the Credit.

(c) Even if the Credit prohibits transhipment, banks will accept an air transport document which indicates that transhipment will or may take place, provided that the entire carriage is covered by one and the same air transport document.

Article 28   Road, Rail or Inland Waterway Transport Documents

(a) If a Credit calls for a road, rail, or inland waterway transport document, banks will, unless otherwise stipulated in the Credit, accept a document of the type called for, however named, which—

(i) appears on its face to indicate the name of the carrier and to have been signed or otherwise authenticated by the carrier or a named agent for or on behalf of the carrier and/or to bear a reception stamp or other indication of receipt by the carrier or a named agent for or on behalf of the carrier.

Any signature, authentication, reception stamp or other indication of receipt of the carrier, must be identified on its face as that of the carrier. An agent signing or authenticating for the carrier, must also indicate the name and the capacity of the party, ie carrier, on whose behalf that agent is acting, and

(ii) indicates that the goods have been received for shipment, dispatch or carriage or wording to this effect. The date of issuance will be deemed to be the date of shipment unless the transport document contains a reception stamp, in which case the date of the reception stamp will be deemed to be the date of shipment, and

(iii) indicates the place of shipment and the place of destination stipulated in the Credit, and

(iv) in all other respects meets the stipulations of the Credit.

(b) In the absence of any indication on the transport document as to the numbers issued, banks will accept the transport document(s) presented as constituting a full set. Banks will accept as original(s) the transport document(s) whether marked as original(s) or not.

(c)   For the purpose of this Article, transhipment means unloading and reloading from one means of conveyance to another means of conveyance, in different modes of transport, during the course of carriage from the place of shipment to the place of destination stipulated in the Credit.

(d)   Even if the Credit prohibits transhipment, banks will accept a road, rail, or inland waterway transport document which indicates that transhipment will or may take place, provided that the entire carriage is covered by one and the same transport document and within the same mode of transport.

Article 29   Courier and Post Receipts

(a)   If a Credit calls for a post receipt or certificate of posting, banks will, unless otherwise stipulated in the Credit, accept a post receipt or certificate of posting which—
  (i)   appears on its face to have been stamped or otherwise authenticated and dated in the place from which the Credit stipulates the goods are to be shipped or dispatched and such date will be deemed to be the date of shipment or dispatch, and
  (ii) in all other respects meets the stipulations of the Credit.

(b)   If a Credit calls for a document issued by a courier or expedited delivery service evidencing receipt of the goods for delivery, banks will, unless otherwise stipulated in the Credit, accept a document, however named, which—
  (i)   appears on its face to indicate the name of the courier/service, and to have been stamped, signed or otherwise authenticated by such named courier/service (unless the Credit specifically calls for a document issued by a named Courier/Service, banks will accept a document issued by any Courier/Service), and
  (ii) indicates a date of pick-up or of receipt or wording to this effect, such date being deemed to be the date of shipment or dispatch, and
  (iii) in all other respects meets the stipulations of the Credit.

Article 30   Transport Documents issued by Freight Forwarders

Unless otherwise authorised in the Credit, banks will only accept a transport document issued by a freight forwarder if it appears on its face to indicate—
  (i)   the name of the freight forwarder as a carrier or multimodal transport operator and to have been signed or otherwise

authenticated by the freight forwarder as carrier or multimodal transport operator, or

(ii) the name of the carrier or multimodal transport operator and to have been signed or otherwise authenticated by the freight forwarder as a named agent for or on behalf of the carrier or multimodal transport operator.

Article 31 "On Deck", "Shipper's Load and Count", Name of Consignor

Unless otherwise stipulated in the Credit, banks will accept a transport document which—

(i) does not indicate, in the case of carriage by sea or by more than one means of conveyance including carriage by sea, that the goods are or will be loaded on deck. Nevertheless, banks will accept a transport document which contains a provision that the goods may be carried on deck, provided that it does not specifically state that they are or will be loaded on deck, and/or

(ii) bears a clause on the face thereof such as "shipper's load and count" or "said by shipper to contain" or words of similar effect, and/or

(iii) indicates as the consignor of the goods a party other than the Beneficiary of the Credit.

Article 32 Clean Transport Documents

(a) A clean transport document is one which bears no clause or notation which expressly declares a defective condition of the goods and/or the packaging.

(b) Banks will not accept transport documents bearing such clauses or notations unless the Credit expressly stipulates the clauses or notations which may be accepted.

(c) Banks will regard a requirement in a Credit for a transport document to bear the clause "clean on board" as complied with if such transport document meets the requirements of this Article and of Articles 23, 24, 25, 26, 27, 28 or 30.

Article 33 Freight Payable/Prepaid Transport Documents

(a) Unless otherwise stipulated in the Credit, or inconsistent with any of the documents presented under the Credit, banks will accept transport documents stating that freight or transportation charges (hereafter referred to as "freight") have still to be paid.

(b)   If a Credit stipulates that the transport document has to indicate that freight has been paid or prepaid, banks will accept a transport document on which words clearly indicating payment or prepayment of freight appear by stamp or otherwise, or on which payment or prepayment of freight is indicated by other means. If the Credit requires courier charges to be paid or prepaid banks will also accept a transport document issued by a courier or expedited delivery service evidencing that courier charges are for the account of a party other than the consignee.

(c)   The words "freight prepayable" or "freight to be prepaid" or words of similar effect, if appearing on transport documents, will not be accepted as constituting evidence of the payment of freight.

(d)   Banks will accept transport documents bearing reference by stamp or otherwise to costs additional to the freight, such as costs of, or disbursements incurred in connection with, loading, unloading or similar operations, unless the conditions of the Credit specifically prohibit such reference.

Article 34   Insurance Documents

(a)   Insurance documents must appear on their face to be issued and signed by insurance companies or underwriters or their agents.

(b)   If the insurance document indicates that it has been issued in more than one original, all the originals must be presented unless otherwise authorised in the Credit.

(c)   Cover notes issued by brokers will not be accepted, unless specifically authorised in the Credit.

(d)   Unless otherwise stipulated in the Credit, banks will accept an insurance certificate or a declaration under an open cover pre-signed by insurance companies or underwriters or their agents. If a Credit specifically calls for an insurance certificate or a declaration under an open cover, banks will accept, in lieu thereof, an insurance policy.

(e)   Unless otherwise stipulated in the Credit, or unless it appears from the insurance document that the cover is effective at the latest from the date of loading on board or dispatch or taking in charge of the goods, banks will not accept an insurance document which bears a date of issuance later than the date of loading on board or dispatch or taking in charge as indicated in such transport document.

(f)(i) Unless otherwise stipulated in the Credit, the insurance document must be expressed in the same currency as the Credit.

(ii) Unless otherwise stipulated in the Credit, the minimum amount for which the insurance document must indicate the insurance cover to have been effected is the CIF (cost, insurance and freight (. . . "named port of destination")) or CIP (carriage and insurance paid to (. . . "named place of destination")) value of the goods, as the case may be, plus 10%, but only when the CIF or CIP value can be determined from the documents on their face. Otherwise, banks will accept as such minimum amount 110% of the amount for which payment, acceptance or negotiation is requested under the Credit, or 110% of the gross amount of the invoice, whichever is the greater.

Article 35   Type of Insurance Cover

(a)   Credits should stipulate the type of insurance required and, if any, the additional risks which are to be covered. Imprecise terms such as "usual risks" or "customary risks" shall not be used; if they are used, banks will accept insurance documents as presented, without responsibility for any risks not being covered.

(b)   Failing specific stipulations in the Credit, banks will accept insurance documents as presented, without responsibility for any risks not being covered.

(c)   Unless otherwise stipulated in the Credit, banks will accept an insurance document which indicates that the cover is subject to a franchise or an excess (deductible).

Article 36   All Risks Insurance Cover

Where a Credit stipulates "insurance against all risks", banks will accept an insurance document which contains any "all risks" notation or clause, whether or not bearing the heading "all risks", even if the insurance document indicates that certain risks are excluded, without responsibility for any risk(s) not being covered.

Article 37   Commercial Invoices

(a)   Unless otherwise stipulated in the Credit, commercial invoices;
  (i)   must appear on their face to be issued by the Beneficiary named in the Credit (except as provided in Article 48), and
  (ii) must be made out in the name of the Applicant (except as provided in sub-Article 48 (h)), and
  (iii) need not be signed.

(b)   Unless otherwise stipulated in the Credit, banks may refuse commercial invoices issued for amounts in excess of the amount permitted by the Credit. Nevertheless, if a bank authorised to pay, incur a deferred payment undertaking, accept Draft(s), or negotiate under a Credit accepts such invoices, its decision will be binding upon all parties, provided that such bank has not paid, incurred a deferred payment undertaking, accepted Draft(s) or negotiated for an amount in excess of that permitted by the Credit.

(c)   The description of the goods in the commercial invoice must correspond with the description in the Credit. In all other documents, the goods may be described in general terms not inconsistent with the description of the goods in the Credit.

Article 38   Other Documents

If a Credit calls for an attestation or certification of weight in the case of transport other than by sea, banks will accept a weight stamp or declaration of weight which appears to have been superimposed on the transport document by the carrier or his agent unless the Credit specifically stipulates that the attestation or certification of weight must be by means of a separate document.

E. MISCELLANEOUS PROVISIONS

Article 39   Allowances in Credit Amount, Quantity and Unit Price

(a)   The words "about" "approximately", "circa" or similar expressions used in connection with the amount of the Credit or the quantity or the unit price stated in the Credit are to be construed as allowing a difference not to exceed 10% more or 10% less than the amount or the quantity or the unit price to which they refer.

(b)   Unless a Credit stipulates that the quantity of the goods specified must not be exceeded or reduced, a tolerance of 5% more or 5% less will be permissible, always provided that the amount of the drawings does not exceed the amount of the Credit. This tolerance does not apply when the Credit stipulates the quantity in terms of a stated number of packing units or individual items.

(c)   Unless a Credit which prohibits partial shipments stipulates otherwise, or unless sub-Article (b) above is applicable, a tolerance of 5% less in the amount of the drawing will be permissible, provided that if the Credit stipulates the quantity of the goods, such quantity of goods is shipped in full, and if the Credit stipulates a unit price, such price is not reduced. This provision does not apply when expressions referred to in sub-Article (a) above are used in the Credit.

Article 40   Partial Shipments/Drawings

(a)   Partial drawings and/or shipments are allowed, unless the Credit stipulates otherwise.

(b)   Transport documents which appear on their face to indicate that shipment has been made on the same means of conveyance and for the same journey, provided they indicate the same destination, will not be regarded as covering partial shipments, even if the transport documents indicate different dates of shipment and/or different ports of loading, places of taking in charge, or despatch.

(c)   Shipments made by post or by courier will not be regarded as partial shipments if the post receipts or certificates of posting or courier's receipts or dispatch notes appear to have been stamped, signed or otherwise authenticated in the place from which the Credit stipulates the goods are to be dispatched, and on the same date.

Article 41   Instalment Shipments/Drawings

If drawings and/or shipments by instalments within given periods are stipulated in the Credit and any instalment is not drawn and/or shipped within the period allowed for that instalment, the Credit ceases to be available for that and any subsequent instalments, unless otherwise stipulated in the Credit.

Article 42   Expiry Date and Place for Presentation of Documents

(a)   All Credits must stipulate an expiry date and a place for presentation of documents for payment, acceptance, or with the exception of freely negotiable Credits, a place for presentation of documents for negotiation. An expiry date stipulated for payment, acceptance or negotiation will be construed to express an expiry date for presentation of documents.

(b)   Except as provided in sub-Article 44(a), documents must be presented on or before such expiry date.

(c)   If an Issuing Bank states that the Credit is to be available "for one month", "for six months", or the like, but does not specify the date from which the time is to run, the date of issuance of the Credit by the Issuing Bank will be deemed to be the first day from which such time is to run. Banks should discourage indication of the expiry date of the Credit in this manner.

Article 43   Limitation on the Expiry Date

(a)   In addition to stipulating an expiry date for presentation of documents, every Credit which calls for a transport document(s) should also stipulate a specified period of time after the date of shipment during which presentation must be made in compliance with the terms and conditions of the Credit. If no such period of time is stipulated, banks will not accept documents presented to them later than 21 days after the date of shipment. In any event, documents must be presented not later than the expiry date of the Credit.

(b)   In cases in which sub-Article 40(b) applies, the date of shipment will be considered to be the latest shipment date on any of the transport documents presented.

Article 44   Extension of Expiry Date

(a)   If the expiry date of the Credit and/or the last day of the period of time for presentation of documents stipulated by the Credit or applicable by virtue of Article 43 falls on a day on which the bank to which presentation has to be made is closed for reasons other than those referred to in Article 17, the stipulated expiry date and/or the last day of the period of time after the date of shipment for presentation of documents, as the case may be, shall be extended to the first following day on which such bank is open.

(b)   The latest date for shipment shall not be extended by reason of the extension of the expiry date and/or the period of time after the date of shipment for presentation of documents in accordance with sub-Article (a) above. If no such latest date for shipment is stipulated in the Credit or amendments thereto, banks will not accept transport documents indicating a date of shipment later than the expiry date stipulated in the Credit or amendments thereto.

(c)   The bank to which presentation is made on such first following business day must provide a statement that the documents were presented within the time limits extended in accordance with sub-Article 44(a) of the Uniform Customs and Practice for Documentary Credits, 1993 Revision, ICC Publication No 500.

Article 45   Hours of Presentation

Banks are under no obligation to accept presentation of documents outside their banking hours.

Article 46   General Expressions as to Dates for Shipment

(a)   Unless otherwise stipulated in the Credit, the expression "shipment" used in stipulating an earliest and/or a latest date for shipment will be understood to include expressions such as, "loading on board", "dispatch", "accepted for carriage", "date of post receipt", "date of pick-up", and the like, and in the case of a Credit calling for a multimodal transport document the expression "taking in charge".

(b)   Expressions such as "prompt", "immediately", "as soon as possible", and the like should not be used. If they are used banks will disregard them.

(c)   If the expression "on or about" or similar expressions are used, banks will interpret them as a stipulation that shipment is to be made during the period from five days before to five days after the specified date, both end days included.

Article 47   Date Terminology for Periods of Shipment

(a)   The words "to", "until", "till", "from" and words of similar import applying to any date or period in the Credit referring to shipment will be understood to include the date mentioned.

(b)   The word "after" will be understood to exclude the date mentioned.

(c)   The terms "first half", "second half" of a month Shall be construed respectively as the 1st to the 15th, and the 16th to the last day of such month, all dates inclusive.

(d)   The terms "beginning", "middle", or "end" of a month shall be construed respectively as the 1st to the 10th, the 11th to the 20th, and the 21st to the last day of such month, all dates inclusive.

F. TRANSFERABLE CREDIT

Article 48   Transferable Credit

(a)   A transferable Credit is a Credit under which the Beneficiary (First Beneficiary) may request the bank authorised to pay, incur a deferred payment undertaking, accept or negotiate (the "Transferring Bank"), or in the case of a freely negotiable Credit, the bank specifically authorised in the Credit as a Transferring Bank, to make the Credit available in whole or in part to one or more other Beneficiary(ies) (Second Beneficiary(ies)).

(b)   A Credit can be transferred only if it is expressly designated as "transferable" by the Issuing Bank. Terms such as "divisible" "fractionable", "assignable", and "transmissible" do not render the Credit transferable. If such terms are used they shall be disregarded.

(c)   The Transferring Bank shall be under no obligation to effect such transfer except to the extent and in the manner expressly consented to by such bank.

(d)   At the time of making a request for transfer and prior to transfer of the Credit, the First Beneficiary must irrevocably instruct the Transferring Bank whether or not he retains the right to refuse to allow the Transferring Bank to advise amendments to the Second Beneficiary(ies). If the Transferring Bank consents to the transfer under these conditions it must, at the time of transfer, advise the Second

Beneficiary(ies) of the First Beneficiary's instructions regarding amendments.

(e)   If a Credit is transferred to more than one Second Beneficiary(ies), refusal of an amendment by one or more Second Beneficiary(ies) does not invalidate the acceptance(s) by the other Second Beneficiary(ies) with respect to whom the Credit will be amended accordingly. With respect to the Second Beneficiary(ies) who rejected the amendment, the Credit will remain unamended.

(f)   Transferring Bank charges in respect of transfers including commissions, fees, costs or expenses are payable by the First Beneficiary, unless otherwise agreed. If the Transferring Bank agrees to transfer the Credit it shall be under no obligation to effect the transfer until such charges are paid.

(g)   Unless otherwise stated in the Credit, a transferable Credit can be transferred once only. Consequently, the Credit cannot be transferred at the request of the Second Beneficiary to any subsequent Third Beneficiary. For the purpose of this Article, a retransfer to the First Beneficiary does not constitute a prohibited transfer.

Fractions of a transferable Credit (not exceeding in the aggregate the amount of the Credit) can be transferred separately, provided partial shipments/drawings are not prohibited, and the aggregate of such transfers will be considered as constituting only one transfer of the Credit.

(h)   The Credit can be transferred only on the terms and conditions specified in the original Credit, with the exception of—
— the amount of the Credit,
— any unit price stated therein,
— the expiry date,
— the last date for presentation of documents in accordance with Article 43,
— the period for shipment,

any or all of which may be reduced or curtailed.

The percentage for which insurance cover must be effected may be increased in such a way as to provide the amount of cover stipulated in the original Credit, or these Articles.

In addition, the name of the First Beneficiary can be substituted for that of the Applicant, but if the name of the Applicant is specifically required by the original Credit to appear in any document(s) other than the invoice, such requirement must be fulfilled.

(i)   The First Beneficiary has the right to substitute his own invoice(s) (and Draft(s)) for those of the Second Beneficiary(ies), for amounts not in excess of the original amount stipulated in the Credit and for the original unit prices if stipulated in the Credit, and upon such substitution of invoice(s) (and Draft(s)) the First Beneficiary can draw under the Credit for the difference, if any, between his invoice(s) and the Second Beneficiary's(ies') invoice(s).

When a Credit has been transferred and the First Beneficiary is to supply his own invoice(s) (and Draft(s)) in exchange for the Second Beneficiary's(ies') invoice(s) (and Draft(s)) but fails to do so on first demand, the Transferring Bank has the right to deliver to the Issuing Bank the documents received under the transferred Credit, including the Second Beneficiary's(ies') invoice(s) (and Draft(s)) without further responsibility to the First Beneficiary.

(j)   The First Beneficiary may request that payment or negotiation be effected to the Second Beneficiary(ies) at the place to which the Credit has been transferred up to and including the expiry date of the Credit, unless the original Credit expressly states that it may not be made available for payment or negotiation at a place other than that stipulated in the Credit. This is without prejudice to the First Beneficiary's right to substitute subsequently his own invoice(s) (and Draft(s)) for those of the Second Beneficiary(ies) and to claim any difference due to him.

G. ASSIGNMENT OF PROCEEDS

Article 49   Assignment of Proceeds

The fact that a Credit is not stated to be transferable shall not affect the Beneficiary's right to assign any proceeds to which he may be, or may become, entitled under such Credit, in accordance with the provisions of the applicable law. This Article relates only to the assignment of proceeds and not to the assignment of the right to perform under the Credit itself.

ICC ARBITRATION

Contracting parties that wish to have the possibility of resorting to ICC Arbitration in the event of a dispute with their contracting partner should specifically and clearly agree upon ICC Arbitration in their contract or, in the event no single contractual document exists, in the exchange of correspondence which constitutes the agreement between them. The fact of issuing a letter of credit subject to the UCP 500 does NOT by itself constitute an agreement to have resort to ICC Arbitration. The following standard arbitration clause is recommended by the ICC—

"All disputes arising in connection with the present contract shall be finally settled under the Rules of Conciliation and Arbitration of the International Chamber of Commerce by one or more arbitrators appointed in accordance with the said Rules".

PART V

CODES

The Banking Code

(1998 Revised Edition)

This Code of Practice is reproduced with the kind permission of the British Bankers' Association, the Building Societies Association and the Association for Payment Clearing Services.

British Bankers' Association
Pinners Hall
105-108 Old Broad Street
London EC2N 1EX
Telephone: 0171 216 8800
Fax: 0171 216 8811

The Building Societies Association
3 Savile Row
London W1X 1AF
Telephone: 0171 437 0655
Fax: 0171 734 6416

Association for Payment Clearing Services
Mercury House
Triton Court
14 Finsbury Square
London EC2A 1BR
Telephone: 0171 711 6200
Fax: 0171 256 5527

This is a voluntary Code followed by banks and building societies in their relations with personal customers in the United Kingdom. It sets standards of good banking practice which are followed as a minimum by banks and building societies subscribing to it. As a voluntary Code, it allows competition and market forces to operate to encourage higher standards for the benefit of customers.

The standards of the Code are encompassed in the 11 key commitments found at the beginning. These commitments apply to the conduct of business for all products and services provided to customers.

Mortgages are covered in more detail in the Council of Mortgage Lenders' Code of Mortgage Lending Practice. Not all subscribers to the Banking Code are members of the Council of Mortgage Lenders.

The Code does not apply to the selling of investments or investment activities as defined by The Financial Services Act 1986.

The Code provides valuable safeguards for customers. It should help them understand how banks and building societies are expected to deal with them. Customers should check who subscribes to it by contacting the Associations shown [above].

The Independent Review Body for the Banking and Mortgage Codes monitors compliance by banks and building societies with the Code and also oversees its review from time to time.

Copies of the Code are available from banks and building societies and the Associations shown [above].

Within the Code, "you" means the customer and "we" means the bank or building society the customer deals with.

This revised edition is effective from 31 March 1999 unless otherwise indicated.

I. KEY COMMITMENTS

1.1 We, the subscribers to this Code, promise that we will—
— act fairly and reasonably in all our dealings with you;
— ensure that all services and products comply with this Code, even if they have their own terms and conditions;
— give you information on our services and products in plain language, and offer help if there is any aspect which you do not understand;
— help you to choose a service or product to fit your needs;
— help you to understand the financial implications of—
    — a mortgage;
    — other borrowing;
    — savings and investment products;
    — card products.
— help you to understand how your accounts work;
— have safe, secure and reliable banking and payment systems;
— ensure that the procedures our staff follow reflect the commitments set out in this Code;
— correct errors and handle complaints speedily;
— consider cases of financial difficulty and mortgage arrears sympathetically and positively;
— ensure that all services and products comply with relevant laws and regulations.

2. INFORMATION

Information Available

2.1   When you become a customer and at any time you ask, we will give you—

Key features
— clear written information explaining the key features of our main services and products;

Your account
— information on how your account works, including—
    — stopping a cheque or other types of payment;
    — when funds can be withdrawn after a credit has been paid into your account and when funds begin to earn interest;
    — unpaid cheques;
    — out of date cheques;
    — when your account details may be passed to credit reference agencies;

Tariff
— a tariff, covering basic account services. This will also be available in branches;

Interest rates
— information on the interest rates which apply to your account(s), when interest will be deducted or paid to you and, on request, a full explanation of how interest is calculated.
— information on where you can get up-to-date details of the interest rates on savings and investment products we offer, including—
    — the newspapers we usually use to notify interest rate changes. These newspapers will reflect the readership of our customers;
    — telephone number(s); and
    — if we have one, our Internet web site address.

ATM charges

2.2   We will give you details of any charges we make for using Automated Teller Machines (ATMs) when we issue the card.

Overdrafts and fixed term products

2.3   We will tell you of any additional charges and interest you may have to pay if—
— your account becomes overdrawn without agreement;
— you exceed your overdraft limit;
— your loan falls into arrears;
— you change your mind about a fixed term product.

Mortgage tariff

2.4   Before you take out a mortgage and at any time you ask, we will give you a tariff covering the operation and repayment of your mortgage, including charges and additional interest costs payable should you fall into arrears.

Other charges

2.5   We will tell you the charges for any other service or product before or when it is provided or at any time you ask.

Helping You To Choose Savings and Investment Accounts

2.6   We will take care to give you clear and appropriate information on the different types of savings and investment accounts available from us to help you to make an informed choice on the product to fit your needs. We will help you understand how your savings and investment accounts work, including any additional charges or loss of interest for withdrawal or cancellation.

2.7   We will give you information on a single savings or investment account if you have already made up your mind.

Cooling-off

2.8   If you are not happy about your choice of savings or investment account(s), (except for a fixed rate account) within 14 days of opening it, we will help you switch accounts or we will give all your money back with interest. We will ignore any notice period and any additional charges.

Terms and Conditions

Plain language

2.9   All written terms and conditions will be fair in substance and will set out your rights and responsibilities clearly and in plain language, with legal and technical language used only where necessary.

Joint accounts

2.10   If you have a joint account, we will give you additional information on your rights and responsibilities.

Closure

2.11   Unless there are exceptional circumstances, eg fraud, we will not close your account without giving you at least 30 days' notice.

Keeping You Informed of Changes

Changes to terms and conditions

2.12   Occasionally terms and conditions may have to be changed. We will tell you how you will be notified of these changes. We will always give you at least 30 days' notice before any change takes effect.

2.13   If the change is clearly to your disadvantage, we will—
— notify you personally; and
— ignore any notice period on your account for at least 60 days starting from the date of the notice so that you can, if you wish, switch your account or close it.

You will not have to pay any additional charges or additional interest as a result of this switch or closure during this 60 day period.

2.14   If there have been significant changes in any one year, we will give or send you a copy of the new terms and conditions or a summary of the changes.

Changes to interest rates are specifically covered by section 2.16.

Charges

2.15   If we increase a charge for basic account services, we will give you at least 30 days' notice.

Interest rates

2.16   The interest rates which will apply to your accounts may change from time to time. When we change the interest rates, we will tell you about the changes for—
  (a) Branch-based accounts
    — within 30 days, by letter, e-mail, or other personal notice; or
    — within 3 working days of the change—
      — by prominent notices in branches; and
      — by placing notices in the newspapers we usually use. To help you compare rates more easily, our notices will state clearly the previous and new interest rates; and
      — by having the previous and new interest rates for your accounts available on our telephone help lines and, if we have one, our Internet web site; and
      — our staff will always be able to help you.
  (b) Non Branch-based accounts
    — Within 30 days by letter, e-mail or other personal notice;
  (c) And for all accounts
    — To help you compare interest rates on all our savings and investment accounts more easily, we will send you, at least once a year, a summary of these products and the current interest rates unless the account is a passbook account with less than £100 in it. This summary will also include—
    — superseded accounts clearly marked;
    — the names of the newspapers we usually use to notify interest rate changes;
    — our telephone help line numbers; and
    — if we have one, our Internet web site address.

In addition, we will also tell you the different interest rates which have applied to the account during the year.

Superseded accounts

2.17   From time to time, we offer new savings and investments accounts. If you have any type of savings and investment account, other than a fixed rate account, which has been 'superseded' because—

— new accounts are no longer opened; or

— the account is not actively promoted;

we will either—

(a) keep the interest rate on the superseded account at the same level as an account with similar features from the current range; or

(b) switch the superseded account to an account with similar features from the current range.

Examples of similar features include notice periods, types of withdrawals, numbers of free withdrawals, how deposits and withdrawals from the account are made.

This means that the interest rate on your account will always be at least as good as the interest rate on an account with similar features from the current range.

2.18 Where there is no account with 'similar features' we will, within 30 days of your account becoming superseded, contact you to—

— tell you that the account is superseded;

— tell you about our other accounts; and

— help you switch accounts without any notice period and without any additional charges.

Marketing of Services

2.19 Occasionally we will bring to your attention additional services and products which may be of benefit to you.

However, when you become a customer, we will give you the opportunity to say that you do not wish to receive this information.

2.20 We will remind you, at least once every three years, that you can ask not to receive this information.

Consent to marketing

2.21 Unless you specifically request it, or give your express consent in writing, we will not pass your name and address to any company, including other companies in our group, for marketing purposes. You will not be asked to give your permission in return for basic banking services.

Host mailing

2.22 We may tell you about another company's services or products and, if you respond positively, you may be contacted directly by that company.

Minors

2.23 We will not send marketing material indiscriminately and, in particular, we will be selective and careful if you are under eighteen years old or where material relates to loans and overdrafts.

Advertising

2.24 We will ensure that all advertising and promotional material is clear, fair, reasonable and not misleading.

Helping You to Choose a Mortgage

2.25 Choosing a mortgage may be your most important financial commitment. There are three levels of service which may be provided and we will tell you which we offer at the outset. These are—
  (a) advice and a recommendation as to which of our mortgages is most suitable for you. When giving advice, we will take care to help you to select a mortgage to fit your needs by asking for relevant information about your circumstances and objectives. Our advice will also depend on your particular needs and requirements and on the market conditions at the time. The reasons for the recommendation will be given to you in writing before you complete your mortgage.
  (b) information on the different types of mortgage products we offer so that you can make an informed choice of which to take;
  (c) information on a single mortgage product only, if we offer only one mortgage product or if you have already made up your mind.

Before you take out your mortgage, we will confirm, in writing, the level of service given.

2.26 Mortgages are covered in more detail in the Council of Mortgage Lenders' Code of Mortgage Lending Practice.

3. ACCOUNT OPERATIONS

Running Your Account

Statements

3.1   To help you manage your account and check entries on it, we will give you regular account statements. These are normally provided monthly, quarterly or as a minimum annually, unless this is not appropriate for the type of account (for example on a passbook account). You may ask for account statements to be sent more frequently than normally available on your type of account.

3.2   If you have a type of account which is accessible by card, and you have a card, we will introduce systems by 1 July 1999 to send you account statements at least quarterly if there have been any card transactions on that account. This does not apply to passbook accounts.

3.3   If your statement or passbook has an entry which seems to be wrong, you should tell us as soon as possible so that we can resolve matters.

Pre-notification

3.4   If charges and/or debit interest accumulate to your current or savings account during a charging period, you will be given at least 14 days' notice of the amount before it is deducted from your account. The 14 days start from the date of posting the notification.

Cheques

3.5   We will keep original cheques paid from your account or copies for at least six years except where these have already been returned to you.

3.6   If, within a reasonable period after the entry has been made, there is a dispute with us about a cheque paid from your account, we will give you the cheque or a copy as evidence (except where the cheque has already been returned to you). If there is an unreasonable delay we will recredit your account until the matter is resolved.

3.7   If you already have your paid cheques returned, we will continue to return your cheques or copies to you and we will tell you our charges for this service.

3.8   When we need to tell you that one of your cheques or other items has been returned unpaid, we will do this either by letter or by other private and confidential means.

Cards and PINs

3.9   We will send you a card only if you request it or to replace one which has already been issued.

3.10   Your PIN (Personal Identification Number) will be advised only to you and will be issued separately from your card.

PIN self selection

3.11   We will tell you if you can select your own PIN and, if so, you will be encouraged to do so carefully. This should make it easier for you to remember your PIN.

We will have systems in place to allow you to select your own PIN by 1 July 2000.

3.12   You can ask not to be issued with a PIN.

Lending

Financial assessment

3.13   All lending will be subject to our assessment of your ability to repay. This assessment may include—
   — taking into account your income and commitments;
   — how you have handled your financial affairs in the past;
   — information obtained from credit reference agencies and, with your consent, others, for example employers, other lenders and landlords;
   — information supplied by you, including verification of your identity and the purpose of the borrowing;
   — credit assessment techniques, for example credit scoring;

— your age;
— any security provided.

Guarantees

3.14   If you want us to accept a guarantee or other security from someone for your liabilities, you may be asked to consent to the disclosure, by us, of your confidential financial information to the person giving the guarantee or other security or to their legal adviser. We will also—
— encourage them to take independent legal advice to make sure that they understand their commitment and the potential consequences of their decision. All the documents they will be asked to sign will contain this recommendation as a clear and prominent notice;
— advise them that by giving the guarantee or other security they may become liable instead of or as well as you;
— advise them of what the limit of their liability will be. An unlimited guarantee will not be taken.

Foreign Exchange Services

3.15   We will give you an explanation of the service, details of the exchange rate and an explanation of the charges which apply to any foreign exchange transactions which you are about to make. Where this is not possible, we will tell you the basis on which these will be worked out.

3.16   If you wish to transfer money abroad, we will tell you how this is done and will give you, at least, the following information—
— a description of the services and how to use them;
— an explanation of when the money you have sent abroad should get there and any reason for potential delays;
— any commission or charges which you will have to pay, including a warning where a foreign bank's charges may also have to be paid by the recipient.

4. PROTECTION

Confidentiality

4.1   We will treat all your personal information as private and confidential (even when you are no longer a customer). Nothing about your accounts

nor your name and address will be disclosed to anyone, including other companies in our group, other than in four exceptional cases permitted by law. These are—
— where we are legally compelled to do so;
— where there is a duty to the public to disclose;
— where our interests require disclosure;
   This will not be used as a reason for disclosing information about you or your accounts (including your name and address) to anyone else including other companies in our group for marketing purposes.
— where disclosure is made at your request or with your consent.

Credit reference agencies

4.2   Information about your personal debts owed to us may be disclosed to credit reference agencies where—
— you have fallen behind with your payments; and
— the amount owed is not in dispute; and
— you have not made proposals satisfactory to us for repayment of your debt following formal demand; and
— you have been given at least 28 days' notice of our intention to disclose.

4.3   We will not give any other information about you to credit reference agencies without your consent.

Data protection

4.4   We will explain that you have a right of access under Data Protection legislation to your personal records held on our computer files.

Bankers' references

4.5   We will tell you if we provide bankers' references. If a banker's reference about you is requested, we will require your written consent before it is given.

Identification

4.6   When you first apply to open an account, we will tell you what identification we need to prove identity. This is important for your security and is required by law. We will also tell you what checks we may make with credit reference agencies.

4.7   If we record telephone conversations, our terms and conditions will explain this.

Taking care

4.8   The care of your cheque book, passbook, cards, electronic purse, PINs, passwords and selected personal information is essential to help prevent fraud and protect your accounts. Please ensure that you—
— do not keep your cheque book and cards together;
— do not allow anyone else to use your card, PIN and/or password;
— always take reasonable steps to keep your card safe and your PIN, password and selected personal information secret at all times;
— never write down or record your PIN on the card or on anything kept with or near it;
— never write down or record your PIN, password or selected personal information without disguising it, for example, never write down or record your PIN using the numbers in the correct order;
— destroy the notification of your PIN and/or password as soon as you receive it.

4.9   It is essential that you tell us as soon as you can if you suspect or discover that—
— your cheque book, passbook, card and/or electronic purse has been lost or stolen;
— someone else knows your PIN, password or your selected personal information.

Loss-what to do

4.10   The fastest method of notifying us is by telephone, using the numbers previously advised or in telephone directories.

4.11   Once you have told us that a cheque book, passbook, card or electronic purse has been lost or stolen or that someone else knows your PIN, password or selected personal information, we will take immediate steps to prevent these from being used to access your accounts.

4.12   We will refund you the amount of any transaction together with any interest and charges—

— where you have not received your card and it is misused by someone else;
— for all transactions not authorised by you after you have told us that someone else knows your PIN, password or selected personal information;
— if additional money is transferred from your account to your electronic purse after you have told us of its loss, theft or that someone else knows your PIN;
— where faults have occurred in the ATMs, or associated systems used, which were not obvious or subject to a warning message or notice at the time of use.

Electronic purse

4.13   You should treat your electronic purse like cash in a wallet. You will lose any money left in the electronic purse at the time it is lost or stolen, in just the same way as if you lost your wallet. However, if your electronic purse is credited by unauthorised withdrawals from your account before you tell us of its loss, theft or misuse, your liability for such amounts will be limited to a maximum of £50, unless you have acted fraudulently or with gross negligence.

Cards

4.14   If your card is misused before you tell us of its loss or theft, or that someone else knows your PIN, your liability will be limited to a maximum of £50, unless you have acted fraudulently or with gross negligence.

4.15   Where a card transaction is disputed, we have the burden of proving fraud or gross negligence or that you have received your card.

In such cases we would expect you to co-operate with us and with the police in any investigation.

Fraud and gross negligence

4.16   If you act fraudulently you will be liable for all losses. If you act with gross negligence which has caused losses you may be liable for them. This may apply if you fail to follow the safeguards set out in section 4.8.

5. DIFFICULTIES

Financial Difficulties

5.1   We will consider cases of financial difficulty sympathetically and positively. Our first step will be to try to contact you to discuss the matter.

How we can help

5.2   If you find yourself in financial difficulties, you should let us know as soon as possible. We will do all we can to help you to overcome your difficulties. The sooner we discuss your problems, the easier it will be for both of us to find a solution. The more you tell us about your full financial circumstances, the more we may be able to help.

5.3   With your co-operation, we will develop a plan with you for dealing with your financial difficulties, consistent with both our interests and yours.

5.4   If you are in difficulties you can also get help and advice from debt counselling organisations. At your request and with your consent, we will liaise, wherever possible, with debt counselling organisations that we recognise, for example—
— Citizens Advice Bureaux; or
— money advice centres; or
— The Consumer Credit Counselling Service.

Complaints

Internal procedures

5.5   We have internal procedures for handling complaints fairly and speedily and we will tell you what these are. These will include establishing a set time for an initial acknowledgement to your complaint. We will tell you how long it might take us to respond more fully.

5.6   If you wish to make a complaint, we will tell you how to do so and what to do if you are not happy about the outcome. Staff will help you with any queries.

Ombudsmen

5.7   Banks and building societies have separate independent ombudsmen or arbitration schemes. The ombudsmen or arbitrators are available to resolve certain complaints made by you if the matter remains unresolved through our internal complaints procedures.

5.8   All building societies must belong to the Building Societies Ombudsman Scheme.

5.9   All banks subscribing to this Code must belong to the Banking Ombudsman Scheme or, where appropriate, to one of the arbitration schemes listed below.

5.10   We will display a notice in a prominent position in all our branches stating which Ombudsman or arbitration scheme we belong to and that copies of the Code are available on request.

5.11   We will give you details about which Ombudsman or arbitration scheme is available to you. You can also get information by contacting the appropriate Ombudsman or arbitration scheme at the addresses listed below—

The Office of the Banking Ombudsman
70 Gray's Inn Road
London WC1X 8NB
Tel: 0171 404 9944
Enquiries only-LO-call Tel: 0345 660902

The Office of the Building Societies Ombudsman
Millbank Tower
Millbank
London SW1P 4XS
Tel: 0171 931 0044

The Finance and Leasing Association Arbitration Scheme
Imperial House,
15-19 Kingsway,
London WC2 6UN
Tel: 0171 836 6511

The Consumer Credit Trade Association Arbitration Scheme
Tennyson House
159/163 Great Portland Street
London W1N 5FD
Tel: 0171 636 7564

Monitoring & Compliance

5.12   We will comply with the law and follow relevant codes of practice or similar documents as members of the British Bankers' Association (BBA), The Building Societies Association (BSA) and the Association for Payment Clearing Services (APACS). The main codes include—
—   BBA, BSA, FLA Code of Practice on the Advertising of Interest Bearing Accounts;
—   BBA Guide to Bankers' References (Status Enquiries);
—   BBA Dormant Accounts Procedure;
—   BSA Code of Practice on Linking of Services;
—   CML Code of Mortgage Lending Practice;
—   CML Statement of Practice on Handling Arrears and Possessions;
—   CML Statement of Practice on the Transfer of Mortgages;
—   Association of British Insurers (ABI) General Business Code of Practice;
—   British Codes of Advertising and Sales Promotion;
—   ITC (Independent Television Commission) Code of Advertising Practice;
—   Guide to Credit Scoring.

5.13   We have a 'Code Compliance Officer' and our internal auditing procedures monitor compliance with the Code.

Review body

5.14   The Code is monitored by the Independent Review Body for the Banking and Mortgage Codes comprised of representatives from the banks and building societies and independent consumers. The address is—

Pinners Hall
105-108 Old Broad Street
London EC2N 1EX
Tel: 0171 216 8800

Complaints concerning the general operation of the Code can be made to them.

5.15   We complete a 'Statement of Compliance' every year which is signed by our Chief Executive and sent to the Independent Review Body for the Banking and Mortgage Codes.

6. HELP SECTION

Sponsoring associations

Enquiries about the Code and requests for copies of it can be addressed to the British Bankers' Association, The Building Societies Association and the Association for Payment Clearing Services. The addresses and telephone numbers are shown at the front of this booklet.

Copies of the Code

All institutions subscribing to the Code will make copies of it available to customers. Copies of the CML Code of Mortgage Lending Practice are available from the Council of Mortgage Lenders (CML) 3 Savile Row, London W1X 1AF, recorded help line telephone number 0171 440 2255.

Additional information

Additional information on a variety of banking and mortgage matters is available in the form of "Bank Facts" from the BBA, "Fact Sheets" and information leaflets from the BSA and the CML and "Pay Points" from APACS. In addition, the Associations operate customer information lines or 'help lines'.

Websites
 Internet sites— www.bba.org.uk
       www.cml.org.uk
       www.bsa.org.uk
       www.apacs.org.uk

Useful Definitions

These definitions explain the meaning of words and terms used in the Code. They are not precise legal or technical definitions.

ATM (Automated Teller Machine)

A cash machine or free standing device dispensing cash and providing other information or services to customers who have a card.

Banker's reference

An opinion about a particular customer's ability to enter into or repay a financial commitment.

Basic banking service

The opening, maintenance and operation of accounts for money transmission by means of cheque and other debit instruments. This would normally be a current account.

Cards

A general term for any plastic card which may be used to pay for goods and services or to withdraw cash. For the purposes of this Code, it excludes electronic purses.

Credit reference agencies

Organisations, licensed under the Consumer Credit Act 1974, which hold information about individuals which is of relevance to lenders. Banks and building societies may refer to these agencies to assist with various decisions, eg whether or not to open an account or provide loans or grant credit. Banks and building societies may give information to or seek information from these agencies.

Credit scoring

A system which banks and building societies use to assist in making decisions about granting consumer credit. Credit scoring uses statistical techniques to measure the likelihood that an application for credit will be a good credit risk.

Electronic purses

Any card or function of a card which contains real value in the form of electronic money which someone has paid for in advance, some of which can be reloaded with further funds and which can be used for a range of purposes.

Guarantee

An undertaking given by a person called the guarantor promising to pay the debts of another if that other person fails to do so.

Notice period

Where notice periods are specified, the notice period starts from the date of posting the notification.

Out of date cheque

A cheque which has not been paid because its date is too old, normally more than six months.

Password

A word or an access code which the customer has selected to permit them access to a telephone or home banking service and which is also used for identification.

Personal customer

A private individual who maintains an account (including a joint account with another private individual or an account held as an executor or trustee, but excluding the accounts of sole traders, partnerships, companies, clubs and societies) or who receives other services from a bank or building society.

PIN (Personal Identification Number)

A number provided on a strictly confidential basis by a bank or building society to a card holder. Use of this number by the customer will allow the card to be used to withdraw cash and access other services from an Automated Teller Machine (ATM).

Security

A word used to describe items of value such as title deeds to houses, share certificates, life policies, etc, which represent assets used as support for a loan. Under a secured loan the lender has the right to sell the security if the loan is not repaid.

Selected personal information

A selection of memorable facts and information of a private and personal nature chosen by the customer (the sequence of which is known only to the customer) which can be used for identification and to verify identification when accessing accounts.

Tariff

A list of charges for services provided by a bank or building society.

Unpaid cheque

This is a term for a cheque which, after being paid into the account of a person to whom it is payable, is subsequently returned 'unpaid' ('bounced') by the bank or building society whose customer issued the cheque. This leaves the person to whom the cheque is payable without the money in his/her account.